Solitons and Nonlinear Wave Equations

Solitons and Nonlinear Wave Equations

R. K. DODD
Department of Mathematics, Trinity College
Dublin, Ireland

J. C. EILBECK
Department of Mathematics, Heriot-Watt University
Edinburgh, Scotland

J. D. GIBBON
Department of Mathematics, Imperial College
London, England

H. C. MORRIS
Department of Mathematics, Trinity College
Dublin, Ireland

1982 ACADEMIC PRESS
A Subsidiary of Harcourt Brace Jovanovich, Publishers

LONDON NEW YORK
PARIS SAN DIEGO SAN FRANCISCO
SÃO PAULO SYDNEY TOKYO TORONTO

ACADEMIC PRESS INC. (LONDON) LTD.
24/28 Oval Road,
London NW1

United States Edition published by
ACADEMIC PRESS INC.
(Harcourt Brace Jovanovich, Inc.)
Orlando, Florida 32887

British Library Cataloguing in Publication Data
Solitons and nonlinear wave equations.
1. Wave-motion, Theory of 2. Nonlinear theories
I. Dodd, R. K.
531.1133 QA927

ISBN 0-12-219120-X
ISBN 0-12-219122-6 (Pbk)
LCCCN 81-66694

Printed in Great Britain by
Whitstable Litho Ltd., Whitstable, Kent

PREFACE

The study of the soliton as a stable particle-like state of nonlinear systems has so caught the imagination of physicists and mathematicians of all descriptions that it can genuinely claim to be one of the few inter-disciplinary subjects of modern day mathematical physics. It may be that the topic of solitons will pass through this wide spectrum of subjects leaving their content unchanged, with only a shift in emphasis! Only time will tell.

The uninitiated reader who wishes to learn about the basic ideas in this subject may have a problem knowing where to start. As the bibliography at the back of the book shows, there is now a variety of books to choose from, each tackling the subject from a different angle. This book has been written for the reader who has no prior knowledge of solitons and nonlinear wave equations and who may find texts of up-to-date research papers too difficult a starting point. One of the unique features of this book is an attempt to combine the ideas of soliton theory with a study of the physical origins of the nonlinear equations concerned. We have also treated the inverse scattering (spectral) transform with a degree of rigour not usually attempted in other texts, although a full grasp of this technical material is not essential to an understanding of the applications of the theory. We have concentrated mainly on the theory and applications of classical solitons with only a brief treatment of quantum mechanical effects which occur in particle physics and quantum field theory. Our treatment should, we hope, appeal to both the research worker and graduate student in applied mathematics, engineering, and theoretical physics.

Our choice of topics is based on two main aims. The first is the study of solutions of certain equations by the inverse scattering (spectral) transform, and the second is the study of general classes of physical systems which give rise to these equations. These two aims are mutually compatible, even though the number of equations solvable by the inverse scattering transform is small. It is precisely the integrable evolution equations such as the KdV, MKdV, sine-Gordon and NLS equations which are the most natural equations to arise on various time and space scales when studying weakly nonlinear dispersive systems of many types. The ubiquitousness of these equations surprises many people who tend to think of their occurrence in terms of 'magic'. One of our aims is to show that this is not so: dispersive nonlinear systems with small or no damping should behave in the same way whether they occur in plasmas, classical fluids, lasers or nonlinear lattices. The evolution of long waves on the one hand, and harmonic wave envelopes in off and on resonant systems on the other hand, along with other nonlinear interactions, are classic phenomena which occur in many areas of physics and applied mathematics. These should now be regarded as standard effects and automatically invoked in any study. A further aim of this book is to encourage the interdisciplinary aspect in those students just beginning in the area. The study of solitons is a good example of an area of research which has brought together workers in different areas ranging from experimental physics to pure mathematics.

The beginnings of this subject is also an example of the value of careful numerical experiment allied with analytical methods. Now that problems in more than one space dimension are being tackled, numerical studies are becoming popular once more. For these reasons, we have thought it worthwhile to include a concise chapter on numerical methods and results.

With these comments in mind, we have tried to choose our chapters carefully. Most students of applied or engineering mathematics in the USA or UK will have studied little quantum mechanics. The basics of this subject, such as the concepts of reflection and transmission coefficients, square integrable wave functions and potential scattering are absolutely essential in studying the inverse scattering transform. For this reason, we have included a chapter on the elementary ideas of quantum mechanics and classical and quantum inverse problems. On the other hand, most students of theoretical physics will have studied very little classical applied mathematics including fluid mechanics or geophysical fluid dynamics. These students will not be so familiar with ideas such as multiple scaling or the method of stretched co-ordinates. We have attempted to work out some of the examples in detail from first principles because we feel that producing equations out of a hat is not a good idea. However, in certain cases, space has precluded a longer derivation. Obvious limitations on the length of the book have prevented us from dealing with other important aspects such as discrete systems, Lie algebraic methods, higher dimensions, perturbation theories and other integrable systems which require more complicated isospectral operators. It would have been nice to have included all these and we hope that colleagues who specialise in these areas are not offended that these aspects of the subject are excluded. We certainly hope we have assigned credit where it is due, although inevitably we will have omitted a number of key contributions through ignorance or oversight. A list of texts and books of edited research articles in which many of these aspects can be found is included at the beginning of the bibliography.

Because this book has been designed as a textbook of basic methods and ideas and not as an up-to-date book of the latest research results, we have in some chapters decided to relegate most of the referencing to chapter notes. Our introduction in Chapter 1, while partially historical, is nevertheless selective and the topics have been chosen in order to lead the reader gently up to the idea of quantum inverse scattering. We could make a list of at least a dozen names who have made major contributions to the subject but we believe that most would agree that M D Kruskal and V E Zakharov deserve the greatest credit for the conception and development of the inverse scattering method.

This book has been produced on the SCRIBE word processing system (Reid and Walker, 1979), including the bibliography and index. Camera ready copy (excluding figures and mathematical symbols) was generated on the Science and Engineering Research Council's Interactive Computing Facility DEC-10 machine at the Edinburgh Regional Computing Centre, and output on a DIABLO daisy-wheel printer. We would especially like to thank Jeff Phillips of ERCC for his valuable assistance with all problems arising during the use of the SERC

ICF facilities. Mathematical symbols and figures were added at the final stage, using a standard typewriter. Many of the figures were generated on the FR80 microfilm recorder at the SERC Rutherford Appleton Laboratory using the ICF network. We are most grateful to Shona McVicar, who undertook the arduous job of inputting the bulk of the text from handwritten or typed copy. Tina Richardson at Imperial College and Mrs M Gardiner at Heriot-Watt University performed the skillful task of typing in the mathematical characters into the text. Thanks are also due to Miss A J Marsland, Mrs J Stewart, and Mrs I Sansonne for useful contributions to proof reading and typing duties. We would also like to thank Anne Trevillion and Anthony Watkinson of Academic Press for their almost infinite patience whilst several deadlines came and went. Several other readers made many useful suggestions, for which we are grateful, although we must accept final responsibility for remaining errors and omissions. We hope the reader will feel encouraged to bring to our notice corrections and suggestions for improvement of the text.

October, 1982
Roger Dodd
Chris Eilbeck
John Gibbon
Hedley Morris

CONTENTS

1. SOLITARY WAVES AND SOLITONS

1.1 The Discovery of the Solitary Wave

A common theme in the development of science is that of an important discovery which is not widely recognised as such when it is first reported. Most often this comes about not through the ignorance or indifference of the scientific community, but because the current state of knowledge of the field is insufficiently developed for the full significance of the result to be realised.

The subject of this book is one such topic. The first conscious observation of what was termed a solitary wave in 1834 was not appreciated until its significance as an important stable state of some nonlinear systems was realised in the mid-1960's. After a span of more than 130 years as a scientific curiosity, the solitary wave has since appeared in many fields of applied mathematics and physics such as meteorology, elementary particle physics, plasma studies, and laser physics.

Because the advances of the last decade have rather overwhelmed the simplicity of the early results, we shall concentrate in this introductory chapter on primary results and observations in order to give the reader some degree of historical perspective. The approach will be heuristic, the aim being to develop key ideas and definitions which can be used later in more detailed and rigorous chapters. In this section we introduce the concept of the "solitary wave": later in the chapter we describe the special features that many types of solitary waves possess. When these features are present the solitary wave is often called a "soliton": the distinction beween these two terms will become clearer in Section 1.4.

The first documented observation of the solitary wave was made in 1834 by the Scottish scientist and engineer, John Scott Russell. Whilst observing the movement of a canal barge, Scott Russell noticed a novel type of water wave on the surface of the canal. Scott Russell's original description (1844) is still worth repeating:

> I was observing the motion of a boat which was rapidly drawn along a narrow channel by a pair of horses, when the boat suddenly stopped - not so the mass of water in the channel which it had put in motion; it accumulated round the prow of the vessel in a state of violent agitation, then suddenly leaving it behind, rolled forward with great velocity, assuming the form of a large solitary elevation, a rounded, smooth and well defined heap of water, which continued its course along the channel apparently without change of form or diminution of speed. I followed it on horseback and overtook it still rolling on at a rate of some eight or nine miles an hour, preserving its

> original figure some thirty feet long and a foot to a foot and a half in height. Its height gradually diminished and after a chase of one or two miles I lost it in the windings of the channel. Such in the month of August 1834 was my first chance interview with that singular and beautiful phenomenon which I have called the Wave of Translation....

This observation was not a purely chance encounter. Scott Russell was engaged in an unpaid study of the design of canal barges for the Union Canal Society of Edinburgh. It seems likely from another of his papers (Scott Russell 1840) that the first sighting of the solitary wave was near the "Hermiston Experimental Station" on the Union Canal, six miles from the centre of Edinburgh.

It is also recorded in some of his reports that he performed various experiments on other canals, rivers and lakes. His researches were spread over a number of years involving tide wave measurements in the Firth of Forth in Scotland and the River Dee in Cheshire. He had a small reservoir in his own back garden built specially for the purpose of studying the resistance of floating bodies. Scott Russell's interests also lay in the very practical problem of ship design, particularly the problem of finding the best shape of a ship's hull which would offer the least resistance to the water. In fact his "wave-line" theory of naval design brought him considerable prestige in his own lifetime.

His researches in general led him to ask a rather more basic question regarding fluid motion. As an object such as a boat moves through the water, it pushes away a hump of fluid, exactly as Scott Russell himself described in the above quotation. Momentum has therefore been transferred from the boat to the fluid. He wondered what happened to this fluid and how it affected the subsequent motion of the boat. Furthermore how does the shape of this hump of fluid evolve? Scott Russell therefore tried to repeat his canal observation in the laboratory in a small scale wave tank in order to study the phenomenon more carefully. He was successful to a large degree in repeating what he had seen in the canal. This is shown in Fig. 1-1, which is one of Scott Russell's original diagrams (Scott Russell 1844).

Our selection from Scott Russell's figures show two experiments (his Figs. 4 and 6). The first experiment, shown in Fig. 1-1a, depicts a raised area of fluid behind a partition. When this partition is suddenly removed, a long bell-shaped wave emerges and propagates down the channel. It was exactly this wave that Scott Russell had observed on the canal. The second experiment (Fig. 1-1b) is identical except that the initial volume of trapped fluid is larger: in this case two solitary waves are formed.

The important point to notice is that once momentum has been imparted to the fluid, this does not result in a rippling motion in which this momentum would be scattering over the whole surface of the water. Instead it remains localised in a stable propagating wave which passes over the fluid and leaves it as it was before the wave had arrived. The third diagram of the sequence in Fig. 1-1a shows that Scott Russell had this in mind. In an idealised situation one should be able to trap the solitary wave once it has reached the end of the channel by slipping down another partition just at the right time. Exactly the same amount of fluid should be trapped at the far end as was released at the beginning, showing that the momentum transfer has remained local. It will be seen later on that this type of behaviour applies to a larger class of nonlinear systems and not just water waves in

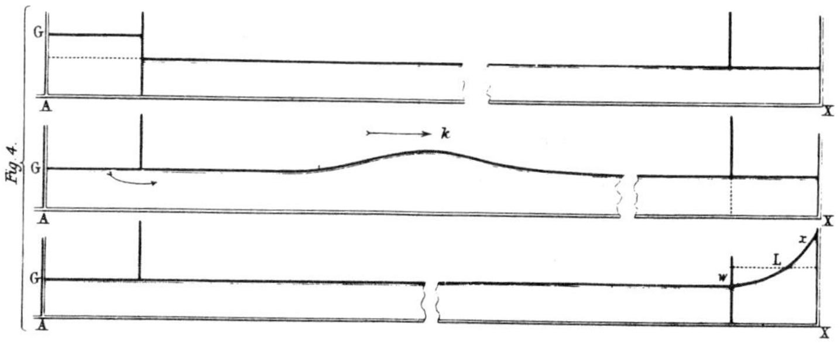

Scott Russell's Figure 4.

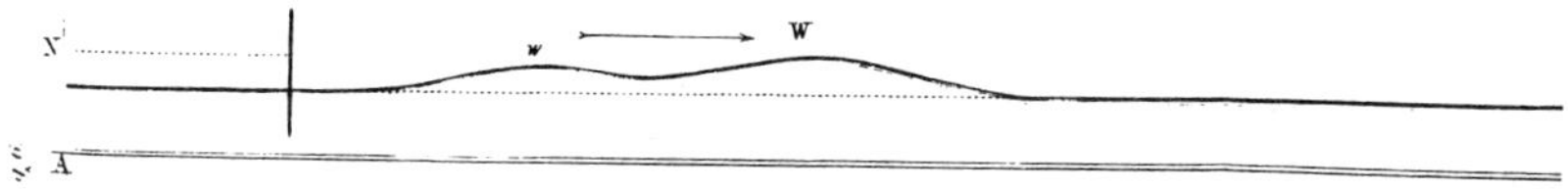

Scott Russell's Figure 6.

Figure 1-1: From Scott Russell's "Report on Waves" (1844).

particular. The words "solitary wave" were coined by Scott Russell himself, mainly because this type of wave motion stands alone and apart from the other types of oscillatory wave motion.

As a conclusion to this Section it would be unfair if we did not at least make a passing mention of some of Scott Russell's other achievements. A recent biography has been written by George Emmerson (1977), and this highly readable book describes well the flavour of early Victorian scientific and engineering practice. Scott Russell was very much a man of his time - bold, hardworking and ready to have a go at anything. Emmerson's biography brings out the ferocious hard work and enthusiasm for every topic which he tackled.

Probably Scott Russell would have been best suited to a university chair, but failure to achieve this led to his involvement in the worlds of business and shipbuilding. It was his yard in London which built the famous ship "The Great Eastern", designed by Brunel, and the tribulations, accidents and failures associated with this venture have unfortunately, and possibly unfairly, tainted Scott Russell's name. It seems from present day evidence that, in a number of incidents, Scott Russell was used as a convenient scapegoat, perhaps because of his humble background and lack of private fortune. It therefore seems fitting that he should get a large measure of posthumous credit for his original researches into solitary waves.

His interest in scientific and engineering problems extended to many other topics, as shown by his extensive list of scientific publications. In particular he made an independent discovery of the Doppler effect. In later life the significance of the solitary wave became almost an obsession, and in his posthumous work "The Wave of Translation" (Scott Russell 1885) he applied a theory of solitary waves in air and ether to predict the depth of the atmosphere (correctly!) and the size of the universe (incorrectly). He would have been delighted, but not surprised, to find his predictions of the wide-ranging importance of the solitary wave verified in many scientific fields.

1.2 Korteweg and de Vries

There was subsequently a gap of more than sixty years between Scott Russell's observation of the shallow water solitary wave and any theoretical treatment of the phenomenon. Despite some attempts by Scott Russell to guess at the analytical formula for the wave profile (Miura 1976), his observation went unexplained in his own lifetime (he died in 1882). In the following decades the solitary wave of translation was briefly mentioned by various mathematicians including Stokes (1847) and Boussinesq (1872). However initial theoretical confirmation of Scott Russell's work had to wait until 1895 when two Dutch scientists , Kortweg and de Vries (1895), derived their now famous equation for the propagation of waves in one direction on the surface of a shallow canal. If the canal has normal depth ℓ and $\ell + \eta$ (η being small) represents the elevation of the surface above the bottom, the partial differential equation which governs the wave motion is:

$$\frac{\partial \eta}{\partial t} = \frac{3}{2}\sqrt{\frac{g}{\ell}}\frac{\partial}{\partial x}\left(\frac{2}{3}\alpha\eta + \frac{1}{2}\eta^2 + \frac{1}{3}\sigma\frac{\partial^2 \eta}{\partial x^2}\right) \tag{1.2-1}$$

where $\sigma = \ell^3/3 - T\ell/\rho g$, is an arbitrary constant, T the surface tension and ρ the density of the fluid. The derivation of (1.2-1) from first principles is not easy and the calculation is explained in Chapter 5. It is equation (1.2-1) which is known as the Korteweg de Vries equation, usually shortened to "KdV" for simplicity. Appropriate scalings will transform it into a more manageable form. If we define

$$\eta = 8\alpha u; \qquad \xi = \left(\frac{2\alpha}{\sigma}\right)^{\frac{1}{2}} x; \qquad \tau = \left(\frac{2\alpha^3 g}{\sigma\ell}\right)^{\frac{1}{2}} t$$

then (1.2-1) becomes

$$u_\tau + u_\xi + 12uu_\xi + u_{\xi\xi\xi} = 0 \tag{1.2-2}$$

The factor of 12 in the nonlinear term in (1.2-2) is purely a matter of choice and can always be rescaled by the transformation $u \to \beta u$. In order to look for a solution similar to Scott Russell's solitary wave of translation we need to look for a travelling wave solution of (1.2-2) which has permanent profile. To do this we look for translationally invariant solutions of the form $u = u(\theta)$, where θ is a linear function of ξ and τ : $\theta = a\xi - \omega t + \delta$. Requiring that u and its derivatives go to zero as $|\theta| \to \infty$, equation (1.2-2) may be integrated to give the solution

$$u = \frac{1}{4}a^2\text{sech}^2\frac{1}{2}[a\xi - (a + a^3)\tau + \delta] \tag{1.2-3}$$

The constants a and δ are arbitrary with the latter playing the role of a phase or "centre shift". This is a wave with a hump-like shape exactly as Scott Russell observed, which moves at a constant velocity without change of shape. The velocity is $1 + a^2$ which is amplitude dependent and hence a taller wave moves faster than a smaller one. A plot of (1.2-3) is given in Fig. 1-2. For smaller values of a the wave is fat and short as in Fig. 1-1 but becomes thinner and sharper the larger a becomes. It was this wave profile which Scott Russell observed in his experiments (see Fig. 1-1) and is therefore his solitary wave of translation.

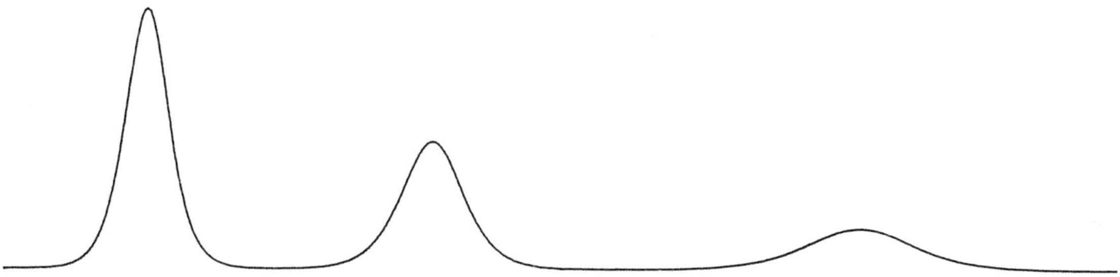

Figure 1-2: Three solitary waves plotted side by side

1.3 The Fermi-Pasta-Ulam Problem

In what seemed at the time a totally unrelated problem, Fermi, Pasta and Ulam in Los Alamos (Fermi et al. 1955) were investigating the behaviour of systems which were primarily linear but in which nonlinearity was introduced as a perturbation. In the absence of such perturbations, the energy in each of the normal modes of the linear system would be constant. It was expected that the nonlinear interactions between the modes would lead to the energy of the system being evenly distributed throughout all of the modes: - a result which would be in accordance with the equipartition theorem. The results they obtained contradicted this idea. The reader might be wondering at this point what this topic has to do with Scott Russell's solitary wave and the KdV equation. It turns out that there is a close connection with the KdV equation, but to see this a little mathematics is needed. The importance of the FPU problem is that the unexpected nature of their results stimulated work on these types of nonlinear systems and much of the modern work on solitons stemmed directly from it. For this reason alone it is important to explain just a few of the details. In their original report FPU gave the results of some extensive numerical calculations which we cannot hope to cover or explain here. Instead we give a brief resume in order to explain how this nonlinear problem connects with the KdV equation.

Consider a dynamical system of N identical particles of unit mass (N large - FPU had N = 64) on a line with fixed end points with forces acting between nearest neighbours. If Q_n denotes the displacement from equilibrium of the nth particle then the equation of motion for this particle can be written in the form

$$\ddot{Q}_n = f(Q_{n+1} - Q_n) - f(Q_n - Q_{n-1}) \tag{1.3-1}$$

where $f(Q)$ is some function which includes both the usual linear nearest neighbour interaction and also some small nonlinear term. Two of the cases which FPU considered are:

$$f(Q) = \gamma Q + \alpha Q^2 \tag{1.3-2}$$

$$f(Q) = \gamma Q + \beta Q^3 \tag{1.3-3}$$

where γ is the linear chain constant and the constants α and β are chosen such that the maximum displacement of Q caused by the nonlinear term is small. Using the nonlinearities given in equations (1.3-2, 1.3-3), if equation (1.3-1) is integrated numerically using initial data in the form of (say) a sine-wave, then it is found that the energy does not spread throughout all the normal modes but remains in the initial mode and a few nearby modes. Furthermore the energy density of those nearby modes has an almost periodic behaviour in time. Figure 1-3 is reprinted from the original report written by FPU and shows a plot of energy against time for the first five modes, with initial data a sine wave.

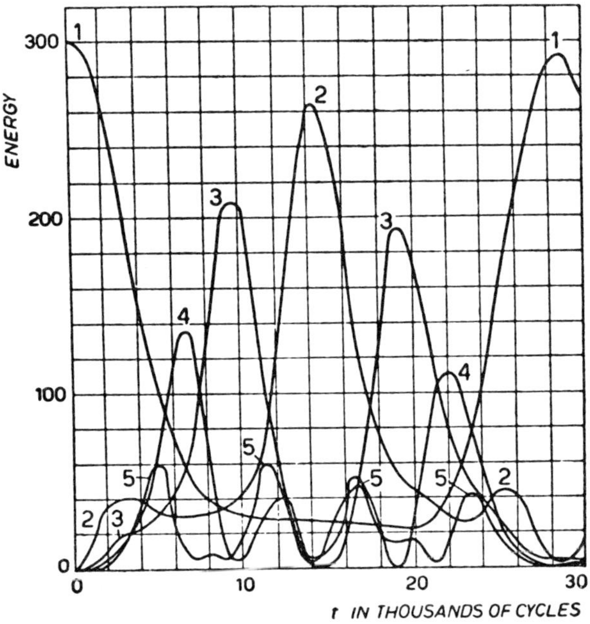

Figure 1-3: Reprinted from FPU (Fermi et al. 1955).

Over a large number of oscillations, the energy in each normal mode is seen to be almost periodic in time, with no loss of energy to higher modes as time increases. The precise explanation of this periodicity stimulated a deeper study of equations such as (1.3-1). Under certain approximations, equation (1.3-1) can be transformed into the KdV equation in the continuum limit.

In order to change terms like Q_n into continuous form we use the Maclaurin expansion

$$\left[\exp\left(a\frac{\partial}{\partial n}\right)\right]F(n) = F(n+a) \tag{1.3-4}$$

where n is now regarded as a continuous variable. Equation (1.3-1) becomes with $Q_n(t) = Q(n,t)$

$$\ddot{Q} = f\{[\exp\left(\frac{\partial}{\partial n}\right) - 1]Q\} - f\{[1 - \exp\left(-\frac{\partial}{\partial n}\right)]Q\} \tag{1.3-5}$$

Expanding the function f as a Maclaurin series we have

$$\ddot{Q} = f'(0)\left[\frac{\partial^2 Q}{\partial n^2} + \frac{1}{12}\frac{\partial^4 Q}{\partial n^4}\ldots\right] + \frac{1}{2!}f''(0)\left[2\frac{\partial Q}{\partial n}\cdot\frac{\partial^2 Q}{\partial n^2} + \ldots\right] + \frac{1}{3!}f'''(0)\left[3\left(\frac{\partial Q}{\partial n}\right)^2\frac{\partial^2 Q}{\partial n^2} + \ldots\right]. \tag{1.3-6}$$

It is difficult at first glance to determine which terms to keep and which to neglect in (1.3-6), and so we shall introduce the lattice spacing ℓ and define the variable $x = \ell n$ as the distance along the lattice. We shall also rescale Q with ℓ by defining $Q = \ell^r P$ where the magnitude of r is to be determined. This quantity will depend on whether we choose the quadratically or cubically nonlinear case. Equation (1.3-6) now becomes

$$\begin{aligned}\ddot{P} &= f'(0)\{\ell^2\frac{\partial^2 P}{\partial x^2} + \frac{\ell^4}{12}\frac{\partial^4 P}{\partial x^4}\ldots\} \\ &+\tfrac{1}{2}f''(0)\{2\ell^{r+3}(\partial P/\partial x)(\partial^2 P/\partial x^2)\ldots\} \\ &+\frac{1}{3!}f'''(0)\{3\ell^{2r+4}(\partial P/\partial x)^2(\partial^2 P/\partial x^2)\ldots.\}\end{aligned} \tag{1.3-7}$$

The choice of f as in (1.3-2) results in $f'(0) = \gamma$: $f''(0) = 2\alpha$: $f'''(0) = 0$. Therefore the choice of $r = 1$ balances the quadratically nonlinear terms with the fourth derivative term. Taking $u(x,t) = \partial P/\partial x$, we have to $O(\ell^4)$:

$$\frac{\partial^2}{\partial x^2}\{\frac{\ell^4}{12}\frac{\partial^2 u}{\partial x^2} + \alpha\ell^4 u + \ell^2\gamma u\} = \frac{\partial^2 u}{\partial t^2} \tag{1.3-8}$$

To $O(\ell^2)$, equation (1.3-8) is none other than the linear wave equation. Equation (1.3-8) has become known as the Boussinesq equation and describes waves which can travel both to the left and the right. Its solitary wave solution for the infinite chain is

$$u = (a^2/8\alpha)\,\text{sech}^2\tfrac{1}{2}(ax - \omega t + \delta) \tag{1.3-9}$$

$$\omega^2 = \gamma a^2\ell^2 + a^4\ell^4/12 \tag{1.3-10}$$

The Boussinesq equation has a very similar structure to the KdV equation, and we would expect it to reduce to the KdV equation for waves travelling in one direction only. This would enable us to follow an initial disturbance in one direction only down the chain. We employ a scaling transformation on x, t and u. Firstly we introduce a small parameter ε and choose new "space" and "time" variables

$$\xi = \varepsilon^p(x - ct) \qquad \tau = \varepsilon^q t \tag{1.3-11}$$

and expand u in powers of ε:

$$u = \varepsilon u^{(1)} + \varepsilon^2 u^{(2)} + \dots \tag{1.3-12}$$

In order to balance the 4th derivative, the time derivative, and the nonlinear terms to the same order in ε, we find that consistency can only be achieved if we take $c = \gamma^{\frac{1}{2}}\ell$, $p = 1/2$ and $q = 3/2$. The resulting equation for $u^{(1)}$ may be integrated with respect to ξ to obtain the KdV equation

$$\frac{\ell^3}{12}\frac{\partial^3 u^{(1)}}{\partial \xi^3} + 2\alpha\ell^3 u^{(1)}\frac{\partial u^{(1)}}{\partial \xi} + 2\gamma^{\frac{1}{2}}\frac{\partial u^{(1)}}{\partial \tau} = 0 \tag{1.3-13}$$

For the cubically nonlinear chain with $f(Q)$ defined as in (1.3-4), we have $f'(0) = \gamma$, $f''(0) = 0$, and $f'''(0) = 6\beta$. We need now a choice of $r = 0$ for a balance between the cubically nonlinear terms and the fourth derivatives. To $O(\ell^4)$ we obtain

$$\frac{\partial^2}{\partial x^2}\left\{ \frac{\ell^4}{12}\frac{\partial^2 u}{\partial x^2} + \beta\ell^4 u^3 + \gamma\ell^2 u \right\} = \frac{\partial^2 u}{\partial t^2} \tag{1.3-14}$$

The same scaling procedure as above now shows that a balance of terms can be achieved if $p = 1$, $q = 3$, and $c = \gamma^{\frac{1}{2}}\ell$. One integration on ξ yields the equation

$$\frac{\ell^3}{12}\frac{\partial^3 u^{(1)}}{\partial \xi^3} + 3\beta\ell^3 (u^{(1)})^2\frac{\partial u^{(1)}}{\partial \xi} + 2\gamma^{\frac{1}{2}}\frac{\partial u^{(1)}}{\partial \tau} = 0 \tag{1.3-15}$$

which has solutions which travel to the right only. Equation (1.3-15) has a cubic, not quadratic, nonlinearity, and is called the Modified KdV equation (MKDV). For an infinite chain, the solitary wave solution of (1.3-15) is

$$\begin{aligned} u^{(1)} &= a(6\beta)^{-\frac{1}{2}}\operatorname{sech}\{a\xi - \omega\tau + \delta\} \\ \omega &= (\ell^3 a^3)/24\gamma^{\frac{1}{2}} \end{aligned} \tag{1.3-16}$$

Physically, what we have achieved is a reduction of the quadratically and cubically nonlinear chains respectively, to two nonlinear partial differential equations, the KdV and MKdV equations on long space and time scales for small amplitude waves travelling unidirectionally. This method is known as reductive perturbation theory, and will be discussed in detail in Chapter 5 for various examples in which the KdV and MKdV equations arise. For the infinite chain the solitary wave profiles represent small amplitude waves carrying momentum down the chain in localised packets in the same way as in the shallow water case as observed by Scott Russell.

1.4 Solitons and the Work of Zabusky and Kruskal

Up until this point we have talked about solitary wave solutions which, using a very loose definition, are no more than waves which propagate without change of form and have some localised shape. Scott Russell's particular interest was in solitary waves in shallow water and as was pointed out in Section 1.1, it was for these waves that he coined the name. However by the above definition there are many equations which have solitary wave solutions. The word "soliton" first appears in the work of Zabusky and Kruskal (1965). Kruskal had been interested for some time in the FPU problem and particulary why recurrence occurred. He had been studying some the nonlinear chain motions mentioned in the last Section. Zabusky and Kruskal describe a numerical study of the KdV equation with a factor δ^2 multiplying the $\partial^3 u/\partial x^3$ term. They chose $\delta = 0.022$, periodic boundary conditions such that $u(x,t) = u(x+2,t)$, and a periodic initial condition $u(x,0) = \cos x$. They noted that initially the wave steepened in regions where it had a negative slope, a consequence of the domination of the nonlinearity over the third derivative due to the smallness of δ^2. As the wave steepens the $\delta^2 u_{xxx}$ term becomes important and balances the nonlinearity. They noted that on the left of the steepening, oscillations developed, each of which grew and finally achieved a steady amplitude and **became in shape almost identical to a solitary wave solution** (1.2-3), each with a different value of a. A remarkable property of these solitary waves was their interaction with one another as they passed through the cycles of evolution forced by the periodic boundary conditions. It was found that they passed through each other without change of form and with only a small change in their phase. This phase shift ensured that the initial state does not quite recur, but nevertheless it came close to recurrence, as in the FPU problem.

Because of their discovery that when two or more KdV solitary waves collide they do **not** break up and disperse, Zabusky and Kruskal called these solitary waves "SOLITONS". The ending "on" is Greek for particle, and the word soliton vividly illustrates the particle-like behaviour of these solitary waves.

This particle-like behaviour does not depend on periodic boundary conditions. We can numerically integrate the KdV equation on boundary conditions $u \to 0$ as $x \to \infty$ and take two well separated solitary wave solutions of the KdV equation as initial data with the taller and therefore faster wave situated on the left. As time increases throughout the numerical integration, the faster wave catches up and collides with the smaller. Figure 1-4 depicts this situation. Each wave profile is a plot of $u(x,t)$ against x for some fixed time t. The complete picture is a superposition of a number of these profiles for a sequence of values of t progressing in equal increments.

The two solitons are well separated before the collision, coalesce in the middle of the picture and separate again afterwards. The fact that the trajectories of each soliton in the (x,t) plane do not coincide before and after collision shows that each one suffers a phase shift. Note also that in the collision region the maximum is smaller than the amplitude of the larger soliton implying that there is no linear superposition in the centre.

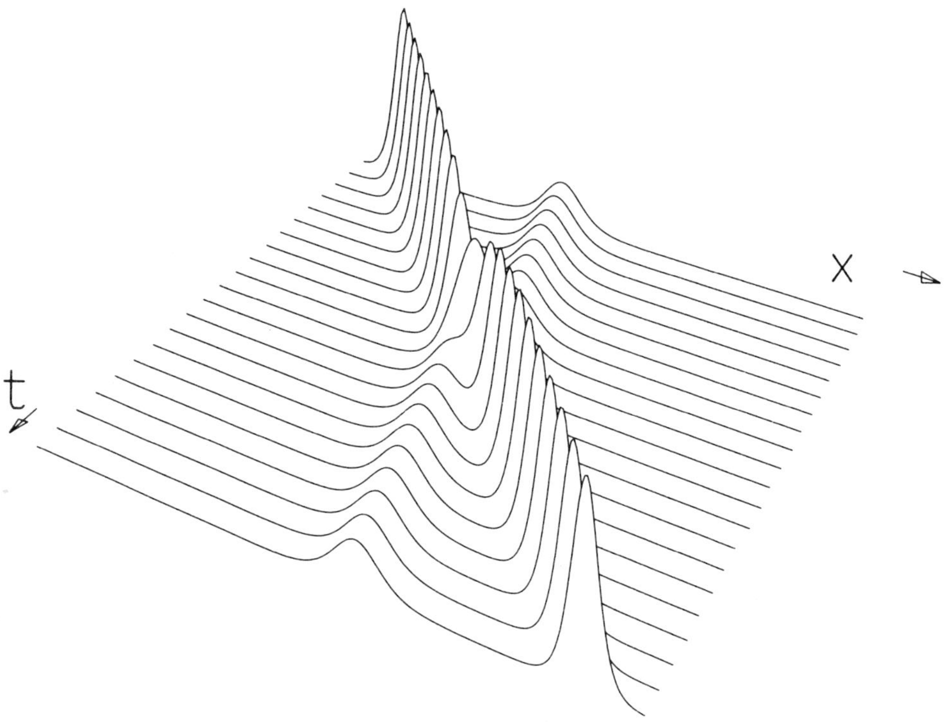

Figure 1-4: Collision of two KdV solitary waves

The fact that the solitary wave re-emerges from the collision with exactly the same shape is surprising since it might be thought that the strong nonlinearity during the collision process would break up the pulses. This property is important because it shows that energy can be progagated in localised stable "packets" without being dispersed. This behaviour is not a property of the KdV equation alone but is shared by the MKdV and Boussinesq equations also, together with many other equations.

It is important to realise that most nonlinear p.d.e.'s which have solitary wave solutions do not exhibit soliton behaviour. Some equations have solutions which approximate soliton behaviour in the sense that when two solitary waves collide they re-emerge after the collision with slight perturbations in shape, leaving a small amount of energy behind in the form of oscillations. Such behaviour is often referred to as soliton-like or the solitary wave collision is said to show inelastic soliton behaviour. However the nomenclature has become somewhat blurred and conventions differ from subject to subject. For instance in particle physics and solid state physics, the transparency of the waves to one another is not so important relative to other particle-like properties such as localisability and finite energy. Consequently some models are used in which the waves are not strictly solitons in the sense defined above (see Section 1.9 of this chapter) but nevertheless they are still called solitons by workers in those fields. However we shall stick to the stricter usage given above.

The wave shape is not an important part of the definition of a soliton.

The KdV soliton is a $(\text{sech})^2$ whereas the modified KdV soliton is a "sech" only - the word soliton refers more to the particle property than the shape. Obviously soliton behaviour has a deep mathematical significance and is more than just an attractive result. The task of Chapters 3,4 and 6 in particular will be to understand this problem and see what characterises those equations which have the soliton property from other equations.

The single soliton wave-form is easy to find by simple integration but this is of no avail if we want an **analytical** formula to describe the collision of two solitons which matches against the numerical solution of Fig. 1-4. Chapters 3 and 4 in particular will be concerned with not only finding analytic solutions which describe the collision of a whole train of solitons but also with the initial value problem which enables $u(x,t)$ to be found from a knowledge of $u(x,0)$ under certain circumstances. This problem is a difficult one and so we shall pre-empt some of the results of those Chapters by calculating a 2-soliton solution by alternative means. This enables us to investigate some of the properties of a soliton collision without becoming enmeshed too deeply in the more difficult mathematics which we will describe in Chapters 3 and 4. To find this solution we first construct a transformation which reduce the KdV equation to a homogeneous equation.

Consider the KdV equation in the form

$$u_{xxx} + 12uu_x + u_t = 0 \tag{1.4-1}$$

which, with $u = w_x$, reduces to

$$w_{xxx} + 6w_x^{\ 2} + w_t = 0 \tag{1.4-2}$$

The transformation

$$w = \frac{\partial}{\partial x} \log f \tag{1.4-3}$$

reduces the KdV equation to a homogeneous equation in $f(x,t)$ (Hirota 1971)

$$ff_{xxxx} - 4f_x f_{xxx} + 3f_{xx}^{\ 2} + ff_{xt} - f_x f_t = 0 \tag{1.4-4}$$

The transformation (1.4-3) is known as the Cole-Hopf transformation (Cole 1951, Hopf 1950) which we discuss further in Section 1.8. The constant outside the derivative in (1.4-3), which is unity in this case, can be adjusted to suit the constant outside the nonlinear term in (1.4-1) such that terms of degree 3 and 4 in f and its derivatives vanish. In trying to solve (1.4-4) we firstly take note of the fact that the single soliton wave form given in (1.2-3) is produced by taking

$$f = 1 + \exp\theta_1 \quad : \quad \theta_i = a_i x - a_i^{\ 3} t + \delta_i \tag{1.4-5}$$

This motivates us to look for a solution in the form

$$f = 1 + \sum_{n=1}^{N} \varepsilon^n f^{(n)} \tag{1.4-6}$$

The factor ε is a convenient expansion parameter. Substitution of this into (1.4-4) yields at different orders of

$$\varepsilon : f^{(1)}_{xxxx} + f^{(1)}_{xt} = 0$$

$$\varepsilon^2 : f^{(2)}_{xxxx} + f^{(2)}_{xt} = -[f^{(1)} f^{(1)}_{xxxx} - 4f^{(1)}_{x} f^{(1)}_{xxx} + 3f^{(1)2}_{xx} + f^{(1)} f^{(1)}_{xt} - f^{(1)}_{x} f^{(1)}_{t}]$$

$$\varepsilon^3 : f^{(3)}_{xxxx} + f^{(3)}_{xt} = -[f^{(1)} f^{(2)}_{xxxx} - 4f^{(1)}_{x} f^{(2)}_{xxx} - 4f^{(1)}_{xxx} f^{(2)}_{x} \tag{1.4-7}$$

$$+ 6f^{(1)}_{xx} f^{(2)}_{xx} + f^{(2)} f^{(1)}_{xxxx} + f^{(2)} f^{(1)}_{xt} - f^{(1)}_{x} f^{(2)}_{t} + f^{(1)} f^{(2)}_{xt}$$

$$- f^{(1)}_{t} f^{(2)}_{x}]$$

At $O(\varepsilon)$ we can easily recover, as an exact solution, the single exponential as in (1.4-5). However, since it is a linear equation we can introduce as many exponentials as we like, although here we shall restrict ourselves to two:

$$f^{(1)} = \exp\theta_1 + \exp\theta_2 \tag{1.4-8}$$

$$\theta_i = a_i x - a_i^3 t + \delta_i \tag{1.4-9}$$

This exact solution for $f^{(1)}$ can be substituted into the right-hand side of the equation at $O(\varepsilon^2)$ to give

$$f^{(2)}_{xxxx} + f^{(2)}_{xt} = 3a_1 a_2 (a_1 - a_2)^2 \exp(\theta_1 + \theta_2) \tag{1.4-10}$$

which integrates to

$$f^{(2)} = \left(\frac{a_1 - a_2}{a_1 + a_2}\right)^2 \exp(\theta_1 + \theta_2) \tag{1.4-11}$$

Most iterative processes result in an infinite series. In this case, the substitution of $f^{(1)}$ and $f^{(2)}$ into the right hand side of the $O(\varepsilon^3)$ equation (1.4-7) shows remarkably that this right-hand side is zero. This result reduces the $O(\varepsilon^3)$ equation to

$$f^{(3)}_{xxxx} + f^{(3)}_{xt} = 0 \tag{1.4-12}$$

We can now take the solution $f^{(3)} = 0$ and we can easily see that with this solution for $f^{(3)}$ all the subsequent $f^{(n)} = 0 (n \geqslant 3)$. This self-truncation of the series for f is absolutely crucial for obtaining an exact solution. The factors of ε can be absorbed into the phase of each θ and we have an **exact** 2-soliton solution:

$$u = \frac{\partial^2}{\partial x^2} \log f(x,t)$$

$$f = 1 + \exp(\theta_1) + \exp\theta_2 + A\exp(\theta_1 + \theta_2) \qquad (1.4\text{-}13)$$

$$A = \left(\frac{a_1 - a_2}{a_1 + a_2}\right)^2$$

The same process will work if we take a 3-parameter solution but the algebra becomes rather forbidding. The KdV equation was solved in this way for an N-soliton solution by Hirota (1971) and the method used above of reducing the equation in question to one or more bilinear equations has become known as Hirota's method (Hirota 1974).

The formula given in (1.4-13) can be analysed for the case when the two solitons are far apart. Let us take $a_1 > a_2 > 0$ and for the first soliton characterised by θ_1 we find that we are in the region of its maximum when $\theta_1 \simeq 0$; that is, in the region of the x-axis where $x \simeq a_1^2 t$. It follows that $\theta_2 \simeq (a_1^2 - a_2^2)t$ and so for

$$\theta_2 \simeq 0 : \theta_1 \to +\infty \qquad t \to +\infty$$
$$\theta_1 \to -\infty \qquad t \to -\infty. \qquad (1.4\text{-}14)$$

Similarly in the region of the second soliton we have

$$\theta_2 \approx 0 : \theta_1 \to \mp\infty \qquad t \to \pm\infty \qquad (1.4\text{-}15)$$

It is now useful to note, before we proceed, that any exponential combination can be factored out of f because of the form of the Cole-Hopf transformation in (1.4-13). Using this property and taking the limits worked out above in f, we find that asymptotically the solution becomes

$$u \sim \sum_{i=1}^{2} \frac{1}{4} a_i^2 \operatorname{sech}^2 \frac{1}{2}(\theta_i + \Delta_i^{\pm}) \qquad t \to \pm\infty \qquad (1.4\text{-}16)$$

where

$$\Delta_1^{(+)} = \log A \qquad \Delta_1^{(-)} = 0$$

$$\Delta_2^{(+)} = 0 \qquad \Delta_2^{(-)} = \log A$$

Hence

$$\Delta_1^{(+)} - \Delta_1^{(-)} = \log A < 0$$
$$\Delta_2^{(+)} - \Delta_2^{(-)} = -\log A > 0 \qquad (1.4\text{-}17)$$

The result implicit in (1.4-17) is that the **larger** soliton (a_1) is shifted **forwards** and the **smaller** (a_2) shifted **backwards** relative to the motion that would have been if no interaction had occurred.

The trajectories of the maxima of the solitons in Fig. 1-4 make this result plain. Whether we think of the soliton collision as a process in which the solitons pass through one another or whether they exchange identities is only a matter of interpretation. Finally we note that equation (1.4-17) shows that the total phase shift within the whole system is zero, indicating that the centre of mass remains fixed.

1.5 The Sine-Gordon Equation

One of the earliest models in field theory was the linear Klein-Gordon equation

$$\phi_{xx} - \phi_{tt} = m^2\phi \qquad (1.5\text{-}1)$$

Equation (1.5-1) derives from the Lagrangian density:

$$L = \frac{1}{2}(\phi_x^{\,2} - \phi_t^{\,2}) + \frac{1}{2}m^2\phi^2 \qquad (1.5\text{-}2)$$

Skyrme (1958) proposed a nonlinear field theory which, for the scalar case and in one space dimension, reduces in simple terms to a nonlinear extension of the Lagrangian density (1.5-2). The $\frac{1}{2}m^2\phi^2$ term is replaced by its simple periodic extension $\frac{1}{2}m^2(1-\cos\phi)$. The field equation now becomes

$$\phi_{xx} - \phi_{tt} = m^2\sin\phi \qquad (1.5\text{-}3)$$

This equation has subsequently become known as the sine-Gordon (SG) equation. Again we look for a wave of permanent profile. Motivated by the Lorenz invariance of (1.5-3) and using boundary conditions $\phi \to 0 \pmod{2\pi}$, two integrations give

$$\phi = 4\tan^{-1}\exp[m\gamma(x - vt) + \delta]$$
$$\gamma^2 = (1 - v^2)^{-1} \qquad (1.5\text{-}4)$$
$$\phi_x = 2m\gamma\,\mathrm{sech}[m\gamma(x - vt) + \delta]$$
$$\sin\tfrac{1}{2}\phi = \mathrm{sech}[m\gamma(x - vt) + \delta] \qquad (1.5\text{-}5)$$

Figure 1-5 shows a plot of $\phi(x,t)$ against x. This solution has been named a "kink" because it represents a twist in the variable ϕ which takes the system from one solution $\phi = 0$ to an adjacent solution with $\phi = 2\pi$. For this plot we have chosen the positive root for γ. The states $\phi = 0 \pmod{2\pi}$ are known as vacuum states as they are constant solutions of zero energy.

In 1962, Perring and Skyrme (1962) found, after numerical experimention, an analytical solution which represents the head-on collision of two kinks of equal but opposite velocity - the kink travelling to the left is called an

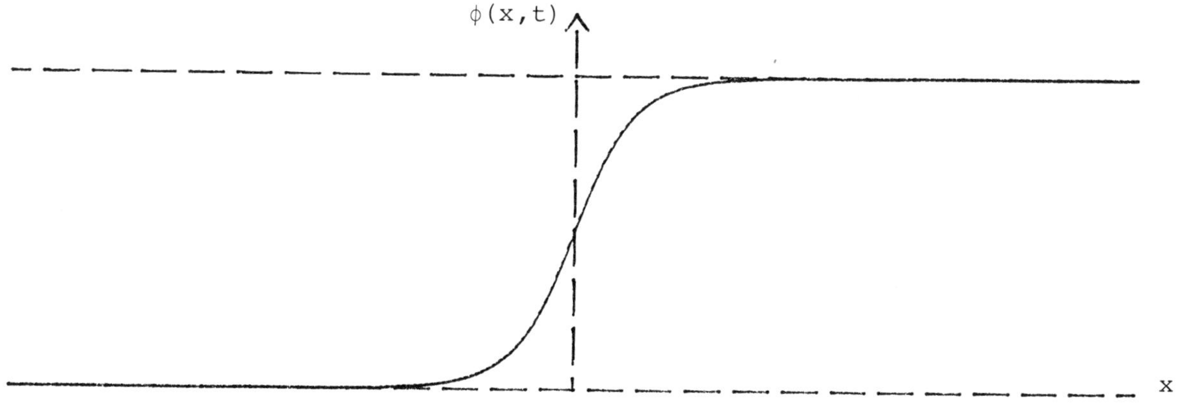

Figure 1-5: A single kink solution of the sine-Gordon equation

anti-kink because it has opposite twist, and corresponds to taking the negative root for γ. This result predated the work of Zabusky and Kruskal but, exactly as the latter found in 1965 when two KdV solitary waves were collided, the collision of these kinks did not produce annihilation or oscillations. This situation is shown in Fig. 1-6 in which a kink and an anti-kink collide and separate. The collision process moves the system from the adjacent vacuum states 0 and 2π to the adjacent states -2π and 0.

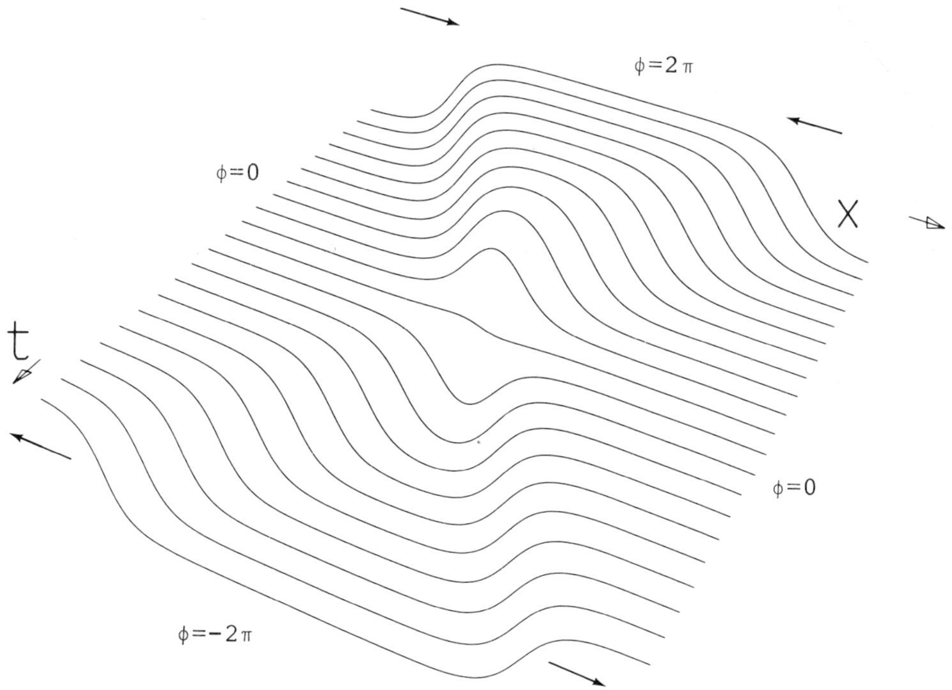

Figure 1-6: Kink-antikink collision in the sine-Gordon equation

In the collision the waves make a transition down to the next two adjacent vacuum states. The waves do not annihilate one another at the centre since ϕ_x is not zero at this point. These waves therefore have the soliton property in that waves of permanent profile when collided will retain their form. However the name "kink" has stuck to the wave shape (1.5-4) and so in future Chapters we shall continue to use this label.

One important point regarding the kink solution is that is has finite energy. This can easily be calculated by integrating the Hamiltonian density of the SG equation

$$H = \tfrac{1}{2}(\phi_x^{\,2} + \phi_t^{\,2}) + m^2(1 - \cos\phi) \tag{1.5-6}$$

over the real axis.The final result is $8m\gamma$, which displays the expected relativistic form for the kink mass. The constant solutions $\phi = \pi (\text{mod } 2\pi)$ cannot be considered as vacuum states since they have infinite energy.

The sine-Gordon equation has an even older history than the KdV equation. Of particular interest are the studies made by the Swedish mathematician Bäcklund in 1875. He was considering a problem in differential geometry involving the theory of surfaces of constant negative curvature (Eisenhart 1960, Forsyth 1959). One contribution that Bäcklund made was to show how it was possible to construct hierarchies of solutions, each of which could be constructed out of the previous ones. The transformations involved are called **Bäcklund Transformations** and have played a major role in the development of present day theory. For this reason we shall illustrate how to find a Bäcklund transformation for the sine-Gordon equation. The method for other equations will arise later, particularly in Chapters 3 and 6.

Consider a general nonlinear Klein-Gordon equation

$$\phi_{\xi\tau} = F(\phi) \tag{1.5-7}$$

expressed in characteristic co-ordinates, $\xi = (x - t)/2$, $\tau = (x + t)/2$.

Let us suppose it is possible to determine a relationship between two **independent** solutions of (1.5-7) in the form of a first order differential system. We take the two independent solutions of (1.5-7) to be $\phi = u + v$ and $\bar{\phi} = u - v$ and then consider the pair of equations

$$\begin{aligned} u_\xi &= f(v) \\ v_\tau &= g(u) \end{aligned} \tag{1.5-8}$$

in the variables $u(\xi,\tau)$ and $v(\xi,\tau)$. As yet we have not specified any functional form for F, g and f, the form of (1.5-8) being motivated by the desire to decouple the ξ and τ derivative parts of (1.5-7) into two separate equations. Differentiating equations (1.5-8) with respect to τ and ξ respectively results in the equations

$$\begin{aligned} u_{\xi\tau} &= g(u)f'(v) \\ v_{\xi\tau} &= g'(u)f(v) \end{aligned} \tag{1.5-9}$$

Although the equations (1.5-9) have a similar structure to (1.5-7), we still cannot take u and v separately as solutions of (1.5-7) because the right hand side of (1.5-9) is comprised of coupled functions of u and v. However a linear combination will suffice so we add and subtract the pair of equations to obtain

$$(u+v)_{\xi\tau} = g(u)f'(v) + g'(u)f(v)$$

$$(u-v)_{\xi\tau} = g(u)f'(v) - g'(u)f(v) \qquad (1.5\text{-}10)$$

Since ϕ and $\bar{\phi}$ are independent solutions of (1.5-7), we find

$$F(u+v) = g(u)f'(v) + g'(u)f(v)$$

$$F(u-v) = g(u)f'(v) - g'(u)f(v) \qquad (1.5\text{-}11)$$

Not F_{uv}!

Differentiating the first equation with respect to u and then v we find very easily that

($F_u = F_v$)

$$\frac{g''(u)}{g(u)} = \frac{f''(v)}{f(v)} = -\lambda \qquad (1.5\text{-}12)$$

The second equation gives the same result also. Since the left hand side of (1.5-12) is a function of u only and the right hand side a function of v only, then λ must be a constant. The functions f and g now satisfy the equations

$$g'' + \lambda g = 0 \qquad f'' + \lambda f = 0 \qquad (1.5\text{-}13)$$

The magnitude of λ is not so important as its sign so we take firstly $\lambda = 1$ and find that

$$g(u) = \beta \sin u \qquad f(v) = \alpha \sin v \qquad (1.5\text{-}14)$$

which in (1.5-11) immediately yields

$$F(\phi) = \sin\phi \qquad \beta = \frac{1}{\alpha} \qquad (1.5\text{-}15)$$

Since $u = \frac{1}{2}(\phi + \bar{\phi})$ and $v = \frac{1}{2}(\phi + \bar{\phi})$ the original equations (1.5-8) are now

$$\tfrac{1}{2}(\phi + \bar{\phi})_{\xi} = \alpha \sin\tfrac{1}{2}(\phi - \bar{\phi})$$

$$\tfrac{1}{2}(\phi - \bar{\phi})_{\tau} = \frac{1}{\alpha} \sin\tfrac{1}{2}(\phi + \bar{\phi}) \qquad (1.5\text{-}16)$$

which are a pair of equations relating two particular solutions ϕ and $\bar{\phi}$ of the sine-Gordon equation $\phi_{\tau\xi} = \sin\phi$. Equations (1.5-16) are the transformations derived by Bäcklund.

The choice of $\lambda = -1$ yields a hyperbolic-sinh in place of the sine functions and the choice $\lambda = 0$ produces the linear Klein-Gordon equation. Hence we have shown that under the restrictions of the choice made in (1.5-8), the sine (sinh)-Gordon equation is the only nonlinear Klein-Gordon equation with a Bäcklund Transformation, a result which makes it a very special equation!

Given one solution ϕ_o of the sine-Gordon equation, the correspondence (1.5-16) enables us to form a 2-parameter family of related solutions. The simplest solution of the sine-Gordon equation is $\phi_o = 0$ and so inserting this into (1.5-16) we find two very simple o.d.e.'s in $\bar{\phi}$, which integrate to

$$\bar{\phi} = 4\tan^{-1}\exp(a\xi + \frac{1}{a}\tau + \delta) \qquad \alpha = -a \tag{1.5-17}$$

This is the single kink solution (1.5-4) expressed in characteristic co-ordinates. From this single kink solution we can construct another solution from (1.5-16). To do this from the differential equations in (1.5-16) by direct integration is difficult and a simpler geometric way is available. Firstly we start with a solution ϕ_o and construct two solutions of the same type, ϕ_1 and ϕ_2, the only difference being that we take $\alpha = -a_1$ in ϕ_1 and $\alpha = -a_2$ in ϕ_2. A third solution can be obtained from these if we suppose that the Bäcklund transformations in in Fig. 1-7 commute

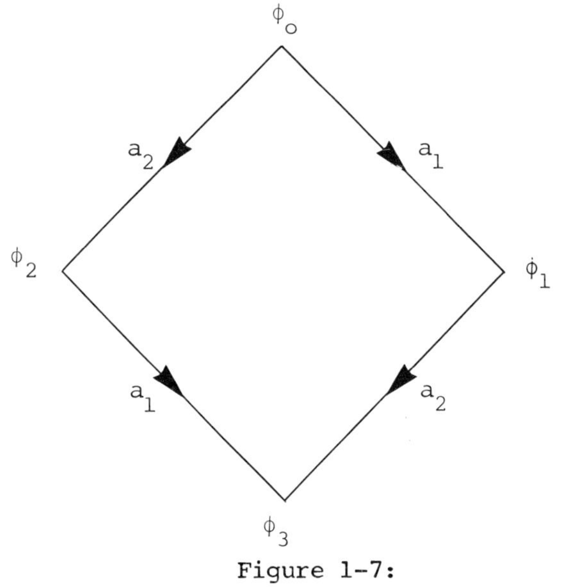

Figure 1-7:

Taking the sequence of pairs ϕ_o, ϕ_1; ϕ_1, ϕ_3; ϕ_2, ϕ_3 and ϕ_o, ϕ_2, and adding the first and third but subtracting the second and fourth, we find that the derivative terms in (1.5-16) cancel and we obtain

$$a_1\sin\frac{1}{4}(\phi_0 - \phi_1 + \phi_2 - \phi_3) = a_2\sin(\phi_0 + \phi_1 - \phi_2 - \phi_3) \tag{1.5-18}$$

which can be rearranged to give

$$\tan\frac{1}{4}(\phi_0 - \phi_3) = \left(\frac{a_1 + a_2}{a_2 - a_1}\right)\tan\frac{1}{4}(\phi_2 - \phi_1) \tag{1.5-19}$$

If we take $\phi_0 = 0$, then ϕ_1 and ϕ_2 will be the single kink solutions (1.5-17) with $a = -a_1$ and $a = -a_2$ respectively. The solution ϕ_3 becomes

$$\phi_3 = 4\tan^{-1}\left[\left(\frac{a_1 + a_2}{a_1 - a_2}\right)\frac{\exp\theta_1 - \exp\theta_2}{1 + \exp(\theta_1 + \theta_2)}\right] \tag{1.5-20}$$

This is a 2-kink solution which, when $a_1, a_2 < 0$, is shown in Fig. 1-6. It is the generalisation of Perring and Skyrme's formula which depicted a kink and anti-kink in head-on collision with opposite but equal velocities. The solution (1.5-20) or other multi-kink solutions can also be found by using the Cole-Hopf transformation method. This is described in the chapter notes.

1.6 The Nonlinear Schrödinger Equation

One further equation which deserves a special mention is the cubic nonlinear Schrödinger (NLS) equation

$$i\frac{\partial\Phi}{\partial t} + \frac{\partial^2\Phi}{\partial x^2} + \beta\Phi|\Phi|^2 = 0 \tag{1.6-1}$$

so-called because it has the structure of the quantum Schrödinger equation with $\beta|\Phi|^2$ as a potential. Φ is now a complex function and for this reason we would expect a travelling wave solution to have an oscillatory modulation. It is not difficult to show that a travelling wave solution of (1.6-1), subject to $\Phi \to 0$ as $|x| \to \infty$, is

$$\Phi = a\sqrt{2/\beta}\,\exp\{i[\tfrac{1}{2}bx - (\tfrac{1}{4}b^2 - a^2)]\}\ \operatorname{sech}[a(x - bt)] \tag{1.6-2}$$

where a and b are arbitrary constants. The sech-shaped wave acts as an envelope to the oscillatory part. The NLS equation plays an extremely important role in the theory of the evolution of slowly varying wave trains in stable weakly nonlinear systems and occurs in a whole series of physical situations including plasma physics and nonlinear optics.

It turns out that the NLS equation shares the same property as the KdV, MKdV and sine-Gordon equations; that is, its travelling wave envelope solutions are solitons. Chapter 8 will contain considerably more detail on solutions of the NLS equation and will contain derivations of some physical models in which the NLS equation plays an important role.

The form of (1.6-2) indicates that we might again seek as before, a transformation of the Cole-Hopf type in order to see if we can calulate a 2-soliton solution.

We define a function f such that the modulus squared of Φ can be written in the form

$$|\Phi|^2 = \frac{2}{3}\frac{\partial^2}{\partial x^2}\log f \tag{1.6-3}$$

and take $\Phi = g/f$ so that

$$gg^* = \frac{2}{\beta}(ff_{xx} - f_x^{\ 2}) \tag{1.6-4}$$

Substitution of both of these into (1.6-1) shows, as in previous examples, that terms higher than degree two vanish leaving

$$fg_{xx} - 2f_x g_x + gf_{xx} + i(g_t f - gf_t) = 0 \tag{1.6-5}$$

A one parameter solution of (1.6-5) is

$$\begin{aligned} g &= 2a(2/\beta)^{\frac{1}{2}}\exp\theta \\ f &= 1 + \exp(\theta+\theta^*) \\ \theta_R &= ax - abt \\ \theta_I &= \tfrac{1}{2}bx + (a^2 - \tfrac{1}{4}b^2)t \end{aligned} \tag{1.6-6}$$

which agrees with (1.6-2). Introduction of a 2-parameter solution into the equations produced by using a similar iteration procedure to that described for the KdV in Section 1.4 shows that self-truncation occurs and an exact 2-parameter solution is obtained. This is left as an exercise to the reader at the end of the chapter.

1.7 Some Basic Principles of Linear Wave Propagation

In the preceding six sections we have concentrated on a phenomenon which occurs during some nonlinear interactions. The properties of the hyperbolic secant shaped solitary wave are very different in character to those of linear harmonic waves. These differences arise not because it is a non-oscillatory wave motion, but also because its velocity is **amplitude dependent**. This contrasts very strongly with the situation for linear waves for which the speed is always independent of the amplitude. We expect that the reader is familiar with linear waves, but we shall summarize some of the elementary principles for reasons of completeness.

For a **linear** system, the most elementary linear wave is the harmonic wave

$$\phi(x,t) = A\exp[i(kx - \omega t)] \tag{1.7-1}$$

where k is called the wave number and related to the wavelength by $k = 2\pi/\lambda$ and A is the amplitude which is k-independent. The requirement that $\phi(x,t)$ satisfies a linear wave equation results in a functional relationship between k and ω

$$\mathcal{L}(k,\omega) = 0 \tag{1.7-2}$$

which is called the dispersion relation. This is the most important characteristic of any linear system since it mirrors the form of the linear differential equation of which ϕ is a solution. For instance, the linear Klein-Gordon equation

$$\phi_{xx} - \phi_{tt} = m^2\phi \tag{1.7-3}$$

has a dispersion relation

$$\omega^2 = m^2 + k^2 \tag{1.7-4}$$

The phase velocity is defined as

$$c_p = \omega/k \tag{1.7-5}$$

and describes how a surface of constant phase moves. Much more important is the concept of group velocity which is defined as

$$c_g = d\omega/dk \tag{1.7-6}$$

and which gives a measure of how fast the bulk of the wave propagates. For a given linear system there may be several group velocities corresponding to different solutions or modes of (1.7-2). Here we are in one dimension only so k is a scalar but in several space dimensions, the wave number is now a vector $\mathbf{k}$. Under such circumstances the phase velocity is still a scalar: $c_p=\omega/k$ whereas the group velocity becomes a vector: $c_g=d\omega/dk$

The word **dispersion** means that for purely real ω, waves of different wave number k will have different phase and group velocities and hence the components of the wave will spread or disperse during the wave propagation. This requires that c_p and c_g are k dependent. This is not true in the case of (1.7-3) when $m = 0$, but is true when $m \neq 0$. All these ideas presume that the waves are propagating in an unbounded, uniform medium. Generally, if $\omega(k)$ is complex

$$\omega = \omega_R + i\omega_I \tag{1.7-7}$$

then the harmonic wave solutions (1.7-1) will grow exponentially and become unstable if $\omega_I>0$ and decay exponentially if $\omega_I<0$. In the latter case the equation is said to be **dissipative** in character because the wave amplitude is decaying as $\exp(-|\omega_I|t)$ as $t\to\infty$. A slightly more complicated situation occurs if the linear equation which models a physical system has in it a parameter (such as ,for example, the Rayleigh number) which is adjustable externally

$$\mathcal{L}\left(\frac{\partial}{\partial t};\frac{\partial}{\partial x};u\right)\phi = 0 \tag{1.7-8}$$

where $\mathcal{L}$ is a linear operator and the dispersion relation is $\mathcal{L}(-i\omega,ik,u)=0$. If $\omega_I>0$ for some values of u but $\omega_I<0$ for other values, then there is a relationship between u and k

$$u = u(k) \tag{1.7-9}$$

resulting from the requirement that $\omega_I=0$ corresponding to marginal stability. For values of u on one side of the curve (1.7-9) $\omega_I>0$ and the waves will be unstable and on the other side where $\omega_I<0$ the waves will be stable.

The quantity ω_I is only zero **on the curve** and the system is said to be **neutrally stable.** The curve (1.7-9) is called a neutral curve because it is a boundary between stable and unstable regions. Many physical systems do not possess a neutral curve but there are a few of importance that do and we shall be returning to this concept in Chapter 9. In that chapter in particular we will consider the case when complex conjugate pairs of roots appear in the dispersion relation.

The above ideas and definitions can be summarised in an example. Consider the equation

$$\phi_t = \delta\phi_{xx} \tag{1.7-10}$$

If $\delta=-i$ then (1.7-10) is

$$i\phi_t = \phi_{xx} \tag{1.7-11}$$

which is just the linearised version of the Schrödinger equation mentioned in the previous Section. This is a purely dispersive equation since $\omega = -k^2$; $c_p = -k$ and $c_g=-2k$. However if δ is real and positive, then (1.7-10) is the heat or diffusion equation which is purely dissipative since $\omega = -i\delta k^2$ The real part of ω is now zero and hence no dispersion occurs and waves decay as $\exp(-\delta k^2 t)$ when $t\to\infty$. Equation (1.7-10) is a suitable example which we can use to discuss the **initial value problem.**

Given some initial data $\phi(x,0) = f(x)$ we want to calculate $\phi(x,t)$ for all t. Even though the initial data $\phi(x,0)$ may not have harmonic form, as in (1.7-1), it may be represented by a Fourier integral

$$\phi(x,0) = \frac{1}{\sqrt{2\pi}}\int_{-\infty}^{\infty} A(k)e^{ikx}dx \tag{1.7-12}$$

where

$$A(k) = \frac{1}{\sqrt{2\pi}}\int_{-\infty}^{\infty} \phi(x,0)e^{-ikx}dx \tag{1.7-13}$$

For a general linear system, $\omega=\omega(k)$ as indicated above and so a solution for all $t \geq 0$ is

$$\phi(x,t) = \frac{1}{\sqrt{2\pi}}\int_{-\infty}^{\infty} A(k)e^{i[kx - \omega(k)t]}dk \tag{1.7-14}$$

Since $A(k)$ is known at least in principle from the initial data $\phi(x,0)$ as a result of equation (1.7-13), the solution for all t can be found in (1.7-14). In practice, neither of the above integrals can be evaluated in terms of elementary functions even when ω is a polynomial in k, except for a few special cases. However the integral representation (1.7-14) can form the basis for the approximation method of steepest descents.

The equation $\phi_t=\delta\phi_{xx}$ is one which will yield to some analysis and so we shall use this as an example. In this case, $\phi(x,t)$ is

$$\phi(x,t) = \frac{1}{\sqrt{2\pi}}\int_{-\infty}^{\infty} dk\left[\int_{-\infty}^{\infty}\phi(\xi,0)e^{-ik\xi}d\xi\right]\exp[i(kx+i\delta k^2t)] \tag{1.7-15}$$

On the assumption that we can reverse the order of integration, (1.7-15) becomes

$$\phi(x,t) = \frac{1}{2\pi}\int_{-\infty}^{\infty}\left[\int_{-\infty}^{\infty}\exp\{i[(x-\xi)k-\delta k^2t]\}dk\right]\phi(\xi,0)d\xi \tag{1.7-16}$$

Writing $\eta = x-\xi$, (1.7-16) indicates that two separate integrals need to be evaluated

$$I_1 = \int_{-\infty}^{\infty}\cos(k\eta)e^{-\delta k^2t}dk \tag{1.7-17}$$

$$I_2 = \int_{-\infty}^{\infty}\sin(k\eta)e^{-\delta k^2t}dk \tag{1.7-18}$$

The integrand of I_2 is odd and hence $I_2 = 0$. I_1 can be evaluated by a simple device. Firstly, let us define $\lambda^2=\delta k^2t$ and $a=\eta/\sqrt{\delta t}$ so that

$$I_1(a) = \frac{1}{\sqrt{\delta t}}\int_{-\infty}^{\infty}\cos(a\lambda)e^{-\lambda^2}d\lambda \tag{1.7-19}$$

Differentiating (1.7-19) with respect to a and integrating by parts with respect to λ gives the simple relation

$$\frac{dI_1}{du} = -\frac{1}{2}aI_1 \qquad I_1(0) = \sqrt{\frac{\pi}{\delta t}} \tag{1.7-20}$$

and hence

$$I_1 = \sqrt{\frac{\pi}{\delta t}}\exp(-\tfrac{1}{4}a^2) \tag{1.7-21}$$

where we have used the result

$$\int_{-\infty}^{\infty}e^{-\lambda^2}d\lambda = \sqrt{\pi} \tag{1.7-22}$$

Finally, the result for $\phi(x,t)$ is now

$$\phi(x,t) = (4\pi\delta t)^{-\frac{1}{2}}\int_{-\infty}^{\infty}\phi(\xi,0)\exp\left[-\frac{(x-\xi)^2}{4\delta t}\right]d\xi \tag{1.7-23}$$

For general initial data, (1.7-23) can only be evaluated by approximate means, the method of steepest descents (Jeffrey and Jeffrey 1946) being particularly suitable when δ is real and small. An account of this technique can be found in the book by Whitham (1974). One class of initial data for which the integral (1.7-23) can be evaluated is the Gaussian function: $\phi(x,0) = A\exp(-\alpha x^2)$. The result is

$$\phi(x,t) = \frac{A}{(1+4\pi\alpha\delta t)^{\frac{1}{2}}}\exp\left[\frac{-\alpha x^2}{1+4\pi\delta t}\right] \qquad (1.7\text{-}24)$$

If δ is real and positive then $\phi \to 0$ as $t \to \infty$ as we expect for a dissipative equation. A plot of this is given in Fig. 1-8.

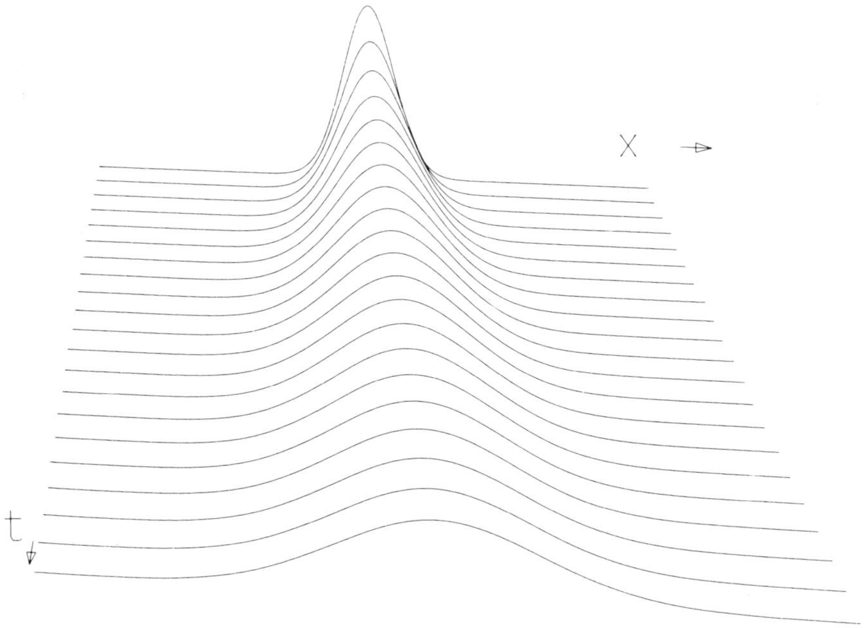

Figure 1-8: Gaussian data decaying (δ=1)

If however δ =-i , then ϕ is now an oscillatory function and Fig. 1-9 shows how the Gaussian initial data starts to disperse. These figures shows the fundamental difference between dissipation and dispersion.

The solitary wave is however a manifestation of both dispersive effects and nonlinear effects. There is a connecting link between nonlinear solitary waves and linear harmonic waves for the KdV equation, if one looks for solutions which are nonlinear in the sense that they have an amplitude dependent velocity but are nevertheless oscillatory. Such a class of solutions can be found if the three integrations of (1.2-2) are performed without imposing the boundary conditions which lead to the soliton solution. Elliptic integrals need to be introduced (Byrd and Friedman 1971) and the final result is expressed in terms of the Jacobian elliptic cnoidal function:

$$u = \frac{1}{4}a^2\text{cn}^2[\frac{1}{2}\theta, K] \qquad (1.7\text{-}25)$$

The modulus of the elliptic function K depends on the value of u and its derivatives at infinity. In the limit when u and its derivatives go to zero at infinity, $K \to 1$:

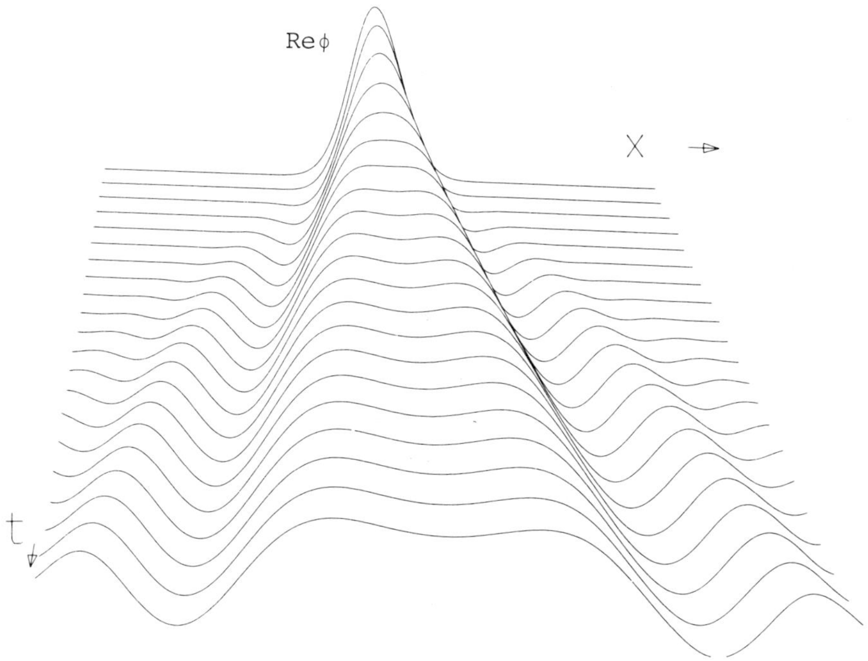

Figure 1-9: Gaussian data dispersing ($\delta = -i$)

$$\mathrm{cn}(\tfrac{1}{2}\,\theta;K) \longrightarrow \mathrm{sech}\,\tfrac{1}{2}\,\theta \qquad K \longrightarrow 1 \tag{1.7-26}$$

For small amplitude waves, when the linearised version of the KdV equation is appropriate, $K \to 0$ and in this limit

$$\mathrm{cn}(\tfrac{1}{2}\,\theta;K) \longrightarrow \cos\tfrac{1}{2}\,\theta \qquad K \longrightarrow 0 \tag{1.7-27}$$

These nonlinear oscillatory solutions are called cnoidal waves and form a bridge between purely linear oscillations on the one hand and pure solitary waves motions on the other.

1.8 Some Elementary Ideas in Nonlinear Wave Propagation

Using the ideas of the previous Section, it can now be seen that the linearised version of the KdV equation written in the form

$$u_t + u_x + u_{xxx} = 0 \tag{1.8-1}$$

is a purely dispersive equation with $\omega = k - k^3$. The u_{xxx} term introduces dispersive effects into the dispersionless equation $u_t + u_x = 0$. Initial data, prescribed for (1.8-1) will therefore disperse or spread as described in Section 1.7.

Let us now concentrate our attention on the nonlinear term uu_x which occurs in the KdV equation and ignore the dispersion term. Consider the problem of the evolution of some initial data

$$u(x,0) = f(x) \tag{1.8-2}$$

prescribed for the dispersionless nonlinear equation

$$u_t + (u+1)u_x = 0 \tag{1.8-3}$$

An equation with this structure would seem difficult to solve but it can be recast into an alternative form by analogy with its linearised problem. The equation $u_t + u_o u_x = 0$ has a solution $u = u(x-u_o t)$ which propagates at speed u_o. If we specify the intial data $u(x,0) = f(x)$, the complete solution is $u(x,t) = f(x-u_o t)$. Although it seems that we are pushing the analogy too far when nonlinearity is included, we see that $(u+1)$ in (1.8-3) is analogous to the wave speed u_o. In parallel with the linear case this would suggest that (1.8-3) might be expressible as the functional equation

$$u = f[x-(u+1)t] \tag{1.8-4}$$

To show that this is possible we can check this by direct calculation.

$$u_x = (1-u_x t)f' \tag{1.8-5}$$

and

$$u_t = -[tu_t + u + 1]f' \tag{1.8-6}$$

giving

$$[u_t + (u+1)u_x](1+tf') = 0 \tag{1.8-7}$$

which is satisfied by solutions of (1.8-3). Surprisingly, (1.8-4) **is** an alternative formulation of (1.8-3) and is furthermore an **implicit** equation in u. The solution (1.8-4) can be derived logically using the method of characteristics. Details are available in Whitham (1974). Using the same ideas, it is easily shown that with initial data $u(x,0) = f(x)$ for the more general nonlinear equation $u_t + c(u)u_x = 0$, the solution can be expressed as $u = f[x-c(u)t]$.

It is clear that for most given functions $f(x)$, (1.8-4) will not be any easier to solve than (1.8-3). There is one choice of initial function which enables us to solve (1.8-4). This is piecewise linear initial data in the form of a triangle

$$f(x) = \begin{cases} u_0 x & 0 < x < 1 \\ u_0(2-x) & 1 < x < 2 \\ 0 & x < 0;\ x > 2 \end{cases} \tag{1.8-8}$$

This triangular shape can be thought of as an approximation to more general humped shaped initial data without running into the problems incurred in solving a transcendental equation and therefore serves as an aid to our intuition concerning the general problem. The functional equation (1.8-4) is now

$$u(\eta,t) = \begin{cases} u_0\,(\eta - ut) & 0 \leqslant \eta - ut \leqslant 1 \\ u_0\,(2 - \eta + ut) & 1 \leqslant \eta - ut \leqslant 2 \\ 0 & \text{otherwise} \end{cases} \tag{1.8-9}$$

where we have simplified (1.8-9) a little by changing the frame of reference by the substitution $x - t = \eta$. Since these equations are intrinsically linear in u we can solve for u to obtain

$$u = \begin{bmatrix} \dfrac{u_0\eta}{1+u_0t} \\ \dfrac{u_0(2-\eta)}{1-u_0t} \\ 0 \end{bmatrix} \quad u_\eta = \begin{bmatrix} \dfrac{u_0}{1+u_0t} \\ \dfrac{-u_0}{1-u_0t} \\ 0 \end{bmatrix} \quad \begin{matrix} 0 \leqslant \eta - ut \leqslant 1 \\ 1 < \eta - ut \leqslant 2 \\ \text{otherwise} \end{matrix} \tag{1.8-10}$$

The expressions for u_η are written in the second column in (1.8-10) and from this it is easy to see that the gradient of the left hand part of the triangle is slowly decreasing as t increases while the gradient of the right hand part of the triangle changes from negative to positive at $t = 1/u_o$, at which point it becomes vertical. The triangle turns over with the bottom right hand angle becoming obtuse. This situation is described by the series of diagrams in Fig. 1-10.

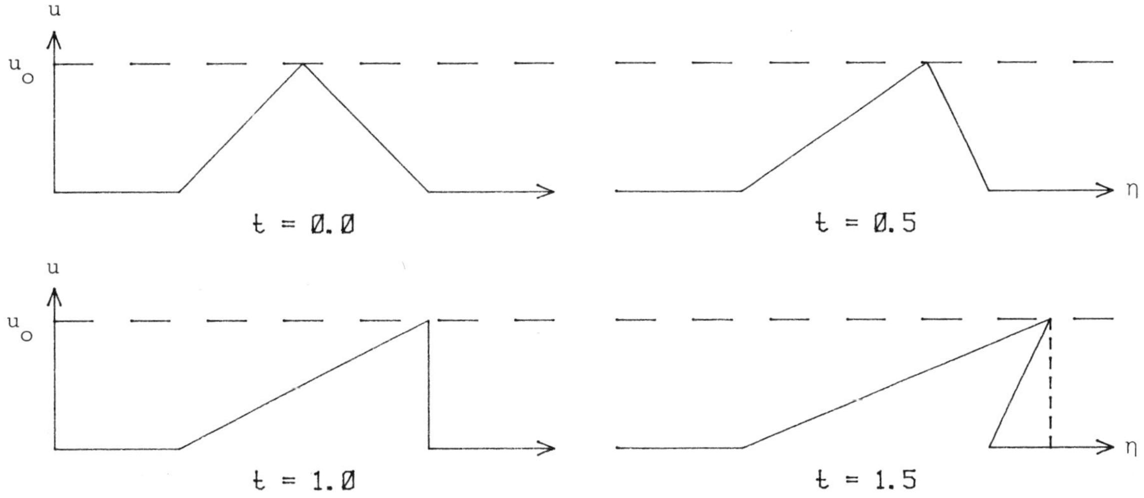

Figure 1-10: The evolution of triangular initial data

An intuitive understanding of why the wave behaves as it does can be seen from the form of (1.8-4) in which we thought of the wave having a velocity of (u + 1). Larger values of u travel faster than smaller values and so the apex of the triangle overtakes all the lower points. Because the wave becomes multi-valued after a certain time ($t = 1/u_o$) we say that the wave breaks and $t = 1/u_0$ is called the minimum breaking time. In the general case, hump-like initial data will become multi-valued and will start to curl very much like a large wave on a beach. This is a very complicated process and is one which it is impossible to describe in a purely analytical fashion as we have done for the simple example of the triangle. A numerical study of how a wave breaks in this fashion has been made by Longuet-Higggins and Cokelet (1976).

If $u(x,t)$ denotes a wave amplitude, then the appearance of multi-valued solutions after the minimum breaking time would be perfectly physical. Equations similar to (1.8-4) appear in inviscid gas dynamics (see execise 5), and in that problem u is a density. Obviously, a multi-valued density is unphysical unless the multi-valuedness can be interpreted as deriving from a discontinuity in the solution for $t \geq u_0^{-1}$. In order to explain this problem we shall, for simplicity, consider equation (1.8-4) in a different frame of reference: $\eta \equiv x' = x - t$; $t' = t$, and from now on we shall drop the primes. This is no more than a Galilean transformation which changes (1.8-4) into the form $u_t + uu_x = 0$. To interpret the multi-valuedness in this way, we replace a multi-valued wave by a discontinuity in (1.8-4). This is shown graphically in Fig. 1-11.

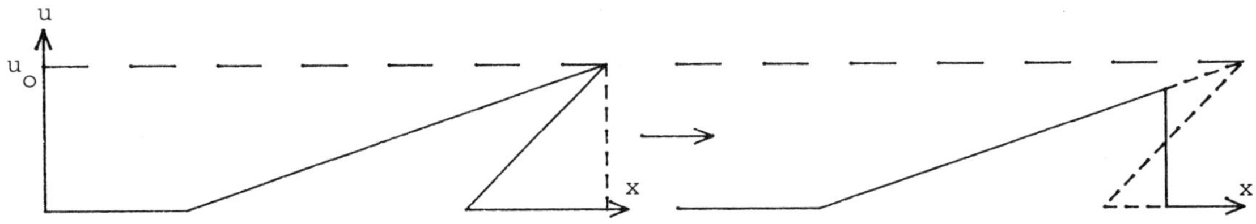

Figure 1-11: Introduction of a discontinuity at $x = x_s$ when $t > 1/u_0$

The problem is to determine where to insert the discontinuity, and to calculate how it evolves in time. We use the fact that the area under the wave must be constant for all time if u is a density since the total mass must be conserved.

The area under the initial data (1.8-10) is u_0 and the area under the triangle with the discontinuity is $x_s h/2$, where h is the height. By similar triangles

$$\frac{h}{x_s} = \frac{u_0}{1 + u_0 t} \tag{1.8-11}$$

giving

$$x_s^2 = 2(1 + u_0 t) \tag{1.8-12}$$

$$h = \sqrt{2u_0}(1 + u_0 t)^{-\frac{1}{2}} \tag{1.8-13}$$

Obviously, x_s and $h \to 0$ as $t \to \infty$. The characteristic plane (Fig. 1-12) shows how the solutions behave before and after the minimum breaking time, where initial data given in (1.8-8) has been used. When the characteristic lines cross, they coalesce into the curve which is the path of the discontinuity.

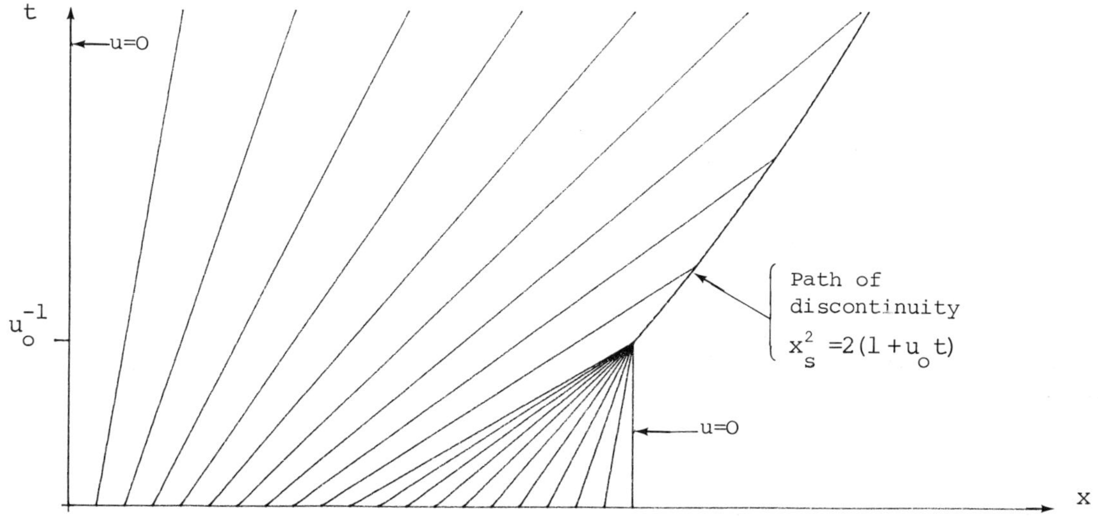

Figure 1-12: Characteristic plane for $u_t + uu_x = 0$

The characteristic lines are given by

$$\frac{dt}{dx} = \frac{1}{u} \qquad \text{on constant } u(x,t) \tag{1.8-14}$$

The line with the minimum gradient occurs at $u = u_o$ causing the lines in the middle of the figure to lean into and meet the lines associated with the smaller values of u which occur when x is large.

The conclusion of these calculations is that if no dispersive term is present the discontinuous (shock) behaviour or multi-valuedness can occur after a finite time. The effect of including the dispersive u_{xxx} term prevents any shock from forming. Zabusky and Kruskal (1965) discussed this problem. They included a $\delta^2 u_{xxx}$ term with $\delta = 0.022$ and found that although steepening occured initially, the wave never turned over. Despite the small values of δ^2, the third derivative is very large in the steep region and shocks can never form. The conclusion is that dispersion counteracts the tendency of nonlinearity to form discontinuities.

The inclusion of dissipation and not dispersion in such systems usually results in the introduction of a δu_{xx} term instead of a u_{xxx} term. This results in a type of nonlinear diffusion equation

$$u_t + uu_x = \delta u_{xx} \tag{1.8-15}$$

which is known as Burgers equation. The quantity δ is the diffusion coefficient and is a real positive constant. Burgers equation is a famous equation for several reasons. Firstly it includes nonlinearity and **dissipation** together in the simplest possible way and may be thought of as a nonlinear version of the heat equation. Secondly, and rather remarkably, it can be **linearised** by a transformation known as the Cole-Hopf transformation, which we have met already in Sections 1.3 and 1.4:

$$u = \alpha \frac{\partial}{\partial x} \log F \tag{1.8-16}$$

The constant α is, as yet, an unknown constant which we will fix below. Using this transformation in (1.8-15) yields

$$\frac{F_t}{F} + \frac{1}{2}\alpha \frac{F_x^2}{F^2} - \delta\left(\frac{F_{xx}}{F} - \frac{F_x^2}{F^2}\right) = c(t) \tag{1.8-17}$$

where $c(t)$ is an integration constant. Fixing $\alpha=-2\delta$ to remove the $(F_x/F)^2$ term we have

$$F_t - F_c(t) = \delta F_{xx} \tag{1.8-18}$$

The function $c(t)$ can be absorbed into F and we shall set it to zero, leaving

$$F_t = \delta F_{xx} \tag{1.8-19}$$

which is the diffusion or heat equation, the initial value problem for which we solved in the previous Section!

The Cole-Hopf transformation in (1.8-16) is exactly that transformation which reduced the KdV equation into a homogeneous equation of degree two (see equation (1.4-4)). One can think of the KdV and Burgers equations as sister equations in some respects. The KdV is the simplest **dispersive** extension of the nonlinear equation $u_t + uu_x=0$ and Burgers is the simplest **dissipative** extension. Obviously they will occur in different physical circumstances. The KdV occurs in dispersive, non-decaying systems while Burgers occurs in viscosity dominated ones.

The simplest solution of Burgers equation which shows decay is the Taylor shock profile solution

$$u = a\delta\{1 - \tanh\tfrac{1}{2}(ax - \delta a^2 t)\} \tag{1.8-20}$$

As $x \to \infty$, $u \to 0$ and as $x \to -\infty$ then $u \to 2a\delta$. A sketch of (1.8-20) for some $t > 0$ is given in Fig. 1-13.

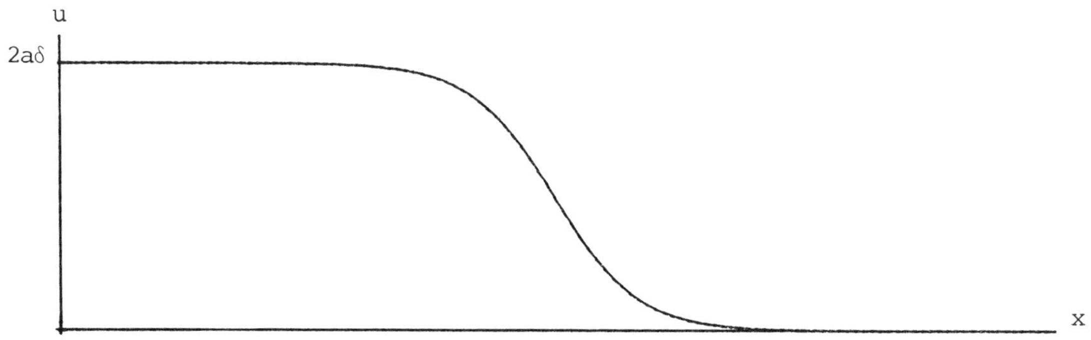

Figure 1-13: The Taylor shock profile solution (1.8-20)

Although this is generally referred to as a "shock" profile, it is a perfectly continuous solution. However, in the limit $\delta \to 0$, the shock front steepens. As in the KdV case, even a term with very small δ will always prevent the wave from breaking because the second derivative is large in the shock region.

1.9 Equations for which Soliton Behaviour Does Not Occur

In order to provide some balance in this Chapter we look at what happens to dispersive nonlinear equations which do **not** display exact soliton behaviour but which do have localised solitary wave solutions. For equations which are amenable to simple numerical integration, a pair of these waves can be used as initial data and collided in some region of the x-axis. The various necessary numerical schemes are discussed in Chapter 10; here we give some brief results in order to fulfil the purpose of this Section.

The first example is the so-called ϕ^4 equation of particle physics:

$$\phi_{xx} - \phi_{tt} = \lambda\phi^3 - m^2\phi \qquad \lambda > 0 \tag{1.9-1}$$

Unlike the sine-Gordon equation, the ϕ^4 equation does not have an infinite number of stable vacuum states but only two; $-m/\lambda^{\frac{1}{2}}$ and $m/\lambda^{\frac{1}{2}}$. A simple travelling wave solution of (1.9-1) is easily found

$$\phi = \pm \tanh[\frac{1}{\sqrt{2}}\gamma(x - vt) + \delta] \tag{1.9-2}$$

The shape of this wave (positive sign) is very like the sine-Gordon kink of Fig. 1-5 in that it takes the system from $\phi=-m/\lambda^{\frac{1}{2}}$ at $x=-\infty$ to $\phi= m/\lambda^{\frac{1}{2}}$ at $x = \infty$. We take a tanh-kink solution (1.9-2) travelling to the right and a tanh-antikink travelling to the left as initial conditions and let them collide in the middle. Remember that there are no other states to flip up to, as in the sine-Gordon case. Figures 1-14 and 1-15 show the result. In Fig. 1-14 the collision energy is quite high. The two kinks bounce out again keeping their structure to some degree but oscillations are also produced which can be thought of as radiation. The collision energy in Fig. 1-15 is much lower and the kinks do not bounce out again but form some sort of a radiating bound state. Equation (1.9-1) will be discussed in greater detail in Chapter 10. Note that these ϕ^4 kinks are **not** solitons in our strict definition, in contrast to the sine-Gordon kinks. This is not surprising because we showed in Section 1.5 that, under the restrictions of the given transformations (equations (1.5-8)), the sine-Gordon was the **only** nonlinear Klein Gordon equation to posses a Bäcklund transformation which generates a whole ladder of exact multi-kink solutions.

The second example is that of a set of equations in nonlinear optics (see Chapter 9) called the Maxwell-Bloch equations:

$$\begin{aligned} E_{xx} - E_{tt} &= -\alpha P_t \\ P_{tt} + \mu^2 P &= (EN)_t \\ N_t &= -EP \end{aligned} \tag{1.9-3}$$

A solitary wave solution of (1.9-2) for boundary conditions $E, P \to 0$; $N \to -1$ as $|x| \to \infty$ has a hyperbolic secant shape like the MkdV equation

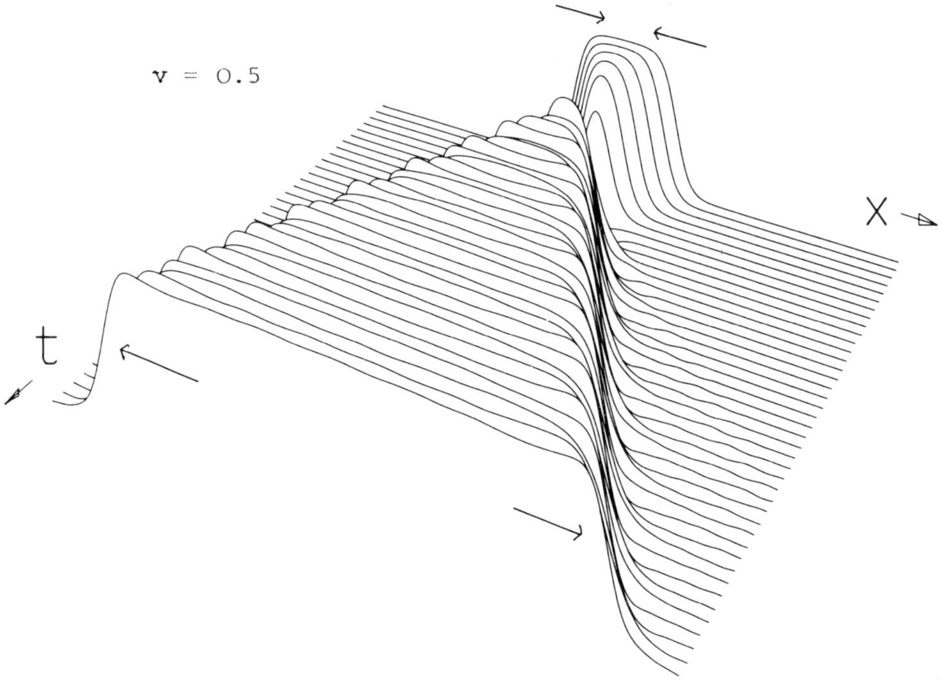

Figure 1-14: High-energy collision of a ϕ^4 kink and antikink

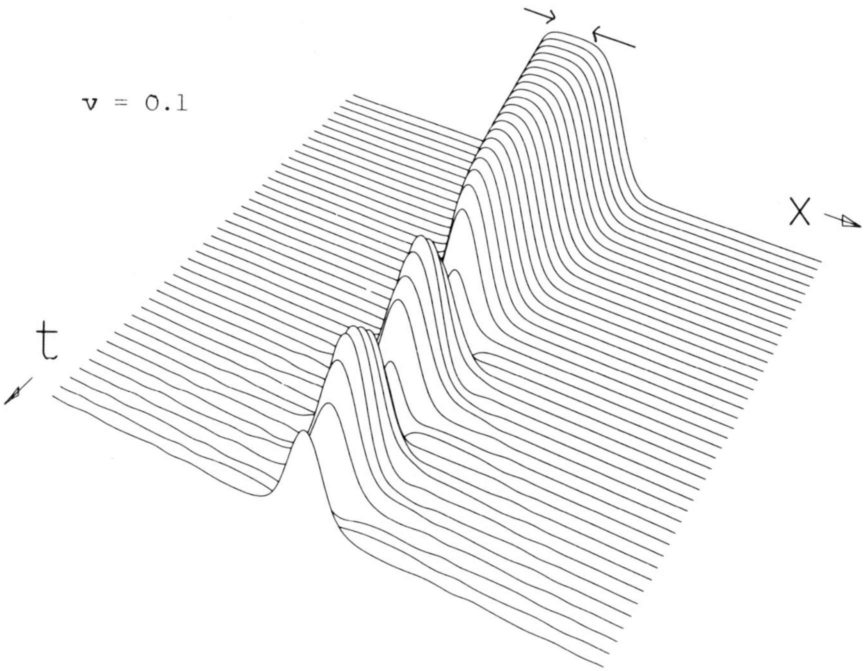

Figure 1-15: Low-energy collision of a ϕ^4 kink and anti-kink

$$E = a\,\mathrm{sech}\,\frac{1}{2}(at - \omega x + \delta) \tag{1.9-4}$$

$$\omega^2 = a^2 + \frac{\alpha a^2}{\frac{1}{4}a^2 + \mu^2} \tag{1.9-5}$$

This solitary wave can move to the left or the right depending on the choice of sign of ω in (1.9-5). The results of a numerical integration of (1.9-3) using the initial conditions of two waves colliding head-on are shown in Fig. 1-16. As in the ϕ^4 example above, ripples are emitted which show that the waves lose energy on collision. The purpose of this chapter is to prove to the reader that one should not get carried away into thinking that all solitary-type wave solutions of any equation behave as solitons. Only a few have this very special property; the majority do not. The task of future chapters is to categorise some that do have this property. Only in Chapter 10 do we discuss numerical studies of equations which do not exhibit this exact soliton property.

1.10 The Connection with Quantum Mechanics

To summarise our results so far, we have discussed methods of studying the interactions of solitary wave solutions of dispersive, nonlinear equations.

The most direct method is numerical integration. If the waves are transparent to one another then this is a good indication (but not a proof!) that soliton behaviour occurs. For some nonlinear problems, stability and accuracy of the numerical integration schemes can be a problem. If ripples are emitted which are larger than the error estimates of the numerical scheme, then particle-like behaviour does not occur. However it is possible that the inelasticity may be very small and the amplitude of ripples emmitted in this case would be so small that they may be missed.

Secondly, reduction to either a single or a pair of homogeneous equations by a Cole-Hopf type transformation. If the series solutions truncate, then exact N-parameter solutions can be found which are the analytical multi-soliton solutions. Lack of truncation indicates (but again does not prove!) that soliton behaviour does not occur as does the impossibility of reducing the equation(s) down to homogeneous degree 2 equations by a Cole-Hopf transformation. Particularly for the nonlinear Klein-Gordon equations, the lack of a Bäcklund transformation indicates the lack of soliton behaviour since these transformations construct higher soliton solutions from other lower solutions.

Note firstly this list is merely a set of points and ideas which have arisen in this chapter. They are only **indications**, but by no means a set of proofs, that some sort of more sophisticated structure does exist behind those equations which apparently have true soliton solutions (KdV, MKdV, sine-Gordon, NLS - there are many others but these are the important ones). Furthermore because such widely differing equations, arising in physical circumstances which apparently have nothing in common, seem to have similar properties, there is an indication again that some generalised structure should exist **which is common to all of them.**

Secondly, because the main property that they all share is the **particle-like** nature of their solutions which either bounce or go through one another elastically, this indicates that they scatter off one another as in

Figure 1-16: Head-on collision of two solitary wave solutions of 1.9-3

one dimensional scattering. This property is therefore a hint that scattering theory of some sort may be applicable. Firstly we do not have any strong mathematical connection as yet betwen any of the equations and any sort of scattering theory, and secondly we do not know whether the scattering theory should be **classical** or **quantum**.

To establish some sort of a connection, we return to the Bäcklund transformation result (1.5-16) which was first derived in 1875 (Bäcklund 1875, Lamb 1976) in differential geometry:

$$\frac{1}{2}(\phi+\bar{\phi})_{\xi} = \alpha\sin\frac{1}{2}(\phi-\bar{\phi}) \tag{1.10-1}$$

$$\frac{1}{2}(\phi-\bar{\phi})_{\tau} = \frac{1}{\alpha}\sin\frac{1}{2}(\phi+\bar{\phi}) \tag{1.10-2}$$

where ϕ and $\bar{\phi}$ are solutions of the sine-Gordon equation. We make the transformation

$$v = \tan\frac{1}{4}(\phi-\bar{\phi}) \tag{1.10-3}$$

and eliminate $\bar{\phi}$ from (1.10-1) to give

$$v_{\xi} = -\alpha v + \frac{1}{2}\phi_{\xi}(1+v^2) \tag{1.10-4}$$

Similarly, in (1.10-2) we find that

$$v_{\tau} = \frac{1}{2\alpha}(1-v^2)\sin\phi - \frac{1}{\alpha}v\cos\phi \tag{1.10-5}$$

The derivation of (1.10-5) requires the use of trigonometric formulae such as

$$\sin(\tan^{-1}v) = \frac{v}{(1+v^2)^{\frac{1}{2}}} \tag{1.10-6}$$

$$\cos(\tan^{-1}v) = \frac{1}{(1+v^2)^{\frac{1}{2}}} \tag{1.10-7}$$

The pair of equations (1.10-4,1.10-5) are separately called Riccati equations because of their structure, having a single derivative with **quadratic** nonlinearity. Riccati equations can always be integrated; to achieve this we introduce two functions ψ_1 and ψ_2 such that $v = \psi_1/\psi_2$. Equation (1.10-4) now becomes

$$\psi_2\frac{\partial\psi_1}{\partial\xi} - \psi_1\frac{\partial\psi_2}{\partial\xi} = -\alpha\psi_1\psi_2 + \frac{1}{2}\phi_{\xi}(\psi_1{}^2+\psi_2{}^2) \tag{1.10-8}$$

We can split (1.10-8) down into two linear equations, which are

$$\begin{aligned}\frac{\partial\psi_1}{\partial\xi} + \frac{1}{2}\alpha\psi_1 &= \frac{1}{2}\phi_{\xi}\psi_2\\ \frac{\partial\psi_2}{\partial\xi} - \frac{1}{2}\alpha\psi_2 &= -\frac{1}{2}\phi_{\xi}\psi_1\end{aligned} \tag{1.10-9}$$

Fquations (1.10-9) are more recognisable if we differentiate them once with respect to ξ to get a pair of second order equations and then set $\alpha = 2i\lambda$.

The first order derivatives can be eliminated and we obtain

$$\left(\frac{\partial^2}{\partial\xi^2} + \lambda^2\right)\begin{pmatrix}\psi_1\\ \psi_2\end{pmatrix} = \begin{pmatrix}-\frac{1}{4}\phi_\xi{}^2 & \frac{1}{2}\phi_{\xi\xi}\\ -\frac{1}{2}\phi_{\xi\xi} & -\frac{1}{4}\phi_\xi{}^2\end{pmatrix}\begin{pmatrix}\psi_1\\ \psi_2\end{pmatrix} \tag{1.10-10}$$

Equation (1.10-10) is a **two channel quantum Schrödinger equation where** λ **is an eigenvalue.** It can be written in the more conducive form

$$\frac{\partial^2}{\partial\xi^2}\underset{\sim}{\psi} + \underset{\approx}{V}\underset{\sim}{\psi} = -\lambda^2\underset{\sim}{\psi} \tag{1.10-11}$$

where V is the potential scattering matrix, λ an eigenvalue and $\tilde{\psi}$ a column vector of wave functions. An important point to note about (1.10-11) is that the potential scattering matrix V is purely a function of the derivatives of ϕ and **that** τ **occurs only as a parameter.**

The solution of this type of scattering problem is given in Chapters 3, 4, and 6. The problem was originally solved 30 years ago by Gel'fand, Levitan, Marchenko and others (see references in Chadan and Sabatier 1977 and Agranovich and Marchenko 1963). The main point about (1.10-11) is that the spectrum of the operator is constant in time. This means that the spectrum of the initial data is the spectrum for all time $t > 0$. The problem which Gel'fand and others solved, known as the inverse scattering problem, is to construct the potential $V(x)$ from a knowledge of the spectrum and the asymptotic behaviour of ψ. In our problem, a construction of the potential is to find the solution of the sine-Gordon equation for all $\tau > 0$.

This connection between a set of transformations first derived in the last century in differential geometry and the quantum Schrödinger equation is rather remarkable, particularly since it provides a method for solving the initial value problem. We should point out here that the Schrödinger equation we have derived here is not the same nonlinear Schrödinger equation (NLS) of Section 1.6 which is a p.d.e. in its own right. The two must not be confused.

Having achieved a mathematical connection between a soliton equation and a problem in quantum physics, we need to pursue it in more detail. Historically, the route we have given above is not the way the connection was first accomplished. This was first achieved by Gardner et al. (1967) who saw how to connect the KdV equation with the scalar Schrödinger scattering problem. That is a different story and will be considered again in Chapter 3. The next chapter is a slight but necessary digression in that it sets up all the necessary framework of quantum scattering theory which will be of great use later on. Those who are familiar with quantum mechanics and/or inverse problems will be able to jump to Chapter 3.

1.11 Notes

Section 1.1

The habit of avidly extracting details of Scott Russell's life from dusty archives has become somewhat of a cult among some scientists engaged in nonlinear studies, and the present authors are not immune to this obsession. Very few subjects have such readily identifiable origins which trace back to one individual. In addition many researchers may have conscious or subconcious romantic longings for that period of Victorian science in which the "gentleman" scientist, often an amateur with private means, played a significant role.

While Emmerson's biography is very entertaining and lucid, not unexpectedly there is relatively little in it with regard to the solitary wave of translation, since this subject only occupied a small period of Scott Russell's early professional life. Lonngren and Scott (1978) and Bullough and Caudrey (1980) both give a short account of Scott Russell's life in their books, and the latter book includes a partial list of Scott Russell's published articles, compiled by one of the present authors (JCE) from sources in the Library of the Royal Society of Edinburgh. Certain publications such as his "Report on Waves" (Scott Russell 1844) are an example of lucid style which many present day workers would do well to emulate.

Section 1.2

The stability of the KdV solitary wave has been proved by Benjamin (1972). See also Chapter 10 for a discussion of alternative long wave equations.

Section 1.3

The calculation showing the reduction of the Boussinesq equation (1.3-8) to the KdV equation, using the ξ, τ coordinates, is one which is explained and repeated in the Notes of Chapter 5. The variables ξ,τ are known as "stretched" coordinates because they change an amount $O(1)$ for changes $O(\varepsilon^{-p})$ and $O(\varepsilon^{-q})$ for x and t. The KdV equation (1.3-10) describes the evolution of weakly nonlinear long waves moving to the right down a chain with quadratic nonlinearity. Similarly, the Modified KdV equation does the same for a chain with cubic nonlinearity.

The choice of $f(Q) = \exp(-Q)$ corresponds to what is known as the Toda lattice (Toda 1970), a very important physical system. This choice for $f(Q)$ is rather special in that this system turns out to be a completely integrable differential-difference equation (Manakov 1975, Flaschka 1974a, b, Henon 1974)

Section 1.4

Zabusky and Kruskal's numerical result that the initial state almost recurred is a general result which has been proved analytically for the KdV equation on periodic boundary conditions. This is an extremely difficult problem to handle, requiring some sophisticated mathematics. The recurrence time depends on how many periodicities occur in the initial data and whether the frequencies are commensurate. See Lax (1975) and Novikov (1974) and Dubrovin and Novikov (1975).

Section 1.5

Prior to the work of Skyrme, the SG equation arose in studies of the motions of Bloch walls in ferromagnetic crystals (Seeger et al. 1953). More recently it has been much studied in connection with Josephson junctions in superconductivity. We return to these topics and other applications of the SG equation in later Chapters.

Hirota's (1971) discovery of the N-soliton solution of the KdV equation followed from his development of a method to reduce the evolution equation in question to a homogeneous equation of degree 2. Following this, the SG equation was a obvious candidate for a similar attack. The work by Lamb

(1971) on the SG equation in nonlinear optics brought the idea of a Bäcklund transformation to the fore since it was through this method that the hierarchy of soliton solutions could be constructed. A direct formula for the full N-soliton solution was found by Hirota (1972) in one form and conjectured by Gibbon and Eilbeck (1972) in another (later to be proved by Caudrey et al. (1973)). Hirota took

$$\phi = 4\tan^{-1}(G/F) \tag{1.11-1}$$

and showed that the SG equation could then be written in the form

$$\begin{aligned} GG_{\xi\tau} - G_\xi G_\tau &= FF_{\xi\tau} - F_\xi F_\tau \\ GF_{\xi\tau} + FG_{\xi\tau} - G_\xi F_\tau - G_\tau F_\xi &= 0 \end{aligned} \tag{1.11-2}$$

The appropiate expressions for G and F taken from (1.5-20) can be seen to satisfy (1.11-2). Hirota provided suitable expressions for G and F which described the collision on N kinks or solitons.

The second set of authors took a slightly different line. They noted that for a one parameter (b) solution of the sine-Gordon equation

$$\begin{aligned} \phi_\xi &= 2b_1 \operatorname{sech}(b_1\xi + \frac{1}{b_1}\tau + \delta_1) \\ \sin\frac{1}{2}\phi &= \operatorname{sech}(b_1\xi + \frac{1}{b_1}\tau + \delta_1) \end{aligned} \tag{1.11-3}$$

could be expressed as

$$\phi_\xi^{\ 2} = 4\frac{\partial^2}{\partial\xi^2}\log(1 + \exp\theta_1) \tag{1.11-4}$$

Taking $\phi_\xi = 2g/f$, where $g^2 = ff_{\xi\xi} - f_\xi^2$, we have now

$$\begin{aligned} \phi_\xi^{\ 2} &= 4\frac{\partial^2}{\partial\xi^2}\log f \\ \cos\phi &= 1 - 2\frac{\partial^2}{\partial\xi\partial\tau}\log f \end{aligned} \tag{1.11-5}$$

subject to the boundary conditions $\phi \to 0(\text{mod } 2\pi)$, $|\xi|\to\infty$. Differentiating $\phi_\xi = 2g/f$ with respect to τ and using (1.11-5) we obtain

$$g_\tau^{\ 2} = f_\tau f_{\xi\xi\tau} - f_{\xi\tau}^{\ 2} + ff_{\xi\tau} - f_\xi f_\tau \tag{1.11-6}$$

A solution for f which can be found is

$$\begin{aligned} M_{ij} &= \frac{2}{b_i + b_j}\cosh\frac{1}{2}(\theta_i + \theta_j) \\ \theta_i &= b_i\xi + \frac{1}{b_i}\tau + \delta_i \end{aligned} \tag{1.11-7}$$

The different forms of the solutions given here are all equivalent after various transformations have been performed. Stage by stage they could have been calculated recursively by expanding g and f (or G and F) in an ε-series

$$g = \varepsilon g^{(1)} + \varepsilon^3 g^{(3)} + \ldots$$
$$f = 1 + \varepsilon^2 f^{(2)} + \varepsilon^4 f^{(4)} \ldots \qquad (1.11\text{-}8)$$

in a similar fashion to (1.4-6) for the KdV equation. In the same fashion as the KdV equation the series would truncate at a certain point leaving the exact solutions displayed here.

This general method of reducing the partial differential equation concerned into one or two homogeneous equations of degree two, and then finding formulae to fit them, has become known as Hirota's method. Hirota has applied it to many equations including the MKdV equation, the Toda lattice, and equations in electrical network theory as well as the KdV and SG equations. In general it is an extremely powerful tool: the word "method" is somewhat inappropriate since in use it relies very heavily on experience and instinct to achieve results. Hirota (1980) gives an exposition of his results and ideas including a list of equations solvable by this method. The object of exercise (1) is to calculate multi-soliton solutions of the Toda lattice by Hirota's method.

Section 1.6

The NLS equation has a special place in the history of soliton theory since it was through this equation that Zakharov and Shabat introduced the very important 2x2 inverse scattering formulation. This formulation can be generalized to solve other equations - see Chapter 6 for details.

Section 1.8

For a much more extensive account of linear and nonlinear wave propagation, the reader is advised to consult the book by Whitham (1974). For an account of nonlinear dispersive wave propagation, see Karpman (1975).

Section 1.9

As noted earlier, a discussion on alternative long wave equations will be found in Chapter 10.

Section 1.10

Along with numerical integration, care must be taken with Hirota's direct method. It is not enough to reduce a partial differential equation to a single or a pair of homogeneous equations of degree two. The series must truncate to give an exact solution. There is a case where this does not happen. Consider the ϕ^4 equation with opposite sign to (1.9-1).

$$\phi_{\xi\tau} = \phi - \beta\phi^3 \qquad (1.11\text{-}9)$$

Writing $\phi = g/f$ and taking

$$\beta g^2 = 2(ff_{\xi\tau} - f_\xi f_\tau) \qquad (1.11\text{-}10)$$

we easily find that the second homogeneous equation is

$$fg_{\xi\tau} + gf_{\xi\tau} - g_\xi f_\tau - g_\tau f_\xi = 0 \tag{1.11-11}$$

It can readily be seen that the series expansion for g and f will only truncate for a 1-parameter solution but not for higher solutions. This is left as an exercise for the reader. On physical grounds, this is not surprising since the solitary wave solution of (1.11-9)

$$\phi = \sqrt{\frac{2}{\beta}}\,\text{sech}\,(a\xi + \frac{1}{a}\tau + \delta) \tag{1.11-12}$$

is unstable. Other equations also exist which can be reduced to homogeneous form but will not truncate. The one great value of Hirota's method is that once reduction to the homogeneous form has been achieved, the calculation showing whether or not a two-parameter solution will truncate the series is only a simple algebraic computation.

1.12 Problems

1.

The Toda lattice or chain is a model for particles on a line, each of unit mass, which undergo nearest neighbour interactions. The equation of motion is

$$\ddot{Q}_n = \exp(Q_{n-1} - Q_n) - \exp(Q_n - Q_{n+1})$$

where $Q_n(t)$ is the displacement from equilibrium of the nth particle. By defining a variable S_n such that

$$Q_n = S_{n-1} - S_n$$

where S_n and all its time derivatives $\to 0$ as $t \to \infty$, show that

$$1 + \ddot{S}_n = \exp(S_{n+1} + S_{n-1} - 2S_n)$$

Find the transformation $S_n(t) = F(f_n(t))$ which reduces the equation to

$$f_{n+1}f_{n-1} - f_n^2 = f_n\ddot{f}_n - \dot{f}_n^2$$

and hence show that one and two parameter solutions of this equation take the form

$$f_n^{(1)} = 1 + \exp\theta_1$$

$$f_n^{(2)} = 1 + \exp\theta_1 + \exp\theta_2 + A_{12}\exp(\theta_1+\theta_2)$$

$$\theta_i = a_i n - \omega_i t + \delta_i$$

$$D(a_i, \omega_i) \equiv \omega_i^2 - 2(\cosh a_i - 1) = 0$$

$$A_{ij} = -\frac{D(a_i - a_j; \omega_i - \omega_j)}{D(a_i + a_j; \omega_i + \omega_j)}$$

Hence show by induction that an N-parameter solution has the form

$$f_n^{(N)}(t) = \log\det \mathbf{M}$$

where the NxN matrix **M** has elements

$$M_{ij} = \delta_{ij} + (1 - A_{ij})^{\frac{1}{2}} \exp\tfrac{1}{2}(\theta_i + \theta_j)$$

2.

The equation

$$\phi_{\xi\tau} = \phi - \phi^3$$

is a form of truncated sine-Gordon equation in characteristic co-ordinates. Taking $\phi = g/f$, show that it can be reduced to two homogeneous equations of degree two in g and f:

$$g^2 = 2(ff_{\xi\tau} - f_\xi f_\tau)$$

$$fg_{\xi\tau} + gf_{\xi\tau} - g_\xi f_\tau - g_\tau f_\xi = gf$$

by expanding g and f as series in a small parameter ε

$$g = \sum_{n=0}^{\infty} \varepsilon^{2n+1} g^{(2n+1)}$$

$$f = 1 + \sum_{n=1}^{\infty} \varepsilon^{2n} f^{(2n)}$$

show that an exact 1-parameter solution can be found

$$g = 2\sqrt{2}\exp\theta_1$$

$$f = 1 + \exp 2\theta_1$$

$$\theta = a_i\xi + \tau/a_i + \delta_i$$

but a 2-parameter solution fails to produce self-truncation in the expansion for g and f.

3.

Given Burger's equation in the form

$$u_t + uu_x = \delta u_{xx}$$

with some smooth initial data $u(x,0) = f(x)$, reduce this equation to the heat equation. By solving the initial value problem for this latter equation, show that in the limit $\delta \to 0$, solutions of Burger's equation can be expressed in the form

$$u = f(x - ut)$$

You will need the two results

(i) $$\int_{-\infty}^{\infty} \exp(-x^2)\,dx = \sqrt{\pi}$$

(ii) $$\int_{-\infty}^{\infty} g(\eta)\exp[-G(\eta)/2\delta] \simeq g(\eta_o)\left[\frac{4\pi\delta}{|G_{\eta\eta}(\eta_o)|}\right]^{\frac{1}{2}} \exp[-G(\eta_o)/2\delta]$$

for small δ. η_o is a stationary point of G.

4.

By taking $u = g/f$, show that the Modified KdV equation

$$u_{xxx} + 6u^2u_x + u_t = 0$$

can be reduced to the homogeneous form

$$g^2 = ff_{xx} - f_x^2$$

$$fg_{xxx} - 3f_xg_{xx} + 3f_{xx}g_x - f_{xxx}g + g_tf - gf_t = 0$$

Bu using the expansion method on g and f, show that a 2-parameter solution is given by

$$g = 2a_1\exp\theta_1 + 2a_2\exp\theta_2 + 2a_2A\exp(2\theta_1+\theta_2) + 2a_1A\exp(\theta_1+2\theta_2)$$

$$f = 1 + \exp 2\theta_1 + \exp 2\theta_2 + 2(1-A)\exp(\theta_1+\theta_2) + A\exp(2\theta_1+2\theta_2)$$

$$\theta_i = a_ix - a_i^3t + \delta_i$$

$$A = \left[\frac{a_i - a_j}{a_i + a_j}\right]^2 .$$

If $a_1 > a_2 > 0$ and the larger soliton is initially to the left, show that the larger soliton is shifted forwards on collision, and has a subsequent separation with a phase shift $\ln A$, while the smaller is shifted backwards with a phase shift $\ln A^{-1}$.

5.

In inviscid gas dynamics, the equations of conservation of mass and momentum are given by

$$\rho_t + (\rho u)_x = 0$$

$$(\rho u)_t + (\rho u^2 + F)_x = 0$$

where ρ is the density and u is the velocity. If F is related to the gas density ρ by the formula

$$F = A(\rho/\rho_o)^\gamma$$

$(1 < \gamma < 2)$ for a polytropic gas molecule), where ρ_o is the eqilibrium density and A is a constant, show that u can be expressed purely as a function of ρ (simple waves) and that these simple wave solutions are governed by the equation

$$\rho_t + c(\rho)\rho_x = 0$$

$$c(\rho) = 2(\gamma A\rho_o)^{\frac{1}{2}}\left[\tfrac{1}{2}(\gamma+1)(\rho/\rho_o)^{(\gamma-1)/2} - 1\right]/(\gamma-1)$$

2. SCATTERING TRANSFORMS

2.1 Inverse Problems and Fourier Analysis

Increased computing power and the ability to make highly accurate experimental measurements has altered the type of problem posed by modern science and engineering. For example, a modern engineer may not be particularly interested in knowing the temperature as a function of time at a point in a body, as such a variable can be accurately measured. However, if there is a mixture of materials and the thermal conductivity of the mixture is unknown, the engineer has no way to measure what the conductivity is. The relevant problem is to determine if, from results on the temperature profile in the body, the thermal conductivity (which is often temperature dependent) can be determined by extrapolating backwards. Problems of this type involving backward extrapolation are known as **Inverse problems.**

The analysis of seismic data is a typical example of such an inverse problem. Apart from its role in monitoring earthquake activity within the earth's crust it also provides a potent geophysical research tool for investigating the detailed structure of the earth's mantle and core. From an analysis of seismic waves reflected and refracted by the layers of rock through which they have passed, information concerning the density and nature of the rock is sought. This is an example of an **Inverse Scattering Problem.** Determine, given the incoming and outgoing waves related to a known source, the nature of an unknown wave scatterer.

Inverse scattering problems are bound to occur in a problem of remote sensing. A typical example is the investigation of the health of arterial walls by means of ultrasonic sound pulses. A weakened region of arterial wall such as an aneurism is detectable by spectral analysis of the pulse data.

Medical science, chemistry, physics and engineering all require detailed information about macromolecules whether they be DNA or a new form of man-made fibre. The most common technique is X-ray crystallography. From an X-ray photograph one must determine the nature of the molecules which scattered the X-rays. Theoretically, this inverse problem is reduced to determining the Fourier transform of the X-ray pattern.

Where X-rays cannot penetrate, neutron and meson beams can. To probe for faults within such inaccessible regions as the heart of a nuclear reactor, engineers can use neutron beams in the same way a geologist would use X-rays. Scanning techniques developed for X-rays can be simply adapted to determine cross sections of the reactor core and thereby identify cracks of potentially lethal nature. Again, the problem requires the resolution of an inverse problem, and in this case the problem reduces to inverting another linear transform known as the Radon transform.

Even more inaccessible than a nuclear fuel rod is the interior of an atom, nucleon or other subnuclear particle. High energy beams of other particles such as electrons or protons are directed onto a target particle in an attempt, by 'destructive testing', to determine its structure. Such a procedure has been likened to the problem of trying to determine the shape of an object by directing a stream of water on it and collecting the water which comes out at various angles. In the atomic scattering case one determines the number of beam particles, and their type, emerging at various angles about the target. From those particle numbers one attempts to determine the nature of the target. This is the quintessential inverse scattering problem.

This book is concerned with showing how an inverse problem of mathematical type has enabled many of the equations mentioned in Chapter 1 to be solved exactly. In this chapter we will start to gather together some of the ideas that have motivated the theory of the so-called **inverse scattering transform** (IST). Its roots lie in quantum mechanics, where most of the descriptive language associated with the technique originated, but spread into various branches of mathematics such as spectral theory. The rigorous development of the method will begin in Chapters 3 and 4 and will be continued in a generalised form in Chapter 6. The aim of this Chapter is to prepare the way for that more abstract theory. This will be done by establishing our notation and introducing the various functions that will arise in the context of specific examples.

Let us start our mathematical considerations by establishing the notation that we will adopt in this and coming chapters. In most of the work that follows we will deal with functions $f:R^2 \to R^n$ of two variables. Often, just one of the two variables will be of particular importance whilst the other remains fixed. To emphasise the dominant variable we will append the other as either a subscript before or a superscript after the function symbol f. To be exact we define the function ${}_kf:R^2 \to R^n$ by

$$ {}_kf(x) = f(x,k) \tag{2.1-1}$$

and the function $f^x:R \to R^n$ by

$$ f^x(k) = f(x,k) \tag{2.1-2}$$

With the exception of numerical subscripts we reserve the notation of a subscript following the function symbol f for the denoting of partial derivatives. For example we will denote $\partial^2f/\partial x^2$ and $\partial^2f/\partial k^2$ by f_{xx} and f_{kk} respectively. On occasions we will switch from one notation to another in accordance with the mode which best expresses the statement we wish to make. On those occasions when we need to consider a function $y:R^3 \to R^n$ of three variables it is also convenient to define the function ${}_kY:R^2 \to R^3$ by ${}_kY(t,x) = Y(t,x,k)$.

In Chapter 1 we solved the linear heat equation by means of Fourier transforms. The methods we will develop in this book can be regarded as an extension of the Fourier transform method to certain classes of special nonlinear equations. Consequently it is worthwhile to pause for a moment and examine the way in which a Fourier transform enables us to solve a linear initial value problem. Consider the model problem

$$\left(\frac{\partial}{\partial t} + c_0 \frac{\partial}{\partial x}\right) q(t,x) = 0 \qquad q(0,x) = q^0(x) \tag{2.1-3}$$

having the exact solution

$$q^t(x) = q(t,x) = q^0(x - c_0 t) \tag{2.1-4}$$

If q is a complex valued absolutely integrable function for $t \geq 0$

$$\int_{-\infty}^{\infty} |q(t,x)|\,dx < \infty \tag{2.1-5}$$

the Fourier transform of q^t is defined by

$$(Fq^t)(k) = \int_{-\infty}^{\infty} e^{-ikx} q(t,x) \frac{dx}{(2\pi)^{\frac{1}{2}}} = B^t(k) \tag{2.1-6}$$

If we denote the space of continuously differentiable absolutely integrable functions on the real line by $A(R)$ then

$$F : A(R) \longrightarrow A(R) \tag{2.1-7}$$

The transformation F is invertible on $A(R)$ and its inverse is given by

$$(F^{-1}B^t)(x) = \int_{-\infty}^{\infty} e^{ikx} B(t,k) \frac{dk}{(2\pi)^{\frac{1}{2}}} = q^t(x) \tag{2.1-8}$$

$$F^{-1} : A(R) \longrightarrow A(R) \tag{2.1-9}$$

Equation (2.1-3) is an equation defining the time evolution of a function $q^t \in A(R)$ which when $t = 0$ starts off as the function q^0. As t varies q^t traces out a path in the function space $A(R)$.

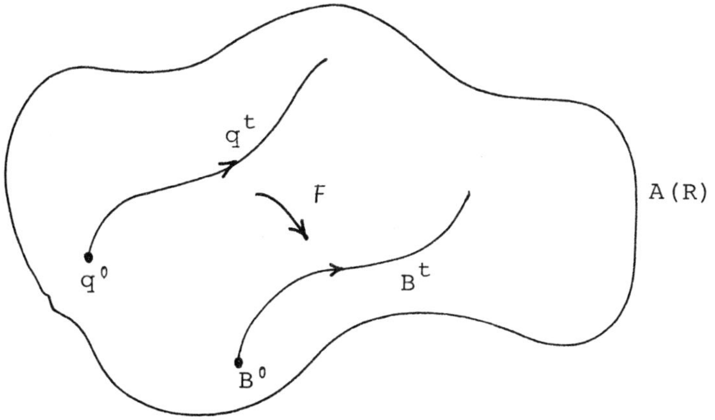

Figure 2-1:

Figure 2-1 shows how this geometric picture is transformed by the mapping F. The transformed picture is also a path B^t in $A(R)$ traced from an initial point B^0. The advantage of applying the Fourier mapping F is that the equation describing the curve in the transformed space is much simpler and may be solved exactly. From (2.1-3) we find that

$$\left(\frac{\partial}{\partial t} + ikc_0\right)B(t,k) = 0 \qquad B(0,k) = B^0(k) \tag{2.1-10}$$

which may be integrated to give

$$B^t(k) = B(t,k) = B^0(k)e^{-ikc_0 t} \tag{2.1-11}$$

If we now apply the inverse transformation F^{-1} we obtain

$$q^t(x) = \int_{-\infty}^{\infty} e^{ik(x - c_0 t)} B^0(k) \frac{dk}{(2\pi)^{\frac{1}{2}}} = q^0(x - c_0 t) \tag{2.1-12}$$

The whole process is illustrated in Fig. 2-2 which serves as a paradigm for the methods we will develop.

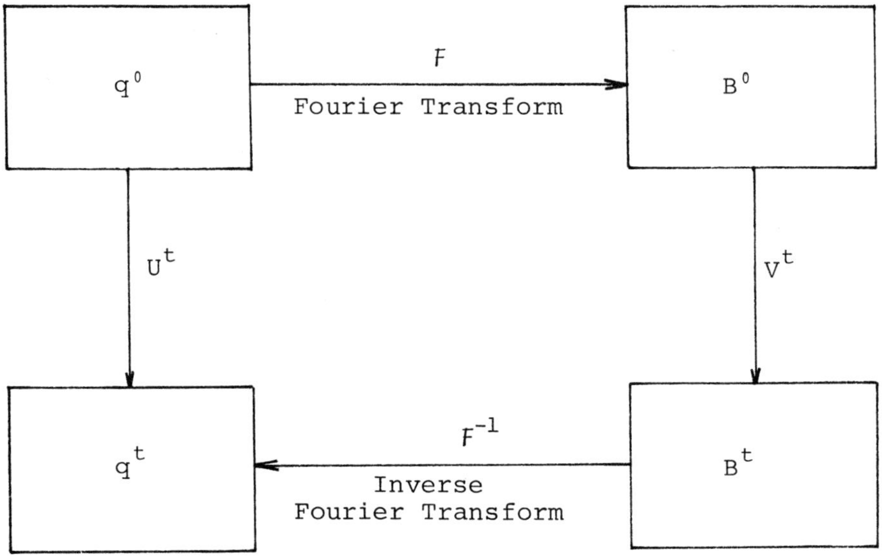

Figure 2-2:

where U^t and V^t are the time evolution operators for equations (2.1-3) and (2.1-10) defined by

$$U^t : q^0 \longrightarrow q^t \quad \text{and} \quad V^t : B^0 \longrightarrow B^t \tag{2.1-13}$$

For equation (2.1-3) there is an alternative way in which a mapping similar to F might be constructed. Consider the eigenvalue equation

$$\left(\frac{d^2}{dx^2} + q(t,x)\right)y(t,x,k) = -k^2 y(t,x,k) \tag{2.1-14}$$

The variable t appears in this equation as a parameter. As t is varied both the allowable values of k^2 (the spectrum of the equation) and the eigenfunctions $_k y$ will vary. If the evolution of q^t is determined by a specific equation such as (2.1-3) then the dependence of $_k y$ on t will also be strongly restricted. Let $_k y$ evolve according to the equation

$$\left(\frac{\partial}{\partial t} + c_0 \frac{\partial}{\partial x}\right)y(t,x,k) = \alpha(k)y(t,x,k) \tag{2.1-15}$$

Equations (2.1-14) and (2.1-15) are then consistent in the sense that

$$\frac{\partial^2}{\partial x \partial t} y(t,x,k) = \frac{\partial^2}{\partial t \partial x} y(t,x,k) \tag{2.1-16}$$

provided that

$$\left(\frac{\partial}{\partial t} + c_0 \frac{\partial}{\partial x}\right)q(t,x) = 0 \tag{2.1-17}$$

and for eigenvalues k^2

$$\frac{\partial}{\partial t}(k^2) = 0 \tag{2.1-18}$$

Consequently the evolution equation (2.1-3) becomes replaced by (2.1-15) and (2.1-18). At this stage it may seem as though we have complicated things as, after all, (2.1-15) is more complicated than (2.1-3) which corresponds to $\alpha = 0$. However, we can make progress with $_k y$ which we cannot make with q^t. Assume that the function q^t satisfies the boundary condition

$$q^t(x) \longrightarrow 0 \qquad \text{for} \quad |x| \longrightarrow \infty \tag{2.1-19}$$

As a result equation (2.1-14) becomes asymptotically

$$\left(\frac{d^2}{dx^2} + k^2\right)y(t,x,k) = 0 \tag{2.1-20}$$

The equation (2.1-20) has a general solution

$$y(t,x,k) = (Ae^{-ikx} + Be^{+ikx}) \tag{2.1-21}$$

where A and B may be functions of t and k. It may be shown that if a solution is specified at $x = -\infty$ then the solution is uniquely fixed at $x = +\infty$. If we chose a solution $_k\Phi$ to (2.1-14) having asymptotic form given by

$$\Phi(t,x,k) \sim \begin{cases} e^{-ikx} & x \to -\infty \\ b^t(k)e^{ikx} + a^t(k)e^{-ikx} & x \to +\infty \end{cases} \tag{2.1-22}$$

then taking the asymptotic limit $x \to -\infty$ and using the asymptotic form (2.1-22) in (2.1-15) gives

$$\alpha(k) = -ikc_0 \tag{2.1-23}$$

If we take the other limit $x \to \infty$ together with (2.1-23) we obtain the evolution of the asymptotic coefficients b^t and a^t to be

$$\frac{\partial}{\partial t} b(t,k) = -2ikc_0 b(t,k) \tag{2.1-24}$$

$$\frac{\partial}{\partial t} a(t,k) = 0 \tag{2.1-25}$$

Equation (2.1-24) is clearly analogous to (2.1-10). Solving (2.1-24) we find that

$$\Phi(t,x,k) \sim \begin{cases} e^{-ikx} & x \to -\infty \\ b^0 e^{ik(x-2c_0t)} + a^0 e^{-ikx}, & x \to +\infty \end{cases} \tag{2.1-26}$$

where the asymptotic solution to (2.1-14) corresponding to $q^0(x)$ is

$$\Phi(0,x,k) \sim \begin{cases} e^{-ikx} & x \to -\infty \\ b^0 e^{ikx} + a^0 e^{-ikx}, & x \to +\infty \end{cases} \tag{2.1-27}$$

consequently the asymptotic form corresponding to $q^0(x-c_0t)$ is given by

$$\Phi(t,x,k) \sim \begin{cases} e^{-ik(x-c_0t)} & x \to -\infty \\ b^0 e^{ik(x-c_0t)} + a^0 e^{-ik(x-c_0t)}, & x \to +\infty \end{cases} \tag{2.1-28}$$

We can normalise the asymptotic coefficient e^{ikc_0} at $x = -\infty$ to unity by using the linearity of (2.1-14). Therefore if a 1-1 mapping T can be established between q^t and the asymptotic data (b^t, a^t) we must have

$$q(t,x) = q^0(x - c_0t) \tag{2.1-29}$$

in agreement with (2.1-12).

We have therefore shown that if T^{-1} can be defined we would be able to solve (2.1-5) by the scheme shown in Fig. 2-3. Here

$$W^t : (b^0, a^0) \longrightarrow (b^t, a^t) \tag{2.1-30}$$

is the time evolution operator for equation (2.1-24)-(2.1-25).

In later sections we will determine the asymptotic data (b^t, a^t) for some specific functions $q(t,x)$. This will help us to decide whether a mapping such as T^{-1} can be defined. In order to have a well defined inverse T^{-1} we need the asymptotic data to be uniquely associated with a function $q(t,x)$. It will turn out that more information than is contained in (b^t, a^t) is needed and that strong constraints are required on $q(t,x)$.

Equation (2.1-14) occurs directly in physical models and the ability to construct the mapping T^{-1} would be immediately useful. Consider the vibrations of a one-dimensional elastic medium of variable density $\rho(y)$. The relevant equation is Newton's equation of motion for the velocity

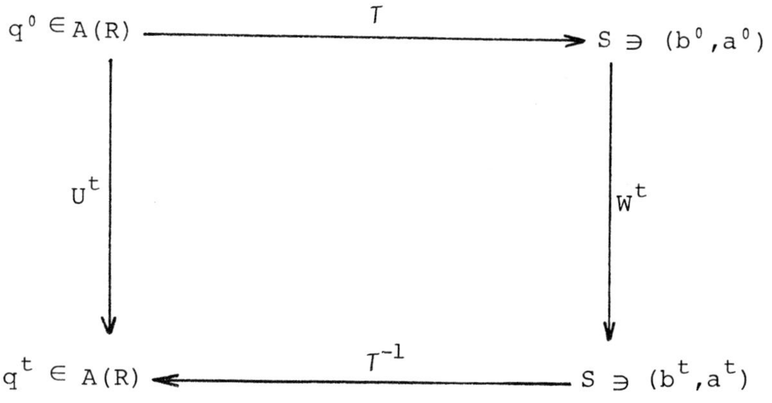

Figure 2-3:

$v(y,t)$.

$$\rho(y)\frac{\partial}{\partial t} v(y,t) = \frac{\partial S}{\partial y}(y,t) \tag{2.1-31}$$

where the stress $S(y,t)$ is given in terms of the displacement field $u(y,t)$ by

$$S(y,t) = \kappa(y)\left(\frac{\partial u}{\partial y}\right)(y,t) \tag{2.1-32}$$

The elastic modulus $\kappa(y)$ is assumed to depend on y and $v(y,t)$ is given in terms of $u(y,t)$ by

$$v(y,t) = \frac{\partial u}{\partial t}(y,t) \tag{2.1-33}$$

If we denote the Fourier transform of a function $f(x,t)$ with respect to t by

$$\tilde{f}(y,k) = \int_{-\infty}^{\infty} e^{-ikt} f(y,t) \frac{dt}{(2\pi)^{\frac{1}{2}}} \tag{2.1-34}$$

equations (2.1-31)-(2.1-33) may be reduced to

$$iw\rho(y)\tilde{v}(y,k) = \frac{\partial \tilde{S}}{\partial y}(y,k) \tag{2.1-35}$$

and

$$ik\tilde{S}(y,k) = \kappa(y)\frac{\partial \tilde{v}}{\partial y}(y,k) \tag{2.1-36}$$

from which $\tilde{v}(x,w)$ can be eliminated to give the Sturm-Liouville equation

$$\frac{\partial}{\partial y}\left(\frac{1}{\rho(y)}\frac{\partial \tilde{S}}{\partial y}(y,k)\right) + \frac{k^2}{\kappa(y)}\tilde{S}(y,k) = 0 \qquad (2.1\text{-}37)$$

Introducing the new independent variable x by

$$x = \int^y \left[\frac{\rho(u)}{k(u)}\right]^{\frac{1}{2}} du \qquad (2.1\text{-}38)$$

and the new dependent variable $Y(y,k)$ by

$$Y(x,k) = \tilde{S}(y,k)\,(\kappa(y)\rho(y))^{-\frac{1}{2}} \qquad (2.1\text{-}39)$$

we find that (2.1-35) may be recast into the canonical form

$$\left[\frac{d^2}{dx^2} + k^2 + Q(y)\right]Y(x,k) = 0 \qquad (2.1\text{-}40)$$

where

$$Q(x) = -\frac{1}{\Phi}\frac{d^2\Phi}{dx^2} \quad \text{and} \quad \Phi(x) = (\kappa(y)\rho(y))^{-1/4} \qquad (2.1\text{-}41)$$

Thus we have returned to an equation of the form (2.1-14).

An important idea that we will need is that of direct and inverse problems. In the above problem the direct problem would be to compute the velocity and displacement fields $v(y,t)$ and $u(y,t)$ given the density $\rho(y)$ and elastic modulus $\kappa(y)$. In an exactly analogous fashion to the thermal conductivity problem mentioned earlier this may not be the relevant problem. It may be possible to make very accurate measurements of the velocity and displacement fields and it may be the elastic modulus which is unknown. Indeed the modulus may depend on the displacement field. A more realistic problem is then the determination of $\kappa(y)$ from the measured displacement field. This is an example of a inverse problem in contrast to the direct problem of determining $u(y,t)$ and $v(y,t)$ given $\rho(y)$ and $\kappa(y)$. To solve the inverse problem of determining $\kappa(y)$ given $\rho(y)$ we would need to determine $Q(x)$ by solving the inverse problem for equation (2.1-35) and then solve (2.1-41) and extract $\kappa(y)$.

The most common occurrence of (2.1-40) is in quantum mechanics where it is known as the Schrödinger equation and the function $Q(x)$ is known as a potential function. In Section 2.3 we will examine how (2.1-40) arises in quantum mechanics. To understand the methods we are going to develop, it is not necessary to know any quantum mechanics. However, as it is the source of many of the ideas we will use, much of the original language has survived.

Before embarking on that study we wish to put it into context by considering a classical problem which will serve to illustrate further the notion of direct and inverse problems.

2.2 Classical Scattering

Let us begin our consideration of direct and inverse scattering problems with the classical case. Consider a particle moving in a plane under the action of a conservative force directed away from a single point O. To such a force field there corresponds a potential $V(r)$, dependent only upon r, the radial variable of polar coordinates located at O. If the radial force on the particle goes to zero as $r \to \infty$ then we may select a corresponding potential $V(r)$ which also goes to zero as $r \to \infty$. At great distances from such a potential the particle moves as though it were effectively free, that is, in a straight line. Figure 2-4 shows a number of particle trajectories in such a potential. In the diagram, the trajectories correspond to particles which start off at $x = -\infty$ travelling in straight lines parallel to, but at varying distances from, the x-axis. The particles are bent as they enter the region of significant variation of the potential, and each penetrates to within a minimum distance R_b of the orgin O dependent upon the initial distance b of the particle from the x-axis. The variable b is called the impact parameter of the particle.

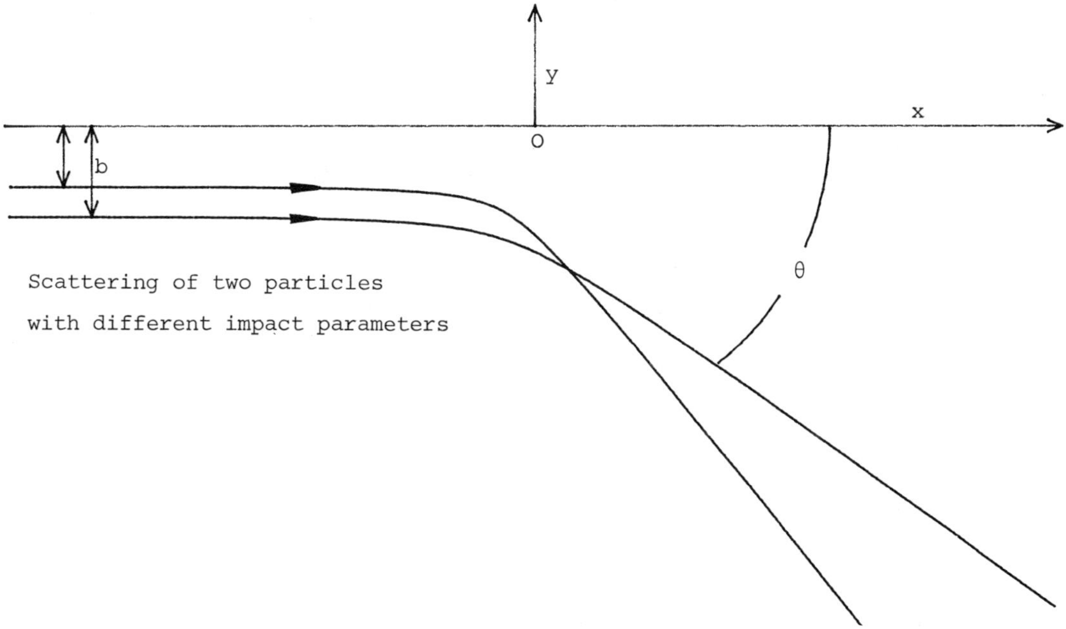

Figure 2-4:

The particle is subsequently repelled by the potential, bent yet further from its initial path along the x-axis, and finally moves off towards a straight line asymptote which passes through the origin O. Such a straight line determines a polar angle $\theta(b)$. The direct problem corresponding to this situation is, given a potential $V(r)$, to determine $\theta(b)$. This problem is physically realisable in the scattering of electrons by thin metal foils. In this case the problem is three-dimensional. However, if the beam is cylindrical about the negative x-axis we can restrict our attention to the plane of a single trajectory and use our previous analysis. Physically, one clearly cannot measure $\theta(b)$. However, a different quantity from which $\theta(b)$ may be determined is measurable. A particle beam is manufactured in the laboratory in which ρ particles of uniform energy E, cross unit cross-sectional area of the beam each second. This beam of particles is directed onto the potential along the negative x-axis. The number scattered into the angular interval θ to $\theta + d\theta$ per second is denoted by

$-2\pi\rho\sigma(\theta)\sin\theta\, d\theta$ and the function $\sigma(\theta)$ is called the differential scattering cross section. The range of θ-values in $(\theta, \theta+d\theta)$ corresponds to a range of b values in $(b, b+db)$. The particle number is a conserved quantity. The number entering each second through the annular area bounded by the impact parameter values b and $b + db$ is $2\pi\rho b db$ and must be equal to the number leaving which is $-2\pi\sigma(\theta)\rho\sin\theta\, d\theta$. Thus we have the relationship

$$\sigma(\theta) = -b \operatorname{cosec}\theta \frac{db}{d\theta} \tag{2.2-1}$$

We may integrate (2.2-1) for monotonically increasing potentials which are singular at $r = 0$ if we note that particles starting out along the x-axis $b = 0$ are reflected back the way they came giving $\sigma(0) = \pi$. Consequently

$$b^2 = 2\int_\theta^\pi \sigma(\theta)\sin\theta\, d\theta \tag{2.2-2}$$

and hence this equation determines θ as a function of the impact parameter b. From a given $V(r)$ the physical direct problem is to calculate the differential scattering cross section $\sigma(\theta)$. The purpose of such experiments is precisely the reverse of the direct problem. The potential $V(r)$ is unknown, the differential scattering cross section is experimentally measurable. Can one, from a knowledge of $\sigma(\theta)$, determine $V(r)$? This constitutes the physical inverse scattering problem.

It is a straightforward application of the orbit theory of classical mechanics to solve the direct problem. There are two conserved quantities, the energy E and the angular momentum J. If we take a particle of unit mass with impact parameter b then these may be expressed in polar coordinates (r,θ) about O as

$$\frac{1}{2}\left[\left(\frac{\partial r}{\partial t}\right)^2 + r^2\left(\frac{d\theta}{dt}\right)\right]^2 + V(r) = E = \frac{1}{2}v_0^2 \tag{2.2-3}$$

$$r^2\frac{d\theta}{dt} = J = bv_0 = b\sqrt{2E} \tag{2.2-4}$$

where v is the initial velocity of the particle.

Using (2.2-4) to eliminate $d\theta/dt$ we obtain

$$\frac{1}{2}\left[\left(\frac{dr}{dt}\right)^2 + \frac{2Eb^2}{r^2}\right] + V(r) = E \tag{2.2-5}$$

or

$$\frac{dr}{dt} = \mp[2(E - V(r)) - \frac{2Eb^2}{r^2}]^{\frac{1}{2}} \tag{2.2-6}$$

where the minus sign applies to the inward leg of the trajectory and the plus to the outward going leg. On the inward portion r decreases until it reaches a minimum where $dr/dt = 0$ at $r = R_b$ and where R_b is the largest zero of the function $(E - V(r) - Eb^2r^{-2})$. Dividing (2.2-4) by (2.2-6) we obtain

$$\frac{d\theta}{dr} = \mp \frac{b\sqrt{E}}{r^2}[E - V(r) - Eb^2/r^2]^{-\frac{1}{2}} \tag{2.2-7}$$

Denoting by ${}_b\phi$ the value of θ which corresponds to $r = R_b$ we may integrate separately for the inward and outward portions to obtain

$$\pi - \phi_b = \int_{R_b}^{\infty} [b^{-2} - r^{-2} - V(r)E^{-1}b^{-2}]^{-\frac{1}{2}} r^{-2}\,dr \tag{2.2-8}$$

$$\phi_b - \theta(b) = \int_{R_b}^{\infty} [b^{-2} - r^{-2} - V(r)b^{-2}E^{-1}]^{-\frac{1}{2}} r^{-2}\,dr \tag{2.2-9}$$

which may be combined to give

$$\theta(b) = \pi - 2\int_{R_b}^{\infty} [b^{-2} - r^{-2} - V(r)E^{-1}b^{-2}]^{-\frac{1}{2}} r^{-2}\,dr \tag{2.2-10}$$

We can, by changing the integration variable from r to $u = r^{-1}$, express (2.2-10) as

$$\theta(b) = \pi - 2\int_0^{U_b} [b^{-2}(1 - \tilde{V}(u)E^{-1}) - u^2]^{-\frac{1}{2}} du \tag{2.2-11}$$

where $U_b = R_b^{-1}$ and $\tilde{V}(u) = V(u^{-1})$.

Equation (2.2-11) solves the direct problem in its neatest form. For example, consider the potential $V(r) = e/r^2$. Substituting this potential into (2.2-11) we obtain,

$$\theta = \pi[1 - (eb^{-2}E^{-1} + 1)^{-\frac{1}{2}}] \tag{2.2-12}$$

which may be inverted to give

$$b^2 = eE^{-1} \frac{(1 - \theta/\pi)^2}{(1 - (1 - \theta/\pi)^2)} \tag{2.2-13}$$

By (2.2-2) the differential scattering cross section is given by

$$\sigma(\theta) = -\frac{1}{2}\operatorname{cosec}\theta \frac{d}{d\theta}(b^2) \tag{2.2-14}$$

By using (2.2-13) in (2.2-14) we now have an expression for the differential scattering cross section in terms of θ :

$$\sigma(\theta) = \frac{e(1 - \theta/\pi)}{\pi E(\theta/\pi(2 - \theta/\pi))^2} \operatorname{cosec}\theta \tag{2.2-15}$$

Let us now consider the inverse problem. To do so we must carefully analyse the structure of (2.2-11). A troublesome aspect of (2.2-11) is that its dependency upon b is twofold. Not only is there explicit b dependence in the integrand but also implicit b dependence in the upper limit of integration U_b which is the smallest zero of the integrand. As a first step towards solving (2.2-11) we must make explicit the implicit dependence in U_b. The integrand can be rewritten as

$$[b^{-2} - u^2(1 - \tilde{V}(u)E^{-1})^{-1}]^{-\frac{1}{2}}[1 - \tilde{V}(u)E^{-1}]^{-\frac{1}{2}}$$

where U_b is the smallest zero of the factor $[b^{-2} - u^2(1-\tilde{V}(u)E^{-1})]^{-1}$. If we define

$$W(u) = u^2(1 - E^{-1}\tilde{V}(u))^{-1} \tag{2.2-16}$$

then because U_b is the smallest zero $W(u)$ has a unique inverse on the range of integration $(0,U_o)$. Consequently, our integration variable can be changed from u to W, the advantage being that the range of W is the interval $W(0) = 0$ up to $W(U_b) = b^{-2}$ and the b dependence is made explicit. Transforming to the variable W equation (2.2-11) becomes

$$\frac{1}{2}(\pi - \theta(b)) = \int_0^{b^{-2}} (b^{-2} - W)^{-\frac{1}{2}}\Gamma(W)\,dW \tag{2.2-17}$$

where

$$\Gamma(W) = (1 - \tilde{V}(u)E^{-1})^{-\frac{1}{2}}\frac{du}{dW} \tag{2.2-18}$$

The structure of (2.2-11) now becomes clear. Defining the integral $A[\psi](s)$ by

$$A[\psi](s) = \int_0^{s} (s - W)^{-\frac{1}{2}}\psi(W)\,dW \tag{2.2-19}$$

we see that (2.2-17) can be expressed as,

$$\frac{1}{2}(\pi - \theta(b)) = A[\Gamma](b^{-2}) \tag{2.2-20}$$

The integral transform (2.2-19) is called an Abel transform and we can now see that the inverse problem is identical to the problem of inverting this Abel transform.

The problem of solving the equation

$$A[\psi](s) = G(s) \tag{2.2-21}$$

is simplified by observing that $A[\psi]$ is an integral operator of Laplace convolution type. The Abel transform of a function ψ is the Laplace convolution of ψ with the function $t^{-\frac{1}{2}}$. Taking the Laplace transform of (2.2-1) and using the Laplace convolution theorem we easily discover that, as the Laplace transform of $t^{-\frac{1}{2}}$ is $\sqrt{\pi}p^{-\frac{1}{2}}$, the Laplace tranform $L[\psi]$ of ψ is given by

$$L[\psi] = \pi^{-\frac{1}{2}}L[G]p^{\frac{1}{2}} \tag{2.2-22}$$

The inverse Laplace transform of $p^{\frac{1}{2}}$ does now exist. However, if we use the elementary properties of Laplace transforms that the Laplace transform of $G'(t) = dG/dt$ is related to the Laplace transform of G by

$$L[G'] = pL[G] - G(0) \tag{2.2-23}$$

we can rewrite (2.2-21) as

$$L[\psi] = \frac{1}{\sqrt{\pi}} p^{-\frac{1}{2}}\{L[G'] + G(0)\} \tag{2.2-24}$$

Hence by means of the convolution theorem we can deduce that

$$\psi(s) = \frac{1}{\pi}\{G(0)s^{-\frac{1}{2}} + \int_0^x G'(t)(s-t)^{-\frac{1}{2}}dt\} \tag{2.2-25}$$

which is the inverse Abel Transform.

The inverse Abel transform above gives,

$$\Gamma(W) = \frac{d}{dW}\frac{1}{2\pi}\int_0^W \frac{(\pi - \theta(s))}{(W-s)^{\frac{1}{2}}}\,ds \tag{2.2-26}$$

where $\tilde{\theta}(s) = \theta(s^{-\frac{1}{2}})$ and $\theta(s) \to \pi$ as $s \to \infty$ because $U \to 0$ as $b \to \infty$. Now let us define

$$v(u) = (1 - \tilde{V}(u)E^{-1}) \tag{2.2-27}$$

By (2.2-16) we then have $v = u^2w^{-1}$ and so differentiating this last expression with respect to W and substituting $du/dw = v^{\frac{1}{2}}(W)$ we obtain, regarding v as a function $\tilde{v}$ of W, the equation

$$\frac{d\tilde{v}}{dW} = [2\Gamma(W)W^{-\frac{1}{2}} - W^{-1}]\tilde{v}(W) \tag{2.2-28}$$

which may be integrated to give

$$\tilde{v}(W) = \exp\int_0^W [\Gamma(W)W^{-\frac{1}{2}} - W^{-1}]\,dW \tag{2.2-29}$$

We can now invert the equation $u^2 = \tilde{v}(W)\,W$ giving W and hence v as a function of u and thereby solving the inverse problem.

The inversion process consists of constructing the sequence of functions

$$\sigma \longrightarrow \theta \longrightarrow \Gamma \longrightarrow \tilde{v} \longrightarrow W \longrightarrow v \longrightarrow V \tag{2.2-30}$$

By way of illustration let us retrace our steps from the differential scattering cross section (2.2-15) to its originating potential. We will construct the sequence (2.2-30) for that case. From (2.2-2) we find that

$$b^2 = \frac{eE^{-1}(1 - \frac{\theta}{\pi})^2}{(1 - (1 - \frac{\theta}{\pi})^2)} \tag{2.2-31}$$

which may be solved to give θ as a function of b

$$\theta(b) = \pi(1 - (eb^{-2}E^{-1} + 1)^{-\frac{1}{2}}) \tag{2.2-32}$$

Consequently we have $\tilde{\theta}(S) = \pi(1 - (eSE^{-1} + 1)^{-\frac{1}{2}})$ and hence substituting this into (2.2-26) gives

$$\Gamma(W) = \frac{d}{dW}\left[\frac{1}{2\pi}\int_0^W \frac{\pi(eSE^{-1}+1)^{-\frac{1}{2}}}{(W-S)^{\frac{1}{2}}}\,dS\right] \tag{2.2-33}$$

Integrating by substitution we have

$$\Gamma(W) = \frac{1}{2}\frac{W^{-\frac{1}{2}}e^{-1}E}{(W+e^{-1}E)} \tag{2.2-34}$$

and further substitution of this into (2.2-32) yields the function $\tilde{v}(W)$ as

$$\tilde{v}(W) = (1+eE^{-1}W)^{-1} \tag{2.2-35}$$

From the equation $u^2 = \tilde{v}(W)W$ we then obtain

$$u^2 = W(1+eE^{-1}W)^{-1} \tag{2.2-36}$$

so that $W(u) = u^2(1 - eE^{-1}u^2)^{-1}$ and

$$v(u) = \tilde{v}(W(u)) = (1-eE^{-1}u^2) \tag{2.2-37}$$

which by the definition of $v(u)$ gives $\tilde{v}(u) = eu^2$. We deduce finally that $V(r) = er^{-2}$.

This problem in classical mechanics exemplifies the general structure of such dual, direct and inverse problems. First the direct problem is solved and determines a transformation between unknown quantities, in this case a single potential function, and some measurable quantities, in this case a differential scattering cross section. Frequently the measurable quantities are of an asymptotic character. This is because in an asymptotic domain the effects of the unknown quantities are assumed to be known and of simple character. Consequently, one is able to interpret the experimental features with some degree of confidence. Transformations set up between an investigated object and asymptotic data by means of a scattering process are called scattering transforms. The inverse problem then becomes the problem of constructing an inverse scattering transform. As we have just seen the construction of such an inverse scattering transform involves the construction of a sequence of auxiliary functions which are generally unimportant in themselves and only steps to the final answer. This will be a common feature of the inverse scattering transforms in this book.

2.3 Scattering in Quantum Mechanics

Classical mechanics is the appropriate description for systems such as the solar system. For such macroscopic systems the measurement process has negligible effect on the outcome of an observation. An important aspect of the classical view of particle motions is that at any moment of time a classical particle can have both a precise position in space as well as precise values for momentum and other observable characteristics such as energy and angular momentum. The classical scattering problem which we considered in the previous section is typical. The equations (2.2-3 - 2.2-4) may be used to model the motion of a comet in the gravitational field of a planet. In a classical model the comet approaches from some infinite point of the universe in a straight line and is deflected by the gravitational field of a fixed planet into a path which becomes asymptotically another straight line. We say that the comet has been scattered by the gravitational field of the planet. At any point in its trajectory the comet has a precise position, momentum, energy and angular momentum. Indeed the last two are constant throughout the motion.

Quantum mechanics is the appropriate description for small scale systems involving the interactions of subatomic particles. Unlike its classical counterpart a quantum particle is not able to possess both a precise position and a precise momentum simultaneously as the measurement process does interfere with systems of subatomic dimensions. However, it is possible to find a set of simultaneously definable observable characteristics. For example, the energy H and angular momentum J^2 of a particle may be simultaneously measured without mutual interference.

In order to illustrate quantum mechanical ideas and in particular the Schrödinger equation we will consider the situation of a quantum particle moving in a potential field $V(x)$ on the one dimensional real line $-\infty < x < \infty$. The problem is very relevant to the techniques which will be developed in later chapters.

A quantum particle can only be described in probabilistic terms. At any time t the state of such a particle is determined by a complex valued function $\psi^t(x)$ of the position x. Such a function $\psi^t(x)$ is called a wave function and has the property that

$$\int_{-\infty}^{\infty} |\psi^t(x)|^2 dx < \infty \tag{2.3-1}$$

and is therefore said to be the square-integrable. The complete linear space of all such functions will be denoted by $L^2(R)$. This requirement is imposed in order that the following interpretation can be made of $\psi^t(x)$. The quantity $|\psi(x)|^2 dx$ is interpreted as giving the relative probability of finding the particle on the interval $(x,x+dx)$. Provided (2.3-1) holds the absolute probability of finding the particle in the interval (a,b) is taken to be

$$\int_a^b |\psi^t(x)|^2 dx \bigg/ \int_{-\infty}^{\infty} |\psi^t(x)|^2 dx \tag{2.3-2}$$

In the transition from classical to quantum mechanics, classical observable quantities such as the position x, momentum p and energy H are replaced by operators $\hat{x}$, $\hat{p}$ and $\hat{H}$ acting on the wave function Ψ. The operators representing position and momentum are usually taken to be

$$\hat{x}\psi^t(x) = x\psi^t(x) \tag{2.3-3}$$

and

$$\hat{p}\psi^t(c) = \frac{\hbar}{i}\frac{\partial\psi^t}{\partial x}(x) \tag{2.3-4}$$

where $\hbar = h/2\pi$ and h is a physical constant with units of action known as Plancks' constant. The numerical value of h is 6.6252×10^{-34} joule sec. The fact that the position and momentum cannot be simultaneously observed is reflected in the fact that the operators $\hat{x}$ and $\hat{p}$ do not commute.

The order in which the momentum and position are measured makes a difference. From (2.3-3),(2.3-4) we find that $[\hat{x},\hat{p}] = \hat{x}\hat{p} - \hat{p}\hat{x} = i\hbar$ and so $\hbar$ is seen to measure the extent to which the individual measurements of position and momentum interfere with each other. In the limit $\hbar \to 0$ it can be shown that quantum mechanics goes across to classical mechanics. This is known as the classical limit.

The classical energy H of a particle of mass m in a potential $V(x)$ is

$$H = \frac{p^2}{2m} + V(x) \tag{2.3-5}$$

and so the corresponding energy or Hamiltonian operator H in quantum mechanics is defined by

$$H = \frac{\hat{p}^2}{2m} + V(\hat{x}) = -\frac{\hbar^2}{2m}\frac{\partial^2}{\partial x^2} + V(x) \tag{2.3-6}$$

The appropriate physical law which determines the time-evolution of a quantum system is expressed in Schrödinger's equation for $\psi^t(x)$

$$i\hbar\frac{\partial \psi^t}{\partial t} = \hat{H}\psi^t \tag{2.3-7}$$

which becomes as a result of (2.3-6)

$$i\hbar\frac{\partial \psi^t}{\partial t}(x) = -\frac{\hbar^2}{2m}\frac{\partial^2\psi^t}{\partial x^2}(x) + V(x)\psi^t(x) \tag{2.3-8}$$

This is known as the time-dependent Schrödinger equation.

One thing that we must check is that this time evolution is consistent with our probabilistic interpretation of ψ^t. If $\psi^t(x)$ is normalised to unity at some time $t = t_0$, that is

$$\int_{-\infty}^{\infty} |\psi^{t_0}(x)|^2 dx = 1 \tag{2.3-9}$$

requiring ψ^t to be normalised for all time. That this is so results from a single conservation law obtainable from (2.3-8). From (2.3-8) it is easily found that

$$i\hbar\frac{\partial}{\partial t}|\psi^t(x)|^2 = -\frac{\hbar^2}{2m}\frac{\partial}{\partial x}\left(\psi^{t*}(x)\frac{\partial\psi^t}{\partial x}(x) - \psi^t(x)\frac{\partial\psi^{t*}}{\partial x}(x)\right) \tag{2.3-10}$$

independent of $V(x)$. Integrating over the real line and assuming that $\psi^t(x) \to 0$ as $x \to \infty$ results in

$$i\hbar\frac{\partial}{\partial t}\int_{-\infty}^{\infty}|\psi^t(x)|^2 dx = -\frac{\hbar^2}{2m}\left[\psi^{t*}(x)\frac{\partial\psi^t}{\partial x}(x) - \psi^t(x)\frac{\partial\psi^{t*}}{\partial x}(x)\right]_{-\infty}^{\infty} = 0 \tag{2.3-11}$$

and so for such functions the norm of ψ^t is a constant of the motion.

Consequently, an asymptotically zero square-integrable wave function once normalised remains normalised and the probabilistic interpretation can be consistently made. Equation (2.3-10) can be written in the form of a conservation law

$$\frac{\partial \rho^t}{\partial t} + \frac{\partial}{\partial x}j^t = 0 \tag{2.3-12}$$

if ρ^t and j^t are defined in the following way

$$\rho^t = |\psi^t|^2 \quad \text{and} \quad j^t = \frac{\hbar}{2im}\left(\psi^{t*}\frac{\partial}{\partial x}\psi^t - \psi^t\frac{\partial}{\partial x}\psi^{t*}\right) \tag{2.3-13}$$

These are interpreted as the probability density and probability current respectively. An alternative way of writing j^t is

$$j^t = \frac{\hbar}{2mi} W(\psi^{t*},\psi^t) \tag{2.3-14}$$

where the bilinear functional $W(\phi_1,\phi_2)$ is defined by

$$W(\phi_1,\phi_2) = \det\begin{bmatrix} \phi_1 & \phi_2 \\ \frac{\partial\phi_1}{\partial x} & \frac{\partial\phi_2}{\partial x} \end{bmatrix} \tag{2.3-15}$$

and is termed the Wronskian of ϕ_1 and ϕ_2. Wronskians will have an important part to play in the theory which will be developed in later chapters.

In quantum mechanics the eigenfunctions of an operator $\hat{A}$ corresponding to the classical observable quantity A are interpreted as representing states of the particle corresponding to precise A values. If we seek solutions $_E\psi^t$ of the time-dependent Schrödinger equation (2.3-7) corresponding to states of precise energy E,

$$H(_E\psi^t) = E(_E\psi^t) = 2\hbar\frac{\partial}{\partial t}(_E\psi^t) \tag{2.3-16}$$

then we can solve equation (2.3-16) to obtain

$$\psi(t,x,E) = e^{-i\frac{E}{\hbar}t}\psi(x,E) \tag{2.3-17}$$

where, by the first equality of (2.3-16), $\Psi(x,E)$ satisfies the equation

$$E\psi(x,E) = -\frac{\hbar^2}{2m}\frac{\partial^2}{\partial x^2}\psi(x,E) + V(x)\psi(x,E) \tag{2.3-18}$$

which is known as the time-independent Schrödinger equation.

We note that as the time-evolution of such energy eigenstates involves only a change of phase in the wave function and that ρ^t and j^t are both time-independent for such states. An important aspect of j^t is its independence from the potential V(x). For energy eigenstates it is possible to find other such quantities. If ϕ_1 and ϕ_2 are two solutions to the time-independent Schrödinger equation (2.3-18), then one easily shows that

$$\frac{\partial}{\partial x}W(\phi_1,\phi_2) = 0 \tag{2.3-19}$$

and that therefore the Wronskian $W(\phi_1,\phi_2)$ of any two constant energy solutions is a constant.

The condition of square-integrability and continuity of $\Psi(x,E)$ and its first derivative together with other boundary conditions that adapt the general equation (2.3-18) to a specific problem place strong restrictions on the possible values of E. If we define

$$Q(x) = -\frac{2m}{\hbar^2}V(x), \quad k^2 = -\frac{2mE}{\hbar^2}, \quad y(x,k) = \psi(x, -\frac{\hbar^2 k^2}{2m}) \tag{2.3-20}$$

then the Schrödinger equation (2.3-18) can be written in the form

$$[\frac{d^2}{dx^2} + Q(x)]y(x,k) = k^2 y(x,k) \tag{2.3-21}$$

which may be interpreted as an eigenvalue problem for the differential operator

$$L = \frac{d^2}{dx^2} + Q(x) \tag{2.3-22}$$

supplemented by suitable boundary conditions.

We will refer to L as the Schrödinger scattering operator. The eigenvalues E represent the possible energy configurations of the system. In the classical case E is continuous and subject only to the natural restriction $E \geqslant V_{min}$ where V_{min} is the minimum value of the potential $V(x)$. However, in the quantum case the possible values may no longer be continuous. Generally there are a discrete number of isolated points together with continuous intervals which form the allowable set of energy values known as the energy spectrum of L. A precise discussion and definition of the spectrum of the Schrödinger operator will be given in chapter three.

Not all the eigenfunctions of a quantum operator are square integrable. Such functions are not wave functions as a probabilistic interpretation cannot be made for them. We will refer to such non-normalisable functions as generalised wave functions. An example of a generalised wave functions is given by the eigenfunction ${}_k y$ of the momentum operator $\hat{p}$

$$\hat{p}({}_k y)(x) = \frac{\hbar}{i}\frac{\partial}{\partial x}({}_k y(x)) = (\hbar k)({}_k y(x)) \tag{2.3-23}$$

corresponding to the momentum eigenvalue $p = hk$. This equation can be integrated to give

$$y(x,k) = b(k)e^{ikx} \tag{2.3-24}$$

where $b(k)$ is a constant.

Although such momentum eigenstates are not physically interpretable wave functions it is a common feature of elementary quantum mechanics to use them as a basis in which to expand square integrable wave functions. This is essentially the technique of Fourier analysis. Mathematically, it is convenient for our immediate discussion to impose a stronger condition on wave functions than membership of $L^2(R)$. The requirement that a wave function be continuously differentiable and absolutely integrable, $\int_{-\infty}^{\infty} | y^t(x) | dx$, is sufficient by the Cauchy-Schwartz inequality to imply square-integrability. Moreover, it allows us to utilise the classical theory of Fourier transforms and generally permits greater manipulative flexibility. From the classical Fourier integral theorem we know that such a continuously differentiable absolutely integrable wave function has a Fourier representation:

$$y^t(x) = \frac{1}{(2\pi)^{\frac{1}{2}}} \int_{-\infty}^{\infty} e^{ikx}\tilde{y}^t(k)dk \qquad (2.3\text{-}25)$$

where, by the Fourier integral theorem

$$y^t(k) = \frac{1}{(2\pi)^{\frac{1}{2}}} \int_{-\infty}^{\infty} e^{-ikx}y^t(x)dx \qquad (2.3\text{-}26)$$

The individual elements of the integrand $e^{ikx}\tilde{y}^t(k)$ are often referred to as the Fourier modes of y^t.

Only in the case $V \equiv 0$ is $_k y$ a solution of the time-independent Schrödinger equation (2.3-18). If $E = \hbar^2k^2/2m$, then the general solution to the potential free time-independent Schrödinger equation

$$[\frac{d^2}{dx^2} + k^2]y(x,k) = 0 \qquad (2.3\text{-}27)$$

A general solution of this equation is given by

$$Y(x,k) = b(k)e^{ikx} + a(k)e^{-ikx} \qquad (k > 0) \qquad (2.3\text{-}28)$$

The case of $V(x) = 0$ corresponds to free motion and so we refer to (2.3-27) as the free Schrödinger equation which describes the quantum dynamics of a particle for which the classical analogue is rectilinear motion. The general solutions of the time-dependent free Schrödinger equation may be expressed by

$$y^t(x) = \int_0^{\infty} dxe^{\frac{ik^2t\hbar}{2m}} Y(x,k) \qquad (2.3\text{-}29)$$

which is a Fourier integral representation in terms of energy-momentum eigenfunctions as those observables may be simultaneously specified in this case.

We see that $y^t(x)$ consists of two parts $y^t_{\pm}(x)$ given by

$$y^t_+(x) = \int_0^{\infty} b(k)e^{i(kx - k^2\frac{t\hbar}{2m})}dk \qquad (2.3\text{-}30)$$

and

$$y^t_-(x) = \int_0^{\infty} a(k)e^{-i(kx + k^2\frac{t\hbar}{2m})}dk \qquad (2.3\text{-}31)$$

The function y^t_+ consists of waves travelling to the right and the function y^t_- consists of waves travelling to the left. When we are dealing with the time independent-Schrödinger equation it should be remembered that the solution

$$y(x,k) = b(k)e^{ikx} + a(k)e^{-ikx} \qquad (2.3\text{-}32)$$

corresponds to a superposition of a rightward travelling wave, which is a momentum eigenfunction of value $k\hbar$ and a leftward travelling wave, which is a momentum eigenfunction of value $-k\hbar$.

Returning to the general case $V \not\equiv 0$ we suppose that $V(x) \to 0$ as $|x| \to \infty$. If a quantum particle is projected with momentum $-k\hbar$ from $x = +\infty$ there is a probability that it will penetrate through the core of the potential and travel on with momentum $-k\hbar$ to $x = -\infty$ where it is again free. There is also a probability that it will be reflected back from the potential and travel back to $x = +\infty$ with momentum $+\hbar k$. These two mutually exclusive possibilities of classical mechanics are both possible in the world of the quantum. In a time-dependent treatment, an initial wave packet travels in from infinity and as it moves into the region of influence of the potential it is generally broken up into its Fourier modes. Part of the initial packet passes through the potential and moves off to $-\infty$ and the remainder is reflected back from the potential. Figure 2-5 shows a sequence of views of a wave packet incident from the right upon the Gaussian potential $Q(x) = 2e^{-x^2}$.

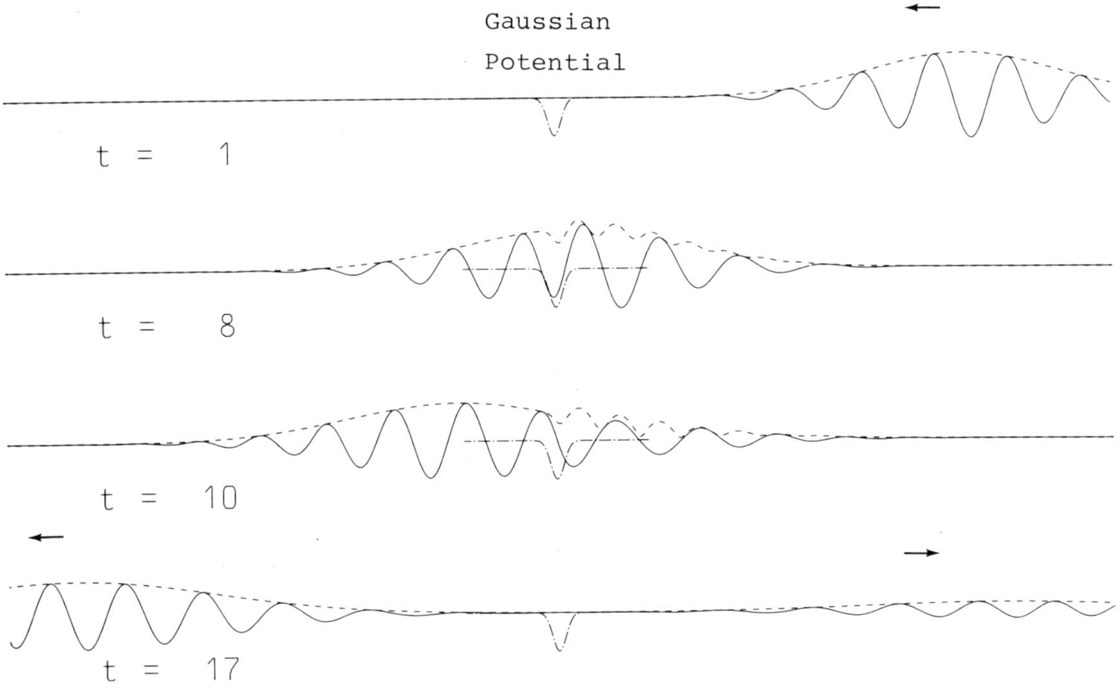

Figure 2-5: -.-.- $Q(x)$, ——— $\mathrm{Re}(y)$, - - - $|y|$

If the initial particle at $x = +\infty$ is represented by the momentum eigenstate e^{-ikx} corresponding to a free leftward travelling particle of momentum $-k\hbar$, we must add to it a reflected wave travelling to the right with momentum $\hbar k$. Therefore, the appropriate asymptotic form of the wave function describing the particle at $x = +\infty$ is given by

$$\phi(x,k) \sim (e^{-ikx} + R(k)e^{+ikx}) \qquad x \longrightarrow +\infty \tag{2.3-33}$$

At the other asymptotic extreme $x \to -\infty$ there is only the rightward travelling wave which has penetrated through the potential. Consequently, the appropriate asymptotic form is given by

$$\tilde{\phi}(x,k) \sim T(k)e^{-ikx} \qquad x \longrightarrow -\infty \tag{2.3-34}$$

The functions $R(k)$ and $T(k)$ are known as reflection and transmission coefficients.

In the analysis that we will make of the above scattering process it is convenient to normalise the wave function $_k\tilde{\phi}$ to a function $_k\phi$ having the asymptotic behaviour $\phi(x,k) \sim e^{-ikx}$ as $x \to -\infty$. This means that we will seek wave functions $_k\phi$ with the asymptotic form

$$\phi(x,k) \sim \begin{cases} b(k)e^{ikx} + a(k)e^{-ikx} & x \longrightarrow +\infty \\ e^{-ikx} & x \longrightarrow -\infty \end{cases} \tag{2.3-35}$$

where $a(k) = 1/T(k)$ and $b(k) = R(k)/T(k)$ and we note that $R(k) = b(k)/a(k)$. One immediate relationship between the functions a and b can be obtained from the constancy of j^t. Since $j^t(+\infty) = j^t(-\infty)$ for real values of k we have

$$|a(k)|^2 - |b(k)|^2 = 1 \tag{2.3-36}$$

The quantities $a(k)$ and $b(k)$ represent asymptotic data concerned with the Schrödinger scattering process in the same way that $\theta(b)$ was concerned with the classical scattering in the previous section. In the remainder of this section and the next we will start to investigate the extent to which a knowledge of $a(k)$ and $b(k)$ can help us to determine $Q(x)$. As a conclusion to this section we will solve two simple potential scattering problems. This will help to make us feel more at home with the functions $a(k)$ and $b(k)$. The examples are chosen for simplicity and exact solubility and should not be thought of as especially typical of any single feature. The examples are (i) the delta potential, and (ii) the square well potential.

2.3.1 The Delta Potential

$$Q(x) = Q_0\delta(x) \tag{2.3-37}$$

where $\delta(x)$ is the Dirac delta distribution and may be thought of as the limiting member of the family of Gaussian potentials $\pi^{-\frac{1}{2}}\alpha\exp(-x^2\alpha^2)$ as $\alpha \to -\infty$.

The Schrödinger equation for the delta potential (2.3-37) takes the form

$$\left[\frac{d^2}{dx^2} + k^2 + Q_0\delta(x)\right]\phi(x,k) = 0 \tag{2.3-38}$$

As $\delta(x)$ is zero for $x \neq 0$ this is essentially a free problem. At $x = 0$ the delta distribution leads to a boundary condition as follows. Integrating equation (2.3-38) on an interval $(-\varepsilon,\varepsilon)$ about $x = 0$ we obtain

$$\left[\frac{d}{dx}\phi(x,k)\right]_{-\varepsilon}^{\varepsilon} + \int_{-\varepsilon}^{\varepsilon}\{Q_0\delta(x) + k^2\}\phi(x,k)\,dx = 0 \tag{2.3-39}$$

Using the continuity of ${}_k\phi$ at $x = 0$ and the basic property of $\delta(x)$ we obtain in the limit $\varepsilon \to 0$

$$\frac{d}{dx}({}_k\phi)(0+) - \frac{d}{dx}({}_k\phi)(0-) = Q_0(\phi_k(0)) \tag{2.3-40}$$

which conflicts with our usual condition of continuity of the first derivative of ${}_k\phi$.

Let us write

$$\phi(x,k) = \begin{cases} b(k)e^{ikx} + a(k)e^{-ikx} & x \longrightarrow +\infty \\ e^{-ikx} & x \longrightarrow -\infty \end{cases} \tag{2.3-41}$$

The continuity of ${}_k\phi$ at $x = 0$ gives

$$a(k) + b(k) = 1 \tag{2.3-42}$$

and the discontinuity condition (2.3-40) gives

$$ik - ik[a(k) - b(k)] = -Q_0 \tag{2.3-43}$$

Solving for $a(k)$ and $b(k)$ we obtain

$$a(k) = [1 + \frac{Q_0}{2ik}], \quad b(k) = -\frac{Q_0}{2ik} \tag{2.3-44}$$

As a check one easily finds that for real k, $|a|^2 - |b|^2 = 1$ as required by the conservation of the probability current j.

If k is allowed to be pure imaginary, corresponding to a state of negative total energy, we see from (2.3-41) that it is possible for ${}_k\phi$ to be normalisable provided $a(k)$ has a zero on the positive imaginary axis of the complex k-plane. Thus we require $Q_0 > 0$ so that the zero of $a(k)$, $k = iQ_0/2$ obtained from (2.3-44) lies on the positive imaginary axis. Equation (2.3-41) then determines the solution to be

$$\phi(x,\frac{iQ_0}{2}) = e^{-Q_0|x|/2} \tag{2.3-45}$$

If we were looking at these wave functions solely within the context of quantum mechanics we would normalise (2.3-45) so that it had unit $L^2(R)$ norm. However, in the context of the techniques that we will develop in this book, a different scaling for wave functions is adopted. We will see in chapters 3 and 4 that the normalising of a wave function $Y(x)$ by requiring $\int_{-\infty}^{\infty} Y^2(x)dx = 1$ rather than $\int_{-\infty}^{\infty}|Y(x)|^2dx = 1$ is more convenient in the context relevant for us. For the delta potential wave function (2.3-45) the normalised wave function is $\sqrt{c}\,\phi(x,\tfrac{1}{2}Q_0 i)$ where

$$c^{-1} = \int_{-\infty}^{\infty}[\phi(x,\frac{iQ_0}{2})]^2dx = \frac{2}{Q_0} \tag{2.3-46}$$

which is the same as the quantum mechanical normalisation as $\phi(x,\tfrac{1}{2}iQ_0)$ is real. The wave-function (2.3-45) describes a particle centred on $x = 0$ the location of the potential. Such a localised state about the potential centre is called a **bound state.** It is the quantum analogue of a bound orbit such as that of the Moon about the Earth. Solution (2.3-45) is referred to as a bound state wave function. From (2.3-30) we see that the energy value of the state is $E = h^2k^2/2m = -Q_o^2h^2/8m$ a negative energy configuration.

2.3.2 The Square Well Potential

In this case $Q(x)$ takes the form

$$Q(x) = \begin{cases} Q_0 & |x| \leqslant \alpha \\ 0 & |x| > \alpha \end{cases} \tag{2.3-47}$$

The Schrödinger equations are

$$\left[\frac{d^2}{dx^2} + k^2 + Q_0\right]\phi(x,k) = 0 \quad \text{for} \quad |x| \leqslant \alpha \tag{2.3-48}$$

and

$$\left[\frac{d^2}{dx^2} + k^2\right]\phi(x,k) = 0 \quad \text{for} \quad |x| > \alpha \tag{2.3-49}$$

to which must be added the subsidiary conditions of continuity of $({}_k\phi)$ and $({}_k\phi)'$ at $x = \pm\alpha$.

We must write down a form for the wave function in each of the regions $x > \alpha$, $|x| < \alpha$ and $x < -\alpha$. An appropriate form for the wave function ${}_k\phi$ is given by

$$\phi(x,k) = \begin{cases} e^{-ikx} & x \leq -\alpha \\ B(k,\zeta)e^{i\zeta x} + A(k,\zeta)e^{-i\zeta x} & |x| \leq \alpha \\ b(k,\zeta)e^{ikx} + a(k,\zeta)e^{-ikx} & x \geq \alpha \end{cases} \tag{2.3-50}$$

where $\zeta^2 = k^2 + Q_o^2$.

Continuity of ${}_k\phi$ at $x = \pm\alpha$ requires that

$$Ae^{i\zeta\alpha} + Be^{-i\zeta\alpha} = e^{ik\alpha} \tag{2.3-51}$$

$$Ae^{-i\zeta\alpha} + Be^{i\zeta\alpha} = ae^{-ik\alpha} + be^{ik\alpha} \tag{2.3-52}$$

and the continuity of $({}_k\phi)'$ at $x = \mp\alpha$ yields

$$Ae^{i\zeta\alpha} - Be^{-i\zeta\alpha} = \frac{k}{\zeta}e^{ik\alpha} \tag{2.3-53}$$

$$Ae^{-i\zeta\alpha} - Be^{i\zeta\alpha} = \frac{k}{\zeta}(ae^{-ik\alpha} - be^{ik\alpha}) \tag{2.3-54}$$

From (2.3-51) and (2.3-53) we obtain

$$A(k,\zeta) = \frac{1}{2}e^{-i\zeta\alpha}(1+\frac{k}{\zeta})e^{ik\alpha}, \qquad B(k,\zeta) = \frac{1}{2}e^{i\zeta\alpha}(1-\frac{k}{\zeta})e^{ik\alpha} \tag{2.3-55}$$

and (2.3-52) together with (2.3-54) determine a and b,

$$a(k,\zeta) = \frac{1}{2}e^{+ik\alpha}(1+\frac{\zeta}{k})Ae^{-i\zeta\alpha} + (1-\frac{\zeta}{k})Be^{i\zeta\alpha} \tag{2.3-56}$$

$$b(k,\zeta) = \frac{1}{2}(1-\frac{\zeta}{k})Ae^{-i\zeta\alpha} + (1+\frac{\zeta}{k})Be^{-ik\alpha} \tag{2.3-57}$$

Combining these last two equations with (2.3-55) produces the final result

$$a(k,\zeta) = e^{2ik\alpha}[\cos 2\zeta\alpha - \frac{i}{2}(\frac{\zeta}{k}+\frac{k}{\zeta})\sin 2\zeta\alpha] \tag{2.3-58}$$

$$b(k,\zeta) = \frac{i}{2}(\frac{\zeta}{k}-\frac{k}{\zeta})\sin 2\zeta\alpha \tag{2.3-59}$$

These satisfy the condition $|a(k,\zeta)|^2 - |b(k,\zeta)|^2 = 1$ for real k and ζ as required by the conservation of the probability current which remains equal to $-k/m$ for all x.

Putting $Q_0 = \bar{Q}_0/2\alpha$ and allowing $\alpha \to 0$ we find that $\zeta\alpha \to 0$ and $\zeta \sin 2\zeta\alpha \to \bar{Q}_0$. In the limit $\alpha \to 0$, equations (2.3-58) and (2.3-59) show that

$$a(k,\zeta) \longrightarrow a(k) = 1 - \frac{i\bar{Q}_0}{2k} \quad \text{as} \quad \alpha \longrightarrow 0 \tag{2.3-60}$$

and

$$b(k,\zeta) \longrightarrow b(k) = \frac{i}{2k}\bar{Q}_0 \quad \text{as} \quad \alpha \longrightarrow 0 \tag{2.3-61}$$

where the limit functions $a(k)$ and $b(k)$ are those of the delta potential (2.3-37). This is to be expected as the potential family

$$Q(x,\alpha) = \begin{cases} -Q_0/2\alpha & |x| \leqslant \alpha \\ 0 & |x| > \alpha \end{cases} \tag{2.3-62}$$

also has the delta potential as its $\alpha \to 0$ limit. This limit is

$$\lim_{\alpha\to 0} Q(x,\alpha) = -\bar{Q}_0\delta(x) \tag{2.3-63}$$

As the delta potential has a bound state $(Q_0 > 0)$ we expect the potential well to have bound states also. To find them it is necessary to investigate the imaginary zeros $k = ik(k > 0)$ of $a(k,\zeta)$. For $k = i\kappa$, $a(k,\zeta)$ takes the form

$$a(i\kappa,\chi) = e^{-\kappa\alpha}(\cos 2\chi\alpha - \frac{1}{2}(\frac{\chi}{\kappa}-\frac{\kappa}{\chi})\sin 2\chi\alpha) \tag{2.3-64}$$

where $\chi^2 = (Q_0 - \kappa^2)^{\frac{1}{2}}$. If $\chi^2 < 0$ it is easy to show that $a(i\kappa,\chi)$ has no zeros. However, if $\chi^2 > 0$ then $k = i\kappa$ is a zero of $a(k,\chi)$ provided that

$$\tan 2\chi\alpha = \frac{2\kappa\chi}{\chi^2 - \kappa^2} \tag{2.3-65}$$

Introducing $\xi = \chi\alpha$ and $\beta^2 = Q_0\alpha^2$, (2.3-65) may be written as

$$\tan 2\xi = \frac{2(\beta^2 - \xi^2)^{\frac{1}{2}}}{(2\xi^2 - \beta^2)} \tag{2.3-66}$$

and we seek solutions for values of $\xi \leqslant \beta$. Equation (2.3-66) may be solved graphically by plotting the curves $y = \tan 2\xi$ and $y = 2(\beta^2 - \xi^2)^{\frac{1}{2}}(2\xi^2 - \beta^2)^{-1}$ and examining where they intersect.

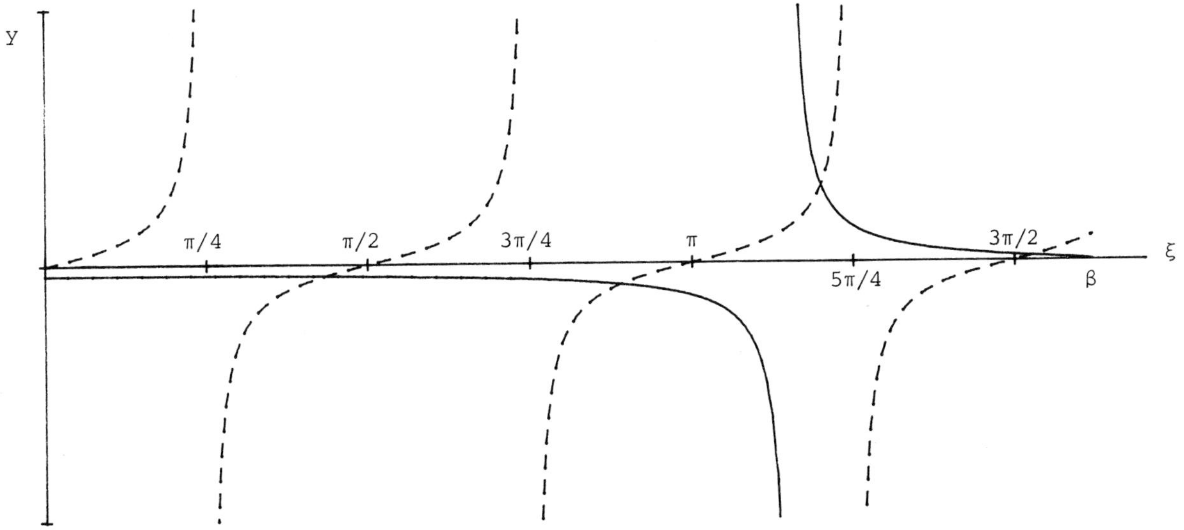

Figure 2-6:

From the graphs in Fig. 2-6 we see that if $N\pi/2 \leqslant \beta \leqslant (N+1)\pi/2$ then there are (N+1) solutions $\{\gamma\}_{n+1}^{N+1}$. If $Q_o = Q_o/2\alpha$ then $\beta \to 0$ as $\alpha \to 0$ and we obtain just one bound state in the $\alpha \to 0$ limit. This is consistent with our previous analysis of the delta potential. A simple example is given by the potential

$$Q(x) = \begin{cases} 2 & |x| \leqslant \pi/4 \\ 0 & |x| > \pi/4 \end{cases} \tag{2.3-67}$$

which has a single bound state of energy $E = -\hbar^2/2m$ with wave function

$$\phi(x,i) = \begin{cases} e^{-x} & x \geqslant \frac{\pi}{4} \\ e^{-\pi/4}\sqrt{2}\cos x & |x| \leqslant \frac{\pi}{4} \\ e^{x} & x \leqslant -\frac{\pi}{4} \end{cases} \tag{2.3-68}$$

This function is shown in Fig. 2-7. The normalised form of this wave function is $\sqrt{C}\phi(x,i)$ where C is given by

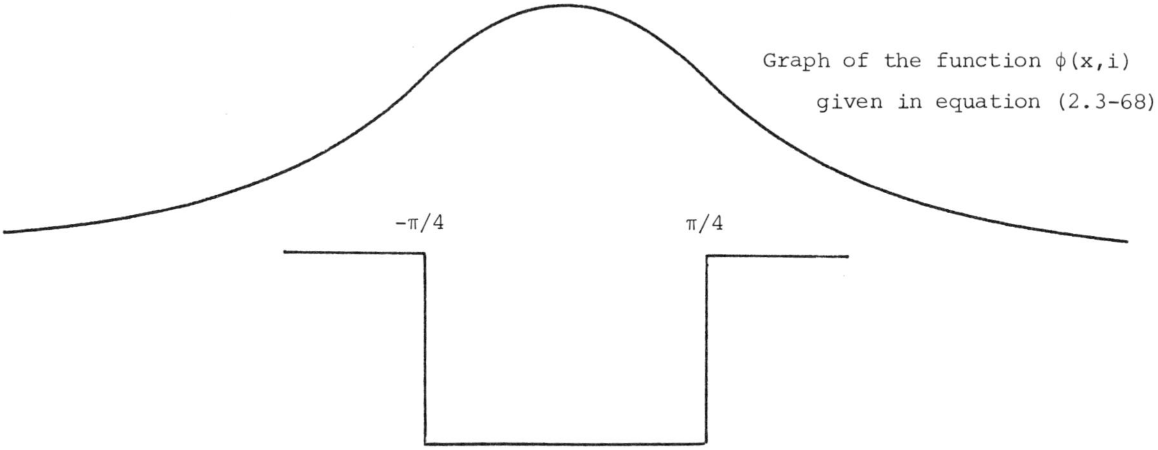

Figure 2-7:

$$C^{-1} = \int_{-\infty}^{\infty} [\phi(x,i)]^2 dx = \int_{\pi/4}^{\infty} e^{-2x} x + 2\int_{-\pi/4}^{\pi/4} e^{-\pi/2} \cos^2 x dx + \qquad (2.3\text{-}69)$$

$$+ \int_{-\infty}^{-\pi/4} e^{2x} \alpha x = (2 + \frac{\pi}{2}) e^{-\pi/2}$$

As the wave function $\phi(x,i)$ is real this normalisation is the same as that of quantum mechanics and consequently the probability of finding the particle in the well is $(2+\pi/4+\pi)$ or approximately 70%.

Taking $k = \sqrt{2}$, then the scattering wave function of positive energy $E = h^2/m$ is

$$\phi(x,2) = \begin{cases} \exp(-i\sqrt{2}x) & x \leq -\pi/4 \\ -\exp(i\pi/2\sqrt{2}).\{\sin 2x + (\cos 2x)/\sqrt{2}\} & x \leq \pi/4 \\ \exp\{-i\sqrt{2}x + i\pi(1 + 1/\sqrt{2})\} & x \geq \pi/4 \end{cases} \qquad (2.3\text{-}70)$$

This describes a particle which is not reflected by the potential but is transmitted through to $+\infty$ with a transmission coefficient $b = \exp[-i\pi(1 + 1/\sqrt{2})]$ which has modulus one. The wave is said to have been phase shifted by an amount $\pi(1 + 1/\sqrt{2})$ by its passage through the potential. Values of k such as $k = \sqrt{2}$ which are zeros of the reflection coefficient $R(k,\zeta)$ are isolated and so a wave packet made up of waves of momentum in an interval about $\sqrt{2}$ will always be affected by transmission. The narrower the band of k values about $\sqrt{2}$ the nearer to perfect transmission will the situation become. Such a narrowly spread pulse will pass almost unaltered through the potential except for a phase shift. In fact as this example shows there are special k values which allow perfect transmission. In the next section we will examine special potentials for which such transparency effects are more general.

2.4 Reflectionless Potentials

There are special potentials which single themselves out for particular attention on account of their remarkable properties. Figures 2-8 - 2-10 shows a sequence of views of a wave packet incident upon the potential

$$Q(x,m) = m(m+1)\operatorname{sech}^2 x \qquad (2.4\text{-}1)$$

for three values of the parameter m.

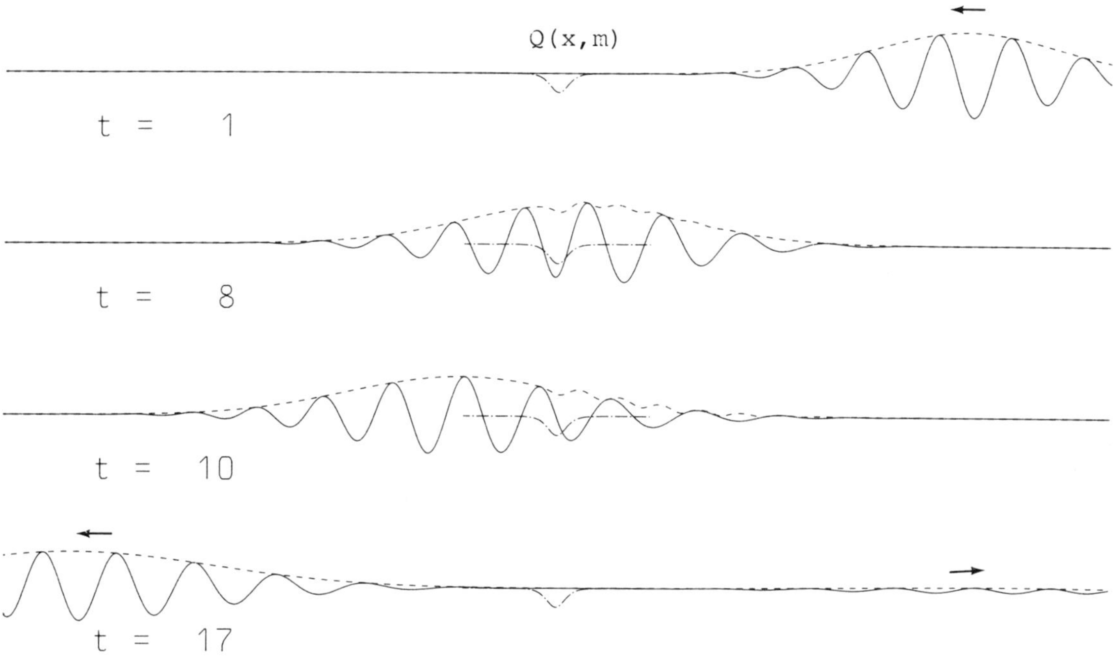

Figure 2-8: -.-.- Q(x), ——— Re(y), - - - |y|, (m < 1).

In Figs. 2-8 and 2-10, corresponding to values of $m < 1$ and $m > 1$, we see the typical behaviour mentioned previously of the packet decomposing into a transmitted and reflected portion. However, in Fig. 2-9, corresponding to $m = 1$, the wavepacket starts to break up into its Fourier modes as soon as it begins to penetrate the well, but as it penetrates further, its individual waves are resynthesised and the packet emerges complete after its passage through the well. Clearly in case Fig. 2-9 when $m = 1$ we have a situation of total transmission with no reflected wave. This is a property of the potential $Q(x,m)$ for any integer value of m and such potentials are known as reflectionless potentials for obvious reasons. The case $m = 1$ can be easily solved using a factorisation method and it is both instructive and useful in what follows for this case to be solved explicitly. The Schrödinger equation to be solved is

$$\left(\frac{d^2}{dx^2} + (2\operatorname{sech}^2 x + k^2)\right) y(x,k) = 0 \qquad (2.4\text{-}2)$$

Defining the operators $L_{\pm}$ by

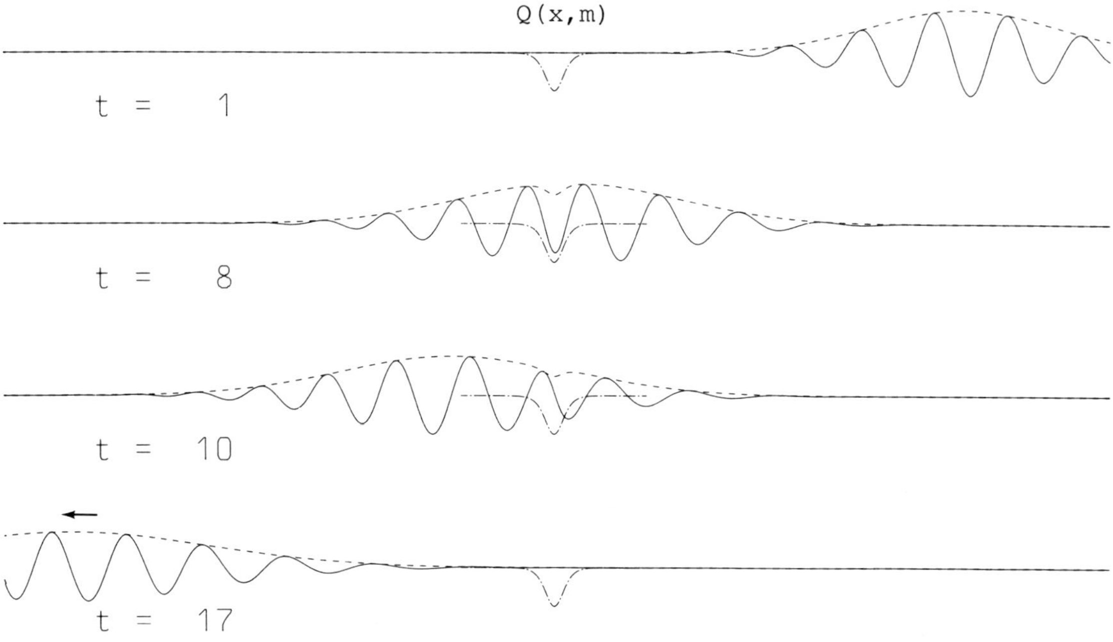

Figure 2-9: -.-.- Q(x), ——— Re(y), - - - |y|, (m = 1).

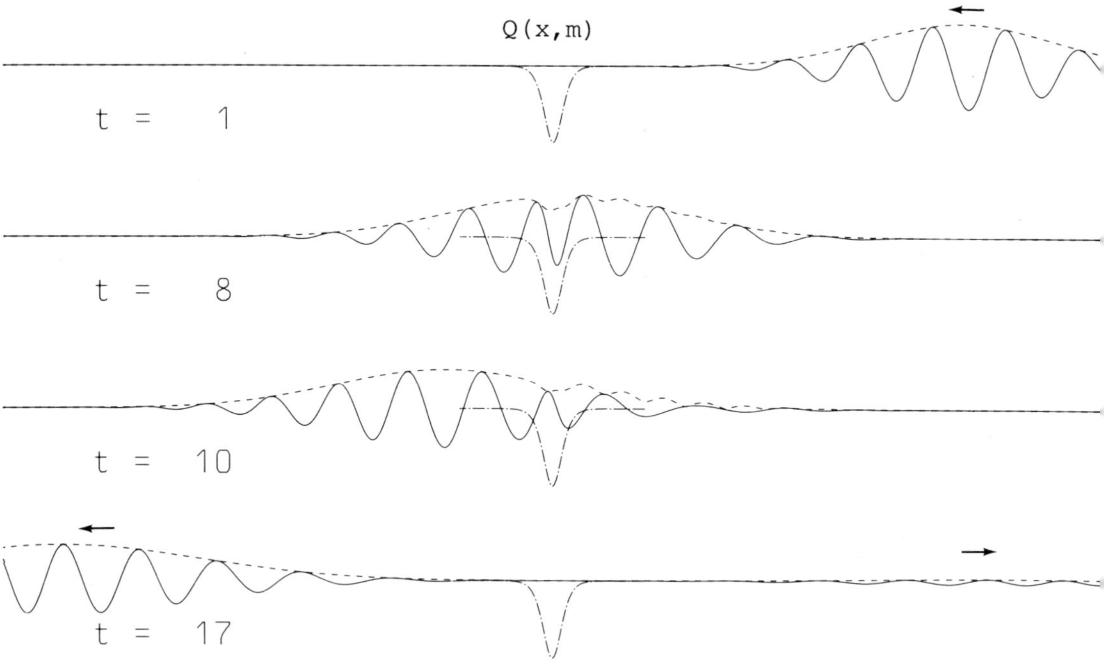

Figure 2-10: -.-.- Q(x), ——— Re(y), - - - |y|, (m > 1).

$$L_{\pm} = (\tanh x \mp \frac{d}{dx}) \tag{2.4-3}$$

We can express (2.4-2) as

$$L_+L_-({}_k y) = (k^2+1)({}_k y) \tag{2.4-4}$$

It is easy to show that ${}_k Y$ defined by

$${}_k Y = L_-({}_k y) \tag{2.4-5}$$

is a solution of the equation

$$\left(\frac{d^2}{dx^2} + k^2\right) Y(x,k) = 0 \tag{2.4-6}$$

Furthermore, for each solution ${}_k Y$ of (2.4-6), the function ${}_k y$, defined by

$${}_k y = L_+({}_k Y) \tag{2.4-7}$$

is a solution of (2.4-2). Using this attractive algebraic structure we can now easily solve the $m = 1$ case. For the scattering problem we seek a solution of (2.4-2) determined by the asymptotic boundary conditions

$$\phi(x,k) \sim \begin{cases} e^{-ikx} & x \to -\infty \\ b(k)e^{ikx} + a(k)e^{-ikx} & x \to +\infty \end{cases} \tag{2.4-8}$$

where $b(k)$ and $a(k)$ are reflection and transmission coefficients. The general solution to (2.4-6) is

$$Y(x,k) = Ae^{-ikx} + Be^{+ikx} \tag{2.4-9}$$

and therefore applying $L_{\pm}$ a general solution of (2.4-2) must be given by

$$y(x,k) = A(\tanh x + ik)e^{-ikx} + B(\tanh x - ik)e^{ikx} \tag{2.4-10}$$

As $x \to \pm\infty$, $\tanh x \to \pm 1$. Consequently we see that

$$a(k) = \frac{ik+1}{ik-1}, \quad b(k) = 0 \tag{2.4-11}$$

corresponding to the exact wave-function ${}_k\phi$ given by

$$\phi(x,k) = \frac{(\tanh x + ik)}{(ik-1)} e^{-ikx} \tag{2.4-12}$$

Bound states of the system correspond to negative values of the energy. For such states $k = i\kappa$ with κ real, and the wave-function is normalisable. Putting $k = i\kappa$ in (2.4-12) gives the wave-function $\phi(x,i\kappa)$ defined by

$$\phi(x,i\kappa) = \frac{(\kappa - \tanh x)}{(1+\kappa)} e^{+\kappa x} \tag{2.4-13}$$

Asymptotically the wave-function $\phi(x,i\kappa)$ has the behaviour

$$\phi(x,i\kappa) \sim \begin{cases} e^{+\kappa x} & \text{for} \quad x \to -\infty \quad (2.4\text{-}14) \\ \left[\frac{\kappa-1}{\kappa+1}\right] e^{+\kappa x} & \text{for} \quad x \to +\infty \quad (2.4\text{-}15) \end{cases}$$

and can only be normalised if the transmission coefficients $(\kappa-1)/(\kappa+1)$ in (2.4-15) is zero. We therefore require $\kappa = 1$ and only one bound state exists with energy $E = -\hbar^2/2m$ and normalisable wave-function

$$\phi(x,i) = \tfrac{1}{2}(1 - \tanh x)e^{x} = \tfrac{1}{2}\operatorname{sech} x \tag{2.4-16}$$

The normalised wave-functions is $\sqrt{2}\,\phi(x,i)$. In general if a wave function y is rescaled to $Y = \sqrt{C}y$ so that $\int_{-\infty}^{\infty} |Y|^2(x)dx = 1$, C is called the normalisation constant. For the potential $Q(x,N)$ where N is an integer one can show in a similar fashion that the wave-function can be given by

$$y(x,k) = \prod_{m=1}^{N} \left(m \tanh x - \frac{d}{dx}\right) Y(x,k) \tag{2.4-17}$$

where, as before, $Y(x,k)$ is a solution of (2.4-6). In this case

$$a(k) = \prod_{m=1}^{N} \frac{ik+m}{ik-m}, \qquad b(k) = 0 \tag{2.4-18}$$

For the bound states we must have the condition

$$a(i\kappa) = 0 \tag{2.4-19}$$

which leads to N bound states with energies E_m given by

$$E_m = -\frac{\hbar^2}{2M} m^2 \qquad m = 1,\ldots,N. \tag{2.4-20}$$

The potential $U(x) = -2\operatorname{sech}^2 x$ has therefore separated itself out for our special attention by virtue of its reflectionless property. By using the scaling symmetries of the Schrödinger equation we can enlarge this single example of a reflectionless potential to a whole two parameter family of potentials $Q(x,\kappa,c)$ defined by

$$Q(x,\kappa,c) = 2\kappa^2 \operatorname{sech}^2(\kappa x + c) \tag{2.4-21}$$

Consider the Schrödinger equation with the potential $Q(z, \ ,c)$

$$\left(\frac{d^2}{dz^2} + (K^2 + 2\kappa^2 \operatorname{sech}^2(\kappa z + c))\right)\omega(x,K) = 0 \tag{2.4-22}$$

Defining x, ${}_k y$ and k by,

$$x = \kappa z + c, \qquad y(x,k) = \omega\left(\frac{x-c}{\kappa}, K\right), \qquad k = K/\kappa \tag{2.4-23}$$

then ${}_k y$ satisfies the Schrödinger equation (2.4-2) and we see that the potential $Q(x,\kappa,c)$ is also reflectionless with transmission coefficient

$a(K)$ given by

$$a(K) = \frac{i\kappa + K}{i\kappa - K} \tag{2.4-24}$$

There are several generalisations of the above factorisation method which are worth exploring. If we have a Schrödinger equation with potential $q(x)$ and wave function $_k y$ given by

$$\left[\frac{d^2}{dx^2} + q(x) + k^2\right] y(x,k) = 0 \tag{2.4-25}$$

can we find a continuous function $u(x)$ such that $_k Y$ defined by

$$\left(\frac{d}{dx} - u(x)\right) y(x,k) = Y(x,k) \tag{2.4-26}$$

satisfies another Schrödinger equation,

$$\left(\frac{d^2}{dx^2} + Q(x) + k^2\right) Y(x,k) = 0 \tag{2.4-27}$$

with k^2 the same in each case?

Substituting (2.4-26) into (2.4-27) we readily discover that for compatibility we must have

$$Q = q + 2\frac{du}{dx} \tag{2.4-28}$$

and

$$-q = \frac{du}{dx} + u^2 + C \tag{2.4-29}$$

where C is an integration constant.

If we define

$$Z(x,\sigma) = \left(\frac{d}{dx} \ln z(x,\sigma)\right) \tag{2.4-30}$$

where $_k z$ is a solution of (2.4-25), then the function $Z(x,\sigma)$ satisfies the equation

$$\frac{dZ}{dx} + Z^2 + \sigma^2 = -q \tag{2.4-31}$$

For (2.4-30) to be well defined everywhere we must have $z(x,\sigma)$ non zero. A necessary and sufficient condition, by the oscillation theorem for Sturm-Liouville systems, is that σ^2 must lie to the left of the continuous and discrete spectrum of (2.4-25). Therefore, if we take $\sigma = iK$ where $K \geq$ (the modulus of the smallest bound state eigenvalue), then a suitable choice for $u(x)$ is the function

$$u(x) = \frac{d}{dx} \log(z(x,iK)) \tag{2.4-32}$$

where $z(x,iK)$ is any wave function for the potential $q(x)$ corresponding to $\sigma = iK$. If we put $Q = -2dW/dx$ and $q = -2dw/dx$ we can eliminate $u(x)$ from (2.4-28)-(2.4-29) and obtain the relationship

$$\frac{dW}{dx} = -\frac{dw}{dx} - K^2 + (W - w)^2 \qquad (2.4\text{-}33)$$

Potentials Q and q related in such a way are said to be Bäcklund Transforms of one another.

As an example of the application of this results we can rederive our reflectionless potential 2K sech (Kx + c) as follows. Setting q = 0, then the most general wave function of (2.4-25) corresponding to k = iK is

$$z(x,iK) = (A \cosh Kx + B \sinh Kx) \qquad (2.4\text{-}34)$$

with A and B nonzero. The corresponding u(x) is given by

$$u(x) = K^2 \tanh(Kx + c) \qquad (2.4\text{-}35)$$

where $c = \tanh^{-1}(B/A)$.

From this we obtain from (2.4-28) the potential

$$Q(x) = 2K^2 \operatorname{sech}^2(Kx + c) \qquad (2.4\text{-}36)$$

As we know u(x) we can construct the wavefunctions corresponding to the potential $2K^2\operatorname{sech}^2(Kx + c)$ explicitly. Selecting any value of $k^2 \le -K^2$ we can construct a new function $u_a(x)$ and derive a new potential and corresponding wavefunctions. Consequently a tower of solutions to various different Schrödinger equations may be constructed.

Sometimes a Bäcklund transformation may be only a symmetry. For example if we choose k= 1 and c = 0 and take the wave function

$$z_1(x,iK) = \left(\frac{K - \tanh x}{k + 1}\right) e^{+Kx} \qquad K^2 > 1 \qquad (2.4\text{-}37)$$

we find that the corresponding u(x), denoted by $u_1(x)$, is given by

$$u_1(x) = \frac{\operatorname{sech}^2 x}{\tanh x - K} + K \qquad (2.4\text{-}38)$$

leading, via (2.4-28), to the potential

$$Q(x) = \frac{2(K^2 - 1)\operatorname{sech}^2 x}{(\tanh x - K)^2} = 2 \operatorname{sech}^2(x + c), \quad c = -\coth^{-1} K \qquad (2.4\text{-}39)$$

The wavefunction ${}_k\phi$ which has the asymptotic behaviour

$$\Phi(x,k) \sim e^{-ikx} \qquad x \to -\infty \qquad (2.4\text{-}40)$$

is easily seen to be

$$\Phi(x,k) = -\frac{1}{(K + ik)}\left[\frac{d}{dx} - K - \frac{\operatorname{sech}^2 x}{(\tanh x - K)}\right] \phi(x,k) \qquad (2.4\text{-}41)$$

Taking the asymptotic limit $x \to -\infty$ we see that we must have

$$a(k) = \frac{(k + i)}{(k - i)}, \qquad b(k) = 0 \qquad (2.4\text{-}42)$$

exactly the same asymptotic data as ${}_k\phi$ which was the wave function corresponding to the potential 2 sech x. It is now obvious that the correspondence between a potential and the pair of functions $a(k)$, $b(k)$ is not one to one. However, there is a difference . The bound state wavefunction is

$$\Phi(x,i) = \frac{1}{(1-K)}[\frac{d}{dx} - u_1]\,\phi(x,i) \qquad (2.4\text{-}43)$$

and the normalisation constant C of $\phi(x,i)$ is given by

$$C^{-1} = \int_{-\infty}^{\infty} \Phi^2(x,i)\,dx = \int_{-\infty}^{\infty}(\frac{d}{dx} - u_1)\phi(x,i)(\frac{d}{dx} - u_1)\phi(x,i)\frac{dx}{(1-K)^2} \qquad (2.4\text{-}44)$$

Integrating the right hand expression by parts gives

$$(K-1)^2C^{-1} = \int_{-\infty}^{\infty} \phi(x,i)(-\frac{d}{dx} - u_1)(\frac{d}{dx} - u_1)\phi(x,i)\,dx$$

$$= \int_{-\infty}^{\infty} \phi(x,i)(-\frac{d^2}{dx^2} + \frac{du_1}{dx} + u_1^2)\phi(x,i)\,dx \qquad (2.4\text{-}45)$$

which by the construction of u_1 gives for (2.4-45),

$$= \int_{-\infty}^{\infty} \phi(x,i)(-\frac{d^2}{dx^2} - 2\,\mathrm{sech}^2 x + K^2)\phi(x,i)\,dx$$

$$= (K^2 - 1)\int_{-\infty}^{\infty} \phi^2(x,i)\,dx \qquad (2.4\text{-}46)$$

$$= \tfrac{1}{2}(K^2 - 1)$$

Consequently, the normalisation factor of the bound state wavefunction has been altered from 2 to $C(K)$ given by

$$C(K) = 2\left(\frac{K+1}{K-1}\right) = 2e^{-2c} \qquad (2.4\text{-}47)$$

Thus, any connection between potentials and the asymptotic behaviour of the wave functions will have to include a specification of normalisation constants as well as the functions $a(k)$, $b(k)$, at the very least. We also note that the construction of the normalisation constant above was more general than the particular potential $2\,\mathrm{sech}^2 x$ used to illustrate it. There is clearly a close interrelation between Bäcklund transformations and symmetries of the Schrödinger equation such as that in (2.4-23). If one generates a Bäcklund transformation from a more general solution than $\phi(x,i)$, potentials outside of the family $Q(x,K,c)$ can be generated, which have an additional bound state.

We may hope that a knowledge of $a(k)$ and $b(k)$ together with the normalisation constants will be sufficient to reconstruct the potential $Q(x)$. However, as our final example will show, strong constraints are going to be needed on the potentials and their associated wavefunctions if this is to be possible. For example the potential

$$Q(x,\alpha) = \frac{2}{(x+\alpha)^2} \tag{2.4-48}$$

has the associated wavefunction,

$$\phi(x,k) = [1 - \frac{1}{2k(x+\alpha)}]e^{-ikx} \tag{2.4-49}$$

This is clearly reflectionless with $b(k) = 0$ and $a(k) = 1$ for all values of k and α. There are no bound states to be normalised and not even a phase shift as the transmission coefficient $T(k)$ is identically unity. The potential (2.4-48) is singular at $x = -\alpha$ and so we may be optimistic about eliminating it by putting constraints such as absolute integrability on our class of potentials. Another aspect which we note at this point is that $(_k\phi)$ has a pole at $k = 0$. This will turn out to be an undesirable feature of which we must be wary. It is interesting to note that the singular potentials $Q(x,\alpha)$ may be obtained by a limiting process applied to an analytic continuation of the family of potentials $Q(x,\gamma,c) = 2\gamma^2\text{sech}^2(\gamma x+c)$ as follows

$$Q(x,\alpha) = \lim_{\gamma\to 0} Q(x,\gamma,\gamma\alpha + \frac{i\pi}{2}) \tag{2.4-50}$$

and suggests a way in which they might be treated.

2.5 Generalistions

The factorisation method used to solve the reflectionless potential problem (2.4-2) has within it a natural generalisation. If our Schrödinger equation can be factored into the form

$$(u(x) - \frac{d}{dx})(u(x) + \frac{d}{dx})\psi(x,k) = (k^2 + A)\phi(x,k) \tag{2.5-1}$$

the the equation (2.4-37) can be reduced to the pair of first order equations below,

$$[\frac{d}{dx} + u(x)]\psi_1 = i\zeta\psi_2 \tag{2.5-2}$$

$$[\frac{d}{dx} - u(x)]\psi_2 = -i\zeta\psi_1 \tag{2.5-3}$$

where $\psi_1 = \psi$ and ψ_2 is defined by (2.5-2) and $\zeta^2 = (k^2 + A)$. Defining new functions v_1 and v_2 by

$$v_1 = \tfrac{1}{2}(\psi_1 + \psi_2), \qquad v_2 = \frac{1}{2i}(\psi_2 - \psi_1) \tag{2.5-4}$$

(2.5-2)-(2.5-3) can be expressed as the two-dimensional eigenvalue problem,

$$\begin{pmatrix} \frac{id}{dx} & u(x) \\ u(x) & -\frac{id}{dx} \end{pmatrix} \begin{pmatrix} v_1 \\ v_2 \end{pmatrix} = \zeta \begin{pmatrix} v_1 \\ v_2 \end{pmatrix} \tag{2.5-5}$$

For this first order scattering problem an asymptotic analysis can be performed similar to that for the Schrödinger equation. The time-dependent system, corresponding to the time-dependent Schrödinger equation, is the Dirac equation given by

$$\left[\begin{bmatrix} i(\partial_x + \partial_t) & 0 \\ 0 & i(\partial_t - \partial_x) \end{bmatrix} + u(x)\right]\begin{bmatrix} v_1 \\ v_2 \end{bmatrix} = \begin{bmatrix} 0 \\ 0 \end{bmatrix} \tag{2.5-6}$$

Taking $u(x) = \tanh x$ which corresponds to the factorisation involved in solving the $2\,\mathrm{sech}^2 x$ reflectionless potential, we can solve the system (2.5-2)-(2.5-3) (see Figs. 2-8-2-10). Using the solution (2.4-12) it is easy to show that after a renormalisation a corresponding solution to (2.5-5) is given by

$$\phi(x,k,\zeta) = \begin{pmatrix} \dfrac{\tanh x + i(\zeta + k)}{i(\zeta + k) - 1} \\ \dfrac{i\tanh x - (k - \zeta)}{i(\zeta + k) - 1} \end{pmatrix} e^{-ikx}, \quad k = \sqrt{\zeta^2 - 1} \tag{2.5-7}$$

In the limit $x \to -\infty$; $\phi(x,k,\zeta)$ $\phi(x,k,\zeta) \sim \begin{pmatrix} 1 \\ k-\zeta \end{pmatrix} e^{-ikx}$ which in turn is a solution to the system

$$\begin{bmatrix} \dfrac{id}{dx} & -1 \\ -1 & -\dfrac{id}{dx} \end{bmatrix}\begin{bmatrix} v_1 \\ v_2 \end{bmatrix} = \xi \begin{bmatrix} v_1 \\ v_2 \end{bmatrix} \tag{2.5-8}$$

This is the asymptotic limit of the system (2.5-5) as $x \to -\infty$. When $x \to +\infty$, (2.5-7) has the asymptotic form

$$\phi(x,k,\zeta) \sim \left(\frac{k + \zeta - i}{k + \zeta + i}\right)\begin{pmatrix} 1 \\ \zeta - k \end{pmatrix} e^{-ikx} \tag{2.5-9}$$

For a general potential $u(x) \to \pm 1$ as $x \to \pm\infty$, a solution $\phi(x,k,\zeta)$ of (2.5-5) which has the asymptotic boundary condition,

$$\phi(x,k,\zeta) \sim \begin{pmatrix} 1 \\ k - \zeta \end{pmatrix} e^{-ikx} \qquad x \to -\infty \tag{2.5-10}$$

will be expressible at $x = +\infty$ in the asymptotic form:

$$\phi(x,k,\zeta) \sim a(\zeta,k)\begin{pmatrix} 1 \\ \zeta - k \end{pmatrix} e^{-ikx} + b(\zeta,k)\begin{pmatrix} \zeta - k \\ 1 \end{pmatrix} e^{ikx} \tag{2.5-11}$$

The vectors $v_- = \begin{pmatrix} 1 \\ \zeta-k \end{pmatrix} e^{ikx}$ and $v_+ = \begin{pmatrix} \zeta-k \\ 1 \end{pmatrix} e^{ikx}$ are, in fact, the eigenfunctions of the linear system

$$\begin{pmatrix} \frac{id}{dx} & +1 \\ +1 & -\frac{id}{dx} \end{pmatrix} \begin{pmatrix} v_1 \\ v_2 \end{pmatrix} = \zeta \begin{pmatrix} v_1 \\ v_2 \end{pmatrix} \tag{2.5-12}$$

which is seen to be the asymptotic form of the system (2.5-5) when $x \to +\infty$. The functions $v_\pm$ correspond to rightward and leftward travelling asymptotic waves of the Dirac system (2.3-6). The $a(\zeta,k)$ and $b(\zeta,k)$ are the natural analogues of the $a(k)$ and $b(k)$ of the previous sections.

Solutions (2.5-7) shows that $u(x) = \tanh x$ is a reflectionless potential for the scattering problem (2.5-5). The asymptotic data $a(\zeta,k)$ and $b(\zeta,k)$ take the form

$$a(\zeta,k) = \left(\frac{k+\zeta-i}{k+\zeta+i}\right), \qquad b(\zeta,k) = 0 \tag{2.5-13}$$

We have been careful to show the dependence of $a(\zeta,k)$ and $b(\zeta,k)$ on both ζ and k since these functions are defined on the Riemann surface of the function $k = \sqrt{\zeta^2-1}$. We showed a similar dependency for the scattering data of the square well but close examination shows that in that case $R(k,\zeta) = b(k,\zeta)/a(k,\zeta) = \tilde{R}(k,\zeta^2)$.

The Wronskian (2.3-15) of the Schrödinger eigenvalue problem may be extended to the first order system (2.5-5) and is defined to be

$$W(V_1,V_2) = \det \begin{pmatrix} V_{11} & V_{12} \\ V_{21} & V_{22} \end{pmatrix} \tag{2.5-14}$$

where $V_1 = \begin{pmatrix} V_{11} \\ V_{21} \end{pmatrix}$ and $V = \begin{pmatrix} V_{12} \\ V_{22} \end{pmatrix}$ are any two solutions of (2.5-2) corresponding to the same ζ value.

As $|u(+\infty)| = |u(-\infty)|$ it may be shown by use of the Wronskian (2.5-14) that $|a(k,\zeta)|^2 - |b(k,\zeta)|^2 = 1$ in an analogous way to the Schrödinger case.

There are two new features to this problem. Firstly, we are dealing with a system written in terms of two first order equations rather than a single second order equation and secondly, in the case $u(x) = \tanh x$, we have a situation in which the potential $u(x)$ does not go to zero as $|x| \to \infty$. In this book we will not pursue further the case in which the potentials do not vanish at $x = \pm\infty$ but instead, we will develop the first of these two possible extensions. Once the case of potentials vanishing at infinity is understood it is a technical rather than a conceptual problem to extend it to the asymptotically non-trivial case.

The most obvious generalisation is to replace the second order Schrödinger eigenvalue problem (2.3-21) by the first order eigenvalue problem

$$\frac{\partial}{\partial x}V_1 + ikV_1 = q(x)V_2 \tag{2.5-15}$$

$$\frac{\partial}{\partial x}V_2 - ikV_2 = r(x)V_1 \tag{2.5-16}$$

which includes (2.5-5) as the special case $q = -r = iu$. We also note that when $r = -1$ we regain the Schrödinger problem for V_2

$$\left(\frac{d^2}{dx^2} + k^2 + q(x)\right) V_2(x,k) = 0 \tag{2.5-17}$$

The functions V_1 and V_2 are functions of x and k but for notational convenience we will not show this dependence except where necessary for additional clarity. As the system was first analysed in full generality by Zakharov and Shabat (1972) and by Ablowitz, Kaup, Newell and Segur (1974) we will refer to it as the ZS-AKNS system. At infinity the system is described by the free equations

$$\frac{\partial}{\partial x} V_1 + ikV_1 = 0 \tag{2.5-18}$$

$$\frac{\partial}{\partial x} V_2 - ikV_2 = 0 \tag{2.5-19}$$

which have the general solution

$$V_1 = ae^{-ikx} \tag{2.5-20}$$

$$V_2 = be^{ikx} \tag{2.5-21}$$

For the general case (2.5-15)-(2.5-16) involving two unknown potentials q and r it proves necessary to specify two wave-functions defined by specific asymptotic behaviours. We can construct solutions ${}_k\phi$ and ${}_k\bar{\phi}$ with the asymptotic behaviours

$$\phi(x,k) \sim \begin{cases} e^{-ikx}\begin{bmatrix}1\\0\end{bmatrix} & x \to -\infty \quad (2.5\text{-}22) \\ \begin{bmatrix}ae^{-ikx}\\ be^{ikx}\end{bmatrix} & x \to +\infty \quad (2.5\text{-}23) \end{cases}$$

and

$$\bar{\phi}(x,k) \sim \begin{cases} \begin{pmatrix}0\\-1\end{pmatrix}e^{ikx} & x \to -\infty \quad (2.5\text{-}24) \\ \begin{bmatrix}\bar{b}e^{-ikx}\\ -\bar{a}e^{ikx}\end{bmatrix} & x \to +\infty \quad (2.5\text{-}25) \end{cases}$$

where the functions a, b, $\bar{a}$, $\bar{b}$ are distinct from the similarly denoted functions of sections 3 and 4.

In a similar study to that carried out in the last two sections we will attempt to take some of the mystery out of the scattering problems (2.5-15)-(2.5-16) by solving some simple illustrative examples. We consider two special cases: (i) the square well, and (ii) the potential $q(x) = -r(x) = -2\text{sech}2x$.

2.5.1 The Square Well Potential

$$q(x) = -r(x) = \begin{cases} 0 & |x| > \gamma \\ -Q_0 & |x| \leqslant \gamma \end{cases} \tag{2.5-26}$$

For the square well potential (2.5-26) the ZS-AKNS system (2.5-15)-(2.5-16) takes the form

$$\begin{pmatrix} \frac{d}{dx} + ik & 0 \\ 0 & \frac{d}{dx} - ik \end{pmatrix} \begin{pmatrix} V_1 \\ V_2 \end{pmatrix} = \begin{pmatrix} 0 \\ 0 \end{pmatrix} \quad \text{for} \quad |x| > \gamma \tag{2.5-27}$$

and

$$\begin{pmatrix} \frac{d}{dx} + ik, & Q_0 \\ -Q_0 & \frac{d}{dx} - ik \end{pmatrix} \begin{pmatrix} V_2 \\ V_2 \end{pmatrix} = \begin{pmatrix} 0 \\ 0 \end{pmatrix} \quad \text{for} \quad |x| \leqslant \gamma \tag{2.5-28}$$

which is similar to that in (2.5-8).

Let us write the solution as

$$\begin{pmatrix} V_1 \\ V_2 \end{pmatrix} = \begin{cases} \begin{bmatrix} Ae^{-ikx} \\ Be^{ikx} \end{bmatrix} & x \geq \gamma \qquad (2.5\text{-}29) \\ \alpha \begin{bmatrix} 1 \\ \frac{k - \zeta}{iQ_0} \end{bmatrix} e^{-i\zeta x} + \beta \begin{bmatrix} \frac{k - \zeta}{iQ_0} \\ 1 \end{bmatrix} e^{i\zeta x}, & |x| \leqslant \gamma \qquad (2.5\text{-}30) \\ \begin{bmatrix} Ce^{-ikx} \\ De^{ikx} \end{bmatrix} & x \leqslant -\gamma \qquad (2.5\text{-}31) \end{cases}$$

where $\zeta^2 = k^2 + Q_0^2$.

Continuity at $x = \pm\gamma$ gives the conditions

$$e^{-ik\gamma}A = \alpha e^{-i\zeta\gamma} + \frac{i(\zeta - k)}{Q_0} e^{i\zeta\gamma}\beta \tag{2.5-32}$$

$$e^{ik\gamma}B = \alpha \frac{i(\zeta - k)}{Q_0} e^{-i\zeta\gamma} + e^{i\zeta\gamma}\beta \tag{2.5-33}$$

$$e^{ik\zeta}C = \alpha e^{i\zeta\gamma} + \frac{i(\zeta - k)}{Q_0} e^{-i\zeta\gamma}\beta \tag{2.5-34}$$

$$e^{-ik\zeta}D = \alpha \frac{i(\zeta - k)}{Q_0} e^{i\zeta\gamma} + e^{-i\zeta\gamma}\beta \tag{2.5-35}$$

If $C = 1$, $D = 0$ we obtain from (2.5-34),(2.5-35)

$$\alpha = e^{ik\gamma - i\zeta\gamma}\left[1 + \left(\frac{\zeta - k}{Q_0}\right)^2\right]^{-1} \tag{2.5-36}$$

$$\beta = -e^{ik\gamma + i\zeta\gamma}\left[\frac{i(\zeta - k)}{Q_0[1 + \left(\frac{\zeta - k}{Q_0}\right)^2]}\right] \tag{2.5-37}$$

and from (2.5-32),(2.5-33)

$$a(k,\zeta) = e^{2ik\gamma}[\cos 2\zeta\gamma - \frac{2k}{\zeta}\sin 2\zeta\gamma] \tag{2.5-38}$$

$$b(k,\zeta) = \frac{Q_0}{\zeta}\sin 2\zeta\gamma \tag{2.5-39}$$

If $C = 0$, $D = -1$

$$\alpha = \frac{i(\zeta - k)e^{-i\zeta\gamma - ik\gamma}}{Q_0[1 + \left(\frac{\zeta - k}{Q_0}\right)^2]} \tag{2.5-40}$$

$$\beta = -e^{-ik\gamma + i\zeta\gamma}[1 + \left(\frac{\zeta - k}{Q_0}\right)^2]^{-1} \tag{2.5-41}$$

and from (2.5-32),(2.5-33)

$$\bar{a}(k,\zeta) = e^{2ik\gamma}[\cos 2\zeta\gamma - \frac{2k}{\zeta}\sin 2\zeta\gamma] = (a(k^*,\zeta^*))^* \tag{2.5-42}$$

$$\bar{b}(k,\zeta) = \frac{Q_0}{\zeta}\sin 2\zeta\gamma = (b(k^*,\zeta^*))^* \tag{2.5-43}$$

By analogy with the Schrödinger case we can see that there are two ways in which bound states can occur. Either the zeros of $a(k,\zeta)$ lie on the positive imaginary k-axis or the zeros of $\bar{a}(k,\bar{\zeta})$ lie on the negative imaginary k-axis. As $\bar{a}(k,\bar{\zeta}) = (a(k^*,\zeta^*))^*$ in this case we need only consider $a(k,\zeta)$.

There will be a zero of $a(k,\zeta)$ at $k = i\kappa$ ($\kappa > 0$ and ζ real) if

$$\cot 2\zeta\gamma = -\frac{\kappa}{\zeta}; \qquad \zeta^2 = Q_0^2 - \kappa^2 \geqslant 0 \tag{2.5-44}$$

Denoting $\zeta\gamma = \xi$ then (2.5-44) can be expressed as an equation for ξ similar to (2.3-65)

$$\cot 2\xi = -\frac{(\delta^2 - \xi^2)^{\frac{1}{2}}}{\xi} \tag{2.5-45}$$

Values of ξ must be in the range $\xi \leq \delta$ where $\delta = \gamma Q_0$. A graphical analysis of equation (2.5-45) is analogous to that of (2.3-66). Figure 2-11 shows that if $N\pi/2 \leq (\delta-\pi)/4 \leq (N+1)\pi/2$, there are (N+1) roots or bound states.

As a specific example let us consider the potential

$$q(x) = \begin{cases} 0 & |x| > \frac{3\pi}{8} \\ -\sqrt{2} & |x| \leqslant \frac{3\pi}{8} \end{cases} \tag{2.5-46}$$

which has a single bound state at $k = i$ with bound state wave function

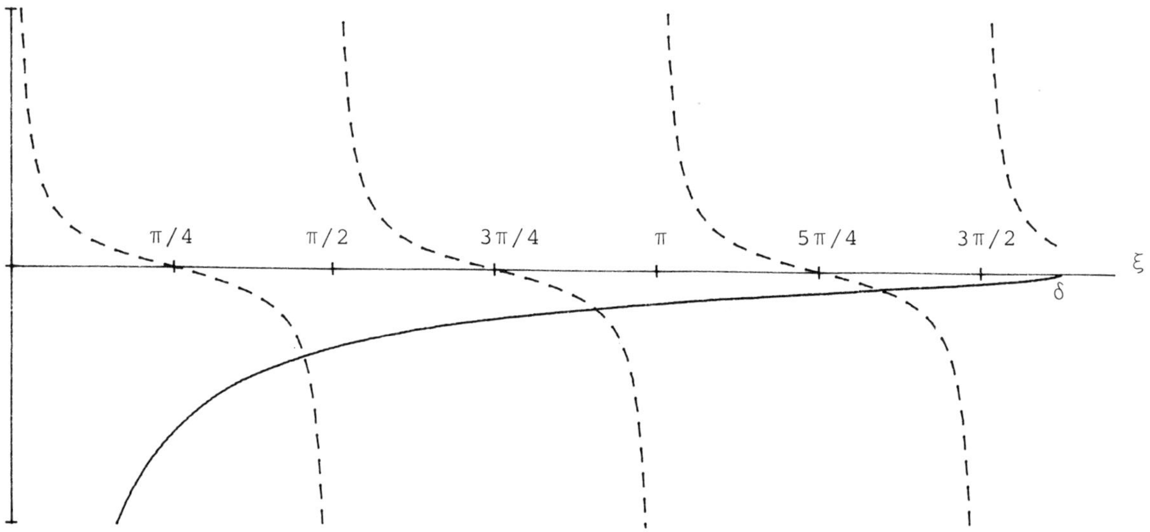

Figure 2-11: δ = 13π/8

$$\phi(x,i) = \begin{bmatrix}\phi_1\\ \phi_2\end{bmatrix} = \begin{cases} \begin{pmatrix}0\\1\end{pmatrix} e^{-x} & x \geqslant \frac{3\pi}{8} \\ \sqrt{2}\, e^{-3\pi/8}\begin{bmatrix}\sin(x+5\pi/8)\\ \sin(x+3\pi/8)\end{bmatrix} & |x| \leqslant \frac{3\pi}{8} \\ \begin{pmatrix}1\\0\end{pmatrix} e^{x} & x < -\frac{3\pi}{8}\end{cases} \tag{2.5-47}$$

If $\chi(x)$ is a bound state wave function of the ZS-AKNS system it is convenient to introduce normalisation constants C defined by

$$C^{-1} = 2\int_{-\infty}^{\infty} \chi_1\chi_2\,dx \tag{2.5-48}$$

We note that this need not be positive, or even real and is not related to quantum mechanical probabilities of finding particles in specific spin states in various positions. For the example (2.5-47) the normalisation constant $C^{-1} = (1 + 3\pi/4\sqrt{2})\exp(-3\pi/4)$.

2.5.2 The Potential q = -r = -2 sech 2x

To solve the potential $q = -2 \operatorname{sech} 2x$ we can use a technique similar to the factorisation method. Defining L_+ by

$$L_+ = \begin{bmatrix} \frac{d}{dx} - \tanh 2x & -\operatorname{sech} 2x \\ \operatorname{sech} 2x & \frac{d}{dx} - \tanh 2x \end{bmatrix} \tag{2.5-49}$$

it is easy to show that ${}_k v = L_+({}_k V)$ is a solution to the ZS-AKNS system

$$\begin{bmatrix} \frac{d}{dx} + ik, & 2\operatorname{sech}2x \\ -2\operatorname{sech}2x, & \frac{d}{dx} - ik \end{bmatrix} v(x,k) = \begin{bmatrix} 0 \\ 0 \end{bmatrix} \tag{2.5-50}$$

whenever ${}_k V$ is a solution to the free system

$$\begin{bmatrix} \frac{d}{dx} + ik, & 0 \\ 0, & \frac{d}{dx} - ik \end{bmatrix} V(x,k) = \begin{bmatrix} 0 \\ 0 \end{bmatrix} \tag{2.5-51}$$

The general solution to (2.5-50) is therefore given by

$$\begin{bmatrix} V_1 \\ V_2 \end{bmatrix} = \begin{bmatrix} -C(ik + \tanh 2x)e^{-ikx} - D\operatorname{sech}2x\, e^{ikx} \\ C\operatorname{sech}2x\, e^{-ikx} + D(ik - \tanh 2x)e^{ikx} \end{bmatrix} \tag{2.5-52}$$

In particular we have the solutions for ${}_k\phi$ and ${}_k\bar{\phi}$, which are

$$\phi(x,k) = \begin{bmatrix} \frac{\tanh 2x + ik}{ik - 1} \\ -\frac{\operatorname{sech}2x}{ik - 1} \end{bmatrix} e^{-ikx} \tag{2.5-53}$$

and

$$\bar{\phi}(x,k) = \begin{bmatrix} \frac{\operatorname{sech}2x}{1 + ik} \\ \frac{\tanh 2x - ik}{1 + ik} \end{bmatrix} e^{ikx} \tag{2.5-54}$$

from which we see that a, b, $\bar{a}$ and $\bar{b}$ are given by

$$a(k) = \left(\frac{1 + ik}{ik - 1}\right), \qquad b(k) = 0 \tag{2.5-55}$$

$$\bar{a}(k) = \left(\frac{ik - 1}{ik + 1}\right), \qquad \bar{b}(k) = 0 \tag{2.5-56}$$

A numerical simulation of a wave packet incident upon the potential $q(x) = -2\operatorname{sech}2x$ would show close similarities with Figs. 2-8 - 2-10. There is a bound state at $k = i$ with corresponding wave function

$$\phi(x,i) = \begin{bmatrix} \phi_1(x,i) \\ \phi_2(x,i) \end{bmatrix} = \begin{bmatrix} \frac{e^{-x}}{2} \\ \frac{e^{x}}{2} \end{bmatrix} \text{sech } 2x \tag{2.5-57}$$

Moreover, the state ϕ normalised according to (2.5-48) is $\tilde{\phi}$ having the asymptotic behaviour $\tilde{\phi}(x,i) \sim \sqrt{C}\, e^{x}$ as $x \to -\infty$, where C is given by

$$C^{-1} = 2\int_{-\infty}^{\infty} \phi_1(x,i)\phi_2(x,i)\,dx = \tfrac{1}{2}\int_{-\infty}^{\infty} \text{sech}^2 2x = \tfrac{1}{2} \tag{2.5-58}$$

There are also bound states corresponding to zeros of $\bar{a}(k)$ on the lower half plane imaginary axis. In this case, $k = -i$ and the corresponding wave function is

$$\bar{\phi}(x,-i) = \begin{bmatrix} \bar{\phi}_1(x,i) \\ \bar{\phi}_2(x,i) \end{bmatrix} = \begin{bmatrix} \frac{e^{x}}{2} \\ -\frac{e^{-x}}{2} \end{bmatrix} \text{sech } 2x \tag{2.5-59}$$

with normalisation constant $\bar{C}^{-1} = -\tfrac{1}{2}\int_{-\infty}^{\infty}\text{sech}^2 2x\,dx = -\tfrac{1}{2}$.

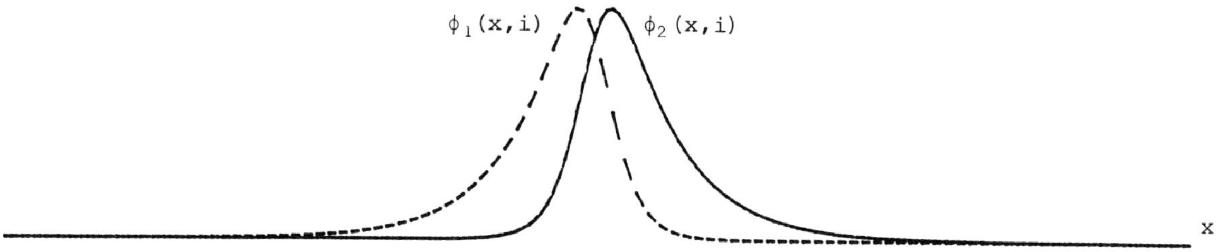

Figure 2-12:

Figure 2-12 shows the two components of the wave function $\phi(x,i)$. As seen in the Schrödinger case, the normalization constants represent a vital part of the scattering data, if we wish to reconstruct a unique potential.

In this chapter we have tried to dispel some of the fears that the reader may experience when he now moves on to tackle the rigorous chapters that follow. We also hope that having a source of examples to which to relate the abstract results will also be beneficial and lead to a greater understanding of the material. Very few of the mathematical problems have been unearthed as yet and we have only attempted to indicate the restrictions on potentials involved in order to produce a well defined scattering transform. By now the reader will have many questions and we hope that the following chapters will answer some of these. We hope also that these chapters will encourage and equip the reader to go on and construct his own solutions.

2.6 Notes

Section 2.1

The application of inverse scattering techniques to problems in remote sensing is discussed by Mittra (1973).

Scanning techniques developed for X-rays are describe in an article by Smith et al. (1977).

The claim that the transformation F is invertible on $A(R)$ is an oversimplification, as the inverse image of an element of $A(R)$ need not be in $A(R)$. It is better to define the image $A(R)$ as a new space $\tilde{A}(R)$ on which F^{-1} is defined. For equations (2.2-11) and (2.2-12) to be valid we then require that $B \in \tilde{A}(R)$ implies $\exp(ikx)B \in \tilde{A}(R)$.

The assumption that $q^t(x)$ satisfies the boundary condition (2.1-19) is necessary because, contrary to intuition, the absolute integrability of a function does not imply that it must go to zero at infinity.

A very good account of classical direct and inverse problems can be found in the article by Keller (1967).

Section 2.2

This section is based on the work of Keller, Kay, and Shmoys (Keller et al., 1956)

Section 2.3

The assumption that $\psi(x) \to 0$ as $x \to \infty$ following equation (2.3-10) is necessary as it does not follow from the integrability properties of the wave function. We impose the conditions $\psi(\pm\infty) = 0$ to make the operator H well defined. The time evolution operator $U = \exp(-iHt)$ is then unitary on the domain of H and can be extended to $L^2(R)$ in a norm preserving manner.

The continuous part of the spectrum of L (equation (2.3-22)) does not exist if we maintain the square integrability condition: they are the generalized eigenvalues obtained when this condition is dropped.

The mathematical reason for requiring the continuity of $d\psi/dx$ is to ensure self-adjoint boundary conditions for the operator $H = -d^2/dx^2 + V(x)$.

Section 2.4

The factorization method is described in the paper by Infeld and Hull (1951).

A useful collection of papers on Bäcklund Transformations is that edited by Miura (1976).

2.7 Problems

1.

The function $f(x)$ is defined by

$$f(x) = \sqrt{2\pi}(1-|x|) \quad \text{for } |x| \leq 1 \text{ and } 0 \text{ otherwise}$$

Show that

$$(Ff)(k) = \frac{2}{k^2}(1 - \cos k)$$

and therefore that

$$F^{-1}(2k^{-2}(1-\cos k)) = f(x)$$

Hence deduce that F^{-1} does not map $A(R)$ onto itself.

2.

Show that the expression

$$F(x) = \sum_{n=1}^{\infty} (2\pi)^{-\frac{1}{2}}\exp(-\tfrac{1}{2}(x-n^2)n^4)$$

defines an infinitely differentiable function which is absolutely integrable. Show that

$$F(N^2) \geqslant (2\pi)^{-\frac{1}{2}} \qquad \text{for all } N \in Z^+$$

and hence deduce that $F(x) \not\to 0$ as $x \to \infty$.

3.

A particle oscillates in a one dimensional symmetric potential well $V(x) = V(-x)$. Given the period of oscillation $T(E)$ for a particle of energy E, find the potential $V(x)$.

4.

Determine the potential which gives rise to the scattering cross section

$$\sigma(\theta) = \frac{e^2}{16E^2\sin^4(\theta/2)}$$

5.

Show that the Schrödinger equation

$$(\frac{d^2}{dx^2} + k^2 - x^2)\psi = 0$$

can be written in the factored form

$$L_+L_-\psi = \lambda\psi$$

with λ and $L_\pm$ defined by

$$\lambda = \tfrac{1}{2}(k^2-1), \qquad L_\pm = (x \mp d/dx)/\sqrt{2}$$

Show that with the normal inner product on $L^2(R)$

$$L_+ = L_-^+$$

and that $L_\pm$ satisfy the commutation relations

$$[L_-, L_+] = 1$$

If ${}_\lambda\phi$ is the square integrable normalised eigenfunction corresponding to eigenvalue λ show that

(i) $\lambda = \| L_-({}_\lambda\phi) \|^2$

(ii) $L_-({}_\lambda\phi) = \sqrt{\lambda}({}_{\lambda-1}\phi)$

(iii) $L_+({}_\lambda\phi) = \sqrt{\lambda+1}({}_{\lambda+1}\phi)$

Deduce from (i) and (ii) that the eigenvalues are $\lambda_n = n$, an integer, and that the corresponding eigenfunctions ${}_n\phi$ are given by

$${}_n\phi = (L_+)^n({}_0\phi)(\sqrt{n!})^{-1}$$

where ${}_0\phi$ satisfies the equation $L_-({}_0\phi) = 0$ and is given by

$${}_0\phi = \frac{1}{\sqrt{\pi}} \exp(-\tfrac{1}{2}x^2)$$

6.

Show that the Schrödinger equation with the potential

$${}_aV(x) = \frac{2}{(x + a)^2}$$

can be exactly solved by the Factorisation method with $u(x)$ chosen to be

$$u(x) = \frac{1}{x + a}$$

Use the exact result to obtain a Bäcklund transform of the potential ${}_aV$.

7.

Use the results of Problem 6 to determine the scattering data $a(k)$ and $b(k)$ for the potential

$$V(x) = \frac{2}{1 + |x|^2}$$

8.

Consider the Schrödinger equation

$$\left[\frac{d^2}{dx^2} + k^2 + W(x)\right]\psi = 0$$

where $W(x)$ is the potential step defined by

$$W(x) = \begin{cases} -2\operatorname{sech}^2 x & x \leq 0 \\ -2 & x \geq 0 \end{cases}$$

Let ψ be defined to be the wave function with the asymptotic behaviour

$$\psi \sim \begin{cases} \exp(ik'x) & x \to +\infty \\ a(k,k')\exp(ikx) + b(k,k')\exp(-ikx) & x \to -\infty \end{cases}$$

with k and k' related by $(k')^2 = k^2 + 2$. Determine the functions $a(k,k')$ and $b(k,k')$.

9.

Solve the AKNS system with $q = -r = W$ where W is the potential step defined by

$$W(x) = \begin{cases} -2 & x \geq 0 \\ -2\operatorname{sech} 2x & x \leq 0 \end{cases}$$

State and determine the scattering data for this problem.

3. THE SCHRÖDINGER EQUATION AND THE KORTEWEG-DE VRIES EQUATION

3.1 The Korteweg-de Vries Equation and Bäcklund Transformations

In 1967, in an amazingly brief and original Physical Review Letter, Gardner, Greene, Kruskal and Miura (1967) exactly solved the Korteweg-de Vries (KdV) equation on the real line

$$q_t + \alpha q q_x + q_{xxx} = 0 \tag{3.1-1}$$

for solutions that tended sufficiently fast to a constant value as $|x| \to \infty$. In their letter they took $\alpha = -6$, but by simply rescaling $q : q \to -\alpha q/6$, α can be chosen as any convenient number. This letter was the culmination of the investigations by these authors into the properties of the KdV (Miura 1968, Miura et al. 1968, Su and Gardner 1969, Kruskal et al. 1970, Gardner 1971, Gardner et al. 1974). These papers derived many of the properties for the KdV which are now known to be characteristic of a whole class of solvable partial differential equations (See Chapter Note 3.1.1).

They showed, for example, that the KdV on the real line possesses an infinite number of conservation laws which we discuss in detail in Section 3.5. Miura, by an examination of the associated conserved densities and the equivalent infinite set for the Modified KdV (MKdV) equation

$$v_t - \beta v^2 v_x + v_{xxx} = 0 \tag{3.1-2}$$

established a transformation which related solutions of the two equations (Miura 1968)

$$q = \frac{1}{\alpha}\,(-\beta v^2 + \varepsilon (6\beta)^{\frac{1}{2}} v_x), \qquad \varepsilon = \pm 1 \tag{3.1-3}$$

Substituting (3.1-3) into (3.1-1) shows that

$$q_t + \alpha q q_x + q_{xxx} \equiv \frac{1}{\alpha}\,(-2\beta v + \varepsilon (6\beta)^{\frac{1}{2}} \frac{\partial}{\partial x})(v_t + \beta v^2 v_x + v_{xxx}) \tag{3.1-4}$$

Clearly if v is a solution of (3.1-2), then q defined by (3.1-3) is a solution of the KdV equation (3.1-1). The transformation (3.1-3) constitutes one half of a Bäcklund Transformation which in general defines a correspondence (rather than a mapping) between solutions either to the same equation or to a pair of equations. The other half of the transformation is obtained by repeatedly substituting (3.1-3) into (3.1-2) until all the x derivatives of v have been eliminated. The complete Bäcklund Transformation so obtained consists of the pair of coupled partial differential equations

$$v_x = \frac{\varepsilon}{(6\beta)^{\frac{1}{2}}}(\alpha q + \beta v^2)$$
$$v_t = -\frac{\alpha\varepsilon}{(6\beta)^{\frac{1}{2}}}\left(\frac{1}{3}\beta v^2 q + q_{xx} + \left(\frac{2\beta}{3}\right)^{\frac{1}{2}}\varepsilon v q_x + \frac{\alpha q^2}{3}\right) \tag{3.1-5}$$

A full treatment of Bäcklund Transformations as they relate to the exactly solvable equations of the ZS-AKNS system is developed in Chapter 6. The important point to note here is that the complete integrability conditions on (3.1-5), $v_{xt} = v_{tx}$, are satisfied provided q is a solution of the KdV equation (3.1-1).

Before demonstrating this, let us generalise the transformation (3.1-5), by exploiting the invariance of (3.1-1) under the Galilean transformation

$$x' = x + \alpha\lambda t$$
$$t' = t \tag{3.1-6}$$
$$q' = q + \lambda$$

Here λ is an arbitrary real number. This symmetry of the KdV, when applied to (3.1-5), widens the scope of the Bäcklund Transformation so that it includes an arbitrary parameter λ. Equations (3.1-5) become

$$v_x = \frac{\varepsilon}{(6\beta)^{\frac{1}{2}}}(\alpha(q-\lambda) + \beta v^2) \tag{3.1-7}$$

$$v_t = \frac{-\alpha\varepsilon}{(6\beta)^{\frac{1}{2}}}\left(\frac{1}{3}(q+2\lambda)(\alpha(q-\lambda) + \beta v^2) + q_{xx} + \left(\frac{2\beta}{3}\right)^{\frac{1}{2}}\varepsilon v q_x\right) \tag{3.1-8}$$

where we have dropped the prime (') notation. The complete integrability condition on equations (3.1-7),(3.1-8) requires that $v_{xt} = v_{tx}$. A brief calculation using (3.1-7),(3.1-8) shows that

$$v_{xt} = \frac{\varepsilon}{(6\beta)^{\frac{1}{2}}}\left(\alpha q_t - \left(\frac{2\beta}{3}\right)^{\frac{1}{2}}\varepsilon v\alpha\left(\frac{1}{3}(q+2\lambda)(\alpha(q-\lambda) + \beta v^2) + q_{xx}\right.\right.$$
$$\left.\left. + \left(\frac{2\beta}{3}\right)^{\frac{1}{2}} v q_x \varepsilon\right)\right)$$
$$v_{tx} = \frac{-\alpha\varepsilon}{(6\beta)^{\frac{1}{2}}}\left(q_x(\alpha(q-\lambda) + \beta v^2) + \frac{\alpha}{3}(q+2\lambda)q_x + q_{xxx}\right. \tag{3.1-9}$$
$$+ \left(\frac{2\beta}{3}\right)^{\frac{1}{2}}\varepsilon v q_{xx} + \frac{\varepsilon}{(6\beta)^{\frac{1}{2}}}(\alpha(q-\lambda) + \beta v^2)\left(\frac{2\beta}{3} v(q+2\lambda)\right.$$
$$\left.\left. + \left(\frac{2\beta}{3}\right)^{\frac{1}{2}} q_x \varepsilon\right)\right)$$

Note that when we differentiate partially with respect to x on functions of x, $\partial/\partial x$ is effectively a total derivative, that is if we denote this operator by $\partial^{\dagger}/\partial x$, then, for this example

$$\frac{\partial^{\dagger}}{\partial x} = \frac{\partial}{\partial x} + q_x \frac{\partial}{\partial q} + q_{xx} \frac{\partial}{\partial q_x} + q_{xxx} \frac{\partial}{\partial q_{xx}} + v_x \frac{\partial}{\partial v} \tag{3.1-10}$$

By equating the two expressions in (3.1-9) we find that the system (3.1-7), (3.1-8) is completely integrable provided q is a solution of the KdV equation (3.1-1). The transformation from the KdV to the equation for which v is a solution is easily achieved by using (3.1-7) to eliminate q from (3.1-8). The resulting equation is

$$v_t + \alpha\lambda v_x - \beta v^2 v_x + v_{xxx} = 0 \tag{3.1-11}$$

which is simply related to the MKdV and is identical to it when $\lambda = 0$. The Bäcklund Transformation (3.1-7),(3.1-8) thus relates the KdV to a one parameter family of MKdV equations (3.1-11) (we assume that α and β are fixed). We have left α and β arbitrary in the text so that the reader may more easily compare in a unified manner the treatments given by various authors in the literature.

If we assume q is a known solution, then equations (3.1-7),(3.1-8) are examples of Riccati equations. That is, they are of first order and the nonlinearity is quadratic. In particular it is possible to linearise them by changing the dependent variable. For the equations (3.1-7),(3.1-8), set

$$v = \gamma \frac{\psi_x}{\psi} \tag{3.1-12}$$

and choose γ so that the quadratic term in (3.1-7) is eliminated. Thus we find that

$$\gamma = -\varepsilon \left(\frac{6}{\beta}\right)^{\frac{1}{2}} \tag{3.1-13}$$

Equation (3.1-8) transforms to the equation

$$\psi G_x - \psi_x G = 0$$

where

$$G \equiv \psi_t - \frac{\alpha}{6} q_x \psi + \frac{\alpha}{3}(q + 2\lambda)\psi_x \tag{3.1-14}$$

After the integration of (3.1-14) the transformed set of equations (3.1-7), (3.1-8) can be written as

$$\psi_{xx} + \frac{\alpha}{6} q\psi = \frac{\alpha k^2}{6}\psi \tag{3.1-15}$$

$$\psi_t + \frac{\alpha}{3}(q + 2k^2)\psi_x + (f(k,t) - \frac{\alpha}{6} q_x)\psi = 0 \tag{3.1-16}$$

where $k^2 = \lambda$ and f is an arbitrary function of t and k.

Equation (3.1-15) is the Schrödinger equation discussed in Section 2.4! However (3.1-15) is now a partial differential equation although t only appears as a parameter. A further comparison with the material of Section 2.4 can be obtained by exploiting the invariance of (3.1-11) under $v \to -v$. If we apply this trivial symmetry to (3.1-7),(3.1-8) and define Q as the corresponding solution to the KdV then

$$-v_x = \frac{\varepsilon}{(6\beta)^{\frac{1}{2}}} (\alpha(Q - \lambda) + \beta v^2) \tag{3.1-17}$$

$$v_t = \frac{\alpha\varepsilon}{(6\beta)^{\frac{1}{2}}} (\frac{1}{3}(Q + 2\lambda)(\alpha(Q - \lambda) + \beta v^2) + Q_{xx} - \left(\frac{2\beta}{3}\right)^{\frac{1}{2}} \varepsilon v Q_x) \tag{3.1-18}$$

It is immediate from (3.1-7) and (3.1-17) that

$$v_x = \frac{\alpha\varepsilon}{2(6\beta)^{\frac{1}{2}}} (q - Q) \tag{3.1-19}$$

If we introduce the functions

$$w = \frac{1}{2}\int_{-w}^{x} q(y)\,dy + \mu, \quad W = \frac{1}{2}\int_{-\infty}^{x} Q(y)\,dy - \mu$$

with

$$\mu = \frac{\varepsilon}{\alpha}\left(\frac{3\beta}{2}\right)^{\frac{1}{2}} v(-\infty) \tag{3.1-20}$$

we obtain from (3.1-19) and (3.1-7),(3.1-8) (or (3.1-17),(3.1-18)) the pair of partial differential equations relating two solutions to the KdV equation

$$(W + w)_x = k^2 - \frac{\alpha}{6}(W - w)^2 \tag{3.1-21}$$

$$(W - w)_t = \frac{\alpha^2}{6}(W - w)^2 (W - w)_x - \alpha k^2 (W - w)_x - (W - w)_{xxx} \tag{3.1-22}$$

In obtaining (3.1-22) we used (3.1-21) so that we could write the final form of the equation symmetrically. Equations (3.1-21),(3.1-22) constitute the auto Bäcklund Transformation for the KdV (3.1-1) (see Chapter Note 3.1.2). That is, it transforms from one solution of an equation to another solution of the same equation. It was first derived by Wahlquist and Estabrook (1973). Equation (3.1-21) is equation (2.4-33) with $k = i\kappa$ and $\alpha = -6$. The use of the transformation (3.1-21),(3.1-22) in obtaining solutions to the KdV equation is treated in the problems at the end of the chapter.

Finally in this section we return to the set of equations (3.1-15), (3.1-16). Eliminating q between the equations we find that ψ is a solution of the equation

$$\psi\psi_t + \alpha k^2 \psi\psi_x - 3\psi_x\psi_{xx} + f\psi^2 + \psi\psi_{xxx} = 0 \tag{3.1-23}$$

The system (3.1-15),(3.1-16) is therefore a Bäcklund Transformation between the KdV equation (3.1-1) and (3.1-23), a point which is not usually emphasized in the literature.

From (3.1-11) and (3.1-19) we see that

$$Q - q = \frac{12}{\alpha}(\log\psi)_{xx} \tag{3.1-24}$$

Since $q = 0$ is a solution to the KdV equation we can use (3.1-24) recursively to generate further solutions to (3.1-1) which all have the form

$$q = \frac{12}{\alpha}(\log \phi)_{xx} \qquad (3.1\text{-}25)$$

ϕ is easily shown to satisfy the equation

$$\phi\phi_{xt} - \phi_x\phi_t + 3\phi_{2x}^2 + \phi\phi_{4x} - 4\phi_x\phi_{3x} = 0 \qquad (3.1\text{-}26)$$

by directly substituting (3.1-25) into (3.1-1). Equation (3.1-26) was derived by Hirota (1971) using the transformation (3.1-25). His analysis of this equation led to a subset of the solutions to the KdV called the N-soliton solutions, as we have already seen in Chapter 1. These solutions emerge as the most important component of the general solution derived by Gardner et al. (1967). In particular notice that if $q = 0$ (c.f. (3.1-15) and (2.4-6)), then the solution Q generated by the Bäcklund Transformation must be the sech^2 reflectionless potential function discussed in Section 2.4.

In this section we have tried to show the interrelation of symmetries and Bäcklund Transformations between the KdV equation and some other nonlinear equations. The point we wish to emphasize is that (3.1-7),(3.1-8) and the related equations, in particular (3.1-15),(3.1-16) are all interpretable as Bäcklund Transformations.

3.2 The KdV Hierarchy of Equations and the Isospectral Schrödinger Equation

If we define the linear operators

$$\mathsf{L} \equiv -\left(\frac{\partial^2}{\partial x^2} + \frac{\alpha}{6} q\right) \qquad (3.2\text{-}1)$$

$$\mathsf{A}_f \equiv -\left(4\frac{\partial^3}{\partial x^3} + q\frac{\partial}{\partial x} + \frac{\alpha}{2} q_x + f(k,t)\right) \qquad (3.2\text{-}2)$$

then upon substituting for k^2 from (3.1-15) into (3.1-16) and rescaling k, equations (3.1-15),(3.1-16) can be written as the linear operator equations

$$\mathsf{L}\psi = k^2\psi \qquad (3.2\text{-}3)$$

$$\psi_t = \mathsf{A}_f\psi \qquad (3.2\text{-}4)$$

The subscript f on A_f is to emphasise that this operator is not unique. It is only defined up to an arbitrary function $f(k,t)$, and thus also depends upon the specific eigenfunction of L in (3.2-3).

In Section 3.1 we derived the fundamental result that the complete integrability conditions on the Bäcklund Transformation (3.1-15),(3.1-16) were satisfied provided q was a solution of the KdV equation (3.1-1). Although we saw there that this implied that $\lambda = k^2$ was independent of time, let us rederive this result from the operator equations (3.2-3), (3.2-4), assuming that this system is completely integrable. With this assumption we can write

$$\frac{\partial}{\partial t}(L\psi) = L_t\psi + L\psi_t \tag{3.2-5}$$

where L_t denotes partial differentiation with respect to t of the **coefficients** of the operator L. Then by (i) partially differentiating (3.2-3) with respect to t and then (ii) substituting for ψ_t from (3.2-4) we have step-wise that

$$\text{(i)}\quad L_t\psi + L\psi_t = (k^2)_t\psi + k^2\psi_t$$

$$\text{(ii)}\quad L_t\psi + LA_f\psi = (k^2)_t\psi + A_f(k^2\psi)$$

and the final result using (3.2-3) again can be written

$$(L_t - [A_f, L])\psi = (k^2)_t\psi \tag{3.2-6}$$

The square bracket in (3.2-6) is the operator commutator $[A_f, L] = A_fL - LA_f$. A direct calculation shows that $[A_f, L]$ is independent of f and that

$$L_t - [A_f, L] = 0 \tag{3.2-7}$$

is the KdV equation (3.1-1). We conclude from (3.2-6) that the eigenvalues $\lambda = k^2$ of the Schrödinger equation (3.2-3) are independent of time provided $q(x,t)$ evolves according to the KdV equation (3.1-1).

When the eigenvalues are independent of time we shall refer to (3.2-3) as the isospectral Schrödinger equation. Let us write $A_1 \equiv A_{f=0}$. Then, without loss of generality, we have that the **nonlinear** KdV equation

$$L_t = [A_1, L] \tag{3.2-8}$$

is **associated** with the **linear** isospectral equation (3.2-3). The presence of the arbitrary function f is then seen to corespond to the fact that A_1 is only defined up to an operator which can commute with L. L and A_1 are called Lax pairs for the KdV equation.

When the equation is written in this form, an obvious question to ask is whether there any other equations which can be associated with the isospectral Schrödinger equation. This is clearly equivalent to an analysis of the possible linear operators A which can replace A_1 in (3.2-8). Lax (1968) found one such family of operators; they have the generic form

$$A_m = -c_m\left(D^{2m+1} + \sum_1^m (b_jD^{2j-1} + D^{2j-1}b_j)\right) \tag{3.2-9}$$

where $D^j = \partial^j/\partial x^j$ and b_j are functionals of the variable q and its x derivatives. The variable q satisfies the nonlinear evolution equation

$$L_t = [A_m, L] \tag{3.2-10}$$

Note that when D acts on b_j it effectively acts as total differentiation with respect to x (c.f. comment before (3.1-10)). The b 's in (3.2-9) are determined by the condition that (3.2-10) should be of zero degree in D.

Thus for $m = 0$ (3.2-10) is the linear wave equation

$$q_t + c_0 q_x = 0 \tag{3.2-11}$$

When $m = 1$ and $c_1 = 4$ we regain the KdV equation (3.1-1) and when $m = 2$, $c_2 = 96$ we obtain the higher order KdV equation

$$q_t + 6q_{5x} + 5\alpha^2 q^2 q_x + 20\alpha q_x q_{2x} + 10\alpha q q_{3x} = 0 \tag{3.2-12}$$

The family of equations generated by (3.2-9) of increasing order and nonlinearity are known as the hierarchy of KdV equations. In the first two sections of this chapter we have seen how a physically important equation, the KdV equation, can be associated with a **linear** eigenvalue problem, the isospectral Schrödinger equation. The material as presented raises a number of technical mathematical problems and suggests several different possible lines of further and deeper research into the significance of this relationship. In which function space is the linear operator L of (3.2-3) acting? Can we categorise the isospectral operators which could replace the Schrödinger operator in a relationship of the form (3.2-10)? Is it possible to determine the complete family of nonlinear equations which can be associated with the Schrödinger operator through an equation analagous to (3.2-10)?

The material of Section 2.4 indicated that when the ordinary Schrödinger equation is defined on the real line and we deal with scattering potentials (that is potentials $q(x)$, such that $q \to 0$ as $|x| \to \infty$), which are reflectionless, the asymptotic data of the linear system, the scattering data uniquely determined the potential. The role of the inverse of the transmission coefficient $a(\zeta)$ seemed particularly important. In Section 3.1 we noted that t only entered as a parameter in (3.2-3) so that for fixed t it is equivalent to (2.-3-18). Is it possible to define uniquely the scattering data for any given function $q(x,t)$ in (3.2-3) which belongs to the class of functions such that $q(x,t) \to 0$ as $|x| \to \infty$? If this is so can $q(x,t)$ be reconstructed from the scattering data? This constitutes the **inverse problem** for the Schrödinger equation which is treated in Chapter 4. What information does the asymptotic form of the evolution of the linear system (3.2-4) tell us about the evolution of the scattering data? The last three questions are also related to the initial value problem for the KdV equation. That is, what **initial data** and constraints need to be prescribed for $q(x,0)$ and its x derivatives so that the solution to the KdV equation exists and is unique for all $t > 0$. This problem is discussed in Section 3.5 and also in Chapters 4 and 6.

The problem of a periodic boundary condition on q, $q(x+c,t) = q(x,t)$ the only other case of physical interest, is not treated in this book.

In the remaining sections we provide answers to all the other questions. The key point is that through the association of the nonlinear KdV with a linear equation, the vast body of knowledge for linear equations is at our disposal to elicit properties of the solutions to the nonlinear equation. That this association is special and not in general possible for an arbitrary nonlinear equation can be seen from the example provided by the Maxwell-Bloch equations discussed in Chapter 1. The fundamental question of just when such an association is possible is currently an area of active research.

3.3 The Schrödinger Scattering Problem

In this section we investigate the mathematical properties of the Schrödinger equation

$$Ly \equiv -\frac{\partial^2 y}{\partial x^2} + Qy = k^2 y \tag{3.3-1}$$

defined on the whole real line, $x \in R$, for fixed t, $t = t_0$. We restrict our considerations to those functions $Q(x,t_0)$ which are summable in every interval $]a,b[, -\infty < a,b < \infty$, and such that $Q \to 0$ as $|x| \to \infty$. (See Chapter Note 3.3.1). This equation defined on the half line has been exhaustedly investigated by both physicists and mathematicians. However there are a number of differences between the two cases. Many of the results contained in this and the subsequent chapter were first given in a paper by Faddeyev (1964) (see Chapter Note 3.3.2). Comparing (3.3-1) with (3.2-1) we see that we have written $Q = -\alpha q/6$ and that we have used y to denote an arbitrary solution to the Schrödinger equation. This is because we wish to reserve the letter ψ for a special class of solutions to be introduced below. We shall suppress the variable t when writing solutions to (3.3-1) in this section, because we only deal with fixed $t = t_0$, but we shall continue to write the partial derivative for differentiation with respect to x to remind ourselves of the t dependency. Notice that in the case of the KdV equation (3.1-6) we can also cover the case when $Q(x,t_0)$ tends to a finite constant value as $|x| \to \infty$.

For the class of function Q which we are considering, the eigenvalue problem (3.3-1) is called a scattering problem. This is because, as discussed in Section 2.2, equation (3.3-1) can then be interpreted as an incident plane wave interacting with a potential well defined by the function $Q(x,t\ )$. The incident and scattered waves can be expressed in terms of certain fundamental solutions which asymptotically, either as $x \to +\infty$ or $x \to -\infty$ have the plane wave character of solutions for the problem $Q(x,t_0) \equiv 0$. This family of solutions parametrised by k are usually called the Jost solutions for (3.3-1); they are defined by the boundary conditions

$$
\begin{aligned}
&\lim_{x\to+\infty} e^{-ikx}\psi(x,k) = 1, && \lim_{x\to+\infty} e^{ikx}\bar{\psi}(x,k) = 1\\
&\lim_{x\to+\infty} e^{-ikx}\psi_x(x,k) = ik, && \lim_{x\to+\infty} e^{ixk}\bar{\psi}_x(x,k) = -ik\\
&\lim_{x\to-\infty} e^{-ikx}\bar{\phi}(x,k) = 1, && \lim_{x\to-\infty} e^{ikx}\phi(x,k) = 1\\
&\lim_{x\to-\infty} e^{-ikx}\bar{\phi}_x(x,k) = ik && \lim_{x\to-\infty} e^{ikx}\phi_x(x,k) = -ik
\end{aligned}
\tag{3.3-2}
$$

We shall investigate the conditions on Q for the existence of the Jost solutions. A natural extension is to enquire whether the Jost solutions exist for complex k. From the boundary conditions (3.3-2) it is clear that the existence of Jost solutions with complex k is connected with the determination of the eigenfunctions of L in $L^2(R)$. This analysis forms a major part of the material in this section.

We now introduce the Hilbert space $L^2(R)$ of complex valued functions which are measurable in the Lebesgue sense and square integrable over the real line. This has associated with it the inner product

$$\langle u,v\rangle = \int_{-\infty}^{\infty} u(x)\,v^*(x)\,dx \qquad (3.3\text{-}3)$$

where $u,v \in L^2(R)$. It is clear that for real k the functions (3.3-2) do not in general belong to $L^2(R)$. They are **improper** or **generalised eigenfunctions** of the linear operator defined by (3.3-1) and which acts in $L^2(R)$. As we shall see there is a way of incorporating them into the Hilbert space framework.

A function defined on the real line which vanishes for sufficiently large $|x|$ is said to have compact support. The set of functions in $L^2(R)$ which have compact support are called the finite functions of $L^2(R)$; they are dense in $L^2(R)$. Consider the set of functions $\{I_a,\ 0 < a < \infty\}$ defined by $I_a(x) = x$ if $|x| < a$ and zero otherwise. Then for any function $Q(x)$ we can associate with it the function $Q_a = Q.I_a$ which has compact support. The derivation of many of the results in this and the next two sections is simplified by assuming that initially Q has compact support $Q_a = Q.I_a$ and then allowing $a \to \infty$. The validity of this technique is discussed in Section 6.1.

The differential expression on the left hand side of (3.3-1) is formally self-adjoint since Q is real. This can be seen by partially integrating twice using the inner product (3.3-3). Formally we have

$$\langle \mathsf{L}u,v\rangle = [uv_x - u_x v]_{-\infty}^{\infty} + \langle u,\mathsf{L}v\rangle \qquad (3.3\text{-}4)$$

However we have to be careful in specifying a domain of definition $\mathcal{D}_{\mathsf{L}}$ for a corresponding self-adjoint operator which acts in $L^2(R)$. Clearly $\mathcal{D}_{\mathsf{L}}$ must be a subspace of $L^2(R)$ consisting of the elements $u \in L^2(R)$ such that $\mathsf{L}u \in L^2(R)$ and also such that the square bracketed expression in (3.3-4) should vanish for arbitrary elements of $\mathcal{D}_{\mathsf{L}}$. Thus we impose the condition that u_x is absolutely continuous on every finite interval if $u \in \mathcal{D}_{\mathsf{L}}$. The condition $\mathsf{L}u \in L^2(R)$ requires that Q should not be too singular. Suitable conditions on Q arise naturally when we consider the existence and uniqueness of the Jost solutions. This self-adjoint operator can also be defined by the closure of the symmetric operator corresponding to (3.3-1) acting on the **core** of functions $C_0^\infty(\mathbb{R})$ which are the infinitely differentiable functions on $\mathbb{R}$ with compact support (Kato 1966). We shall henceforth denote the resulting self adjoint operator by L.

The standard method in the theory of ordinary differential equations in proving the existence and uniqueness of a solution to a given differential equation with boundary or initial conditions given at a single point is to incorporate the condition into the equivalent integral representation. Picard iteration, that is the method of successive approximations, then enables conditions for the existence of the solution to be determined. The integral representation for (3.3-1) can be obtained directly using the method of variation of constants. Alternatively by introducing auxillary variables $y_1 \equiv y$, $y_2 = y_x$ we can convert (3.3-1) into an equivalent system of first order equations

$$\frac{\partial Y}{\partial x} + B(k)y = Q\sigma Y \tag{3.3-5}$$

where

$$Y = (y_1, y_2)^t, \quad B(k) = \begin{pmatrix} 0 & -1 \\ k^2 & 0 \end{pmatrix} \quad \text{and} \quad \sigma = \begin{pmatrix} 0 & 0 \\ 1 & 0 \end{pmatrix}$$

Then since $\exp(B(k)x)$ is analytic in k, the equivalent integral formulation for the solutions of (3.3-1) which satisfy the initial or one point boundary condition

$$Y(x_0, k) = Y_0(k) \equiv \begin{pmatrix} Y_{01}(k) \\ Y_{02}(k) \end{pmatrix} \tag{3.3-6}$$

is

$$Y(x,k) = e^{B(k)(x_0 - x)} Y_0(k) + \int_{x_0}^{x} e^{B(k)(v-x)} Q(v)\sigma Y(v)\, dv \tag{3.3-7}$$

Expanding $\exp(B(k)x)$ in a power series we obtain

$$e^{(k)x} = I\cos kx + k^{-1}B(k)\sin kx \tag{3.3-8}$$

where I is the 2×2 identity matrix. Substituting (3.3-8) into (3.3-7) results in the following equation for $y \equiv y_1$

$$y(x,k) = \cos k(x_0 - x)y_{01}(k) - k^{-1}\sin k(x_0 - x)y_{02}(k) \tag{3.3-9}$$

$$-k^{-1}\int_{x_0}^{x} Q(v)\sin k(v - x)y(v,k)$$

In order to derive the integral representations of the Jost solutions we see from (3.3-6) that we require the asymptotic behaviour of the functions and their first x-derivatives. Thus the single point boundary conditions (3.3-2) are sufficient to obtain, for example, the integral representations

$$\psi(x,k) = e^{ikx} + k^{-1}\int_{x}^{\infty} Q(v)\sin k(v - x)\psi(v,k)\,dv \tag{3.3-10}$$

$$\phi(x,k) = e^{-ikx} - k^{-1}\int_{-\infty}^{x} Q(v)\sin k(v - x)\phi(v,k)\,dv \tag{3.3-11}$$

Of course there are many ways in which (3.3-1) can be converted into a first order system of differential equations. The choice $y_1 = iky - y_x$, $y_2 \equiv -y$ has the advantage that the first order system so defined is an eigenvalue problem

$$LY \equiv i\begin{pmatrix} \frac{\partial}{\partial x} & -Q \\ 1 & -\frac{\partial}{\partial x} \end{pmatrix} Y = kY \tag{3.3-12}$$

It is the generalisation of this problem obtained by replacing the first entry in the second row of L, 1, by an arbitrary function $R(x,t)$ which constitutes the ZS-AKNS scattering problem investigated in Chapter 6.

Let us now use the method of successive approximations to establish the conditions under which Jost solutions exist for (3.3-1). We begin by continuing (3.3-10),(3.3-11) into the complex k-plane, $k = \xi + i\eta$. Then

$$|e^{-ikx}||\sin kx| = \frac{e^{\eta x}}{2}|e^{ikx} - e^{-ikx}| \leqslant e^{\eta x}\max\{e^{-\eta x}, e^{\eta x}\}$$

so that for $x \leqslant 0$

$$|e^{-ikx}||\sin kx| \leqslant e^{(\eta - |\eta|)x} \tag{3.3-13}$$

Taking the modulus of (3.3-11) and using (3.3-13) yields the inequality

$$|\phi(x,k)e^{ikx}| \leqslant 1 + |k|^{-1}\int_{-\infty}^{x} |Q(v)|e^{(\eta - |\eta|)(v-x)}|\phi(v,k)e^{ikv}|dv \tag{3.3-14}$$

We now show that a solution of (3.3-11) can be found in the form

$$\phi(x,k)e^{ikx} = \sum_{i=0}^{\infty} h_i(x,k) \tag{3.3-15}$$

where

$$h_0(x,k) = 1,\quad h_{j+1}(x,k) = -k^{-1}\int_{-\infty}^{x} Q(v)e^{-ik(v-x)}\sin k(v-x)h_j(v,k)dv$$

when $\text{Im}\, k \geqslant 0$. The iteration of (3.3-14) in $\text{Im}\, k \geqslant 0$ leads to the expansion

$$|\phi(x,k)e^{ikx}| \leqslant 1 + |k|^{-1}P_0(x) + \frac{1}{2!}|k|^{-2}P_0^2(x) + \ldots + \\ + \frac{1}{r!}|k|^{-r}P_0^r(x) + \ldots \tag{3.3-16}$$

where

$$P_0(x) = \int_{\infty}^{x} |Q(v)|dv \quad \text{and} \quad |h_j(x,k)| \leqslant \frac{P_0^j}{j!}(x)|k|^{-j} \tag{3.3-17}$$

The formula (3.3-16) is established with the help of the relationship

$$\frac{d}{dx} P_0^r(x) = r P_0^{r-1}(x)|Q(x)|, \qquad r = 1,.... \tag{3.3-18}$$

The series on the right hand side of (3.3-16) constitutes a **majorant** for the series (3.3-15), as follows from formula (3.3-17). This series converges uniformly for $x \in \mathbb{R}$ provided the formal sum, $\exp(P_0(x)|k|^{-1})$ exists. A sufficient condition for the existence of the sum is that Q is absolutely integrable on $\mathbb{R}$, $Q \in L^1(\mathbb{R})$, which implies that

$$P_0(\infty) = \int_{-\infty}^{\infty} |Q(v)|dv < \infty \tag{3.3-19}$$

Since sine and Q are summable in every finite interval, it follows at once that the series Σh_n converges absolutely and uniformly for $x \in \mathbb{R}$ and fixed k, $\operatorname{Im} k \geqslant 0$, $k \neq 0$, and so defines a solution $\phi(x,k)$ of (3.3-11). We also have the estimate

$$|\phi(x,k)e^{ikx}| < \exp(|k|^{-1}P_0(x)) \qquad k \neq 0 \tag{3.3-20}$$

The point $k = 0$ appears to be exceptional in the analysis because the inequality (3.3-14) is not valid for this value. However directly from (3.3-10),(3.3-11) we obtain the integral representations

$$\begin{aligned} \psi(x,0) &= 1 + \int_x^{\infty} (v - x)Q(v)\psi(v,0)dv \\ \phi(x,0) &= 1 - \int_{-\infty}^{x} (v - x)Q(v)\phi(v,0)dv \end{aligned} \tag{3.3-21}$$

for the Jost solutions when $k = 0$.

Put

$$\begin{aligned} h_0(x) &= 1 \\ h_{j+1}(x) &= - \int_{-\infty}^{x} (v - x)Q(v)h_j(v)dv \qquad j = 0,1... \end{aligned} \tag{3.3-22}$$

From this recurrence relation one can deduce the estimate

$$|h_j(x)| \leqslant \frac{1}{j!}(M(x))^j \tag{3.3-23}$$

where

$$M(x) = \int_{-\infty}^{x} (x - v)|Q(v)|dv$$

If we define

$$P_j(x) = \int_{-\infty}^{x} |v|^j |Q(v)|dv \qquad j = 0,1,2... \tag{3.3-24}$$

then it follows that the Jost solutions are defined for $k = 0$ provided that $P_0(\infty) < \infty$ and $P_1(\infty) < \infty$ and $x < a < \infty$. In fact it is easy to establish due to the inequality

$$\left|\frac{e^{-ikx}\sin kx}{k}\right| = \frac{1}{2}\left|\frac{1-e^{-2ikx}}{k}\right| = \left|\int_x^0 e^{-2iky}dy\right| < -x \qquad \begin{array}{l} x \le 0 \\ \operatorname{Im}k > 0 \end{array} \tag{3.3-25}$$

a similar result when $\operatorname{Im} k \geqslant 0$ and $k \neq 0$

$$|\phi(x,k)e^{ikx} - 1| \leqslant M(x)\exp M(x) \qquad \operatorname{Im}k \ge 0 \tag{3.3-26}$$

Note that the bound blows up exponentially as $x \to \infty$. Consequently $\phi(x,k)\exp(ikx)$ is uniformly bounded in $\operatorname{Im} k \geqslant 0$ on any interval $x \in]-\infty,a]$, $a < \infty$. We can conclude from this that $\phi(x,k)\exp(ikx)$ is continuous in $\operatorname{Im} k \geqslant 0$ and analytic in $\operatorname{Im} k > 0$ since the iterates possess these properties. Another bound is obtained by observing that with $h(x,k) = \phi(x,k)\exp(ikx)$

$$|h(x,k) - 1| \leqslant x\int_{-\infty}^{x} |Q(v)|\,|h(v,k)|\,dv + \int_{-\infty}^{x} (-v)\,|Q(v)|\,|h(v,k)|\,dv$$

$$\leqslant K_1 x\int_{-\infty}^{x} (1+|v|)\,|Q(v)|\,dv + K_2\int_{-\infty}^{0} (-v)\,|Q(v)|\,dv$$

$$\leqslant \begin{cases} K_3(1+x)\displaystyle\int_{-\infty}^{x} (1+|v|)\,|Q(v)|\,dv & x \geqslant 0 \\ K_4\displaystyle\int_{-\infty}^{x} (-v)\,|Q(v)|\,dv & x < 0 \end{cases}$$

Hence for $x \in \mathbb{R}$ we have the inequality

$$|h(x,k) - 1| \leqslant K_5(1+\max(x,0))\int_{-\infty}^{x} (1+|v|)\,|Q(v)|\,dv \tag{3.3-27}$$

Combining this with the "good" behavoir of $h(x,k)$ away from $k = 0$, (3.3-20) we obtain

$$|\phi(x,k)e^{ikx} - 1| \leqslant \frac{K(1+\max(x,0))}{(1+|k|)}\int_{-\infty}^{x} (1+|v|)\,|Q(v)|\,dv \tag{3.3-28}$$

(c.f. Fadeeyev (1974) and the paper by Deift and Trubowitz (1979) where this result was first derived). More directly, analyticity of $\phi(x,k)$ in $\operatorname{Im} k > 0$ requires $\phi_k(x,k)$ to exist. From (3.3-11) we easily obtain

$$h_k(x,k) = -\int_{-\infty}^{x} Q(v)\,(k^{-1}\sin k(v-x)\,e^{ik(x-v)})_k\,h(v,k)\,dv$$
$$-\int_{-\infty}^{x} Q(v)\,(k^{-1}\sin k(v-x)\,e^{ik(x-v)})\,h_k(v,k)\,dv \qquad (3.3\text{-}29)$$

which satisfies the differential equation

$$-h_{xxk} + 2i(kh_{kx} + h_x) + Qh_k = 0 \qquad (3.3\text{-}30)$$

Consideration of $\{k^{-1}\sin k(v-x)\exp[ik(x-v)]\}_k$ shows that

$$|\{k^{-1}\sin(kx)\exp(-ikx)\}_k| = |(\int_x^0 \exp(-2iky)\,dy)_k| \leqslant x^2 \qquad \begin{matrix}\mathrm{Im}k\geq 0\\ x\leq 0\end{matrix} \qquad (3.3\text{-}31)$$

Finally using the estimate (3.3-28) one arrives at the inequality

$$|h_k(x,k)| \leqslant K_5(1 + x\max(x,0)) + \int_{-\infty}^{x} (x-v)\,|Q(v)|\,|h_k(v,k)|\,dv \qquad (3.3\text{-}32)$$

since

$$\int_{-\infty}^{x} |Q(v)|\,(v-x)^2\,|h(x,v)|\,dv \leqslant K_5(1 + x\max(x,0))$$

provided $P_2(\infty) < \infty$. Henceforth we will assume that $P_i(\infty) < \infty$ for $i = 0,1,2$. We can combine these requirements into the single condition that

$$\int_{-\infty}^{\infty} |Q(v)|\,(1+v^2)\,dv < \infty \qquad (3.3\text{-}33)$$

Iteration of (3.3-32) leads one to the conclusion that h_k is uniformly bounded for all $\mathrm{Im}\,k \geqslant 0$ in any half line $x \leqslant a$. Consequently ϕ_k exists and is continuous in $\mathrm{Im}\,k \geqslant 0$ and analytic in $\mathrm{Im}\,k > 0$. It is also simple to show from (3.3-11) and (3.3-27) that h_x is uniformly bounded. Explicitly these bounds are given by the expressions

$$|h_k(x,k)| \leqslant K(1 + x\max(x,0)) \qquad \mathrm{Im}k \geqslant 0$$
$$|h_x(x,k)| \leqslant \frac{K}{1+|k|} \qquad \mathrm{Im}k \geqslant 0 \qquad (3.3\text{-}34)$$

When $k = 0$ we can also define solutions which satisfy the integral equations

$$\tilde{\psi}(x) = x + \int_x^{\infty} (v - x)Q(v)\tilde{\psi}(v)\,dv \tag{3.3-35}$$

$$\tilde{\phi}(x) = x - \int_{-\infty}^{x} (v - x)Q(v)\tilde{\phi}(v)\,dv \tag{3.3-36}$$

By means of the substitution $h(x) = x^{-1}\tilde{\phi}(x)$ equation (3.3-36) is transformed into

$$h(x) = 1 - x^{-1}\int_{-\infty}^{x} (v - x)vQ(v)h(v)\,dv \tag{3.3-37}$$

In this case the method of successive approximations leads to the recursion formula

$$h_0(x) = 1, \quad h_{j+1}(x) = -x^{-1}\int_{-\infty}^{x} (v - x)vQ(v)h_j(v)\,dv \qquad j = 0,1\ldots \tag{3.3-38}$$

and the estimates

$$|h_j(x)| \leqslant \frac{1}{j!}(S(x,\delta))^j \qquad x \notin \,]-\delta,\delta[\tag{3.3-39}$$

where

$$S(x,\delta) = \int_{-\infty}^{x} (\delta^{-1}v^2 + |v|)\,|Q(v)|\,dv$$

A bound on $\tilde{\phi}_x$ is easily established in the same manner. Analogous results are produced when the procedures are repeated for the other Jost solutions and for the function ψ which are conveniently summarised in the following theorem.

<u>**Theorem 3-1:**</u> The Schrödinger equation

$$[-\partial^2/\partial x^2 + Q(x)]y = k^2 y \qquad \text{on } \mathbb{R}$$

has solutions $\psi(x,k)$, $\phi(x,k)$, $\overline{\psi}(x,k)$, $\overline{\phi}(x,k)$ which satisfy the integral equations

$$\psi(x,k) = e^{ikx} + k^{-1}\int_x^{\infty} Q(v)\sin k(v - x)\psi(v,k)\,dv$$

$$\phi(x,k) = e^{-ikx} - k^{-1}\int_{-\infty}^{x} Q(v)\sin k(v - x)\phi(v,k)\,dv$$

$$\bar{\psi}(x,k) = e^{ikx} + k^{-1}\int_x^{\infty} Q(v)\sin k(v-x)\bar{\psi}(v,k)\,dv$$

$$\bar{\phi}(x,k) = e^{-ikx} - k^{-1}\int_{-\infty}^{x} Q(v)\sin k(v-x)\bar{\phi}(v,k)\,dv$$

provided $\int_{-\infty}^{\infty}(1+|v|)|Q(v)|\,dv < \infty$. For every $x \in \mathbb{R}$ the solutions $\phi(x,k)$, $\psi(x,k)$ and their x-derivatives $\phi_x(x,k)$, $\psi_x(x,k)$ are continuous with respect to k, $\text{Im}\,k \geqslant 0$ and analytic with respect to k in $\text{Im}\,k > 0$. If $\int_{-\infty}^{\infty}(1+v^2)|Q(v)|\,dv < \infty$ then $\phi_k(x,k)$ and $\psi_k(x,k)$ are analytic in $\text{Im}\,k > 0$ and continuous in $\text{Im}\,k \geqslant 0$. The analogous properties hold for the functions $\bar{\phi}(x,k)$, $\bar{\psi}(x,k)$, $\bar{\phi}_x(x,k)$, $\bar{\psi}_x(x,k)$, $\bar{\phi}_k(x,k)$, $\bar{\psi}_k(x,k)$ in $\text{Im}\,k \leqslant 0$.

When $k = 0$ then in addition to the Jost solutions there are solutions given by the integral equations

$$\tilde{\psi}(x) = x + \int_x^{\infty}(v-x)Q(v)\tilde{\psi}(v)\,dv$$

$$\tilde{\phi}(x) = x - \int_{-\infty}^{x}(v-x)Q(v)\tilde{\phi}(v)\,dv$$

which can be defined and are continuous outside any neighbourhood of the origin provided that $\int_{-\infty}^{\infty}(1+v^2)|Q(v)|\,dv \leqslant \infty$.

These solutions have the uniform asymtotic behaviour

$$\psi(x,k) = e^{ikx}(1+o(1)), \qquad \bar{\psi}(x,k) = e^{-ikx}(1+o(1)),$$

$$\psi_x(x,k) = e^{ikx}(ik+o(1)), \qquad \bar{\psi}_x(x,k) = e^{-ikx}(-ik+o(1)), \qquad \text{as } x \to +\infty$$

$$\phi(x,k) = e^{-ikx}(1+o(1)), \qquad \bar{\phi}(x,k) = e^{ikx}(1+o(1)),$$

$$\phi_x(x,k) = e^{-ikx}(-ik+o(1)), \qquad \bar{\phi}_x(x,k) = e^{ikx}(ik+o(1)), \qquad \text{as } x \to -\infty$$

$$\tilde{\psi}(x) = x(1+o(1)), \qquad \tilde{\psi}_x(x) = 1 + o(1), \qquad \text{as } x \to +\infty$$

$$\tilde{\phi}(x) = x(1+o(1)), \qquad \tilde{\phi}_x(x) = 1 + o(1)) \qquad \text{as } x \to -\infty$$

The proof of the theorem also establishes the following bounds on the Jost solutions.

Corollary 3-1-1: For all $x \in \mathbb{R}$ and every k such that $\text{Im}\, k > 0$, and $|k| \geqslant \delta$ there exists a real number C_δ such that

$$|\phi(x,k)| \leqslant C_\delta \exp \eta x, \qquad |\psi(x,k)| \leqslant C_\delta \exp(-\eta x)$$

If we require the Jost solutions to be analytic for real k, the function Q has to satisfy very strong conditions. Thus we certainly require them to be infinitely differentiable with respect to k. In general the Jost solutions will be infinitely differentiable with respect to k provided

$$P_j(\infty) \equiv \int_{-\infty}^{\infty} |x|^j |Q(x)| dx < \infty, \quad j = 0,1... \tag{3.3-40}$$

Thus for the Jost solutions to be analytic for real k it is necessary that the function Q decays faster than any power of $|x|$ as $|x| \to \infty$.

Concerning the uniqueness of the Jost functions we have the following theorem.

Theorem 3-2: The Jost solutions $\psi(x,k), \phi(x,k), \bar{\psi}(x,k), \bar{\phi}(x,k)$ are unique under the conditions of Theorem 3-1.

Proof: Let

$$\phi_j(x,k) = e^{-ikx} - k^{-1} \int_{-\infty}^{x} Q(v) \sin k(v-x) \phi_j(v,k) dv \qquad j = 1,2$$

be two solutions satisfying the conditions of Theorem 3-1. Then if $w(x,k) = |(\phi_1(x,k) - \phi_2(x,k)) \exp(ikx)|$ we have that

$$w(x,k) \leqslant s(x,k) \qquad \text{Im}k \geqslant 0 \tag{3.3-41}$$

where

$$s(x,k) = \int_{-\infty}^{x} (x-v) |Q(v)| w(v,k) dv$$

From (3.3-41) it follows that

$$s_x = \int_{-\infty}^{x} |Q(v)|w(v)\,dv \leqslant s(x) \int_{-\infty}^{x} |Q(v)|dv$$

so that

$$\frac{\partial}{\partial x}\left\{s(x,k)\exp - \int_{-\infty}^{x} (x-v)\,|Q(v)|\,dv\right\} \leqslant 0$$

and upon integrating this relationship between $-\infty$ and x we obtain

$$s(x,k) \leqslant 0 \qquad \text{Im}k \geqslant 0 \tag{3.3-42}$$

Equations (3.3-41) and (3.3-42) give immediately that $w(x,k) \equiv 0$. Thus we have for $\text{Im}\,k \geqslant 0$ that the Jost solution $\phi(x,k)$ is unique provided $\int_{-\infty}^{\infty}(1+|v|)|Q(v)|dv$ is finite. Uniqueness for the other Jost solutions is established in the same way.

An examination of the defining boundary conditions (3.3-2) for the Jost solutions leads to the following corollary to Theorem (3-2).

Corollary 3-2-1: The Jost solutions $\phi,\bar{\phi}$ and $\psi,\bar{\psi}$ are related by the relationships

$$\bar{\phi}(x,k) = \phi(x,-k^*) = \phi^*(x,k^*)$$

$$\bar{\psi}(x,k) = \psi(x,-k^*) = \psi^*(x,k^*)$$

where * denotes the operation of complex conjugation.

Let us now consider the space of solutions to the second order linear equation (3.3-1) or to the equivalent first order system (3.3-5) for fixed k. We shall simply write $y(x)$,(or $Y(x)$) to denote such a solution. The x-dependency has been included to indicate that solutions might have, at this stage, different domains of definition. In both the scalar and vector equations it is easy to check that the sum of any two solutions on their common domain of definition and the product of a solutions by an arbitrary complex number is also a solution. The aggregate of solutions is therefore a linear function space on their common domain of definition over the complex numbers. Since $Q(x) \to 0$ as $|x| \to \infty$ it follows for fixed k that any solution to (3.3-1) or (3.3-5) is defined by an initial condition at $x = +\infty$ or $x = -\infty$ which has the form of a linear combination of exponential functions $\exp(\pm ikx)$

The next step is to prove that any such boundary condition uniquely defines a solution to (3.3-1) or (3.3-5). This can be established in the same manner as we proved the existence and uniqueness of the Jost solutions (see Naimark, 1968). The proof also shows that any solution is defined almost everwhere in x. It follows that the asymptotic relationship which exists between an arbitrary solution and a linear combination of Jost solutions is valid almost everwhere. We can therefore represent any solution unambiguously by the function y or Y alone.

If Y_1, Y_2 are distinct solutions of (3.3-5) then they are **linearly independent** provided

$$\alpha Y_1 + \beta Y_2 = 0 \qquad \alpha, \beta \in C \tag{3.3-43}$$

implies $\alpha = \beta = 0$.

That is, one solution is not a multiple of the other, if (3.3-43) is true. We next observe that the order of the scalar equation (3.3-1) or the dimension of the vector space in which solutions to (3.3-5) take their values is the maximal dimension of the space of solutions (k fixed) for (3.3-1), (3.3-5) respectively. Thus let Y_1, Y_2, Y_3 be three distinct solutions to (3.3-5). Then if

$$\alpha Y_1 + \beta Y_2 + \gamma Y_3 = 0$$

it follows that

$$Y_3(x_0) = -\gamma^{-1}(\alpha Y_1(x_0) + \beta Y_2(x_0)) \tag{3.3-44}$$

for any $x_0 \in \mathbb{R}$ and that since solutions are uniquely determined by initial conditions, Y_3 is linearly dependent upon Y_1 and Y_2.

Define the **Wronskian** of two solutions Y_1, Y_2, $W(Y_1,Y_2)$ by

$$W(Y_1,Y_2) \equiv \det\begin{vmatrix} y_{11} & y_{21} \\ y_{12} & y_{22} \end{vmatrix} = (y_{11}y_{22} - y_{21}y_{12}) \tag{3.3-45}$$

From (3.3-43) we see that Y_1,Y_2 are linearly independent if their Wronskian $W(Y_1,Y_2)$ is non-zero.

If we now allow k to vary, then we can denote a member of the resulting one (complex) parameter family of solutions by ${}_kY$ where ${}_kY(x) = Y(x,k)$. It should be clear to the reader that since any solution is expressible in terms of the Jost solutions, an arbitrary family of solutions $Y(x,k)$ may be only defined for real k, $\text{Im}\, k \leqslant 0$ or $\text{Im}\, k \geqslant 0$.

<u>Lemma 3-3:</u>

The solutions ${}_kY_1, {}_kY_2$ to (3.3-5) are linearly independent if and only if their Wronskian $W({}_kY_1, {}_kY_2)$ is non-zero.

The solutions ${}_ky_1$, ${}_ky_2$ to (3.3-1) are linearly independent if and only if their Wronskian

$$W({}_ky_1, {}_ky_2) \equiv ({}_ky_1\, {}_ky_{2x} - {}_ky_2\, {}_ky_{1x})$$

is nonzero.

Either from (3.3-5) or (3.3-1) by eliminating Q from between the two equations satisfied by the solutions ${}_kY_1, {}_kY_2$ we obtain the following relationship for their Wronskian

$$W({}_k y_1, {}_\kappa y_2)(x) - W({}_k y_1, {}_\kappa y_2)(x_0) = (k^2 - \kappa^2) \int_{x_0}^{x} {}_k y_1(v)\, {}_\kappa y_2(v)\, dv \tag{3.3-46}$$

From this important relationship we can establish immediately that for $\kappa = k$ the Wronskian $W({}_k y_1, {}_k y_2)$ is independent of x and is a function of k. In the case of the Jost solutions we obtain

$$W({}_k\phi, {}_k\psi) \equiv 2ik\, a(k), \qquad W({}_k\bar{\phi}, {}_k\bar{\psi}) \equiv -2ik\, \bar{a}(k) \tag{3.3-47}$$

$$\begin{aligned} &W({}_k\psi, {}_k\bar{\psi}) = -2ik,\ k \neq 0, \qquad && W({}_k\phi, {}_k\bar{\phi}) = 2ik,\ k \neq 0 \\ &W({}_0\psi, \tilde{\psi}) = 1,\ k = 0, && W({}_0\phi, \tilde{\phi}) = 1,\ k = 0 \end{aligned} \tag{3.3-48}$$

$$W({}_k\phi, {}_k\bar{\psi}) = -2ikb(k), \qquad W({}_k\bar{\phi}, {}_k\psi) = 2ik\bar{b}(k) \tag{3.3-49}$$

Equation (3.3-48) is established using the asymptotic behaviour of the Jost solutions as given in Theorem 3-1. Equations (3.3-47) and (3.3-49) are **definitions** for the functions occuring on the right hand side of the identity signs. As we shall show, unless Q has compact support, the Wronskians in (3.3-47)-(3.3-49) are not defined everywhere. Provided the conditions of Theorem 3-1 are satisfied, we have that $a(k)$ is analytic for $\mathrm{Im}\, k > 0$, $\bar{a}(k)$ is analytic for $\mathrm{Im}\, k < 0$, and both functions are continuous for $\mathrm{Im}\, k = 0$, $k \neq 0$. In fact it is clear from (3.3-47)-(3.3-49) and Theorem 3-1 that $ka(k)$ and $kb(k)$ are continuous on the real axis for all k. Since the maximal number of linearly independent solutions to (3.3-1) or (3.3-5) is two, we have the following theorem from Lemma 3-3 and equations (3.3-47)-(3.3-49).

<u>**Theorem 3-4:**</u> For fixed k, under the conditions of Theorem 3-1, the Jost solutions ${}_k\phi, {}_k\psi, ({}_k\bar{\phi}, {}_k\bar{\psi})$ form a basis for the solution space of $\partial^2/\partial x^2 - Q\, y = k^2 y$ if $\mathrm{Im}\, k \geqslant 0$ ($\mathrm{Im}\, k \leqslant 0$) and $k \neq 0$, unless $a(k) = W({}_k\phi, {}_k\psi)/2ik$ is zero ($\bar{a}(k) = -W({}_k\bar{\phi}, {}_k\bar{\psi})/2ik$ is zero). If $\mathrm{Im}\, k = 0$ and $k \neq 0$ then the Jost solutions ${}_k\psi, {}_k\bar{\psi}$ or ${}_k\phi, {}_k\bar{\phi}$ are always a basis. In particular, if $b(k) = W({}_k\bar{\phi}, {}_k\psi)/2ik$ and $\bar{b}(k) = -W({}_k\phi, {}_k\bar{\psi})/2ik$ and $a(k)$, $\bar{a}(k)$ are non-zero, then any pair of the Jost solutions ${}_k\phi, {}_k\psi, {}_k\bar{\phi}, {}_k\bar{\psi}$ can be used as a basis to the solution space if $\mathrm{Im}\, k = 0$ and $k \neq 0$. For $k = 0$ a basis is provided by the pairs of functions $({}_0\psi, \tilde{\psi})$ or $({}_0\phi, \tilde{\phi})$ almost everwhere.

<u>**Corollary 3-4-1:**</u>
The function $a(k)$ ($\bar{a}(k)$) is analytic in the upper (lower) k half-plane and continuous down (up) to the real k-axis ($k \neq 0$). The functions $b(k)$ and $\bar{b}(k)$ are continuous on the real k-axis ($k \neq 0$). Furthermore it follows from the relationship between the Jost solutions that

$$a(k) = a(-k) = a^*(k)$$

and

$$\bar{b}(k) = b(-k) = b^*(k)$$

A basis for the solution space to (3.3-1) for fixed k, for example $({}_k\phi, {}_k\psi)$ if $\operatorname{Im} k > 0$, is called a fundamental system of solutions. The fundamental system of solutions to (3.3-5) written in the form of a matrix, for example

$$_k\Phi = \begin{pmatrix} {}_k\phi & {}_k\bar{\phi} \\ {}_k\phi_x & {}_k\bar{\phi}_x \end{pmatrix} \quad \text{or} \quad {}_k\Psi = \begin{pmatrix} {}_k\psi & {}_k\bar{\psi} \\ {}_k\psi_x & {}_k\bar{\psi}_x \end{pmatrix} \tag{3.3-50}$$

if $\operatorname{Im} k = 0,\ k \neq 0$, are called the **fundamental matrix solutions** to (3.3-5). Immediately from (3.3-5) we have that

$$_k\Phi_x = (Q\sigma - B(k))\ {}_k\Phi \tag{3.3-51}$$

With a harmless abuse of language we shall often simply say that Φ and Ψ are fundamental matrix solutions to (3.3-5). For real k any solution $_ky$ to (3.3-1) can be written as a linear combination of the fundamental solutions $_k\psi, {}_k\bar{\psi}$

$$_ky = {}_kc_1\ {}_k\bar{\psi} + {}_kc_2\ {}_k\psi \tag{3.3-52}$$

where $_kc_1$ and $_kc_2$ are complex **numbers.** The associated k-parameter family of solutions y can therefore be expressed in terms of the Jost solutions $\bar{\psi}, \psi$ and a pair of complex **functions** $c_1(k), c_2(k)$

$$y(x,k) = c_1(k)\bar{\psi}(x,k) + c_2(k)\psi(x,k) \tag{3.3-53}$$

In particular for the Jost solutions ϕ and $\bar{\phi}$ we find using (3.3-47)-(3.3-49) that

$$\phi(x,k) = a(k)\bar{\psi}(x,k) + b(k)\psi(x,k) \tag{3.3-54}$$

$$\bar{\phi}(x,k) = \bar{b}(k)\bar{\psi}(x,k) + \bar{a}(k)\psi(x,k) \tag{3.3-55}$$

In terms of the fundamental matrix solutions to (3.3-5), equations (3.3-47)-(3.3-49) can be written as

$$\Phi(x,k) = \Psi(x,k)A(k) \tag{3.3-56}$$

where

$$A(k) = \begin{pmatrix} b(k) & \bar{a}(k) \\ a(k) & \bar{b}(k) \end{pmatrix}$$

Directly from (3.3-47)-(3.3-49) we find by expanding the determinants that

$$W({}_k\phi, {}_k\psi)W({}_k\bar{\phi}, {}_k\bar{\psi}) - W({}_k\phi, {}_k\bar{\psi})W({}_k\bar{\phi}, {}_k\psi) = W(\phi, \bar{\phi})W(\psi, \bar{\psi}) \tag{3.3-57}$$

This implies the following relationship between the functions a, $\bar{a}$, b and $\bar{b}$ for real $k \neq 0$

$$a(k)\bar{a}(k) - b(k)\bar{b}(k) = 1 \tag{3.3-58}$$

At this point it is probably worthwhile to pause in the mathematical development and to consider the relationship between this section and the more physical treatment of the same material given in Section 2.2. Let us assume that we can define bounded continuous functions $T_+(k) = a^{-1}(k)$ and $R_+(k) = b(k)a^{-1}(k)$ for real k. Then (3.3-54) can be written as

$$T_+(k)\phi(x,k) = \bar{\psi}(x,k) + R_+(k)\psi(x,k) \tag{3.3-59}$$

The asymptotic behaviour of the Jost solutions given in Theorem 3-1 enables the following interpretation to be given to (3.3-59). $\bar{\psi}(x,k)$ represents an incident plane wave from the right $(\exp(-ikx))$ which interacts with the potential Q. As a result of the interaction part of the wave is transmitted $(T_+(k)\exp(-ikx))$ and the other part is reflected $(R_+(k)\exp(ikx))$. The interaction is represented diagrammatically below.

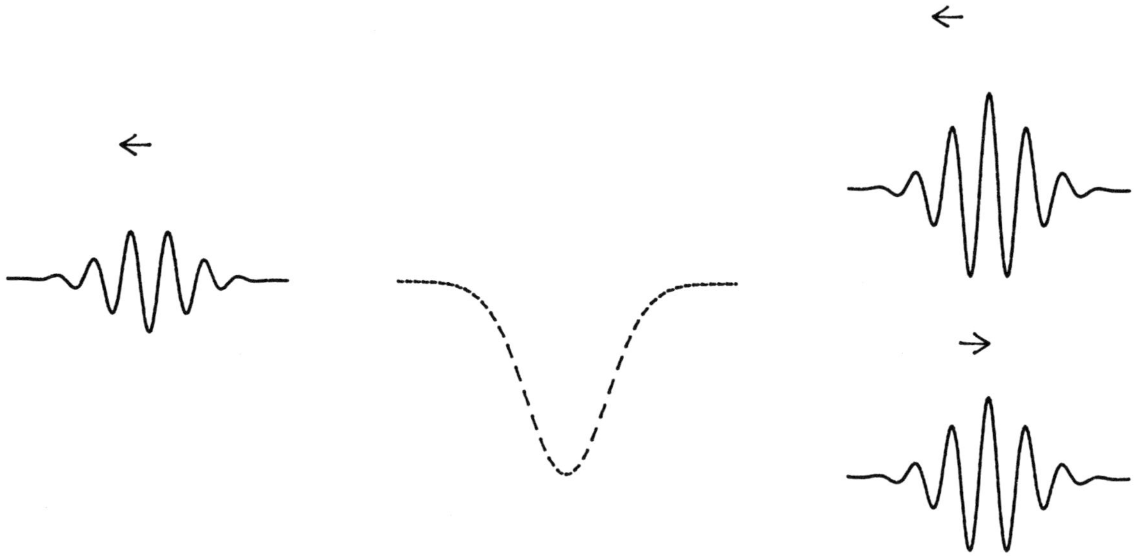

Figure 3-1: Representation of the solution (3.3-59) as a scattering process

The "collision" is perfectly elastic because from (3.3-58) and Corollary 3-4-1

$$|T_+(k)|^2 + |R_+(k)|^2 = 1 \tag{3.3-60}$$

This is a "snapshot" of the interaction at the fixed time $t = t_0$. As t varies then so will the Jost solutions and the transmission and reflection coefficients, but the overall picture of the interaction will remain unchanged. The precise relationship between the scattering coefficients and the potential is presented in Chapter 4, and in the following sections of this chapter we develop formulae which give the time evolution of the scattering coefficients R_+ and T_+. If we define

$T_-(k) = a^{-1}(k)$ $(\equiv T_+(k))$ and $R_-(k) = -\bar{b}(k)a^{-1}(k) \equiv -b^*(k)a^{-1}(k)$ for real k then we can establish an analogous picture for a plane wave $(\exp(ikx))$ incident from the left interacting with the potential Q by using (3.3-55). The matrix

$$\tilde{S}(k) = \begin{pmatrix} T_+(k) & R_-(k) \\ R_+(k) & T_-(k) \end{pmatrix} \tag{3.3-61}$$

is called the scattering matrix. A discussion of the role of $\tilde{S}$ in scattering theory is given in Section 3.4. It is clear from Corollary 3-1-1 that $\tilde{S}(k)$ is continuous for $k \neq 0$. In fact as we now establish, it is continuous for all real k provided $\int_{-\infty}^{\infty}(1 + v^2)|Q(v)|dv < \infty$.

Using the properties of the Jost solutions as given in Theorem 3-1 we obtain

$$\phi(x,k)e^{ikx} = e^{2ikx}\left[\frac{1}{2ik}\int_{-\infty}^{\infty} Q(v)^{-ikv}\phi(v,k)\,dv\right] + \left[1 - \frac{1}{2ik}\int_{-\infty}^{\infty} Q(v)e^{ikv}\phi(v,k)\,dv\right] + o(1) \quad \text{as} \quad x \to +\infty \tag{3.3-62}$$

but from (3.3-54)

$$\phi(x,k)e^{ikx} = b(k)e^{2ikx} + a(k) + o(1) \quad \text{as} \quad x \to +\infty \tag{3.3-63}$$

so that

$$\begin{aligned} a(k) &= 1 - \frac{1}{2ik}\int_{-\infty}^{\infty} Q(v)e^{ikv}\phi(v,k)\,dv \\ b(k) &= \frac{1}{2ik}\int_{-\infty}^{\infty} Q(v)e^{-ikx}\phi(v,k)\,dv \end{aligned} \tag{3.3-64}$$

The estimate (3.3-27) for $\phi\exp(ikx)$ and the representations (3.3-64) also establish that a is analytic in $\operatorname{Im} k > 0$ and that both a and b are continuous for real $k \neq 0$. From (3.3-64) putting $m = \int_{-\infty}^{\infty}Q(v)\phi(v,0)dv$ (which requires $\int_{-\infty}^{\infty}(1+v^2)|Q(v)|dv < \infty$) we have

$$a(k) = \left(1 - \frac{m}{2ik}\right) - \frac{1}{2i}\int_{-\infty}^{\infty} Q(v)[\phi(v,k)\exp(ikv) - \phi(v,0)]k^{-1}\,dv \tag{3.3-65}$$

Thus taking the limit $k \to 0$ and using (3.3-34) for the two cases $m = 0$ and $m \neq 0$ we find that although the limit exists $a(0) = \text{const.} \neq 0$ and a is continuous at $k = 0$ in the $m = 0$ case, when $m \neq 0$, a is not continuous at $k = 0$. However, T_+ is continuous at $k = 0$, in fact for the $m \neq 0$ case

$$T_+(k) = k(\text{const.}) + o(1) \quad \text{as} \quad k \to 0 \tag{3.3-66}$$

Similarly b is not continuous at $k = 0$ but (3.3-64) and (3.3-34) can be used to establish the continuity of R_+ at $k = 0$. We conclude that the relationship (3.3-59) is valid for all real k and that $\tilde{S}(k)$ is continuous. If $m \neq 0$ then from (3.3-64)

$$\begin{aligned}(1 + R_+(k))a(k) &= 1 + \int_{-\infty}^{\infty} Q(v) \frac{e^{ikv} - e^{-ikv}}{2ik} \phi(v,k)\,dv \\ &= 1 + \int_{-\infty}^{\infty} Q(v) v\phi(v,k)\,dv + o(1) \quad \text{as} \quad k \to 0\end{aligned} \tag{3.3-67}$$

From which we conclude that

$$R_+(0) = -1 \quad \text{provided} \quad \int_{-\infty}^{\infty} Q(v)\phi(v,0)\,dv \neq 0 \tag{3.3-68}$$

It is clear from (3.3-64) that T_+ and R_+ are analytic in the complex k-plane when Q has compact support.

So far we have not considered the bound states for this interaction process. By definition these solutions belong to $L(R)$ and from our earlier comments on solutions to (3.3-1) we see that the only candidates are solutions which have the asymptotic behaviour $\exp(\pm ikx)$ as $x \to \pm\infty$, $\text{Im}\, k > 0$ or $\exp(\pm ikx)$ as $x \to \mp\infty$, $\text{Im}\, k < 0$. Thus the bound state solutions are the eigenfunctions of L which satisfy relations of the type

$${}_k\phi = {}_kc\,{}_k\psi \quad \text{Im}k>0 \;; \quad {}_k\bar{\phi} = {}_k\bar{c}\,{}_k\bar{\psi} \qquad \text{Im}k<0 \tag{3.3-69}$$

If we assume $\int_{-\infty}^{\infty}|Q(v)|(1 + v^2)dv < \infty$, then it is not possible for $\lambda = k^2 = 0$ to be an eigenvalue of L, because the general solution of (3.3-1) is then

$$\begin{aligned}y(x,0) &= c_1\psi(x,0) + c_2\tilde{\psi}(x) \\ &\sim c_1 + c_2 x(1 + O(1)) \quad \text{as} \quad x \to +\infty\end{aligned} \tag{3.3-70}$$

which does not belong to $L^2(\mathbb{R})$.

In the remainder of this section we shall show that there is a finite number of eigenfunctions satisfying (3.3-69) provided $\int_{-\infty}^{\infty} |Q(v)|(1+v^2)dv < \infty$.

From the definitions (3.3-47) we see that relations of the type (3.3-69) require either $a(k) = 0$, $\text{Im}\, k > 0$, or $\bar{a}(k) = 0$, $\text{Im}\, k < 0$. It follows that an analysis of the zeros of a will establish whether there is an infinite or finite number of eigenfunctions for L.

Two approaches are open to us. The first is to continue working within the theory of ordinary differential equations and the second is to develop the spectral theory of the self-adjoint operator L in $L^2(R)$. We shall adopt the first method for the remainder of this section. The spectral theory for L is the subject of the next section. Why do we want such a rigorous development anyway? The reason lies at the heart of the inverse method. We shall establish in the next chapter the conditions under which the function Q can be uniquely constructed from the scattering data of the associated linear scattering Schrödinger equation (3.3-1). This is formed from the normalisation constants for the bound states, the eigenvalues of L and one of the reflection coefficients.

Directly from (3.3-46) we have

$$W({}_{k^*}\phi^*, {}_k\phi) = (k^{*2} - k^2)\int_{-\infty}^{x} |{}_k\phi|^2 dx, \quad \text{Im} k > 0 \tag{3.3-71}$$

If ${}_k\phi$ is a bound state solution, that is an eigenfunction of L, then the relationship (3.3-69) and the boundary condition for ϕ results in the following formula from equation (3.3-71)

$$\xi\eta \, \| {}_k\phi \|^2 = 0, \quad \text{where} \quad k = \xi + i\eta, \; \eta > 0 \tag{3.3-72}$$

Here $||.||$ is the $L^2(R)$ norm and we obtain immediately that $\text{Re } k = \xi = 0$. Since as we have just shown the case $a(0) = 0$ is excluded, we conclude that the zeros of a lie on the positive imaginary axis of the k-plane. From the relation $\bar{a}(k^*) = a^*(k^*)$, $\text{Im } k > 0$, we see that the zeros of $\bar{a}$ are the complex conjugates of the zeros of a and consequently we deduce that the eigenvalues of L are strictly negative. Thus if $\lambda = -\eta^2$, $\eta > 0$ is an eigenvalue of L then $k = i\eta$ is a zero of a. It might appear that there is a paradox here; namely that corresponding to a single eigenvalue $\lambda = -\eta^2$ of L there are two eigenfunctions $\{{}_{i\eta}\phi, {}_{-i\eta}\bar{\phi}\}$. However, from Corollary 3-2-1, it follows that ${}_{i\eta}\phi = {}_{-i\eta}\bar{\phi} = ({}_{i\eta}\phi)^*$ so that the eigenfunctions are real.

We next show that a has only a finite number of zeros. In order to prove this result we require the asymptotic behaviour of a as $|k| \to \infty$, $\text{Im } k \geqslant 0$. It seems appropriate at this point to also present the large k behaviour of b and the Jost solutions which we shall require in later sections. Immediately from (3.3-64) and (3.3-20) we get

$$a(k) = 1 - \frac{1}{2ik}\int_{-\infty}^{\infty} Q(v)dv + O\left(\frac{1}{|k|^2}\right) \quad \text{Im} k \geqslant 0 \quad \text{as} \quad |k| \to \infty \tag{3.3-73}$$

$$b(k) = \frac{1}{2ik}\int_{-\infty}^{\infty} Q(v)\exp(2ikv)dv + O(|k|^{-2}) \quad \text{as} \quad |k| \to \infty \tag{3.3-74}$$

The Riemann-Lebesgue theorem improves on the last result, $b(k) = o(1/|k|)$ as $|k| \to \infty$. Similarly the asymptotic behaviour of the Jost solutions is obtained from iterating their integral representations and using (3.3-20).

Lemma 3-5: The Jost solutions have the uniformly valid asymptotic expansions,

$$\phi(x,k)e^{ikx} = 1 + \frac{1}{(-2ik)}\int_{-\infty}^{x} Q dv + O(|k|^{-2})$$

$$\psi(x,k)e^{-ikx} = 1 + \frac{1}{(2ik)}\int_{x}^{\infty} Q dv + O(|k|^{-2})$$

as $|k| \to \infty$, $\text{Im } k \geq 0$ if $\int_{-\infty}^{\infty}|Q(v)|\,(1+|v|)dv < \infty$. The expansions for $\bar{\phi},\bar{\psi}$ are obtained by taking the complex conjugates of the expansions for ϕ and ψ.

It follows from (3.3-73) that, since the zeros of an analytic function are isolated and a is non-zero on the real axis, a has only a finite number of zeros. In addition they are also **simple** as we now prove. Differentiate (3.3-1) with respect to k and multiply by ϕ

$$-\phi_{xxk}\,\phi + Q\phi_k\,\phi = 2k\phi^2 + k^2\phi_k\,\phi \qquad \text{Im}k \geq 0 \tag{3.3-75}$$

Multiplying (3.3-1) by ϕ_k, subtracting from (3.3-75), and integrating once results in the formula

$$W(\phi,\phi_k)(\infty) - W(\phi,\phi_k)(-\infty) = -2k\int_{-\infty}^{\infty}\phi^2 dx \quad \text{Im}k > 0 \tag{3.3-76}$$

Upon differentiating (3.3-47) once with respect to k and evaluating the Wronskians at $x = +\infty$, we obtain

$$W(\phi_k,\psi)(\infty) + W(\phi,\psi_k)(\infty) = 2ia + 2ika_k \qquad \text{Im}k \geq 0 \tag{3.3-77}$$

Consequently, at a zero of a, $k_j = i\eta_j$, $\eta_j > 0$, $\phi_j = c_j\psi_j$ where $u_j \equiv {}_{k_j}u$ (3.3-76) and (3.3-77) can be combined to form

$$c_j^2 W(\psi_j,\dot{\psi}_j)(\infty) - W(\phi_j,\dot{\phi}_j)(-\infty) - 2ik_j\dot{a}_j c_j = -2k_j\int_{-\infty}^{\infty}\phi_j^2 dx \tag{3.3-78}$$

In (3.3-78) we have used the notation $\dot{f}_j \equiv f_k|_{k=k_j}$ to improve the readability of the formula. Evaluating the Wronskians in (3.3-78) produces the result,

$$D_{-j}^{-1} \equiv ic_j\dot{a}_j = \langle\phi_j,\phi_j\rangle\,,\; D_{+j}^{-1} \equiv ic_j^{-1}\dot{a} = \langle\psi_j,\psi_j\rangle \tag{3.3-79}$$

It follows immediately from (3.3-79) together with our earlier considerations that the eigenvalues of L are simple and that $D_{-j}(D_{+j})$ is interpretable as the normalisation constant for the eigenfunction ϕ_j (ψ_j).

It is possible to show that the number of zeros of $\phi(x,0)$ for fixed x_0 is the same as the numb•r of eigenvalues of the Dirichlet operator defined by the differential expression (3.3-1) on $L^2(-\infty, x_0)$ (Coddington and Levison, 1955). This implies using the minimax principle that L has a discrete spectrum if and only if $\phi(x,0)$ vanishes for some x (Deift and Trubowitz, 1979). Further one can show that if M is the number of eigenvalues of L that

$$M \leqslant \int_{-\infty}^{\infty} |v|\,|Q(v)|\,dv \tag{3.3-80}$$

Finally notice that the $m = 0$ case in (3.3-65) requires that $\phi(x,0)$ and $\psi(x,0)$ should be linearly dependent

$$0 = m \equiv \int_{-\infty}^{\infty} Q(v)\phi(v,0)\,dv = W({}_0\phi, {}_0\psi) \tag{3.3-81}$$

This condition is highly unstable and only occurs as we show in Section 4.3 when another eigenvalue is about to be added to the spectrum of L. We give now a summary of the results obtained in the last few pages of this section.

Lemma 3-6: If $\int_{-\infty}^{\infty}|Q(v)|(1 + v^2)dv < \infty$ then the function a is uniquely defined and continuous in $\text{Im}\, k \geqslant 0,\ k \neq 0$, analytic in $\text{Im}\, k > 0$ and has a finite number of zeros $k_j = i\eta_j,\ \eta_j > 0$. The function b is uniquely defined and continuous for real $k \neq 0$. The functions ka, kb are continuous for all real k. For real $k \neq 0$ a and b satisfy the condition

$$|a(k)|^2 - |b(k)|^2 = 1$$

They have the asymptotic behaviour

$$a(k) = 1 + O(|k|^{-1}) \quad \text{as} \quad |k| \to \infty \qquad b(k) = O(|k|^{-1}) \quad \text{as} \quad |k| \to \infty$$

Theorem 3-7: If $\int_{-\infty}^{\infty}|Q(v)|(1 + v^2)dx < \infty$ then the scattering data

$S_\pm = \{R_\pm, D_{\pm j}, \lambda_j, j=1,\ldots,M\}$ is uniquely defined.

(i) The eigenvalues λ_j are all distinct, strictly negative, and simple.

(ii) The normalization constants $D_{\pm j}$ are positive.

(iii) The scattering matrix

$$\tilde{S}(k) = \begin{pmatrix} T_+(k) & R_-(k) \\ R_+(k) & T_-(k) \end{pmatrix}$$

where $T_+ = T_- = T$ is uniquely defined and has the properties

(a) $T(k) = T^*(-k),\ R_\pm(k) = R_\pm^*(-k)$ (reality)

(b) $T(k)R_+^*(k) + R_-(k)T^*(k) = 0$

$$|T(k)|^2 + |R_+(k)|^2 = 1 = |T(k)|^2 + |R_-(k)|^2 \quad \text{(unitarity)}$$

(c) $\tilde{S}(k)$ is continuous

(d) $T(k) = 1 + O(1/|k|)$ as $|k| \to \infty$, k real

(e) Either $T(k) = \alpha k + o(k)$ $\alpha \neq 0$ as $|k| \to 0$

or $|T(k)| > \text{const} > 0$ as $|k| \to 0$

What have we actually accomplished in this section? The main object has been to establish the analytic properties of certain special solutions, the Jost solutions, to the Schrödinger equation and the analytic properties of the function $a = W(\phi,\psi)/2ik$. We have also investigated the asymptotic behaviour of these functions in the complex k-plane. In the course of this work we showed that L has only a finite number of strictly negative, simple eigenvalues. Furthermore, Theorem 3-7 gives a partial answer to the properties of the scattering data for a given potential Q.

In the next section we will show that there is a resolution of the identity in terms of the principal functions of L. That is we determine the **spectral family** for L. The principal functions of L prove to be the eigenfunctions and the Jost solutions for which $\text{Im}\, k = 0$. These latter functions do not belong to $L^2(\mathbb{R})$ and are associated with the **continuous spectrum** of L. In order to obtain this result we use complex analysis in the λ-plane so that a knowledge of the analytic and asymptotic properties of the Jost solutions is essential. The beauty of this approach is that by studying the (complex) k-parameter family of Jost solutions we have neatly covered the properties of both principal functions required for the resolution.

The function a plays a central role in this analysis; it represents an average of the spectral properties of L. Indeed the logarithm of a is defined by the regularised trace of the resolvent operator. Again in the Hamiltonian formulation of the inverse it is the coefficients in the large k asymptotic expansion of a which determine some of the Hamiltonians of the solvable nonlinear equations as we briefly outline in Section 3.5.

All the lemmas and theorems of this section have their analog for the first order system of equations (3.3-5) equivalent to the scalar Schrödinger equation (3.3-1). Some of the exercises at the end of this chapter develop this alternative approach to a study of the Schrödinger equation.

3.4 Spectral Theory for the Schrödinger Operator

In this section we consider the spectral theory of the self-adjoint operator L in $L^2(\mathbb{R})$. Our first aim will be to obtain the resolution of the identity for L. This defines the spectral family for L, the family of projection operators in $L^2(R)$ such that an arbitrary element of $L^2(R)$ has a unique decomposition in terms of elements in the subspaces defined by the spectral family. We discuss briefly the relationship with Fourier analysis and the expansion of an arbitrary element of $L^2(R)$ in terms of the improper generating basis or generalised eigenfunctions so defined. Throughout the section, unless otherwise stated, we assume that $\int_{-\infty}^{\infty} |Q(v)|(1+v^2)dv < \infty$.

The method we use to obtain the resolution of the identity starts from the resolvent operator, ${}_\lambda R$ of L. Some of the elements of the basis do not belong to $L^2(R)$, namely those eigenfunctions associated with the continuous spectrum of L. However it is possible to incorporate them into the Hilbert space framework by introducing the uniquely defined matrix spectral distribution function $F(k)$ of L. Specifying $F(k)$ is equivalent to defining the scattering data $S_\pm$ for L. We deduce the important result that F uniquely determines $S_\pm$ and vice-versa. This in effect completes the direct or forward scattering problem, the specification of $S_\pm(t_0)$ along with its properties from the given initial $Q(x,t_0)$.

We also define a scalar spectral function $\sigma(k)$ for which we introduce the Hilbert space $L_2^2(\sigma,R)$ of complex C^2-valued functions which are square integrable on R with respect to the measure σ. We show that there exists isometric mappings of $L^2(R)$ onto $L_2^2(\sigma,R)$ and vice-versa in which L is transformed into an operator $\tilde{L}$ which simply multiplies elements in $L_2^2(\sigma,R)$ by $\lambda = k^2$.

We then set up a general theory of perturbation operators whereby L is obtained from a reference operator L_0 through an intertwining operator. In this section this is used to give an operator theoretic interpretation of the scattering matrix and to introduce transformation operators in preparation for Chapter 4.

The general theory given here is used again in Section 3.5 to indicate an alternative method for obtaining the time evolution of the scattering data.

If λ is not a eigenvalue of L then the operator ${}_\lambda R = (L - \lambda I)^{-1}$ exists and is called the resolvent of L. The number λ belongs to the resolvent set or the continuous spectrum of L according to whether ${}_\lambda R$ is bounded or unbounded on its domain of definition.

In the theory of ordinary differential equations the kernel A^{-1} of the inverse operator A^{-1} of a linear differential operator A defined on $L^2(R)$ is called a Green's function. Thus if

$$Au = f \tag{3.4-1}$$

then

$$(A^{-1}f)(x) = \int_{-\infty}^{\infty} A^{-1}(x,y)\, f(y)dy \tag{3.4-2}$$

Compatability between (3.4-1) and (3.4-2) requires that

$$A\, A^{-1}(x,y) = \delta(x-y) \tag{3.4-3}$$

where δ is the Dirac delta function or distribution. For the resolvent operator ${}_\lambda R$ of L these formulae become for $u \in D_L$

$$({}_{\lambda}R\, u)\,(x) = \int_{-\infty}^{\infty} R(x,y,\lambda)\, u(y)dy \tag{3.4-4}$$

and

$$(L-\lambda I)\, R(x,y,\lambda) = \delta(x-y) \tag{3.4-5}$$

or

$$\left(-\frac{\partial^2}{\partial x^2} + Q(x) - \lambda\right) R(x,y,\lambda) = \delta(x-y) \tag{3.4-6}$$

From (3.4-4) we deduce that the boundary conditions on R are

$$\lim_{|x|\to\infty} R(x,y,\lambda) = 0, \quad \lim_{|y|\to\infty} R(x,y,\lambda) = 0 \tag{3.4-7}$$

If $\text{Im}\, k > 0$, $\lambda = k^2$, then from (3.4-7) the existence and uniqueness theorems of Section 3.3 require that for fixed $y = y_0$

$$\begin{aligned} R(x,y_o,k^2) &= f(y_o,k)\ \psi(x,k) \quad \text{as} \quad x \to +\infty \\ &= g(y_o,k)\ \phi(x,k) \quad \text{as} \quad x \to -\infty \end{aligned} \tag{3.4-8}$$

Similarly as a function of x and y, (3.4-7) and (3.4-8) require by continuity that

$$\begin{aligned} R(x,y,k^2) &= h(k)\ \phi(y,k)\ \psi(x,k) \quad x \geq y \\ &= h(k)\ \psi(y,k)\ \phi(x,k) \quad y \geq x \end{aligned} \tag{3.4-9}$$

To determine the function h integrate (3.4-6) over the interval $]y-\varepsilon, y+\varepsilon[$ to obtain

$$-\lim_{\varepsilon\to 0} \frac{\partial}{\partial x} R(x,y,\lambda)\Big|_{x=y-\varepsilon}^{x=y+\varepsilon} = 1 \tag{3.4-10}$$

which upon using (3.4-9) becomes

$$-h(k)\ W({}_k\phi, {}_k\psi) = 1 \tag{3.4-11}$$

where W is the Wronskian functional defined in Section 3.3. From equations (3.3-47) we find that $W({}_k\phi, {}_k\psi) = 2ika(k)$ so that we have finally

$$R(x,y,k^2) = \begin{cases} \dfrac{i}{2ka(k)}\ \phi(y,k)\ \psi(x,k) & x \geq y \\[2ex] \dfrac{i}{2ka(k)}\ \psi(y,k)\ \phi(x,k) & y > x \end{cases} \qquad \text{Im}\ k \geq 0 \tag{3.4-12}$$

We now show that ${}_{\lambda}R$ is bounded on $\{\lambda : \lambda \notin \sigma(L),\ \lambda \in C\}$, where $\sigma(L)$ is the spectrum of L^{λ} and that $\{\lambda : \lambda \geqslant 0\}$ constitutes the continuous spectrum of L.

Lemma 3-8:
For every $\eta > 0$ the formulae

$$A_o u(x) = \exp(-\eta x)\int_{-\infty}^{x} \exp(\eta y)\,u(y)\,dy$$

$$B_o u(x) = \exp(\eta x)\int_{x}^{\infty} \exp(-\eta y)\,u(y)\,dy$$

define in $L^2(R)$ continuous linear operators and

$$\|A_o\| \leqslant \eta^{-1}, \quad \|B_o\| \leqslant \eta^{-1}.$$

Proof: Let $u \in L(R)$. Then

$$|(A_o u)(x)| \leq \exp(-\eta x)\{\int_{-\infty}^{x/2} \exp(\eta y)\,|u(y)|\,dy + \int_{\frac{x}{2}}^{x} \exp(\eta y)\,|u(y)|\,dy\}$$

$$\leq \frac{\exp-\eta x}{(2\eta)^{\frac{1}{2}}}\{e^{\eta x/2}\left(\int_{-\infty}^{x} |u(y)|^2 dy\right)^{\frac{1}{2}} + (e^{2\eta x} - e^{-\eta x})^{\frac{1}{2}}\left[\int_{\frac{x}{2}}^{x} |u(y)|^2 dy\right]^{\frac{1}{2}}\}$$

and consequently $|A_0\, u(x)| \to 0$ as $x \to \infty$. Since

$$\frac{d[(A_o u)(x)]}{dx} = -\eta(A_o u)(x) + u(x)$$

we obtain

$$\frac{d}{dx}|(A_o u)(x)|^2 = -2\eta|(A_o u)(x)|^2 + (A_o u)(x)\,.\,u^*(x) + u(x).(A_o u)^*(x)$$

Then imposing the boundary conditions after integrating over the real line we find that

$$0 = -2\eta\|A_o u\|^2 + \langle A_o u, u\rangle + \langle u, A_o u\rangle$$

and the Cauchy-Schwarz inequality results in

$$\|A_o u\| \leq \frac{1}{\eta}\|u\|$$

The second part of the lemma is proved in an analogous fashion.

Lemma 3-8 and the formula for the kernel of the resolvent operator (3.4-12) are used in the proof of the following theorem.

Theorem 3-9: The set of complex numbers $\{\lambda: \lambda = k^2, a(k) \neq 0, \text{Im } k > 0\}$ belong to the resolvent set of L. The resolvent operator $_\lambda R = (L-\lambda I)^{-1}$ is the integral operator

$$_\lambda R\, v(x) = \int_{-\infty}^{\infty} R(x,y,\lambda)\, v(y)dy$$

$v \in D_l$ with the kernel

$$R(x,y,\lambda) = \begin{cases} \dfrac{i}{2\lambda^{\frac{1}{2}} a(\lambda^{\frac{1}{2}})} \phi(y,\lambda^{\frac{1}{2}})\, \psi(x,\lambda^{\frac{1}{2}}) & x \geq y \\ \dfrac{i}{2\lambda^{\frac{1}{2}} a(\lambda^{\frac{1}{2}})} \psi(y,\lambda^{\frac{1}{2}})\, \phi(x,\lambda^{\frac{1}{2}}) & y > x \end{cases} \qquad \text{Im}\lambda^{\frac{1}{2}} \geq 0$$

$$= \quad (R(x,y,\lambda^*))^* \qquad \text{Im}\lambda^{\frac{1}{2}} < 0$$

which is defined if $\int_{-\infty}^{\infty} |Q(v)|(1+|v|)dv < \infty$. For every $\delta > 0$ there is a number $C_\delta > 0$ such that

$$\| _\lambda R \| \leq \frac{C_\delta}{|\lambda^{\frac{1}{2}}|\; |a(\lambda^{\frac{1}{2}})|\; \text{Im }\lambda^{\frac{1}{2}}} \qquad \text{for} \quad \text{Im }\lambda^{\frac{1}{2}} > 0, \quad |\lambda| \geq \delta$$

Every number $\lambda \geqslant 0$ belongs to the continuous spectrum of the operator L.

Proof: Let $u \in D_L$ then

$$_{k^2}Ru(x) = \frac{i\; a(k)^{-1}}{2k} \{\psi(x,k)\int_{-\infty}^{x} \phi(y,k)u(y)dy + \phi(x,k)\int_{x}^{\infty} \psi(y,k)u(y)dy\} \qquad \text{Im } k > 0$$

and an analogous expression obtains if $\text{Im } k < 0$. The resolvent operator is defined provided $\int_{-\infty}^{\infty} |Q(v)|(1+|v|dv < \infty$. Corollary 3-1-1 gives

$$|\psi(x,k)| \leq C_\delta \exp \eta x, \qquad |\phi(x,k)| \leq C_\delta \exp -\eta x$$

so that

$$|_{k^2}R(u(x)| \leq \frac{|a(k)|^{-1}}{2|k|} C_\delta \{\exp\eta x \int_{-\infty}^{x} \exp -\eta y\; u(y)dy + \exp -\eta x \int_{x}^{\infty} \exp\eta y\; u(y)dy\} \qquad \text{Im } k > 0$$

Using Lemma 3-8 we obtain the inequality

$$\| _{k^2}R \| \leq \frac{C_\delta}{|k|\,|a(k)|\,\text{Im } k} \qquad \text{Im } k > 0$$

Introduce the finite functions $_a\phi^*(x,k) = \phi^*(x,k)$, $|x| \leqslant a$ and 0 otherwise. Then

$$ {}_{k^2}R_a\phi(x) = \frac{i}{2ka(k)}\,\psi(x,k)\|{}_a\phi\|^2 \qquad |x| \geq a $$

which implies that

$$ \|{}_{k^2}R_a\phi\|^2 \geq \int_a^\infty |{}_{k^2}R_a\phi|^2\,dx = \frac{\|{}_a\phi\|}{4|ka(k)|^2}\int_a^\infty |\psi(x,k)|^2\,dx \qquad \text{Im } k \geq 0 $$

Since $\psi(x,k) = \exp(ikx)(1 + o(1))$ as $x \to +\infty$ there exists a sufficiently large a such that for $a < x < \infty$, Im $k \geqslant 0$, $k \neq 0$

$$ |\psi(x,k)| \geq \tfrac{1}{2}\exp{-\eta x} \qquad \eta = \text{Im } k $$

and

$$ \|{}_{k^2}R_a\phi\| \geq \frac{\|{}_a\phi\|\exp{-\eta a}}{4|k||a(k)|\sqrt{2\eta}} $$

It follows that $\|{}_{k^2}R\| \to \infty$ as Im $k \to 0$ and so $\lambda \geqslant 0$ belongs to the spectrum of L .

Let $R(L-\lambda I)$ be the range of $(L-\lambda I)$. Then we now have to show that for $\lambda \geqslant 0$, $R(L-\lambda I)$ is dense in $L^2(R)$ so that the inverse can be defined. A condition equivalent to this is that the orthogonal complement of $R(L-\lambda I)$ is the zero element. But since the space of solutions of $L^A u = \lambda u$ coincides with the orthogonal complement it follows from $L^A = L$ that this has only the zero element as a solution. Thus the set of numbers $\lambda \geqslant 0$ constitutes the continuous spectrum of L .

The principal functions are therefore of the following type. Corresponding to the **discrete or point spectrum** $\{\lambda_j = -\eta_j^2 \; , \; \eta_j > 0\}$ which is finite and simple under the conditions of Lemma 3-8 are the **eigenfunctions** $\{\phi_j \equiv i\eta_j\phi \; , \; j=1,\ldots,M\}$. The functions $\{{}_k\phi : k^2 = \lambda \geqslant 0\}$ are the **generalised eigenfunctions** associated with the **continuous spectrum**. They do not belong to $L^2(R)$ and so are not eigenfunctions in the strict sense.

We now proceed to derive the spectral family for L . As a concrete example of a spectral family consider a Hermitian operator A acting in a finite dimensional vector space over the complex field. Let $\{E_i \; , \; i = 1,\ldots,N\}$ be a decomposition of the identity operator into projection operators which are invariant on the eigenspaces of A with eigenvalues $\{\lambda_i : i = 1,\ldots,N\}$. Then

$$ E_iE_j = \delta_{ij} \quad \text{and} \quad \sum_{i=1}^{N} E_i = I \tag{3.4-13} $$

Define a "step-function" by

$$ {}_\lambda P = \sum_{\lambda_j<\lambda}^{N} E_j \tag{3.4-14} $$

Then since

$$A = \sum_{j=1}^{N} \lambda_j E_j \tag{3.4-15}$$

it follows from (3.4-14) that

$$A = \int_{-\infty}^{\infty} \lambda d_\lambda P \tag{3.4-16}$$

Thus (3.4-16) is a way of representing the self-adjoint operator A in diagonal form and $\{_\lambda P\}$ is called the **spectral family** associated with A. In a Hilbert space (3.4-16) is still a valid way of representing the diagonalisation of a self-adjoint operator, (Smirnov 1964, Kato 1966, Naimark 1964) but in this case $\{_\lambda P\}$ need no longer be a step-function. In books on functional analysis, what is meant by the resolution of the identity is precisely the spectral family $\{_\lambda P\}$. However, we shall use it for the Hilbert space version of (3.4-13). The spectral family leads to the notion of a spectral distribution function.

Consider the contour integral

$$I_{R,r} = \frac{1}{2\pi i} \int_{\Gamma_{R,r}} \frac{R(x,y,\lambda)d\lambda}{(\lambda - z)} \tag{3.4-17}$$

where $\Gamma_{R,r}$ is the contour depicted in the Fig. 3-2.

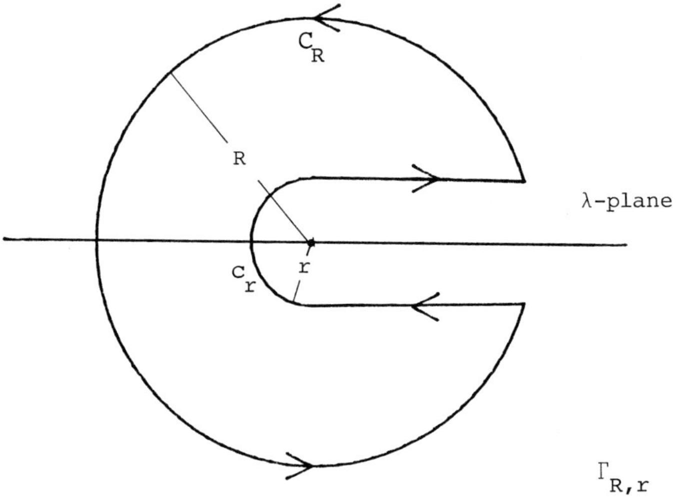

Figure 3-2: The contour for the resolvent operator

The contour consists of the circular arc of radius R, two line segments parallel to the real axis, and a smaller arc of radius r. The radii R and r are chosen so that $\{z, \lambda_1, \ldots, \lambda_2\}$ lie within the contour $\Gamma_{R,r}$. Then since $R(x,y,\lambda)$ is analytic within $\Gamma_{R,r}$ by Theorem 3-9 it follows from Cauchy's theorem that

$$I_{R,r} = R(x,y,z) + \sum_{j=1}^{M} \operatorname{Res}_{\lambda=\lambda_j} \left[\frac{R(x,y,\lambda)}{(\lambda - z)} \right] \tag{3.4-18}$$

We also have from (3.4-12) and Lemmas 3-5 and 3-6 that as

$$|R(x,y,\lambda)| \le \frac{1}{2|\lambda^{\frac{1}{2}}|} e^{-\eta(x-y)} + o(1) \quad \text{as} \quad |\lambda| \to \infty, \eta = \operatorname{Im}\lambda^{\frac{1}{2}} > 0 \tag{3.4-19}$$

Hence in the limit

$$\operatorname*{Lt}_{R\to\infty} \int \frac{R(x,y,\lambda)\,d\lambda}{(\lambda - z} = 0 \tag{3.4-20}$$

and

$$\operatorname*{Lt}_{R\to\infty} I_{R,r} = \frac{1}{2\pi i} \int_{\Gamma_r} \frac{R(x,y,\lambda)\,d\lambda}{(\lambda - z)} \tag{3.4-21}$$

where Γ_r is the contour $\operatorname*{Lt}_{R\to\infty} (\Gamma_{R,r} - C_R)$ shown in Fig. 3-3.

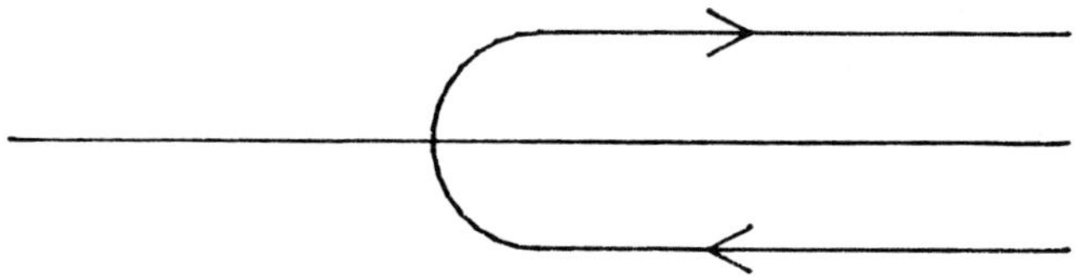

Figure 3-3: The contour Γ_r .

Finally we use the fact that $T_+(0)$ is finite as estabished in Section 3.3 to deduce that

$$\operatorname*{Lt}_{r\to 0_+} \int_{C_r} \frac{R(x,y,\lambda)\,d\lambda}{(\lambda - z} = 0 \tag{3.4-22}$$

If we combine (3.4-21) and (3.4-22) with (3.4-18) in the limit $R \to \infty$, $r \to 0_+$, then we obtain an expression for the resolvent kernel

$$R(x,y,\lambda) = J(x,y) - \sum_{j=1}^{M} \operatorname{Res}_{\lambda=\lambda_j} \left\{ \frac{R(x,y,\lambda)}{(\lambda - z)} \right\}$$
$$J(x,y) \equiv \frac{1}{2\pi i} \int_0^\infty (R(x,y,\lambda_+) - R(x,y,\lambda_-)) \frac{d\lambda}{(\lambda - z)} \tag{3.4-23}$$

where $\lambda_\pm = \lambda \pm i0$. The next step is to rewrite the right hand side of (3.4-23) in terms of the principal functions. If $\operatorname{Im} \lambda^{\frac{1}{2}} < 0$ the resolvent kernel is defined by

$$R(x,y,\lambda) = (R(x,y,\lambda^*))^* \qquad \operatorname{Im}\lambda^{\frac{1}{2}} < 0 \tag{3.4-24}$$

so that

$$J(x,y,) = J_1(x,y) + J_1^*(x,y) \tag{3.4-25}$$

and

$$J_1(x,y) \equiv \frac{1}{4\pi}\int_0^\infty [\frac{b}{a}(\lambda^{\frac{1}{2}})\psi(x,\lambda^{\frac{1}{2}})\psi(y,\lambda^{\frac{1}{2}}) + \psi^*(x,\lambda^{\frac{1}{2}})\psi(y,\lambda^{\frac{1}{2}})]\frac{d\lambda}{\lambda^{\frac{1}{2}}(\lambda-z)} \tag{3.4-26}$$

It is simpler to express J in terms of functions defined on the k-plane ($k = \lambda^{\frac{1}{2}}$)

$$J(x,y) = \frac{1}{2\pi}\int_{-\infty}^{\infty} (\frac{b}{a}(k)\psi(x,k)\psi(y,k) + \psi^*(x,k)\psi(y,k))\frac{dk}{(k^2-z)} \tag{3.4-27}$$

Under the conditions on Q given in Theorem 3-9, the poles of $R(x,y)$ are in one to one correspondence with the simple zeros of a. It is therefore straight forward to evaluate the residues

$$\operatorname{Res}_{\lambda=\lambda_j}\left[\frac{R(x,y,\lambda)}{(\lambda-z)}\right] = \frac{ic_j\psi_j(x)\psi_j(y)}{(\lambda_j-z)\dot{a}_j} \tag{3.4-28}$$

Using the normalisation constants

$$D_{+j}^{-1} \equiv i\,c_j^{-1}\,\dot{a}_j = \langle\psi_j,\psi_j\rangle \tag{3.4-29}$$

the resolvent kernel can be written in terms of the principal functions as

$$R(x,y,z) = \frac{1}{2\pi}\int_{-\infty}^{\infty} [\frac{b}{a}(k)\psi(x,k)\psi(y,k) + \psi^*(x,k)\psi(y,k)]\frac{dk}{(k^2-z)} + \sum_{j=1}^{M} D_{+j}\frac{\psi_j(x)\psi_j(y)}{(\lambda_j-z)} \tag{3.4-30}$$

Then upon operating with $(L-zI)$ we obtain **the resolution of the identity in terms of the principal functions of** L

$$\delta(x-y) = \frac{1}{2\pi}\int_{-\infty}^{\infty} [\frac{b}{a}(k)\psi(x,k)\psi(y,k) + \psi^*(x,k)\psi(y,k)]dk + \sum_{j=1}^{M} D_{+j}\psi_j(x)\psi_j(y) \tag{3.4-31}$$

The appearance of both the improper eigenfunctions $_k\psi$ and $_k\psi^*$ in the resolution corresponds to the fact that **the continuous spectrum is a 2-fold degenerate Lebesgue spectrum.** By treating the improper eigenfunctions as generalised functions one can show that they are not orthonormal in the generalised sense. If we extend the definition of <,> to include generalised functions then we find that

$$
\begin{aligned}
\langle {}_k\phi, {}_{\bar{k}}\phi \rangle &= 2\pi(|a(k)|^2\delta(k-\bar{k}) + a(k)b(k)\delta(k+\bar{k})) \\
\langle {}_k\psi, {}_{\bar{k}}\psi \rangle &= 2\pi(|a(k)|^2\delta(k-\bar{k}) - a(k)b(k)\delta(k+\bar{k})) \\
\langle {}_k\phi, {}_{\bar{k}}\psi \rangle &= 2\pi a(k)\delta(k+\bar{k}) \\
\langle {}_k\phi, \psi_j \rangle &= 0 \\
\langle {}_k\psi, \psi_j \rangle &= 0 \\
\langle \psi_j, \psi_j \rangle &= D_j^{-1}
\end{aligned}
\tag{3.4-32}
$$

In obtaining these results extensive use is made of the relationship (3.3-46), the asymptotic form $|x| \to \infty$ of the Jost solutions, and the formula

$$\underset{x\to\infty}{\mathrm{Lt}}\left[\frac{e^{\pm ikx}}{k}\right] = \pm i\pi\delta(k) \tag{3.4-33}$$

which occurs in the theory of generalised functions.

If we put $g_1(x,k) \equiv \psi(x,k)$, $g_2(x,k) \equiv \psi(x,-k)$ then (3.4-32) can be written

$$\delta(x-y) = \int_{-\infty}^{\infty} \sum_{i,j=1}^{2} g_i(x,\lambda^{\frac{1}{2}})\, g_j^*(y,\lambda^{\frac{1}{2}})\, dF_{ij}(\lambda) \tag{3.4-34}$$

where

$$dF_{ij}(\lambda) = H_{ij}(\lambda)\,d\lambda \tag{3.4-35}$$

and

$$H(\lambda) = \begin{cases} \dfrac{1}{8\pi\lambda^{\frac{1}{2}}}\begin{pmatrix} 1 & \dfrac{b}{a}(\lambda)^{\frac{1}{2}} \\ \dfrac{b^*}{a^*}(\lambda)^{\frac{1}{2}} & 1 \end{pmatrix} & 0 \le \lambda < \infty \\[2ex] \dfrac{1}{4\lambda^{\frac{1}{2}}}\begin{pmatrix} 0 & 1 \\ 1 & 0 \end{pmatrix}\displaystyle\sum_{j=1}^{M} D_j\,\delta(\lambda-\lambda_j) & -\infty<\lambda<0 \end{cases} \tag{3.4-36}$$

The corresponding Hermitian measure $F(\lambda)$ is called the (matrix) **spectral distribution function** for L corresponding to the resolution of the identity in terms of the eigenfunctions $\psi(x,k)$ (see Chapter Note 3.4.1). In order to define it, it is clear that we need to specify the set of **scattering data**

$$S_+ = \{D_j, R_+ = \frac{b}{a}, \lambda_j \quad j = 1,\ldots,M\}$$

If we had used the eigenfunctions $g_1(x,k) \equiv \phi(x,k)$, $g_2(x,k) \equiv \phi(x,-k)$ then for $\lambda \geqslant 0$ the matrix H in (3.4-36) would have been

$$H(\lambda) = \frac{1}{8\pi\lambda^{\frac{1}{2}}} \begin{pmatrix} 1 & -\frac{b^*}{a}(\lambda^{\frac{1}{2}}) \\ -\frac{b}{a^*}(\lambda^{\frac{1}{2}}) & 1 \end{pmatrix} \qquad \lambda \geqslant 0 \tag{3.4-37}$$

The $\lambda < 0$ components have the same form as (3.4-36), but D_{-j} , (Section 3.3 equation (3.3-79)) replaces D_{+j}. In this case the scattering data is the set $S_- = \{D_{-j}, R_- = -b^*/a, \lambda_j , j = 1,\ldots,M\}$. In the next chapter we show that provided the components satisfy certain constraints the potential Q can be uniquely constructed from either S_+ or S_-.

Physically the scattering process naturally selects the eigenfunctions ϕ/a and ψ/a (see Chapter Note 3.4.2). In order to pursue this approach we rewrite (3.4-31) as

$$\delta(x-y) = \frac{1}{4\pi} \int_{-\infty}^{\infty} \frac{1}{|a|^2} \quad \phi(x,y)\phi^*(y,k) + \psi(x,k)\psi^*(y,k) \; dk + \sum_{j=1}^{M} D_{+j}\psi_j(x)\psi_j(y) \tag{3.4-38}$$

Consider now the eigenfunction expansion of an arbitrary $u \in L^2(R)$ in terms of the eigenfunctions of L . Using (3.4-38) this is given by

$$u(x) = \frac{1}{4\pi} \int_{-\infty}^{\infty} \frac{1}{|a|^2} (\langle u, {}_k\phi\rangle\phi(x,k) + \langle u, {}_k\psi\rangle\psi(x,k))dk + \sum_{j=1}^{M} D_{+j} \langle u, \psi_j\rangle\psi_j(x) \tag{3.4-39}$$

Introducing the notation

$$p^-(x,k) = \frac{\psi(x,k)}{a(k)}, \qquad m^-(x,k) = \frac{\phi(x,k)}{a(k)} \tag{3.4-40}$$

and assuming for simplicity that there is no discrete spectrum, (3.4-39) can be written as

$$u(x) = \int_{-\infty}^{\infty} (u_1(k)m^-(x,k) + u_2(k)p^-(x,k))d\sigma(k) \tag{3.4-41}$$

where $\tilde{u} = (u_1,u_2)^T$, $d\sigma(k) = dk/4\pi$ and

$$u_1(k) = \int_{-\infty}^{\infty} u(x)m^-(x,k)\,dx, \quad u_2(k) = \int_{-\infty}^{\infty} u(x)p^-(x,k)\,dx \tag{3.4-42}$$

$\sigma(k)$ is a spectral distribution function for L (when the discrete spectrum is absent) and it can be shown that $\tilde{u} \in L^2_{(2)}(\sigma,R)$ where $L^2_{(2)}(\sigma,R)$ is the Hilbert space of C^2-valued functions which are square integrable on R with respect to the σ measure. In future we shall write L_x and L_σ for $L^2(R)$ and $L^2_{(2)}(\sigma,R)$ respectively. Norms and inner products in the different spaces shall also be distinguished by an x or σ suffix. The functionals in (3.4-42) are generalisations of the ordinary Fourier transformations which are called Krein direction functionals or L-Fourier transformations. Notice that these functionals lead to the incorporation of the improper eigenfunctions into the Hilbert space framework as kernels of the map, $T : u \mapsto u$, $u \in L_x$. It follows from (3.4-41) and (3.4-42) that T is an **isometry which is unitary,** T^*T = Identity operator of L_x and TT^* = Identity operator of L_σ.

$$\|u\|_x^2 = \int_{-\infty}^{\infty} (u_1(k)u_1^*(k) + u_2(k)u_2^*(k))\,d\sigma(k) = \|u\|_\sigma^2 \tag{3.4-43}$$

At the same time the T-transformed operator corresponding to L is extremely simple

$$\|u\|_\sigma^2 = \langle \lambda u,u\rangle_x = \langle Lu,u\rangle_x = \langle TLT^* u,u\rangle_\sigma \tag{3.4-44}$$

so that

$$\tilde{L} \equiv TLT^* = k^2 I \tag{3.4-45}$$

is just multiplication by the square of the independent variable. Thus L furnishes a trivial example of an operator on L_x which is **diagonal** in the L spectral representation. In general an operator A on L_x is **diagonal** in the spectral representation of L if for arbitrary $u \in L_x$, $\widetilde{(Au)} = \tilde{A}\tilde{u}$.

Consider now the particular case of two operators L, L_0 on L_x such that L is **unitarily equivalent** to L_0. That is, there exists a unitary intertwining operator U, such that

$$LU = UL_o \tag{3.4-46}$$

In this case L can be spectrally represented on the same space as L_0. Let $T_0 : u \mapsto \tilde{u}_0$ be the isometry introduced earlier, which defines the L_0 -Fourier transformations.

Then

$$\langle L_o v, u\rangle_x = \langle T_o(L_o v), T_o u\rangle_{\sigma_o} = \langle (T_oU^*)L(T_oU^*)^*(T_oU^*)Uv, (T_oU^*)Uu\rangle_{\sigma_o}$$

and

$$\langle L_o v,u\rangle_x = \langle LUv,Uu\rangle_x = \langle (TLT^*)T(Uv),T(Uu)\rangle_\sigma$$

so that

$$L_\sigma \equiv L_{\sigma_o} \quad \text{and} \quad T = T_o U^* \tag{3.4-47}$$

There may exist more than one intertwining operator defining the unitary equivalence between L and L_0. If V is another such operator then consistency with (3.4-46) requires that

$$[L_o, N] = 0 \quad \text{where} \quad N = U^* V \tag{3.4-48}$$

is the unitary **normalisation** operator. Notice that N is necessarily diagonal in the L_0 spectral representation since it commutes with L, and is bounded.

For the direct or forward scattering problem discussed in the previous section $L_0 \equiv L\ (Q=0)$. The scattering matrix can be obtained from this theory by introducing the Møller operators $U_\pm$ as the intertwining operators. In this case, still assuming that there is not discrete spectrum, the L_0-Fourier transformations are defined by inner products of arbitrary elements of L_x with the eigenfunctions

$$m_o(x,k) = e^{-ikx} \qquad p_o(x,k) = e^{ikx} \tag{3.4-49}$$

Then the map

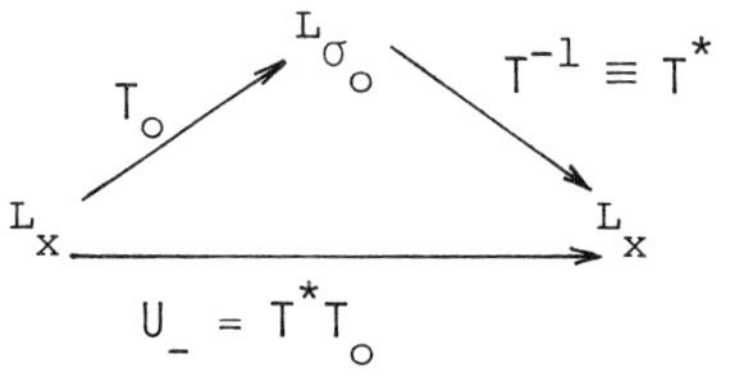

$$u \xrightarrow{T_o} u_o \xrightarrow{T^*} v \qquad u,v \in L_x\,, \quad u_o \in L_{\sigma_o} \tag{3.4-50}$$

defines one of the Møller wave operators. As the reader has probably guessed, the + eigenfunctions are defined by

$$m^+(x,k) = \frac{\phi^*(x,k)}{a^*(k)}\,, \quad p^+(x,k) = \frac{\psi^*(x,k)}{a^*(k)} \tag{3.4-51}$$

Then from relations (3.3-54),(3.3-55) and their inverses we have

$$\begin{aligned} \psi^* &= b^* \psi^* + a^* \psi \\ \psi^* &= a^* \phi - b\phi^* \end{aligned} \tag{3.4-52}$$

so that

$$p^- = \frac{1}{a} m^+ - \frac{b^*}{a} p^+$$
$$m^- = \frac{1}{a} p^+ + \frac{b}{a} m^+ \tag{3.4-53}$$

and (3.4-50) can be rewritten using (3.4-53) as

$$\begin{aligned} U_- u(x) &= \int_{-\infty}^{\infty} [(u_{01}(k) m^-(x,k) + u_{02}(k) p^-(x,k)] d\sigma(k) \\ &= \int_{-\infty}^{\infty} [\{\frac{1}{a(k)} u_{01}(k) - \frac{b^*}{a}(k) u_{02}(k)\} p^+(x,k) + \\ &\qquad \{\frac{b}{a}(k) u_{01}(k) + \frac{1}{a(k)} u_{02}(k)\} m^+(x,k)] d\sigma(k) \\ &= U_+ w(x) \end{aligned} \tag{3.4-54}$$

w is the element of L_x whose L_0 representative is formed from the coefficients of p^+ and m^+ in (3.4-54). It follows that the normalisation operator for this case, called the scattering operator

$$S = U_+^* U_- \tag{3.4-55}$$

has the L_0 representative

$$\tilde{S} = \begin{pmatrix} \frac{1}{a} & -\frac{b^*}{a} \\ \frac{b}{a} & \frac{1}{a} \end{pmatrix} \tag{3.4-56}$$

which is the unitary matrix $\tilde{S}$ called the scattering matrix introduced in Section 3.3.

In general, knowledge of the discrete spectrum of L together with either the function $R_+(k) = b(k)/a(k)$ or the function $R_-(k) = -b^*(k)/a(k)$, $k \in R$ is sufficient to completely determine $\tilde{S}$ when Q satisfies the condition $\int_{-\infty}^{\infty} |Q(v)|(1 + v^2)dv < \infty$. To prove this we introduce the function

$$h(k) = \prod_{j=1}^{M} \frac{(k - i\eta_j)}{(k + i\eta_j)} a^{-1}(k) \qquad \text{Im } k \geq 0 \tag{3.4-57}$$

where $\{\lambda_j = -\eta_j^2,\ \eta_j > 0 : j = 1, \ldots, M\}$ is the discrete spectrum for L and

$$a^{-1}(k) = 1 + o(1) \text{ as } |k| \to \infty, \quad \text{Im } k \geq 0 \tag{3.4-58}$$

Then since $h(k)$ is analytic in $\text{Im } k > 0$ and continuous in $\text{Im } k \geq 0$ it follows from the general form of Cauchy's theorem that

$$\operatorname{Log} h(k) = \frac{1}{2\pi i}\int_{\gamma_1} \frac{\operatorname{Log} h(z)}{(z-k)}\,dz = \frac{1}{2\pi i}\int_{C_1} \frac{\operatorname{Log} h(z)\,dz}{(z-k)} + $$

$$+ \frac{1}{2\pi i}\int_{-\infty}^{\infty} \frac{\operatorname{Log} h(x)\,dx}{(x-k)} \qquad \operatorname{Im} k > 0 \tag{3.4-59}$$

$$0 = \frac{1}{2\pi i}\int_{\gamma_2} \frac{\operatorname{Log} h^*(z)\,dz}{(z-k)} = \frac{1}{2\pi i}\int_{C_2} \frac{\operatorname{Log} h^*(z)\,dz}{(z-k)} \tag{3.4-60}$$

$$- \frac{1}{2\pi i}\int_{-\infty}^{\infty} \frac{\operatorname{Log} h^*(x)\,dx}{(x-k)} \qquad \operatorname{Im} k > 0$$

The contours γ_1, γ_2, C_1 and C_2 are defined in Fig. 3-3.

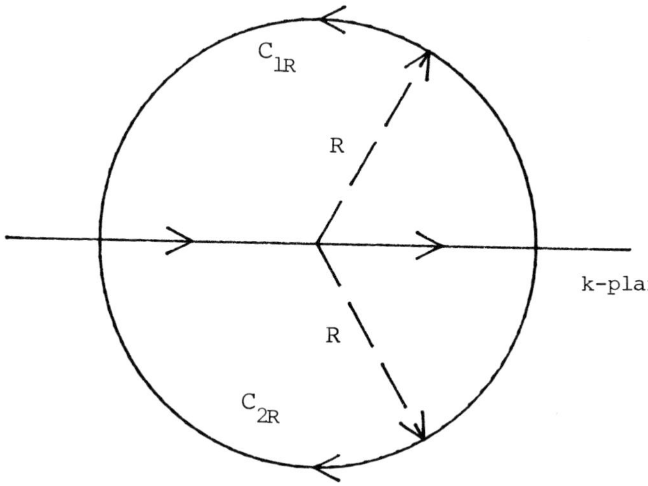

Figure 3-4: The contours γ_1, γ_2, C_1 and C_2

The contours γ_1, γ_2 are the limits of the contours $\gamma_{1R} = C_{1R} + [-R,R]$ $\gamma_{2R} = C_{2R} + [-R,R]$ as $R \to \infty$. C_1 and C_2 are the bounding semicircles in this limit for the upper and lower half k-plane respectively. Since the integrals along the contours C_1 and C_2 are zero, combining the two expressions results in the formula

$$a^{-1}(k) = \prod_{j=1}^{M} \frac{k+i\eta_j}{k-i\eta_j} \exp \frac{1}{2\pi i}\int_{-\infty}^{\infty} \frac{\log|a(x)|^{-2}}{(k-x)}\,dx \qquad \operatorname{Im} k > 0 \tag{3.4-61}$$

This formula can be continued down to the real k axis by replacing k with $y + i\varepsilon$ and taking the limit $\varepsilon \to 0$, whereupon the integral becomes principally valued. The relationship (3.3-58) can be written as

$$|a|^{-2} = 1 - \frac{bb^*}{aa^*} \tag{3.4-62}$$

Consequently if R_+ or R_- is given then $|a|^{-2}$ is determined from (3.4-49) and a^{-1} from (3.4-61). It also follows from (3.4-62) that

$$|a|^{-2} = 1 - R_+R_-^*|a|^{-2} \tag{3.4-63}$$

and from (3.3-79) and (3.4-29) that

$$D_{-j}^{-1}D_{+j}^{-1} = -\dot{a}_j^2 \tag{3.4-64}$$

so that S_+ determines S_- and vice-versa. We summarise these results on the forward scattering problem in the following theorem.

Theorem 3-10: If $\int_{-\infty}^{\infty}|Q(v)|(1 + v^2)dv < \infty$ then either of the sets

> S_+, S_- of scattering data are uniquely defined. The scattering matrix $\tilde{S}$ which is unitary and continuous and is uniquely determined by either of the sets of scattering data. The functions $T_\pm \equiv a^{-1}$ are determined from
>
> $$a^{-1}(k) = \prod_{j=1}^{M}\left(\frac{k+i\eta_j}{k-i\eta_j}\right)\exp\frac{1}{2\pi i}\int_{-\infty}^{\infty}\frac{\log|a(x)|^{-2}}{(k-x)}dx \qquad \mathrm{Im}\,k > 0$$
>
> $$|a(x)|^{-2} = 1-|R_+(x)|^2 = 1 - |R_-(x)|^2$$
>
> which can be continued down to the real axis whereupon the integral becomes principal valued.

Of particular relevance to the inverse method are the intertwining operators which transform between the eigenfunctions of L and $L_0 \equiv L(Q=0)$. They are called **transformation operators** in the literature (Agranovich and Marchenko 1963, Kay and Moses 1955,56) although Faddeyev (1963) reserves this name for intertwining operators associated with eigenfunctions defined by regular boundary conditions. For the isospectral Schrödinger operator a representation is easily obtained from the properties of the Jost solutions established in Section 3.3. Thus defining $\ell(x,k) \equiv \phi(x,k) - \exp(-ikx)$ we have that $_x\ell$ is analytic in $\mathrm{Im}\,k > 0$ and satisfies

$$|_x\ell(k)| < \frac{Ke^{-\eta x}}{|k|}, \qquad k = \xi + i\eta,\ \eta > 0 \tag{3.4-65}$$

so that $_x\ell \in L^2(R)$ and

$$\int_{-\infty}^{\infty}|_x\ell(\xi+i\eta)|^2 d\xi = O(e^{-2\eta x}) \tag{3.4-66}$$

Therefore by a theorem of Titchmarsh (1948), (Theorem 96, pg 129) the Fourier transforms

$$K_+(x,y) = \frac{1}{2\pi}\int_{-\infty}^{\infty}(\psi(x,k)-e^{ikx})e^{-iky}dk \qquad y \geqslant x$$
$$K_-(x,y) = \frac{1}{2\pi}\int_{-\infty}^{\infty}(\phi(x,k)-e^{-ikx})e^{iky}\,dk \qquad x \leqslant y \tag{3.4-67}$$

exist. Inverting (3.4-67) we obtain the transformation operators

$$\psi(x,k) = U_\psi e^{ikx} \equiv e^{ikx} + \int_x^{\infty} K_+(x,y)e^{iky}dy \tag{3.4-68}$$

$$\phi(x,k) = U_\phi e^{-ikx} \equiv e^{-ikx} + \int_{-\infty}^{x} K_-(x,y)e^{-iky}dy \tag{3.4-69}$$

which are valid in $\operatorname{Im} k > 0$ and ${}_xK_+, {}_xK_-$ are L^2 functions for $y \geqslant x$ and $y \leqslant x$ respectively. Notice that (3.4-67) implies that the kernels are real.

The existence and uniqueness of the transformation operators (3.4-68), (3.4-69) as well as detailed properties of the kernels can be established by assuming that the Jost solution ψ for example, has the form (3.4-68). Then following the method developed by Marchenko and Agranovich (1963), one can obtain a Volterra integral equation for the kernel K_+ and use the method of successive approximations to establish existence. From (3.3-10) and (3.4-68) we get

$$\int_x^{\infty} K_+(x,y)e^{iky}dy = \int_x^{\infty} Q(v)\,\frac{\sin k(v-x)}{k}\,e^{ikv}dv$$
$$+\int_x^{\infty} Q(v)dv\int_x^{\infty}\frac{\sin k(v-x)}{k}\,e^{iku}K_+(x,u)du \equiv J_1+J_2 \tag{3.4-70}$$

Using the formulae

$$\frac{\sin k(v-x)}{\lambda}e^{iku} = \tfrac{1}{2}\int_x^{2v-x} e^{iks}ds$$
$$\frac{\sin k(v-x)e^{iku}}{k} = \tfrac{1}{2}\int_{x-v+u}^{v-x+u} e^{iks}ds \tag{3.4-71}$$

and changing the order of integration in J_1, J_2 we obtain

$$J_1 = \tfrac{1}{2}\int_x^{\infty} Q(v)dv\int_x^{2v-x} e^{iks}ds = \{\tfrac{1}{2}\int_{\frac{1}{2}}^{\infty} e^{iks}ds\}\{\int_{\frac{1}{2}(x+s)}^{\infty} Q(v)dv\}$$
$$J_2 = \int_x^{\infty} Q(v)dv\int_v^{\infty} K_+(v,u)du\,\tfrac{1}{2}\int_{x-v+u}^{v-x+u} e^{iks}ds \tag{3.4-72}$$

$$= \int_x^\infty e^{iks}\left[\tfrac{1}{2}\int_x^{\frac{1}{2}(x+s)} Q(s)ds \int_{s+v-x}^{s+x-v} K_+(v,u)du \right.$$

$$\left. + \tfrac{1}{2}\int_{\frac{1}{2}(x+s)}^{\infty} Q(v)dv \int_v^{s+v-x} K_+(v,u)du\right]$$

Since we are assuming that $P_0(x) < \infty$ the first change in the order of integration is easily justified. The second change is established a posteriori to the proof of existence. Substituting the expressions (3.4-72) for J_1, J_2 and taking into account the uniqueness of the Fourier integral representation, we arrive at the following integral equation for the kernel $K_+(x,y)$

$$K_+(x,y) = \tfrac{1}{2}\int_{\frac{1}{2}(x+y)}^{\infty} Q(v)dv + \tfrac{1}{2}\int_x^{\frac{1}{2}(x+y)} Q(v)dv \int_{y+x-v}^{y+v-x} K_+(v,u)du$$

$$+ \tfrac{1}{2}\int_{\frac{1}{2}(x+y)}^{\infty} Q(v)dv \int_v^{y+v-x} K_+(v,u)du \qquad \infty < x \le y < \infty \tag{3.4-73}$$

Assuming a solution in the form

$$K_+(x,y) = \sum_{m=0}^{\infty} K_m(x,y) \tag{3.4-74}$$

where

$$K_0(x,y) = \tfrac{1}{2}\int_{\frac{1}{2}(x+y)}^{\infty} Q(v)dv$$

and

$$K_{m+1}(x,y) = \tfrac{1}{2}\int^{\frac{1}{2}(x+y)} Q(v)dv \int_{y+x-v}^{y+v-x} K_m(v,u)du$$

$$+ \tfrac{1}{2}\int_{\frac{1}{2}(x+y)}^{\infty} Q(v)dv \int_v^{y+v-x} K_m(v,u)du \qquad m = 0,1,\ldots$$

We now show by induction that

$$|K_m(x,y)| \leqslant R_0[\tfrac{1}{2}(x+y)]\frac{1}{m!} N^m(x) \qquad m = 0,1,..$$

where

$$N(t) = \int_t^\infty (v-x)|Q(v)|dv \quad \text{and} \quad R_i(x) = \int_x^\infty |v|^i |Q(v)|dv \tag{3.4-75}$$

From (3.4-74) we have

$$|K_{m+1}(x,y) \leqslant \frac{1}{2m!} R_o[\tfrac{1}{2}(x+y)] \{\int_x^{\frac{1}{2}(x+y)} (N(v))^m |Q(v)|(v-x)dv\}$$

$$+ \int_{\frac{1}{2}(x+y)}^\infty \{(N(v))^m |Q(v)| \tfrac{1}{2}(y-x)dv\}$$

$$\leqslant \frac{1}{2m!} R_o[\tfrac{1}{2}(x+y)] \int_x^\infty (N(v))^m |Q(v)|(v-x)dv \tag{3.4-76}$$

$$= \frac{1}{(2m+1)!} R_o[\tfrac{1}{2}(x+y)] |V^{m+1}(x)|$$

The result shows that

$$|K_+(x,y)| \leqslant \tfrac{1}{2} R_o[\tfrac{1}{2}(x+y)] \exp N(x) \tag{3.4-77}$$

It is clear from this result that K_+ is unbounded as $x \to -\infty$.

The second change in the order of integration is justified by Fubini's theorem since for $x > a > -\infty$

$$\int_x^\infty |K_+(x,y)|dy \leqslant \tfrac{1}{2} \exp N(x) \int_x^\infty dy \int_{\frac{1}{2}(x+y)}^\infty |Q(s)|ds$$

$$= \exp N(x) \int_x^\infty \int_y^\infty |Q(x)|dsdy$$

$$= \exp N(x) \int_x^\infty (y-x)|Q(y)|dy$$

$$= N(x) \exp N(x)$$

and so

$$\int_x^\infty dy \int_y^\infty |Q(s)|ds \int_s^\infty |K_+(s,v)|dv \leqslant \int_x^\infty dy \int_y^\infty |Q(s)|N(s)| \exp[N($$

$$= \int_x^\infty (s-x)|Q(s)|N(s)\exp(N(s))ds \tag{3.4-78}$$

$$= \exp[N(x)]\{N(x)-1\} < \infty$$

One then shows that the function defined by the right hand side of (3.4-68) satisfies the Schrödinger equation with the correct boundary condition for ψ. The function ψ is the unique function of Theorem 3-1.

The continuous differentiability of K_+ follows from (3.4-73). In fact we immediately obtain

$$\frac{d}{dx} K_+(x,x) = -\tfrac{1}{2} Q(x)$$
$$K_{+\,2x}(x,y) - Q(x)\,K_+(x,y) - K_{+\,2y}(x,y) = 0 \qquad y < x \tag{3.4-79}$$

provided Q is differentiable. From the equation satisfied by K_+ we obtain by differentiating the boundary conditions

$$\lim_{x+y\to\infty} K_{+x}(x,y) = 0,\ \lim_{x+y\to\infty} K_{+t}(x,y) = 0, \tag{3.4-80}$$

The initial value problem posed by (3.4-79) is a characteristic or **Goursat problem.** It is straightforward to obtain the solution by using Riemann's method (for the existence and uniqueness of a solution to a Goursat problem as well as Riemann's method see Garabedian (1964)).

Faddeyev (1964) and more recent authors on the inverse scattering or spectral transform use the representation

$$\psi(x,k) = e^{ikx}(1 + \int_0^\infty B_+(x,y)e^{2iky}dy) \tag{3.4-81}$$

and an analogous representation for ϕ. The connection with the Levin representation (3.4-68) is given by $K_+(x,y) = \tfrac{1}{2}B_+(x,\tfrac{1}{2}(x-y))$. We use the Levin representation for consistency with the material of Chapter 6.

3.5 Nonlinear Equations Associated with the Isospectral Schrödinger Equation

In the last two sections we restricted our attention to a single function Q and considered only the x behaviour of the function and of the corresponding Jost solutions of the Schrödinger equation. Of course our principal interest in the linear Schrödinger equation is to solve those nonlinear equations which can be associated with it in the sense of Section 3.2. We showed there that the Lax heirarchy of KdV equations was one such family. In this section we shall obtain a very large family of equations which can be associated with the Schrödinger equation and which moreover are solvable. In fact these two criteria are related; the solvable equations are always associated.

The detailed discussion of the inverse method is delayed until the next chapter. However, some outline of the method is required in order to understand how a large class of solvable equations may be obtained. Essentially the method consists of obtaining equations which govern the evolution of the scattering data from its initial specification $S_+(t_0) = \{\lambda_j(t_0), D_{+j}(t_0), R_+(k,t_0), j = 1,...M, k \in R\}$ for a given function $Q(x,t_0)$. Then because, as we shall see, provided S_+ obeys certain conditions the function Q is uniquely determined by S_+ for arbitrary values of the parameter t. If the evolution equations of the scattering data can be solved then the function Q is uniquely determined at subsequent times. Thus we will have solved the initial value problem for the equation satisfied by Q with initial function $Q(x,t_0)$. The evolution equations for the scattering data are first order and involve bilinear functionals of the Jost solutions. In general we cannot expect any simplication because Q could satisfy any nonlinear equation with the prescribed boundary conditions that Q along with its x- and t-derivatives vanish "sufficiently fast" as $|x| \to \infty$. However, if we restrict ourselves to a certain class of equations for which the evolution equation of the scattering data are linear, then they can be solved with the consequent implied solution of Q for arbitrary time.

As Ablowitz et al. (1974) have stressed, this procedure can be viewed as a nonlinear analogue of Fourier analysis. In fact the ansatz one makes in order to render the evolution equations for the scattering data linear involves the specification of a function which is essentially the dispersion relation of the linearised equation satisfied by Q. This material is discussed in detail in this section. We also use the theory of intertwining operators outlined in the previous section to develop an operator theoretic approach to obtaining the evolution of the scattering data from the Lax pair associated with a given solvable equation.

In general Q will depend upon x, t and possible some further spatial variables $y = (y_1,...,y_n)$ and similarly the Jost solutions are also functions of these additional variables. It is now convenient to consider k to be a **function** of the variables (t,y). We shall assume initially that the dependency of Q and the Jost solutions upon these variables is C^1. That is we require the existence and continuity of the partial derivatives with respect to t and y. In addition we shall require that Q is C^∞ with respect to x and that Q and its derivatives tend to zero "sufficiently fast" as $|x| \to \infty$. The reasons for these conditions will become apparent from the text. We adopt the notation that if $p \in R^{n+2}$ its local coordinates are (x,t,y). Let $C: R \to R^{n+1}$ define a curve in R^{n+1}, $t = t(u)$, $y = y(u)$.

This curve lifts to a curve $C^*_{x_0}$, $x = x_0$, $t = t(u)$, $y = y(u)$, in R^{n+2}. Let $C^* = \{C^*_{x_0} : x_0 \in R, C$ any curve in $R^{n+1}\}$. If $f \in C^1(R^{n+1},R)$ then the derivative of f along any member of C^* is given by

$$\frac{df}{du} = \frac{dt(u)}{du}\frac{\partial f}{\partial t} + \frac{d\underline{y}(u)}{du} \cdot \frac{\partial f}{\partial \underline{y}} \qquad (3.5\text{-}1$$

where $t = t(u)$, $y = y(u)$ is a parametrisation for the chosen curve. The dot, "." in (3.5-1) is just the ordinary scalar product. Thu

$(0, dt(u)/du|_{u=0}, dy(u)/du|_{u=0})$ are the components of the tangent vector to the curve $C^*_{x_0}$ at $p_0 = (x_0, t_0 = t(0), y_0 = y(0))$. The derivative of f along **any** curve of C^{*0} at p_0 will clearly have the form (3.5-1) and so we can write

$$X_{p_0}(f) = (\pi^* X^t(x_0, t_0, y_0)\frac{\partial}{\partial t} + \pi^* X^y(x_0, t_0, y_0) . \frac{\partial}{\partial y}) f \qquad (3.5\text{-}2)$$

where $\pi^* X^t = X^{t_0}\pi$, X^t and X^{y_i} are arbitrary functions on R^{n+1} and π is the projection map $\pi(x,t,y) = (t,y)$. It follows that

$$X = \pi^* X^t \frac{\partial}{\partial t} + \pi^* X^y . \frac{\partial}{\partial y} \qquad (3.5\text{-}3)$$

defines a **vector field** on R^{n+2} if X^t and X^y are specified functions on R^{n+1}. Solving

$$\frac{dt}{du} = X^t \qquad \frac{dy}{du} = X^y \qquad (3.5\text{-}4)$$

would enable the x-parameter family of **orbits** or curves to be obtained which have (3.5-3) defining their tangent vectors. Consider now the graphs of functions $k : R^{n+1} \to C$. Introduce the manifold $J^0(R^{n+1}, C)$ with local coordinates (t,y,k). Then curves on $J^0(R^{n+1}, C)$ can be lifted to curves on $J^0(R^{n+1}, C)xR$ and in this case we obtain vector fields

$$X = \pi^* X^t (\frac{\partial}{\partial t} + k_t \frac{\partial}{\partial k}) + \pi^* X^y . (\frac{\partial}{\partial y} + k_y \frac{\partial}{\partial k}) \qquad (3.5\text{-}5)$$

$(\pi(x,t,y,k,k_t,k_y) = (t,y))$ on $J^1(R^{n+1}, C)xR$ corresponding to functions X^t, X^{y_i} defined on $R^{(n+1)}$.

Classically one usually deals with the calculus of variations when one wishes to indicate an arbitrary directional derivative. Consequently we introduce the notation δ and Δ to denote the differential operators which map functions on $J^0(R^{n+1}, C)xR$ into functions on $J^1(R^{n+1}, C)xR$

$$\delta = \pi^* X^t (\frac{\partial}{\partial t} + k_t \frac{\partial}{\partial k}) + \pi^* X^y . (\frac{\partial}{\partial y} + k_y \frac{\partial}{\partial k})$$

$$\Delta = \delta - \delta k \frac{\partial}{\partial k} \qquad (3.5\text{-}6)$$

It is usual practice to omit the π^* map from (3.5-6). We shall also use δ to denote the corresponding operator which acts on functions on R^{n+1}, that is the operator

$$X^t \frac{\partial}{\partial t} + X^y . \frac{\partial}{\partial y} \tag{3.5-7}$$

It follows from their definition that the δ and Δ operators commute with $\partial/\partial x$.

In order to obtain the Δ variation of the scattering data we adopt the procedure due to Flaschka and Newell (1975). The fundamental matrix solution

$$\Phi = \begin{pmatrix} \phi & \bar{\phi} \\ \phi_x & \bar{\phi}_x \end{pmatrix}$$

as shown in Section 3.3 satisfies the equation

$$\Phi_x = P\Phi \equiv (Q\sigma - B(k))\Phi \qquad \text{Im } k = 0$$

where

$$B(k) = \begin{pmatrix} 0 & -1 \\ k^2 & 0 \end{pmatrix} \quad \text{and} \quad \sigma = \begin{pmatrix} 0 & 0 \\ 1 & 0 \end{pmatrix} \tag{3.5-8}$$

The inverse of Φ is found to be upon using the Wronskian $W({}_k\phi, {}_k\bar{\phi}) = 2ik$

$$\Phi^{-1} = \frac{1}{2ik} \begin{pmatrix} \bar{\phi}_x & -\bar{\phi} \\ -\phi_x & \phi \end{pmatrix} \tag{3.5-9}$$

Making a Δ variation of (3.5-8) and multiplying on the left by Φ^{-1} produces

$$\Phi^{-1} \Delta \Phi_x = \Phi^{-1} \Delta P \Phi + \Phi^{-1} P \Delta \Phi \tag{3.5-10}$$

By differentiating $\Phi\Phi^{-1} = 1$ with respect to x, solving for Φ_x^{-1}, and using (3.5-8) we also have

$$\Phi_x^{-1} \Delta \Phi = -\Phi^{-1} \Phi_x \Phi^{-1} \Delta \Phi = -\Phi^{-1} P \Delta \Phi \tag{3.5-11}$$

Combining equations (3.5-10),(3.5-11) and integrating with respect to x over R results in the expression

$$\Phi^{-1} \Delta \Phi \Big|_{x=-\infty}^{x=\infty} = \int_{-\infty}^{\infty} \Delta Q (\Phi^{-1} \sigma \Phi) dx \tag{3.5-12}$$

Since the fundamental matrix solutions are linearly dependent for real k

$$\Phi = \Psi A \,, \qquad A = \begin{pmatrix} b & \bar{a} \\ a & \bar{b} \end{pmatrix} \tag{3.5-13}$$

where A is the matrix introduced in Section 3.3. Consequently (3.5-12) becomes

$$\begin{aligned} A^{-1}\Delta A + A^{-1}(\Psi^{-1}\Delta\Psi)(x{=}+\infty)A - (\Phi^{-1}\Delta\Phi)(x{=}-\infty) \\ = \int_{-\infty}^{\infty} \Delta Q(\Phi^{-1}\sigma\Phi)\,dx \end{aligned} \tag{3.5-14}$$

To ensure the existence of the integrals in (3.5-12) and (3.5-14) we impose the condition that $\Delta Q \to 0$ "sufficiently fast" as $|x| \to \infty$, which is met by our earlier assumptions. Since the Δ variation is independent of x we can interchange differentiation by Δ and the limit and then use the fact that $\Delta \exp(\pm ikx) = 0$ to obtain

$$A^{-1}\Delta A = \int_{-\infty}^{\infty} \Delta Q(\Phi^{-1}\sigma\Phi)\,dx \tag{3.5-15}$$

If we introduce the bilinear functional

$$I_\Delta(u,v) = \int_{-\infty}^{\infty} \Delta Q(u,v)\,dx \tag{3.5-16}$$

Then equations (3.5-15) are equivalent to

$$\Delta a = -\frac{1}{2ik} I_\Delta(\phi,\psi) \tag{3.5-17}$$

$$\Delta b = \frac{1}{2ik} I_\Delta(\phi,\bar{\psi}) \tag{3.5-18}$$

To obtain the form given in (3.5-17),(3.5-18) one can either use the Wronskian relations (3.3-47)-(3.3-49) or more directly exploit the relation-ships which exist between the fundamental matrix solutions Φ and Ψ and the matrix A given by equation (3.5-13).

The Δ variations of $\bar{a}$ and $\bar{b}$ have been omitted from (3.5-12) as they are just the complex conjugates of Δa and Δb. In Section 3.4 we defined the scattering data S_+ to be the set

$$S_+ = \{\lambda_j\,,\ D_{+j}\,,\ R_+(k)\,,\quad k\in R,\ j = 1,\ldots,M\}$$

That is, it consists of the eigenvalues and the normalisation "constants" of the eigenvectors of L together with the function R_+ defined on the real line. It is straightforward to obtain from (3.5-17),(3.5-18) the Δ variation of R

$$\Delta R_+ = \frac{\Delta b}{a} - \frac{b}{a^2}\,\Delta a = R_+\left[\frac{1}{2ik}\ \frac{1}{ab}\ I_\Delta(\phi,\phi)\right] \qquad (3.5\text{-}19)$$

Equation (3.5-19) is valid for any function Q which satisfies $\int_{-\infty}^{\infty}|Q(x)|(1 + x^2)dx < \infty$ and which vanishes along with its derivatives as $|x| \to \infty$, for arbitrary values of t and y. Therefore in general (3.5-19) is a very complicated expression which we could not hope to be able to solve **exactly**, that is find an integrating factor for the equation. However, if we impose the **constraint**

$$I_\Delta(\phi,\phi) - 2ikab\Omega = 0$$
$$\Omega = \Omega(k,t,y) \qquad (3.5\text{-}20)$$

then this will select from the class of all possible equations formed from functions Q which meet the conditions, those for which the Δ variation of R_+ is given by a **linear** equation

$$\Delta R_+ = \Omega R_+ \qquad (3.5\text{-}21)$$

In particular, certain variations will render (3.5-21) exactly solvable. As we noted briefly at the beginning of this section, if we can solve the equations governing the evolution of S_+ then we have effectively solved the initial value problem for the nonlinear equation satisfied by Q. The question then is, can we obtain from (3.5-20) the family of equations such that for any member of the family the evolution of the R_+ function is given by (3.5-21)?

From the Schrödinger equation (3.3-1) we obtain for improper eigenfunctions u, v with the same eigenvalue the relations

$$-u_{xx}v = (k^2-Q)uv \qquad (3.5\text{-}22)$$

$$-(u_{xx}v)_x + Q_x uv = (k^2-Q)(uv)_x \qquad (3.5\text{-}23)$$

$$(uv)_{xx} = u_{xx}v + 2u_x v_x + uv_{xx} = 2(Q-k^2)uv + 2u_x v_x \qquad (3.5\text{-}24)$$

$$-u_{xx}v_x = (k^2-Q)uv_x \qquad (3.5\text{-}25)$$

$$-v_{xx}u_x = (k^2-Q)u_x v \qquad (3.5\text{-}26)$$

Using (3.5-23), (3.5-25) and (3.5-26) we can form

$$(u_{xx}v - u_x v_x)(x) = \int_{-\infty}^{x} Q_s uv \, ds + (u_{xx}v - u_x v_x)(-\infty) \tag{3.5-27}$$

and combining this expression with (3.5-24) and (3.5-22) results in the equation

$$-\frac{1}{4}\left(\frac{\partial^2}{\partial x^2} + 2\int_{-\infty}^{x} ds\, Q_s - 4Q\right)uv = k^2 uv + \tfrac{1}{2}(u_{xx}v - u_x v_x)(-\infty) \tag{3.5-28}$$

In (3.5-28) $\int_{-\infty}^{x} dsQ$ is the integral operator defined by

$$\left(\int_{-\infty}^{x} ds\, Q_s\right)u(x) = \int_{\infty}^{x} Q_s(s)\, u(s)\, ds \tag{3.5-29}$$

Equations (3.5-27) and (3.5-28) are the fundamental relationships which we shall require. In particular if $u = v = \phi$ then (3.5-27) becomes

$$ab = \frac{1}{4k^2}\int_{-\infty}^{\infty} Q_x \phi^2 dx \tag{3.5-30}$$

and (3.5-28) becomes an **eigenvalue problem,**

$$L_1 \phi^2 = k^2 \phi^2$$

where

$$L_1 \equiv -\frac{1}{4}\left(\frac{\partial^2}{\partial x^2} + 2\int_{-\infty}^{x} ds\, Q_s - 4Q\right) \tag{3.5-31}$$

Equation (3.5-30) enables the constraint equation (3.5-20) to be written as

$$\int_{-\infty}^{\infty} (\Delta Q + Q_x C)\phi^2 dx = 0$$
$$C(k,t,y) = \frac{i}{2k}\Omega(k,t,y) \tag{3.5-32}$$

Remember that our aim is to force (3.5-32) to yield the solvable evolution equations. If C and the coefficients of Δ are functions of t and y only, then requiring

$$\Delta Q + CQ_x = 0 \tag{3.5-33}$$

for all t and y is sufficient to generate a nontrivial evolution equation involving only Q and its derivatives and which trivially satisfies the constraint equation. How can be generalise this example to the situation when C and Δ depend arbitrarily upon k? First consider the case when

$C(k) = -4k^2$ and Δ is independent of k. Then (3.5-32) can be written after using (3.5-31) as

$$\int_{-\infty}^{\infty} (\Delta Q - 4Q_x L_1)\phi^2 dx = 0 \qquad (3.5\text{-}34)$$

This has the right form, but L_1 acts on ϕ^2 so we cannot simply equate the bracket to zero. Thus we need to transfer the action of L_1 from ϕ^2 onto Q_x. If we can do this then the bracket in (3.5-34) is again independent of ϕ and so equating it to zero is a sufficient nontrivial condition on Q for the validity of (3.5-20). To lift the action onto Q_x we form the **adjoint** L_1^A of L (see Chapter Note 3.5.1). Define

$$< L_1 u, Q_x v> = \int_{-\infty}^{\infty} L_1 u \, Q_x v^* dx \qquad (3.5\text{-}35)$$

where u and v are arbitrarily elements of $L^2(R)$. Notice that (3.5-35) is well defined even if $u(x,k) = \phi^2(x,k)$, k real, because of the asymptotic behaviour of the function Q_x. Integrating by parts shows that

$$-\frac{1}{4}\int_{-\infty}^{\infty}\left(\frac{\partial^2}{\partial x^2} + 2\int_{-\infty}^{x} dy\, Q_x - 4Q\right)u\, Q_x v^* dx$$

$$= -\frac{1}{4}\left[u_x Q_x v^* - u(Q_x v^*)_x - 2\int_x^{\infty} Q_s v^* ds \int_{-}^{} Q_s u \, ds\right]_{-\infty}^{\infty}$$

$$-\frac{1}{4}\int_{-\infty}^{\infty} u\left[\frac{\partial^2}{\partial x^2} - 4Q + 2Q_x\int_x^{\infty} ds\right](Q_x v^*)dx \qquad (3.5\text{-}36)$$

The form of integration which results in the integral operator $\int_x^{\infty} dy$ in (3.5-36) has been chosen so that upon evaluation the square bracketed term vanishes and we obtain

$$< L_1 u, Q_x v> = <u, L_1^A Q_x v>$$

$$L_1^A \equiv -\frac{1}{4}\left(\frac{\partial^2}{\partial x^2} - 4Q + 2Q_x\int_x^{\infty} dy\right) \qquad (3.5\text{-}37)$$

In particular (3.5-37) is valid when $v \equiv 1$ and $u = \phi^2$ so that (3.5-34) can be written

$$\int_{-\infty}^{\infty} (\Delta Q - 4L_1^A Q_x)\phi^2 dx = 0 \qquad (3.5\text{-}38)$$

This constraint is satisfied trivially when Q evolves according to

$$\Delta Q - 4L_1^A Q_x = 0 \tag{3.5-39}$$

Reverting to the notation of Sections 3.1 and 3.2 we put $q = -6\alpha^{-1}Q$ and choose $\Delta = \partial/\partial t$ then simplifying (3.5-39) we find that it becomes

$$q_t + \alpha q q_x + q_{xxx} = 0 \tag{3.5-40}$$

which is the KdV equation (3.1-1). Clearly the same method will enable us to obtain the nonlinear evolution equations corresponding to when C is an arbitrary real polynomial in k^2. Thus if

$$C(t,y,k^2) = \sum_{i=0}^{n} g_i(t,y)k^{2i} \tag{3.5-41}$$

we have that

$$C(t,y,k^2)\phi^2 = \sum_{i=0}^{n} g\ (t,y)k^{2i}\phi^2 = \sum_{i=0}^{n} g_i(t,y)(L_1)^{2i}\phi^2$$
$$= C(t,y,L_1)\phi^2 \tag{3.5-42}$$

In obtaining equation (3.5-42) we have implied conditions on the differentiability of Q. The basic requirement needed here and in what follows is that Q is C^∞ with respect to x and C^1 with respect to (t,y). In addition we require that Q and its derivatives tend to 0 as $|x| \to \infty$ "sufficiently fast" so that the singular integrals are defined.

Let us use this information to obtain the general nonlinear equation defined by (3.5-41). From (3.5-32) and (3.5-41),(3.5-42) we have

$$\int_{-\infty}^{\infty} (\Delta[L_1^A)Q(x,t,y) + C(t,y,L_1^A)\ Q_x(x,t,y)]\ \phi^2(x,t,y)dx = 0 \tag{3.5-43}$$

where

$$\Delta(L_1^A) \equiv X^t(t,y,L_1^A)\frac{\partial}{\partial t} + X^y(t,y,L_1^A)\cdot\frac{\partial}{\partial y} \tag{3.5-44}$$

and so the constraint equation is trivially satisfied by the nonlinear evolution equation

$$\Delta(L_1^A)Q(x,t,y) + C(t,y,L_1^A)Q_x(x,t,y) = 0 \tag{3.5-45}$$

A larger class of nonlinear equations which satisfy the constraint equation are derived by allowing C to be a rational function of k^2, $C = C_2/C_1$ where C_1 and C_2 are of the form (3.5-41). The corresponding evolution equation is then

$$C_1(t,y,L_1^A)\Delta(L_1^A)Q(x,t,y) + C_2(t,y,L_1^A)Q_x(x,t,y) = 0 \tag{3.5-46}$$

It follows that the family of nonlinear evolution equations for which the Δ-evolution of the R_+ function obeys a linear equation is generated by the set of arbitrary real polynomials of k^2, C_1, C_2. More formally C_1, C_2 could be taken as any real analytic functions of k.

The equations can be written in an alternative form which introduces an operator which plays an important role in the Hamiltonian structure of this particular inverse method. Let $u, v \in L^2(R)$, then

$$\begin{aligned}
\langle u, L_1^A v_x\rangle &= -\frac{1}{4}\int_{-\infty}^{\infty} u\left(\frac{\partial^2}{\partial x^2} - 4Q + 2Q_x\int_x^{\infty} ds\right)\frac{\partial v^*}{\partial x}\,dx \\
&= -\frac{1}{4}\int_{-\infty}^{\infty} u\left(\frac{\partial}{\partial x}(v^*_{xx} - 4Qv^*) + 2Q_x v^*\right)dx \\
&= \langle u, \frac{\partial}{\partial x} L_2 v\rangle
\end{aligned}$$

where

$$L_2 \equiv -\frac{1}{4}\left(\frac{\partial^2}{\partial x^2} - 4Q - 2\int_x^{\infty} ds Q_s\right) \tag{3.5-47}$$

Equation (3.5-46) can now be written

$$C_1(t,y,L_1^A)\Delta(L_1^A)\ Q(x,t,y) + \frac{\partial}{\partial x} C_2(t,y,L_2)\ Q(x,t,y) = 0 \tag{3.5-48}$$

This follows by induction since for integral r

$$\langle u, (L_1^A)^r v_x\rangle = \langle u, (L_1^A)^{r-1}\frac{\partial}{\partial x}(L_2 v)\rangle \tag{3.5-49}$$

After taking into account the boundary conditions on Q equation (3.5-48) can also be written as

$$\int_{-\infty}^{x} ds C_1(t,y,L_2)\ \Delta(L_2)\ Q(s,t,y) + C_2(t,y,L_2)\ Q(x,t,y) = 0 \tag{3.5-50}$$

Before considering in detail examples of these equations let us return to the consideration of the evolution of S_+. We have just shown that solvable equations (3.5-50) are associated with the linear evolution of R_+

$$\Delta R_+ = \Omega R_+ \equiv -2ik\, C_1(k^2)/C_2(k^2)R_+ \tag{3.5-51}$$

We were forced to make Δ-variations of R because the evaluations in (3.5-14) are undefined for a δ-variation. However, for Im k > 0 or Im k < 0, the most general variation we can perform is a δ variation of the scattering data a, b or $\bar{a}$, $\bar{b}$ respectively. In this case analyticity of b and $\bar{b}$ requires that Q has compact support. Equations (3.5-17), (3.5-18) are now replaced by

$$\begin{aligned} \delta a + a\delta k &= -\frac{1}{2ik}\cdot I_\Delta(\phi,\psi) \\ \delta b + b\delta k &= \frac{1}{2ik}\cdot I_\Delta(\phi,\bar{\psi}) \qquad \text{Im } k > 0 \end{aligned} \tag{3.5-52}$$

so that

$$\delta\left(\frac{b}{a}\right) = \frac{1}{2ika^2} I_\Delta(\phi,\phi) \equiv \Omega ba^{-1} \qquad \text{Im}k > 0 \tag{3.5-53}$$

The δ-variation of the barred scattering data is found by taking the complex conjugate of equations (3.5-52). If $k \in C(R^{n+1},C)$ then corresponding to any orbit of δ, $t = t(u)$, $y = y(u)$ there is an induced curve $k(u) \equiv k(t(u),y(u))$ in the complex k-plane. In particular let k_j correspond to the eigenvalue $\lambda_j = k_j^2$ of L. Then the functional

$$D_+(k) = -i\, b(k)(k-k_j)/a(k) \tag{3.5-54}$$

where we have supressed the t and y dependency is well defined, differentiable, and has the value

$$D_{+j} \equiv D_+(k_j) = -\, i\, b_j/\dot{a}_j \tag{3.5-55}$$

at $k = k_j$ which is the normalisation "constant" for the corresponding eigenfunction ψ_j as we showed in Sections 3.3 and 3.4. From (3.5-53) and (3.5-54) we obtain

$$(\delta D_+(k) - \Omega(k)D_+(k)).(k-k_j) - D_+(k)\,\delta(k-k_j) = 0 \tag{3.5-56}$$

is really a k-parameter family of operators in the notation of Section 3.3

$$k^\delta = {}_kX^t(t,y)\left\{\frac{\partial}{\partial t} + k_t\frac{\partial}{\partial k}\right\} + {}_kX^y(t,y).\left\{\frac{\partial}{\partial y} + k_y\frac{\partial}{\partial k}\right\} \tag{3.5-57}$$

Consequently assuming that the coefficients of ${}_k\delta$ are analytic

$$_k\delta = {}_{k_j}\delta + (k-k_j)_{k_j}\delta_k + O(|k-k_j|^2) \tag{3.5-58}$$

and (3.5-56) can be written

$$\begin{aligned}({}_k\delta D_+(k) - \Omega(k)D(k) - D_+(k)\ {}_{k_j}\delta_k(k-k_j))(k-k_j)\\ + {}_{k_j}\delta(k-k_j) + O(|k-k_j|^2) = 0\end{aligned} \tag{3.5-59}$$

We can evaluate the limit as $k \to k_j$ in two ways. First note that ${}_{k_j}\delta k_j = 0$ and then take the limit to obtain

$$_{k_j}\delta k|_{k=k_j} = (\delta k)|_{k=k_j} \equiv X^t(t,y,k_j)k_{jt} + X^y(t,y,k_j).k_{jy} = 0 \quad j=1,\ldots,M \tag{3.5-60}$$

Secondly divide (3.5-59) by $(k - k_j)$ and then take the limit using (3.5-60) which gives

$$_{k_j}\Delta D_j(t,y) - \Omega(t,y,k_j)D_{+j}(t,y) - D_{+j}(t,y)_{k_j}\Delta_k k_j(t,y) = 0 \quad j=1,\ldots,M \tag{3.5-61}$$

In producing this formula we have made use of the facts that $\delta k(t,y)|_{k=k_j} \equiv \Delta k(t,y)|_{k=k_j}$ and $\delta_k\ k(t,y)|_{k=k_j} \equiv \Delta_k k(t,y)_{k=k_j}$. When Q does not have compact support, (3.5-61) is still the required equation. For this case the integral in (3.5-53) converges only at an eigenvalue of L. The method of proof proceeds by introducing an auxillary analytic function $\ell(k)$ such that $\ell(k_j) = b_j$ and $\ell_k(k_j) = b_{kj}$. The existence of b_{kj} follows from the integral representation (3.3-64) and Theorem 3-1. In general (3.5-60) is a nonlinear equation for k_j. However, once k_j has been determined (3.5-61) is linear. We shall not consider the general family of solvable equations because of the difficulty of analysing (3.5-60). Special solutions can be obtained however, but for the purposes of this book we restrict ourselves in the remainder of this section and the next chapter to the case of principal interest $\Delta \equiv \partial/\partial t$, and C_1, C_2 are functions of k^2 alone. In this case $S_+(t)$ is uniquely defined by $S_+(t_0)$. Furthermore we observe the following interesting property of the corresponding solvable equations. Linearising (3.5-46) we obtain

$$C_1\left(-\frac{1}{4}\frac{\partial^2}{\partial x^2}\right) Q_t(x,t) + C_2\left(-\frac{1}{4}\frac{\partial^2}{\partial x^2}\right) Q_x(x,t) = 0 \tag{3.5-62}$$

and putting $Q(x,t) = \exp\{i(\omega t-kx)\}$ we get the dispersion relation

$$\omega(k) = k\ C_2(\tfrac{1}{4}k^2)/C_1(\tfrac{1}{4}k^2) \tag{3.5-63}$$

from which we deduce that the phase speed of the elementary solution ω/k determines an equivalence class of solvable nonlinear equations. Two equations belong to the same equivalence class if $C_2^1 = E.C$ and $C_1^1 = E.C_1$ where E is a real analytic function of k^2. To summarise the results we have obtained so far we make the "sufficiently fast" criterion explicit. Q is a function of **rapid decrease** or belongs to the Schwartz class if it is C^∞ and

$$\sup_{(x,t)\varepsilon R^2} \left| x^{\alpha_1} t^{\alpha_2} \frac{\partial^{\beta_1+\beta_2} Q}{\partial x^{\beta_1} \partial t^{\beta_2}} \right| < \infty \tag{3.5-64}$$

where α_i, β_i are non-negative integers. If we invert the process whereby we obtained the solvable equations, then we see that provided $Q(x,t)$ is uniquely reconstructable from $S_+(t)$ the initial value problem is solved if $Q(x,t_0)$ satisfies a general Schwartz condition

$$\sup_{(x,t)\in R^2} \left| t^{\alpha_2} x^{\alpha_1} \frac{\partial^{\beta_1+\beta_2} Q}{\partial x^{\beta_1} \partial t^{\beta_2}} \right| < \infty \quad \begin{array}{l} \beta_1 \text{ an integer} \\ \beta_2 \text{ a non-negative integer} \end{array} \tag{3.5-65}$$

Thus if we interpret negative indices as anti-derivatives this will ensure that the integrals of $Q(x,t_0)$ and $Q_t(x,t_0)$ also have the rapid decrease property. In fact we only require β_2 to take the values 0 and 1, but for simplicity we will use the more general statement. This is, of course, a very strong condition on $Q(x,t_o)$. Much weaker conditions are sufficient for specific solvable equations (see Section 4.1), but this is the simplest condition to impose to ensure that Q is a classical solution for an arbitrary member of the family of solvable equations given by (3.5-46).

Theorem 3-11: If $\int_{-\infty}^{\infty} |Q(x,t_0)|(1 + x^2)dx < \infty$ then the scattering data $S_+(t_0)$ is uniquely determined from $Q(x,t_0)$.

The function $\Omega(k)$ defines an equivalence class of solvable nonlinear evolution equations

$$C_1(L_1^A)Q_t(x,t) + C_2(L_1^A)Q_x(x,t) = 0$$

each member of which has the same $S_+(t)$

$$R_+(t) = R_+(t_o)\exp\Omega(k)(t-t_o)$$
$$D_{+j}(t) = D_{+j}(t_o)\exp\Omega_j(t-t_o),\ \Omega_j = \Omega(k_j)$$
$$k_j(t) = k_j(t_o)$$

$\Omega(k) = -2ikC_2/C_1 \equiv -2i\omega(2k)$ and $\omega = \omega(k)$ is the dispersion relation of the corresponding linearised equation.

A sufficient condition for Q to be a solution to the initial value problem is that the initial data should be so defined as to ensure that the reconstructed Q belongs to the general Schwartz class.

One of the problems here, as we shall see in the next chapter, is in defining the initial data for an arbitrary solvable equation. As examples of nonlinear equations solvable by this inverse method we have the following

KdV

$$C_1(k^2) = 1, \quad C_2(k^2) = -4k^2$$

$$Q_t - 6QQ_x + Q_{xxx} = 0$$

KdV Hierarchy

$$C_1(k^2) = 1, \quad C_2(k^2) = a_i k^{2i} \quad a_i \varepsilon R.$$

$$Q_t + a_i (L_1^A)^{2i} Q_x = 0$$

This is the same set of equations as derived in Section 3.1 using a Lax pair, since both sets of equations have the same linearised dispersion relation and $C_1 \equiv 1$

A long wave equation

$$C_1(k^2) = (1+k^2), \quad C_2(k^2) = -4$$

$$Q_{xxt} - 4Q_t - 4QQ_t + 2Q_x \int_x^\infty Q_t dy + Q_x = 0$$

This equation reduces to the KdV equation in the long wave, small amplitude limit and has the good stability property that it responds feebly to short waves.

Finally we consider the role played by a in this inverse method. From equation (3.5-22)-(3.5-26) we obtain in a similar fashion to equation (3.5-28) that

$$L_2\, uv = k^2 uv + \frac{1}{2}(u_{xx}v - u_x v_x)(\infty) \tag{3.5-66}$$

where L_2 is the operator defined by (3.5-47). In particular

$$L_2 \phi\psi = k^2 \phi\psi - k^2 a \tag{3.5-67}$$

and since

$$L_2 1 = \tfrac{1}{2}Q$$

$$(L_2 - k^2 I)\left(\frac{\phi\psi}{a} - 1\right) = -\tfrac{1}{2}Q \tag{3.5-68}$$

Although the trace of the resolvent operator is not defined it is possible to define a regularised trace with respect to the reference operator $L_0 \equiv L(Q=0)$,

$$d(\lambda) \equiv \mathrm{Tr}\left((L-\lambda I)^{-1} - (L_0-\lambda I)^{-1}\right) \tag{3.5-69}$$

For the isospectral Schrödinger operator using the definition of the resolvent kernel given by (3.4-12) we find that

$$d(k^2) = -\frac{1}{2k}\frac{d}{dk}\log a(k) = \frac{i}{2k}\int_{-\infty}^{\infty}\left(\frac{\phi\psi}{a} - 1\right)dx \tag{3.5-70}$$

so that using (3.5-68) we have formally that

$$\frac{d}{dk}\log a(k) = \frac{i}{2}\int_{-\infty}^{\infty}(L_2-k^2)^{-1}Q\,dx \tag{3.5-71}$$

It is also possible to rewrite (3.5-17) in terms of the resolvent operator for 2

$$\Delta a = -\frac{1}{2ik}I_\Delta(\phi,\psi) = -\frac{a}{2ik}\int_{-\infty}^{\infty}\Delta Q\left(\frac{\phi\psi}{a} - 1\right)dx \tag{3.5-72}$$

under the boundary conditions on ΔQ. Then using (3.5-68) we find that

$$\Delta \log a(k) = \frac{1}{4ik}\int_{-\infty}^{\infty}\Delta Q(L_2-k^2)^{-1}Q\,dx \tag{3.5-73}$$

Equations (3.5-72) and (3.5-73) are extremely interesting and important equations. Assuming it is valid to expand $(I-L_2/k^2)^{-1}$ for larger k we get from (3.5-71)

$$\frac{d}{dk}\log a(k) \sim \frac{-i}{2k^2}\int_{-\infty}^{\infty}\sum_{n=0}^{\infty}\left(\frac{L_2}{k^2}\right)^n Q\,dx \quad \text{as } |k| \to \infty \tag{3.5-74}$$

and formally integrating we obtain

$$\log a(k) \sim \frac{i}{2}\int_{-\infty}^{\infty}\sum_{n=0}^{\infty}\frac{1}{(2n+1)}\frac{1}{k^{2n+1}}(L_2)^n Q\,dx \quad \text{as } |k| \to \infty \tag{3.5-75}$$

Defining

$$c_{2n+1} \equiv \frac{i}{2}\int_{-\infty}^{\infty}\frac{L_2{}^n Q}{(2n+1)}\,dx$$

we obtain the trace formulae due to Zakharov and Faddeev (1971). The first three terms in the expansion are

$$c_1 = \frac{i}{2}\int_{-\infty}^{\infty} Q\,dx, \quad c_3 = \frac{i}{8}\int_{-\infty}^{\infty} Q^2\,dx, \quad c = \frac{i}{32}\int_{-\infty}^{\infty}(Q_x^2 + 2Q^3)\,dx \tag{3.5-76}$$

The validity of this exapansion can be justified by obtaining an asymptotic expansion for $\log a$ starting from the isospectral Schrödinger equation. It is not hard to see that $\Delta a = 0$. From (3.3-60) and (3.5-21) we have

$$\Delta|R_+|^2 + \Delta|T_+|^2 = \Delta|T_+|^2 = 0 \tag{3.5-77}$$

since Ω is imaginary. Then from (3.4-61) it follows that $\Delta a = 0$. It is also possible to show directly that the integral on the right hand side of (3.5-73) is identically zero for the solvable equations. However, the equation (3.5-73) is non-trivial when Δ is interpreted as a functional derivative.

The **functional** or **Frechet derivative** is defined by

$$\int_{-\infty}^{\infty}\frac{\Delta F(Q)}{\Delta Q}\cdot v\,dx \equiv \lim_{\varepsilon\to 0}\frac{d}{d\varepsilon}F(Q+\varepsilon v)$$

where

$$F(Q) = \int_{-\infty}^{\infty} F(Q, Q_x, \ldots, Q_{nx})\,dx \tag{3.5-78}$$

In (3.5-78), n is a positive integer.

As an example, if $F(Q) = \int_{-\infty}^{\infty}(Q_x^2 + 2Q^3)dx$ then

$$\lim_{\varepsilon\to 0}\frac{d}{d\varepsilon}\int_{-\infty}^{\infty}\{(Q_x+\varepsilon v_x)^2 + 2(Q+\varepsilon v)^3\}dx = \int_{-\infty}^{\infty}(6Q^2 - 2Q_{xx})\,v\,dx \tag{3.5-79}$$

after an integration by parts so that

$$\frac{\Delta F(Q)}{\Delta Q} = 6Q^2 - 2Q_{xx} \tag{3.5-80}$$

In the above v is assumed to belong to the same space as Q so, for example, we can assume the space of functions are of general Schwartz type. A simple generalisation of the definition enables the Frechet derivative to be defined for this case also (simply allow n to be an integer in (3.5-78) with negative indices interpreted as integrals). It is easy to see from the definition that the Frechet derivative, when restricted to functions of the Schwartzclass, is just the Euler-Lagrange operator in this function space

$$\frac{\Delta}{\Delta Q} = \frac{\partial}{\partial Q} - \frac{d}{dx}\frac{\partial}{\partial Q_x} + \frac{d^2}{dx^2}\frac{\partial}{\partial Q_{xx}} - \ldots. \tag{3.5-81}$$

A check with the derivation shows that (3.5-73) can be intepreted as this type of variational derivative. This is because the operation of total differentiation and the Frechet derivative commute so that the assumptions made in deriving (3.5-73) are valid in this case also. That is (3.5-81) can be taken as defining the variation of $\log a$ when Q varies in an arbitrary manner (not necessarily as a solution) in the function space. Asymptotically as $|k| \to \infty$, (3.5-73) becomes

$$\Delta \log a \sim -\frac{1}{4ik^3}\int_{-\infty}^{\infty}\sum_{n=0}^{\infty} \Delta Q\left(\frac{L_2}{k^2}\right)^n Q\,dx \quad \text{as } |k| \to \infty \tag{3.5-82}$$

but

$$\log a \sim \sum_{n=0}^{\infty}\left(\frac{c_{2n+1}}{k^{2n+1}}\right) \qquad \text{as } |k| \to \infty \tag{3.5-83}$$

and consequently

$$\frac{\Delta c_{2n+3}}{\Delta Q} = \frac{1}{4i}(L_2)^n Q \tag{3.5-84}$$

Putting $C_2(L_2) = \sum_{j=1}^{n} a_j (L_2)^j$, $c_1 \equiv 1$ and then using (3.5-84) shows that the corresponding solvable equation can be written in **Hamiltonian form**

$$Q_t + \frac{\partial}{\partial x}\left\{\frac{\Delta H}{\Delta Q}\right\} = 0 \tag{3.5-85}$$

with Hamiltonian

$$H = \sum_{j=1}^{n} 4ia_j\, c_{2j+3} \tag{3.5-86}$$

It follows that there is a Hamiltonian structure associated with the hierarchy of Lax equations. Thus for example taking $H = -16ic_5$ we find upon using (3.5-81) that we obtain the KdV equation (3.1-1) after rescaling

$q = -6\alpha^{-1}Q$. This aspect of the equations solvable by the inverse scattering transform method as well as the Hamiltonian form of the equations for rational linearised dispersion relations takes us outside the scope of the present volume. Further details are contained in the paper of Zakharov and Faddeev (1971), the papers of Gel'fand and Dikii (1975-78) and the articles by Flaschka and Newell (1975) and Dodd and Bullough (1979).

In Section 3.2, we derived the hierarchy of KdV equations through the existence of a Lax pair (A, L) (now often written as (P, L) after Peter Lax), L being the isospectral Schrödinger operator. We referred to the KdV hierarchy as being associated with the operator L. In this section we have obtained a large class of solvable nonlinear equations which we should also like to be able to show were associated, that is to determine the A in the Lax pair. In general this is a very involved process. However, for the Lax hierarchy there is a fairly direct route which we shall now describe. The diagonal of the resolvent kernel is defined by

$$R(x,x,k^2) = \frac{i}{2ka(k)}\,\psi(x,k)\phi(x,k) \qquad \text{Im } k > 0 \tag{3.5-87}$$

which is immediate from the definition of resolvent kernel given in Section 3.4. Clearly from (3.5-72) we have that

$$\frac{\Delta \log a}{\Delta Q} = R \tag{3.5-88}$$

so that in particular it follows from the asymptotic expansion for large k of this relationship from (3.5-85),(3.5-86) and (3.5-84) that

$$R_{j+1} = L_2 R_j \tag{3.5-89}$$

where R has the asymptotic expansion

$$R \sim \sum_{j=0}^{\infty} \frac{R_j}{k^{2j+1}} \qquad \text{as } |k| \to \infty \tag{3.5-90}$$

Equivalently from differentiating (3.5-68) we see that R satisfies the third order linear equation

$$-R_{xxx} + 2Q_x R + 4(Q-k^2)R_x = 0 \tag{3.5-91}$$

In this notation the solvable equations of this type can be written as

$$Q_t + \frac{\partial}{\partial x}\sum_{j=0}^{n} b_j R_j = 0 \tag{3.5-92}$$

It is therefore clear that the commutator of L and A_j must give rise to some multiple of R_{jx}. If we interpret R given by (3.5-90) as a generating function then we can obtain a generating operator for the operators corresponding to the R_j functions entering into (3.5-92) in the following way

$$-R_x \equiv [L,A] \tag{3.5-93}$$

where A is the generating operator for the A_j's.

$$\begin{aligned} -R_x &= -R_x(L-k^2)(L-k^2)^{-1} = (R_x \frac{\partial^2}{\partial x^2} - QR_x + k^2R_x)(L-k^2)^{-1} \\ &= (R_x \frac{\partial^2}{\partial x^2} - \frac{1}{4} R_{xxx} + \frac{1}{2} Q_x R)(L-k^2)^{-1} \end{aligned} \tag{3.5-94}$$

The last expression results upon substituting for k from (3.5-91). We also have that

$$\begin{aligned} [L,A] &= (La-aL)(L-k^2)^{-1} \\ &= (-a_{xx} - 2a_x \frac{\partial}{\partial x} + Qa-aQ)(L-k^2)^{-1} \end{aligned} \tag{3.5-95}$$

if we put

$$A = a(L-k^2)^{-1} \tag{3.5-96}$$

Then comparing (3.5-95) with (3.5-94) and assuming that a is a first order operator leads to the required representation

$$A = \left(-\frac{1}{2} R \frac{\partial}{\partial x} + \frac{1}{4} R_x \right) (L-k^2)^{-1} \tag{3.5-97}$$

Specific A_j's are obtained from the asymptotic expansion of the generating operator

$$A = \sum_{j=1}^{\infty} \frac{A_j}{k^{2j+1}} \qquad \text{as } |k| \to \infty \tag{3.5-98}$$

The operator for the KdV equation is obtained by taking an appropriate multiple of A_2. The annoying factor i which appears in this analysis can be transformed away by a change of variable $k \mapsto ik$.

Having established an operator theoretic method for obtaining the solvable equations it seems reasonable to ask whether the time evolution of the scattering data may also be obtained in this way without recourse to the properties of the isospectral Schrödinger equation itself. Unfortunately the details of such an approach are yet to be completed. However, we can apply some of the ideas of Section 3.4 on intertwining operators to give an outline of the essence of such a method. Put $L_o \equiv L(Q=0)$ and $L \equiv L(Q(t))$ then defining $U_\pm$ as we did in section 4 we find from the definition of an intertwining operator (3.4-46) that

$$L_t = [A_\pm, L] \quad \text{where} \quad A_\pm = U_{\pm t} U_\pm^* \tag{3.5-99}$$

and

$$S_t = (U_+^* U_-)_t = U_{+t}^* U_- + U_+^* U_{-t} = U_+^* A_+^* U_- + U_+^* A_- U_- = U_+^* (A_- - A_+) U_+ S \tag{3.5-100}$$

Use has been made of the fact that $A_\pm$ are skew-adjoint which is easily proved by differentiating with respect to t the unitary condition on $U_\pm$. Transforming (3.5-100) to the spectral representation of L_0 gives

$$\tilde{S}_t = \tilde{B}\tilde{S} \quad \text{where } \tilde{B} = T_o U_+^* (A_- - A_+) U_+ T_o^* \tag{3.5-101}$$

$\tilde{B}$ is assumed to be diagonal (since $\tilde{B}$ is unbounded this has to be added as an extra condition). In this way we have obtained an equation for the evolution of the scattering matrix. An example is given in Flaschka and Newell (1975).

3.6 Notes

Section 3.1

1. We use the work **solvable** in the same sense as many authors would use the word integrable. That is, there exists a unique invertable transformation of the p.d.e. to a system of variables, in which the transformed p.d.e. can be explicitly integrated. However, it is not possible, in general, to recover the solution of the p.d.e. in closed form by inverting the transformation, although we know that it exists. When the initial conditions are such that this is possible we shall call the solution an **exact solution** of the p.d.e..

Confusion arises through the use of the term integrable by some authors to denote that the p.d.e., intepreted as an infinite dimensional Hamiltonian system, has an infinite number of conserved quantities and so **by analogy** with the finite dimensional case is integrable.

2. An alternative approach to Bäcklund transformations is to investigate the **factorisation** of the linear operators associated with the equation. Although this was not their objective the results of this type of **differential algebra** were given in a paper by Burchnall and Chaundy (1922). A useful summary is contained in Ince (1926), Chapter 5. For the Schrödinger equation (3.1-15) to be factorisable requires that

$$\left(\frac{\partial^2}{\partial x^2} + \frac{\alpha q}{6}\right) \equiv \left(\frac{\partial}{\partial x} + u\right)\left(\frac{\partial}{\partial x} + v\right)$$

which is satisfied provided $u = -v$

$$v_x - v^2 = \frac{q}{6}$$

which is the Miura transformation. The permutation of this factorisation is allowed provided there exists a function Q such that

$$\left(\frac{\partial^2}{\partial x^2} + \frac{\alpha Q}{6}\right) \equiv \left(\frac{\partial}{\partial x} + v\right)\left(\frac{\partial}{\partial x} - v\right)$$

which together with the previous factorisation and the Miura transformation leads to the auto-Backlund transformation for the KdV.

Section 3.3

1. The best review of the Schrödinger equation on the half line is probably the article by Faddeyev (1963). The book by Agranovich and Marchenko (1963) contains the most complete analysis of this and the n-component generalisation. The review is also an interesting historical account of the development of the ideas and techniques used in tackling the problem.

In dealing with the Schrödinger equation on the real line $x \in R$, the paper by Kay and Moses (1956) has precedence, but the authors work in a more general context applying techniques developed in their earlier papers to the real line case which makes the paper rather difficult to read. Faddeyev's paper (1964) paper contains the criteria the scattering data must satisfy if the potential is to be of decrease type (in our case we require

$\int_{-\infty}^{\infty}(1 + x^2)|Q(x)|dx < \infty$), and this paper was used as the basis for this and the next chapter. There is a mistake in the paper, Faddeyev misses the fact that the second moment is required besides $\int_{-\infty}^{\infty}(1 + |x|)|Q(x)|dx < \infty$ in order that the scattering matrix should be continuous when $k = 0$. This condition was first given by Chadan and Sabatier (1977) and more recently by Deift and Trubowitz (1979). This last paper contains the most thorough and detailed analysis of inverse scattering on the real line.

Section 3.4

1. The spectral family in Hilbert space (page 124) is defined as follows For $u \in L^2(R)$ define ${}_\lambda P$ by

$$
{}_{\Delta}Pu(x) = \frac{1}{4\pi}\int_{\Delta\cap(0,\infty)} \frac{1}{\lambda^{\frac{1}{2}}}\Big(\frac{b}{a}(\lambda^{\frac{1}{2}})\psi(x,\lambda^{\frac{1}{2}})\psi(u,\lambda^{\frac{1}{2}}) + \psi^{*}(x,\lambda^{\frac{1}{2}})\psi(u,\lambda^{\frac{1}{2}})
$$

$$
+ \frac{b}{a^{*}}(\lambda^{\frac{1}{2}})\psi^{*}(x,\lambda^{\frac{1}{2}})\psi^{*}(u,\lambda^{\frac{1}{2}}) + \psi(x,\lambda^{\frac{1}{2}})\psi^{*}(u,\lambda^{\frac{1}{2}})\Big)d\lambda
$$

$$
+ \sum_{\lambda_j\in\Delta} D_j\psi_j(x)\psi_j(u) \qquad \text{where } \psi(u,\lambda^{\frac{1}{2}}) = \int_{-\infty}^{\infty} u(x)(x,\lambda^{\frac{1}{2}})dx
$$

and Δ is an interval in R. The operator

$$
A = \int_{-\infty}^{\infty} A(\lambda)\,d_{\lambda}P
$$

is defined as the limit of the sum, $\sum_{i=1}^{n} A(\lambda_i)_{\Delta_i}P$ where λ_i is in eachcase an arbitrary point of the interval Δ_i and $\{\cup\Delta_i\} = R$. In the case when the discrete spectrum is absent then $h_1(x,\lambda) = \psi(x,\lambda^{\frac{1}{2}})$, $h_2(x,\lambda) = (x,-\lambda^{\frac{1}{2}})$ or $h_2(x,\lambda) = \phi(x,\lambda^{\frac{1}{2}})$ serve as an **improper generating** basis and the matrix function $H_{jk}(\Delta) = (h_j(\Delta), h_k(\Delta))$ is the matrix distribution function for this basis. They correspond in the two cases (up to a factor) to the functions F and σI introduced in Section 3.4.

2. In scattering theory the Møller operators are defined by

$$
U_{+} = \underset{k\to\pm\infty}{\text{strong-lim}}\; \exp(ikL)\exp(-ikL_{o})
$$

The corresponding eigenfunctions are obtained from the Lippman-Schwinger equation

$$
u_{\pm}(x,k) = u_{o}(x,k) - \int_{-\infty}^{\infty} R_{o}(x,y,k^{2}\pm i0)Q(y)\,u_{\pm}(y,k)dy
$$

where $R(x,y,k^2\pm i0)$ is the limit of the Green's function for $Q = 0$ as k^2 approaches the real axis from above or from below. The vector $u_0 = (\exp(ikx),\exp(-ikx))$. In particular

$$
u_{+}(x,k) = u_{o}(x,k) - \frac{i}{2k}\int_{-\infty}^{\infty} e^{ik|x-y|}Q(y)\,u_{+}(y,k)dy
$$

and from the asymptotic form of this equation we find that

$$
u_{+}(x,k) = \Big(\frac{\psi^{*}}{a^{*}}(x,k),\ \frac{\phi^{*}}{a^{*}}(x,k)\Big)
$$

and similarly

$$u_-(x,k) = (\frac{\phi}{a}(x,k), \frac{\psi}{a}(x,k))$$

Section 3.5

1. We use the notation P^A to denote the formal adjoint of an operator P with respect to some naturally defined bilinear form associated with the operator L. The bilinear form required here is

$$(u,v) = \int_{-\infty}^{\infty} u(x)v(x)\,dx$$

since L_1^A is the required operator even if Q is complex.

3.7 Problems

Section 3.1

1. If (ψ_n, q_n, k_n), $n = 0,1,\ldots,$ satisfy the Schrödinger equation

$$\psi_{xx} + \frac{\alpha}{6} q\psi = \frac{\alpha}{6} k^2\psi$$

show that the recursion formula

$$q_{n+1} - q_n = \frac{12}{\alpha} (\log \psi_n)_{xx}$$

leads to the recursion relationship

$$q_{n+1} = \frac{12}{\alpha} \left[\log\left(\prod_{i=0}^{n} \psi_i\right)\right]_{xx}$$

In particular show that if $q_0 = 0$ that $q_1(x,t) = 2k_0^2 \operatorname{sech}^2((\alpha/6)^{\frac{1}{2}} k_0 x + g(t))$ and determine the function $g(t)$ for the KdV equation. Is it possible to determine q_2 for $q_0 = 0$ from this formula?

2. Consider the Miura transformation

$$v_x = q + v^2$$

between

$$q_t + 6qq_x + q_{xxx} = 0 \quad \text{(KdV)}$$

and

$$v_t - 6v^2 v_x + v_{xxx} = 0 \quad \text{(MKdV)}$$

tart with $q = 0$ then we generate a **one parameter** family of rational solu...ns to the MKdV $v = -(x + c)^{-1}$ where C is an arbitrary constant. It is clear however that the transformation does define a mapping from the MKdV to the KdV. This rational solution can be made time dependent by adjusting the coefficient. Thus $v = 2(x + C(t))^{-1}$ is a solution of the MKdV provided,

$$C_t = 18(x + C)^{-2}$$

Obtain the corresponding rational solution of the KdV. Consider n such solutions characterised by the functions C_i, $i = 1,\ldots,n$. Show that the discretisation of the corresponding evolution equations for the C_j obtained by replacing the continuous variable x by a sum over the C_i's (excluding C_j for the jth equation of motion) can be interpreted as a finite dimensional Hamiltonian system.

3. By looking for a solitary wave solution of the MKdV

$$v_t + 6\varepsilon v^2 v_x + v_{xxx} = 0 \qquad \varepsilon = \pm 1$$

or otherwise show that the equation only has a soliton solution,

$$v(x,t) = k \operatorname{sech}(kx + k^3 t)$$

when $\varepsilon = 1$. Under the Miura transformation the corresponding solution of the KdV

$$q_t + 6qq_x + q_{xxx} = 0$$

is the singular solution

$$q(x,t) = K^2(\sec^2\theta + \sec\theta\tan\theta) \quad , \quad \theta = Kx + K^3 t$$

However, if we start from the soliton solution of the KdV

$$q(x,t) = 2k^2 \operatorname{sech}^2 k(x - 4k^2 t)$$

we obtain the one (c) parameter family of solutions in the case $\varepsilon = -1$

$$V(x,t) = -2k \operatorname{cosech}\phi + \coth^2\phi(c + x - k^{-1}\coth\phi)^{-1} \quad ,$$

$$\phi = k(x - 4k^2 t)$$

which is non-singular if c is finite.

4. For the KdV equation,

$$q_t + 6qq_x + q_{xxx} = 0$$

the auto-Bäcklund transformation is, in terms of the potential function $w_x = q/2$

$$(W - w)_x = k^2 - (W - w)^2$$
$$(W - w)_t = 6(W - w)^2(W - w)_x - 6k^2(W - w)_x - (W - w)_{xxx}$$

Starting from $w = 0$ show that a one (c) parameter family of nonsingular solutions of the KdV is

$$q(x,t,c) = 2k^2\text{sech}^2(k(x - 4k^2t) + c)$$

and deduce by a symmetry argument applied to the transformation that a singular family of solutions is

$$q^*(x,t,c,) = -2k^2\text{cosech}^2(k(x - 4k^2t) + c)$$

Consider now three solutions of the KdV which are Bäcklund related,

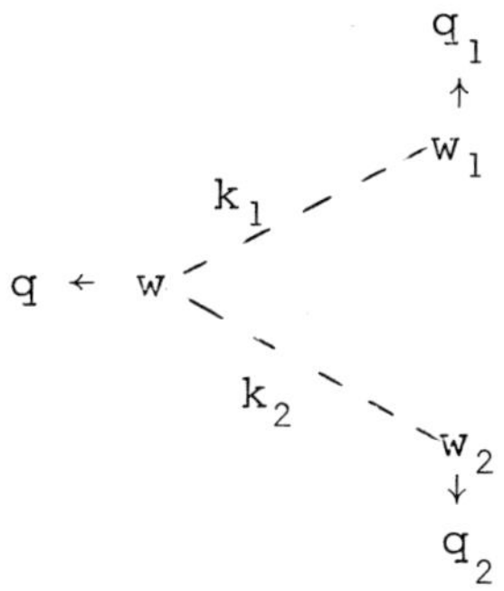

The dotted lines in the diagram indicate a Bäcklund correspondence whereas the solid arrows are maps. The x-part of the transformations are

$$(w_1 + w)_x = k_1^2 - (w_1 - w)^2$$

$$(w_2 + w)_x = k_2^2 - (w_2 - w)^2$$

By considering the further transformations

$$(w_{12} + w_1)_x = k_2^2 - (w_{12} - w_1)^2$$

$$(w_{21} + w_2)_x = k_1^2 - (w_{21} - w_2)^2$$

show that the condition $w_{12} = w_{21}$ is satisfied provided

$$w_{12} = w + \frac{(k_2^2 - k_1^2)}{(w_2 + w_1)}$$

from which we can deduce q_{12}. This commutability of the Bäcklund transformation enables solutions to the KdV to be obtained by purely algebraic means using the nonlinear superposition relationship just presented. In this way one can construct a tower or ladder of solutions. If one starts from the zero solution the soliton tower is generated. One has to be careful to obtain a regular solution at each step.

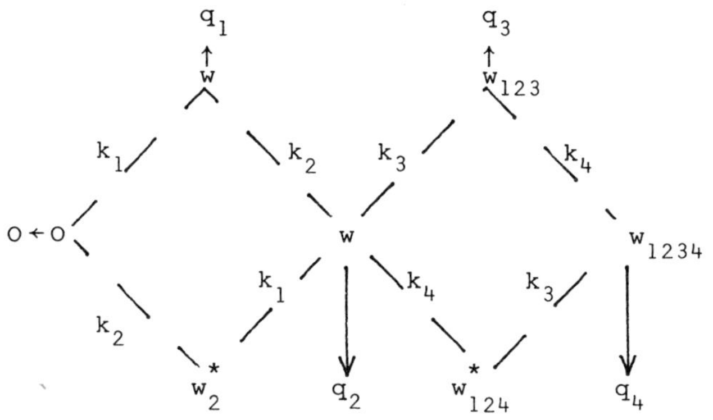

$k_1 < k_2 < \ldots$, q_n is the n-soliton solution.

5. By generalising the Miura transformation it is straightforward to relate the KdV to a family of equations which contain the KdV and the MKdV as particular members,

$$q_t + \alpha q q_x + q_{xxx} = 0$$

$$v_t + \mu v_x + \phi v^2 v_x + \theta v v_x + v_{xxx} = 0$$

One does this by rescaling and translating the v variable,

$$(6\phi)^{\frac{1}{2}} i\varepsilon v_x + \mu + (\phi v^2 + 2\theta v) - \alpha q = 0$$

In particular for $\phi^{\frac{1}{2}} = 6^{\frac{1}{2}}\delta$, $\theta = \frac{1}{2}\alpha$, $\mu = 0$ and $\alpha = 6$ the transformation can be written

$$i\delta v_x + \delta^2 v^2 + v - q = 0$$

and gives a correspondence between solutions of the equations

$$q_t + (3q^2 + q_{xx})_x = 0$$

$$v_t + (3\delta^2 v^3 + \frac{3}{2} v^2 + v_{xx})_x = 0$$

The equations have been written in **conservation form**,

$$H_t + F_x = 0$$

where H is the **conserved density** and F the associated **flux.** A characteristic property of the solvable equations we are considering is the existence of an ifinite number of integrals (constants) of the motion $I = \int_{-\infty}^{\infty} H\, dx$. This follows directly from the conservation law provided q and v along with their derivatives all vanish on the boundary (here chosen to be $\pm\infty$). We can obtain a formal power series for v from the Miura transformation by putting $v = \Sigma_0^\infty L_i(q)\,\delta^i$, where L_i is a functional of q and its x-derivatives. By substituting for v into the corresponding conservation law and equating coefficients of δ^i we obtain an infinite number of conserved densitites H for the KdV. Show that the first few densities with $\alpha = 6$ are, up to a scaling factor,

$$H_0 = q,\ H_2 = q^2,\ H_4 = 2q^3 - q_x^2\ ,\ H_6 = 9q^4 - 18qq_x^2 + 9q_x^2 15$$

Notice that the corresponding odd densitites are absent because they are perfect x-derivatives and therefore trivial. Construct a proof to show that all the odd densities are trivial.

Section 3.1

1. Show that if $\int_{-\infty}^{\infty}(1 + x^2)|Q(x)|dx < \infty$ then $\int_{-\infty}^{\infty}(1 + |x|)|Q(x)|dx < \infty$.

2. Starting from $h(x,k) = \phi(x,k)\exp(ikx)$

$$|h(x,k)| \leqslant 1 + x\int_{-\infty}^{x}|Q(v)|\,|h(v,k)|dv + \int_{-\infty}^{x}(-v)\,|Q(v)|\,|h(v,k)|dv$$

justify the estimate $|h(x,k) - 1| \leqslant K_5(1 + \max(x,0))\int_{-\infty}^{\infty}(1 + |v|)|Q(v)|\,dv$.

3. Derive the inequality (3-3-32)). It is necessary to consider the cases $x \leqslant 0$, $x \geqslant 0$ separately. Iterate (3.3-32) to show that

$$|h_k(x,k)| \leqslant K(1 + \max(x,0))\exp M(x)$$

and thereby deduce the first estimate in (3.3-34).

4. Show that

$$|h_x(x,k)| \leqslant K(1+|k|)^{-1}\int_{-\infty}^{x}(1+|v|)|Q(v)|dv \qquad x \in R$$

and so confirm the second estimate in (3.3-34).

5. Consider a solution of the form $\phi^1(x) = f(x)\phi(x,0)$ to the equation $-y_{xx} + Qy = 0$. Show that a possible solution is

$$\phi^1(x) = \left\{\int^{x}\phi^{-2}(v,0)\,dv\right\}\phi(x,0), \qquad \phi^1(x) = x + o(1) \text{ as } x \to \infty$$

and conclude that $\phi^1 \equiv \tilde{\phi}$.

6. Improve upon the estimate given in Lemma (3.3-2),

$$\phi(x,k)e^{ikx} = 1 + \frac{1}{2ik}\int_{-\infty}^{x}(e^{2(ik(x-v)}-1)Q(v)\,dv + \frac{1}{2(2ik)^2}\left\{\int_{-\infty}^{x}Q(v)\,dv\right.$$

$$+\, o(1/|k|^2) \quad \text{as} \quad |k| \to \infty$$

This formula can be simplified if $Q_x \in L^2(R)$.

7. Show that if Q has n derivatives in $L^1(R)$ then

$$R_+(k) = o(1/|k|^{n+1}) \qquad \text{as} \quad |k| \to \infty$$

8. Consider the Schrödinger operator $L_D = -\partial^2/\partial x^2 + Q$ acting in $L^2(-\infty, x_0)$ and satisfying a Dirichelet condition at x_0. Show that $\phi(x,k)$ Im $k > 0$ has a finite number of simple zeros $k_j = i\eta_j$, $\eta_j > 0$ and that there are no real zeros apart from possibly $k = 0$ which is called a virtual eigenvalue.

9. Using the operator defined in problem (3.3-8) show that the number of eigenvalues of L_D is equivalent to the number of zeros of $\phi(x,0)$. Further by consideration of a pair of consecutive zeros of $\phi(x,0)$ say $a < b$ show that

$$\phi(x,o) = \phi_x(a,0)(a-x) + \int_a^x Q(v)(v-x)\phi(v,0)\,dv, \qquad a < x < b$$

and deduce by consideration of the case $\phi(x,0) \leqslant 0$, $0 \leqslant a \leqslant x \leqslant b$ that if n is the number of eigenvalues of L_D then,

$$n \leqslant \int_{-\infty}^{x_0} (x_0 - v)\,|Q(v)|\,dv < \infty$$

10. Consider the potential barrier

$$Q(x) = \begin{cases} 0 & x < 0 \\ 1 & 0 \leqslant x \leqslant 1 \\ 0 & x > 1 \end{cases}$$

and show that for fixed $k \neq 0$, that $\phi(x,k)e^{ikx}$ is a bounded function whereas for $k = 0$ $\phi(x,0)$ grows linearly as $x \to +\infty$. Compare this with the general behaviour of ϕ as displayed by (3.3-27).

11. For the first order system (3.3-12) define the Jost solutions for real k by the boundary conditions,

$$\underset{x \to -\infty}{\mathrm{Lt}} \quad \phi e^{ikx} = \begin{pmatrix} 2ik \\ 1 \end{pmatrix}, \quad \underset{x \to -\infty}{\mathrm{Lt}} \quad \bar{\phi} e^{-ikx} = \begin{pmatrix} 0 \\ 1 \end{pmatrix}$$

$$\underset{x \to +\infty}{\mathrm{Lt}} \quad \psi e^{-ikx} = \begin{pmatrix} 2ik \\ 1 \end{pmatrix}, \quad \underset{x \to +\infty}{\mathrm{Lt}} \quad \bar{\psi} e^{ikx} = \begin{pmatrix} 0 \\ 1 \end{pmatrix}$$

Establish conditions on Q for the existence and uniqueness of these solutions and their analytic continuation into the complex k-plane. Introduce the functions $a(k), \bar{a}(k), b(k)$ and $\bar{b}(k)$ through the relations

$$\phi(k) = a(k)\bar{\psi}(k) + b(k)\psi(k)$$

$$\bar{\phi}(k) = \bar{a}(k)\psi(k) + \bar{b}(K)\bar{\psi}(k)$$

and investigate their analytic properties using analogous technique to those of Section 3.3 In particular show that a has only a finite number of zeros.

12. Derive the first few terms in the large k expansion of a,

$$a(k) = 1 - \frac{1}{2ik}\int_{-\infty}^{\infty} Q(v)\,dv + \frac{1}{2}\,\frac{1}{2ik^2}\int_{-\infty}^{\infty} Q(v)\,dv + o(|k|^{-2})$$

$$\text{as } |k| \to \infty$$

Hence conclude that if $\lambda_0 = -\eta_0^2$ is the eigenvalue with the greatest magnitude that

$$|\lambda_0| \approx \left(\int_{-\infty}^{\infty} Q(v)\,dv \right)^2$$

and that there are no eigenvalues if

$$\int_{-\infty}^{\infty} Q(v)\,dv < 0$$

13. From problem (3.3-9) we observe that the Dirichelet operators on $L^2(-\infty,0)$ and $L^2(0,\infty)$ have respectively the bounds

$$n_{-\infty} \leqslant -\int_{-\infty}^{\infty} v|Q(v)|\,dv \quad , \quad n_{+\infty} \leqslant \int_{-\infty}^{\infty} v|Q(v)|\,dv$$

on the number of eigenvalues in their spectra. Conclude that the number of eigenvalues for the Dirichelet operator on $L^2(-\infty,0) \oplus L^2(0,\infty)$ has the bound

$$n_D \leqslant \int_{-\infty}^{\infty} |v|\,|Q(v)|\,dv$$

However, from the minimax principle and the spectral mapping theorem one can show that $M = n + 1$ so that

$$M \leqslant 1 + \int_{-\infty}^{\infty} |v|\,|Q(v)|\,dv$$

14. Starting from the representations (3.3-64) investigate the behaviour of a and b as $|k| \to 0$. In particular obtain equation (3.3-66)

15. Consider the one parameter family of potentials

$${}_aQ = \begin{cases} a, & -2 \leqslant x \leqslant -1 \\ -1, & -1 < x \leqslant -1 \\ 0, & \text{otherwise} \end{cases}$$

and find the Jost solution $\phi(x,k)$. For $a = -4\frac{2}{\pi}$ determine the number of eigenvalues of the Schrödinger operator by calculating the number of zeros of $\phi(x,0)$. Compare with the bound given in problem (3.3-14).

Show that there is a unique $a = a_0$ such that $\phi(x,0) = \text{constant}$ for $x > 0$ and so $\phi(x,0)$ and $\psi(x,0)$ are linearly dependent. Conclude that ${}_aR_{\mp}(0) = -1$ for $a \neq a_0$, but that ${}_{a_0}R_{+}(0) = -1$, so that no matter how ${}_aQ$

converges the corresponding sequence of reflection coefficients ${}_aR_+(0)$ do not converge uniformly in the neighbourhood of zero.

16. Show from equations (3.3-64) that when Q is on compact support R_+ and T_+ are analytic throughout the complex k-plane. If Q is supported on a half line show from (3.3-64) and the analogous equations for the ψ representation that T_+ is analytic throughout the complex k-plane whereas R_+ is analytic in the upper half k-plane.

Section 3.4

1. Consider the **complex** isospectral Schrödinger equation,

$$-\frac{\partial^2 y}{\partial x^2} + Qy = k^2 y$$

and repeat the analysis of this chapter for this equation. In particular investigate the resolution of the identity for this problem. Notice that the presence of **spectral singularities** that is real k such that $a(k) = 0$ cause convergence problems in the expansion. Either restrict the class of functions belonging to $L^2(R)$ or introduce a suitable regularisation of the divergent integrals to that the expansion of $v \in L^2(R)$ in terms of the eigenfunctions is well defined. Are the eigenvalues simple in this problem and where are they located?

2. Establish the formulae (3.4-32). By assuming that $v \in L^2(R)$ has an expansion

$$v = v_\psi \psi + v_{\psi^*}\psi^* + \sum_{j=1}^{M} v_j \psi_j$$

Use the formulae to determine the coefficients in the expansion and deduce the resolution of the identity from it.

3. Prove the existence and uniqueness of the solution to the Goursat problem posed by (3.4-74).

4. Introduce a set of solutions to the isospectral Schrödinger equation (3.3-1) defined by the boundary conditions,

$$g_1(0) = 1 \quad , \quad g_{1x}(0) = ik$$

$$g_2(0) = 1 \quad , \quad g_{2x}(0) = -ik$$

and show that the row vector $G = (g_1, g_2)$ satisfies the integral equation

$$G(x,k) = G_0(x,k) + k^{-1}\int_0^x \sinh(x-y)Q(y)G(y,k)\,dy$$

where $G_0(x,k) = (\exp(ikx), \exp(-ikx))$. Prove that the components of G are entire functions of k. Put

$$G(x,k) = H(x,k)B(k) \quad \text{where} \quad H = (\psi,\phi)$$

and determine B. Define the Jost matrix by

$$J(k) = T^{-1}(k)B(k)$$

and show that $\det J(k) = T^{-1}(k)$ so that the bound states are determined by

$$\det J(k) = 0$$

Interpret the eigenfunctions of J.

Section 3.5

1. Define operators

$$B_{\bar{\psi}} \equiv B_{\phi} = -\left(4\frac{\partial^3}{\partial x^3} + q\frac{\partial}{\partial x} + \frac{\alpha}{2}q_x - 4ik^3\right)$$

$$B_{\bar{\phi}} \equiv B_{\psi} = -\left(4\frac{\partial^3}{\partial x^3} + q\frac{\partial}{\partial x} + \frac{\alpha}{2}q_x + 4ik^3\right)$$

These operators form Lax pairs with the isospectral Schrödinger operator for the KdV as shown in Section 3.2. In particular notice that $\bar{\psi}_t = B_{\bar{\psi}}\bar{\psi}$, $\bar{\phi}_t = B_{\bar{\phi}}\bar{\phi}$, $\psi_t = B_{\psi}\psi$ and $\phi_t = B_{\phi}\phi$ ensure that the boundary conditions defining the Jost solutions are valid for all t. Show that since for real k

$$\phi = a\bar{\psi} + b\psi$$

that $R_{+t} = 8ik^3 R_+$ and obtain the evolution equation for the normalisation constant of the eigenfunctions for this scattering problem.

2. Show that

$$\frac{d}{dk}\log a(k) = -i\int_{\infty}^{\infty}\left(\frac{\psi\phi}{a} - 1\right)dx$$

by starting from the Schrödinger equation.

3. Using the defininti on of a in terms of the Wronskian of ϕ and ψ, and the Schrödinger equation, obtain a recursion formula for the coefficients in the asymtotic expansion $\log a$ as $|k| \to \infty$. Calcuate the fist three non-trivial terms in this expansion and compare them with (3.5-76).

5. Derive the representation of the Frechet derivative for functionals $F(Q)$ which involve in their definition both derivatives and integrals of Q given that $Q : R \to R$ and is of general Schwartz type.

5. Starting for the definition

$$\mathrm{Tr}((L - k^2)^{-1} - (L_0 - k^2)^{-1}) = -\frac{1}{2k}\frac{d}{dk}\log a(k)$$

and taking Δ variations formally derive (3.5-52) (you will need to use the representation of the resolvent kernel given in Theorem 3-9).

4. THE INVERSE METHOD FOR THE ISOSPECTRAL SCHRÖDINGER EQUATION AND THE GENERAL SOLUTION OF THE SOLVABLE NONLINEAR EQUATIONS

4.1 The Inverse Scattering Problem and the Marchenko Equation for the Iso-spectral Schrödinger Equation

The inverse scattering problem for the Schrödinger equation is to reconstruct the potential from the asymptotic data which results from the scattering process. Examples of this procedure were given in Section 2.2. The first attempts at solving the inverse problem were undertaken by Fröberg (1948) and Hylleraas (1948). However, Bargmann (1949) showed by constructing counter examples that their solutions were not unique. Their error lay in using only the scattering matrix as their asymptotic data rather than the scattering data (S_+ or S_-). The scattering data completely determines the (matrix) spectral distribution function which can be associated with the Schrödinger equation for given boundary conditions (see Section 3.4). As a result of the researches of several Russian mathematicians it was established that the specification of the spectral distribution function, subject to its satisfying certain conditions, was both necessary and sufficient to ensure the unique reconstruction of the potential Q.

Our principal interest in this work is that Q can be recovered from solving a linear integral equation which involves the scattering data so that if $S_{\pm}(t)$ is determined in the manner indicated in Section 3.5 from the initial data $S_{\pm}(0)$ at $t = 0$ then we can in principal construct $Q(x,t)$ at arbitrary later times from the initial $Q(x,0)$. This linear integral equation is called either the Gelfand-Levitan equation (1951) or the Marchenko equation (Marchenko and Agranovich, 1963). The distinction depends upon whether the scattering problem has regular or irregular boundary conditions. Since the fundamental solutions for our problem are the Jost solutions, the corresponding integral equation is a Marchenko equation.

The main purpose of this section is to derive the Marchenko equation for the Schrödinger equation on the real line and to find the constraints on the scattering data which ensure that it has a unique solution Q satisfying

$$\int_{-\infty}^{\infty} (1 + x^2)\,|Q(x)|\,dx < \infty.$$

The most direct derivation starts from the fundamental relation (3.3-59)

$$T_+(k)\,\phi(x,k) = \psi^*(x,k) + R_+(k)\,\psi(x,k) \tag{4.1-1}$$

where

$$T_+ = a^{-1} \text{ and } R_+ = ba^{-1}$$

are continuous bounded functions, and from Lemma 3-6

$$R_+(k) = O(1) \text{ as } |k| \to \infty$$

and (4.1-2)

$$T_+(k) = O(|k|^{-1}) \text{ as } k \to \infty$$

These properties of the scattering matrix entries, together with the boundary conditions defining the Jost solutions (3.3-2) ensure the existence of the integrals in the Fourier transform of (4.1-1)

$$\frac{1}{2\pi}\int_{-\infty}^{\infty} (T_+(k)-1)\phi(x,k)e^{iky}dk = \frac{1}{2\pi}\int_{-\infty}^{\infty} R_+(k)(\psi(x,k)-e^{ikx})e^{iky}dk$$

$$+\frac{1}{2\pi}\int_{-\infty}^{\infty} R_+(k)e^{ik(x+y)}k+\frac{1}{2\pi}\int_{-\infty}^{\infty}(\psi^*(x,k)-e^{-ikx})e^{iky}dk+\frac{1}{2\pi}\int_{-\infty}^{\infty}(e^{-ikx}-\phi(x,k))e^{iky}dk \quad (4.1\text{-}3)$$

Define the Fourier transforms of $f \in L^{12}(R) = L^1(R) \cap L^2(R)$ by

$$\hat{f}(x) = (2\pi)^{-\frac{1}{2}}Tf(x) \equiv \frac{1}{2\pi}\int f(k)e^{ikx}dk$$

and (4.1-4)

$$\check{f}(k) = (2\pi)^{\frac{1}{2}} T^*f(k) \equiv \int^{\infty} f(x)e^{-ikx}dx$$

where T is a unitary operator. Then since

$$\langle f,g\rangle = \langle T^*Tf,g\rangle = \langle Tf,Tg\rangle$$

and (4.1-5)

$$(2\pi)^{-\frac{1}{2}} T(e^{iky}f(k)) = \frac{1}{2\pi}\int_{-\infty}^{\infty} f(k)e^{ik(x+y)}dk = \hat{f}(x+y)$$

it follows that the first integral on the right hand side of (4.1-3) can be written as

$$\frac{1}{2\pi}\int_{-\infty}^{\infty} R_+(k)(\psi(x,k)-e^{ikx})e^{iky}dk = \frac{1}{2\pi}\langle T(e^{iky}R_+(k)),T(\psi(x,k)-e^{ikx})^*\rangle$$

$$=(2\pi)^{-\frac{1}{2}}\int_{-\infty}^{\infty}\hat{R}_+(u+y)T^*(\psi(x,k)-e^{ikx})(u)\,du$$

(4.1.6)

In Section 3.4, we defined the transformation operators U_+, U_-, which have real valued kernels

$$\psi(x,k) = U_+ e^{ikx} \equiv e^{ikx} + \int_{-\infty}^{\infty} K_+(x,y) e^{iky} dy \tag{4.1-7}$$

$$\phi(x,k) = U_- e^{-ikx} \equiv e^{-ikx} + \int_{-\infty}^{\infty} K_-(x,y) e^{-iky} dy \tag{4.1-8}$$

Using (4.1-7) we have

$$\begin{aligned} T^*(\psi(x,k) - e^{ikx})(u) &= T^* \int_{-\infty}^{\infty} K_+(x,y) e^{iky} dy \ (u) \\ &= (2\pi)^{-\frac{1}{2}} \int_{-\infty}^{\infty} K_+(x,y) \int_{-\infty}^{\infty} e^{ik(y-u)} dk dy = (2\pi)^{\frac{1}{2}} K_+(x,u) \end{aligned} \tag{4.1-9}$$

The transformations (4.1-7),(4.1-8) also enable the other terms on the right hand side of (4.1-3) to be written in terms of the kernels K_+, K_- of the transformation operators

$$\begin{aligned} \frac{1}{2\pi} \int_{-\infty}^{\infty} & (T_+(k) - 1)\phi(x,k) e^{iky} dk \\ &= \int_x^{\infty} \hat{R}_+(u+y) K_+(x,u) du + \hat{R}_+(x+y) + K_+(x,y) - K_-(x,y) \end{aligned} \tag{4.1-10}$$

The integral on the left hand side of (4.1-10) is evaluated by integrating around the contour Γ_R consisting of the line segment $(-R,R)$ and the semi-circle C_R of radius R in the upper half k-plane, described in an anti-clockwise sense. Then in the limit $R \to \infty$ we obtain

$$\begin{aligned} \lim_{R\to\infty} \left\{ \frac{1}{2\pi i} \oint_{\Gamma_R} (T_+(k)-1)\phi(x,k) e^{ikx} dk \right\} &= \frac{1}{2\pi i} \int_{-\infty}^{\infty} (T_+(k)-1)\phi(x,k) e^{iky} dk \\ &+ \lim_{R\to\infty} \left\{ \frac{1}{2\pi i} \int_{C_R} (T_+(k)-1)\phi(x,k) e^{iky} dk \right\} \end{aligned} \tag{4.1-11}$$

The second integral on the right hand side of (4.1-11) vanishes in the limit and so using the residue theorem produces the result

$$\frac{1}{2\pi i} \int_{-\infty}^{\infty} (T_+(k)-1)\phi(x,k) e^{iky} dk = \sum_{j=1}^{M} \operatorname{Res}_{k=i\eta_j} \left\{ T_+(k)\phi(x,k) e^{iky} \right\} \tag{4.1-12}$$

where $\lambda_j = -\eta_j^2 \in \sigma(L)$, the spectrum of the Schrödinger operator L. In Section 3.4, we introduced the normalisation constants $D_{+j}^{-1} \equiv ic_j^{-1} a_j = \langle\psi_j,\psi_j\rangle$, where $\phi_j = c_j\psi_j$ are the eigenfunctions of L. We can now derive from (4.1-7) the expression

$$\frac{1}{2\pi i} \int_{-\infty}^{\infty} (T_+(k)-1)\phi(x,k) e^{iky} dk = i \sum_{j=1}^{M} D_{+j} \left(e^{-\eta_j(x+y)} + \int_{-\infty}^{\infty} K_+(x,u) e^{-\eta_j(y+u)} du \right) \tag{4.1.13}$$

If we substitute (4.1-13) into (4.1-10) then after some rearrangement we obtain the time-dependent Marchenko equation

$$K_+(x,y,t) + \Omega_+(x+y,t) + \int_x^\infty K_+(x,u,t)\,\Omega_+(u+y,t)\,du = 0\,, \qquad y \geq x$$

$$\Omega_+(x,t) = \hat{R}_+(x,t) + \sum_{j=1}^{M} D_{+j}(t)e^{-\eta_j x} \tag{4.1-14}$$

The time dependency, t, which enters as a parameter in the above equation, will continue to be suppressed throughout the rest of this section since all the calculations are made at one fixed time.

An analogous equation can be obtained for the K_- kernel

$$K_-(x,y,t) + \Omega_-(x+y,t) + \int_{-\infty}^{x} K_-(x,u,t)\,\Omega_-(u+y,t)du = 0 \qquad y \leq x \tag{4.1-15}$$

$$\Omega_-(x,t) = \hat{R}_-(x,t) + \sum_{j=1}^{M} D_{-j}e^{\eta_j x}$$

and

$$D_{-j}^{-1} = ic_j\dot{a}_j \qquad \text{so that} \quad D_{-j}D_{+j} = -\dot{a}_j^{\,2}$$

Notice that Ω_+ and Ω_- are defined once the scattering data S_+ or S_- is given and that from (3.4-79)

$$Q(x,t) = -2\frac{d}{dx}K_+(x,x,t) = 2\frac{d}{dx}K_-(x,x,t) \tag{4.1-16}$$

Thus once the solution of either of the Marchenko equations has been determined, provided of course that it exists and is unique, the potential Q is completely specified.

We shall now obtain the conditions which the scattering data must obey if Q is to satisfy the condition $\int_{-\infty}^{\infty}|Q(x)|(1+x^2)dx < \infty$. Conditions on the scattering data such that the reconstructed Q should satisfy the condition $\int_{-\infty}^{\infty}|Q(x)|(1+|x|)dx < \infty$ are unknown at present (Deift and Trubowitz, 1979). Of course the scattering data will be subjected to further conditions if Q is to be a solution of a solvable equation. These extra conditions are derived in Section 4.2. Readers who are principally interested in solutions to the solvable equations can omit a detailed study of the material of this section. The results are summarised in Theorem 4-3.

We work from now on entirely with the Marchenko equation (4.1-14) and quote the analogous results for (4.1-15). We begin by investigating the conditions the scattering data must satisfy if Q satisfies the condition $\int_{-\infty}^{\infty}|Q(x)|(1+x^2)dx < \infty$.

Putting $y = x$ in (4.1-14) we obtain

$$K_+(x,x) + \Omega_+(2x) + \int_x^\infty K_+(x,u)\,\Omega_+(x+u)\,du = 0 \tag{4.1-17}$$

which we consider as an equation for Ω_+. For clarity of notation we shall omit the suffix + from the functions of this equation in the following calculation. Differentiating (4.1-17) with respect to x we get

$$\Omega_x(2x) - \tfrac{1}{2}Q(x) + \int_x^\infty K_x(x,u)\,\Omega(x+u)\,du + \int_x^\infty K(x,u)\,\Omega_x(x+u)\,du - K(x,x)\,\Omega(2x) = 0 \tag{4.1-18}$$

If we define

$$I(x) = \int_x^\infty K_x(x,u)\,\Omega(x+u)\,du \tag{4.1-19}$$

then using (3.4-79) it is not too hard to show that

$$\begin{aligned}\frac{d}{dx} I(x) &= Q(x)\int_x^\infty K(x,u)\,\Omega(x+u)\,du + \int_x^\infty K_{uu}(x,u)\,\Omega(x+u)\,du \\ &\quad + \int_x^\infty K_x(x,u)\,\Omega_x(x+u)\,du - K_x(x,x)\,\Omega(2x) \\ &= Q(x)\left(\int_x^\infty K(x,u)\,\Omega(x+u)\,du + K(x,x) + \Omega(2x)\right) \\ &\quad - Q(x)K(x,x) - \frac{d}{dx}\left(\int_x^\infty \Omega(x+u)\,K_x(x,u)\,du\right) \\ &= \frac{d}{dx}\left(K^2(x,x) - \int_x^\infty \Omega(x+u)\,K_u(x,u)\,du\right)\end{aligned} \tag{4.1-20}$$

Equation (4.1-18) can now be written as

$$\Omega_x(2x) - \tfrac{1}{2}Q(x) + 2\int_x^\infty K(x,u)\,\Omega_x(x+u)\,du + K^2(x,x) = 0 \tag{4.1-21}$$

From the properties of the kernel $K(x,u)$ we can define the convolution operation " $*$ " by

$$({}_xK *{}_xK)(u) = \int_{-\infty}^\infty K(x,2u-y)\;K(x,y)\,dy \qquad u \ge x \tag{4.1-22}$$

Then

$$\int_x^\infty \Omega_u(2u)({}_xK*{}_xK)(u) = \int_x^\infty K(x,y)\int_{\frac{1}{2}(x+y)}^\infty K(x,2u-y)\Omega_u(2u)\,du\,dy$$

$$= -\tfrac{1}{2}\int_x^\infty K(x,y)(\Omega_y(x+y) + K_y(x,y))\,dy$$

$$= \tfrac{1}{4}K^2(x,x) - \frac{1}{2}\int_x^\infty K(x,y)\Omega_x(x+y)\,dy \qquad (4.1\text{-}23)$$

where we have used the y-derivative of the Marchenko equation. Combining (4.1-21) and (4.1-23) gives an equation which we can iterate for $\Omega_+(2x)$

$$\Omega_x(2x) = \tfrac{1}{2}Q(x) + \int_x^\infty P(x,y)\Omega_y(2y)\,dy \qquad (4.1\text{-}24)$$

where

$$P(x,y) = 4(K(x,2y-x) + ({}_xK*{}_xK)(y))$$

Since the iterates of a Volterra integral equation always converge, we can use the estimate

$$K(x,y) \le C(x)R_o(\tfrac{1}{2}(x+y)) \le C(a)\,R_o(\tfrac{1}{2}(x+y)) \qquad -\infty<a \leq x \leq y \qquad (4.1\text{-}25)$$

derived from the inequality (3.4-61) to obtain a bound on $\Omega_+(2x)$. Here and in the following section, $C_j(x)$ is a monotonic function bounded as $x \to +\infty$ and generally increasing as $x \to -\infty$. Since one can show that $k\,R_+T_+^{-1} \in L^1(\mathbf{R})$, $(T_+-1) \in L^2(\mathbf{R})$ and

$$kR_+ = kR_+T_+^{-1} + kR_+(T_+-1)T_+^{-1} \qquad (4.1\text{-}26)$$

it follows that

$$\frac{d}{dx}\hat{R}_+(x)^{\vee} = \check{g}_1 + \check{g}_2 \qquad (4.1\text{-}27)$$

where $g_1 \in L^1(R)$ and $g_2 \in L^2(R)$. Iterating (4.1-24) and using the estimate (4.1-25) together with the decomposition given by (4.1-27) enables one to derive the estimate

$$\left|\tfrac{1}{2}Q(x) - \Omega_x(2x)\right| < C_1(x)R_o^2(x) \qquad (4.1\text{-}28)$$

Since $\Omega_+(2x)$ is locally L^1 we deduce that $\Omega_+(2x)$ is absolutely continuous.

Suppose that $\int_{-\infty}^{\infty}|Q(x)|\ (1 + |x|^n)dx < \infty$ then since

$$|\Omega_x(2x)| = |(\tfrac{1}{2}Q(x) - \Omega_x(2x)) - \tfrac{1}{2}Q(x)| \leqslant |\tfrac{1}{2}Q(x) - \Omega_x(2x)| + |\tfrac{1}{2}Q(x)| \quad (4.1\text{-}29)$$

it follows that for $-\infty < a < x$

$$\begin{aligned}\int_x^\infty |\Omega_y(2y)|(1+|y|^n)dy &\leq \frac{1}{2}\int_x^\infty |Q(y)|(1+|y|^n)\,dy + \int_x^\infty C_1(y)R_o^{\,2}(y)(1+|y|^n)dy \\ &\leq \frac{1}{2}\int_x^\infty |Q(y)|(1+|y|n)\,dy + C_1(a)\int_x^\infty R_o(y)dy\int_y^\infty |Q(s)|(1+|s|^n)ds \\ &\leq \int_x^\infty |Q(y)|(1+|y|^n)\,dy\,(\frac{1}{2}+C_1(a)\int_x^\infty (x-a)|Q(x)|dx) < C_2(x) < \infty\end{aligned}$$

(4.1-30)

provided that in addition $P_1(\infty) < \infty$. Thus if $\int_{-\infty}^{\infty}|Q(x)|(1 + x^2)dx < \infty$ then Ω_x satisfies the condition $\int_a^\infty |\Omega_x(x)|(1 + x^2)dx < \infty$, $-\infty < a \leqslant x$. Since Ω is absolutely continuous with boundary condition $\Omega(\infty) = 0$ we also have as a result of this condition that

$$\int_a^\infty |\Omega(x)|dx \leq \int_a^\infty dx\int_a^\infty |\Omega_y(y)|dy = \int_a^\infty (x-a)|\Omega_x(x)|\,dx \leq C_3(a) < \infty \quad (4.1\text{-}31)$$

In the case of Ω_- the functions C_i and R_o are replaced by functions D_i and P_o in relations similar to (4.1-28), (4.1-30) and (4.1-31). In this case, however, the D_i are monotonically non-decreasing functions bounded as $x \to +\infty$ and generally increasing as $x \to +\infty$.

The functions $\hat{R}_+$, $\hat{R}_-$ differ only from Ω_+, Ω_- by exponentially decreasing terms as $x \to +\infty$, $x \to -\infty$ respectively. Thus, besides Theorem 3-7 the scattering data must satisfy the further conditions given in Theorem 4-1 if Q satisfies $\int_{-\infty}^{\infty}|Q(x)|(1 + x^2)dx < \infty$.

Theorem 4-1: The Fourier transforms $\hat{R}_+$, $\hat{R}_-$ of the scattering coefficients $R_+ = ba^{-1}$, $R_- = -b^*a^{-1}$ are absolutely continuous and satisfy the conditions

$$\int_a^\infty |\hat{R}_{+x}(x)|(1+x^2)\,dx < C_3(a) < \infty$$

$$\int_{-\infty}^b |\hat{R}_{-x}(x)|(1+x^2)\,dx < D_3(a) < \infty$$

for all $a > -\infty$, $b < \infty$ if $\int_{-\infty}^{\infty}|Q(x)|(1 + x^2)dx < \infty$.

We now see that the necessary conditions given by Theorems 3-7 and 4-1 on the scattering data are also sufficient for the unique reconstruction of a scattering potential function Q for the Schrödinger equation and which, moreover, obeys the condition $\int_{-\infty}^{\infty}|Q(x)|(1 + x^2)dx < \infty$. This result can be established by proving the validity of the statements in the following five steps, for which we assume without loss of generality, that S_+ is given. We first outline the steps involved and then return to discuss each of the steps in more detail.

Step 1

Given that S_+ satisfies

$$|R_+(k)| \le 1, \quad R_+(-k) = R_+^*(k)$$

and

$$\int_a^{\infty} |\hat{R}_{+x}(x)|(1+x^2)dx < C_3(a)$$

$a > -\infty$ where C_3 is a non-decreasing function, bounded as $x \to +\infty$ and increasing as $x \to -\infty$, form

$$\Omega_+(x) = \hat{R}_+(x) + \sum_{j=1}^{M} D_{+j} e^{-\eta_j x}$$

Here M is the number of "eigenvalues" $\lambda_j = -\eta_j^2, \eta_j > 0$ and the positive real numbers D_{+j}, $j = 1,\ldots,M$ are the corresponding "normalisation constants". Form the unitary scattering matrix $\tilde{S}(k)$ and check that it is a continuous function of k and that either

(i) $T_+(k) = \alpha k + O(|k|^2)$ as $|k| \to 0$

(ii) $T_+(k) \geqslant C \ne 0$ as $|k| \to 0$

Prove that under these conditions the solution $K_+(x,y)$ of the Marchenko equation (4.1-14) exists and is unique. Prove that the derivatives of $K_+(x,y)$ exist and deduce estimates for K_+ and its derivatives.

Step 2

Form
$$u_1(x,k) = e^{ikx} + \int_x^{\infty} K_+(x,y)e^{iky}dy$$

and
$$v_1(x,k) = u_1(x,-k) + R_+(k)u_1(x,k) \qquad \text{Im}k = 0$$

and show that provided $R_+(k) = o(1/|k|)$ as $|k| \to \infty$ Step 1 implies that u_1 and v_1 are the unique functions such that

(a) u_1 and v_1 can be analytically continued into $\mathrm{Im}k > 0$ with $\exp(-ikx)u\,(x,k)$ bounded for $\mathrm{Im}\,k \geqslant 0$, $k \neq 0$ and v_1 having simple poles at the points $k_j = i\eta_j$, $j = 1,\ldots,M$.

(b) The residues of v_1 are related to the values of u_1 for $k = i\eta_j$ by the relation

$$\mathrm{Res}_{k=i\eta_j}(v_1(x,k)) = -iD_{+j}u_1(x,i\eta_j), \quad j = 1,\ldots,M.$$

(c) On the real axis

$$u_1(x,-k) = u_1^*(x,k), \quad v_1(x,-k) = v_1^*(x,k)$$

(d) For $|k| \to \infty$, $\mathrm{Im}\,k \geqslant 0$

$$u_1(x,k)e^{-ikx} = 1 + O(\frac{1}{|k|}), \quad v_1(x,k)e^{ikx} = 1 + O(\frac{1}{|k|})$$

Step 3

From the scattering matrix $\tilde{S}$ check that

$$\int_{-\infty}^{b} |\hat{R}_{-x}(x)|\,(1+x^2)dx < D_3(b)$$

where D_3 is a non-decreasing function bounded as $x \to -\infty$. Calculate $D_{-j} = -a_j D_{+j}$, $j = 1,\ldots,M$ where $a(k) \equiv T_+^{-1} = T_-^{-1}$. Form

$$\Omega_-(x) = \hat{R}_-(x) + \sum_{j=1}^{M} D_{-j}e^{\eta_j x}$$

and repeat Steps 1 and 2 using the Marchenko equation (4.1-15) for the kernel $K_-(x,y)$ and the functions

$$u_2(x,k) = e^{-ikx} + \int_{-\infty}^{x} K_-(x,y)e^{-iky}dy$$

$$v_2(x,k) = R_-(k)\,u_2(x,k) + u_2(x,-k) \quad \mathrm{Im}k = 0$$

In Step 2 u_2, v_2, R_- replace u_1, v_1, R_+ in the calculation and in part (b) we have the relation

$$\operatorname{Res}_{k=i\eta_j}(v_2(x,k)) = iD_{-j}u_2(x,i\eta_j) \qquad j = 1,\ldots,M$$

Step 4

Establish that the functions u_i are solutions of the Schrödinger equation

$$-u_{ixx}(x,k) + Q_i(x)u_i(x,k) = k^2u_i(x,k) \qquad i = 1,2$$

where $Q_1(x) = -2\frac{d}{dx}K_+(x,x)$, $Q_2(x) = 2\frac{d}{dx}K_-(x,x)$ and, moreover

$$\int_a^\infty |Q_1(x)|(1+x^2)\,dx < \infty, \qquad \int_{-\infty}^b |Q_2(x)|(1+x^2)\,dx < \infty$$

Step 5

The functions u_i satisfy the relations

$$T_-u_1(x,k) = R_-(k)u_2(x,k) + u_2(x,-k)$$
$$T_+u_2(x,k) = R_+(k)u_1(x,k) + u_1(x,-k)$$

where $T_+ = T_-$. From this we conclude that $Q \equiv Q_1 = Q_2$ and that the inverse problem has a unique solution Q which satisfies the constraint

$\int_{-\infty}^\infty |Q(x)|(1 + x^2)dx < \infty$. The functions u_1, u_2 are the Jost solutions

ψ, ϕ. Having outlined the five steps, we now return to discuss each in more detail.

Step 1

$T_+^{-1}(k)$ is formed using (3.4-61) (see, however, the chapter notes for this section). The conditions under Step 1 are checked for the scattering matrix

$$\tilde{S}(k) = \begin{pmatrix} T_+(k) & R_-(k) \\ R_+(k) & T_-(k) \end{pmatrix}$$

where $|T_+|^2 + |R_+|^2 = 1$ and $R_+ = ba^{-1}$, $R_- = -b^*a^{-1}$. Rewrite (4.1-14) after suppressing the t-dependency as the operator equation

$${}_xh(y) + {}_xg(y) + {}_xN_xh(x) = 0 \qquad (4.1\text{-}32)$$

where $\qquad {}_xh(y) \equiv K_+(x,y), \quad y \geq x; \quad {}_xg(y) \equiv \Omega_x(x+y) \quad y \geq x$

and $$ {}_xNf(y) = \int_x^{\infty} \Omega_+(y+y)f(u)\,du \quad y \geq x, \quad f \in L^1(x,\infty) $$

Equation (4.1-32) is a **Fredholm equation of the second kind.** The function ${}_xg$ is absolutely integrable and bounded on $[x,\infty[$ (see equation (4.1-31)). It follows that it is also quadratically integrable, i.e. ${}_xg \in L^{12}(x,\infty)$. In order to prove the existence and uniqueness of the solution to (4.1-32) which also belongs to $L^{12}(x,\infty)$ we shall apply the **Fredholm alternative theorems.** We start by proving that ${}_xN$ is a completely continuous operator in $L^{12}(x,\infty)$. Put

$$ p(x) = \int_x^{\infty} |\Omega_{+u}(u)|\,du \tag{4.1-33} $$

Using the conditions on S, we see that $p(x)$ is absolutely integrable on $[a,\infty[$, $a > -\infty$ and

$$ |\Omega_+(x)| \leq \int_x^{\infty} |\Omega_{+u}(u)|du = p(x) \tag{4.1-34} $$

so that

$$ \int_x^{\infty} du \int_x^{\infty} dy\ |\Omega_+(u+y)|^2 \leq \left(\int_{2x}^{\infty} p(y)dy\right)^2 < \infty $$

and consequently ${}_xN$ is completely continuous in $L^2(x,\infty)$. Similarly one can show that ${}_xN$ is completely continuous in $L^1(x,\infty)$ as a consequence of the absolute integrability of p.

In order to establish existence and uniqueness of the solution in $L^{12}(x,\infty)$ we need only show that the homogeneous equation

$$ {}_xh(y) + {}_xN\,{}_xh(y) = 0 \tag{4.1-35} $$

has only a trivial solution in $L^{12}(x,\infty)$. Suppose that this is not so and that ${}_xh(y) \in L^2(x,\infty)$ is non-trivial. Since $\Omega_+(x)$ is real, without loss of generality we assume that ${}_xh(y)$ is also real. Define ${}_xh(y) = 0$ if $y < x$, multiply (4.1-35) by ${}_xh$, then integrate over $]-\infty,\infty[$

$$ \int_{-\infty}^{\infty} {}_xh^2(y)dy + \int_{-\infty}^{\infty} dy \int_{-\infty}^{\infty} du\ {}_xh(y)\left\{\sum_{j=1}^{M} D_{+j}e^{-\eta_j(u+y)} + R_+(u+y)\right\}{}_xh(u) = 0 \tag{4.1-36} $$

Since $\langle T_xh, T_xh\rangle = \langle {}_xh, {}_xh\rangle$, it follows that this can be written in terms of the Fourier transforms as

$$\frac{1}{2\pi}\int_{-\infty}^{\infty}({}_x\hat{h}(k)\,{}_x\hat{h}(-k)+R_+(k)\,{}_x\hat{h}^2(k))dk + \sum_{j=1}^{M} D_{+j}\,{}_x\hat{h}^2(i\eta_j)e^{-2\eta_j x}=0 \tag{4.1-37}$$

By putting

$${}_xf = {}_x\hat{h}(-k) + R_+(k)\,{}_x\hat{h}(k)$$

and (4.1-38)

$${}_xg = |T_+(k)|\,{}_x\hat{h}(k)$$

we find that (4.1-37) can be written as

$$\frac{1}{4\pi}\int_{-\infty}^{\infty}(|{}_xf(k)|^2+|{}_xg(k)|^2)dk + \sum_{j=1}^{M} D_{+j}\,{}_x\hat{h}^2(i\eta_j)e^{-2\eta_j x}=0 \tag{4.1-39}$$

Since ${}_xh(y)$ is real, ${}_xh(i\eta)$ is also real, thus the left hand side of (4.1-39) contains only positive terms. Hence we conclude that

$${}_xf(k) = {}_xg(k) = 0 \tag{4.1-40}$$

From (4.1-38) and (4.1-40) we deduce that either ${}_x\hat{h}(k)=0$ ($T_+(k) > C$ as $|k|\to 0$) or else ${}_x\hat{h}(k) = 0$ except at $k = 0$ ($T_+(k) = \alpha k + 0(|k|)$, as $|k|\to 0$). In the last case $R_+(0) = -1$ and the first relation of (4.1-38) then gives that ${}_x\hat{h}(k) = 0$ for all k. Hence the homogeneous equation (4.1-35) only has the trivial solution. So from the complete continuity of ${}_xN$ in $L^2(x,\infty)$ and the Fredholm alternative theorem, the solution of (4.1-32) exists and is unique in $L^2(x,\infty)$. Moreover, it is also solvable in $L^1(x,\infty)$, since

$$|{}_xh(y)| \le \int_x^{\infty} |\Omega_+(u+y)|\,|{}_xh(u)|\,du \le p(y+x)\int_x^{\infty} |{}_xh(u)|\,du \tag{4.1-41}$$

and $p(x)$ is bounded on $[x,\infty[$ with $p(\infty) = 0$. Thus any solution in $L^1(x,\infty)$ also belongs to $L^2(x,\infty)$ since (4.1-41) shows that it is quadratically integrable.

To complete Step 1 we now derive estimates for the kernel $K_+(x,y)$ and its derivatives. Since

$$\int_x^{\infty} |{}_xNf(y)|dy \le \int_x^{\infty} dy \int_x^{\infty} |f(u)|\,|\Omega_+(u+y)|\,du$$
$$\le \left(\int_x^{\infty}|f(y)|dy\right)\left(\int_x^{\infty}|\Omega_+(y)|dy\right), \quad -\infty < a < x < \infty,\ f\in L^1(a,\infty) \tag{4.1-42}$$

we have the estimate

$$\| {}_x N \|_1 \le \int_x^\infty |\Omega_+(y)|\,dy \le C_3(x) \tag{4.1-43}$$

where $\| \ \|_1$ is the operator norm derived from $L^1(x,\infty)$, $x \geqslant a > -\infty$. Equation (4.1-35) and the corresponding inhomogeneous equation has a unique solution so that $(I + {}_x N)^{-1}$ exists and is bounded on $[a,\infty[$. Following Agranovich and Marchenko (1964) one then shows that $\| (I + {}_x N)^{-1} \|_1$ is a continuous function of x. Since $\| {}_x N \| \to 0$ as $x \to \infty$

$$\max_{-\infty < a \leqslant x < \infty} \| (I + {}_x N)^{-1} \|_1 = C(a) \tag{4.1-44}$$

From (4.1-14) we get

$$|K_+(x,y)| \le p(x+y) \ (1 + \int_x^\infty |K_+(x,u)|\,du) \tag{4.1-45}$$

and

$$\int_x^\infty |K_+(x,y)|\,du \le \| (I + {}_x N)^{-1} \|_1 \int_x^\infty |\Omega_+(x+u)|\,du \le C(a)\ C_3(2x) \tag{4.1-46}$$

so that

$$|K_+(x,y)| \le C_4(x)\,p(x+y) \tag{4.1-47}$$

Starting from (4.1-14) it is possible to establish the existence of the partial derivatives of $K_+(x,y)$. To obtain estimates on the derivatives, first note that

$$\int_x^\infty |K_+(x,u)|\,|\Omega_{+y}(u+y)|\,du \le C_4(x) \int_x^\infty p(x+u)\,|\Omega_{+y}(u+y)| \le C_4(x)\,p(x+y)\,p(2x) \tag{4.1-48}$$

$$|K_{+y}(x,y)| + |\Omega_{+y}(x+y)| \le C_4(x)\,p(2x)\,p(x+y) \tag{4.1-49}$$

Similarly from the equation satisfied by the x-derivative of K_+ we obtain

$$\int_x^\infty |K_{+x}(x,y)|\,dy \le \| (I + {}_x N)^{-1} \|_1 \left(\int_x^\infty |\Omega_{+x}(x+y)|\,dy + |K_+(x,x)| \int_x^\infty |\Omega(x+y)|\,dy \right) \le C(a)\,p(2x)\,(1 + C_3(x)\,C_4(x)) \tag{4.1-50}$$

so that

$$|K_{+x}(x,y) + \Omega_x(x+y)| \leq C_5(x)p(2x)p(x+y) \tag{4.1-51}$$

and consequently

$$\left|\frac{d}{dx} K_+(x,x) + \frac{d}{dx} \Omega_+(2x)\right| \leq C_6(x)p^2(2x) \tag{4.1-52}$$

Finally we have the result that

$$\int_a^\infty (1+x^2)\left|\frac{d}{dx} K_+(x,x)\right| dx \leq \int_a^\infty (1+x^2)\left|\frac{d}{dx} K_+(x,x) + \frac{d}{dx}\Omega_+(2x)\right| dx + \int_{2a}^\infty \left(1 + \frac{x^2}{4}\right)\left|\frac{d}{dx}\Omega_+(x)\right| dx \tag{4.1-53}$$

The second integral on the right hand side of (4.1-53) exists by hypothesis. For the first integral we have

$$\begin{aligned}\int_a^\infty (1+x^2)\left|\frac{d}{dx} K_+(x,x) + \frac{d}{dx}\Omega_+(2x)\right| dx &\leq C_6(a)\int_a^\infty (1+x^2)p^2(2x)dx \\ &\leq C_7(a)\int_a^\infty p(2x)\,dx < \infty\end{aligned} \tag{4.1-54}$$

The last inequality follows from dividing the range of integration into the intervals $[a,-1[$, $[-1,1]$, $]1,\infty[$. We summarise the results of Step 1 in the following lemma.

<u>**Lemma 4-2:**</u>
If Ω_+ constructed from S_+ satisfies the condition

$$\int_a^\infty |\Omega_{+x}(x)|(1+x^2)\,dx < \infty, \qquad a > -\infty$$

and if $\tilde{S}(k)$ constructed from S_+ is continuous and unitary with the properties

(i) $R_+(-k) = R_+^*(k)$

(ii) either (a) $T_+(k) = T_-(k) = \alpha k + o(|k|)$ as $|k| \to 0$ $\quad \mathrm{Im}\, k \geqslant 0$

$R_\pm(k) + 1 = \beta_\pm k + o(|k|)$ as $|k| \to 0$ $\quad k$ real

or (b) $T_+(k) = T_-(k) = \text{constant} > 0$ as $|k| \to 0$ $\quad \mathrm{Im}\, k \geqslant 0$

then the solution to the Marchenko equation (4.1-14), $K_+(x,y)$ is unique. Moreover the first derivatives of $K_+(x,y)$ exist and the following estimates hold

$$|K_+(x,y)| \leq C_6(x)\; p(x+y)$$

$$|K_{+y}(x,y) + \Omega_{+y}(x+y)| \leq C_4(x)\; p(x+y)p(2x)$$

$$|K_{+x}(x,y) + \Omega_{+x}(x+y)| \leq C_5(x)\; p(2x)\; p(x+y)$$

$$\left|\frac{d}{dx} K_+(x,x) + \frac{d}{dx}\Omega_+(2x)\right| \leq C_6(x)p^2(2x)$$

for $-\infty \leqslant a < x < \infty$ and

$$\int_a^{\infty} \left|\frac{d}{dx}K_+(x,x)\right|(1+x^2)\,dx < \infty$$

tep 2

$$u_1(x,k) = e^{ikx} + \int_x^{\infty} K_+(x,y)e^{iky}dy$$

$$v_1(x,k) = u_1(x,-k) + R_+(k)u_1(x,k), \qquad \text{Im}k = 0$$

ote that the definition of $u_1(x,k)$ implies that $u_1(x,-k) = u_1^*(x,k)$, Im $k = 0$.
From the properties of $K_+(x,y)$ we have immediately that $u_1(x,k)$ is analytic in Im $k > 0$ and for $|k| \to \infty$ we have

$$u_1(x,k)e^{-ikx} = 1 + O(\frac{1}{|k|}) \text{ as } |k| \to \infty \tag{4.1-55}$$

his follows since $K_{+y}(x,y)$ exists. Integrating the expression for $u_1(x,k)$ by arts and using the estimate in Lemma 4-2 we obtain (4.1-55).
From (4.1-14) we get

$$K_+(x,y) + \hat{R}_+(x+y) + \int_x^{\infty} K_+(x,u)\hat{R}_+(u+y)du = \tilde{K}_+(x,-y) \quad y \geq x, \tag{4.1-56}$$

here

$$\tilde{K}_+(x,y) = -\left(\sum_{j=1}^{M} D_{+j}\left(e^{-\eta_j(x-y)} + \int_x^{\infty} K_+(x,u)e^{-\eta_j(u-y)}du\right)\right) \quad y \leq x \tag{4.1-57}$$

aking the Fourier transform of (4.1-56) we obtain the representation

$$v_1(x,k) = e^{-ikx} + \int_{-\infty}^{x} \tilde{K}_+(x,y)e^{iky}\,dy \tag{4.1-58}$$

or $v_1(x,k)$. We conclude that $v_1(x,k)$ is meromorphic in Im $k > 0$ nd provided that

$$R_+(k) = O(\frac{1}{|k|}) \text{ as } |k| \to \infty \tag{4.1-59}$$

hen

$$v_1(x,k)e^{ikx} = 1 + O(\frac{1}{|k|}) \text{ as } |k| \to \infty \tag{4.1-60}$$

Cauchy's theorem applied to (4.1-58) and (4.1-57) then gives that

$$\operatorname{Res}_{k=i\eta_j} v_1(x,k) = -iD_{+j}u_1(x,i\eta_j) \tag{4.1-61}$$

Since the numbers η_j are distinct. From equation (4.1-58) we also have

$$v_1(x,-k) = v_1^*(x,k), \qquad \operatorname{Im}k = 0.$$

Steps 3 and 4

Step 3 is a straightforward repetition of the calculations given under Steps 1 an 2. For Step 4, assume for simplicity that

$$\int_a^\infty (1+x^2)\,|\Omega_{+xx}(x)|\,dx < \infty, \qquad -\infty < a \quad . \tag{4.1-62}$$

Then we can prove the existence of K_{+xx} and K_{+yy} which satisfy the equations

$$K_{+yy}(x,y) = -\Omega_{+yy}(x+y) - \int_x^\infty K_+(x,u)\,\Omega_{+yy}(u+y)\,du \tag{4.1-63}$$

$$\begin{aligned} K_{+xx}(x,y) &= -\Omega_{+yy}(x+y) + \frac{d}{dx}K_+(x,x)\Omega_+(x+y) \\ &\quad + K_+(x,x)\Omega_{+x}(x+u)\Big|_{u=x}\Omega_+(x+y) \\ &\quad - \int_a^\infty K_{+xx}(x,u)\,\Omega_+(u+y)\,du \end{aligned} \tag{4.1-64}$$

Integrating (4.1-63) by parts and subtracting (4.1-64) leads to the equation

$$\begin{aligned} &K_{+xx}(x,y) - K_{+yy}(x,y) - Q_1(x)\,K_+(x,y) \\ &+ \int_x^\infty (K_{+xx}(x,u) - K_{+uu}(x,u) - Q_1(x)K_+(x,u))\Omega_+(u+y)\,du = 0 \end{aligned} \tag{4.1-6}$$

where

$$Q_1(x) = -2\frac{d}{dx}K_+(x,x)$$

Since this equation has only the trivial solution it follows that

$$K_{+xx}(x,y) - K_{+yy}(x,y) - K_+(x,y) - Q_1(x)K_+(x,y) = 0 \tag{4.1-6}$$

Furthermore, under our assumptions

$$\lim_{(x,y)\to\infty} K_+(x,y) = 0, \qquad \lim_{(x,y)\to\infty} K_{+x}(x,y) = 0 \tag{4.1-6}$$

By taking the Fourier transform of (4.1-66), it is straightforward to obtain

$$-u_{1xx}(x,k) + Q_1(x)u_1(x,k) = k^2u_1(x,k) \tag{4.1-6}$$

and from Lemma 4-2

$$\int_a^\infty |Q_1(x)|(1+x^2)\,dx < \infty \quad , \qquad -\infty < a \tag{4.1-69}$$

n fact, as Agranovich and Marchenko (1964) show, u_1 satisfies (4.1-68) even when t is not assumed that Ω_+ has a second derivative. In the same way we can show hat

$$-u_{2xx}(x,k) + Q_2(x)u_2(x,k) = k^2u_2(x,k) \tag{4.1-70}$$

here

$$\int_{-\infty}^b |Q_2(x)|(1+x^2)\,dx < \infty \quad , \qquad b < \infty$$

nd

$$Q_2(x) = 2\frac{d}{dx}K_-(x,x)$$

tep 5

From Step 3, we have the unique functions u_2, v_2 which for real k satisfy the elation

$$v_2(x,k) = R_-(k)u_2(x,k) + u_2(x,-k) \tag{4.1-71}$$

ultiplying (4.1-71) by $R_-(-k)$ the relationship can be rewritten as

$$|R_-(k)|^2u_2(x,k) + R_-(-k)u_2(x,-k) = R_-(-k)\,v_2(x,k) \tag{4.1-72}$$

hen from the unitary property of $\tilde{S}$ we obtain

$$-|T_+(k)|^2u_2(x,k) + v_2(x,-k) = R_-(-k)\,v_2(x,k) \tag{4.1-73}$$

Define functions

$$u(x,k) = v_2(x,k)T_+^{-1}(k)\,; \quad v(x,k) = u_2(x,k)\,T_+(k) \tag{4.1-74}$$

hen (4.1-73) can be written as

$$v(x,k) = u(x,-k) + R_+(k)u(x,k) \tag{4.1-75}$$

rom their definitions(4.1-74) it is straightforward to establish that u and v atisfy the defining properties of u_1 and v_1 whence by uniqueness $u \equiv u_1$, $v \equiv v_1$ and we conclude that

$$Q = Q_1 \equiv Q_2 \tag{4.1-76}$$

nd

(4.1-77

Steps 1 - 5 solve the inverse scattering problem.

Theorem 4-3: The set $S_+ = \{R_+, \lambda_j = (i\eta_j)^2,\ D_{+j}\ j = 1,\dots,M\}$ constitutes the scattering data for a real potential Q, $\int_{-\infty}^{\infty} |Q(x)|(1 + x^2)dx < \infty$ provided

(i) The numbers λ_j are all distinct and strictly negative

(ii) The scattering matrix $\tilde{S} = \begin{pmatrix} T_+ & R_- \\ R_+ & T_- \end{pmatrix}$

where $T_+(k) = T_-(k) \equiv T(k)$

$$T(k) = \begin{cases} \prod_{j=1}^{M} \left(\frac{k+i\eta_j}{k-i\eta_j}\right) \exp\frac{1}{2\pi i} \int_{-\infty}^{\infty} (\xi-k)^{-1} \log[1-|R_+(\xi)|^2]\,d\xi & \text{Im}k>0 \\ \lim_{\varepsilon\to 0} T(k+i\varepsilon) & \text{Im}k=0 \end{cases}$$

has the properties

(a) $T(k) = T^*(-k)$, $R_\pm(k) = R^*_\pm(-k)$ (reality)

(b) $T(k)R^*_+(k) + R_-(k)T^*(k) = 0$

$|T(k)|^2 + |R_+(k)|^2 = 1 = |T(k)|^2 + |R_-(k)|^2$ (unitarity)

(c) $\tilde{S}(k)$ is continuous

(d) $T(k) = 1 + O(1/|k|)$ as $|k| \to \infty$

$R_\pm(k) = O(1/|k|)$ as $|k| \to \infty$, k real

(e) Either $T(k) = \alpha k + o(k)$, $\alpha \neq 0$ as $|k| \to 0$ $\text{Im } k \geqslant 0$

and $R_\pm(k) = \beta_\pm k + o(k)$, $\beta_\pm \neq 0$ as $|k| \to 0$ k real

or $|T(k)| \geqslant$ const. > 0 as $|k| \to 0$ $\text{Im } k \geqslant 0$

(iii) $\hat{R}_\pm$ are absolutely continuous and

$$\int_a^{\infty} |R_{+x}(x)|(1+x^2)\,dx < \infty, \qquad \int_{-\infty}^{b} |R_{-x}(x)|(1+x^2)\,dx < \infty$$

for $-\infty < a;\ b < \infty$

(iv) The numbers D_{+j} are positive

Theorem 4-3 is the main result of this section. We conclude this section with discussion of other methods for solving the inverse scattering problem. We begin by deriving, from the completeness relationship established in Section 3.4, a Marchenko equation which relates the scattering data for two different scattering potential Q_1, Q_2. Let $\psi^{(1)}$, $\psi^{(2)}$ be the Jost solutions corresponding to these potentials defined by conditions analogous to (3.3-2). Then we have the transformation operators

$$\psi^{(2)}(x,k) = \psi^{(1)}(x,k) + \int_x^\infty K(x,k)\,\psi^{(1)}(y,k)\,dy \tag{4.1-78}$$

$$\psi^{(1)}(y,k) = \psi^{(2)}(y,k) + \int_y^\infty \tilde{K}(v,y)\,\psi^{(2)}(v,k)\,dv \tag{4.1-79}$$

We shall omit the properties of the kernels of these operators and concentrate on a brief derivation of the result. From the completeness relation (3.4-31) for the Jost solutions $\psi^{(1)}$, $\psi^{(2)}$ we have

$$\delta(x-y) = \frac{1}{2\pi}\int_{-\infty}^{\infty}\left(R_+^{(i)}(k)\psi^{(i)}(x,k)\psi^{(i)}(y,k) + \psi^{*(i)}(x,k)\psi^{(i)}(y,k)\right)dk + \sum_{j=1}^{M^{(i)}} D_{+j}^{(i)}\psi_j^{(i)}(x)\,\psi_j^{(i)}(y) \qquad i = 1,2 \tag{4.1-80}$$

Multiply (4.1-79) by $(R_+^{(2)}\psi^{(2)}(x,k) + \psi^{*(2)}(x,k))$ and integrate with respect to k; then use (4.1-80) for $i = 2$ to obtain

$$\frac{1}{2\pi}\int_{-\infty}^{\infty} dk\,\psi^{(1)}(y,k)(R_+^{(2)}(k)(R_+^{(2)}(k)\,\psi^{(2)}(x,k) + \psi^{*(2)}(x,k)) + \sum_{j=1}^{M^{(2)}} D_{+j}^{(2)}\psi_j^{(2)}(x)\,\psi^{(1)}(y,i\eta_j^{(2)}) = \int_y^\infty dv\;\delta(x-v)\,\tilde{K}(v,y) + \delta(x-y) \tag{4.1-81}$$

Repeating this operation on (4.1-78) and using (4.1-81) we obtain

$$\int_x^\infty dv\;K(x,v)\left\{\frac{1}{2\pi}\int_{-\infty}^{\infty} dk\,\psi^{(1)}(y,k)\;R_+^{(2)}(k) + \sum_{j=1}^{M^{(2)}} D_{+j}^{(2)}\psi^{(1)}(v,i\eta_j^{(2)})\right\}$$
$$+\frac{1}{2\pi}\int_{-\infty}^{\infty}\psi^{(1)}(x,k)\psi^{(1)}(y,k)R_+^{(2)}(k)\,dk + \sum_{j=1}^{M^{(2)}} D_{+j}^{(2)}\psi^{(2)}(x,i\eta_j^{(2)})\,\psi^{(2)}(y,i\eta_j^{(2)})$$
$$+\frac{1}{2\pi}\int_{-\infty}^{\infty} dk\;\psi^{(1)}(y,k)\psi^{*(2)}(x,k) - \int_{-\infty}^{\infty} dv\;\delta(x-v)\,\tilde{K}(v,y) - \delta(x-y) = 0 \tag{4.1.82}$$

Then equation (4.1-80) for $i = 1$ provides the Marchenko equation

$$K(x,y) + \Omega(x,y) + \int_x^\infty K(x,v)\,\Omega(v,y)\,dv = 0 \qquad y \geq x \tag{4.1-83}$$

where

$$\Omega(x,y) = \frac{1}{2\pi}\int_{-\infty}^{\infty} dk\;\psi^{(1)}(x,k)\;\psi^{(1)}(y,k)\;(R_+^{(2)}(k) - R_+^{(1)}(k)) + \sum_{j=1}^{M^{(2)}} D_{+j}^{(2)}\psi^{(1)}(x,i\eta_j^{(2)})\psi^{(1)}(y,i\eta_j^{(2)}) - \sum_{j=1}^{M^{(2)}} D_{+j}^{(1)}\psi_j^{(1)}(x)\,\psi_j^{(1)} \tag{4.1.84}$$

The potentials are related to the Marchenko equation through the relationship

$$Q_2(x) - Q_1(x) = -2\frac{d}{dx}K(x,x) \tag{4.1-85}$$

Provided $Q_{ix} \in L^1(R)$, K satisfies the equation

$$K_{xx}(x,y) - K_{yy}(x,y) + (Q_1(x) - Q_2(y))K(x,y) = 0 \tag{4.1-86}$$

with the boundary conditions

$$\lim_{(x,y)} K_x(x,y) = 0, \quad \lim_{(x,y)} K_y(x,y) = 0 \tag{4.1-87}$$

Clearly the Marchenko equation (4.1-83) is identical to (4.1-14) when $Q_1 \equiv 0$. Equation (4.1-83) is the fundamental equation used in proving that the KdV is an integrable Hamiltonian system in the approach due to Zakharov and Faddeev (1971). One of the crucial steps in the proof is to show that the inverse scattering or spectral transformation is a symplectic transformation for the (infinite dimensional) manifolds defined with local coordinates (Q) and (S_+). They show that the symplectic form

$$\begin{aligned}\omega(Q,P) &= \int_{-\infty}^{\infty} (\delta_1 Q(x)\delta_2 P(x) - \delta_1 P(x)\delta_2 Q(x))dx \\ P(x) &= \int_{-\infty}^{x} Q(y)dy\end{aligned} \tag{4.1-88}$$

is mapped onto the symplectic form $\omega(S_+)$, thus proving the result. The $\delta_i Q$ are arbitrary tangent vectors in the infinite dimensional space. From (4.1-83) and (4.1-85) it is easy to see that

$$\delta Q = -2\left\{\frac{1}{2\pi}\int_{-\infty}^{\infty} dk\, \psi^2(x,k)\,\delta R_+(x) + \sum_{j=1}^{M}(\delta D_{+j}\psi_j^{\,2}(x) + 2iD_{+j}\psi_j\dot{\psi}_j\delta\eta_j)\right\} \tag{4.1-89}$$

where $Q = Q_1$ and $\psi = \psi^{(1)}$.

Other methods for reconstructing Q from the scattering data are due to Newton (1980) and Deift and Trubowitz (1979).

In this section we have obtained a precise statement for the reconstruction of a Q from the scattering data S_+ so that $\int_{-\infty}^{\infty}|Q(x)|(1 + x^2)dx < \infty$, (Theorem 4-3). In the next section we use the techniques of this section and Theorem 4-3 to solve the initial value problem for a solvable evolution equation. Clearly, further conditions of differentiability have to be imposed on the initial data, but it turns out that these conditions (at least in the simplest of cases when the initial data is fairly smooth) lead to only slight modifications of the theory given above in order to ensure the existence and uniqueness of the solution to the initial value problem.

4.2 The Initial Value Problem for Solvable Equations

From the results of Section 3.5, we see that we are concerned with the **initial value problem**

$$C_1(L_1{}^A)Q_t + C_2(L_1{}^A)Q_x = 0 \qquad x \in \mathbb{R}, \quad t \geq 0$$
$$Q(x,0) = F(x) \tag{4.2-1}$$

where $C_1(k^2)$, $C_2(k^2)$ are real analytic functions. Amongst the family of equations (4.2-1) is a class of evolution equations which have the form $Q_t = K(Q)$ depending only upon Q and its x-derivatives and defined on some appropriate function space. In this case (4.2-1) can be written

$$Q_t + C(L_1{}^A)Q_x = 0 \qquad x \in \mathbb{R}, \quad t \geq 0$$
$$Q(x,0) = F(x) \tag{4.2-2}$$

where C is real analytic and $K(Q) \equiv C(L_1^A)Q$. The KdV is the most important and most studied member of this class

$$Q_t + 6Q\,Q_x + Q_{xxx} = 0 \qquad t \geq 0, \quad x \in \mathbb{R}$$
$$Q(x,0) = F(x) \tag{4.2-3}$$

Notice that this is an example of a **characteristic initial value problem** so that the Cauchy-Kowalewski theorem is not applicable to obtain even a local analytic result. However, it is quite easy to show that the solutions, if they exist, to (4.2-3) are unique if we impose the conditions that $Q_t \in C^3(R)$ and tends to zero along with its first two derivatives as $|x| \to \infty$ (Lax, 1968). In this case, let $P(x,t)$ be another solution and form $W = Q - P$. Then it is quite easy to show upon imposing these conditions that

$$E_t(t) \leq m\,E\,(t) \tag{4.2-4}$$

where

$$E(t) = \tfrac{1}{2}\int_{-\infty}^{\infty} t\; W\,(x)\,dx \quad \text{and} \quad \frac{m}{6} = \max_{x,t \in \mathbb{R}} |2P_x - Q_x| \tag{4.2-5}$$

Consequently $E(t) \leqslant E(0)\exp(mt)$ and if $E(0) = 0$ then $W(t)$ is identically zero for all $t > 0$. The existence of a solution to the initial value problem for the KdV has been determined for a number of increasingly restrictive conditions on the initial data. This material is discussed at the end of this section.

We shall now prove the existence of a solution to (4.2-2) when $F(x)$ is $C^n(R)$ and rapidly decreases along with its derivatives as $|x| \to \infty$ and satisfies $\int_{-\infty}^{\infty}|F(x)|\,(1 + x^2)dx < \infty$. The number n of course depends upon the particular function C in (4.2-2). Before entering into technicalities, we shall give an outline of the method which is called the **inverse scattering method** or the **inverse spectral transform method**).

The method not only proves existence but is also constructive. Thus a large class of solutions called the N-soliton solutions can be explicitly obtained and in addition the asymptotic properties of the general solution can also be derived. These are the topics of the final section of this chapter.

The inverse method is most easily described by referring to Fig. 4-1 below.

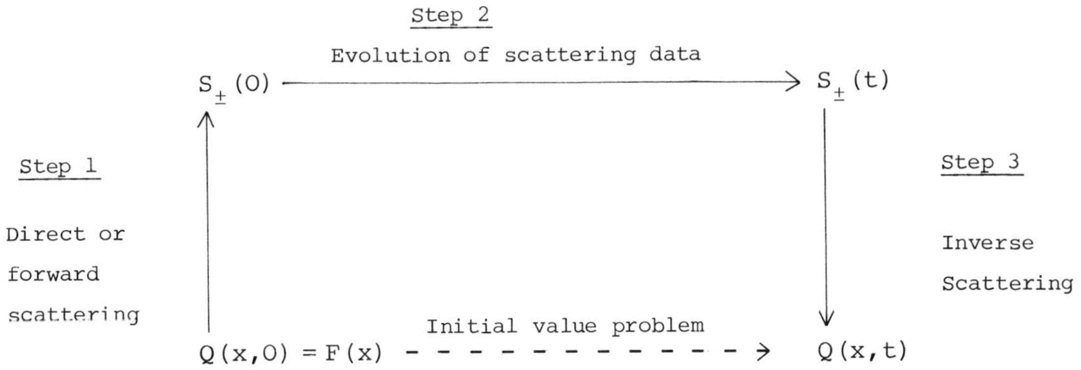

Figure 4-1: The Inverse Method

The goal is to solve the initial value problem for equations (2.2-1) represented by the dotted line on the Fig. 4-1. The solid lines in the figure indicate how this can be achieved by a series of three steps which involve **linear processes** only.

Step 1 - Direct scattering

The scattering data $S_{\pm}(0)$ corresponding to the given initial function $F(x)$ is obtained by solving the Schrödinger equation for the Jost solutions. The results of Chapter 3, Theorem 3-10, establish that $S_{\pm}(0)$ is unique.

Step 2 - Evolution of the scattering data

For a given equation (4.2-1), the real analytic function C uniquely determines the evolution of the scattering data, equation (3.5-2), and can be solved to obtain the scattering data at time $t > 0$, $S_{\pm}(t)$.

Step 3 - Inverse spectral transform

The scattering data $S_{\pm}(t)$ satisfies the requirements of Theorem 4-3 and consequently the potential $Q(x,t)$ can be uniquely constructed from it and, moreover, $\int_{-\infty}^{\infty} |{}_tQ(x)| (1 + x^2)dx < \infty$, $t > 0$.

Of course, we have omitted differentiability and integrability conditions on $F(x)$ which are required to ensure that the reconstructed $Q(x)$ is a solution of the equation (4.2-1). These modifications are considered below.

It is in this way that we solve the initial value problem (4.2-2). We emphasize why the method works. **This is because Step 2 involves the solution of linear equations which only requires knowledge of the boundary value of $Q(x,t)$ and its derivatives as $|x| \to \infty$ for $t > 0$.** The boundary value here is zero, but other boundary conditions, such as Q periodic, or Q tending to a constant non-zero value as $|x| \to \infty$, also lead to inverse methods from which Q can be uniquely reconstructed from the initial data.

The inverse method can be interpreted as a generalisation of Fourier analysis techniques to nonlinear equations. To see this consider equation (4.2-2) when Q is "small"; that is it satisfies the (non-local) condition

$$P_o(\infty,t) << 1 \tag{4.2-6}$$

In this case the equation (4.2-2) can be written as

$$Q_t + C\left(-\frac{1}{4}\frac{\partial^2}{\partial x^2}\right)Q_x = 0 \qquad Q(x,0) = F(x) \tag{4.2-7}$$

in which as we saw in Section 3.5, C is interpreted as the phase speed of an elementary solution. Using the equation (3.3-64) we have the result for direct scattering.

Direct scattering

$$R_+(k,0) \approx \frac{1}{2ik}\int_{-\infty}^{\infty} F(x)e^{-ikx}\,dx \tag{4.2-8}$$

There is no discrete spectrum which is a characteristic feature of the nonlinear solvable systems.

Evolution of the scattering data

From equation (3.5-51) one can obtain the evolution of the scattering data

$$R_+(k,t) \approx R_+(k,0)e^{-2ik\,C(k^2)t} \tag{4.2-9}$$

Inverse spectral or scattering transformation

Solving the Marchenko equation (4.1-14) under condition (4.2-6) we get

$$Q(x,t) \approx \frac{2i}{\pi}\int_{-\infty}^{\infty} kR_+(k)e^{2ik(x-C(k^2)t)}\,dk \tag{4.2-10}$$

It is now easy to check using (4.2-10) that $P_o(\infty,t) << 1$ as a consequence of $P_o(\infty,0) << 1$.

To solve the initial value problem (4.2-2) we follow closely the ideas of Tanaka (1974) for the KdV, except in proving that the constructed $Q(x,t)$ satisfies the equation (4.2-2) where we use an approach more appropriate to our methods. We adopt the notation

$$h^{(i,j,\ell)}(x,y,t) \equiv \frac{\partial^{i+j+\ell}}{\partial x^i \partial x^j \partial x^\ell} h(x,y,t)$$

with the obvious contraction for functions of one or two variables. The existence of a solution to (4.2-2) can be deduced by first proving the following lemmas:

Lemma 4-4: If $F(x)$ is C^n and

$$P_o^{(j)}(x) = \int_{-\infty}^{x} |F^{(j)}(y)|dy, \quad R_o^{(j)}(x) = \int_{x}^{\infty} |F^{(j)}(y)|dy \qquad j \le n$$

are finite for any x then $B_\pm^{(j,\ell)}(x,y)$ exist for $j+\ell \leqslant n+1$ and the following estimates hold (where $B_\pm(x,y) = K_\pm(x,x+2y)$)

$$|B_+^{(j,\ell)}(x,y) + F^{(j+\ell-1)}(x+y)| \leqslant C(x) \sum_{m=0}^{j+\ell-1} R_o^{(m)}(\tfrac{1}{2}(x+y))$$

$$|B_-^{(j,\ell)}(x,y) - F^{(j+\ell-1)}(x+y)| \leqslant D(x) \sum_{m=0}^{j+\ell-1} P_o^{(m)}(\tfrac{1}{2}(x+y))$$

Proof: We consider only B_+; the result for B_- is established in an analogous fashion. For $j=\ell=0$ the result is given by (3.4-77). Differentiating (3.4-73) with respect to x we obtain

$$B_{+x}(x,y) - F(x+y) = -\int_0^y F(x+y-z)B\;(x+y-z)dz \tag{4.2-11}$$

The existence and estimates of $B_+^{(j,o)}$ are obtained by differentiating (4.2-11 with respect to x. Suppose in general the statement has been proved for $j+\ell \leqslant m$ and for $j+\ell \quad m+1$, $\ell \leqslant \ell'$. The existence of $B_+^{(m-k',k'+1)}$ and its estimates follow from

$$B_{+x}(x,y) - B_{+y}(x,y) = -\int_x^\infty F(z)B_+(z,y)dz \tag{4.2-12}$$

and its repeated differentiations.

We next need the regularity of the scattering data ($S_+(0)$ or $S_-(0)$) induced by the regulatiry of $F(x)$. This can be derived from the Fourier representations of the components of $S(k,0)$ obtained by substituting (4.1-7)-(4.1-8) into the defining relations for a and b, equations (3.3-47)-(3.3-49). This gives

$$2ika(k) = 2ik - \int_{\infty}^{\infty} F(y)\,dy + \int_0^{\infty} \pi_2(y) e^{2iky} dy, \quad \text{Im}k \geq 0$$
$$2ikb(k) = e^{-2ikx} \int_{-\infty}^{\infty} {}_x\pi_1(y) e^{-2iky} dy, \quad \text{Im}k = 0 \tag{4.2-13}$$

where
$$ {}_x\pi_1(y) = -B_{+x}(x,y) + B_{-x}(x,y) + \int_{-\infty}^{\min(y,0)} B_{-x}(x,z) B_+(x,y-z)\,dz - \int_{\max(y,0)}^{\infty} B_{+x}(x,z)\, B_-(x,y-z)\,dz$$
$$\pi_2(y) = B_{+x}(x,y) - B_{-x}(x,-y) - B_{+y}(x,y) + B_{-y}(x,-y) - B_+(x,0)\, B_-(x,-y) + \int_{z=0}^{y} (B_{+x}(x,z) - B_y(x,z))\, B_-(x,z-y)\,dz - \int_{z=0}^{y} B_+(x,z)\, B_{-x}(x,z-y)\,dz$$

and

$$B_\pm(x,y) \equiv 2K_\pm(x,2y+x)$$

Tanaka proves the following lemma

Lemma 4-5:
If $F(x)$ is C^n then $\{{}_x\pi_1^{(m)}(y), \pi_2^{(m)}(y), m \leqslant n\}$ are continuous functions.

Proof: From Lemma 4-4 it follows that ${}_x\pi_1(y)$ is differentiable except at $y = x$. For $y \gtrless x$ we have

$$ {}_x\pi^{(m)}(y) = \mp B_\pm^{(1,m)}(x,y) + \sum_{j=0}^{m-1} B_\pm^{(i,j)}(x,y)\, B_\mp^{(0,m-j)}(x,\mp 0) + \int_{-\infty}^{\min(y,0)} B_{-x}(x,z) B_+^{(0,m)}(x,y-z)\,dz - \int_{\max(y,0)}^{\infty} B_{+x}(x,z) B_-^{(0,m)}(x,y-z)\,dz \tag{4.2-14}$$

Thus we need only show that

$$\mp B_\pm^{(1,m)}(x,\pm 0) + \sum_{j=0}^{m-1} B_\pm^{(1,j)}(x,\pm 0)\, B_\mp^{(0,m-j-1)}(x,\mp 0)$$

does not depend upon sign. For $m > 1$ we can use

$$\frac{\partial}{\partial x}\left(\frac{\partial}{\partial x}-\frac{\partial}{\partial y}\right) B_{\pm}(x,y) - F(x)\, B_{\pm}(x,y) = 0 \tag{4.2-15}$$

$$\mp \frac{\partial}{\partial x} B_{\pm}(x,\pm 0) = F(x)$$

to simplify (4.2-14). This is equation (3.4-79) written in terms of the kernels $B_{\pm}$ instead of $K_{\pm}$. Thus for $m = 1$ we get

$$\mp B_{\pm}^{(1,)}(x,\pm 0) + B_{\pm}^{(1,0)}(x, 0)\, B_{\mp}(x,\mp 0)$$

$$= \mp (B_{\pm}^{(2,0)}(x,\pm 0) - F(x)\, B_{\pm}(x,\pm 0)) + B_{\pm}^{(1,0)}(x,\pm 0) B_{\mp}(x,\mp 0)$$

$$= \mp(\mp F_x(x) - F(x) B_{\pm}(x,\pm 0)) \mp F_{(x)} B_{\mp}(x,\mp 0) \tag{4.2-16}$$

and the result for $m > 1$ is proved in a similar fashion.

Lemma 3-6 shows that $T_+(k) = a^{-1}(k)$ is continuous for all k provided $\int_{-\infty}^{\infty}(1 + x^2)|F(x)|dx < \infty$. In addition the Jost solutions are m-times differentiable with respect to k provided $P_j(\infty) < \infty$, $j = 0,\ldots,m$. This information is sufficient to ensure the existence of $a^{(m)}$ and $b^{(m)}$ using their definitions in terms of the Wronskians of Jost solutions, (3.3-47) and (3.3-49). However we require more than this, we need to know conditions under which $k^p R_+^{(m)}$ is square integrable. In order to establish some conditions for this we prove the following lemmas.

Lemma 4-6:
If $F(x)$ is C^n and $F^{(j)}(x) = O(|x|^{-\ell})$ as $|x|\to\infty$, then ${}_x\pi_1$, π_2 are class C^n and furthermore,

$${}_x\pi_1^{(j)}(y) = O(|y|^{1-\ell}) \text{ as } |y| \to \infty$$

$$\pi_2^{(j)}(y) = O(y^{1-\ell}) \text{ as } y \to \infty$$

if $0 \leqslant j \leqslant n$.

Proof: Lemma 4-5 establishes that ${}_x\pi_1$ is $C^n(R)$ and π_2 is $C^n(R^+)$ from its definition. From the bounds on $B_{\pm}^{(i,j)}$, Lemma 4-4 we obtain

$$|B_{\pm}^{(i,j)}(x,y)| \le C_{\pm}(x)\,|x+y|^{1-\ell} \quad i + j \leqslant n + 1, \text{ as } y \to \infty \tag{4.2-17}$$

The result then follows from the definitions of ${}_x\pi_1$ and π_2.

Lemma 4-7:
$a(k)$ is of class $C^{\ell-2}$ except possibly at $k = 0$. If $j \leqslant \ell - 2$ then $(ka(k))^{(j)}$, $\text{Im}k = 0$ is bounded.
$b(k)$ is of class $C^{\ell-2}$ except possibly at $k = 0$. If $j \leqslant \ell - 2$ then

$b^{(j)}(k) = O(|k|^{-(n+1)})$ as $|k| \to \infty$.

Proof: Differentiating (4.2-13) m times, $1 \leqslant m \quad \ell - 2$ we obtain,

$$2i(k(a(k)-1))^{(m)} = \int_0^\infty (2iy)^m \pi_2(y) e^{2iky} dy \qquad \text{Im} k \geq 0 \tag{4.2-18}$$

$$2i(kb(k))^{(m)} = \int_{-\infty}^\infty (-2iy)^m {}_o\pi_1(y) e^{-2iky} dy \qquad \text{Im} k = 0 \tag{4.2-19}$$

Thus $ka(k)$ is of class $C^{\ell-2}(R)$ so $a(k)$ is class $C^{\ell-2}$ except perhaps at $k = 0$. The function $k(a(k) - 1)$ is bounded $\text{Im}k \geqslant 0$, and $(ka(k))^{(m)}$, $1 \leqslant m \leqslant \ell - 2$ is bounded. From (4.2-19) we get similarly that $kb(k)$ is class $C^{\ell-2}$ so that $b(k)$ is class $C^{\ell-2}$ except perhaps at $k = 0$. We also have that

$$(2ik)^p (2ikb(k))^{(m)} = \int_{-\infty}^\infty \frac{\partial^p}{\partial y^p} ((-2iy)^m {}_o\pi_1(y)) e^{-2iky} dy \tag{4.2-20}$$

from which it follows that $b^{(m)}(k) = o(1/|k|^{n+1})$; $m \leqslant \ell - 2$ as $|k| \to \infty$.

Lemma 4-8:
If $F(x)$ is C^n and $F^{(j)}(x) = 0(|x|^{-\ell})$ as $|x| \to \infty$ then $R_+ \in C^M$ and $R_+^{(m)}(k) = O(|k|^{-(n+1)})$ for all $m \leqslant M$ where $M = \ell - 2$ if $\lim_{k \to 0} ka(k) \neq 0$ or $M = \ell - 3$ if $\lim_{k \to 0} ka(k) = 0$.

This follows immediately from the previous two lemmas since $R_+ = ba^{-1}$. A similar result is valid for R_-, but we shall deal solely with S_+ although of course one could equally well work with S_-.

In the case under consideration Theorem 3-12 gives

$$R_+(k,t) = R_+(k,0) \exp(-2ikC(k^2)t) \tag{4.2-21}$$

where C is real analytic. Since $R_+(k,0)$ has the properties given in Lemma 4-8 it follows that $R_+(k,t)$ shares these properties. Furthermore, $k^m R_+^{(p,o)}(k,0)$ is square integrable $m \leqslant n$, $p \leqslant M$. Form the function,

$$\hat{R}_+(x,t) = \frac{1}{2\pi} \int_{-\infty}^\infty R_+(k,0) \exp(ik(x-2C(k^2)t)) dk \tag{4.2-22}$$

then from the preceeding considerations and the fact that if a function and its first derivative is square integrable, then its Fourier transform is integrable, we have,

Lemma 4-9:
Let $\deg C(k^2) = s$ and let $F(x)$ be of class C^n, $n > \max\{4, 2s\}$, then $\hat{R}_+^{(j,\ell)}(x,t)$ exist and are integrable for $j + (2s+1)\ell \leqslant n$. Furthermore $x\hat{R}_{+x}$, $x^2\hat{R}_{+x}$, $x\hat{R}_{+xx}$, $x^2\hat{R}_{+xx}$ are also integrable. We assume throughout that $F^{(j)}(x) = O(|x|^{-m})$ as $|x| \to \infty$, where $m \geqslant 6$, and that the coefficients of $C(k^2)$ are strictly negative.

As outlined in section 1 one can determine $S_+(0) = \{R_+(k,0), D_{+j}(0), \eta_j;\ j = 1,\ldots,M\}$ from $F(x)$. Form the function,

$$\Omega_+(x,t) = \sum_{j=1}^{m} D_{+j}(t)\exp -\eta_j x + \hat{R}_+(x,t) \tag{4.2-23}$$

where

$$D_{+j}(t) = D_{+j}(0)\exp(2\eta_j C(-\eta_j^2)t) \tag{4.2-24}$$

Then Theorem 4-3 and Lemma 4-9 ensures the existence of a function $Q(x,t) = -2d/dx\, K_+(x,x,t))$ where $K_+(x,y,t)$ is the unique solution of the Marchenko equation (4.1-14) and $Q(x,t)$ satisfies the isospectral Schrödinger equation,

$$-y_{xx} + Q(x,t)y = k^2 y \tag{4.2-25}$$

We now require to know the differentiability of $Q(x,t)$ and the Jost solutions to (4.2-25).

Lemma 4-10:
Suppose that $\Omega_\pm^{(i,j)}(x,t)$, $j + (2s+1)\ell \leqslant n$ exist where $\deg C(k^2) = s$ and $F(x)$ is C^n. Then the ith x and jth t derivatives of the Jost solutions and the function $Q(x,t)$ exist and are continuous for $i + (2s+1)j \leqslant n$, $i + (2s+1)j \leqslant n-1$ respectively, and $Q^{(i,j)} \to 0$ as $|x| \to \infty$.

Proof: Using the results of Section 4.1, we get by differentiating the Marchenko equation (4.1-14) written in terms of the B_+ kernel,

$$\begin{aligned} &B_+^{(i,0,0)}(x,y,t) + \int_0^\infty H_+(x+y+u,t)B_+^{(i,0,0)}(x,u,t)\,du \\ &= -H_+^{(i,0)}(x+y,t) - \sum_{m=1}^{i} C_m^i \int_0^\infty H_+^{(m,0)}(x+y+u,t)B_+^{(i-m,0)}(x,u,t) \end{aligned} \tag{4.2-26}$$

where $H_+(x,t) = 2\Omega_+(2x,t)$ from which we deduce that

$$|B_+^{(i,j,0)}(x,y,t) + H_+^{(i,j,0)}(x,y,t)| \leq C_+(x) \sum_{m=j}^{i+j} p_{+m}(x+y,t) \tag{4.2-27}$$

where

$$p_{+m}(x,t) = \int_x^\infty |H_+^{(m,0)}(y,t)|\,dy$$

Then from Lemma 4-9 we see that $B_{+(x,y,t)}^{(i,j,o)}$ exist for $i + j \leqslant n$. If we now consider t-derivatives of (4.1-14) we obtain,

$$B_{+t}(x,y,t) + \int_0^\infty H_+(x+y+u)B_{+t}(x,u,t)\,du + D_+(x,y,t) = 0 \tag{4.2-28}$$

where

$$D_+(x,y,t) = -\sum_{j=0}^{s} a_j(-1)^j \left\{ H_+^{(2j+1,0)}(x+y,t) + \int_0^\infty B_+(x,u,t)\, H_+^{(2j+1,0)}(x+y+u,t)\,du \right\}$$

This last expression is obtained from the fact that if $C(k^2) = \sum_{j=0}^{s} a_j k^{2j}$ then

$$\Omega_{+t}(x,t) + 2\sum_{j=0}^{s} a_j(-1)^j \Omega_+^{(2j+1,0)}(\mathbf{x},t) = 0 \tag{4.2-29}$$

Using (4.2-26) and (4.2-27) and a similar argument for B_-, we deduce that $B_{\pm(x,y,t)}^{(i,j,\ell)}$ exist and are continuous for $(i + j + (2s + 1)\ell) \leqslant n$ and that from (3.4-68)-(3.4-69) the ith x and the jth t, $(i + (2s + 1)j) \leqslant n$, derivatives of the Jost solutions exist. It follows from the definition of Q that $Q_{(x,t)}^{(i,j)}$ exists and is continuous for $(i + (2s + 1)j) \leqslant n - 1$. Moreover, from Lemma 4-9 and equations (4.2-27), (4.2-28), $|B_{+(x,y,t)}^{(i,j,\ell)}| \to 0$ as $x, y, t \to +\infty$ and from the analogous equations for B_-, $|B_{-(x,y,t)}^{(i,j,\ell)}| \to 0$ as $x, y, t \to -\infty$. Consequently $Q_{(x,t)}^{(i,j)} \to 0$ as $|x|, t \to \infty$ for the given index range. The precise asymptotic behaviour of $Q_{(x,t)}^{(i,j)}$ can be obtained from an analysis of R_+ although we do not present this here.

We now establish the main result of this section.

Theorem 4-11: The initial value problem

$$Q_t + C(L_1^A)Q_x = 0$$
$$Q(x,0) = F(x)$$

where $C(k^2)$ is a polynomial of degree s has a solution $Q(x,t)$ if F is of class C^n where $n = \max\{6, 2s+2\}$ and $|F^{(j)}(x)| = O(|x|^{-\ell})$, $\ell > 5$, as $|x| \to \infty$.

<u>**Proof:**</u> From the definition of L_1^A and equation (4.2-2) we see that if $\deg C(k^2) = s$ then we require the existence of $Q^{(j,o)}_{(x,t)}$, $j = 2s + 1$ and $Q^{(o,1)}_{(x,t)}$. Thus put $n = 2s + 2$, then $C(L_1)\phi^2$ exists and is continuous so that

$$R_{+t}(k,t) + C(k^2)\, R_+(k,t) = \int_{-\infty}^{\infty} (Q_t + Q_x C(L_1))\phi^2(x,k)\,dx \tag{4.2-30}$$

By definition (4.2-21) the left hand side of (4.2-30) is identically zero and so

$$Q_t + C(L_1^A)Q_x = 0 \tag{4.2-31}$$

A number of other results have been obtained for the initial value problem associated with the KdV which essentially either refine or extend the above analysis. Some of these, notably those due to Murray (1978) are capable of being also applied to the general class of evolution equations. Bona and Smith (1975) require for existence F to belong to the Sobolev space $H^{s+1}(R)$, $(s + 1) \geqslant 1$. Cohen (1978) establishes the existence of a solution of the KdV for a "box-shaped" initial function which lies outside the results of Bona and Smith. By the imposition of explicit decay rates on the initial data Murray (1978,79) also proves the existence of classical (as opposed to weak solutions) solutions when $F \in C^s(R)$, any s, (compare with Theorem 4-11). Another interesting result, due to Deift and Trubowitz (1979), is the application to the KdV on the line of the analogue of the statement that periodic potentials are real analytic if and only if the widths of the forbidden bands are exponentially decreasing (Trubowitz, 1977). This theorem establishes the reciprocity between analyticity in a strip, $Q(z)$ analytic in $|\mathrm{Im} z| < a$ and exponential decay of R_+. They use this theorem to prove the existence of a solution $Q(x,t)$ to the KdV which is meromorphic in $|\mathrm{Im} z| < a$ with n poles defined by the bound states of the initial function $F(x)$.

If we consider the equations (4.2-1) next then the problem is how to interpret the solution obtained from the inverse method when the corresponding initial value problem appears to require more information than we are given. The resolution to this paradox lies in the comment made after equation (3.5-63), that is one does not solve (4.2-1), but rather one obtains solutions which are common to an equivalence class of equations. It follows from this interpretation that one is really solving the initial values problem which we write formally as

$$\begin{aligned} &Q_t + C_1^{-1}(L_1^A)C_2(L_1^A)Q_x = 0 \\ &Q(x,0) = F(x) \end{aligned} \tag{4.2-32}$$

Consequently we have to investigate the conditions on $F(x)$ for such a solution to exist. First we observe that the formal equation in (4.2-32) is equivalent to the system

$$
\begin{aligned}
Q_t &= f_1 \\
(L_1^A - k_j^2 I) f_j &= f_{j+1} \qquad j = 1, \ldots, n-1 \\
(L_1^A - k_n^2 I) f_n &= -C_2 (L_1^A) Q_x
\end{aligned}
\tag{4.2-33}
$$

when $C_1(k^2) = \prod_{j=1}^{n} (k^2 - k_j^2)$. The jth equation can be written as

$$
-\tfrac{1}{4} h_{jxxx} + (Q - k_j^2) h_{jx} + \tfrac{1}{2} Q_x h_j = f_{j+1}
$$

with

$$
h_j = \int_x^\infty f_j \, dy \tag{4.2-34}
$$

We now show that the system (4.2-33) can be written in terms of the Jost solutions and that the associated initial value problem is well posed. To show this consider an alternative representation of the same system

$$
\Phi_x = P\Phi \qquad \Phi_t = M\Phi \tag{4.2-35}
$$

where P is the operator (3.5-8) and M is a matrix function (it is **not** a differential operator). Then complete integrability of this system requires that

$$
P_t - M_x + [P, M] = 0 \tag{4.2-36}
$$

from which we obtain,

$$
-\tfrac{1}{4} B_{xxx} + (Q - k^2) B_x + \tfrac{1}{2} Q_x B = \tfrac{1}{2} Q_t \tag{4.2-37}
$$

where

$$
M = \begin{pmatrix} -\tfrac{1}{2} B_x & B \\ -\tfrac{1}{2} B_{xx} + (Q - k^2) B_x & \tfrac{1}{2} B_x \end{pmatrix} \tag{4.2-38}
$$

Thus (4.2-34) is (4.2-37)-(4.2-38) with $h_j = B$ and $f_{j+1} = \frac{1}{2} Q_t$.
Solving (4.2-35) for M we obtain

$$M = \Phi\left(N + \int_{-\infty}^{x} \Phi^{-1} P_t \Phi \, dy\right) \Phi^{-1} \tag{4.2-39}$$

with

$$N = \begin{pmatrix} 0 & K \\ H & 0 \end{pmatrix} \qquad K, H \text{ are constants} \tag{4.2-40}$$

It is now straightforward to obtain the relationship

$$h_j = \frac{1}{2ik_j} \left(-2\alpha_j \phi\bar{\phi} + \beta_j \phi^2 - \gamma_j \bar{\phi}^2\right) \tag{4.2-41}$$

with

$$\alpha_j = -k_j^{-1} I(f_{j+1}, \phi, \bar{\phi}), \quad \beta_j = -k_j^{-1} I(f_{j+1}, \bar{\phi}, \bar{\phi}) + K_j$$

$$\gamma_j = k_j^{-1} I(f_{j+1}, \phi, \phi) + H_j \quad \text{and } I(f,u,v) \equiv -i\int_{-\infty}^{x} fuv\,dy \tag{4.2-42}$$

Where the Jost functions are evaluated at $k = k_j$. As a simple example consider the long wave equation given in Section 4.5.

$$Q_{xxt} - 4Q_t - 4QQ_t + 2Q_x \int_x^{\infty} Q_t \, dy + Q_x = 0 \tag{4.2-43}$$

We can find some solutions to this equation by solving by the inverse method the evolution equation which belongs to the same equivalence class as (4.2-36),

$$\begin{aligned} Q_t &= f \\ (L_1^A + 1) f &= \tfrac{1}{4} Q_x \end{aligned} \tag{4.2-44}$$

where f is given by

$$f \equiv h_x = -\tfrac{1}{2}\left(-2\alpha\phi\bar{\phi} + \beta\phi^2 - \gamma\phi^2\right)_x \tag{4.2-45}$$

and

$$\alpha = \frac{i}{4}\, I(Q_y,\phi,\bar{\phi}) = \frac{1}{8}\,(\bar{\phi}_{xx}\phi+\phi_{xx}\bar{\phi}-2\bar{\phi}_x\phi_x-4)$$

$$\beta = \frac{i}{4}\, I(Q_y,\bar{\phi},\bar{\phi}) + K = \tfrac{1}{4}(\bar{\phi}_{xx}\bar{\phi}-\bar{\phi}_x{}^2)+K$$

$$\gamma = \frac{-i}{4}\, I(Q_y,\phi,\phi) + H = -\tfrac{1}{4}(\phi_{xx}\phi-\phi_x{}^2)+H$$

In obtaining expressions (4.2-45) we have assumed that $k_1 = +i$ and for simplicity let us assume that Q is on compact support so that the scattering functions and $\tilde{S}$ are defined throughout the complex k-plane. Then simplifying the expression for f we obtain

$$\begin{aligned} f &= \frac{1}{8}\,(-2\phi\phi_x\bar{\phi}\bar{\phi}_x-4\phi\bar{\phi}+\phi^2\bar{\phi}_x{}^2+\phi^2\phi_x{}^2)_x - \tfrac{1}{2}K(\phi^2-\bar{\phi}^2)_x \\ &= -\tfrac{1}{2}(\phi\bar{\phi})_x - \tfrac{1}{2}(K\phi^2-H\bar{\phi}^2)_x \end{aligned} \qquad (4.2\text{-}46)$$

The boundary conditions on Q require that $f \to 0$ as $|x| \to \infty$. Consequently from (4.2-46) we find that $H = 0$ and either (i) $a(i) = 0$ or (ii) $K = \overline{b(i)/a(i)}$. Condition (i) is ruled out by hypothesis. Condition (ii) gives that

$$Q_t = -\tfrac{1}{2}\left(\frac{\phi\psi}{a}\right)_x \qquad (4.2\text{-}47)$$

$$(\phi\psi)_{xxx} = 4(Q+1)(\phi\psi)_x + 2Q_x(\phi\psi) \qquad (4.2\text{-}48)$$

If we now impose the initial condition $Q(x,0) = F(x)$ then the mixed problem (4.2-47) is well posed since we can solve (4.2-48) at $t = 0$ for the solution $\phi\psi$ such that $\phi\psi \to a(k,0)$ as $|x| \to \infty$. Then since $a(k,t) = a(k,0)$ this is the required boundary condition for all subsequent times. If we had chosen $k_1 = -i$ then $\bar{\phi}$, $\bar{\psi}$ and $\bar{a}$ would replace ϕ, ψ and a in (4.2-47)-(4.2-48). Notice that compact support is not required in establishing the validity of (4.2-48) since one can proceed directly from the equations themselves. If we had considered the case with $C_2(k^2) = k^2 - 1$ then we obtain the conditions (i) $a(\pm 1) = \bar{a}(\pm 1) = 0$ or (ii) $b(\ 1) = 0$. The first condition is ruled out by the properties of the function a. The second condition then gives for the form of the equation

$$Q_t = -\tfrac{1}{2}(\phi\bar{\phi})_x \qquad (4.2\text{-}49)$$

$$(\phi\bar{\phi})_{xxx} = 4(Q-1)(\phi\bar{\phi})_x + 2Q_x(\phi\bar{\phi}) \qquad (4.2\text{-}50)$$

which is again well posed. The general principle to be stated here can be easily derived from a consideration of the general equation using the formal methods developed in Section 3.5. It was shown there that the differential operator $\partial_x \equiv \partial/\partial x$ intertwines the action of the operators L_1^A, L_2,

$$\partial_x L_2 = L_1^A \partial_x \tag{4.2-51}$$

Thus if ∂_x^{-1} is the integral operator, $\partial_x^{-1}\partial_x = \partial_x\partial_x^{-1} = I$ then

$$(L_1^A)^r = \partial_x L_2^r \partial_x^{-1} \quad \text{for integral } r \tag{4.2-52}$$

It follows that we can formally write the equation (4.2-1) as

$$Q_t + \partial_x (C_1^{-1}(L_2) C_2(L_2)) Q = 0 \tag{4.2-53}$$

Then from (3.5-68) we have that

$$-2(\frac{\phi\psi}{a} - 1) = (L_2 - k^2 I)^{-1} Q \tag{4.2-54}$$

so that in particular if $C_2 = \alpha$ and $C_1 = (k^2 - k_1^2)^n$ we see from (4.2-54) that

$$(L_2 - k^2 I)^{-n} Q = \frac{-2}{(n-1)!} \frac{\partial^{n-1}}{\partial (k^2)^{n-1}} (\frac{\phi\psi}{a} - 1) \Big|_{k=k_1} \tag{4.2-55}$$

and that the evolution equation is

$$Q_t + \alpha\partial_x \left(\frac{-2}{(n-1)!} \frac{\partial^{n-1}}{\partial (k^2)^{n-1}} (\frac{\phi\psi}{a} - 1)\Big|_{k=k_1} \right) = 0 \tag{4.2-56}$$

together with the equation (4.2-50) and its derivatives up to the $2(n-1)$th order evaluated at $k = k_1$. We require that k_1 is not an eigenvalue of L, $\text{Im}k > 0$. If $\text{Im}k_1 < 0$ then $\bar{a}, \overline{\phi\psi}$ replace $\phi\psi$, a in (4.2-56). If $\text{Im}k_1 = 0$ then from the asymptotic form of (4.2-56) as $|x| \to \infty$ in addition we require that

$$\frac{\partial^m b(k)}{\partial k^m}\Big|_{k=k_1} = 0, \quad m=0,..,2(n-1).$$

Thus certainly in this case further conditions have to be imposed on the initial data if the equation is to be solved by the inverse method. As far as we are aware, the initial value problem for equation (4.2-56) has not been thoroughly investigated in the literature.

4.3 N-Soliton Solutions for the Solvable Equations

In the case when the initial data is such that the reflection coefficient (R_+ or R_-) is zero the inverse method is exactly solvable and gives the N-soliton solutions of the solvable equations (4.2-1) if the spectrum of the Schrödinger operator initially has N eigenvalues. The Marchenko equation ((4.1-14) or (4.1-15)) is then an equation with a degenerate kernel and can be solved using the techniques of linear algebra.

Assume that the kernel (4.1-14) can be written in the separable form

$$K(x,y,t) = h(x,t).f(y) \tag{4.3-1}$$

and that

$$\Omega(x+y,t) = g(x,t).f(y) \tag{4.3-2}$$

where h is an unknown N component column vector and g, f are respectively the N component column vectors whose jth entries are $D_j(t)\exp(-\eta_j x)$ and $\exp(-\eta_j y)$. This if we consider the Marchenko equation (4.1-14) for example, which can be written as

$$(A(x,t)h(x,t) + g(x,t)).f(y) = 0 \qquad y \geq x \tag{4.3-3}$$

where A is the invertible matrix function

$$A(x,t) = I + E(x,t)$$

$$E(x,t)_{ij} = \left(\int_x^\infty g(u,t)f^T(u)\,du\right)_{ij} = \frac{D_{+i}(t)e^{-i(\eta_i+\eta_j)x}}{(\eta_i+\eta_j)}$$

Invertibility of A follows from the existence and uniqueness of the solution to the Marchenko equation. Solving (4.3-3) we obtain

$$K(x,y,t) = -(A^{-1}(x,t)g(x,t)).f(y) \tag{4.3-4}$$

which is therefore the unique solution to the Marchenko equation thus justifying the ansatz (4.3-1). It follows that

$$\begin{aligned} K(x,x,t) &= \mathrm{Tr}\left(A^{-1}\frac{\partial}{\partial x}A\right) \\ &= (\det A)^{-1}\,\mathrm{Tr}\left(\mathrm{adj}\,A\frac{\partial}{\partial x}A\right) \\ &= \frac{\partial}{\partial x}\log(\det A(x,t)) \end{aligned} \tag{4.3-5}$$

and that the solution to the initial value problem (4.1-1) is

$$Q(x,t) = -2 \frac{\partial^2}{\partial x^2} \log(\det(I+E(x,t))) \tag{4.3-6}$$

Either proceeding directly from the Marchenko equation on examining the properties of $\det(I + E(x,t))$ we see that we can replace the definition of E by the more symmetrical

$$E_{ij}(x,t) = \frac{D_{+i}^{\frac{1}{2}}(t) D_{+j}^{\frac{1}{2}}(t) \exp-(\eta_i+\eta_j)x}{(\eta_i+\eta_j)} \tag{4.3-7}$$

From the results of Section 3.5 we have

$$D_{+i}(t) - D_{+i}(0) \exp\left[2\eta_i \int_0^t C(t', -\eta_i^2)dt'\right] \tag{4.3-8}$$

Thus the N-soliton solutions of all the solvable equations have the same functional form and the difference arises in the time evolution of the normalisation constants. This, as we have seen, is simply due to the different dispersion relations for the linearised solvable systems.

As examples we give the explicit formulae for the one-soliton and two-soliton solutions of the general solvable equation (4.2-1).

<u>One-soliton Solution</u>

$$Q(x,t) = -2\eta^2 \text{sech}^2 \left(\eta x - \eta \int_a^t C(t\ ,-\eta^2)dt' - \delta\right) \tag{4.3-9}$$

where

$$\delta = \tfrac{1}{2}\log \frac{1}{2\eta} D_+(0)$$

<u>Two-soliton Solution</u>

$$Q(x,t) = \frac{-4\eta_1\eta_2\left[4\sinh^2\tfrac{1}{2}(\delta_2-\delta_1)+\tanh\tfrac{1}{2}(\delta_2-\delta_1)\left\{e^{(\delta_1-\delta_2)}\cosh(\theta_1+\delta_1)+e^{(\delta_2-\delta_1)}\cosh(\theta_2+\delta_2)\right\}\right]}{\left(\tan\tfrac{1}{2}(\delta_2-\delta_1)\cosh\left[\tfrac{1}{2}(\theta_1+\theta_2)+\tfrac{1}{2}(\delta_1+\delta_2)\right]+\cosh\left[\tfrac{1}{2}(\theta_1-\theta_2)+\tfrac{1}{2}(\delta_1-\delta_2)\right]\right)^2} \tag{4.3.10}$$

where

$$e^{\delta_1} = \frac{(\eta_1-\eta_2)}{2\eta_1(\eta_1+\eta_2)} \qquad e^{\delta_2} = \frac{(\eta_1-\eta_2)}{2\eta_2(\eta_1+\eta_2)}$$

and

$$\theta_i = \log D_{+i}(0) + 2\eta_i \int_0^t C(t',-\eta_i^2)dt' - 2\eta_i x$$

It is interesting to consider in detail the asymptotic behaviour as $|t| \to \infty$ of the N-soliton solution (4.3-6). A brief discussion of this was given in Section 1.4 for the KdV equation. We assume here that $C = C(k^2)$ (if C depends explicity upon t the following analysis may break down). Since the η_j are distinct we can arrange them in an ordered sequence $\{\eta_i : \eta_i < \eta_j \text{ if } i < j\}$. We use a suffix to denote the components of a vector, thus (4.3-1) can be written as

$$K(x,y,t) = \sum_{j=1}^{N} K_j(x,y,t) \text{ where } K_j(x,y,t) = h_j(x,t)e^{-\eta_j y} \tag{4.3-11}$$

and (4.3-3) as the system of linear equations

$$K_j(x,x,t)e^{2\theta_j\eta_j} + \sum_{\ell=1}^{N} \frac{K_\ell(x,x,t)}{(\eta_j+\eta_\ell)} + 1 = 0 \tag{4.3-12}$$

where

$$\theta_j = x + C(-\eta_j^{\,2})t - \delta_j$$

and

$$\delta_j = \frac{1}{\eta_j} \log D_{+j}^{\frac{1}{2}}(0)$$

Either $C(-\eta_j^2) > C(-\eta_k^2)$, $j < k$ or $C(-\eta_j^2) < C(-\eta_k^2)$, $j < k$, but since the second case is mapped onto the first by the transformation $t \to -t$, we shall only consider the first possibility. In the vicinity of the line $x = \delta_j - C(-\eta_j^2)t$, we have

$$\begin{aligned} \theta_k &\to \pm\infty \text{ as } t \to \pm\infty \quad k < j \\ \theta_k &\to \mp\infty \text{ as } t \to \pm\infty \quad k > j \\ \theta_j &\approx 0 \end{aligned} \tag{4.3-13}$$

Using Cramer's rule to solve (4.3-12) we observe that K_j is a ratio of two functions p_j and q. The function q is common to all the K's and depends upon all the functions $e^{\eta_k\theta_k}$, $k = 1,\ldots,N$, whereas p differs for each K and is independent of $e^{\eta_j\theta_j}$.

It follows that

$$
\left.\begin{aligned}
K_m &= 0+O(e^{-\lambda t}) \quad m < j \\
K_m &= K_m^{\,o} + O(e^{-\lambda t}) \quad m \geqslant j
\end{aligned}\right\} \quad \text{as } t \to \infty \tag{4.3-14}
$$

$$
\left.\begin{aligned}
K_m &= K_m^{o} + O(e^{\lambda t}) \quad m \leqslant j \\
K_m &= 0+O(e^{\lambda t}) \quad m > j
\end{aligned}\right\} \quad \text{as } t \to -\infty \tag{4.3-15}
$$

where λ is a positive constant. The leading order terms in the asymptotic expansion of the functions $K_j(x,x,t)$ therefore satisfy

$\underline{t \to +\infty}$

$$
\begin{aligned}
&K_j(x,x,t)\,(e^{2\theta_j\eta_j} + \frac{1}{2\eta_j}) + \sum_{\ell=j+1}^{N} \frac{K_\ell^{o}(x,x,t)}{(\eta_j+\eta_\ell)} + 1 = 0 \\
&\sum_{\ell=j+1}^{N} \frac{K_\ell^{o}(x,x,t)}{(\eta_k+\eta_\ell)} + \frac{K_j^{o}(x,x,t)}{(\eta_k+\eta_j)} + 1 = 0 \qquad k \gtrless j
\end{aligned} \tag{4.3-16}
$$

$\underline{t \to -\infty}$

$$
\begin{aligned}
&K_j^{o}(x,x,t)\,(e^{2\theta_j\eta_j} + \frac{1}{2\eta_j}) + \sum_{\ell=1}^{j-1} \frac{K_\ell^{o}(x,x,t)}{(\eta_j+\eta_\ell)} + 1 = 0 \\
&\sum_{\ell=1}^{j-1} \frac{K_\ell^{o}(x,x,t)}{(\eta_k+\eta_\ell)} + \frac{K_j^{o}(x,x,t)}{(\eta_k+\eta_j)} + 1 = 0 \qquad k \gtrless j
\end{aligned} \tag{4.3-17}
$$

When $|x| \to \infty$ all the K_ℓ^0 in (4.3-16) and (4.3-17) tend to constant values so that the corresponding Q obeys the boundary condition $Q \to 0$ as $|x| \to \infty$. The solutions defined by (4.3-16) and (4.3-17) are therefore solitons, since Q is solely a function of $\theta_j = x + C(-\eta_j^2) + \delta_j$ and $Q \to 0$ as $|\theta_j| \to \infty$. Thus the N-soliton solution breaks up into N one soliton solutions as $|t| \to \infty$ with the jth soliton centred about $x = \delta_j - C(-\eta_j^2)t$ and moving with velocity $-C(-\eta_j^2)$.

Because of the properties of $C(-\eta_j^{\,2})$ we see that as $t \to -\infty$, the solitons are ordered with the fastest at the back and the slowest at the front of the group. This ordering is reversed for $t \to +\infty$ and so the N-soliton solution can be interpreted as the interaction of N one soliton solutions. The remarkable feature of this interaction is that the solitons re-emerge after the collison, the sole change due to the collision being a change of phase (as we saw in the two soliton collision for the KdV in Chapter 1). In terms of the notation employed here, a two soliton collision results in a change of phase

$$\delta_1 = -\frac{1}{\eta_1}\log\left|\frac{\eta_1+\eta_2}{\eta_1-\eta_2}\right| \qquad \delta_2 = \frac{1}{\eta_2}\log\left|\frac{\eta_1+\eta_2}{\eta_1-\eta_2}\right| \tag{4.3-18}$$

the slower soliton acquiring a negative phase shift and the faster soliton a positive one. Clearly equations (4.3-16) and (4.3-17) only involve the velocities of the other solitons emerging from the N-soliton solution as $|t| \to \infty$ and not their phases. Thus the total phase shift acquired by a particular soliton through collisions with the slower members of the group will be the sum of the phase shift due to pairwise collisions.

This remarkable stability of the soliton-soliton collision leads to the consideration of the stability of the soliton in general. One such method due to Benjamin (1972) utilises the conserved densities associated with the solvable equation to define a measure of the difference between a general solution Q and a soliton solution Q_s

$$\Delta C_3(Q,Q_s) = C_3(Q) - C_3(Q_s) \tag{4.3-19}$$

where $C_3(Q)$ is the third density in the infinite hierarchy of conserved densities for the equation. This material takes us outside the scope of this book. It suffices to remark that Benjamin demonstrates the stability of the KdV soliton. Another approach to this question is to work directly from the Marchenko equation itself. However, this presents many technical problems, although a partial result was obtained by Zakharov (1971).

One of the characteristic features of solvable nonlinear equations is the existence of **Bäcklund transformations** which relate solutions of the same equations or even solutions of two different solvable equations. A mathematical outline of Bäcklund transformations is given in Section 6.3. Here we content ourselves with producing a large family of such transformations. The material of Section 3.2 is also relevant, and furnishes particular examples of Bäcklund transformations for the KdV.

We begin by considering two potentials, Q, Q' and their associated fundamental matrix solutions Φ, Φ' which satisfy the equations

$$\Phi_x = P\Phi \qquad \Phi'_x = P'\Phi \tag{4.3-20}$$

where

$$P \equiv (Q\sigma - F(k))$$

$$F(k) = \begin{bmatrix} 0 & -1 \\ k^2 & 0 \end{bmatrix} \quad \text{and} \quad \sigma = \begin{bmatrix} 0 & 0 \\ 1 & 0 \end{bmatrix}$$

(compare with equations (3.5-8)). The assumption here is that the continuous spectra for the two Schrödinger operators $L \equiv L(Q)$, $L' \equiv L(Q')$ agree, but their discrete spectra differ. Then proceeding in the same manner which produced (3.5-15) we find that

$$(A')^{-1}\,\Delta A = \int_{-\infty}^{\infty} \Delta Q\,((\Phi')^{-1}\,\sigma\Phi)\,dx \tag{4.3-21}$$

where

$$\Delta Q \equiv Q'-Q, \quad \Delta A \equiv A'-A$$

and A', A are the scattering function matrices (3.5-13) associated with the system (4.3-20). The right hand side of (4.3-21) involves products of the eigenfunctions for the two operators L, L'. Just as in the case when we considered the square eigenfunctions one can determine an eigenvalue problem for the function $\phi\,\phi'$. The procedure is to write down equations (3.5-22)-(3.5-26) for eigenfunctions ${}_ku$, ${}_kv'$ of L, L' and then to apply the operations which results in (3.5-27)-(3.5-28). This gives the relations

$$\begin{aligned}
(u_{xx}v' + v'_{xx}u - 2u_xv'_x)(x) &= \int_{-\infty}^{x} (Q_s+Q_s')\,uv'\,ds + \int_{-\infty}^{x} (Q-Q')(uv'_s - v'u_s)\,ds \\
&\quad + (u_{xx}v'+v'_{xx}u-2u_xv'_x)(-\infty) \\
\frac{\partial^2}{\partial x^2} - (Q+Q')\,uv' &= -2k^2uv'+2u_xv'_x \\
(u_xv'-v_x'u)(x) &= \int_{-\infty}^{x} (Q-Q')\,uv'\,ds+(u_xv'-v'_xu)(-\infty)
\end{aligned} \tag{4.3-22}$$

Combining these equations together we obtain

$$\begin{aligned}
B_1uv' &= k\;uv' + \frac{1}{4}\,(u_{xx}v'+v'_{xx}u-2u_xv'_x)(-\infty) \\
&\quad + \frac{1}{4}\,(v'_xu-u_xv')(-\infty)\int_{-\infty}^{x} (Q-Q')\,ds
\end{aligned} \tag{4.3-23}$$

where

$$B_1 \equiv -\frac{1}{4}\left\{\frac{\partial^2}{\partial x^2} - 2(Q'+Q) + \int_{-\infty}^{x} ds\,(Q'_s+Q_s) + \int_{-\infty}^{x} ds\,(Q'-Q)\int_{-\infty}^{x} dy\,(Q'-Q)\right\}$$

Notice that when $Q' = Q$, B_1 is equivalent to L_1, (3.5-31). Equations (4.3-20) can be written

$$\Delta R_+ = \frac{1}{2ikaa'} I_\Delta(\phi',\phi) \tag{4.3-24}$$

$$1 + \bar{b}'b - a\bar{a}' = -\frac{1}{2ik} I_\Delta(\bar{\phi}',\phi)$$

where

$$I_\Delta(u,v) \equiv \int_{-\infty}^{\infty} \Delta Q\, uv\, dx$$

If we evaluate (4.3-23) in the limit $x \to +\infty$ for the function $\phi.\phi'$ we obtain the relation

$$(R_+' + R_+) = -\frac{1}{4k^2aa'} \int_{-\infty}^{\infty} \left[(Q_x' + Q_x) + (Q'-Q)\int_{-\infty}^{x} ds\,(Q'-Q)\right]\phi\phi'\,dx \tag{4.3-25}$$

If we introduce $G(k^2)$, which is either real analytic or a ratio of real analytic functions, we can combine (4.3-24) and (4.3-25) to give the equation

$$\begin{aligned}(R_+' - R_+) &+ 2ikG(k^2)\,(R_+' + R_+) \\ &= \frac{1}{2ikaa} \int_{-\infty}^{\infty} \left[(Q'-Q) + G(k^2)\,((Q_x' + Q_x) \right. \\ &\qquad \left. + (Q'-Q)\int_{-\infty}^{x} ds\,(Q'-Q)\right]\phi\phi'\,dx\end{aligned} \tag{4.3-26}$$

All this is reminiscent of Section 3.5 when we were determining the solvable equations.

We also have that

$$G(B_1)\phi\phi' = G(k^2)\phi\phi' \tag{4.3-27}$$

so that we repeat the trick used there to transfer the action of the operator B_1 onto the potential functions by forming the adjoint

$$\begin{aligned}B_1^A \equiv -\frac{1}{4}\Big\{\frac{\partial^2}{\partial x^2} &- 2(Q'+Q) + (Q_x' + Q_x)\int_x^{\infty} ds \\ &+ (Q'-Q)\int_x^{\infty}(Q'-Q)\int_s^{\infty} dy\Big\}\end{aligned} \tag{4.3-28}$$

Then a nontrivial relationship is established between the functions Q', Q by requiring that

$$R'_+ = \frac{(1-2ikG(k^2))}{(1+2ikG(k^2))} R_+ \tag{4.3-29}$$

The functions Q', Q are then related by

$$(Q'-Q)+G(B_1^A)\left\{(Q_x'+Q_x) + (Q'-Q)\int_x^\infty (Q'-Q)\,ds\right\} = 0 \tag{4.3-30}$$

From (3.5-51)

$$\begin{aligned} R'_{+t} &= \frac{(1-2ikG(k^2))}{(1+2ikG(k^2))} R_{+t} \\ &= -2ikC(k^2)\, R'_+ \end{aligned} \tag{4.3-31}$$

where $C(k^2)$ is the ratio of real analytic functions defining the solvable equation satisfied by Q, (3.5-45). Since R'_+ satisfies the same evolution equation as R_+ it follows that Q' satisfies the same equation and that consequently (4.3-30) defines a correspondence between certain solutions to a solvable nonlinear equation.

Equation (4.3-30) is the **generalised Bäcklund transformation,** or rather one half of it, which we set out to determine and (4.3-29) represents the associated change in the reflection coefficient. It was first obtained by Calogero and Degasperis (1976). The method of deriving it used here is due to Dodd (1978). Note that (4.3-30) is common to **all** the solvable equations as it does not depend upon the phase speed of the linearised equation, the function $C(k^2)$. The other half of the transformation has to be obtained by substituting into (4.3-30) the equations satisfied by Q' and Q and rearranging to obtain a symmetrical expression in Q'_t, Q_t and the x-derivatives of these functions. The simplest transformation given by (4.3-30) corresponds to the first half of the Bäcklund transformation given by (4.3-30) corresponds to the first half of the Bäcklund transformation (3.1-21) determined for the KdV. To obtain it put $G(k^2)$ equal to a constant, $G(k^2) = 1/2p$, $p \neq 0$ and we find that

$$(Q'_x+Q_x) + (Q'-Q)\left\{2p + \int_x^\infty (Q'-Q)\,ds\right\} = 0 \tag{4.3-32}$$

If we introduce the potential function

$$w' = \tfrac{1}{2}\int_x^\infty Q'\;ds + \alpha \qquad w = \tfrac{1}{2}\int_x^\infty Q\,ds + \beta \tag{4.3-33}$$

where $\alpha - \beta = p$, then this can be written

$$(w'+w)_x = p^2 - (w'-w)^2$$

which is the half of the Bäcklund transformation common to all solvable equations. For this case the reflection coefficients are related by

$$R'_+ = - \frac{(k+ip)}{(k-ip)} R_+ \qquad (4.3\text{-}34)$$

If one starts from the zero solution $Q = 0$ then it is not hard to check that (4.3-33) integrates to given the soliton solution (4.3-41). The coefficient of the t-variable in the solution is of course fixed by the phase speed $C(k^2)$ of the linearised version of the solvable equation being considered. This in fact gives a clue to the form of Q' for arbitrary Q. The transformation (4.3-33) "adds a soliton to Q". This is clear from (4.3-34) if we interpret $-p^2$ as an additional eigenvalue which is present in $\sigma(L(Q'))$ but absent from $\sigma(L(Q))$. As one sees from the first calculation the soliton solution calculated in this way is not unique. The Bäcklund transformation is not therefore a transformation (that is a mapping) in the usual sense because the integration introduces a one parameter family of solutions (p fixed) and therefore establishes a **correspondence** between solutions. In the case of $Q = 0$ the parameter enters into the phase shift of the soliton solution generated by the Bäcklund transformation.

As a further example the choice $G(k^2) = \frac{1}{2}(-k^2/abc + (ab+ac+bc)/abc)$ with $a + b + c = 0$ corresponds to a composite Bäcklund transformation for which

$$R'_+ = - \frac{(k+ia)(k+ib)(k+ic)}{(k-ia)(k-ib)(k-ic)} R_+ \, , \quad a + b + c = 0 \qquad (4.3\text{-}35)$$

and thus $\sigma(L(Q'))$ has an additional **three eigenvalues** $-a^2, -b^2, -(a+b)$. In general $G(k^2) = \sum_{i=0} a_i k^{2i}$ will result in the change,

$$R'_+ = - \prod_{j=1}^{2s+1} \frac{(k+i\alpha_i)}{(k-i\alpha_i)} R_+ \qquad (4.3\text{-}36)$$

where the α_i satisfy $(2s + 1) - (s + 1) = s$ conditions, and $\sigma(L(Q'))$ has an additional $2s + 1$ eigenvalues. The reason for calling the transformations corresponding to (4.3-24), (4.3-36) composite are that they are compositions of the transformations corresponding to (4.3-34).

In this regard the commutation relation or "theorem of permutivity" for the Bäcklund transformation (4.3-34) discussed in the exercises to Chapter 3 is seen to be a consequence of the fact that

$$R_{3+} = \frac{(k+ip_2)}{(k-ip_2)} R_{1+} = \frac{(k+ip_2)}{(k-ip_2)} \frac{(k+ip_1)}{(k-ip_1)} R_{0+}$$

$$= \frac{(k+ip_1)(k+ip_2)}{(k-ip_1)(k-ip_2)} R_{0+} = - \frac{(k+ip_1)}{(k-ip_1)} R_{2+} \qquad (4.3\text{-}37)$$

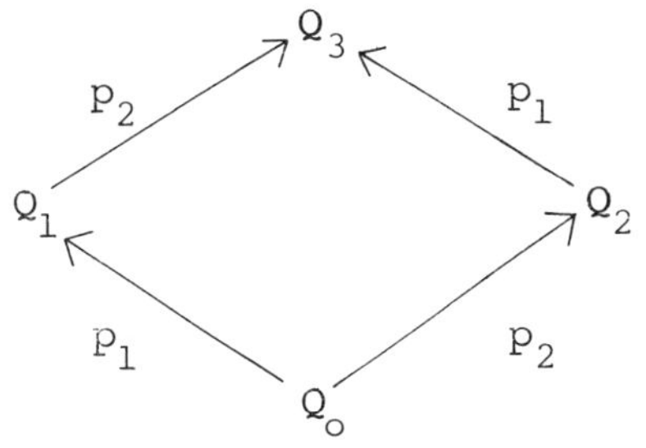

Figure 4-2: The algebraic superposition principle for solvable equations

This is conveniently represented by the commuting diagram, Fig. 4-2 where Q_1, Q_2, Q_3 are specific solutions obtained from Q_0 by applications of (4.3-32) to Q_0 with parameters p_1 and p_2. It is easy to obtain from (4.3-33) **the algebraic superposition principle** which results from the commutation relation (4.3-37),

$$w_3 = \frac{(p_1^{\,2}-p_2^{\,2}) + w_o(w_1-w_2)}{(w_1 - w_2)} \tag{4.3-38}$$

w_3 is a new solution which is obtained by algebraic means from a given solution w_o & two particular transforms of it w_1 and w_2 with parameters p_1 and p_2. Further solutions can be obtained by combining families of solutions arising in this way. The "tower" of solitons can be generated in this fashion.

If the Bäcklund transformation (4.3-30) is extended by allowing G to be also a function of t, then **formal** transformations may be defined which transform between solutions to different solvable equations. Thus let Q', Q be solutions of solvable equations characterised by functions C' and C respectively. In this case,

$$R_{+t}{}' = -2ikC\ (k^2)R_+{}' \tag{4.3-39}$$

and we assume that R_+' and R_+ are related by (4.3-29) so that integrating we obtain,

$$G(k^2,t) = -\frac{1}{2ik}\tan\left\{k(C(k^2) - C'(k^2))(t-t_o) + F(k^2))\right\}$$

where

$$\tan(kF(k^2)) = -2ikG(k^2,t_o) \tag{4.3-40}$$

The formal transformation is then defined by (4.3-30) with the G defined by equation (4.3-40).

At this point it is worth deriving the effect of a Bäcklund transformation upon the rest of the scattering data as well as the transmission coefficient T_+. It is immediate from (4.3-29) and the work of section 5, chapter 3 that

$$D_{+j}{}' = \frac{(1-2ikG(k^2))}{(1+2ikG(k^2))} D_{+j} \quad j = 1,\ldots,M \tag{4.3-41}$$

It is much harder to derive the effect of the Bäcklund transformation upon the transmission coefficient T_+. Here we give the formula for $G(k^2) = 1/2p$ and deduce the general result. From (4.3-30) we get upon integrating that

$$\int_{-\infty}^{\infty} (Q'-Q)\,dx = -\,4p \tag{4.3-42}$$

One of equations (4.3-24) can be written

$$a'-a = -\frac{1}{2ik} I_\Delta(\psi',\phi) \tag{4.3-43}$$

which becomes, upon using the equation (4.3-30) with $G(k^2) = 1/2p$,

$$a'-a = \frac{1}{4ikp}\int_{-\infty}^{\infty}\left[(Q_x' + Q_x) + (Q'-Q)\int_x^{\infty}(Q'-Q)\,ds\right]\psi'\phi\,dx \tag{4.3-44}$$

It is not hard to obtain from (4.3-22)

$$\int_{-\infty}^{\infty} (Q_x{}'+Q_x)+(Q'-Q)\int_x^{\infty}(Q'-Q)\,ds)\,\psi'\phi\,dx = 2k[2a'k-2ak-ia\int_{-\infty}^{\infty}(Q'-Q)dx] \tag{4.3-45}$$

Then (4.3-44) can be written using (4.3-45) and (4.3-42) as

$$T_+'(k) = \frac{(k-ip)}{(k+ip)} T_+(k) \tag{4.3-46}$$

It follows that if $\sigma(L(Q'))$ has an N additional eigenvalues to $\sigma(L(Q))$ then by repeated application of (4.3-32) (with different parameters p_j) we get the formulae

$$R_+{}'(k) = (-1)^N \prod_{j=1}^{N} \frac{(k+ip_j)}{(k-ip_j)} R_+(k) \tag{4.3-47}$$

$$T_+{}'(k) = \prod_{j=1}^{N} \frac{(k+ip_j)}{(k-ip_j)} T_+(k) \tag{4.3-48}$$

We now consider the relationship between the eigenfunctions of Bäcklund related isospectral Schrödinger operators. This material provides an alternative method for establishing some of the results of Section 4.1 and uses the factorisation technique of Section 2.2.

Suppose that $L(Q) \equiv -\partial^2/\partial x^2 + Q$ has bound states $\{-k_n^2 < \ldots < -k_1^2\}$ and let $\beta > k_1$ with g a solution of

$$(L(Q) + \beta^2 I)g = 0 \tag{4.3-49}$$

Note that g is **not** an eigenfunction of $L(Q)$. It is easy to show that $L(Q) + \beta^2 I$ can be factorised into a product of first order operators,

$$L(Q) + \beta^2 I = C^A C \qquad \text{where } C \equiv \frac{\partial}{\partial x} + u \tag{4.3-50}$$

and

$$u^2 - u_x = Q + \beta^2 \tag{4.3-51}$$

From this factorisation it follows that we can select the particular function u defined by

$$u = -g_x g^{-1} \tag{4.3-52}$$

so that

$$C = g \frac{\partial}{\partial x} g^{-1} \tag{4.3-53}$$

A calculation then shows that

$$CC^A = L(Q') + \beta^2 I \tag{4.3-54}$$

where

$$Q'(x) = Q(x) - 2 \frac{\partial^2}{\partial x^2} g(x)$$

The formal equivalence defined by (4.3-54) can be rigorously justified within the Hilbert space framework of $L^2(R)$. Next observe that

$$(L(Q') + \beta^2 I)g^{-1} = -(g \frac{\partial}{\partial x} g^{-1})(g^{-1} \frac{\partial}{\partial g} g)g^{-1} = 0 \tag{4.3-55}$$

Then if

$$g_\alpha(x) = \alpha\phi(x,i\beta) + \psi(x,i\beta) \tag{4.3-56}$$

we have

$$g_\alpha(x) = \begin{cases} a(i\beta)\alpha e^{\beta x}(1+O(1)) & \text{as } x \to -\infty \\ a(i\beta)e^{-\beta x}(1+O(1)) & \text{as } x \to -\infty \end{cases} \tag{4.3-57}$$

β is not a zero of a so that g_α^{-1} is continuous and $g_\alpha^{-1} \in L^2(\mathbb{R})$. Consequently g_α^{-1} is the eigenfunction of $L(Q')$ for the eigenvalue $-\beta^2$. The result (4.3-57) which we have obtained by analytic continuation of the boundary conditions of the Jost solutions (Q assumed on compact support then this condition is relaxed) is directly established by Deift and Trubowitz (1979). The obvious conclusion that $\sigma(CC^A) = \{\sigma(C^AC), -\beta^2\}$ is justified by a basic fact in spectral theory (Deift 1978) that if B is a closed operator then $\sigma(BB^A)\backslash\{0\} = \sigma(B^AB)\backslash\{0\}$. We simply apply this result to C and then add in the additional eigenvalue.

The relationship between the Jost solutions of $L(Q)$ and $L(Q')$ is easily obtained from

$$C^ACy = 0 \Rightarrow (CC^A)Cy = 0 \tag{4.3-58}$$

We also have directly from the definition that

$$\int_{\pm a}^{\infty} |x|^j |Q'(\pm x)|\,dx = \int_{\pm a}^{\infty} |x|^j \left|-2\beta^2 + 2\left(\frac{g_{\alpha x}}{g_\alpha}\right)^2\right| dx + \int_{\pm \alpha}^{\infty} |x|^j |Q(\pm x)|\,dx \tag{4.3-59}$$

$$j = 1,2$$

$$\frac{g_{\alpha x}}{g_\alpha} = \begin{cases} \dfrac{\phi_x(i\beta)}{\phi(i\beta)} & \text{as } x \to +\infty \\[2ex] \dfrac{\psi_x(i\beta)}{\psi(i\beta)} & \text{as } x \to -\infty \end{cases} \tag{4.3-60}$$

Then we use Theorem 4-11 to get the result that

$$\left|-2\beta^2 + 2\left(\frac{g_{\alpha x}}{\alpha}\right)^2\right| = O\left(\frac{1}{|x|^4}\right) \quad \text{as} \quad |x| \to \infty \tag{4.3-61}$$

so that $\int_{-\infty}^{\infty}(1 + x^2)|Q'(x)|\,dx < \infty$. Finally we notice that the above arguments can be reversed to **remove** an eigenfunction from $L(Q)$.

<u>**Theorem 4-12:**</u> Let

(i) $\beta > k_j$, where $-k_j^2 \in \sigma(L)$, $j=1,\ldots,N$ and $\alpha > 0$ (adding a bound state) or

(ii) $\beta = k_N$ (removing a bound state).

For (i) define $g_\alpha(x) = \alpha\phi(x,i\beta) + \psi(x,i\beta)$
and $Q' = Q - 2\partial^2/\partial x^2 \ln g_\alpha$
then $\int_{-\infty}^{\infty}(1 + x^2)|Q'(x)|\,dx < \infty$ and $\sigma(L(Q')) = \{\sigma(L(Q)), -\beta^2\}$.

Furthermore, the scattering data for the two operators is related by

$$D'_{+j} = \left(\frac{\beta+k_j}{\beta-k_j}\right) D_{+j} \qquad j = 1,\ldots,N$$

$$D'_{+M+1} = \frac{2\beta}{\alpha} T_+(i\beta)$$

$$R'_+ = -\left(\frac{k+i\beta}{k-i\beta}\right) R_+$$

and the transmission coefficients by

$$T'_+ = \left(\frac{k+i\beta}{k-i\beta}\right) T_+$$

The function g_α^{-1} is the eigenfunction corresponding to $-\beta^2$ and the Jost solutions for $L(Q')$ are defined by

$$\phi'(x,k) = \frac{-1}{(ik-\beta)} g_\alpha \frac{\partial}{\partial x} g_\alpha^{-1} \phi(x,k)$$

$$\psi'(x,k) = \frac{1}{(ik-\beta)} g_\alpha \frac{\partial}{\partial x} g_\alpha^{-1} \psi(x,k)$$

For case (ii)

$$D'_{+j} = \frac{\beta - k_j}{\beta + k_j} D_{+j} \qquad j = 1,\ldots,N-1$$

$$R'_+ = -\left(\frac{k-i\beta}{k+i\beta}\right) R_+$$

and

$$T'_+ = \left(\frac{k-i\beta}{k+i\beta}\right) T_+$$

The Jost solutions in this case are related by

$$\phi'(x,k) = \frac{-1}{(ik+\beta)} \psi(x,ik\beta) \frac{\partial}{\partial x}\left(\frac{\phi(x,k)}{\psi(x,i\beta)}\right)$$

$$\psi'(x,k) = \frac{1}{(ik+\beta)} \psi(x,i\beta) \frac{\partial}{\partial x}\left(\frac{\psi(x,k)}{\psi(x,i\beta)}\right)$$

and

$$Q' = Q - 2\frac{\partial^2}{\partial x^2} \ln\psi(x,i\beta) \quad \text{with} \int_{-\infty}^{\infty} (1+x^2)\,|Q'|\,dx < \infty$$

$$\sigma(L(Q')) = \sigma(L(Q))\backslash -\beta^2$$

Proof:

All that remains to check is that the scattering data is related in the given manner.

For case (i) we have that

$$\phi'(x,k) = -(ik-\beta)\ g_\alpha \frac{\partial}{\partial x} g_\alpha^{-1}\phi(x,k)$$
$$\sim \frac{(\beta+ik)}{(ik-\beta)} a(k)e^{-ikx} - b(k)e^{+ikx} \text{ as } x \to -\infty \tag{4.3-62}$$

The result for the transmission coefficients follows directly from this and the result for R_+ is obtained from the definition.

The eigenfunction g_α^{-1} has the asymptotic behaviour,

$$g_\alpha^{-1} \sim \begin{cases} (\alpha a)^{-1}e^{-\beta x} & x \to +\infty \\ a^{-1}e^{\beta x} & x \to -\infty \end{cases} \tag{4.3-63}$$

When Q' is on compact support $\phi_j' = b_j'\psi_j'$ at a bound state, where $b_j' = b'(ik_j)$. From (4.3-63) it follows that $b_{N+1}' = -b_{N+1} = \alpha^{-1}$ and so

$$D_{+N+1} = \frac{b'_{N+1}}{i\dot{a}'_{N+1}} = \lim_{k\to i\beta}\left\{\frac{1}{i} R'_+(k-i\beta)\right\}$$
$$= -2\beta R_+(i\beta) = 2\beta\alpha^{-1}T_+(i\beta)$$

which is therefore also the result when Q' is not on compact support. The proofs for case (ii) are similar.

By repeated application of Theorem 4-12 one can obtain the results given earlier deduced from the generalised Bäcklund transformation. The fact that eigenvalues can be removed from $L(Q)$ enables the construction of a unique $Q^{(\text{reduced})}$ such that $\sigma(L(Q^{(\text{reduced})}))$ is the empty set.

The existence of a unique Q^{reduced} simplifies the existence and uniqueness proof of the solution to the Marchenko equation as indicated in Sections 4.1 and 4.2.

If the initial data is smooth then the results of this chapter indicate that the N-soliton evolves from initial data which has precisely the form (4.3-6) where $t = t$ is some fixed number. However as we indicated in Section 4.2, the situation is much more complex when the smoothness condition is relaxed.

A useful question to ask for arbitrary initial data is "how many solitons will emerge?" In Chapter 3 we gave as an estimate for the number of solitons

$$N \leqslant 1 + \int_{-\infty}^{\infty} |v|\ |F(v)|dv \tag{4.3-64}$$

where $F(v) = Q(x,0)$ is piecewise continuous. In the case of specific initial functions other estimates can be derived. Those most often given in the literature usually omit the effect of the continuous spectrum. Thus Berezin and Karpman (1966) and Karpman (1968) have used a finite number of the conserved densities $\{C_n\}$ to approximate the discrete spectrum. A simpler estimate on the number of eigenvalues can be obtained using the Bohr-Sommerfield quantisation rule. This works well provided the initial disturbance is large so that the approximation is valid. In this case we obtain (Landau and Lifshitz, 1965; pg.162).

$$\oint p\,dx = \oint (\lambda - F(x))^{\frac{1}{2}} dx = 2\pi(N + \tfrac{1}{2}) \tag{4.3-65}$$

where the integral is over a complete cycle of the periodic classical motion. We deduce that

$$N \approx \frac{1}{\pi} \int_{-\infty}^{\infty} |F(x)|^{\frac{1}{2}}\, dx \tag{4.3-66}$$

If $\lambda_o = -\eta_o^2$ is the eigenvalue of largest magnitude then in this case

$$|\lambda_0| \approx \max\{|F(x)|\} \tag{4.3-67}$$

An alternative more general expression for λ_o is obtained from the asymptotic expansion of $a(k)$ given in (3.3-73). If we assume that $|\lambda_o|$ is sufficiently large we obtain

$$\eta_0 \approx -\tfrac{1}{2}\int_{-\infty}^{\infty} F(v)\,dv \tag{4.3-68}$$

Better estimates are obtained by including further terms in the asymptotic expansion equating the left hand side to zero and solving the resulting algebraic equation. Equation (4.3-68) indicates that a condition for the absence of solitons from the initial data is that

$$\int_{-\infty}^{\infty} F(v)\,dv > 0 \tag{4.3-69}$$

Although the solution to the Marchenko equation for arbitrary initial data cannot be given in closed form it is possible to obtain the general features of a solution to a solvable equation in the long time asymptotic region when solitons are absent from the initial data. For the equations considered in this book it turns out that there is an asymptotic region where the long time behaviour is similar to that of linear dispensive systems. That is the asymptotic form of the solution is locally periodic with a slowly varying amplitude and phase. Solutions of this type can be obtained from a direct asymptotic analysis of the equation using multiple scales for example. However asymptotic methods alone cannot prove the validity of the solution or give the precise dependence upon the initial data. For this one must use some properties associated with the inverse method for the solvable equation. Two approaches, which overlap to some extent, appear in the literature. The first starts from the isospectral problem itself, and the second utilises the conservation laws associated with the equation. For the KdV the isospectral problem has been used by Zakharov and Manakov (1976), and the conservation laws by Ablowitz and Segur (1977). Because the first method works directly from the isospectral Schrödinger equation, it appears that one can obtain results for the whole solvable family. Indeed this is the case, but the results apply only to part of the asymptotic region for almost all initial data. The reason for this is that for the isospectral Schrödinger equation $R_+(0) = -1$ except in the special cases when an soliton is about to be added to the spectrum of L. This gives rise to an internal layer being included in the asymptotic region which can only be obtained from a separate analysis of each solvable equation. This situation does not arise for the solvable equations of the AKNS system which we also consider in Section 6.3 of this book.

For this reason and because of the complexity and length of the analysis we shall only give here the results for the KdV equation obtained by Ablowitz and Segur (1977). Their results are given for the KdV in the form

$$Q_t + 6QQ_x + Q_{xxx} = 0 \tag{4.3-70}$$

that is $Q \to -Q$ in our formulation (3.3-1).

The features of the solution are depicted in the Fig. 4-3 for $R_+(0) = -1$.

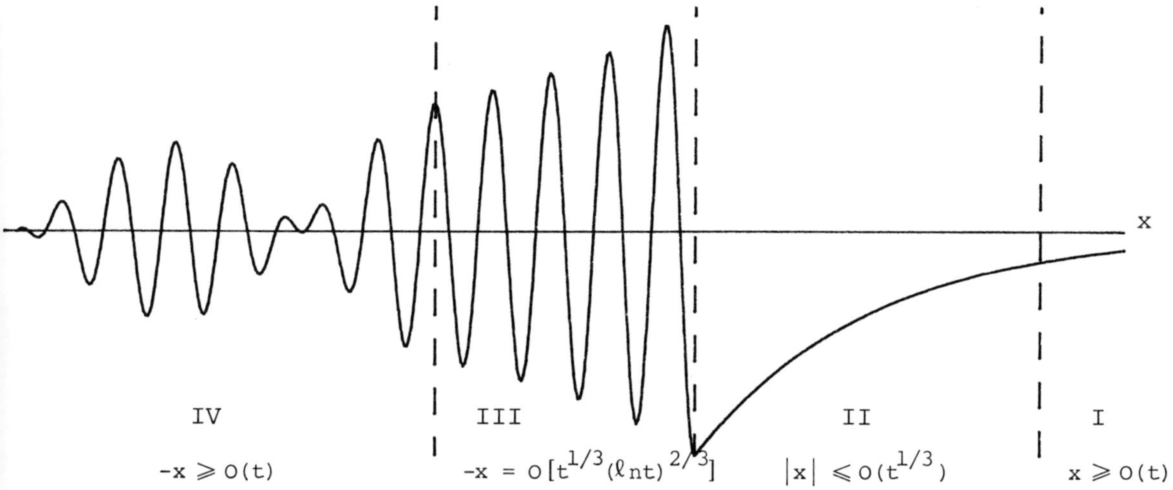

Figure 4-3: Long-time behaviour of solutions with soliton free initial data

The analysis leads to the recognition of four distinct asymptotic regions. Undetermined constants and functions are evaluated by matching the solutions on the overlap regions (see the chapter on matched asymptotic expansions in Nayfeh (1973) for example). Within each region the solution has the following character

I. $x \geqslant O(t)$

The solution is exponentially small and can be obtained by the method of steepest descents from the Marchenko equation.

$$Q(x,t) = \frac{1}{2(3\pi t)^{\frac{1}{2}}} R_+\left(\frac{i}{2}\left(\frac{x}{3t}\right)^{\frac{1}{2}}\right)\left(\frac{x}{3t}\right)^{\frac{1}{4}} e^{-2(x/3t)^{3/2}t}(1 + O(t^{-1}))$$

Here $X \equiv x/t = O(1)$ and t orders the expansion

II. $|x| \leqslant O(t^{1/3})$

The solution is self-similar and grows linearly as $(x/t^{1/3}) \to \infty$.

$$Q(x,t) = (3t)^{-2/3}F(\eta,t), \quad \eta = (3t)^{-1/3}$$

$$F(\eta,t) = f(\eta) + (3t)^{-1/3}f_1(\eta) + (3t)^{-2/3}f_2(\eta) + \ldots$$

where $f(\eta)$ satisfies the self-similar equation for the KdV

$$f_{\eta\eta\eta} + 6ff_{\eta} - (2f + \eta f_{\eta}) = 0$$

Two branches are of interest

(a) $|R_+(0)| < 1$, f oscillates as $\eta \to -\infty$.

$$f(\eta) \approx \frac{i}{\pi^{\frac{1}{2}}} \ln\{1 - |R_+(0)|^2\}^{\frac{1}{4}} \cos\theta$$
$$+ \frac{1}{2\pi} \ln\{1 - |R_+(0)|^2\}^{\frac{1}{2}} (-\eta)^{-\frac{1}{2}} (1 - \cos 2\theta)$$

where

$$\theta \approx \frac{2}{3}(-\eta)^{3/2} + \frac{3}{4\pi} \ln\{-\eta(|-|R_+(0)|^2)\} + \theta_0$$

(b) $|R_+(0)| = 1$ as $\eta \to \infty$ $f(\eta) = \frac{1}{2}\eta$

and as $\eta \to -\infty$

$$f(\eta) = \tfrac{1}{2}\eta - \tfrac{1}{2}(-2\eta)^{-\frac{1}{2}} + \tfrac{1}{2}(-2\eta)^{-2} - \frac{5}{2}(-2\eta)^{-7/2}$$
$$+ O((-2\eta)^{-5})$$

Case (b) is the situation which nearly always arises.

III. $(-x) = O\{t^{1/3}(\ln t)^{2/3}\}$

The ordering of the asymptotic series for F in II, Case (b) breaks down in the $\eta \to -\infty$ limit. The solution in this region corresponds to a collisionless shock. That is a thin region across which the asymptotic solution changes smoothly from one type of behaviour to another without any dissipative mechanism.

$$Q(x,t) = (3t)^{-2/3}(-2\eta)\, g(Z)$$

where

$$g(Z) \approx a(Z) + b(Z)\,\mathrm{cn}^2(2K(\nu)\theta + \theta_0, \nu(Z))$$

$$a = \frac{-\nu}{4(2-\nu)}, \qquad b = \frac{\nu}{2(2-\nu)}$$

$$Z = \ln(1 + \varepsilon\xi), \quad \varepsilon^{-1} = \ln 3t + \frac{5}{4}\ln(2\ln 3t)$$

$$2\xi = (-2\eta)^{3/2} - 2\ln 3t - \frac{15}{4}\ln(-2\eta), \quad \theta = \varepsilon^{-1}\int_0^Z w(z)e^z\,dz$$

$$w^2 = \frac{1}{(6K)^2(2-\nu)}$$

and

$$\frac{d\nu}{dZ} = -\frac{8}{3}\frac{(1-\nu/2)}{\nu}\left[(1-\frac{\nu}{2})\frac{E}{K} - 1 + \nu\right]$$

The range of the variable Z is $]0,\infty[$ and $K(\nu)$, $E(\nu)$ are respectively the complete elliptic integrals of the first and second kind.

IV. $(-x) \geqslant O(t)$

The solution in this region has the form

$$Q(x,t) = \tau^{-\frac{1}{2}}(2d)\,X^{\frac{1}{4}}\cos\theta$$

$$\theta = \tau(2/3\ X^{3/2} - (2d^2)\frac{\ln\tau}{\tau} + \frac{1}{\tau}(\theta_0 - 3d^2\ \ln X))$$

$$d^2(X) = -1/4\pi\ \ln(1 - |R_+(X^{\frac{1}{2}})|^2)$$

$$X = -x/3t,\quad \tau = 3t$$

The function θ_o which was not determined by Ablowitz and Segur can be determined by the Zakharov and Manakov method (see Section 6.3).

4.4 Notes

Section 4.1

1. After Bargmann's counter examples to the work of Fröberg and Hylleraas, Levinson (1949) showed that the lack of uniqueness was due to the presence of a discrete spectrum. Marchenko (1950,1952) showed that it was the spectral distribution function which determined the potential and thus related the problem to the inverse Sturm-Liouville problem which had already been treated in the literature. Analogous results were derived by Borg (1949), Jost and Kohn (1952) and Holmberg (1952). The procedure of reconstructing the potential from the spectral function by solving a linear integral equation is due to Gel'fand and Levitan (1951). The form of this integral equation which we use is due to Marchenko (1955). Necessary and sufficient conditions which the spectral function must satisfy were formulated by Krein (1953).

All of this work was concerned with the Schrödinger equation on the half-line. The extension of these results to the Schrödinger equation on the whole line was investigated by Kay and Moses (1956). A detailed analysis for the case that the reconstructed potential should be of the type

$\int_{-\infty}^{\infty} |Q(v)|(1+|v|)dv < \infty$ was given by Faddeyev (1964). Errors in this analysis were pointed out by Chadan and Sabatier (1977) and by Trubowitz and Deift (1979).

2. Newton presents two methods for the reconstruction of Q; one corresponds to solving a Gelfand-Levitan equation and the other is a Marchenko theory, which we omit from this brief resume. Essentially his method is based upon introducing functions $g_{\pm}(x,k)$ which are solutions to (3.3-1) and which moreover satisfy **regular**

boundary conditions at $x = 0$

$$g_+(0,k) = 1 \qquad g_{+x}(0,k) = ik$$

$$g_-(0,k) = 1 \qquad g_{-x}(0,k) = -ik$$

The theory then proceeds in a similar way as for the half-line case (Fadeeyev, 1963) by the introduction, in this case of a Jost matrix function J which relates the Jost solutions to the regular solutions

$$(g_+, g_-) = (\psi, \phi) J$$

There is a Povsner-Levitan representation for the $g_\pm$'s

$$g_\pm(x) = \exp(\pm ikx) - \int_{-x}^{x} dy\, H(x,y)\exp(\pm iky)$$

where H satisfies a wave equation analogous to that satisfied by $K_\pm$ when $Q_x \in L^1(R)$. The Gelfand-Levitan equation which he obtains is

$$H(x,y) - \Theta(x,y) + \int_{-x}^{x} H(x,u)\Theta(u,y)\,du = 0$$

where

$$\Theta(x,y) = \sum_{j=1}^{M} P_j(x)P_j(y) + \frac{1}{2\pi}\int_{-\infty}^{\infty} S(x,k)\exp(iky)\,dk$$

and

$$P_j(x) = (r_{j1}\exp(-\eta_j x) + r_{j2}\exp(\eta_j x))n_j^{-1}$$

$$S(x,t) = M_{11}(k)\exp(ikx) + M_{12}(k)\exp(-ikx)$$

with

$$M(k) = (J^A(k)J(k))^{-1} - I$$

$r_j = (r_{j1}, r_{j2})$ the row vector such that

$$_{i\eta_j}J\bar{r}_j = 0$$

and

$$n_j^2 = c_j^{-1}\{i(1,c_j)_{i\eta_j}\dot{J}\bar{r}_j\}\{-i\mathrm{Res}_{k=i\eta_j}T_+(k)\} \tag{1}$$

The uniqueness of the reconstructed Q is guaranteed by the uniqueness to the solution of (1) under the conditions of Theorem 4-3) . H is uniquely reconstructable from S_+. However, Newton`s approach is to show that J is uniquely reconstructable from $S(k)$ (red.) where S (reduced) is the scattering matrix constructed from S_+ with the omission of the eigenvalues from T_+. Thus solving the Gelfand-Levitan equation results in a potential Q(reduced). The eigenvalues are added back in by the Bäcklund transformation described in section 4.3 to recover the complete potential $Q(x)$.

Deift and Trubowitz adopt a completely different approach. They show that Q is defined in terms of the Jost solutions and the scattering data by a so-called trace formula

$$Q(x) = \frac{2i}{\pi} \int_{\infty}^{\infty} kR_+(k)\psi^2(x,k)\,dk + \sum_{j=1}^{M} \{2c_j \exp(-2\eta_j x)\}_x \psi_j^2(x)$$

They deal with the reduced Q and then add in the bound state solutions using the Bäcklund transformation. Thus using

$$\underset{(\text{reduced})}{Q(x)} = \frac{2i}{\pi} \int_{\infty}^{\infty} kR_+(k)\exp(2ikx)h^2(x,k)\,dk \tag{2}$$

where $h(x,k) = \exp(-ikx)\,\psi(x,k)$. The Schrödinger equation can be written as

$$\begin{pmatrix} h(x,k) \\ \ell(x,k) \end{pmatrix}_x = \begin{pmatrix} \exp(-2ikx)\ell(x,k) \\ \exp(2ikx)h(x,k)Q_+(x,h) \end{pmatrix} \qquad x \in R_{(3)} \tag{3}$$

where $Q_+(x,h)$ is defined as the expression on the right hand side of (2).

Thus the inverse problem in their method is to solve (3) with the initial (singular) data

$$\underset{x \to \infty}{\text{Lt}} \begin{pmatrix} h(x,k) \\ \ell(x,k) \end{pmatrix} = \begin{pmatrix} 1 \\ 0 \end{pmatrix} \tag{4}$$

That is to show that for a suitably defined R_+ the global solution to (3),(4) exists and is unique. Then one shows that Q(reduced) defined by (2) has reflection coefficient R_+.

Section 4.2

1. There is an extensive literature on the periodic and conditionally periodic problems. The article by Krichever (1977) probably provides the best survey and bibliography on these problems.

2. To establish Lemma 4-9 it is necessary to investigate the asymptotic behaviour of

$$\hat{R}_+(x,t) = \frac{1}{2\pi} \int_{-\infty}^{\infty} R_+(k,0) \exp(i\theta(k))\,dk$$

where $\theta = k(x - 2C(k^2)t)$. Throughout we assume that $C(k) = -\sum^{s} a_i R^{2i}$ and $a_i > 0$. Consider the case when $x > 1$. Introduce the operator D defined by

$$D[f(k)] = \frac{d}{dk}\left[\frac{f(k)}{x+\omega^{(1)}t}\right]$$

where $\omega = -2kC(k^2)$. Then integrating by parts we find that

$$\hat{R}_+(x,t) = \frac{1}{2\pi}(-i)^{-m} \int_{-\infty}^{\infty} D^m\{R_+(k,0)\}\exp(i\theta(k))\,dk$$

which is defined provided (a) $x + \omega^{(1)}$ is never zero on the range (this is satisfied since $x > 1$ and $0 \leqslant t$); (b) $D^m(R(k,0))$ exists and decays at least as fast as $O(1/|k|)$ as $|k| \to \infty$. Condition (b) is satisfied provided $m \leqslant \ell - 3$ (see Lemma 4-8). An analysis of the integrand in the last equation shows that the dominant term T_{hij} in the expansion for $D^m(R_+(k,0))$ which has $(x+\omega^{(1)}t)^{-(h+m)}$ as its denominator is

$$R_+^{(i)}(k,0)(\omega^{(2)})^j t^h \{x+\omega^{(1)}t\}^{-(h+m)}$$

The dominant part of this expansion has the form

$$R_+^{(i)}(k,0)k^{(2s-1)j}t^h\{x+\omega^{(1)}t\}^{-(h+m)}$$

where $i + h = m$, $0 \leqslant j \leqslant h \leqslant m$. Thus we get

$$\|T_{ijk}\| \leqslant \|R_+^{(i)}(k,0)\|_{L^1} \sup_k \{k^{(2s-1)j} t^h (x+\omega^{(1)}t)^{-(h+m)}\}$$

$$\leqslant (\text{constant})\,|x|^{-m-j/2s} t^{\{h-(2s-1)j/2s\}}$$

Finally from the inequality

$$|\hat{R}_+(x,t)| \leqslant (\text{constant})t^n x^{-m}$$

we obtain the behaviour of $\hat{R}_+(x,t)$ as $x,\ t \to +\infty$. One can easily establish the existence of

$$\hat{R}_{+x}(x,t) = \frac{1}{2\pi}\int_{-\infty}^{\infty} (ik)R_+(k,0)\exp(i\theta)\,dk$$

$$\hat{R}_{+xx}(x,t) = \frac{1}{2\pi}\int_{-\infty}^{\infty} (-k^2)R_+(k,0)\exp(i\theta)\,dk$$

from which we obtain using the method of the previous result, $0 < t < T$

$$|\hat{R}_+^{(i,o)}(x,t)| \leqslant (\text{constant})x^{-m+i/2s}$$

for $x > 1$. It follows that $|x^2\hat{R}_{+xx}(x,t)| \leqslant \text{const}\ x^{-\ell+5+1/s}$. The behaviour as $x \to -\infty$ is obtained by the stationary phase method. To leading order one finds that

$$\hat{R}_+^{(i,o)}(x,t) = o\{|x|^{-(j+n+1)/2s}\} \qquad \text{as } x \to -\infty$$

Section 4.3

1. This result was derived by Ablowitz, Kruskal and Segur (1979). The outline of their argument is as follows. For Q such that $\int_{-\infty}^{\infty}|Q|(1+v^2)dv < \infty$ define the potential Q_L which has compact support by $Q_L = Q.\ I_L,\ I_L = 1, |x| \leqslant L,\ 0$ otherwise. The function $\mu = \phi_x/\phi$ satisfies

$$\mu_x + \mu^2 - Q + k^2 = 0$$

In particular

$$\mu_L(-L,k) = -ik$$

$$\mu_L(L,k) = \frac{ik(e^{ikL} + R_{L+}(k)e^{ikL})}{(e^{-ikL} + R_{L+}(k)e^{ikL})}$$

from which we deduce that

$$e^{2ikL} R_{L+}(k) = \frac{ik + \mu_L(L,k)}{ik - \mu_L(L,k)}$$

Since we know from Section 3.3 that either

(i) $R_+(k) = -1 + O(k)$ as $k \to 0$

or

(ii) $R_+(0) = \text{const}, \quad (\text{const}) < 1$

it follows that there exists a perturbation expansion of μ for small k

$$\mu = \mu^0 + k\mu^1 + O(k^2) \quad \text{as} \quad k \to 0$$

We obtain from this that

(i) $\mu_L^0(L) \neq 0$

or

(ii) $\mu_L^0(L) = 0$

Since $\mu_L^0(L)$ is a complicated expression dependent upon Q_L, (ii) is a special case and case (i) is the usual or generic condition. Since Q can be approximated arbitrarily closely by Q_L by choosing L sufficiently large and R is continuous at $k = 0$ we can show that the convergence is uniform.

4.5 Problems

Section 4.1

1. Show that iteration of the equation (4.1-24)

$$\Omega_x(2x) = \tfrac{1}{2}Q(x) + \int_x^\infty P(x,y)\Omega_y(2y)dy$$

where

$$P(x,y) = 4(K(x,2y-x) + (xK^* xK)(y)$$

leads to the expression

$$\Omega_x(2x) = \tfrac{1}{2}Q(x) + \tfrac{1}{2}\int_x^\infty dx_1 P(x,x_1)Q(x_1)$$

$$+ \dots + \tfrac{1}{2}\left(\int_x^\infty dx_1 P(x,x_1)\int_{x_1}^\infty dx_2 P(x_1,x_2) \dots \int_{x_{n-1}}^\infty dx_n P(x_{n-1},x_n)Q(x_n)\right)$$

$$+ \left(\int_x^\infty dx_1 P(x,x_1)\int_{x_1}^\infty dx_2 \dots \int_{x_{n-1}}^\infty dx_n P(x_{n-1},x_n) \int_{x_n}^\infty \Omega_{x_{n+1}}(2x_{n+1}) \quad P(x_n,x_{n-1})dx_{n+1}\right)$$

Since $\frac{d}{dx}\hat{R}_+(x) = g_1 + g_2, \qquad g_1 \in L^1(R);\; g_2 \in L^2(R)$

show that $\int_{x_n}^{\infty} \Omega_{x_{n+1}}(2x_{n+1})P(x_n,x_{n+1})dx_{n+1} \;\varepsilon\; L^{\infty}(a,\infty), -\infty < a$

and that consequently the second bracket in (*) converges uniformly to 0 in $x \geq a$. Conclude that

$$|\Omega_x(2x) - \tfrac{1}{2}Q(x)| \leqslant C_1 R_0^{\,2}(x) \exp\left\{C_2\int_x^{\infty} R_0(y)dy\right\}$$

and thus establish the inequality (4.1-28).

2. Derive equation (4.1-89) from (4.1-83)-(4.1-84) and evaluate

$$\int_{-\infty}^{\infty}(\delta_1 Q(x)\delta_2 P(x) - \delta_1 P(x)\delta_2 Q(x))dx \quad \text{with } P(x) = \int_{-\infty}^{x} Q(y)dy$$

using the Wronskian relation

$$(v_x y - y_x v)_x = (k_1^{\,2} - k_2^{\,2})vy$$

where $y(k_1)$ and $v(k_2)$ are eigenfunctions of the Schrödinger equation (3.3-1). It is necessary to interpret the functions as generalised functions so that the integrals are defined. Simplifications of the resulting formulae are achieved through extensive use of the result from generalised function theory that

$$\underset{x\to\pm\infty}{P}\left\{\frac{e^{ikx}}{k}\right\} = \pm i\pi\delta(k)$$

Section 4.2

1. Derive the expressions for the functions $_x\pi_1(y)$, $\pi_2(y)$ given in (4.2-13) from (4.1-8) and (3.3-47)-(3.3-49)

2. Investigate the initial value problem for the KdV with the initial data

$$Q(x,0) = \begin{cases} 1 & |x| \leqslant a \\ 0 & \text{otherwise} \end{cases}$$

Section 4.3

1. Derive the two soliton solution for the second member in the KdV hierarchy of equations for which $\Omega(k) = -2ik^5$ from the superposition principle (4.3-33) and compare with the result directly obtained from the Marchenko equation (4.3-10).

2. Obtain an explicit formula for the asymptotic $(t\to\infty)$ break down of a two-soliton solution from the formula (4.3-10).

3. The KdV and the MKdV are scale invariant. Thus if $Q(x,t)$, $P(x,t)$ solve

$$Q_t - 6QQ_x + Q_{xxx} = 0$$

$$P_t - 6P^2P_x + P_{xxx} = 0$$

respectively then so do $Q'(x,t) = \lambda Q(\lambda^{\frac{1}{2}}x, \lambda^{3/2}t)$; $P'(x,t) = \gamma P(\gamma x, \gamma^3 t)$; λ, γ constants. It follows that the self-similar solutions which are invariant under this scaling have the form

$$Q(x,t) = (3t)^{-2/3}F(x(3t)^{-1/3}), P(x,t) = (3t)^{-2/3}G(x(3t)^{-1/3})$$

where F and G satisfy the equations

$$\dddot{F} = 6F\dot{F} + 2F + \theta\dot{F}$$

$$\ddot{G} = 2G^3 + \theta G + \nu$$

and $F = F(\theta)$, $G = G(\theta)$, $\theta = x(3t)^{-1/3}$, ν a constant. The equation satisfied by G is the Painleve II equation (Ince 1956). The relation between the F and G equations is easy to obtain from the Miura transformation (3.1-3),

$$Q_{\pm} = P^2 \pm P_x$$

which gives

$$F_{\pm} = G^2 \pm \dot{G}$$

The KdV is also Galilean invariant, $Q'(x,t) = Q(x + 6\beta t, t) + \beta$.

4. The Bäcklund transformation for the KdV equation of (4.3-4) is given in terms of the potential function $W_x = \frac{1}{2}Q$ by

$$(W + w)_x = k^2 + (W - w)^2$$

$$(W - w)_t = 6(W - w)^2(W - w)_x + 6k^2(W - w)_x - (W - w)_{xxx}$$

where W and w are solutions of the equation

$$W_t - 6W_x^2 + W_{xxx} = 0$$

Show that the parameter k in the transformation arises from the Galilean invariance of the KdV. The equation satisfied by W also possesses self-similar solutions. In fact they have the form $W = \frac{1}{2}(3t)^{-1/3}\Phi(\theta)$; $\theta = x(3t)^{-1/3}$ where Φ satisfies the equation

$$\dddot{\Phi} = 3\Phi^2 + \Phi + \theta\dot{\Phi}$$

We also have $\dot{\Phi}_{\pm} = G^2 \pm \dot{G}$. If we make the change of dependent variable $\Phi = -\frac{1}{4}\theta^2 + \Omega$ then the equation for Ω can be integrated once

$$\overset{\circ\circ\circ}{\Omega} = -2\theta\dot{\Omega} + \Omega + 3\dot{\Omega}^2$$

$$\ddot{\Omega}^2 = -2\theta\dot{\Omega}^2 + 2\Omega\dot{\Omega} + 2\dot{\Omega}^3 + n^2$$

where n is a constant of integration. Eliminating Ω between these equations leads to the conclusion that

$$\dot{\Omega}(\theta,n^2) = G^2(\theta, \pm n) + \dot{G}(\theta,\pm n) + \tfrac{1}{2}\theta$$

where G satisfies the PII equation with $\nu_{\pm} = \pm n - \frac{1}{2}$.

5. Using the relationships derived in question 4.3.4. show that

$$G(\theta, \pm n) = \frac{1}{2\dot{\Omega}(\theta,n^2)} (\ddot{\Omega}(\theta,n^2) \pm n)$$

$$\Omega(\theta,n^2) = \{(G^2(\theta, \pm n) + \tfrac{1}{2}\theta)^2 - \dot{G}^2(\theta, \pm n) \pm 2nG(\theta, \pm n)\}$$

provided $\Omega \neq 0$. If this is the case show that $n = 0$ so that $G(\theta,0)$ satisfies the Riccati equation

$$\dot{G} + G^2 + \tfrac{1}{2}\theta = 0$$

The Galilean transformation in terms of the function W is given by

$$W'(x,t) = W(x + 6\beta t,t) + \tfrac{1}{2}\beta x + \frac{3}{2}\beta^2 t$$

The function $\Omega(\theta)$ does not transform into a solution of the equation

$$\overset{\circ\circ\circ}{\Omega} = -2\theta\Omega + \Omega + 3\dot{\Omega}^2$$

under this transformation. It follows that the required form of the Bäcklund transformation for this equation is

$$\dot{\Omega} + \dot{\omega} = \theta + \tfrac{1}{2}(\Omega - \omega)^2$$

$$\dddot{\Omega} - \dddot{\omega} = \frac{3}{2}(\Omega-\omega)^2(\Omega-\omega) + (\Omega-\omega) + \theta(\dot{\Omega}-\dot{\omega})$$

Show that these equations imply

$$\omega(\theta,m^2) = \Omega(\theta,n^2) - \frac{1}{\Omega(\theta,n^2)} (\ddot{\Omega}(\theta,n^2) + n^2)$$

where $m^2 = (n-1)^2$. Deduce from this relation that

$$G(\theta, 1-n) = -G(\theta, n)$$

and furthermore from the expression

$$G(\theta, \pm n) = \{2\mathring{\Omega}(\theta, n^2)\}^{-1} \, (\mathring{\mathring{\Omega}}(\theta, n^2) \pm n)$$

that

$$G(\theta, -n) = G(\theta, n) + \frac{n}{\dot{G}(\theta, n) + G^2(\theta, n) + \tfrac{1}{2}\theta}$$

6. Use the Bäcklund transformation

$$G(\theta, 1-n) = -G(\theta, n)$$

$$G(\theta, -n) = G(\theta, n) + \frac{n}{\dot{G}(\theta, n) + G^2(\theta, n) + \tfrac{1}{2}\theta}$$

for the Painleve' II equation

$$\ddot{G} = 2G^3 + \theta G + (n - \tfrac{1}{2})$$

to generate a family of solutions starting from $G = 0$ for $n = \tfrac{1}{2}$. Thus $G(\tfrac{3}{4}, -\tfrac{1}{2}) = \theta^{-1}$, $G(\theta, -3/2) = -\theta^{-1} + \frac{3\theta^2}{4+\theta^2}$ are the first two members of this family.

7. Obtain the leading order asymptotic form of the solution to an arbitrary member of the KdV hierarchy by the stationary phase method and its region of validity for initial data from which solitons are absent.

8. As was shown in Question 4.3.3. the MKdV

$$P_t - 6P^2P_x + P_{xxx} = 0$$

possesses a similarity solution $P(x,t) = (3t)^{-1/3}G(\theta); \theta = x(3t)^{-1/3}$. For initial data from which solitons are absent and which decays exponentially as $|x| \to \infty$, show that the MKdV possesses a "slowly varying similarity solution" of the form

$$P \approx \frac{(-\theta)^{\frac{1}{4}} d}{(3t)^{1/3}} \sin\left(\frac{2}{3}(-\theta)^{3/2} - \frac{3}{4} d^2 \ln(-\theta) + \phi_0\right)$$

for $\theta \ll -1$.

Is it possible to account for the presence of the $\ln$ term in the phase by a general argument? Show that this solution matches onto the similarity solution which is valid for $\theta = O(1)$.

9. Use Miura's transformation to obtain the asymptotic form of a solution without solitons in the region $\theta \leqslant O(1)$ to the KdV from the results of Question 4.3.8. In

articular show that the slowly varying similarity solution is

$$Q \approx \frac{d(\theta)^{\frac{1}{4}}}{(3t)^{2/3}} \cos\left(\frac{2}{3}(\theta)^{3/2} - \frac{3d^2}{4}\ln(\theta) + \phi_0\right)$$

here $Q_t - 6QQ_x + Q_{xxx} = 0$.

Compare with the results of Section 4.3. (after rescaling the equation). Explain he difference in the results (hint: the generic case is $R_+(0) = -1$).

. ISOLATING THE KORTEWEG-DE VRIES EQUATION IN SOME PHYSICAL EXAMPLES

.1 Introduction

The last two chapters have concentrated on the techniques necessary in olving the KdV equation, showing in particular the correspondence between olitons and the discrete eigenvalues of the potential of the quantum chrödinger equation. We now turn our attention to investigating a few cases here the KdV equation arises as a realistic model governing the evolution of aves in media in which weak nonlinear effects are important. We shall oncentrate solely on the KdV equation and its generalisations in this hapter, in order to provide some continuity with the previous three. arallel chapters are included later which will deal with the sine-Gordon, onlinear Schrödinger and other associated equations which have similar roperties.

The examples given here will be restricted to four: partly because of lack f space and partly because the ones with which we shall deal are fairly ypical of the many cases which may be reduced to the KdV equation. Similar ethods can be used to analyse other parallel examples which will be eferenced at the end of this chapter.

The first example is a fairly straightforward one, which occurs in plasma hysics where the KdV equation governs the evolution of long compressive aves in a plasma of cold ions and hot electrons. We begin with this example s it is probably the simplest on which to demonstrate how the perturbation ethods work. The shallow water wave problem is placed second because of its ength: it takes some space to set up the problem before the Korteweg and de ries result of 1895 can be established. Thirdly, an interesting case occurs n meteorology in studies of the propagation of nonlinear Rossby waves in a omogeneous rotating fluid. This latter case is slightly different and has ome appealing features mathematically in that a second space dimension (y) ccurs in the original equations and the coefficients of the final KdV quation are found to be integrals over y. The fourth and final example is aken from electric circuit theory in which a nonlinear capacitance is used. ne final result is a generalized nth order KdV equation, the nonlinearity epending on the capacitance. We have chosen this example to illustrate how ne modified KdV equation can arise in certain circumstances.

The KdV equation itself is a single scalar equation involving one ependent variable and two independent variables and as such, is quite simple n structure. However, the original equations of motion of most physical ystems are not so simple and generally contain several dependent variables. or instance, in fluid mechanics the governing equations are the Navier-tokes equations which contain the fluid density $\rho(x,t)$ and the fluid velocity vector $\underline{v}(x,t)$ and perhaps several othervariables and equations o state as well, depending on whether thermodynamic considerations are take into account. We need a procedure which, in a systematic fashion, wil reduce such sets of equations to simpler forms. Such procedures are usuall perturbative in nature and so the title of reductive perturbation theory i usually applied. One of the useful features of this form of perturbatio theory is that it enables us to look in a natural way for long waves; tha is, waves whose wavelengths are long compared to a typical length scale. Fo instance, in the quote given in Chapter 1, Scott Russell noted that th solitary wave which evolved from the mass of water at the prow of the boa was about thirty feet long and a foot to a foot and one half in height - typical long wave for a wave in a canal. On the mathematical level, in orde to build this length scale into the original equations of motion, we need t rescale both space and time in order to introduce space and time variable which are appropriate for the description of long wavelength phenomena. Thi rescaling enables us to isolate from the system the relevant equations o motion which describe how the system reacts on the new space and time scales It turns out that for a large class of dispersive (not dissipative) systems the KdV equation is the equation which governs such weak nonlinear long wav behaviour. The reduction process is slightly ill-defined in that it rests o experience in knowing how to pick the relevant scales. Before starting, i makes the process easier if all the variables in the problem are scaled t dimensionless form which enables us to remove some of the awkward physica constants from the equations. Next we expand all the dependent variables i terms of a perturbation parameter ε. For example

$$\rho = \rho^{(0)} + \varepsilon\rho^{(1)} + \varepsilon^2\rho^{(2)} \tag{5.1-1}$$

$$\underline{v} = \varepsilon\underline{v}^{(1)} + \varepsilon^2\underline{v}^{(2)} \tag{5.1-2}$$

The presence or absence of a first term can usually be determined by th boundary conditions. In most cases, for example, the density is normall perturbed about its equilibrium value and so $\rho \to \rho_0$ as $x \to \infty$ wherea $\underline{v} \to 0$. How the choice is made depends very much on the physica circumstances. The next thing to determine is the scaling of x and t Given that x is scaled in a certain way, then it is the dispersio relation which gives the information of how the time part of the syste reacts. The dispersion relation for harmonic waves can easily be found fro the linearised version of the original set of equations. Solutions will b in the form $\exp(i\theta)$ where $\theta = kx - \omega(k)t$. The function $\omega(k)$ satisfie the dispersion relation and k is the wave number. If we are looking fo long waves however, then this corresponds to waves with small wave number k and hence long wave length. Even though such waves are not harmonic waves we can nevertheless use the limiting form of the dispersion relation in th long wave limit. Consequently we write k as $k = \varepsilon^p\kappa$ where κ is a ne wave number of $O(1)$ and p is, as yet, some unknown number which is to b determined later. Hence $\theta(x,t)$ can now be written $\theta(x,t) = \kappa\varepsilon^p x - \omega(\varepsilon^p\kappa)t$. We are considering here only dispersive and no dissipative systems and hence a Taylor expansion of $\omega(k)$ will yield eithe all even or all odd terms in k. Purely dispersive systems cannot have mixture of both odd and even terms. The examples in this chapter are of the sort in which only odd terms in k appear, and so the first two terms in a Taylor expansion for $\omega(k)$ gives $\omega(k) = \omega'(0)\varepsilon^p\kappa + \omega'(0)\varepsilon^{3p}\kappa^3$. The function θ can now be written as

$$\theta = \kappa\varepsilon^p(x-\omega'(0)t) - \kappa^3\varepsilon^{3p}\omega'''(0)t \qquad (5.1\text{-}3)$$

Since the first and third derivative terms in ω are constants, one of them can be scaled out and (5.1-3) then yields a natural scaling for x and t

$$\xi = \varepsilon^p(x-at); \qquad \tau = \varepsilon^{3p}t \qquad (5.1\text{-}4)$$

The new variables ξ and τ are long in the sense that it needs a large change in x and t in order to change ξ and τ appreciably. In order to determine the value of p (which will not necessarily be an integer) a plausibility argument is needed. When the basic equations are expanded in powers of ε and space and time are also rescaled as in (5.1-4), then a suitable choice for the value of p often becomes apparent. Intuitively it can be seen that if p is chosen too large then derivatives in τ will not occur until higher orders of ε, which would imply that many of the dependent variables (such as some of the lower order $\rho^{(n)}$ and $v^{(n)}$) would appear to be independent. This is undesirable since in order to obtain an evolution equation it would be necessary to consider higher order terms in the perturbation expansions. It often turns out that when the KdV equation occurs then p usually takes the value of $\frac{1}{2}$.

The procedures explained above are very much a rule of thumb and as we do not intend to spend too much time on this topic the interested reader is recommended to read some specialised texts on the subject such as Nayfeh (1974), Cole (1968) and Bender and Orszag (1978). Many more complicated examples arise in other areas of applied mathematics where different types of asymptotic expansions are used in different regions of space. The scalings of the space and time variables may also change from region to region which gives rise to the need to match across the boundaries. However, the four examples in this chapter are much simpler and are purely restricted to obtaining the KdV equation as a model equation which governs long wave behaviour.

5.2 Ion Acoustic Waves

Consider a one-dimensional sea of electrons each of mass m_e and charge $-e$ with density n_e per unit volume and ions each of mass m_i and charge $+e$ with density n_i per unit volume. This is technically known as a plasma of electrons and ions. Since the electron mass is much lighter than any ion mass, the electron inertia can be neglected but the electrostatic effect of the electron charge cannot be neglected. For this reason we need some means of expressing this effect and the usual method is to treat the electrons as a "gas" (see Pines 1968). An electron gas can be thought of as a many-body problem and in an idealised situation may be described by an equation of state

$$P = k_B T_e n_e \tag{5.2-1}$$

where k_B is known as Boltzmann's constant and P is the pressure. The value of the electron temperature T_e gives a measure of how energetic (hot) the electrons in the gas are. For the type of situation we are considering here, the electron temperature is usually greater than the ion temperature. Hence we have a gas of charged energetic (hot) electrons superimposed on a background of much more massive but less energetic (cold) ions $(T_i << T_e)$. In order to model this situation, we require some equation which relates the electron density n_e to the electrostatic potential ϕ. To obtain this relationship we note that the electric force on the electrons due to the potential $\phi(x,t)$ is $en_{e,x}$ and in the electron gas this electrostatic force is balanced by the pressure gradient

$$en_e \frac{\partial \phi}{\partial x} = k_B T_e \frac{\partial n_e}{\partial x} \tag{5.2-2}$$

which integrates to

$$n_e = n_o \exp\left[\frac{e}{k_B T_e} \phi\right] \tag{5.2-3}$$

where n_0 is the equilibrium background density. It is not necessary to have any other equations for the state of the electrons since (5.2-3) provides the necessary coupling betwen the ions and electrons through the potential ϕ.

For the ions the equations of conservation of mass and momentum are

$$\frac{\partial n_i}{\partial t} + \frac{\partial}{\partial x}(n_i v_i) = 0 \tag{5.2-4}$$

$$m_i \left(\frac{D}{Dt} v_i\right) = - e \frac{\partial \phi}{\partial x} \tag{5.2-5}$$

where the total derivative is given by

$$\frac{D}{Dt} = \frac{\partial}{\partial t} + v_i \frac{\partial}{\partial x} \tag{5.2-6}$$

For the electrostatic potential ϕ, Poisson's equation is

$$\frac{\partial^2 \phi}{\partial x^2} = 4\pi e(n_e - n_i) \tag{5.2-7}$$

The type of perturbation scheme mentioned in Section 5.1 is much simpler if all the unpleasant constants are scaled out and new variables introduced. The process is quite easy if applied in successive stages. Equations (5.2-3) and (5.2-7) indicate that ϕ and n_i can be rescaled as

$$\Phi = \frac{e}{k_B T_e} \phi \qquad n = n_i/n_o \tag{5.2-8}$$

Using these rescaled variables again in (5.2-7) it is easy to show that a new dimensionless x-variable can be introduced

$$\bar{x} = \lambda^{-1} x \text{ where } \lambda = \left(\frac{4\pi n_o e^2}{k_B T_e} \right)^{-\frac{1}{2}} \tag{5.2-9}$$

The constant λ is known as the Debye length for the plasma. Using the expressions given above for Φ, $\bar{n}$ and $\bar{x}$ in (5.2-4) and (5.2-5) it is easy to find dimensionless velocity and time variables

$$\bar{t} = \omega_p t; \quad v = (\lambda\omega_p)^{-1} v_i \; ; \; \omega_p = \left(\frac{4\pi n_o e^2}{k_B T_e} \right)^{\frac{1}{2}} \tag{5.2-10}$$

where ω_p is known as the plasma frequency, and $\lambda\omega_p$ as the ion sound speed. Equations (5.2-4) and (5.2-5) now become the much simpler set of dimensionless equations of motion

$$\begin{aligned} n_{\bar{t}} + (nv)_{\bar{x}} &= 0 \\ v_{\bar{t}} + vv_{\bar{x}} &= -\Phi_{\bar{x}} \\ \Phi_{\bar{x}\bar{x}} &= \exp\Phi - n \end{aligned} \tag{5.2-11}$$

Boundary conditions are now $n \to 1$; $v, \Phi \to 0$ as $|x| \to \infty$.

We now apply the type of procedure outlined in Section 5.1 and make asymptotic expansions for n, Φ and v:

$$\begin{aligned} n &= 1 + \varepsilon n^{(1)} + \varepsilon^2 n^{(2)} \ldots & n^{(i)} &\to 0 \text{ as } |x| \to \infty \\ \Phi &= \varepsilon\Phi^{(1)} + \varepsilon^2\Phi^{(2)} \ldots & \Phi^{(i)} &\to 0 \text{ as } |x| \to \infty \\ v &= \varepsilon v^{(1)} + \varepsilon^2 v^{(2)} \ldots & v^{(i)} &\to 0 \text{ as } |x| \to \infty \end{aligned} \tag{5.2-12}$$

The expansions in (5.2-12) can be used to linearise (5.2-11). Eliminating $v^{(1)}$ and $n^{(1)}$ yields for $\Phi^{(1)}$

$$\Phi^{(1)}_{\bar{x}\bar{x}\bar{t}\bar{t}} + \Phi^{(1)}_{\bar{x}\bar{x}} - \Phi^{(1)}_{\bar{t}\bar{t}} = 0 \tag{5.2-13}$$

which has the dispersion relation: $\omega^2 = k^2(1 + k^2)^{-1}$. Therefore for small k $(k = \varepsilon^p \kappa\, ;\; p > 0)$ the first two terms in the expansion of $\omega(k)$ are the k and k^3 terms. Using the arguments of Section 5.1, we therefore rescale x and t by defining ξ and τ to be

$$\xi = \varepsilon^p(x - a\bar{t}) \qquad \tau = \varepsilon^{3p}\bar{t} \tag{5.2-14}$$

Equations (5.2-11) written out in full are now

$$0 = \left(\varepsilon^{3p}\frac{\partial}{\partial\tau} - a\varepsilon^{p}\frac{\partial}{\partial\xi}\right)[1 + \varepsilon n^{(1)} + \varepsilon^2 n^{(2)} \ldots] + \varepsilon^{p}\frac{\partial}{\partial\xi}[\varepsilon v^{(1)} + \varepsilon^2(v^{(2)} + n^{(1)}v^{(1)})\ldots] \tag{5.2-15}$$

$$0 = \left(\varepsilon^{3p}\frac{\partial}{\partial\tau} - a\varepsilon^{p}\frac{\partial}{\partial\xi}\right)\left(\varepsilon v^{(1)} + \varepsilon^2 v^{(2)} \ldots\right) + \tfrac{1}{2}\varepsilon^{p}\frac{\partial}{\partial\xi}[\varepsilon^2(v^{(1)})^2] + \varepsilon^{p}\frac{\partial}{\partial\xi}[\varepsilon\phi^{(1)} + \varepsilon^2\phi^{(2)} \ldots] \tag{5.2-16}$$

$$0 = \varepsilon^{2p}\frac{\partial^2}{\partial\xi^2}(\varepsilon\Phi^{(1)} + \varepsilon^2\Phi^{(2)} \ldots) - \varepsilon(\Phi^{(1)} - n^{(1)}) - \varepsilon^2[\Phi^{(2)} - n^{(2)} + \tfrac{1}{2}(\Phi^{(1)})^2]\ldots \tag{5.2-17}$$

For the moment we will gather terms and write down the coefficients of the various powers of ε without, as yet, setting any to zero, using the convention that subscripts refer to partial derivatives. For equation (5.2-15)

$$\varepsilon^{p+1}: \qquad -an^{(1)}_{\xi} + v^{(1)}_{\xi} \tag{5.2-18}$$

$$\varepsilon^{p+2}: \qquad -an^{(2)}_{\xi} + v^{(2)}_{\xi} + (n^{(1)}v^{(1)})_{\xi} \tag{5.2-19}$$

$$\vdots$$

$$\varepsilon^{3p+1}: \qquad n^{(1)}_{\tau} \tag{5.2-20}$$

For equation (5.2-16)

$$\varepsilon^{p+1}: \qquad -av^{(1)}_{\xi} + \Phi^{(1)}_{\xi} \tag{5.2-21}$$

$$\varepsilon^{p+2}: \qquad -av^{(2)}_{\xi} + \Phi^{(2)}_{\xi} + v^{(1)}v^{(1)}_{\xi} \tag{5.2-22}$$

$$\vdots$$

$$\varepsilon^{3p+1}: \qquad v^{(1)}_{\tau} \tag{5.2-23}$$

For equation (5.2-17)

$$\varepsilon: \qquad -(\Phi^{(1)} - n^{(1)}) \tag{5.2-24}$$

$$\varepsilon^{2}: \qquad -[\Phi^{(2)} - n^{(2)} + \tfrac{1}{2}\Phi^{(1)^2}] \tag{5.2-25}$$

$$\varepsilon^{2p+1} \qquad \Phi^{(1)}_{\xi\xi} \tag{5.2-26}$$

$$\varepsilon^{2p+2} \qquad \Phi^{(2)}_{\xi\xi} \tag{5.2-27}$$

We naturally need p to be positive and so using the boundary conditions the lowest order terms (i.e. ε^{p+1} and ε) just give

$$n^{(1)} = v^{(1)} = \Phi^{(1)} \tag{5.2-28}$$

for $a = +1$. For the moment we shall only consider $a = 1$ and mention $a = -1$ later. The result (5.2-28) is independent of the value of p $(p > 0)$ so to determine p we need to go to the next order. Equations (5.2-18) to (5.2-23) indicate that if $3p + 1 > p + 2$ then no τ-derivative occurs at all at order $p + 2$. This is unsatisfactory because $3p + 1 > p + 2$ implies that $p > \frac{1}{2}$ which means that in (5.2-26) the second derivative on $\Phi^{(1)}$ at order ε^{2p+1} is of a higher order than ε^2. From the three equations, $\Phi^{(2)}, n^{(2)}$ and $v^{(2)}$ can easily be eliminated to show that $n^{(1)} = 0$. If we accept this result then it would be necessary to go to higher orders of perturbation theory to obtain an evolution equation for $n^{(1)}$. However, if we set $3p + 1 = p + 2$, then $p = \frac{1}{2}$ and the $n^{(1)}$ and $v^{(1)}$ terms are of the same order as the ε^{p+2} terms where quadratic nonlinearities in $n^{(1)}$ occur. This result in the third set of equations is also satisfactory since it matches the quadratically nonlinear term $(\Phi^{(1)})^2$ with the $\Phi^{(1)}_{\xi\xi}$ dispersive term.

Therefore setting $p = \frac{1}{2}$, replacing $v^{(1)}$ and $\Phi^{(1)}$ with $n^{(1)}$ and setting terms at each order to zero we obtain the equations:-

$$v^{(2)}_{\xi} - n^{(2)}_{\xi} + 2n^{(1)}n^{(1)}_{\xi} + n^{(1)}_{\tau} = 0 \tag{5.2-29}$$

$$\Phi^{(2)}_{\xi} - v^{(2)}_{\xi} + n^{(1)}n^{(1)}_{\xi} + n^{(1)}_{\tau} = 0 \tag{5.2-30}$$

$$n^{(1)}_{\xi\xi} - \tfrac{1}{2}[n^{(1)}]^2 = \Phi^{(2)} - n^{(2)} \tag{5.2-31}$$

Eliminating $v^{(2)}$ from (5.2-29),(5.2-30) by addition we have the ξ-derivative of $\Phi^{(2)} - n^{(2)}$ which, fortunately, we have also in (5.2-31). The three equations together give

$$\tfrac{1}{2}n^{(1)}_{\xi\xi\xi} + n^{(1)}n^{(1)}_{\xi} + n^{(1)}_{\tau} = 0 \tag{5.2-32}$$

which is exactly the KdV equation. The solitons are compressive waves because the scaling given in (5.2-14) implies that we are already in a frame of reference of speed unity (the ion sound speed) and the soliton velocities are positive for the form of the KdV equation given in (5.2-32). The choice of $a = -1$ alters the result slightly in that $v^{(1)} = -n^{(1)} = -\Phi^{(1)}$ and the resulting equation is

$$\tfrac{1}{2}n^{(1)}_{\xi\xi\xi} + n^{(1)}n^{(1)}_{\xi} - n^{(1)}_{\tau} = 0 \tag{5.2-33}$$

This form of the KdV equation shows that the solitons move to the left as (5.2-14) also implies when $a = -1$ (time reversal).

The physical interpretation of this result in the light of the previous chapters is fairly simple. If, by means of an electrostatic probe, a disturbance is introduced into an initially uniform plasma (which constitutes an initial value problem for the KdV equation) the number of solitons which will emerge is exactly the number of bound states of the initial disturbance. For instance, initial data in the form of a square wave, which is easy to produce experimentally, will break down into as many solitons as it has discrete eigenvalues. This has been tested by Hershkowitz, Romesser and Montgomery (1972) who calculated theoretically the number and energy of the bound states of a given square wave input and then proceeded with this same input experimentally. The calculation of the discrete eigenvalues of square wave initial data is given in the chapter exercises (see also the calculations and exercises of Chapter 2).

5.3 Long Waves in Shallow Water

The problem of analysing surface water waves is a complicated one since as well as being a moving boundary problem, different types of wave motion can occur, depending on whether the waves in question are large or small in amplitude in ratio to the depth. The contents of this section are such that we restrict ourselves to long surface waves in a shallow fluid while ignoring frictional effects; a situation which is relevant to Scott Russell's experiments and one which was investigated theoretically by de Boussinesq (1877), Korteweg and de Vries (1895) and others. Here we are only sampling a very extensive subject and the interested reader is recommended to look at other more advanced texts which deal with a variety of water wave motions, such as the relevant chapters of such books as Lamb (1932), Stoker (1957), Whitham (1974) and Lighthill (1978).

The subject has a somewhat long and distinguished history in that it goes back to the time of Sir George Stokes and before, to the early middle part of the last century. Scott Russell's solitary wave observations pre-date the Navier-Stokes equations by a short time and as we saw in Chapter 1, his results remained dormant for a long time before their real significance was understood. The fact that some quantum mechanics is necessary to solve the problem in full makes it a very modern one despite its long pedigree.

It will take some time to set the problem up in a suitable form but our main aim, as we mentioned in Section 5.1, is to obtain the simplest set of equations of motion, having scaled all the variables into dimensionless form. Once this has been achieved, the perturbation procedure is easier to handle.

Consider an inviscid, incompressible fluid of density ρ with velocity vector $v = (v_1, v_2, v_3)$ in a co-ordinate system (x,y,z). The bottom of the fluid is at $z = -h$ and the undisturbed surface at $z = 0$. We shall ignore frictional forces and so gravity is the only external force acting on the fluid: $F = -\rho g\hat{k}$, where $\hat{\underline{i}}, \hat{\underline{j}}, \hat{\underline{k}}$ are the usual unit vectors. Only two basic equations of motion are necessary. Because of incompressibility

$$\text{div } \underline{v} = 0 \tag{5.3-1}$$

and the momentum equation is

$$\rho\left[\frac{\partial}{\partial t} + \underset{\sim}{v}.\underline{\nabla}\right]\underline{v} = -\underline{\nabla}P - g\rho\hat{\underline{k}} \tag{5.3-2}$$

where P is the pressure in the fluid. For irrotational motion $\text{curl}\,\underline{v} = 0$ which implies that a potential function ϕ exists such that $\underline{v} = \nabla\phi$. Furthermore, because $\underline{v}\text{x curl } \underline{v} = 0$ we have

$$(\underline{v}.\underset{\sim}{\nabla})\underline{v} = \underline{\nabla}[\tfrac{1}{2}(\nabla\phi)^2] \tag{5.3-3}$$

Using this result in (5.3-2) and integrating we find that our two equations of motion are now

$$\nabla^2\phi = 0 \tag{5.3-4}$$

$$P - P_o = -\rho\{\phi_t + \tfrac{1}{2}(\underset{\sim}{\nabla}\phi)^2 + gz\} \tag{5.3-5}$$

where P_o is the atmospheric pressure acting on the surface of the fluid. Clearly we should also have an arbitrary function of time as an integration constant in (5.3-5) but this can be removed by absorbing it into ϕ so we shall ignore it. The main complication with this problem is that we are really more interested in what happens at the surface and less interested in the value of ϕ or P somewhere inside the fluid. It is the motion of the upper boundary which will give the evolution equation for long surface waves and consequently we are in the position of wanting the evolution of the boundary given some solution in the fluid and not the other way round. We therefore specify the upper free surface (which is unknown at present) by the equation

$$\Gamma(x,y,z,t) = z - S(x,y,t) = 0 \tag{5.3-6}$$

Our task is to determine the equation for S. Any deformation of the surface will move those particles which form the surface but by the very fact that they are the surface, they will return to their original positions when the effect of the wave has disappeared. Intuitively we would therefore expect the total derivative $D\Gamma/Dt$ of Γ to be zero. To prove this mathematically is quite easy. Given the surface $z = S(x,y,t)$ then $\hat{\underline{n}}$ is an inwarding pointing unit normal where

$$\hat{\underline{n}} = \frac{\hat{\underline{i}}\, S_x + \hat{\underline{j}}\, S_y - \hat{\underline{k}}}{[1+S_x^2 + S_y^2]^{\frac{1}{2}}} \tag{5.3-7}$$

The normal fluid velocity is $\underline{v}.\hat{\underline{n}}$. The velocity of the surface $z = S(x,y,t)$ is

$$\frac{-S_t}{[1+S_x^2+S_y^2]^{\frac{1}{2}}} \tag{5.3-8}$$

and so equating the two yields the result

$$v_1 S_x + v_2 S_y + S_t = v_3 \tag{5.3-9}$$

which is $D\Gamma/Dt = 0$.

Since $\underline{v} = \nabla\phi$, equations (5.3-9) can be expressed as

$$\phi_x S_x + \phi_y S_y + S_t = \phi_z \qquad \text{on } z = S(x,y,t) \tag{5.3-10}$$

Note that this is a boundary condition and is only valid on $z = S$ but not in the whole fluid. Bernoulli's equation, (5.3-5), again only gives us any interesting information at the surface. Ignoring changes in air pressure, which are negligible, the pressure terms **at the surface** can be written

$$P - P_o = - T(S_{xx} + S_{yy}) \tag{5.3-11}$$

where the term on the right of (5.3-11) is a surface tension term (Korteweg and de Vries 1895; Whitham 1974, Chapter 13). For fluids such as water, surface tension forces are quite small so we shall neglect them for the moment in order to keep the calculation as simple as possible and return to them later on in this section. At the free surface, equation (5.3-5) is now

$$0 = \phi_t + \tfrac{1}{2}(\underset{\sim}{\nabla}\phi)^2 + gS \qquad \text{when} \quad z = S(x,y,t) \tag{5.3-12}$$

The final boundary condition requires that the normal component of $\underline{v}$ must be zero on the bottom of the fluid: i.e. $\phi_z = 0$ on $z = -h$.

Finally, the complete set of equations is

$$\nabla^2\phi = 0 \qquad \text{in the fluid } (-h < z < 0) \tag{5.3-13}$$

$$\phi_x S_x + \phi_y S_y + S_t = \phi_z \tag{5.3-14}$$

$$\phi_t + \tfrac{1}{2}[\phi_x^{\,2} + \phi_y^{\,2} + \phi_z^{\,2}] + gS = 0 \qquad \text{when } z = S(x,y,t) \tag{5.3-15}$$

$$\phi_z = 0 \qquad z = -h \tag{5.3-16}$$

The aim is now to solve Laplace's equation in the fluid and then apply the three boundary conditions. In their original paper in the Philosophical Magazine, Korteweg & de Vries solved Laplace's equation by finding a solution for ϕ in terms of a rapidly convergent series in Z where $Z = h + z$ (Z is the fluid depth, measured from the bottom). They state that they followed the same method used by Lord Rayleigh in an earlier paper (1876). Before we do this it is desirable if all the variables, including the fluid depth, are scaled into dimensionless form. This is essential in determining which terms dominate when the wavelength is large or small in comparison to the depth. Consequently we need to introduce a typical wavelength λ into the problem and also a typical time scale. The time scaling can be deduced by looking for the phase velocity of the waves which is obtainable from the linearized version of (5.3-14),(5.3-15). The result is

$$\begin{aligned} \nabla^2\phi &= 0 \\ \phi_{tt} + g\phi_z &= 0 \qquad \text{on } z = 0 \\ \phi_z &= 0 \qquad \text{on } z = -h \end{aligned} \tag{5.3-17}$$

Looking for separable solutions of (5.3-17) in the form

$$\phi(x,y,z,t) = R(z)\exp[i(k_1x + k_2y - \omega t)] \qquad (5.3\text{-}18)$$

Laplace's equation yields

$$R'' = (k_1^2 + k_2^2)R \qquad (5.3\text{-}19)$$

Taking the solution of (5.3-19) in the form

$$R(z) = C\cosh[k(z+\delta)] \qquad k^2 = k_1^2 + k_2^2 \qquad (5.3\text{-}20)$$

and applying $\phi_z = 0$ at $z = -h$, we find that $\delta = h$. Using this result in the boundary condition at the free surface $z = 0$ yields the dispersion relation

$$\omega^2 = gk\tanh(kh) \qquad (5.3\text{-}21)$$

which is an even function in k. The dispersion relation (5.3-21) gives an expression for the phase speed c which is

$$c^2 = gh.\left[\frac{\tanh(kh)}{(kh)}\right] \qquad (5.3\text{-}22)$$

Equation (5.3-22) shows that for waves **having a wavelength** λ **much greater than the fluid depth** with $kh << 1$ we have $c^2 \simeq c_0^2 = gh$. This is a useful result because it gives the relevant measure of time for such long waves as $t_0 = \lambda / c_0$. We therefore introduce new dimensionless variables

$$\bar{x} = x/\lambda \ ; \quad \bar{t} = t/t_0 \ ; \quad \bar{z} = z/h \qquad (5.3\text{-}23)$$

The only variable which remains to be transformed into dimensionless form is S. We therefore introduce a typical amplitude "a" and define a dimensionless form for S

$$u = S/a \qquad (5.3\text{-}24)$$

From now on we shall drop the y variation and consider the wave motion purely in the x-direction, a situation relevant to canal motion. Following Korteweg and de Vries, we adopt the notation $Z = z + h$ which in dimensionless form is $\bar{Z} = 1 + \bar{z}$. The Z co-ordinate now measures distances from the bottom of the fluid. Our fluid equations now yield an appropriate scaling for ϕ. For instance, (5.3-15) is

$$u + \left(\frac{c}{ga\lambda}\right)\left(\frac{\partial\phi}{\partial\bar{t}}\right) + \frac{1}{2gah^2}\cdot\frac{h^2}{\lambda^2}\left(\frac{\partial\phi}{\partial\bar{x}}\right)^2 + \left(\frac{\partial\phi}{\partial\bar{Z}}\right)^2 = 0$$

This suggests that a suitable scaled form for ϕ is $\Phi = c\phi/ga\lambda$. It is easy to check that Φ is dimensionless. The total set of equations in dimensionless form is now

$$0 = u + \frac{\partial\Phi}{\partial\bar{t}} + \tfrac{1}{2}\nu\left(\frac{\partial\Phi}{\partial\bar{x}}\right)^2 + \frac{\nu}{\mu}\left(\frac{\partial\Phi}{\partial\bar{Z}}\right)^2 \qquad (5.3\text{-}25)$$

$$\mu\left(\frac{\partial u}{\partial\bar{t}}\right) + \mu\nu\left(\frac{\partial\Phi}{\partial\bar{x}}\right)\left(\frac{\partial u}{\partial\bar{x}}\right) = \frac{\partial\Phi}{\partial\bar{Z}} \qquad \text{on } \bar{Z} = 1 + \nu u \qquad (5.3\text{-}26)$$

$$\mu\frac{\partial^2\Phi}{\partial\bar{x}^2} + \frac{\partial^2\Phi}{\partial\bar{Z}^2} = 0 \qquad 0 < Z < 1 + \nu u \qquad (5.3\text{-}27)$$

and

$$\frac{\partial\Phi}{\partial\bar{Z}} = 0 \qquad \text{on } \bar{Z} = 0 \qquad (5.3\text{-}28)$$

where the dimensionless constants ν and μ are given by

$$\nu = a/h \qquad \text{and} \qquad \mu = (h/\lambda)^2 \qquad (5.3\text{-}29)$$

Although we have taken some time in obtaining them, equations (5.3-25)-(5.3-28) have a definite adantage over (5.3-13)-(5.3-16) in that they have built into them the dimensionless constants ν and μ, the magnitude of which give a measure of whether the waves are long or short in comparison to the depth or small or large in amplitude. We are now in a position to solve Laplace's equation as a series in $\bar{Z}$ and then apply the three boundary conditions.

Taking Φ in the form

$$\Phi(\bar{x},\bar{Z},\bar{t}) = \sum_{n=0}^{\infty} \bar{Z}^n \rho_n(\bar{x},\bar{t})$$

and substituting in Laplace's equation yields a recurrence relation

$$\frac{\partial^2\rho_n}{\partial\bar{x}^2} + (n+2)(n+1)\rho_{n+2} = 0 \qquad (5.3\text{-}30)$$

The boundary condition $\partial\Phi/\partial\bar{Z} = 0$ on $Z = 0$ immediately shows that $\rho_1 = 0$ and hence all odd $\rho_{(n)}$ are also zero. Writing $\rho = \rho_o$, the left hand column yields a general formula for ρ_{2j} which gives an expression for Φ in the form

$$\Phi = \sum_{j=0}^{\infty} (-1)^j \mu^j . \frac{\bar{Z}^{2j}}{(2j)!} . \frac{\partial^{2j}\rho}{\partial\bar{x}^{2j}} \qquad (5.3\text{-}31)$$

This is now a series solution for Φ throughout the fluid, for the region $0 \le Z \le 1 + \nu u$. We are more interested in what happens at the surface at $Z = 1 + \nu u$, and so we substitute (5.3-31) into the two free surface conditions (5.3-25),(5.3-26) and gather terms in powers of μ. The equations are simpler if the variable $w = \rho_{\bar{x}}$ is introduced and to facilitate this (5.3-25) is differentiated with respect to $\bar{x}$ after the substitution has been made. The pair of upper surface boundary conditions become

$$0 = u_{\bar{t}} + [(1+\nu u)w]_{\bar{x}} - \mu[\tfrac{1}{6} w_{\bar{x}\bar{x}\bar{x}}(1+\nu u)^3 + \tfrac{1}{2}(1+\nu u)^2 u_{\bar{x}} w_{\bar{x}\bar{x}}] + O(\mu^2) \quad (5.3\text{-}32)$$

$$0 = u_{\bar{x}} + w_{\bar{t}} + \nu w w_{\bar{x}} - \tfrac{1}{2}\mu[(w_{\bar{x}\bar{x}\bar{t}} + \nu w w_{\bar{x}\bar{x}\bar{x}} - \nu w_{\bar{x}} w_{\bar{x}\bar{x}})(1+\nu u)^2$$
$$+ 2\nu u_{\bar{x}}(1+\nu u)(w_{\bar{x}\bar{t}} + \nu w w_{\bar{x}\bar{x}} - \nu w_{\bar{x}}^2)] + O(\mu^2) \quad (5.3\text{-}33)$$

In the long wave approximation, if all the terms in ν and μ are neglected ($a << h$, $\lambda >> h$) then all we are left with is $u_{\bar{t}\bar{t}} - u_{\bar{x}\bar{x}} = 0$, which is the linear wave equation. Note that it is only valid to neglect higher terms in μ if $h << \lambda$; that is, the wavelength of the waves is much greater than the depth. If, however, terms in ν are retained but terms in μ are neglected then we obtain what are known as the nonlinear shallow water equations.

$$0 = u_{\bar{t}} + w_{\bar{t}} + \nu(uw)_{\bar{x}} \quad (5.3\text{-}34)$$

$$0 = u_{\bar{t}} + w_{\bar{t}} + \nu w w_{\bar{x}} \quad (5.3\text{-}35)$$

These now have nonlinear terms but as yet no dispersive terms. In order to extract from (5.3-32),(5.3-33) their leading order terms in the long wave approximation, it is not enough to take just the lowest order μ terms because we also need to re-scale $\bar{x}$ and $\bar{t}$. Following the procedure sketched out in Section 5.1, we expand u and w as

$$u = \varepsilon u^{(1)} + \varepsilon^2 u^{(2)} \ldots\ldots$$
$$w = \varepsilon w^{(1)} + \varepsilon^2 w^{(2)} \ldots\ldots \quad (5.3\text{-}36)$$

Boundary conditions on u and w are $u, w \to 0$ as $x \to \infty$ so there is no ε^0 term in (5.3-36). The dispersion relation has already been given (equation (5.3-22)) and for waves much longer than the depth h ($kh << 1$) then the expansion of the hyperbolic tangent function will give a k and a k^3 term in ω as expected. As in Section 5.1, the appropriate space and time variables are now

$$\xi = \varepsilon^p(x-at) \qquad \tau = \varepsilon^{3p} t \quad (5.3\text{-}37)$$

Using these substitutions in (5.3-32) gives

$$(\varepsilon^{3p}\frac{\partial}{\partial\tau} - a\varepsilon^p\frac{\partial}{\partial\xi})(\varepsilon u^{(1)} + \varepsilon^2 u^{(2)} \ldots) + \varepsilon^p\frac{\partial}{\partial\xi}[(\varepsilon w^{(1)} + \varepsilon^2 w^{(2)} \ldots)$$
$$(1 + \nu(\varepsilon u^{(1)} + \varepsilon^2 u^{(2)} \ldots))] - \tfrac{1}{6}\mu[1 + \nu(\varepsilon u^{(1)} + \varepsilon^2 u^{(2)} \ldots)]^3 \quad (5.3\text{-}38)$$
$$\varepsilon^{3p}\frac{\partial^3}{\partial\xi^3}(\varepsilon w^{(1)} + \varepsilon^2 w^{(2)} \ldots) \;\ldots\ldots\ldots = 0$$

We have ignored the quadratic terms at $O(\varepsilon)$ because they will be of order ε^{3p+2} at least and it will not be necessary to go to such a high order. In a similar fashion, equation (5.3-33) becomes

$$\left(\varepsilon^{3p}\frac{\partial}{\partial\tau}-a\varepsilon^{p}\frac{\partial}{\partial\xi}\right)(\varepsilon w^{(1)}+\varepsilon^{2}w^{(2)}\ldots)+\varepsilon^{p}\frac{\partial}{\partial\xi}(\varepsilon u^{(1)}+\varepsilon^{2}u^{(2)}\ldots)$$
$$+\tfrac{1}{2}\nu\varepsilon^{p}\frac{\partial}{\partial\xi}[\varepsilon^{2}w^{(1)}]^{2}\ldots-\tfrac{1}{2}\mu[1+\nu(\varepsilon u^{(1)}+\varepsilon^{2}u^{(2)}\ldots)]^{2}\varepsilon^{2}\frac{\partial^{2}}{\partial\xi^{2}} \tag{5.3-39}$$
$$\times\left[\left(\varepsilon^{3p}\frac{\partial}{\partial\tau}-a\varepsilon^{p}\frac{\partial}{\partial\xi}\right)(\varepsilon w^{(1)}+\varepsilon^{2}w^{(2)})\right]\ldots\ldots=0$$

Taking each of these two equations in turn it can be shown in a similar way to the ion acoustic wave problem of Section 5.2 that $p = \frac{1}{2}$. In each equation the lowest order of ε is $\varepsilon^{3/2}$ and the next is $\varepsilon^{5/2}$. For equation (5.3-38)

$$\varepsilon^{3/2}:\ -au_{\xi}^{(1)}+w_{\xi}^{(1)}=0$$
$$\varepsilon^{5/2}:\ -au_{\xi}^{(2)}+w_{\xi}^{(2)}+u_{\tau}^{(1)}+[w^{(1)}n^{(1)}]_{\xi}-\frac{1}{6}\mu w_{\xi\xi\xi}^{(1)}=0$$

For equation (5.3-39)

$$\varepsilon^{3/2}:\ -aw_{\xi}^{(1)}+u_{\xi}^{(1)}=0$$
$$\varepsilon^{5/2}:\ -aw_{\xi}^{(2)}+u_{\xi}^{(2)}+w_{\tau}^{(1)}+\nu w_{\xi}^{(1)}w^{(1)}+\tfrac{1}{2}a\mu w_{\xi\xi\xi}^{(1)}=0$$

The $\varepsilon^{3/2}$ terms imply that $a^2 = 1$ so taking $a = +1$ we have at this order

$$u^{(1)}=w^{(1)} \tag{5.3-40}$$

Substituting this into the two equations at order $\varepsilon^{5/2}$ we see, by addition, that the $u^{(2)}$ and $w^{(2)}$ terms cancel leaving

$$u_{\tau}^{(1)}+\frac{3}{2}\nu u^{(1)}u_{\xi}^{(1)}+\frac{1}{6}\mu u_{\xi\xi\xi}^{(1)}=0 \tag{5.3-41}$$

which is the KdV equation. The introduction of the variables ξ and τ correctly introduces a dispersion term into the problem which balances the nonlinear term. However small the value of μ, the dispersive term will always prevent the breaking of any wave since this term will always become significant when the wave becomes steep enough despite the smallness of its coefficient. If surface tension is included (see equation (5.3-11) the extra term $T.u_{\bar{x}\bar{x}}$ in (5.3-25) will become a $u_{\xi\xi\xi}$ term in (5.3-33). This will be included in the final equation since it will contribute at order $\varepsilon^{5/2}$ but all it does is alter the coefficient of the $u_{\xi\xi\xi}^{(1)}$ term in the final KdV equation. Inspection of equation (1.2-1) in Chapter 1 in which Korteweg and de Vries's result is stated shows that the surface tension term is included in the coefficient of the third derivative. Our equation looks different to theirs because our amplitude, space and time variables are dimensionless and the 1895 result is not written in dimensionless form.

Scott Russell's observations can now easily be explained. In Chapter 1 we showed a diagram of one of his experiments (Fig. 1-1) in which water held behind a partition at a level which was higher than the rest of the trough. Removal of the partition caused this water to flow forwards and, depending on the height and width, solitons were produced. This, of course, is equivalent to having square-wave initial data for the KdV equation

$$u = \begin{cases} u_0 & 0 < x < L \\ 0 & x < 0; \quad x > L \end{cases} \tag{5.3-42}$$

The number of discrete eigenvalues of this data gives the number of solitons which will emerge and flow down the trough. (See the calculations of Chapters 2 to 4).

We note that for the Schrödinger equation problem only the square **well** and not the square potential **barrier** has any bound states, but in the form that the Schrödinger problem was set up in Chapters 2 and 3, the KdV equation had a negative sign in the nonlinear term (we took $\alpha = -6$) so a negative potential in that problem means a positive potential here where the KdV equation has a positive sign in the nonlinear term.

5.4 A Problem in Geophysical Fluid Dynamics

The third example with which we shall be concerned is somewhat different from the previous two in that it occurs in Geophysical Fluid Dynamics which deals with the study of atmospheres and oceans. The treatment here will be deliberately naive and incomplete because we will be trying to reduce a very large and complicated subject down to its essentials. Our concern will be with the development of large scale wave motions in a rotating, homogeneous, shallow fluid layer in the inviscid limit. This situation, as we shall see, approximately models the evolution of waves in the atmosphere (Rossby waves). The task of the first parts of this section will be to outline briefly some of the ideas and approximations which are necessary for an understanding of the vorticity equation which we will derive and then reduce to the KdV equation by the reductive perturbation method.

Large scale motions are typically defined to be those which are significantly affected by the Earth's rotation. A measure of this is a dimensionless number called the Rossby number, which is defined as follows. If L and U are typical horizontal length and velocity scales and Ω is the planetary rotation frequency, then the Rossby number ε (not to be confused with the ε in the rest of the chapter used as a perturbation parameter) is defined as

$$\varepsilon = U/L\Omega \tag{5.4-1}$$

Consequently for rotation to be dominant, a necessary condition is that $\varepsilon \ll 1$. Typical values of L and U determined from observed weather patterns are of the order of 1000 km and 10 m/sec respectively. Corresponding values of the Rossby number ε based on a value of $\Omega = 7.3X10 \quad sec^{-1}$ for the Earth range from 0.1 to 0.5. A similar analysis for the oceans gives values of the Rossby number of order 10^{-3}.

In formulating the problem it is much easier to work in the Earth's rotating frame since this is the frame where experimental measurements are made. This is an accelerating frame and both centrifugal and Coriolis forces must be taken into account. The centrifugal force is small and can be absorbed into an effective gravitational potential. The Coriolis force is consequently the dominant rotational effect.

In the rotating frame of the Earth whose angular velocity vector is Ω, the Navier Stokes equations for an inviscid fluid are

$$\rho \frac{D\underline{q}}{Dt} = - \nabla P + \rho \nabla\phi - \rho \, 2\underline{\Omega} X \underline{q} \tag{5.4-2}$$

where ρ and P are the fluid density and pressure, $\underline{q}$ is the velocity vector in the Earth's frame and ϕ is the effective gravitational potential. Other equations are necessary to specify the system, an example being the equation of conservation of mass

$$\frac{D\rho}{Dt} = - \rho\nabla.\underline{q} \tag{5.4-3}$$

The **vorticity** $\underline{\omega}$ is defined to be

$$\underline{\omega} = \text{curl } \underline{q} \tag{5.4-4}$$

and therefore taking the curl of (5.4-2) shows that the governing equation for the vorticity in the Earth's frame is

$$\frac{D\omega}{Dt} = [(\underline{\omega}+2\underline{\Omega}).\nabla]\underline{q} - (\underline{\omega}+2\underline{\Omega})(\nabla.\underline{q}) + (\nabla\rho \, X \, \nabla P)\rho^{-2} \tag{5.4-5}$$

The correction term 2Ω to the vorticity is called the **planetary vorticity.** Using (5.4-3) in combination with (5.4-5) we obtain

$$\frac{D}{Dt}(\omega/\rho) = \rho^{-1}[(\underline{\omega}+2\underline{\Omega}).\nabla]\underline{q} + \rho^{-3}(\nabla\rho \, X \, \nabla P) \tag{5.4-6}$$

Let us now assume that some conserved quantity λ already exists,

$$\frac{D\lambda}{Dt} = 0 \tag{5.4-7}$$

It is now easy to show that

$$(\underline{\omega}+2\underline{\Omega}).\frac{D}{Dt}(\nabla\lambda) = - (\nabla\lambda).[(\underline{\omega}+2\underline{\Omega}).\nabla]\underline{q} \tag{5.4-8}$$

Taking the scalar product of (5.4-6) with $\nabla\lambda$ and combining with (5.4-8) yields

$$\frac{D}{Dt}[\rho^{-1}(\underline{\omega}+2\underline{\Omega}).\nabla\lambda] = (\nabla\lambda).\rho^{-3}(\nabla\rho \, X \, \nabla P) \tag{5.4-9}$$

The right hand side of (5.4-9) is zero under two possible circumstances. Either λ is a function of ρ and P only or $\nabla\rho \times \nabla P = 0$. The last condition is satisfied if the fluid is **barotropic** which means that the density ρ is constant on surfaces of constant pressure: $\rho = \rho(P)$ and therefore $\nabla\rho \times \nabla P = 0$. Under these circumstances (5.4-9) implies that the scalar quantity Π, called the **potential vorticity**

$$\Pi = \rho^{-1}(\underline{\omega}+2\underline{\Omega}).\nabla\lambda \tag{5.4-10}$$

is conserved.

5.4.1 The Geostrophic Approximation and the Taylor-Proudman Theorem

An estimate of the ratio of the relative acceleration and the Coriolis force is given by

$$(U^2/L)/(2U\Omega) = (U/2L\Omega) = \varepsilon \tag{5.4-11}$$

which is the Rossby number. As we have seen, the Rossby number is small for large scale atmospheric and oceanic circulations. We therefore assume the primary balance of forces in the Navier Stokes equations is between the pressure gradient, the Coriolis force and the gravitational force. This is written mathematically as

$$2\underline{\Omega} \times \underline{q} = -\rho^{-1}\nabla P + \nabla\phi \tag{5.4-12}$$

If the components of q are given in a local frame by (u,v,w) then in component form (5.4-12) can be written approximately as

$$fu = -\rho^{-1}\frac{\partial P}{\partial y} \tag{5.4-13}$$

$$fv = \rho^{-1}\frac{\partial P}{\partial x} \tag{5.4-14}$$

$$0 = -\frac{\partial P}{\partial z} - \rho g \tag{5.4-15}$$

where $f = 2\Omega$.

In deriving these equations we have ignored vertical fluid motions. This is reasonable since motions occur in a thin spherical shell in which the horizontal scale is several thousand kilometers but the vertical scale is of the order of a single kilometer. Equations (5.4-13)-(5.4-15) are derived in the approximation that the north-south motions are small so that a local coordinate system can be used. We are using x as the west to east coordinate, y as the south to north coordinate and z as the upward coordinate.

Let us now differentiate equation (5.4-13) with respect to z and combine the result with (5.4-15) to obtain

$$f\frac{\partial u}{\partial z} = g\rho^{-1}(\partial\rho/\partial y)+\rho^{-2}\frac{\partial\rho}{\partial z}(\partial P/\partial y) \tag{5.4-16}$$

We now note that

$$\frac{\partial P}{\partial y} = -\frac{\partial P}{\partial z}(\partial z/\partial y)_P = \rho g(\partial z/\partial y)_P$$

giving

$$f\frac{\partial u}{\partial z} = g\rho^{-1}[(\partial\rho/\partial y) + (\partial z/\partial y)_P\frac{\partial\rho}{\partial z}] = \rho^{-1}(\partial\rho/\partial y)_P g \qquad (5.4\text{-}17)$$

If the fluid is barotropic then $(\partial\rho/\partial y)_P = 0$ and u is independent of z. An equivalent result is also valid for v. This result is the **Taylor Proudman Theorem:** If a fluid is in geostrophic balance and also barotropic, then the horizontal velocity components are independent of the vertical coordinate.

5.4.2 The equations of motion for a shallow fluid layer

An appropriate model for the earth's atmosphere is to consider a shallow layer of incompressible rotating fluid of constant density confined by a rigid horizontal plane above and a variable rigid surface below. A schematic representation of this model is given in Fig. 5-1 below.

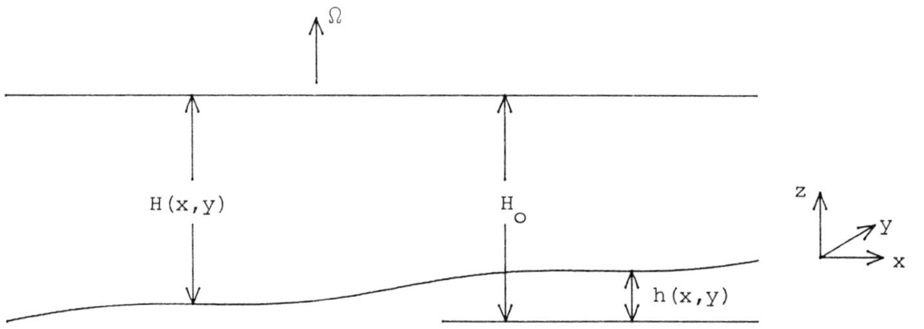

Figure 5-1:

The fluid is automatically barotropic because it has constant density and therefore the Taylor-Proudman theorem applies. That result together with the continuity equation can be shown to lead to the result

$$\frac{D}{Dt}\left[\frac{z-h(x,y)}{H(x,y)}\right] = 0 \qquad (5.4\text{-}18)$$

The function $(z - h)/H$ may be used as the conserved quantity λ in the construction of a potential vorticity Π defined by (5.4-10) and which will be a second conserved scalar function. The dominant contribution to the potential vorticity Π arises from the vertical component ζ of the vorticity vector $\underline{\omega}$ and is, in local coordinates

$$\rho\Pi = (\zeta+2\Omega)\ \hat{\underline{k}}.\underline{\nabla}\ [z-h(x,y))/H] = (\zeta+2\Omega)/H \qquad (5.4\text{-}19)$$

and by equations (5.4-13),(5.4-14)

$$\zeta = \hat{\underline{k}}.\text{curl}\ \underline{q} = (\rho f)^{-1}\nabla^2 P = \nabla^2\psi \qquad (5.4\text{-}20)$$

where $\psi = P(\rho f)^{-1}$ is termed the **geostrophic stream function.**

The potential vorticity Π is conserved and so for an incompressible fluid

$$\frac{D(\rho\Pi)}{Dt} = 0 \qquad (5.4\text{-}21)$$

and we have

$$\zeta_t + u\zeta_x + v\zeta_y - (\zeta+f)(uh_x+vh_y)/H = 0 \qquad (5.4\text{-}22)$$

Equation (5.4-22) can be considerably simplified. Replacing $(\zeta + f)$ by f (since $\zeta << f$) and H by its averaged value H_o we find that (5.4-22) becomes

$$\frac{\partial}{\partial t}(\nabla^2\psi) + \frac{\partial\,[\psi;\nabla^2\psi(fh/H_o)]}{\partial(x,y)} = 0 \qquad (5.4\text{-}23)$$

Equation (5.4-23) is called the **quasi-geostrophic potential vorticity equation.**

5.4.3 Rossby Waves

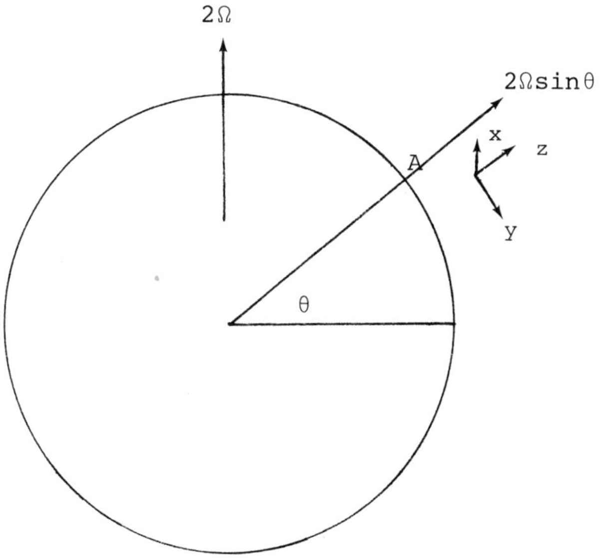

Figure 5-2: To account for the earth's curvature at A, local cartesian co-ordinates are employed.

We can see from Fig. 5-2 that because of the curvature of the Earth's surface we can choose **local** coordinates such that $h(x,y)$ may be approximately represented by the linear function

$$h(x,y) = \alpha y \qquad (5.4\text{-}24)$$

Equation (5.4-23) can now be written as

$$\left(\frac{\partial}{\partial t} + \frac{\partial\psi}{\partial x}\frac{\partial}{\partial y} - \frac{\partial\psi}{\partial y}\frac{\partial}{\partial x}\right)(\nabla^2\psi) + \beta\,\frac{\partial\psi}{\partial x} = 0 \qquad (5.4\text{-}25)$$

where

$$\beta = (\alpha f/H_o) \tag{5.4-26}$$

An important feature of this correction is that it introduces dispersion into the nonlinear vortictiy equation. Furthermore, (5.4-25) possesses an **exact** plane wave solution even though it is a nonlinear equation:

$$\psi = A\cos(kx+ly-\omega t) \tag{5.4-27}$$

$$\omega = -(\beta k/(k^2+l^2)) \tag{5.4-28}$$

It is clear from (5.4-28) that provided $\beta \neq 0$ there are dispersive waves called **Rossby waves.**

The above ideas have taken some space to come to their final form but we are now almost ready to embark on our usual scaling procedures for isolating long waves. The above ideas have been condensed in a deliberately simplified fashion for those readers who are unfamiliar with rotating fluid mechanics but who are nevertheless concerned about the origins of the potential vorticity equation (5.4-25).

5.4.4 Solitary Rossby Waves

Let us now look for long wave solutions of (5.4-25) instead of harmonic Rossby waves. The approximations made so far mean that it is only valid for us to examine long wave propogation in the x-direction but not the y-direction.

From (5.4-13) we notice that the horizontal velocity component u and the geostrophic stream function ψ are related by

$$u(x,y) = -(\partial\psi/\partial y) \tag{5.4-29}$$

A special solution of (5.4-25) is therefore seen to be

$$\psi^U(y) = -\int^y U(s)\,ds \tag{5.4-30}$$

where $U(y)$ is an arbitrary velocity profile dependent only on y. Hence the corresponding flow is purely in the x-direction. Let us now look for solutions which are perturbations about the flow (5.4-30)

$$\psi(x,y,t) = \psi^U(y) + \Psi(x,y,t) \tag{5.4-31}$$

which gives for $\Psi(x,y,t)$

$$\left[\frac{\partial}{\partial t} + U\frac{\partial}{\partial x} + \frac{\partial\Psi}{\partial x}\frac{\partial}{\partial y} - \frac{\partial\Psi}{\partial y}\frac{\partial}{\partial x}\right](\nabla^2\Psi) + (\beta - U'')\frac{\partial\Psi}{\partial x} = 0 \tag{5.4-32}$$

Clearly for $U \neq$ constant, there are no harmonic wave solutions of (5.4-32), but for $U = U_o$, constant harmonic wave solutions exist with a dispersion relation

$$\omega = kU_o - \beta k(k^2+1^2)^{-1} \tag{5.4-33}$$

which for small k, corresponding to long wave motions, may be approximated by its Taylor expansion

$$\omega = \omega_1 k + \omega_3 k^3 + \ldots \tag{5.4-34}$$

This indicates, as in Section 5.1, that in this case we would introduce stretched coordinates

$$\xi = \varepsilon^{\frac{1}{2}}(x-at) \qquad \tau = \varepsilon^{3/2}t \tag{5.4-35}$$

and apply standard reductive perturbation theory. In the general case $U \neq$ a constant we can still try to apply this technique with the same form of stretched coordinates but we must anticipate that modifications will be required.

We expand Ψ as

$$\Psi(x,y,t) = \sum_{m=1}^{\infty} \varepsilon^m \psi^{(m)}(\xi,y,\tau) \tag{5.4-36}$$

and substituting (5.4-35) and (5.4-36) into (5.4-32) we find that

$$[\varepsilon^{3/2}\frac{\partial}{\partial\tau} + \varepsilon^{\frac{1}{2}}\bar{U}\frac{\partial}{\partial\xi} + \varepsilon^{\frac{1}{2}}\frac{\partial}{\partial\xi}(\varepsilon\Psi^{(1)}+.)\frac{\partial}{\partial y} - \varepsilon^{\frac{1}{2}}\frac{\partial}{\partial y}(\varepsilon\Psi^{(1)}+\varepsilon^2\psi^{(2)}+.)\frac{\partial}{\partial\xi}] \tag{5.4-37}$$

$$\times[(\varepsilon\frac{\partial^2}{\partial\xi^2} + \frac{\partial^2}{\partial y^2})(\varepsilon\Psi^{(1)}+\varepsilon^2\psi^{(2)}+.)] + \beta\varepsilon^{\frac{1}{2}}(\beta-\bar{U}'')\frac{\partial}{\partial\xi}(\varepsilon\Psi^{(1)}+\varepsilon^2\psi^{(2)}+.) = 0$$

$$\bar{u} = u - a \tag{5.4-38}$$

Taking the two lowest orders of ε, we find that

$$\varepsilon^{3/2}: \quad \frac{\partial}{\partial\xi}[\bar{U}\frac{\partial^2\psi^{(1)}}{\partial y^2} + (\beta-\bar{U}'')\psi^{(1)}] = 0 \tag{5.4-39}$$

$$\varepsilon^{5/2}: \quad \frac{\partial}{\partial\xi}[\bar{U}\frac{\partial^2\psi^{(2)}}{\partial y^2} + (\beta-\bar{U}'')\Psi^{(2)}] + \frac{\partial}{\partial\tau}\frac{\partial^2\psi^{(1)}}{\partial y^2} + \bar{U}\frac{\partial^3\psi^{(1)}}{\partial\xi^3} \tag{5.4-40}$$

$$+ \frac{\partial\Psi^{(1)}}{\partial\xi}\frac{\partial^3\Psi^{(1)}}{\partial y^3} - \frac{\partial\Psi^{(1)}}{\partial y}\frac{\partial^3\psi^{(1)}}{\partial\xi\partial y^2} = 0$$

We immediately note from equation (5.4-39) that integrating with respect to ξ and setting the integration constant to zero shows that $\psi^{(1)}$ satisfies a linear equation in y. We can therefore introduce an arbitrary function of ξ and τ such that

$$\psi^{(1)} = A(\xi,\tau)Y(y) \tag{5.4-41}$$

where $Y(y)$ satisfies

$$Y''(y) + \nu(y)Y(y) = 0 \tag{5.4-42}$$

$$\nu(y) = (\beta-\bar{U}'')/\bar{U} \tag{5.4-43}$$

subject to the side wall boundary conditions $Y(y_1) = Y(y_2) = 0$. We now see where "a" plays a role, since it is obviously an eigenvalue in the eigenvalue problem (5.4-42) once a given shear profile $U(y)$ has been specified.

We will assume for simplicity that no **critical layers** occur in which some value y_c of y exists $(y_1 < y_c < y_2)$ for which $U(y_c) = a$. If this occurs then the analysis is not consistent as $\nu \to \infty$ when $y \to y_c$.

Examining terms of second highest order we now find, on using (5.4-41) that $\psi^{(2)}$ satisfies the equation

$$-\frac{\partial}{\partial\xi}\left[\bar{U}\frac{\partial^2\psi^{(2)}}{\partial y^2} + (\beta-\bar{U}'')\psi^{(2)}\right]=\bar{U}YA_{\xi\xi\xi}+(YY'''-Y'Y'')AA_{\xi}+Y''A_{\tau} \tag{5.4-44}$$

Multiplying the left hand side by $Y/\bar{U}$ and integrating over y from y_1 to y_2 we obtain for the left hand side

$$-\frac{\partial}{\partial\xi}\int_{y_2}^{y_1} Y(s)\left[\frac{\partial^2\psi^{(2)}}{\partial s^2}(s) + \nu(s)\psi^{(2)}(s)\right]ds \tag{5.4-45}$$

Integrating by parts twice gives

$$-\frac{\partial}{\partial\xi}\left[\left(Y\frac{\partial\psi^{(2)}}{\partial s} - \psi^{(2)}\frac{\partial Y}{\partial s}\right)\Big|_{y_2}^{y_1}\right]+\int_{y_2}^{y_1}\psi^{(2)}\left(\frac{\partial^2 Y}{\partial s^2} + \nu Y\right)ds \tag{5.4-46}$$

Appealing to the boundary conditions and using equation (5.4-39) shows that this expression is zero. Consequently for (5.4-44) to be consistent the same operation carried out on the right hand side must also produce zero. This leads to the following consistency requirement on the previously arbitrary function $A(\xi, \tau)$:

$$A_\tau + \mu AA_\xi + \gamma A_{\xi\xi\xi} = 0 \tag{5.4-47}$$

where

$$\mu = \left(\int_{y_2}^{y_1}(\nu Y^3/\bar{U})\,ds\right)/\left(\int_{y_2}^{y_1}(\nu Y^2/\bar{U})\,ds\right) \tag{5.4-48}$$

and

$$\gamma = \left(\int_{y_2}^{y_1}(Y^2)\,ds\right)/\left(\int_{y_2}^{y_1}(\nu Y^2/\bar{U})\,ds\right) \tag{5.4-49}$$

Equation (5.4-47) is, of course, the KdV equation with constant coefficients. As mentioned above if critical layers exist then the whole analysis breaks down. This is manifest in the fact that the integrals defining μ and γ become singular in that situation.

We have spent a lot of time giving some of the elements of Geophysical Fluid Dynamics in this chapter in order to give a degree of background, not only for this example, but also for another example in Chapter 10 which will build on the work of this section. In particular we will consider a variation of the potential vorticity equation (5.4-30) which will yield the sine-Gordon equation.

.5 The Modified and Generalised Korteweg-de Vries Equation

In the previous three examples, we have concentrated primarily on the KdV quation for which $p = 1/2$. The modified KdV equation also occurs in many roblems by using exactly the same stretched co-ordinate method except that he value of p usually takes on the value of unity. The method however is dentical to the one described in the earlier section of this chapter so it s not worth devoting a whole chapter to the MKdV equation on its own. The inal example of this chapter is one in which the degree of the nonlinearity s left at an arbitrary degree n (+ve integer) at the beginning of the roblem. Needless to say, the value of p therefore depends on n and the inal equation turns out to be

$$u_\tau + au^n u_\xi + bu_{\xi\xi\xi} = 0 \tag{5.5-1}$$

hich is a KdV equation with an $(n + 1)$th degree of nonlinearity. This quation incorporates the KdV $(n = 1)$ and MKdV $(n = 2)$ equations, which are he only two which are integrable. The advantage of this example is that it hows that the scaling method for both equations can be incorporated into a eneral framework.

The example we shall use is a nonlinear transmission line in electronics hich has no loss effects. This effectively means that no resistances are ncorporated in the circuit and the resistance of the components are so small hat they can be taken to be zero. The model we shall use can be found in cott (1970).

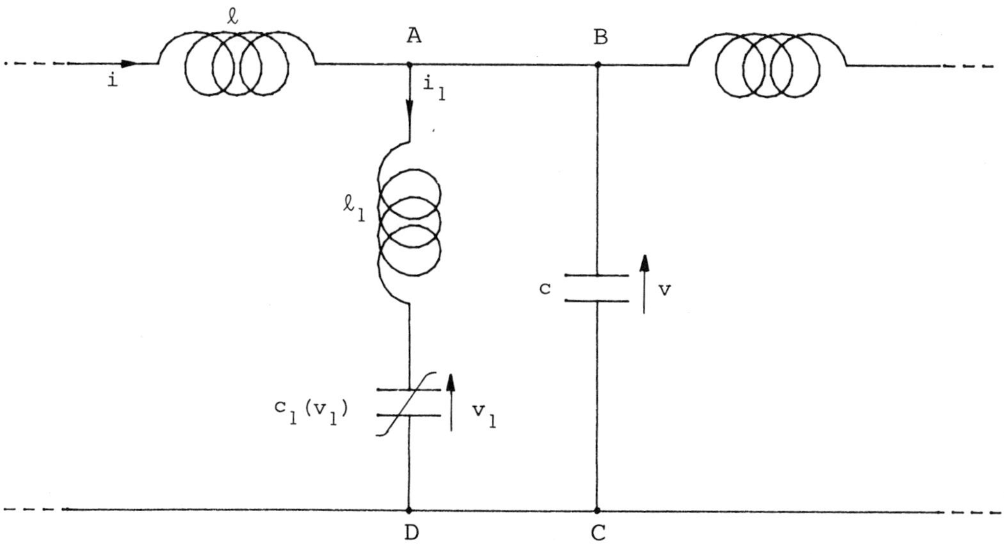

Figure 5-3:

o inductances of values ℓ and ℓ_1 are put together with two capacitors, in Fig. 5-3, one of constant capacitance c but the other with a pacitance $c_1(v_1)$ which varies with the voltage across it. This voltage labelled v_1, and the current in that part of the circuit is called i_1.

Remembering that the potential drop across an inductance ℓ is $-\ell\partial i/\partial t$ where i is the current, we need to apply Kirchoff's laws around the circuit. The first equation is the equivalent of conservation of mass:

$$\ell \frac{\partial i}{\partial t} + \frac{\partial v}{\partial x} = 0 \tag{5.5-2}$$

That is, the potential drop across the inductance ℓ must be balanced by the change of voltage per unit length across the whole circuit. Secondly, the total potential around ABCD must be zero, which gives

$$v_1 - v + \ell_1 \frac{\partial i_1}{\partial t} = 0 \tag{5.5-3}$$

Across the capacitor c_1, we have immediately

$$c_1(v_1) \frac{\partial v_1}{\partial t} = i_1 \tag{5.5-4}$$

and finally the current balance around ABCD gives

$$c \frac{\partial v}{\partial t} + \frac{\partial i}{\partial x} + i_1 = 0 \tag{5.5-5}$$

These equations have no loss terms and so the system is dispersive and non dissipative. This can easily be shown by finding the dispersion relation of the linearised set of equations. It is more convenient to eliminiate i_1 immediately to obtain

$$v_1 - v = \ell_1 c \cdot \frac{\partial^2 v}{\partial t^2} + \ell_1 \frac{\partial^2 i}{\partial x \partial t} \tag{5.5-6}$$

$$c \frac{\partial v}{\partial t} + \frac{\partial i}{\partial x} + c_1(v_1) \cdot \frac{\partial v_1}{\partial t} = 0 \tag{5.5-7}$$

$$\ell \frac{\partial i}{\partial t} + \frac{\partial v}{\partial x} = 0 \tag{5.5-8}$$

Linearising around $i = v = v_1 = 0$, $c_1 = c_o$ the dispersion relation turns out to be purely real and as in the previous sections we take stretched co-ordinates $\xi = \varepsilon^p(x - \alpha t)$; $\tau = \varepsilon^{3p} t$. We have not yet specified the variable capacitance $c_1(v_1)$. We shall now take $c_1(v_1)$ to have the form

$$c_1(v_1) = c_0(1 - a v_1^{\,n}) \tag{5.5-9}$$

where n is a positive integer and c_o and a are also positive. This physically means that the capacitance is nonlinear and varies as the potential across it varies. We shall now apply the usual method to equations ((5.5-6)-(5.5-8) with i, v, and v_1 expanded in the form:

$$i = \varepsilon i^{(1)} + \varepsilon^2 i^{(2)} \; \ldots . \tag{5.5-10}$$

It is not necessary to put in all the working as this has been done before. Instead we shall just list the terms obtained at each order of ε .

For (5.5-6)

$$\begin{aligned}
&O(\varepsilon): && v_1^{(1)} - v^{(1)} \\
&O(\varepsilon^2): && v_1^{(2)} - v^{(2)} \\
&O(\varepsilon^m): && v_1^{(m)} - v^{(m)} \\
&O(\varepsilon^{2p+1}): && -[\alpha^2 \ell_1 c\, v_{\xi\xi}^{(1)} - \alpha \ell_1 i_{\xi\xi}^{(1)}]
\end{aligned} \tag{5.5-11}$$

For (5.5-7)

$$\begin{aligned}
&O(\varepsilon^{p+1}): && -\alpha c v_\xi^{(1)} + i_\xi^{(1)} - \alpha c_0 v_{1\xi}^{(1)} \\
&O(\varepsilon^{p+m}): && -\alpha c v_\xi^{(m)} + i_\xi^{(m)} - \alpha c_0 v_{1\xi}^{(m)} \\
&O(\varepsilon^{3p+1}): && c v_\tau^{(1)} + c_0 v_{1\tau}^{(1)} \\
&O(\varepsilon^{n+p+1}): && \alpha a c_0 [v_1^{(1)}]^n v_{1\xi}^{(1)}
\end{aligned} \tag{5.5-12}$$

For (5.5-8)

$$\begin{aligned}
&O(\varepsilon^{p+1}): && -\alpha \ell\, i_\xi^{(1)} + v_\xi^{(1)} \\
&O(\varepsilon^{p+m}): && -\alpha \ell\, i_\xi^{(m)} + v_\xi^{(m)} \\
&O(\varepsilon^{3p+1}): && \ell\, i_\tau^{(1)}
\end{aligned} \tag{5.5-13}$$

Considering equations (5.5-11) and (5.5-13) first, both of which are linear, we do not yet know at which order of ε we get a balance between terms at $O(\varepsilon^m)$ and the double derivative terms on the one hand and the τ-derivative terms on the other. However, in (5.5-11), taking the $O(\varepsilon^m)$ terms to balance the $O(\varepsilon^{2p+1})$ terms gives $m = 2p + 1$. In (5.5-13) we have the $O(\varepsilon^{p+m})$ terms to balance the $O(\varepsilon^{3p+1})$ terms, which also gives $m = 2p + 1$. We have equivalently the same result in (5.5-12) but we also need to account for the nonlinearity which occurs at $O(\varepsilon^{n+p+1})$. We take therefore $p + m = 3p + 1 = n + p + 1$ and so $p = \frac{1}{2}n$ and $m = n + 1$. Equations (5.5-11)-(5.5-13) now become:

$$\begin{aligned}
&O(\varepsilon^j): && v_1^{(j)} - v^{(j)} \qquad 1 \le j < n+1 \\
&O(\varepsilon^{n+1}): && v_1^{(n+1)} - v^{(n+1)} = \alpha^2 \ell_1 c v_{\xi\xi}^{(1)} - \alpha \ell_1 i_{\xi\xi}^{(1)}
\end{aligned} \tag{5.5-14}$$

$$\begin{aligned}
&O(\varepsilon^j): && -\alpha c v_\xi^{(j)} + i_\xi^{(j)} - \alpha c_0 v_{1\xi}^{(j)} = 0 \qquad 1 \le j < n+1 \\
&O(\varepsilon^{\frac{3}{2}n+1}): && -\alpha c v_\xi^{(n+1)} + i_\xi^{(n+1)} - \alpha c_0 v_{1\xi}^{(n+1)} + c v_\tau^{(1)} + c_0 v_{1\tau}^{(1)} \\
& && + c_0 \alpha a [v_1^{(1)}]^n v_{1\xi}^{(1)} = 0
\end{aligned} \tag{5.5-15}$$

$$O(\varepsilon^{j}): \qquad -\alpha\ell\, i_\xi^{(j)} + v_\xi^{(j)} = 0 \qquad 1 \le j < n+1$$
$$O(\varepsilon^{\frac{3}{2}n+1}): \quad -\alpha\ell\, i_\xi^{(n+1)} + v_\xi^{(n+1)} + \ell i_\tau^{(1)} = 0 \tag{5.5-16}$$

The three linear equations at $O(\varepsilon^j)$ $(1 \le j \le n+1)$ are easily solved to give

$$v_1^{(j)} = v^{(j)} = \alpha\ell i^{(j)} \qquad 1 \le j < n+1 \tag{5.5-17}$$

provided that

$$\alpha^2 = [\ell(c+c_0)]^{-1} \tag{5.5-18}$$

The positive root of (5.5-18) fixes the value of α in the stretched co-ordinates ξ and τ for right-ward travelling waves. Finally, we have three equations:

$$v_1^{(n+1)} - v^{(n+1)} = -\alpha^3\ell_1\ell c_0 i_{\xi\xi}^{(1)} \tag{5.5-19}$$

$$0 = i_\xi^{(n+1)} - \alpha c v_\xi^{(n+1)} - \alpha c_0 v_{1\xi}^{(n+1)} + \alpha^{-1} i_\tau^{(1)} \tag{5.5-20}$$
$$+ \alpha a c_0 (\alpha\ell)^{n+1}[i^{(1)}]^n\, i_\xi^{(1)}$$

$$0 = -\alpha\ell i_\xi^{(n+1)} + v_\xi^{(n+1)} + \ell i_\tau^{(1)} \tag{5.5.21}$$

Multiplying (5.5-20) by $\alpha\ell$ and adding to (5.5-21) we find that

$$\alpha^2\ell c_0[v_\xi^{(n+1)} - v_{1\xi}^{(n+1)}] + 2\ell i_\tau^{(1)} + \alpha a c_0(\alpha\ell)^{n+2}[i^{(1)}]^n i_\xi^{(1)} = 0 \tag{5.5-22}$$

The first term in (5.5-22) can be eliminated by using (5.5-19) and we finally arrive at

$$(\alpha^5\ell\ell_1 c_0^2)\, i_{\xi\xi\xi}^{(1)} + \alpha^2 a c_0(\alpha\ell)^{n+1}[i^{(1)}]^n i_\xi^{(1)} + 2i_\tau^{(1)} = 0 \tag{5.5-23}$$

This is a KdV equation with $(n + 1)$th degree nonlinearity. The stretched co-ordinates are $\xi = \varepsilon^{n/2}x$; $\tau = \varepsilon^{3n/2}(x - \alpha t)$. If the capacitor c_1 varies linearly with the voltage $(n = 1;\ p = 1/2)$ we have the KdV equation with ξ and τ exactly as in the other examples. If it varies quadratically however, then we have the MKdV equation $(n = 2;\ p = 1)$.

5.6 Notes

Sections 1 and 2

The general idea of using stretched co-ordinates to analyse long waves in both dispersive systems (governed by the KdV) and in dissipative systems (governed by Burgers equation - see Section 1.8)

$$u_t + uu_x = \delta u_{xx} \tag{5.6-1}$$

has been around a long time. Cole use this method to obtain Burgers equation (5.6-1) for dissipative shocks in gas dynamics in 1951. We shall not repeat this calculation here as it is set as one of the exercises for this chapter. Regarding a general derivation of both the KdV and Burgers equation, Su and Gardner (1969) showed that a general class of dispersive systems would give the KdV and a class of dissipative systems would give Burgers equation respectively on a long wave scaling. Taniuti and Wei (1968) have given systematic derivation of the KdV equation, for a general system, using reductive perturbation theory. Kodama and Taniuti (1978) and Ichikawa et al. (1976) have taken into account higher order approximations in the reductive perturbation method. A general list of references on the general reductive perturbation method developed by Taniuti and co-workers can be found in this last reference. Further places where a discussion of various perturbation methods can be found are Nayfeh (1973) and Bender and Orszag (1978). The perceptive reader may ask about the value of ε. In each of the problems we have considered, the equations of motion have been rescaled into dimensionless form and thereby already have intrinsic length and time scales built into them. The magnitude of ε, which is used as a measure of the amplitude, is given in each problem by the magnitude of the initial data, and on these length and time scales must be small enough such that ε can be used as an expansion parameter. If it is too large, then the weakly nonlinear approximation is invalid.

In terms of specific applications the KdV is ubiquitous and the references we give in the rest of these notes, although large in number, are nevertheless selective. Beginning with Korteweg and de Vries (1895) the occurrence of the KdV and modified KdV equations seems to have been the most frequent in the theory of waves in nonlinear chains and lattices and in plasmas. Gardner and Morikawa (1960) in an unpublished report noted the similarity between a result they had derived for weak long hydromagnetic waves in plasma and water waves, the result being the KdV equation. Washimi and Taniuti (1966) obtained the KdV for weakly nonlinear compressive ion acoustic waves in a plasma, and Section 5.2 follows their calculation. Berezin and Karpman (1964, 1967) and Karpman (1967) both numerically and analytically obtained similar results independently - see the book by Karpman (1975) for references on this early work. The review by Jeffrey and Kakutani (1969) gives a survey of the earlier results and references regarding derivatives of the KdV equations for ion acoustic and hydromagnetic plasma waves. For a general book on methods in plasma physics see Davidson (1972). Further important papers, a list of which is by no means exhaustive, are to be found on ion acoustic solitons by Tappert and Zabusky (1971), Tappert and Judice (1972), Tappert (1972) and Konno et al. (1977) and on hydromagnetic waves by Kawahara (1969), Kever and Morikawa (1969), Morton (1964), Taniuti and Washimi (1968) and Kakutani et al (1968). A fairly comprehensive list of references on long wave KdV solitons in plasmas can be found in Ichikawa (1979) and the review by Scott, Chu and McLaughlin (1973). Experimental work on ion acoustic solitary waves has been performed by Ikezi, Taylor and Baker (1970) and by Cohn and MacKenzie (1973). As mentioned in the text,

Hershkowitz, Romesser and Montgomery (1972) calculated the number and magnitude of discrete eigenvalues of some chosen initial data from the Schrodinger equation and then showed there was quite good experimental agreement with theory.

All the above reductive perturbation theory calculations were done in one spatial dimension. Kadomtsev and Petviashivili (1970) considered the evolution of weakly nonlinear long waves in dispersive media in which they took into account the transverse co-ordinate y. They obtained a 2-dimensional version of the KdV equation

$$(u_{xxx} + 6u^2 + u_x + u_t)_x - u_{yy} = 0 \tag{5.6-2}$$

which Kako and Rowlands (1976) have also obtained for the 2-dimensional propagation of ion acoustic solitons.

In parallel with the KdV arising in plasma physics in various areas, it also arose as a model equation at about the same time in studies in nonlinear lattices where the general concern was how a noninear system shared energy among the various Fourier modes in the lattice. We introduced the FPU problem in Section 1.3 and most of the subsequent work on nonlinear lattices and chains stemmed from this, the work of Zabusky, Kruskal and others being fundamental. For references see Zabusky (1967, 1968, 1969, 1973), Zabusky and Kruskal (1965). Zabusky, Deem and Kruskal (1968) and one of the present authors (JCE) have made films available on loan regarding the propagation and interaction of solitons.

As Su and Gardner (1969) and Taniuti and Wei (1968) have shown, the KdV is the evolution equation one would expect to obtain for certain types of weakly nonlinar dispersive systems. We have mainly referenced the various cases where it occurs in plasma physics and nonlinear lattices. Obviously other areas exist where other workers have derived it as a model equation other than in the water wave example of the next section. Tappert and Varma (1970) and Narayanamurti and Varma (1970) have noted that it occurs in studies of the propagation of heat pulses in solids. Nairboli (1970) has derived it for the propagation of longitudinal dispersive waves in elastic rods. Leibovitch (1970) and Wijngaarden (1968) have also derived the KdV equation for weakly nonlinear waves in rotating fluids and liquid/gas bubble mixtures respectively. Various other references can be found in the books of articles edited by Leibovitch and Seebass (1968) (see the article by R. Miura) Bullough and Caudrey (1980), Lonngren and Scott (1978), and Calogero (1978). See also the books by Karpman (1975), Whitham (1974), and Bhatnagar (1980) for various discussions of weakly nonlinear waves in various types of media.

Section 3

The calculations of this section although generally following Korteweg and de Vries (1895) are slightly different in that we have used the reductive perturbation scaling arguments of this chapter to order the magnitude of the various terms in the expansions. Freeman and Johnson (1970) have shown that the same calculation goes through with a shear flow in the fluid; the coefficients of the resulting KdV equation being integrals over the shear.

Madsen and Mei (1969) took the Euler equations with a sloping bottom for the fluid chanel and discovered numerically that although a solitary wave resulted, it broke up as it evolved over the sloping bottom. Using this numerical result Johnson (1973) showed that a type of KdV equation with a modified nonlinear term arose. See also Zabusky and Galvin (1971) for a discussion of the KdV equation and shallow water waves. If the transverse spatial dimension is also taken into account for waves in a shallow canal then a 2-dimensional KdV equation arises similar to the Kadomtsev-Petviashvili result for plasmas but with a change of sign on the u_{yy} term. Freeman and Davey (1975) have undertaken a careful analysis of the limiting cases for 2-dimensional water waves regarding the various possible scalings concerning wave amplitude, depth and wavelength and for long waves on a shallow fluid they showed that (5.6-2) with a positive sign on the u_{yy} term is the appropriate equation. In another limit, they obtained the so-called Davey-Stewartson equations which generalise to two space dimensional the NLS equation of Chapter 1 - see section notes of Chapter 8. Hammack (1973) and Hammack and Segur (1974) have attempted to match theory and experiment by considering the evolution of initial data in wave tank experiments and comparing it with the predictions made by inverse scattering.

Section 4

(a) Geophysical fluid dynamics is a difficult subject and requires careful study if it is to be pursued in any depth. Part of the problem lies in the fact that present methods in applied mathematics only allow us to solve the Navier-Stokes equations for highly idealised and very simplified situations. The various approximations and scaling methods used must therfore be treated with care. Pedlosky's (1980) recent book on geophysical fluid dynamics and Greenspan's book on rotating fluids (1968) are two texts which can be read without any prior knowledge of the subject. A series of lectures also by Pedlosky (1971) are very useful for the beginner. The book by Drazin and Reid (1981) on hydrodynamic stability is a good text for those wanting to study the stability properties of and wave propagation in both rotating and parallel flows.

(b) The final vorticity equation (5.4-25) contains the term $\beta\partial\Psi/\partial x$ which in turn arises equivalently from a βy term added onto $\nabla^2\Psi$. The root of this term was the slope of the channel which took into account the curvature of the Earth's surface. This term is mathematically equivalent to what is known as Rossby's beta-plane approximation (Pedlosky 1971) which very simply takes account locally of the planetary sphericity. This is usually derived in the following way: for a fluid channel in the mid-latitudes, Fig. 5-2 makes it clear that equation (5.4-19) for the potential velocity should really read

$$\rho\Pi = (\zeta+2\Omega\sin\theta)/H \tag{5.6-3}$$

where $2\Omega\sin\theta$ has replaced the 2Ω term. Expanding θ around some angle θ_0 we may rewrite (5.6-3) approximately as

$$\rho\,\Pi = (\zeta+f_0+\beta y)/H \tag{5.6-4}$$

where

$$f_0 = 2\Omega\sin\theta_0; \quad \beta = 2\Omega R^{-1}\cos\theta_0; \quad y = R\delta\theta \tag{5.6-5}$$

The form of equation (5.6-4) can be recovered from (5.4-12) if spherical co-ordinates had been used. In this case, $f = 2\Omega$ in equations (5.4-13)-(5.4-15) is replaced by $f = 2\Omega\sin\theta$.

(c) The derivation of the KdV equation given in this context was performed by Redekopp (1977) who also obtained the MKdV equation in a similar manner for a stratified fluid. In the same paper Redekopp also showed how to recover the KdV result when critical layers are present.

Section 5

The modified KdV equation which arises in this section, as we have shown, arises in much the same way as the KdV equation but using a slightly different scaling in the stretched co-ordinates. It arises for example in certain anharmonic lattices (Zabusky 1967), Alven waves in plasmas (Kakutani and Ono 1973), long waves in stratified rotating fluids (Redekopp 1977) as well as in the electric circuit example given in this section.

With regard to the anharmonic lattice, in Section 1.3, we considered a nonlinear chain with quadratic and cubic nonlinearities given in equations (1.3-2),(1.3-3) with an equation of motion of the form

$$\ddot{Q}_n = f(Q_{n+1}-Q_n) - f(Q_n-Q_{n-1}) \tag{5.6-6}$$

In the same way as the electric circuit in which we considered a capacitance which had a dependence on the voltage of the form $c_1(v_1) = c\,[1 - av_1]$, we shall consider nonlinearities in $f(Q)$ of the form

$$f(Q) = Q + \beta_n Q^n \tag{5.6-7}$$

Following the calculations of Section 1.3 through in the same way, we would arrive at

$$\frac{\partial^2}{\partial n^2}\left[\frac{1}{12}\frac{\partial^2 u}{\partial n^2} + \beta_n u^n + u\right] - \frac{\partial^2 u}{\partial t^2} = 0 \tag{5.6-8}$$

In the same fashion as before, we take "stretched" co-ordinates $\xi = \varepsilon^p(n - \alpha t)$; $\tau = \varepsilon^{3p}t$ and take u as

$$u = \varepsilon u^{(1)} + \varepsilon^2 u^{(2)} \ \ldots . \tag{5.6-9}$$

Equation (5.6-8) now becomes

$$\varepsilon^{2p} \frac{\partial^2}{\partial \xi^2} [\varepsilon^{2p} \frac{\partial^2}{\partial \xi^2} (\varepsilon u^{(1)} + \varepsilon^2 u^{(2)} + \ldots \varepsilon^m u^{(m)} \ldots) + \beta_m (\varepsilon u^{(i)} \ldots)^n +$$

$$(\varepsilon u^{(1)} + \ldots \varepsilon^m u^{(m)} \ldots)] = (\varepsilon^{3p} \frac{\partial}{\partial \tau} - \alpha \varepsilon^p \frac{\partial}{\partial \xi})^2 (\varepsilon u^{(1)} + \ldots \varepsilon^m u^{(m)} \ldots)$$

$$\tag{5.6-10}$$

At various orders of ε we obtain

$$\begin{array}{ll} O(\varepsilon^{2p+1}) & u_{\xi\xi}^{(1)} - \alpha^2 u_{\xi\xi}^{(1)} \\ O(\varepsilon^{2p+m}) & u_{\xi\xi}^{(m)} - \alpha^2 u_{\xi\xi}^{(m)} \\ O(\varepsilon^{4p+m}) & u_{\xi\xi\xi\xi}^{(m)} + 2\alpha u_{\xi\tau}^{(m)} \\ O(\varepsilon^{2p+m}) & \beta_n [(u^{(1)})^n]_{\xi\xi} \end{array} \tag{5.6-11}$$

To obtain a balance between dispersion and nonlinearity we need $4p + m = 2p + n$. Furthermore, it is obvious that it is not necessary to go to values of m any higher than unity, so finally we obtain $\alpha = 1$ and $p = \frac{1}{2}(n - 1)$. This gives the final equation

$$u_{\xi\xi\xi}^{(1)} + \beta_n [(u^{(1)})^n]_\xi + 2u_\tau^{(1)} = 0 \tag{5.6-12}$$

which is the same generalised KdV equation as (5.5-23).

5.7 Problems

1. In viscous gas dynamics, the equations of conservation of mass and momentum are given by

$$\rho_t + (\rho u)_x = 0$$

$$(\rho u)_t + (\rho u^2 + P - \mu u_x)_x = 0$$

where ρ and u are the density and velocity respectively, and μ is a positive constant. The variable P, which is the pressure, is related to the density in the following way: $P = A(\rho/\rho_o)^\gamma$ where ρ_o is the equilibrium density and A a constant.

By considering the linearised version of the above equation, show that the stretched co-ordinates needed for finding the evolution of weakly nonlinear long waves are

$$\xi = \varepsilon^{p}(x - \alpha t) \; ; \; \tau = \varepsilon^{2p}t$$

When the density ρ and the velocity u are expanded as a series in ε, show that $p = 1$ is the value of the index which gives Burger's equation for $\rho^{(1)}$ in the form

$$2\rho_{\tau}^{(1)} + \alpha\rho_{o}^{-1}(1 + \gamma)\rho^{(1)}\rho_{\xi}^{(1)} = \rho_{o}^{-1}\mu\rho_{\xi\xi}^{(1)}$$

where $\alpha^{2} = A\gamma\rho_{o}^{-1}$.

2. Consider the coupled pair of equations

$$N_{x} = -\tfrac{1}{2}(E^{2})_{t}$$

$$E_{xxt} + E_{t} = (EN)_{x}$$

where $N \to -1$, $E \to 0$ as $|x| \to \infty$. By expanding E as

$$E = \varepsilon E^{(1)} + \varepsilon^{2}E^{(2)} + \ldots$$

find the expansion for N and show that $\xi = \varepsilon(x-t)$; $\tau = \varepsilon^{3}t$ are the appropriate stretched co-ordinates. In these co-ordinates show that the evolution equation for E is the MKdV equation

$$E_{\xi\xi\xi}^{(1)} + [\tfrac{1}{2}(E^{(1)})^{3}]_{\xi} = E_{\tau}^{(1)}$$

3(a). The equation of Kac and van Moerbeke (KvM) (1975) is given by

$$\dot{u}_{n} = \tfrac{1}{2}[\exp(u_{n+1}) - \exp(u_{n-1})]$$

By transferring into the continuum approximation $u_{n}(t) \to u(n,t)$, show that when the lowest order dispersive and nonlinear terms are included, the KvM equation is approximated by

$$\frac{\partial u}{\partial t} = \frac{1}{6}\frac{\partial^{3}u}{\partial n^{3}} + u\frac{\partial u}{\partial n} + \frac{\partial u}{\partial n}$$

(b). By defining a new variable x_{n} such that

$$u_n = x_n - x_{n+1}$$

Show that the x_n satisfy a half-Toda lattice taking only even points ($\xi_j = x_{2n}$) (Moser 1975)

$$\ddot{\xi}_j = \exp(\xi_{j-1} - \xi_j) - \exp(\xi_j - \xi_{j+1})$$

4. In many matrix problems involving stretched co-ordinates in one spatial dimension for KdV type problems but also involving a further variable y as in Section 5.4, equations arise of the form

$$\underline{\underline{L}}\underline{\phi} = \underline{f}_1(y)A_\tau + \underline{f}_2(y)A^m A_\xi + \underline{f}_3(y)A_{\xi\xi\xi} \equiv R$$

$\underline{\underline{L}}$ is a differential operator in y. If $\underline{\phi} = \underline{u}$ is a solution of the homogeneous problem $\underline{\underline{L}}\underline{\phi} = 0$ and $\underline{v}$ is a solution of the adjoint problem $\underline{\underline{L}}^* \underline{v} = 0$, show that a solvability condition is given by

$$\alpha_1 A_\tau + \alpha_2 A^m A_\xi + \alpha_3 A_{\xi\xi\xi} = 0$$

where

$$\alpha_i = \langle \underline{v}, \underline{f}_i(y) \rangle$$

and the brackets $\langle\ \rangle$ denote an inner product.

5. Apply the method described in Problem 4 to recover equation (5.4-47) from (5.4-44).

6. The equation

$$v_{tt} - c^2 v_{xx} + \alpha L(v^2)_{xx} + L^2 v_{xxtt} = 0$$

arises in longitudinal waves in elastic rods (Ostrovskii and Sutin 1977) where $L \gg 1$ (α, $c \sim O(1)$). Use L^{-1} as a pertubation parameter to show that with stretched co-ordinates $\xi = (x-ct)L^{-1}$ and $\tau = ctL^{-1}$ and $v = L^{-1}u$, after ignoring derivatives in τ greater than one, we obtain

$$u_\tau = \alpha c^{-2} u u_\xi + \tfrac{1}{2} u_{\xi\xi\xi} - u_{\xi\xi\tau}$$

7. For the partial differential equation

$$\frac{\partial^n u}{\partial x^n} + \frac{\partial^r}{\partial x^r}(u^m) + \frac{\partial^2 u}{\partial x^2} - \frac{\partial^2 u}{\partial t^2} = 0$$

$n > r$; $m > 1$, show that, with co-ordinates $\xi = \varepsilon^p(x-t)$, $\tau = \varepsilon^{(n-1)p}t$ and $u = \varepsilon u^{(1)} + \ldots$, a balance between terms is obtained at the lowest order of ε, if p is chosen as $p = (m-1)/(n-r)$.

8. A model which generalises the work in Section 5.4 to two fluid layers is given by

$$\left(\frac{\partial}{\partial t} + \frac{\partial\psi_i}{\partial x}\frac{\partial}{\partial y} - \frac{\partial\psi_i}{\partial y}\frac{\partial}{\partial x}\right)\left[\nabla^2\psi_i + (-1)^i(\psi_i - \psi_j)F + \beta y\right] = 0$$

$$i \neq j = 1,2.$$

A double eigenvalue problem occurs in this case and two phase speeds a_1 and a_2 are needed. Consequently we must alter the scaling slightly to $\xi = \varepsilon^{\frac{1}{2}}x$; $\tau = \varepsilon^{3/2}t$ and take

$$\psi_i = -\int \bar{U}(y)\,dy + \Psi_i \qquad \bar{U} = U_i(y) - a_i$$

$$\Psi_i = \varepsilon\Psi^{(1)} + \varepsilon^2\Psi^{(2)} + \ldots \qquad i = 1,2.$$

Show that with the same boundary conditions as in Section 5.4 ($Y_i = 0$ at $y = y_1$ and $y = y_2$), the eigenvalue problem is given by

$$Y_i'' + \nu_i(y)Y_i = 0$$

where

$$\nu_i = [\beta - \bar{U}_i'' - (-1)^i F(\bar{U}_i - \bar{U}_j)]\bar{U}_i^{-1} \qquad i \neq j = 1,2.$$

$$\Psi_i^{(1)} = A(\xi,\tau)Y_i(y)$$

Apply the method of Problem 4 or otherwise to show that A must satisfy the KdV equation

$$A_\tau + \mu AA_\xi + \gamma A_{\xi\xi\xi} = 0$$

with μ and γ given by

$$\mu = \sum_{i=1}^{2}\int_{y_2}^{y_1}\nu_i Y_i^3/\bar{U}_i\,ds \Big/ \sum_{i=1}^{2}\int_{y_2}^{y_1}\nu_i Y_i^2/\bar{U}_i\,ds$$

$$\gamma = \sum_{i=1}^{2}\int_{y_2}^{y_1}Y_i^2\,ds \Big/ \sum_{i=1}^{2}\int_{y_2}^{y_1}\nu_i Y_i^2/\bar{U}_i\,ds$$

6. THE ZAKHAROV-SHABAT/AKNS INVERSE METHOD

6.1 The Forward Problem for the Zakharov-Shabat scattering problem and a Class of Solvable Equations

The inverse method and the structure of the KdV equation would have remained a mathematical curiosity, if further important physical equations had not been found that were solvable in this way. However, in 1971, in a paper of fundamental importance, Zakharov and Shabat showed that the nonlinear Schrödinger equation

$$iQ_t + 2Q|Q|^2 + Q_{xx} = 0 \qquad (6.1\text{-}1)$$

could also be solved by the inverse method for initial data which decayed sufficiently fast as $|x| \to \infty$. The Lax pair found by them is given by the formulae

$$L_1 \equiv i\begin{pmatrix} 1+p & 0 \\ 0 & 1+p \end{pmatrix}\frac{\partial}{\partial x} - i(1-p^2)^{\frac{1}{2}}\begin{pmatrix} 0 & -Q^* \\ Q & 0 \end{pmatrix}$$

$$A_1 \equiv ip\begin{pmatrix} 1 & 0 \\ 0 & 1 \end{pmatrix}\frac{\partial^2}{\partial x^2} + i\begin{pmatrix} -(1-p)|Q|^2 & (1-p^2)^{\frac{1}{2}}Q_x^* \\ (1-p^2)^{\frac{1}{2}}Q_x & (1+p)|Q|^2 \end{pmatrix} \qquad (6.1\text{-}2)$$

where $p > 0$. L_1 is an isospectral operator, that is the eigenvalues of L_1 are time independent, provided (L_1, A) satisfy the Lax equation

$$L_{1t} = [A_1, L_1] \qquad (6.1\text{-}3)$$

A calculation shows that in this case (6.1-3) is equivalent to the nonlinear Schrödinger equation (6.1-1).

By transforming the eigenfunctions (Y_1) of L_1

$$Y_1 = \exp(-ikp^{-1}x)\begin{pmatrix} 0 & (1-p)^{\frac{1}{2}} \\ (1+p)^{\frac{1}{2}} & 0 \end{pmatrix} Y_2 \qquad (6.1\text{-}4)$$

and rescaling the eigenvalues (λ) of L_1

$$k = \lambda p(1-p^2)^{-1} \tag{6.1-5}$$

the Lax pair (6.1-2) can be replaced by the operators

$$L_2 \equiv i \begin{pmatrix} \partial/\partial x & -Q \\ -Q^* & -\partial/\partial x \end{pmatrix} \tag{6.1-6}$$

$$_fA_2 \equiv \begin{pmatrix} i|Q|^2 - 2ik^2 & iQ_x + 2Qk \\ iQ^*_x - 2Q^*k & -i|Q|^2 + 2ik^2 \end{pmatrix} + f(k) \begin{pmatrix} 1 & 0 \\ 0 & 1 \end{pmatrix} \tag{6.1-7}$$

where $f(k) = ik(p + p^{-1})$. In obtaining the final form of (6.1-7), the eigenvalue problem $L_2y = ky$ was used repeatedly to eliminate the x-derivatives of the eigenfunctions. It is clear that we have a family of operators $_fA_2$ which will formally serve as the evolution operator in the Lax equation since L_2 commutes with an x-independent diagonal matrix. The Lax equation is now only formally valid because $_fA_2$ depends upon the eigenfunction. However, we can easily reverse the last step made in obtaining (6.1-7) and rewrite it as a differential operator or require the Lax equation to be independent. Another point to note is that L_2 is no longer self-adjoint so that the analysis in this and the succeeding sections shows considerable differences when compared with the inverse method involving the isospectral Schrödinger equation.

The key observation of Ablowitz, Kaup, Newell, and Segur (AKNS) (Ablowitz et al. 1973, 1974) was that the inverse problem posed by (6.1-6),(6.1-7) could be generalised to the problem represented by

$$L \equiv i \begin{pmatrix} \partial/\partial x & -Q(x,t) \\ R(x,t) & -\partial/\partial x \end{pmatrix} \tag{6.1-8}$$

$$_fA \equiv \begin{pmatrix} A(x,t,k) & B(x,t,k) \\ C(x,t,k) & -A(x,t,k) \end{pmatrix} + f(k)I \tag{6.1-9}$$

The corresponding nonlinear equations arising from the formal Lax equation are

$$\begin{aligned} A_x - QC + RB &= 0 \\ Q_t - B_x - 2AQ + 2ikB &= 0 \\ R_t - C_x + 2AR + 2ikC &= 0 \end{aligned} \tag{6.1-10}$$

A technique for obtaining some of the associated equations is to substitute polynomials in k for A, B, C and solve (6.1-10) recursively. In this way one quickly establishes that, besides the nonlinear Schrödinger equation, the following important equations can also be associated with the isospectral operator (6.1-8).

The modified KdV equation: $Q_t \pm 6Q^2Q_x + Q_{xxx} = 0$

$$A = -4ik^3 \pm 2iQ^2;$$
$$B = 4Qk^2 + 2iQ_xk \pm 2Q^3 - Q_{xx};$$
$$C = \pm 4Qk^2 \pm 2iQ_xk + 2Q^3 \pm Q_{xx};$$
$$R = \pm Q.$$

The KdV equation: $Q_t + 6QQ_x + Q_{xxx} = 0$

$$A = -4ik^3 + 2iQ - Q_x;$$
$$B = 4Qk^2 + 2iQ_xk - 2Q^2 - Q_{xx};$$
$$C = -4k^2 + 2Q;$$
$$R = -1.$$

In addition, inverse powers of k also generate solvable equations.

The sine-Gordon equation: $U_{xt} = \pm \sin U$

$$A = \pm \frac{i}{4k} \cos U; \quad B = \pm \frac{i}{4k} \sin U; \quad C = \pm \frac{i}{4k} \sin U;$$
$$R = -Q = \tfrac{1}{2}U_x$$

For this case the corresponding evolution operator in the Lax equation is an integral operator. In fact it is not hard to show that it is defined by

$$A = \pm \frac{1}{4} \begin{pmatrix} I_1 & I_2 \\ I_2 & I_1 \end{pmatrix} \tag{6.1-11}$$

where $\quad I_1 v(x) = \int^x \cos\tfrac{1}{2}(U(x) + U(y))v(y)dy$

and $\quad I_2 v(x) = \int^x \sin\tfrac{1}{2}(U(x) + U(y))v(y)dy$

For an arbitrary formal A operator the corresponding Lax operator will in general be an integro-differential operator. For most of this section we shall omit the explicit t-dependency of all functions since the operations we use only involve differentiation and integration with respect to x and k at the same time.

We now introduce the Hilbert space of functions $L^2_{(2)}(R)$ of complex $\mathbb{C}^2$-valued functions which are square integrable on R with respect to the Lebesgue measure. The inner produce on $L^2_{(2)}(R)$ is defined by

$$< V,U > = \int_{-\infty}^{\infty} (v_1(x)u_1^*(x) + v_2(x)u_2^*(x))dx \tag{6.1-12}$$
$$V,U \in L^2_{(2)}(R)$$

We can define an operator which we shall also denote by L which acts in $L^2_{(2)}(R)$ using the differential operator (6.1-8). We require that if $v \in D$ then $L\,v \in D_L$. It is necessary, as in the isospectral Schrödinger operator case, that Q and R should not be too singular. We certainly assume that they are summable in every interval $]a,b[$, $-\infty < a < b < \infty$. The precise conditions on Q and R used in this book are given in Theorems 6-1 and 6-6.

In the remainder of this section we shall treat the forward or direct scattering problem for the operator L in the case when neither R nor Q are constant functions. The case R = constant reduces to the isospectral Schrödinger operator for which the forward problem with L a 2×2 first order differential matrix operator was analysed in the exercises to Chapter 3. The material and the development which we follow in this section roughly corresponds to Sections 3.3, 3.4 and 3.5 for the isospectral Schrödinger operator. The reader is advised to bear this material in mind because of the terser exposition given in this section and also in order to compare the two methods.

The Jost solutions for

$$LY = kY \tag{6.1-13}$$

are defined by the boundary conditions

$$\begin{aligned} &\mathop{\mathrm{Lt}}_{x\to\infty} e^{-ikx}\psi(x,k) = \begin{pmatrix}0\\1\end{pmatrix} \qquad \mathop{\mathrm{Lt}}_{x\to\infty} e^{ikx}\bar{\psi}(x,k) = \begin{pmatrix}1\\0\end{pmatrix} \\ &\mathop{\mathrm{Lt}}_{x\to-\infty} e^{ikx}\phi(x,k) = \begin{pmatrix}1\\0\end{pmatrix} \qquad \mathop{\mathrm{Lt}}_{x\to-\infty} e^{-ikx}\bar{\phi}(x,k) = \begin{pmatrix}0\\1\end{pmatrix} \end{aligned} \tag{6.1-14}$$

That is they have the same asymptotic behaviour as $|x| \to \infty$ as the eigenfunctions of (6.1-13) with $R \equiv 0$, $Q \equiv 0$. To establish conditions under which these solutions exist and are unique as well as to determine whether they admit analytic continuations for complex k, we apply the method of successive approximations to the integral equation equivalent to (6.1-13) with boundary conditions (6.1-14). In particular we obtain the representation

$$e^{-ikx}\psi(x,k) = \begin{pmatrix}0\\1\end{pmatrix} - \int_x^\infty P(x,y,k)e^{-iky}\psi(y,k)\,dy \tag{6.1-15}$$

where

$$P(x,y,k) \equiv e^{ik(y-x)}\begin{pmatrix} 0 & e^{ik(y-x)}Q(y) \\ e^{-ik(y-x)}R(y) & 0 \end{pmatrix}$$

If we define $H(x,k) = \exp(-ikx)\psi(x,k)$ then we require to show that

$$H(x,k) = \sum_{j=0}^{\infty} H^j(x,k) \tag{6.1-16}$$

that is that the sum in (6.1-16) converges uniformly onto $H(x,k)$. The terms in the series are defined by

$$H^0 \equiv \begin{pmatrix}0\\1\end{pmatrix} \tag{6.1-17}$$

and

$$H^j(x,k) = \int_x^\infty P(x,y,k)H^{j-1}(y,k)\,dy \qquad j \geqslant 1 \tag{6.1-18}$$

We use the norms

$$|Y(x_0)| = |y_1(x_0)| + |y_2(x_0)|$$

and

$$|P(x_1,x_0,k)| = \sum_{i,j}^{2} |P_{ij}(x_1,x_0,k)| \qquad x_1,x_0 \in R \tag{6.1-19}$$

Consider $V \in L^1_{(2)}(R)$ so that $\int_x^\infty |V(y)|dy$ is defined but V does not necessarily belong to $L^2_{(2)}(R)$. Then

$$\begin{aligned} |\int_x^\infty V(y)\,dy| &\leqslant |\int_x^\infty v_1(y)\,dy| + |\int_x^\infty v_2(y)\,dy| \\ &\leqslant \int_x^\infty |v_1(y)|dy + \int_x^\infty |v_2(y)|dy \\ &= \int_x^\infty |V(y)|dy \end{aligned} \tag{6.1-20}$$

We also have that

$$\begin{aligned} &|P(x_1,x_0,k)\;Y(x_0)| \leqslant |P(x_1,x_0,k)||Y(x_0)| \\ &= e^{\eta(x_1-x_0)}\{e^{\eta(x_1-x_0)}|Q(x_0)| + e^{-\eta(x_1-x_0)}|R(x_0)|\}|Y(x_0)| \end{aligned} \tag{6.1-21}$$

where $\eta = \text{Im}k$. Define $P = (|Q| + |R|)$ so that we obtain from (6.1-18) that

$$|H^j(x,k)| \leqslant \int_x^\infty P(y_j)e^{(\eta-|\eta|)(x-y_j)} \cdots \int_{y_2}^\infty P(y_1)e^{(\eta-|\eta|)(y_2-y_1)}\,dy_1\ldots dy_j \tag{6.1-22}$$

Evaluating the right hand side of (6.1-22) we find that for $\text{Im}k \geqslant 0$

$$|H^j(x,k)| \leqslant \frac{1}{j!}\left(\int_x^\infty P(y)\,dy\right)^j \tag{6.1-23}$$

and consequently

$$|H(x,k)| \leqslant \exp\{\int_x^\infty P(y)dy\} \tag{6.1-24}$$

Thus the series is absolutely and uniformly convergent for $x \in R$ and $\mathrm{Im}\, k \geqslant 0$ provided $S_0(\infty) < \infty$ where

$$S_j^{\pm}(x) = \pm\int_x^{\pm\infty} |y|^j (|Q(y)| + |R(y)|dy \quad \text{and } S_j(\infty) = S_j^{\pm}(\infty) \tag{6.1-25}$$

The uniqueness proof is standard and the uniform convergence of the series for H proves the analyticity of $\psi(x,k)$ in $\mathrm{Im}\, k > 0$ and the continuity in $\mathrm{Im}\, k \geqslant 0$. From (6.1-15) it follows that H_k satisfies the equation

$$H_k(x,k) = -\int_x^\infty P(x,y,k)\, H_k(y,k)dy - \int_x^\infty P_k(x,y,k)H(y,k)dy \tag{6.1-26}$$

Since

$$|P_k| \leqslant 2(y-x)(|Q| + |R|) \qquad \mathrm{Im}k \geqslant 0, \quad y \geqslant x \tag{6.1-27}$$

if $\{S_0(\infty), S_1(\infty)\} < \infty$ it is easy to see that

$$|\int_x^\infty P_k(x,y,k)H(y,k)dy| \leqslant K\{1-\min(x,0)\} \tag{6.1-28}$$

for some constant K. If we substitute this result into the norm of (6.1-26) and iterate we obtain

$$|H_k(x,k)| \leqslant K\{1-\min(x,0)\}\exp S_0^+(x) \tag{6.1-29}$$

Then we deduce from the local uniform convergence of the series for $H_k(x,k)$ that $H_k(x,k)$ is analytic $\mathrm{Im}\, k > 0$ and continuous in $\mathrm{Im}\, k \geqslant 0$ provided $S_0(\infty) < \infty$ and $S_1(\infty) < \infty$. We also have from repeating this process that the rth derivative of ψ exists and is continuous for real k provided $S_j(\infty) < \infty$, $j = 0,\ldots,r$. A sufficient condition for ψ to be analytic for real k is therefore that $S_j(\infty) < \infty$ for all j. It is also clear from the proof for the existence of ψ that if Q and R have compact support they are analytic throughout the complex k-plane. Analogous results are easily proved for the other Jost solutions.

Theorem 6-1: Provided $S_0(\infty) < \infty$ the isospectral operator equation $Ly = ky$ has unique solutions, called the Jost solutions, which satisfy the integral equations

$$\psi(x,k) = \binom{0}{1} e^{ikx} - \int_x^{\infty} M(x,y,k)\ \psi(y,k)dy$$

$$\phi(x,k) = \binom{1}{0} e^{-ikx} + \int_{-\infty}^{x} M(x,y,k)\phi(y,k)dy$$

$$\bar{\psi}(x,k) = \binom{1}{0} e^{-ikx} - \int_x^{\infty} M(x,y,k)\ \bar{\psi}(y,k)dy$$

$$\bar{\phi}(x,k) = \binom{0}{1} e^{-ikx} + \int_{-\infty}^{x} M(x,y,k)\ \bar{\phi}(y,k)dy$$

where
$$M(x,y,k) = \begin{pmatrix} 0 & e^{-ik(x-y)}Q(y) \\ e^{ik(x-y)}R(y) & 0 \end{pmatrix}$$

For every $x \in R$ the solutions $\phi(x,k)$, $\psi(x,k)$ are continuous with respect to k, $\text{Im}\, k \geqslant 0$ and analytic with respect to k in $\text{Im}\, k > 0$. If $S_1(\infty) < \infty$ then $\phi_k(x,k)$ and $\psi_k(x,k)$ are analytic in $\text{Im}\, k > 0$ and continuous in $\text{Im}\, k \geqslant 0$. The analogous properties hold for the functions $\bar{\phi}(x,k), \bar{\psi}(x,k), \bar{\phi}_k(x,k)$ and $\bar{\psi}_k(x,k)$ in $\text{Im}\, k \leqslant 0$. The solutions have the uniform asymptotic behaviour

$$\psi(x,k) = e^{ikx}(\binom{0}{1} + o(1)), \quad \bar{\psi}(x,k) = e^{-ikx}(\binom{1}{0} + o(1) \quad \text{as } x \to +\infty$$

$$\phi(x,k) = e^{-ikx}(\binom{1}{0} + o(1)), \quad \bar{\phi}(x,k) = e^{ikx}(\binom{0}{1} + o(1) \quad \text{as } x \to -\infty$$

Corollary 6-1-1: For all $x \in R$ and every k such that $\text{Im}\, k > 0$ there exists a real positive number C such that

$$|\phi(x,k)| \leqslant C \exp \eta x, \quad |\psi(x,k)| \leqslant C \exp -\eta x$$

Corollary 6-1-2: If Q and R have compact support then the Jost solutions are analytic throughout the complex k-plane.

For this case we can derive from (6.1-13) for solutions ${}_{k_1}U$, ${}_{k_2}V$ the Wronskian relationship

$$\begin{aligned} &W({}_{k_1}U,{}_{k_2}V)(x_0) - W({}_{k_1}U,{}_{k_2}V)(x) \\ &= i(k_2-k_1)\int_x^{x_0}(u_1(y,k_1)\ v_2(y,k_2) + u_2(y,k_1)\ v_1(y,k_2))dy \end{aligned} \tag{6.1-30}$$

From this relationship we see in particular that the Wronskians of the Jost solutions are **independent** of x if evaluated at the same value of k. By analogy with the material of Section 3.3, we can introduce the scattering functions a, $\bar{a}$, b, $\bar{b}$ through the Wronskians

$$W(\phi,\psi) \equiv a \quad \mathrm{Im}k \geqslant 0 \qquad W(\bar{\psi},\phi) \equiv b \quad \mathrm{Im}k = 0 \tag{6.1-31}$$

$$W(\bar{\psi},\bar{\phi}) \equiv \bar{a} \quad \mathrm{Im}k \leqslant 0 \qquad W(\bar{\phi},\psi) \equiv \bar{b} \quad \mathrm{Im}k = 0 \tag{6.1-32}$$

$$W(\psi,\bar{\psi}) \equiv -1 \quad \mathrm{Im}k = 0 \qquad W(\phi,\bar{\phi}) \equiv 1 \quad \mathrm{Im}k = 0 \tag{6.1-33}$$

The Wronskians in (6.1-33) are evaluated from the asymptotic properties of the Jost solutions as $|x| \to \infty$. Under the conditions of Theorem 6-1 the scattering functions are not defined throughout the complex k-plane; their domain of existence is given to the right hand side of the definitions in (6.1-31)-(6.1-33). For fixed k, any pair of Jost solutions which are defined for this value may be used as a basis of the solution space to (6.1-13) (k fixed) provided their Wronskian is non-zero. It is straightforward to show from (6.1-31)-(6.1-33) that

$$\begin{aligned} \phi(x,k) &= a(k)\bar{\psi}(x,k) + b(k)\psi(x,k) \\ \bar{\phi}(x,k) &= \bar{a}(k)\psi(x,k) + \bar{b}(k)\bar{\psi}(x,k) \end{aligned} \tag{6.1-34}$$

It is convenient for later work to introduce the fundamental matrix solutions $\Phi = (\phi,\bar{\phi})$, $\Psi = (\psi,\bar{\psi})$ whose columns are the Jost solutions. Then (6.1-34) can be written as

$$\Phi(x,k) = \Psi(x,k)\ A(k) \tag{6.1-35}$$

where

$$A(k) = \begin{pmatrix} b(k) & \bar{a}(k) \\ a(k) & \bar{b}(k) \end{pmatrix}$$

By expanding determinants we find that

$$\begin{aligned} &W({}_k\Phi,{}_k\Psi)\ W({}_k\bar{\Phi},{}_k\bar{\Psi}) - W({}_k\Phi,{}_k\bar{\Psi}) \,.\, W({}_k\bar{\Phi},{}_k\Psi) \\ &= W({}_k\Phi,{}_k\bar{\Phi}) \,.\, W({}_k\Psi,{}_k\bar{\Psi}) \end{aligned} \tag{6.1-36}$$

which can be written using the definitions (6.1-31)-(6.1-33) as

$$a(k)\bar{a}(k) - b(k)\bar{b}(k) = 1 \tag{6.1-37}$$

The properties of the scattering functions are easily obtained from their integral representations. These are found by equating the asymptotic expansions as $x \to \infty$ for the Jost solutions ϕ and $\bar{\phi}$ using (6.1-34) and the integral representations given in Theorem 6-1. This leads to the formulae

$$a(k) = 1 + \int_{-\infty}^{\infty} Q(y)e^{iky}\phi_2(y,k)\,dy \tag{6.1-38}$$

$$b(k) = \int_{-\infty}^{\infty} R(y)e^{-iky}\phi_1(y,k)\,dy \tag{6.1-39}$$

$$\bar{a}(k) = 1 + \int_{-\infty}^{\infty} R(y)e^{-iky}\bar{\phi}_1(y,k)\,dy \tag{6.1-40}$$

$$\bar{b}(k) = \int_{-\infty}^{\infty} Q(y)e^{iky}\bar{\phi}_2(y,k)\,dy \tag{6.1-41}$$

from which we deduce using Theorem 6-1 that all the scattering functions are uniquely defined and continuous for $\text{Im } k = 0$. The functions $a, \bar{a}$ admit unique analytic continuations into $\text{Im } k > 0$, $\text{Im } k < 0$ respectively. In general the functions $b, \bar{b}$ are not defined for complex k unless Q and R have compact support or obey some stronger conditions than those given in the theorem. However, b and $\bar{b}$ **are defined** at a zero of $a, \bar{a}$ respectively since in this case the integrals in (6.1-39) or (6.1-41) converge. From (6.1-31)-(6.1-33), this is seen to correspond to the cases when either ϕ, ψ are linearly dependent (at a zero of a) or when $\bar{\phi}$, $\bar{\psi}$ are linearly dependent (at a zero of $\bar{a}$). Letting k_j, $\bar{k}_j$ be zeros of $a, \bar{a}$ respectively, we have

$$\begin{aligned} \phi_j &= b_j\psi_j \\ \bar{\phi}_j &= \bar{b}_j\bar{\psi}_j \end{aligned} \tag{6.1-42}$$

where $V_j \equi {}_{k_j}V$ and $b_j, \bar{b}_j$ are the evaluations of $b, \bar{b}$ defined by (6.1-39), (6.1-41) at $k_j, \bar{k}_j$ respectively.

From Theorem 6-1 we see from the asymptotic properties $|x| \to \infty$ of the Jost solutions that if $\text{Im } k_j > 0$ or $\text{Im } \bar{k}_j < 0$ then (6.1-42) define eigenfunctions of L. In the case $\text{Im } k_j = 0$ or $\text{Im } \bar{k}_j = 0$, the relations (6.1-42) do not define proper eigenfunctions of L. In this case $\{k_j, \bar{k}_j\}$ are called **spectral singularities** of L.

Another contrast with the Schrödinger operator is that L can have multiple eigenvalues. However, before considering the consequences of these properties for the spectral expansion of an element $V \in L^2_{(2)}(\mathbb{R})$ let us obtain some more information about the number of eigenvalues. To this end we derive the asymptotic expansion for $a, \bar{a}$ as $|k| \to \infty$. For completeness we also include the large k properties of the Jost solutions and the other scattering functions. The results are easily obtained from the integral representations of Theorem 6-1, and equations (6.1-38)-(6.1-41).

<u>**Lemma 6-2:**</u> The Jost solutions have the uniformly valid asymptotic expansions

$$\phi e^{ikx} = \binom{1}{0} - \frac{1}{2ik}\begin{pmatrix} \int_{-\infty}^{x} R(y)Q(y)\,dy \\ R(x) \end{pmatrix} + o(|k|^{-1})$$

$$\psi e^{-ikx} = \binom{0}{1} + \frac{1}{2ik}\begin{pmatrix} Q(x) \\ -\int_{x}^{\infty} R(y)Q(y)\,dy \end{pmatrix} + o(|k|^{-1})$$

$$\bar{\phi} e^{-ikx} = \binom{0}{1} + \frac{1}{2ik}\begin{pmatrix} Q(x) \\ \int_{-\infty}^{x} R(y)Q(y)\,dy \end{pmatrix} + o(|k|^{-1})$$

$$\bar{\psi} e^{ikx} = \binom{1}{0} + \frac{1}{2ik}\begin{pmatrix} \int_{x}^{\infty} R(y)Q(y)\,dy \\ -R(x) \end{pmatrix} + o(|k|^{-1})$$

as $|k| \to \infty$ within their domains of definition provided $S_0(\infty) < \infty$, Q and R are differentiable, and $Q_x, R_x \in L^1(R)$.

Lemma 6-3: Under the conditions of Theorem 6-1 the scattering functions $a(k)$, $\bar{a}(k)$, $b(k)$, $\bar{b}(k)$ exist and are uniquely defined and continuous in $\text{Im } k \leqslant 0$, $\text{Im } k \leqslant 0$, $\text{Im } k = 0$, $\text{Im } k = 0$ respectively. The functions $a(k)$, $\bar{a}(k)$ are analytic in $\text{Im } k > 0$, $\text{Im } k < 0$ respectively. A zero k_0 of a or $\bar{a}$ is either an eigenvalue ($\text{Im } k_0 > 0$ or $\text{Im } k_0 < 0$) or a spectral singularity ($\text{Im } k_0 = 0$) of L. If Q and R are differentiable and $Q_x, R_x \in L^1(R)$ then the scattering functions have the asymptotic behaviour

$$a(k) = 1 - \frac{1}{2ik}\int_{-\infty}^{\infty} R(y)Q(y)\,dy + o(|k|^{-1})$$

$$\bar{a}(k) = 1 + \frac{1}{2ik}\int_{-\infty}^{\infty} R(y)Q(y)\,dy + o(|k|^{-1})$$

$$b(k) = o(|k|^{-1})$$

$$\bar{b}(k) = o(|k|^{-1})$$

as $|k| \to \infty$.

From the results of Lemma 6-3 we observe that a, $\bar{a}$ are analytic and bounded in the upper half k-plane, respectively the lower half k-plane. It follows that the zeros of $a(k)$, $(\bar{a}(k))$ are bounded in the upper (lower) half k-plane. Since $a(k)$ $(\bar{a}(k))$ is analytic in the upper (lower) half k-plane, the set of roots is no more than countable and can have limit points only on the real axis.

We now define the resolvent operator $_kR$ for L in order to determine the continuous spectrum for L and also so that we can obtain the resolution of the identity in terms of the principal functions associated with L. The resolvent operator $_kR = (L - kI)^{-1}$ is defined by

$$ {}_k RV = \int_{-\infty}^{\infty} R(x,y,k)V(y)\,dy \qquad V \in L^2_{(2)}(R) \tag{6.1-43}$$

where the resolvent kernel or Green's function $R(x,y,k)$ satisfies the conditions

$$ (L-kI)\ R(x,y,k) = \delta(x-y)I \tag{6.1-44}$$

$$ \lim_{|x|\to\infty} R(x,y,k) = 0 \qquad \lim_{|y|\to\infty} R(x,y,k) = 0 \tag{6.1-45}$$

Clearly considering $R(x,y,k)$ as a function of x with y and k ($\mathrm{Im}\,k > 0$) fixed we must have

$$ R(x,y,k) = {}_k\phi(x)\,{}_kU^T(y) + {}_k\psi(x)\,{}_kV^T(y) \tag{6.1-46}$$

where T is the transpose operation. Imposing the conditions (6.1-45) and their transpose then gives

$$ R(x,y,k) = \begin{cases} {}_k\psi(x)\ {}_k\phi^T(y)\ {}_kM & x > y \\ {}_k\phi(x)\ {}_k\psi^T(y)\ {}_kM & x < y \end{cases} \qquad \mathrm{Im}k > 0 \tag{6.1-47}$$

The matrix ${}_kM$ is determined from the jump condition

$$ \lim_{\varepsilon\to 0} R(x,y,k)\Big|_{x=y-\varepsilon}^{x=y+\varepsilon} = i\begin{pmatrix} -1 & 0 \\ 0 & 1 \end{pmatrix} \tag{6.1-48}$$

which yields upon using the Wronskian relationship (6.1-31) that

$$ {}_kM = \frac{i}{a(k)}\sigma_1, \text{ where } \sigma_1 = \begin{pmatrix} 0 & 1 \\ 1 & 0 \end{pmatrix}$$

If we define

$$ V^A \equiv V^T\sigma_1 = (v_2, v_1) \tag{6.1-49}$$

then repeating the above calculation for $\mathrm{Im}\,k < 0$ we finally arrive at

$$ R(x,y,k) = \begin{cases} \dfrac{i}{a(k)}\ \psi(x,k)\,\phi^A(y,k) & x > y \\ \dfrac{i}{a(k)}\ \phi(x,k)\,\psi^A(y,k) & x < y \end{cases} \quad \mathrm{Im}k > 0$$

$$ R(x,y,k) = \begin{cases} \dfrac{-i}{\bar{a}(k)}\ \bar{\psi}(x,k)\,\bar{\phi}^A(y,k) & x > y \\ \dfrac{-i}{\bar{a}(k)}\ \bar{\phi}(x,k)\,\bar{\psi}^A(y,k) & x < y \end{cases} \quad \mathrm{Im}k < 0 \tag{6.1-50}$$

Now the resolvent operator is only defined if k does not belong to the spectrum of L. The set of such regular points is called the resolvent set of (R_L) of L. The set of non-regular points in the spectrum of L which are not eigenvalues form the continuous spectrum of L. We now show that the continuous spectrum of L is the real axis $\text{Im } k = 0$. We first prove the lemma

Lemma 6-4: For every $\eta = \text{Im } k > 0$

$$A_0 V(x) = \exp(\eta x) \int_x^{\infty} \exp(-\eta y) V(y)\, dy$$

$$B_0 V(x) = \exp(-\eta x) \int_{-\infty}^{x} \exp(\eta y) V(y)\, dy$$

define in $L^2_{(2)}(R)$ linear continuous operators such that

$$\| A_0 \| \leqslant \frac{1}{|\eta|}, \qquad \| B_0 \| \leqslant \frac{1}{|\eta|}$$

The norms for the operators defined by

$$C_0 V(x) = \exp(-\eta x) \int_x^{\infty} \exp(\eta y) V(y)\, dy$$

$$D_0 V(x) = \exp(\eta x) \int_{-\infty}^{x} \exp(-\eta y) V(y)\, dy$$

$\eta < 0$, satisfy the same condition.

Proof: Let $V \in L^2_{(2)}(R)$ and put $U = A_0 V$.

$$|U(x)| \leqslant \exp(\eta x) \int_x^{\infty} |\exp(-\eta y)\ V(y)|\, dy$$

$$= \exp(\eta x) \int_x^{\infty} \begin{pmatrix} \exp(-\eta y) \\ \exp(-\eta y) \end{pmatrix} \cdot \begin{pmatrix} |v_1(y)| \\ |v_2(y)| \end{pmatrix} dy$$

$$\leqslant \exp(\eta x) [\{ \int_{\frac{3x}{4}}^{\infty} \exp(-2\eta y)\, dy\}^{\frac{1}{2}} \{ \int_{\frac{3x}{4}}^{\infty} |V(y)|_E^2\, dy\}^{\frac{1}{2}}$$

$$+ \{ | \int_x^{\frac{3x}{4}} \exp(-2\eta y)\, dy | \}^{\frac{1}{2}} \{ | \int_x^{\frac{3x}{4}} |V(y)|_E^2\, dy | \}^{\frac{1}{2}}]$$

$$= 2\exp(\eta x) \{ \frac{\exp-(3\eta x/2)}{2\eta} \}^{\frac{1}{2}} \{ \int_{\frac{3x}{4}}^{\infty} |V(y)|_E^2\, dy \}^{\frac{1}{2}}$$

$$+ 2\{ | \frac{1-\exp(\eta x/2)}{2\eta} | \}^{\frac{1}{2}} \{ | \int_x^{\frac{3x}{4}} |V(y)|_E^2 dy | \}^{\frac{1}{2}}$$

where $|V(y)|_E^2 = |v_1(y)|^2 + |v_2(y)|^2$. Hence $U(x)$ has the boundary conditions $U(x) \to 0$ as $|x| \to \infty$. Since $dU(x)/dx = \eta U(x) - V(x)$

$$\frac{d}{dx} |U(x)|_E^2 = \eta |U(x)|_E^2 - U^{\dagger}_{(x)} V_{(x)} - V^{\dagger}_{(x)} U_{(x)}$$

where $V^{\dagger} = (V^T)^*$. It then follows from the Cauchy-Schwartz inequality that

$\|A_0V\| \leqslant 1/|\eta|\,(\|V\|)$. Hence since $\|A_0V\| = \sup_{\|V\|=1} \|A_0 V\|$ we have $\|A_0\| \leqslant 1/\eta$. Similar bounds are obtained for B_0, C_0 and D_0 in an analogous fashion.

Theorem 6-5: Under the conditions of Theorem 6-1 the resolvent operator ${}_kR = (L - kI)^{-1}$ is the integral operator

$$ {}_kRV(x) = \int_{-\infty}^{\infty} R(x,y,k)V(y)\,dy $$

with kernel defined by (6.1-50). There exists a number $C > 0$ such that

$$ \|{}_kR\| \leqslant \begin{cases} \dfrac{C}{|a(k)|\,|\eta|} & \text{Im}k > 0 \\[2ex] \dfrac{C}{|\bar{a}(k)|\,|\eta|} & \text{Im}k < 0 \end{cases} $$

and for every $r > 0$ there is a number $c_r > 0$ such that

$$ \|{}_kR\| \geqslant \begin{cases} \dfrac{c_r}{|a(k)|\,|\eta|^{\frac{1}{2}}} & \text{Im}k > 0 \\[2ex] \dfrac{c_r}{|\bar{a}(k)|\,|\eta|^{\frac{1}{2}}} & \text{Im}k < 0 \end{cases} $$

for all k in the domain $\text{Im}\,k \gtrless 0$, $|k| \geqslant r$.

Proof: Let $V \in L^2_{(2)}(R)$ and represent ${}_kR$ as

$$ {}_kRV = (A + B + C + D)V $$

where

$$ AV = \int_x^{\infty} R(x,y,k)V(y)\,dy \qquad \text{Im}k > 0 $$

$$ BV = \int_{-\infty}^{x} R(x,y,k)V(y)\,dy \qquad \text{Im}k > 0 $$

with analogous definition for C and D in $\text{Im}\,k < 0$. Then using Corollary 6-1-1 we obtain

$$ \begin{aligned} |AV(x)| &\leqslant \left|\frac{\phi(x,k)}{a(k)}\right| \left|\int_x^{\infty} \psi^A(y,k)V(y)\,dy\right| \\ &\leqslant \left|\frac{\phi(x,k)}{a(k)}\right| \int_x^{\infty} |\psi^A(y,k)|\,|V(y)|\,dy \end{aligned} $$

$$\leqslant \frac{C}{|a(k)|} e^{\eta x} \int_x^{\infty} e^{-\eta y} |V(y)| dy$$

$$= \frac{C}{|a(k)|} e^{\eta x} \int_x^{\infty} \begin{pmatrix} e^{-\eta y} \\ e^{-\eta y} \end{pmatrix} \begin{pmatrix} |v_1(y)| \\ |v\ (y)| \end{pmatrix} dy$$

Then using Lemma 6-4 we have upon putting $\bar{V} = (|v_1|, |v_2|)^T$

$$|AV(x)| \leqslant \frac{C}{|a(k)|} A_0(\bar{V}(x))$$

$$\leqslant \frac{C}{|a(k)|} |A_0(\bar{V}(x))|$$

$$\leqslant \frac{C}{|a(k)|} \| A_0 \| \| V \|$$

so that

$$\| A \| \leqslant \frac{C}{|a(k)| |\eta|}$$

For the second part of the theorem, introduce the function

$$F_b(x) = \begin{cases} \begin{pmatrix} \psi_2^* \\ \psi_1^* \end{pmatrix} (x, k^*) & b < x < \infty \\ 0 & -\infty < x < b \end{cases} \qquad \text{Im}k > 0$$

Then $F_b \in L^2_{(2)}(R)$ and from (6.1-50)

$${}_kRF_b(x) = \frac{i}{a(k)} \phi(x,k) \| F_b \|^2 \quad \text{for} \quad -\infty < x < b$$

Hence

$$\| {}_kRF_b \|^2 \geqslant \int_{-\infty}^{b} |{}_kRF_b(x,k)|_E^2 \, dx$$

$$= \frac{\| F_b \|^4}{|a|^2} \int_{-\infty}^{b} |\phi|_E^2 \, dx$$

We now choose b sufficiently large and negative so that

$$|\phi|_E > \tfrac{1}{2} \exp \eta x$$

Then $\int_{-\infty}^{b} |\phi|_E^2 \, dx \geq \frac{1}{8\eta} e^{2\eta b}$

and $\| {}_kRF_b \| \geq \frac{\|F_b\|^2 e^{2\eta b}}{|a(k)| \, 2(2\eta)^{\frac{1}{2}}}$

Then since $\|F_b\|^2 = \int_{-\infty}^{b} |\phi|_E^2 dx \leq c_r$ for $|k| \leq r$ the result follows.

Corollary 6-5-1: The resolvent set R_L consists of all k such that $\mathrm{Im}\, k > 0$ or $\mathrm{Im}\, k < 0$ and $a(k) \neq 0$, $\bar{a}(k) \neq 0$.

Corollary 6-5-2: The spectral singularities belong to the continuous spectrum of L which is the real axis $\mathrm{Im}\, k = 0$.

Corollary 6-5-1 follows immediately from the bound on R. Corollary 6-5-2 results from the fact that $\|{}_kR\| \to \infty$ as $k \to k_0$ where $\mathrm{Im}\, k_0 = 0$.

Before considering the spectral expansion we impose a stronger condition on the potential functions Q and R so that we can treat the spectral singularities.

Theorem 6-6: If the functions Q and R satisfy the conditions

$$|Q(x)| \leq C \exp -2\varepsilon|x|$$

$$|R(x)| \leq C_\varepsilon \exp -2\varepsilon|x|$$

then the Jost solutions ϕ, ψ are analytic functions of k when $\mathrm{Im}\, k > -\varepsilon$ and $\bar{\phi}$, $\bar{\psi}$ are analytic functions of k when $\mathrm{Im}\, k < \varepsilon$.

The proof is similar to that used for Theorem 6-1. It is also straightforward to prove the corollary.

Corollary 6-6-1: The scattering functions a, $\bar{a}$ are analytic in $\mathrm{Im}\, k > -\varepsilon$, $\mathrm{Im}\, k < \varepsilon$ respectively. The functions b, $\bar{b}$ are analytic in $-\varepsilon < \mathrm{Im}\, k < \varepsilon$.

For the remainder of this section we shall assume that Theorem 6-6 applies. As a consequence of Corollary 6-6-1 and Lemma 6-3 we see that the set of zeros of a and $\bar{a}$ is finite. Let M, N be the number of real, complex zeros of a respectively, and let the barred quantities denote the real and complex zeros of $\bar{a}$. The multiplicity m_j ($\bar{m}_j$) of a zero k_j of a ($\bar{k}_j$ of $\bar{a}$) is called the multiplicity of the **singular value** of L.

It is necessary at this point to introduce the transformation operators associated with the Jost solutions of the scattering problem (6.1-13). This will enable us to investigate the k-derivatives of these solutions fairly easily.

Theorem 6-7: Under the conditions of Theorem 6-1 the transformation operators for the scattering problem (6.1-13) have the triangular representations

$$\tilde{\Psi}(x,k) = \tilde{\Psi}_\infty(x,k) + \int_x^\infty K_+(x,y)\tilde{\Psi}_\infty(y,k)\,dy$$

$$\Phi(x,k) = \Phi_{-\infty}(x,k) + \int_{-\infty}^x K_-(x,y)\Phi_{-\infty}(y,k)\,dy$$

where $\tilde{\Psi} = (\bar{\psi},\psi)$, $K_+ = (\bar{K}_+,K_+)$, $K_- = (K_-,\bar{K}_-)$,

and

$$\tilde{\Psi}_\infty = \Phi_{-\infty}(x,k) = \mathrm{diag}(e^{-ikx},e^{ikx}).$$

If in addition R and Q are differentiable then K_+, K_- satisfy the partial differential equations

$$\sigma K_{\pm x}(x,y) + K_{\pm y}(x,y)\sigma - \sigma P(x)K_\pm(x,y) = 0$$

$$K_\pm(x,x)\sigma - \sigma K_\pm(x,x) \mp \sigma P(x,x) = 0$$

$$\lim_{x,y\to\infty} K_+(x,y) = 0;\ \lim_{x,y\to-\infty} K_-(x,y) = 0$$

where $\sigma = \begin{pmatrix}-1 & 0\\ 0 & 1\end{pmatrix}$; $P = \begin{pmatrix}0 & Q\\ R & 0\end{pmatrix}$; $\pm\sigma\{K_\pm(x,x)\sigma-\sigma K_\pm(x,x)\} = P(x)$

In this case $K_\pm$ exist and are unique. If Q and R are bounded then

$$|K_+(x,y)| \leqslant |Q(\tfrac{1}{2}(x+y))|$$
$$+ \tfrac{1}{2}R_M T_Q^+(\tfrac{1}{2}(x+y))[T_Q^-(\tfrac{1}{2}(x+y))+1]\exp(T_Q^-(\infty)T_R^+(x))$$

$$|K(x,y)| \leqslant \tfrac{1}{2}|R(\tfrac{1}{2}(x+y))|$$
$$+ \tfrac{1}{2}Q_M T_R^-(\tfrac{1}{2}(x+y))[T_R^+(\tfrac{1}{2}(x+y))+1]\exp(T_R^+(-\infty)T_Q^-(x))$$

where $T_Q^\pm(x) = \pm\int_x^{\pm\infty}|Q(y)|\,dy,\qquad T_R^\pm(x) = \pm\int_x^{\pm\infty}|R(y)|\,dy$

and

$$R_M = \sup_{x\in R} R(x),\qquad Q_M = \sup_{x\in R} Q(x)$$

The estimates for $\bar{K}_\pm$ are obtained by replacing $K_\pm$ by $\bar{K}_\pm$, Q by R and

R by Q in these expressions.

Proof: We only consider the proof of existence and uniqueness for the kernel K_+. The proofs are similar for the other kernels. The integral representation of ψ, Theorem 6-1 and the definition of Ψ given above in terms of the transformation operator acting on $\begin{pmatrix}0\\1\end{pmatrix}\cdot\exp(ikx)$ lead to the relationship

$$\int_x^\infty K_+(x,y)e^{iky}dy = -\binom{1}{0}\int_x^\infty Q(y)e^{-ik(x-2y)}dy - \int_x^\infty\int_y^\infty M(x,y,k)K_+(y,u)e^{iku}dudy \tag{6.1-51}$$

Assuming for the moment that it is permissable to interchange the order of integration in the double integral on the right hand side of (6.1-51), this integral can be written as

$$\int_x^\infty\int_y^\infty \begin{pmatrix} K_{+2}(y,u)e^{-ik(x-y-u)}Q(y) \\ K_{+1}(y,u)e^{ik(x-y+u)}R(y)\end{pmatrix} dudy = \int_x^\infty e^{iky}dy \begin{pmatrix} \int_x^{\frac{1}{2}(x+y)} Q(s)K_{+2}(s,y+x-s)ds \\ \int_x^\infty R(s)K_{+1}(s,y-x+s)ds \end{pmatrix} \tag{6.1-52}$$

whence by uniqueness of the Fourier representation we obtain

$$\begin{aligned} K_{+1}(x,y) &= -\tfrac{1}{2}Q\{\tfrac{1}{2}(x+y)\} - \int_x^{\frac{1}{2}(x+y)} Q(s)K_{+2}(s,y+x-s)ds \\ K_{+2}(x,y) &= -\int_x^\infty K_{+1}(s,y-x+s)R(s)ds \end{aligned} \tag{6.1-53}$$

If Q and R are bounded then we can establish the existence of K_+ by iterating equations (6.1-53). This gives

$$\begin{aligned} |K_{+1}(x,y)| &\leqslant \tfrac{1}{2}|Q(\tfrac{1}{2}(x+y))| + \tfrac{1}{2}R_M T_Q^+(\tfrac{1}{2}(x+y))T_Q^-(\tfrac{1}{2}(x+y)) \times \\ &\qquad \exp(T_Q^-(\infty)T_R^+(x)) \\ |K_{+2}(x,y)| &\leqslant \tfrac{1}{2}R_M T_Q^+(\tfrac{1}{2}(x+y))\exp(T_Q^-(\infty)T_R^+(x)) \end{aligned} \tag{6.1-54}$$

from which the result follows. If we assume Q and R are differentiable then directly from (6.1-53) we obtain the partial differential equations and boundary conditions in the theorem. If we transform these equations to characteristic coordinates then there exists a unique solution to the resulting Goursat problem as can be shown by Riemann's method.

Corollary 6-7-1: When R and Q satisfy the requirements of Theorem 6-6 then

$$|K_+(x,y)| \leqslant C_\varepsilon^+ \exp(-\varepsilon(x+y))$$

$$|K_-(x,y)| \leqslant C_\varepsilon^- \exp(\varepsilon(x+y))$$

Corollary 6-7-2: If Q and R are differentiable and satisfy the requirements of Theorem 6-6 and

$$|R_x(x)| \leqslant C_\varepsilon^1 \exp(-2\varepsilon|x|) \quad ; \quad |Q_x(x)| \leqslant C_\varepsilon^1 \exp(-2\varepsilon|x|)$$

then

$$|K_{\pm x}(x,y)| \leqslant C_\varepsilon^{1\pm} \exp\{\mp\varepsilon(x+y)\}$$

Proof: We get by differentiating (6.1-53) with respect to y that

$$K_{+1y}(x,y) = -G[\tfrac{1}{2}(x+y)] - \int_x^{\frac{1}{2}(x+y)} Q(s) K_{+2y}(s,y+x-s)\,ds$$

$$K_{+2y}(x,y) = -\int_x^\infty K_{+1y}(s,y-x+s) R(s)\,ds$$

with

$$G[\tfrac{1}{2}(x+y)] = \tfrac{1}{4}\dot{Q}[\tfrac{1}{2}(x+y)] + \tfrac{1}{4}Q[\tfrac{1}{2}(x+y)] \int_{\frac{1}{2}(x+y)}^\infty Q(s)R(s)\,ds$$

If $\dot{Q}$ and $\dot{R}$ have the properties stated in the corollary then

$$|G[\tfrac{1}{2}(x+y)]| \leqslant \tilde{C}_\varepsilon \exp(-\varepsilon|x+y|)$$

Here $\tilde{C}_\varepsilon = \max(C_\varepsilon^1, C_\varepsilon)$ then since the integral equation for K_{+y} have the same form as (6.1-53) the result follows using the method of Corollary (6-7-1) but in which $\tilde{C}_\varepsilon$ replaces C_ε. The estimate for K_{+x} follows from the estimates for $K_+(x,y)$, $K_{+y}(x,y)$ and the equations

$$K_{+2x}(x,y) = \int_x^{\infty} K_{+1y}(s,y-x+s)R(s)ds + K_{+1}(x,y)R(x)$$

$$K_{+1x}(x,y) = -\tfrac{1}{4}Q[\tfrac{1}{2}(x+y)] + Q(x)K_{+2}(x,y)$$

$$-\tfrac{1}{4}Q[\tfrac{1}{2}(x+y)]\int_{\frac{1}{2}(x+y)}^{\infty} Q(s)R(s)ds - \int_x^{\frac{1}{2}(x+y)} Q(s)K_{+2y}(s,y+x-s)ds$$

The estimates for the other derivatives of $K_{\pm}$ are proved in an analogous fashion.

Notice that if we use the estimate given in Corollary 6-7-1 then we can deduce immediately from the transformation operators that the Jost solutions are analytic in half planes bounded by $|\text{Im } k| > \varepsilon$, the same result as that obtained in Theorem 6-6 by direct calculation. However we shall use Theorem 6-7 to elucidate the properties of the functions

$$J^{(o,r)}_{(x,k_0)} \equiv \frac{\partial}{\partial k^r} J(x,k)\Big|_{k=k_0} \qquad r = 0,1,2.,m-1$$

where J is a Jost solution and $k = k_0$ is a singular value of L of multiplicity m (either real or complex). The functions $J^{(o,r)}(x,k_0)$, $r = 1,\ldots,m-1$ are the functions **associated** with $J(x,k)$. The existence of these functions follows immediately from the transformation operator representation of the Jost solutions and the Corollary 6-7-1 6-7. If we require them to exist for weaker conditions on the potentials, then as we saw in the proof of Theorem 6-1, we would require $\{S_i, i = 0,\ldots,m\} < \infty$. The eigenfunctions and the associated functions satisfy the equations

$$LJ = kJ$$
$$LJ^{(O,j)} = kJ^{(O,j)} + jJ^{(o,j-1)} \qquad j = 1,..,m-1 \tag{6.1-55}$$

If $J(x,k_0)$, $\bar{J}(x,k_0)$ are any pair of eigenfunctions ($\text{Im } k_0 > 0$, $\text{Im } k_0 = 0$, or $\text{Im } k_0 < 0$) then $\bar{J}$ and its associated functions must also satisfy the relations (6.1-55). It follows that there exist complex numbers d^s_i such that

$$\bar{J}^{(o,s)}_{(x,k_0)} = \sum_{i=0}^{s} d^s_i \, J^{(o,i)}_{(x,k_0)} \qquad s = 0,\ldots,m-1 \tag{6.1-56}$$

This can be written in a simpler form if we introduce the analytic function $d(k)$ defined in a neighbourhood of $k = k_0$ such that $d^{s-i}(k_0) = d^s_i(C^s_i)^{-1}$. Then (6.1-56) can be written as

$$\bar{J}^{(o,s)}_{(x,k)} = \frac{\partial^s}{\partial k^s}(d(k)J(x,k))\Big|_{\mu=k_o} \tag{6.1-57}$$

From the properties of the eigenfunctions and the associated functions we have the following lemma.

Lemma 6-8: The functions $\psi^{(o,r)}(x,k_0)$, $\phi^{(o,r)}(x,k_0)$, Im $k_0 > 0$ or $\bar{\psi}^{(o,r)}(x,k_0)$, $\bar{\phi}^{(o,r)}(x,k_0)$, Im $k_0 \leqslant 0$ where $k_0 \in \sigma(L)$ of multiplicity m $(r = 0,\ldots, m-1)$ belong to $L^2_{(2)}(R)$.

The proper eigenfunctions and their associated functions are called the **principal functions for the discrete spectrum.** The Jost solutions for real k are called the principal functions **for the continuous spectrum.** We shall also need the **principal functions for the spectral singularities.** These are the Jost solutions and their k-derivatives evaluated at a real zero of a or $\bar{a}$. Corollary 6-7-1 and the transformation operators can easily be used to establish the following lemma.

Lemma 6-9: The principal functions for the spectral singularities obey

$$\sup_{x \in R} \frac{|J^{(o,r)}(x,k)|}{(1+|x|)^r} < \infty$$

where $J(x,k_0)$ is a Jost solution and k_0 (Im $k_0 = 0$) is a spectral singularity $(r = 0,1,\ldots)$.

We saw from the integral representations of the scattering functions (6.1-38)-(6.1-41) that the evaluation of the functions b and $\bar{b}$ exists at an eigenvalue $k_0 \in \sigma(L)$ (Im $k_0 > 0$ or Im $k_0 < 0$ for b, $\bar{b}$ respectively). As a direct consequence of Lemma 6-8 we see that this is also true of their k-derivatives to order m-1 if k_0 is of multiplicity m. It follows that formally b ($\bar{b}$) can replace d in a relation of the type (6.1-57) since the function and its k-derivatives are only evaluated at the eigenvalue.

Definition

$$\frac{\partial^s}{\partial k^s}(b(k)f(k))\Big|_{k=k_0} \equiv \sum_{i=0}^{s} C_i^s\, b^{(s-i)}(k_0)\, f^{(i)}(k_0)$$

where f is analytic, $k \in \sigma(L)$, Im $k_0 > 0$, and k_0 has multiplicity $m \geqslant s+1$.

A similar definition of course will also be used when $\bar{b}$ replaces b. We shall use this definition extensively throughout the rest of the chapter. An important point to note is that the definition is only valid for Theorem 6-7. If weaker definitions are adopted then derivatives of b or $\bar{b}$ may not exist. Thus if $\{S_i,\ i = 0,\ldots,n\} < \infty$ is used in Theorem 6-1, derivatives of b only exist to order n. In this case if $k \in \sigma(L)$ has multiplicity $m > n$, the last $m - n$ terms in formulae of the type (6.1-56) do not involve extensions of b and its derivatives.

We now expand the kernel of the resolvent operator in terms of the principal functions and then deduce the resolution of the identity from it. Consider the integral

$$I_R = \frac{1}{2\pi i}\int_{\Gamma_R} \frac{R(x,y,k)}{(k-z)}\,dk \tag{6.1-58}$$

where Γ_R is the contour depicted in the Fig. 6-1, and $z \in R_L$. In this figure, the

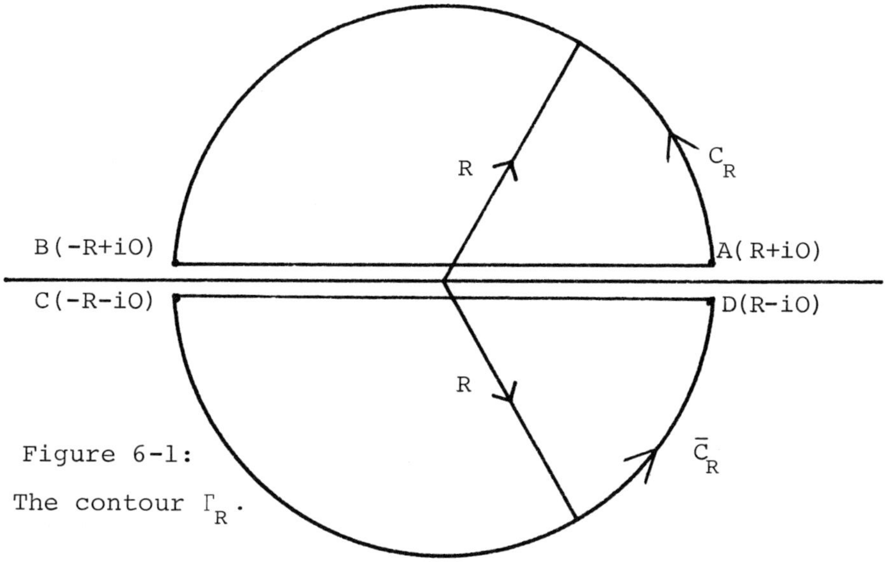

Figure 6-1:
The contour Γ_R.

contour Γ_R is composed of the arc C_R and segment [B,A] with arc $\bar{C}_R$ and segment [D,C]. R is chosen large enough so that all the eigenvalues of L and z lie within Γ_R.

From Cauchy's Theorem we get

$$I_R(x,y,z) = R(x,y,z) + \sum_{j=1}^{N} \text{Res}_{k=k_j}\left\{\frac{R(x,y,k)}{(k-z)}\right\} + \sum_{j=1}^{\bar{N}} \text{Res}_{k=\bar{k}_j}\left\{\frac{R(x,y,k)}{(k-z)}\right\} \tag{6.1-59}$$

where $z \in R_L$. Introduce the functions

$$\begin{aligned} p_j(k) &= -\frac{i(k-k_j)^{m_j} b(k)}{(m_j-1)!\,a(k)} \qquad & j = 1,\dots,N \\ \bar{p}_j(k) &= \frac{i(k-\bar{k}_j)^{\bar{m}_j} \bar{b}(k)}{(\bar{m}_j-1)!\,\bar{a}(k)} \qquad & j = 1,\dots,\bar{N} \end{aligned} \tag{6.1-60}$$

then (6.1-59) can be written as

$$\begin{aligned} I_R(x,y,z) = R(x,y,z) &- \sum_{j=1}^{N} \left\{ \left(\frac{d}{dk}\right)^{m_j-1} \frac{p_j(k)\psi(x,k)\psi^A(y,k)}{(k-z)} \right\}_{k=k_j} \\ &- \sum_{j=1}^{\bar{N}} \left\{ \left(\frac{d}{dk}\right)^{\bar{m}_j-1} \frac{\bar{p}_j(k)\bar{\psi}(x,k)\bar{\psi}^A(y,k)}{(k-z)} \right\}_{k-\bar{k}_j} \end{aligned} \tag{6.1-61}$$

Consider now a cut along the real k-axis and define a spectral singularity k_j as belonging to the upper or lower edge of this cut according to whether $a(k_j) = 0$ or $\bar{a}(k_j) = 0$ respectively. Notice that this includes the case when

$a(k_j) = \bar{a}(k_j) = 0$. Around each k_j on the upper (lower) edge describe a semi-circle situated in $\mathrm{Im}\, k \geqslant 0$ ($\mathrm{Im}\, k \leqslant 0$), the radii ($\delta$) of the semi-circles being so chosen that they are mutually disjoint and $\delta < \varepsilon_0$.

Definition

$$\varepsilon_0 = \min\{|\mathrm{Im} k_0|, \varepsilon : k_0 \in \sigma(L)\}$$

Define γ, ($\bar{\gamma}$) as the contour on the upper (lower) edge of the cut which is obtained from the real axis by replacing the horizontal diameter of the constructed semi-circles by the semi-circles themselves in the upper (lower) half k-planes.

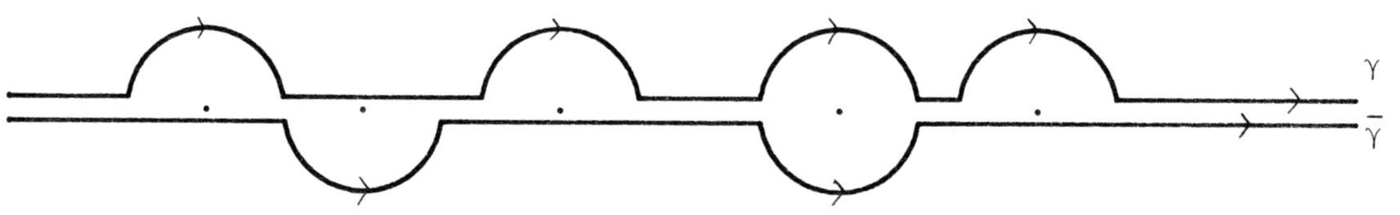

Figure 6-2: The contours $\gamma, \bar{\gamma}$ and $\Gamma = \gamma + \bar{\gamma}$.

Using Lemmas 6-2, 6-3 and equation (6.1-50) we get the estimate

$$|R(x,y,k)| \leqslant C + O\left(\frac{1}{|k|}\right) \quad \text{as} \quad |k| \to \infty \tag{6.1-62}$$

Hence as $R \to \infty$

$$I_\infty(x,y,k) = \frac{1}{2\pi i}\int_\gamma \frac{R(x,y,k)}{(k-z)}dk - \frac{1}{2\pi i}\int_\gamma \frac{R(x,y,k)}{(k-z)}dk \tag{6.1-63}$$

Combining (6.1-61) and (6.1-63) we obtain the equation for the resolvent

$$\begin{aligned} R(x,y,z) &= \frac{1}{2\pi i}\int_\gamma \frac{R(x,y,k)}{(k-z)}dk - \frac{1}{2\pi i}\int_\gamma \frac{R(x,y,k)}{(k-z)}dk \\ &+ \sum_{j=1}^{N}\left\{\left(\frac{d}{dk}\right)^{m_j-1} \frac{p_j(k)\,\psi(x,k)\,\psi^A(y,k)}{(k-z)}\right\}_{k=k_j} \\ &+ \sum_{j=1}^{\bar{N}}\left\{\left(\frac{d}{dk}\right)^{\bar{m}_j-1} \frac{\bar{p}_j(k)\,\bar{\psi}(x,k)\,\bar{\psi}^A(y,k)}{(k-z)}\right\}_{k=\bar{k}_j} \end{aligned} \tag{6.1-64}$$

It is preferable to write (6.1-64) as an integral along the real axis. However, in this case the integrals will diverge. A way round this is to regularise the integrals in the following way. Introduce the functions

$$
\begin{aligned}
f_{ij}(k) &= \begin{cases} \dfrac{(k-k_i)^j}{j!} & |k-k_i| < \delta \qquad \operatorname{Im} k \geqslant 0 \\[2mm] 0 & |k-k_j| > \delta \quad \text{or} \quad \operatorname{Im} k < 0 \end{cases} \\
\bar{f}_{ij}(k) &= \begin{cases} \dfrac{(k-\bar{k}_i)^j}{j!} & |k-\bar{k}_i| < \delta \qquad \operatorname{Im} k \leqslant 0 \\[2mm] 0 & |k-\bar{k}_i| > \delta \quad \text{or} \quad \operatorname{Im} k > 0 \end{cases}
\end{aligned}
\tag{6.1-65}
$$

where δ is the radius of the semi-circles comprising Γ. Clearly the function

$$
f(g(k)) = g(k) - \sum_{i=N+1}^{M+N} \sum_{j=0}^{m_i-1} f_{ij}(k) g^{(j)}(k_i) - \sum_{i=\bar{N}+1}^{\bar{M}+\bar{N}} \sum_{j=0}^{\bar{m}_i-1} \bar{f}_{ij}(k) g^{(j)}(\bar{k}_i) \tag{6.1-66}
$$

where $g(k)$ is sufficiently differentiable, will have $k = k_i(\bar{k}_i)$ as a root of multiplicity $m_i(\bar{m}_i)$, since, for example

$$
\left(\frac{d}{dk}\right)^j (f(g(k)))_{k=k_i} = g^j(k)\big|_{k=k_i} - g^j(k_i) = 0 \qquad j=0,1.,m_i-1 \tag{6.1-67}
$$

If we use (6.1-34), then since the contour integral of an analytic function is independent of the contour, we can re-write the γ-contour integral in (6.1-64) as

$$
\begin{aligned}
\frac{1}{2\pi i} \int_\gamma \frac{R(x,y,k)}{(k-z)} dk = & \int_{-\infty}^{\infty} dk \{ \frac{b(k)}{a(k)} f_1[\psi(x,k)\psi^A(y,k)/(k-z)] + \\
& + (k-z)^{-1}[\psi(x,k)\bar{\psi}^A(y,k)h(x-y) + \\
& + \bar{\psi}(x,k)\psi^A(y,k)h(y-x)] \} + \frac{1}{2\pi} \sum_{\ell=N+1}^{M+N} \sum_{j=0}^{m_\ell-1} \int_\gamma \frac{b(k)dk}{a(k)} f_{\ell j}(k) \{ \left(\frac{d}{dk}\right)^j \frac{\psi(x,k)\psi^A(y,k)}{(k-z)}\bigg|_{k=k_\ell} \}
\end{aligned}
\tag{6.1-68}
$$

where $f_1(g(k)) \equiv f(g(k)) \quad \operatorname{Im} k > 0, \qquad f_2(g(k) \equiv f(g(k)) \quad \operatorname{Im} k < 0$

and

$$
h(x) = \frac{\frac{1}{2}(x+|x|)}{|x|} \qquad x \neq 0 \tag{6.1-69}
$$

Repeating the calculation for the integral around the $\bar{\gamma}$ contour in (6.1-64) we finally arrive at the formula

$$R(x,y,z) = \frac{1}{2\pi}\int_{-\infty}^{\infty} dk \, \{ (\frac{b}{a}(k) f_1 \left(\frac{\psi(x,k)\psi^A(y,k)}{k-z} \right) + \frac{\bar{b}}{\bar{a}}(k) f_2 \left(\frac{\bar{\psi}(x,k)\bar{\psi}^A(y,k)}{k-z} \right) +$$
$$+ [\bar{\psi}(x,k)\psi^A(y,k) + \psi(x,k)\bar{\psi}^A(y,k)](k-z)^{-1}] \}$$
$$+ \sum_{i=1}^{M+N} \{ (\frac{d}{dk})^{m_i - 1} \frac{p_i(k)\psi(x,k)\psi^A(y,k)}{(k-z)} \}_{k=k_i} \tag{6.1-70}$$
$$+ \sum_{i=1}^{\bar{M}+\bar{N}} \{ (\frac{d}{dk})^{\bar{m}_i - 1} \frac{\bar{p}_i(k)\bar{\psi}(x,k)\bar{\psi}^A(y,k)}{(k-z)} \}_{k=\bar{k}_i}$$

where the p_i, $\bar{p}_j$ for $i = M+1, \ldots, M+N$, $j = \bar{N}+1, \ldots, \overline{M+N}$ are any functions such that

$$p_i^{(m_i-1-j)}(k_i) = \frac{1}{2\pi} \cdot \frac{j!\,(m_i-1-j)!}{(m_i-1)!} \int_{\gamma} \frac{b(k)}{a(k)} f_{ij}(k)\,dk$$

$$\bar{p}_i^{(\bar{m}_i-1-j)}(k_i) = \frac{1}{2\pi} \frac{j!\,(\bar{m}_i-1-j)!}{(\bar{m}_i-1)!} \int_{\bar{\gamma}} \frac{\bar{b}(k)}{\bar{a}(k)} \bar{f}_{ij}(k)\,dk \tag{6.1-71}$$

The functions f_{ij}, $\bar{f}_{ij}$ can be replaced by any other functions which have the same properties in the neighbourhood of a spectral singularity. The method of regularising the divergent integral used here is an adaptation of the method due to Lyantse (1965) for the non-self-adjoint Schrödinger operator. Formally we can obtain the resolution of the identity from (6.1-70) by operating with $(L - z\bar{L})$.

$$\delta(x-y)I = \frac{1}{2\pi}\int_{-\infty}^{\infty} dk \, (\frac{b}{a}(k) f_1(\psi(x,k)\psi^A(y,k))$$
$$+ \frac{\bar{b}}{\bar{a}}(k) f_2(\bar{\psi}(x,k)\bar{\psi}^A(y,k)) + \bar{\psi}(x,k)\psi^A(y,k) + \psi(x,k)\bar{\psi}^A(x,k))$$
$$+ \sum_{i=1}^{M+N} \{ (\frac{d}{dk})^{m_i-1} p_i(k)\psi(x,k)\psi^A(y,k) \}_{k=k_i} \tag{6.1-72}$$
$$+ \sum_{i=1}^{\bar{M}+\bar{N}} \{ (\frac{d}{dk})^{\bar{m}_i-1} \bar{p}_i(k)\bar{\psi}(x,k)\bar{\psi}^A(y,k) \}_{k=\bar{k}_i}$$

The next question to answer is to determine the class of functions which can be expanded in the basis (6.1-72). Let $H^m(R)$ be the Hilbert space of functions F on R which take their values in C^2 with the norm

$$\| F \|_m^2 = \int_{-\infty}^{\infty} (1+|x|)^{2m} |F(x)|_E^2 \, dx \tag{6.1-73}$$

so that in particular $H^0(R) \equiv L^2_{(2)}(R)$. Then define

$$m_0 = \max\{m_{N+1},\ldots,m_{M+N}, \bar{m}_{\bar{N}+1}, \ldots, \bar{m}_{\bar{M}+\bar{N}}\} \tag{6.1-74}$$

and put $H_+ = H^{m_0+1}(R)$, $H_- = H^{-(m_0+1)}(R)$. Then $H_+ \subset L^2_{(2)}(R) \subset H_-$ and for $F \in H_+$ we have the relation

$$\| F \|_+ \geqslant \| F \| \geqslant \| F \|_- \tag{6.1-75}$$

between corresponding norms. It is clear from Lemma 6-9 that H_- contains the principal functions for the spectral singularities. The L-Fourier transforms of $F \in L^2_{(2)}(R)$ are defined by

$$J^m(F,k) = \int_{-\infty}^{\infty} J^{(o,m)A}_{(x,k)}(x,k)F(x)\,dx \equiv (J^{(o,m)}, F) \qquad \operatorname{Im} k \neq 0 \tag{6.1-76}$$

where J is a Jost solution. The L-Fourier transforms exist for $k \in \sigma(L)$ and $m = 0,\ldots, m_k - 1$ where m is the multiplicity of k. If $F \in H_+$ we also have

$$\int_{-\infty}^{\infty} | (\tfrac{d}{dk})^{m_0+1} J(F,k) |_E^2 \, dk \leqslant C_{m_0} \int_{-\infty}^{\infty} (1+|x|)^{2m_0+2} |F(x)|_E^2 \, dx \qquad \operatorname{Im} k = 0 \tag{6.1-77}$$

It follows that if $F \in H_+$ then the L-Fourier transforms exist in this case also and that

$$\begin{aligned} F(x) = {} & \frac{1}{2\pi} \int_{-\infty}^{\infty} dk \left(\frac{b}{a}(k) f_1 (\psi(x,k)\psi(F,k)) + \frac{\bar{b}}{\bar{a}}(k) f_2 (\bar{\psi}(x,k)\bar{\psi}(F,k))\right. \\ & \left. + \bar{\psi}(x,k)\psi(F,k) + \psi(x,k)\bar{\psi}(F,k)\right) \\ & + \sum_{i=1}^{M+N} \left\{ \left(\frac{d}{dk}\right)^{m_i-1} p_i(k)\psi(x,k)\psi(F,k) \right\}_{k=k_i} + \sum_{i=1}^{\bar{M}+\bar{N}} \left\{ \left(\frac{d}{dk}\right)^{\bar{m}_i-1} \bar{p}_i(k)\bar{\psi}(x,k)\bar{\psi}(F,k) \right\}_{k=\bar{k}_i} \end{aligned} \tag{6.1-78}$$

where the integral converges in the norm of H_-. In this formula $(d/dk)^{m_i-1}$ is interpreted symbolically, since $(d/dk)^{m_i-1} J(F,k)$ will diverge except at $k = k_i$. Therefore we make the interpretation

<u>Definition</u>

$$(\tfrac{d}{dk})^m J(F,k) \equiv J^m(F,k) \tag{6.1-79}$$

For $F \in L^2_{(2)}(R)$ one can show that there exist functionals $p_{ij}(F)$, $\bar{p}_{ij}(F)$ on $L^2_{(2)}(R)$ such that

$$F(x) = \frac{1}{2\pi}\int_{-\infty}^{\infty} dk\Big(\frac{b}{a} f_1(\psi(x,k))\psi(F,k) + \frac{\bar{b}}{\bar{a}} f_2(\bar{\psi}(x,k))\bar{\psi}(F,k)$$
$$+ \bar{\psi}(x,k)\psi(F,k) + \psi(x,k)\bar{\psi}(F,k)\Big) \tag{6.1-80}$$
$$+ \sum_{i=1}^{M+N}\sum_{j=0}^{m_i-1} p_{ij}(F)\psi^{(o,j)}(x,k_i) + \sum_{\ell=1}^{\bar{M}+\bar{N}}\sum_{j=0}^{\bar{m}_\ell-1} \bar{p}_{\ell j}(F)\bar{\psi}^{(oj)}(x,\bar{k}_\ell)$$

where the integral converges in $H_{-m_0}(R)$.

For $i = 1,\ldots, N$, $\ell = 1,\ldots, \bar{N}$ a comparison of (6.1-80) with (6.1-78) shows that $p_{ij}(F)$, $\bar{p}_{\ell j}(F)$ can be expressed in terms of the functionals $\psi_j(F,k_i)$, $\bar{\psi}_j(F,k_j)$ explicitly. For $i = N+1,\ldots, M+N$, $\ell = \bar{N}+1,\ldots, M+N$, $\psi_j(x,k_i)$, $\bar{\psi}_j(x,\bar{k}_\ell)$ are linearly independent modulo $L^2_{(2)}(R)$ which follows from their asymptotic properties as $x \to \infty$. The requirement that the right hand side of (6.1-80) should belong to $L^2_{(2)}(R)$ uniquely determines the numbers $p_{ij}(F)$, $\bar{p}_{\ell s}(F)$, $i = N+1,\ldots, M+N$, $= \bar{N}+1,\ldots, \bar{M}+\bar{N}$. $j = 0,\ldots,m-1$ and $s = 0,\ldots,\bar{m}_\ell-1$.

These results take a simpler form if we consider for example a more restricted class of function in $L^2_{(2)}(R)$ for which

$$\int_{-\infty}^{\infty} \left|\frac{\psi(F,k)}{a(k)}\right|_E^2 dk < \infty \quad ; \quad \int_{-\infty}^{\infty} \left|\frac{\bar{\psi}(F,k)}{\bar{a}(k)}\right|_E^2 dk < \infty \tag{6.1-81}$$

Alternatively we could impose conditions on Q and R so that spectral singularities are absent. Of course the problem here is that in general conditions on Q and R which lead to this result are not known. The simplest case to consider is when the spectrum of L contains only simple eigenvalues. In this case the resolution of the identity can be written as

$$\delta(x-y)I = \frac{1}{2\pi}\int_{-\infty}^{\infty} dk\Big\{\frac{b}{a}(k)\psi(x,k)\psi^A(y,k)$$
$$+ \frac{\bar{b}}{\bar{a}}(k)\bar{\psi}(x,k)\bar{\psi}^A(y,k) + \bar{\psi}(x,k)\psi^A(y,k) + \psi(x,k)\bar{\psi}^A(y,k)\Big\} \tag{6.1-82}$$
$$+ \sum_{j=1}^{M} p_j\psi(x,k_j)\psi^A(y,k_j) + \sum_{j=1}^{M} \bar{p}_j\bar{\psi}(x,k_j)\bar{\psi}^A(y,k_j)$$

where

$$p_j = -\frac{ib(k_j)}{a_k(k_j)} \qquad \bar{p}_j = \frac{i\bar{b}(\bar{k}_j)}{\bar{a}_k(\bar{k}_j)}$$

If we extend the definition of the bilinear form $(U,V) = \int_{-\infty}^{\infty} U^A(x)V(x)dx$ introduced in (6.1-76) to generalized functions, then using the result from generalised functions that

$$\lim_{x\to\infty} P\left(\frac{e^{\pm ikx}}{ik}\right) = \pm\,\pi\delta(k)$$

we can evaluate all the inner products for the principal functions using (6.1-30). For the eigenfunctions we obtain

$$
\begin{aligned}
({}_{k_1}\phi, {}_{k_2}\phi) &= 2\pi a(k_1) b(k_1)\delta(k_1-k_2) \\
({}_{k_1}\psi, {}_{k_2}\phi) &= 2\pi a(k_1)\delta(k_1-k_2) \\
({}_{k_1}\bar{\psi}, {}_{k_2}\phi) &= 0 \\
({}_{k_1}\phi, {}_{k_2}\bar{\phi}) &= 2\pi a(k_1)\bar{a}(k_1)\delta(k_1-k_2) \\
({}_{k_1}\psi, {}_{k_2}\psi) &= -2\pi a(k_1)\bar{b}(k_1)\delta(k_1-k_2) \\
({}_{k_1}\psi, {}_{k_2}\bar{\psi}) &= 2\pi a(k_1)\bar{a}(k_1)\delta(k_1-k_2)
\end{aligned}
\tag{6.1-83}
$$

In equations (6.1-83) $\text{Im}\, k_1 = \text{Im}\, k_2 = 0$. All the other non-zero products are obtained from (6.1-83) by symmetry or by the "-" operation defined by $(\bar{f}) = \bar{f}$, $(\bar{f}) = f$. The only other possible non-zero products arise from the eigenfunctions

$$
\begin{aligned}
({}_{k_j}\psi, {}_{k_\ell}\phi) &= i\delta_{j\ell} a_k(k_j) \\
({}_{\bar{k}_j}\bar{\psi}, {}_{\bar{k}_\ell}\bar{\phi}) &= -i\delta_{j\ell}\bar{a}_k(\bar{k}_\ell)
\end{aligned}
\qquad k_j, k_\ell, \bar{k}_j, \bar{k}_\ell, \in \sigma(L)
\tag{6.1-84}
$$

To obtain the brackets for the associated functions we use the result that

$$
(U^m(k_1),\ V^{(r)}(k_2)) = \frac{r(U^{(m)}(k_1), V^{(r-1)}(k_2)) - m(U^{(m-1)}(k_1), V^{(r)}(k_2))}{(k_1-k_2)} - \left[\frac{W(U^{(m)}(k_1), V^{(r)}(k_2))}{i(k_1-k_2)}\right]_{-\infty}^{\infty}
\tag{6.1-85}
$$

However in general the bilinear form so defined is singular except in the case when L has only simple eigenvalues. When Q and R have compact support, Theorem 6-6 leads to the conclusion **that the scattering functions a, b, $\bar{a}$ and $\bar{b}$ are analytic throughout the k-plane.** It follows from (6.1-82) (it is easiest to see from (6.1-72)) that the resolution of the identity can then be written in the concise form

$$
\begin{aligned}
\delta(x-y)\, I = -\frac{1}{2\pi}\int_C dk\{&\frac{b}{a}(k)\psi(x,k)\psi^A(y,k) \\
&+ \bar{\psi}(x,k)\psi^A(y,k)\} + \frac{1}{2\pi}\int_{\bar{C}} dk\{\frac{\bar{b}}{\bar{a}}(k)\bar{\psi}(x,k)\bar{\psi}^A(y,k) \\
&+ \psi(x,k)\bar{\psi}^A(y,k)\}
\end{aligned}
\tag{6.1-86}
$$

where C and $\bar{C}$ are the limits of C_R and $\bar{C}_R$ as $R \to \infty$. The expression (6.1-72) can then be obtained for the non-compact support case by deforming the contours C, $\bar{C}$ into the real axis if we interpret $b(k)$, $\bar{b}(\bar{k})$ for complex k, $\bar{k}$ in the same formal fashion as earlier in this section. Thus (6.1-86) provides a very convenient notation in which to write the resolution of the identity. We shall use it extensively in the next section.

So far we have not considered the degeneracy which occurs in the problem through relationships between the two scattering potentials. Some of the possibilities of this type are given in the following lemma, which is established from an examination of (6.1-13) and the defining relations for the scattering functions (6.1-31)-(6.1-33)

Lemma 6-10:

(i) If $R = \alpha Q$, α a complex number, then $\bar{M} = M$, $\bar{N} = N$ and $\bar{\psi}(x,k) = E\psi(x,-k)$, $\bar{\phi}(x,k) = \frac{1}{\alpha}E\phi(x,-k)$, $\bar{a}(k) = a(-k)$, $\bar{b}(k) = \frac{1}{\alpha}b(-k)$, where $E = \begin{pmatrix} 0 & 1 \\ \alpha & 0 \end{pmatrix}$.

The eigenvalues and spectral singularities in $\text{Im}k \geqslant 0$, $\text{Im}k \leqslant 0$ are related by $\bar{k}_j = -k_j$.

(ii) If $R = \alpha Q^*$, α a real number then $\bar{M} = M$, $\bar{N} = N$, and $\bar{\psi}(x,k) = E\psi^*(x,k^*)$, $\bar{\phi}(x,k) = \frac{1}{\alpha}E\phi^*(x,k^*)$, $\bar{a}(k) = a^*(k^*)$, $\bar{b}(k) = \frac{1}{\alpha}b^*(k^*)$. The eigenvalues and spectral singularities in $\text{Im}k \geqslant 0$, $\text{Im}k \leqslant 0$ are related by $\bar{k}_j = k_j^*$.

(iii) If $R = \alpha Q = \alpha Q^*$, α real then $\bar{M} = M$, $\bar{N} = N$ and the sets of eigenvalues and spectral singularities decompose into the sets $\{\pm k_j, \mp k_j^*\}$, $j = 1,\dots,p$, $\{\pm i\eta_j\}$, $j = 1,\dots,q$ where $N = 2p + q$ and $\{k_j, -k_j\}$, $j = 1,\dots,M/2$ respectively.

(iv) In particular when $R = Q^*$, the operator L is self-adjoint and there is no discrete spectrum.

Just as in the case of the isospectral operator defined by the Schrödinger equation one can introduce concepts analogous to **spectral family** and **spectral function** for the non-self adjoint isospectral operator considered here. The required generalisations are called the **generalised spectral measure** and the **generalised spectral function** respectively. The latter is of prime importance theoretically because it enables the **Parseval equality**, that is the main fact in spectral theory, to be written in the non-self-adjoint case. This extension is due to Marchenko (1960) and Foias (1960). The reconstruction of Q and R from the generalised spectral function is then analogous to the self-adjoint case of Chapers 3 and 4. Some of the exercises at the end of this chapter establish this connection in the self-adjoint case. We proceed now to define the scattering data for this problem. This as in the self-adjoint case is defined by the asymptotic properties of the principal functions at infinity. The scattering data $S_\pm$ are defined to be the sets,

$$S_+ = \{R_+ = \frac{b}{a},\ \bar{R}_+ = \frac{\bar{b}}{\bar{a}},\ k_j,\ P_{+j}(x),\ \bar{k}_\ell,\ \bar{P}_{+\ell}(x),$$

$$j=1,\ldots,N,\quad \ell=1,\ldots,\bar{N}\} \tag{6.1-87}$$

$$P_{+j}(x) = -ie^{-ik_j x}\ \text{Res}_{k=k_j}\{\frac{b(k)}{a(k)}\ e^{ikx}\}$$

$$\bar{P}_{+\ell}(x) = ie^{i\bar{k}_\ell x}\ \text{Res}_{k=\bar{k}_\ell}\ \{\frac{\bar{b}(k)}{\bar{a}(k)}\}\ e^{-ikx}\}$$

$$S_- = \{R_- = \frac{\bar{b}}{a},\ \bar{R}_- = \frac{b}{\bar{a}},\ k_j,\ P_{-j}(x),\ \bar{k}_\ell,\ \bar{P}_{-\ell}(x),$$

$$j=1,\ldots,N,\quad \ell=1,\ldots,\bar{N}\} \tag{6.1-88}$$

$$P_{-j}(x) = -ie^{-ik_j x}\ \text{Res}_{k=k_j}\{\frac{1}{b(k)a(k)}\ e^{ikx}\}$$

$$\bar{P}_{-\ell}(x) = ie^{-i\bar{k}_\ell x}\ \text{Res}_{k=\bar{k}_\ell}\ \{\frac{1}{\bar{b}(k)\bar{a}(k)}\ e^{-ikx}\}$$

The set S_- corresponds to using the Jost solutions ϕ and $\bar{\phi}$ to define the generalised spectral measure. The functions $P_{\pm j}(x)$, $\bar{P}_{\pm j}(x)$ are called the **normalisation polynomials.** When there are only simple zeros they reduce to the normalisation constants for the eigenfunctions.

The reconstruction of a and $\bar{a}$ from $S_\pm$ should more properly be considered in the next section, since in the forward problem a and $\bar{a}$ are determined by the initial data. However it is interesting to consider it at this point because it leads to the conclusion that the zeros of a and $\bar{a}$ must be related in some way if a and $\bar{a}$ are to be uniquely determined by $S_\pm$. Define functions

$$f(k) = a(k)\prod_{j=1}^{N}\left(\frac{k-\bar{K}_j}{k-k_j}\right)^{m_j} \qquad \text{Im}k \geqslant 0 \tag{6.1-89}$$

$$\bar{f}(k) = \bar{a}(k)\prod_{j=1}^{\bar{N}}\left(\frac{k-K_j}{k-\bar{k}_j}\right)^{\bar{m}_j} \qquad \text{Im}k \leqslant 0 \tag{6.1-90}$$

In these formulae $\{k_j,\ \bar{k}_\ell,\ j=1,\ldots,N,\ \ell=1,\ldots,\bar{N}\}$, are the set of discrete eigenvalues of L with multiplicities m_j and $\bar{m}_\ell$. The numbers $\{\bar{K}_j,\ K_\ell,\ \text{Im}K_\ell > 0,\ \text{Im}\,\bar{K}_j < 0,\ j=1,\ldots,N,\ \ell=1,\ldots,\bar{N}\}$ are chosen so that the set $\{\bar{K}_j\}$ is distinct from the set $\{\bar{k}_\ell\}$ and similarly the set $\{K_\ell\}$ is distinct from the set $\{k_j\}$. Then since the zeros of a and $\bar{a}$ are isolated the functions f and $\bar{f}$ are analytic within the upper and lower half k-plane respectively. Furthermore the functions $f,\ \bar{f}$ have the properties that as $|k| \to \infty$, $f(k) = 1 + O(1/|k|)$, $\bar{f}(k) = 1 + O(1/|k|)$. Then by Cauchy's theorem we get

$$\ell n f = \frac{1}{2\pi i}\int_\gamma \frac{\ell n f(z)}{(z-k)}\,dz \qquad \text{Im}k > 0$$

$$0 = \frac{1}{2\pi i}\int_{\bar{\gamma}} \frac{\ell n \bar{f}(z)}{(z-k)}\,dz \qquad \text{Im}k > 0 \tag{6.1-91}$$

where we have assumed that ℓn is restricted to a principal value. The conditions $\ell n\bar{f}(z) \to 0$, $\ell n f(z) \to 0$ as $|z| \to \infty$, z real, impose certain boundary conditions on $\ell n a(z)$ and $\ell n \bar{a}(z)$. Choosing the principal branches of the functions

$$h(z) = \sum_{j=1}^{N} \ell n \left(\frac{z-\bar{K}_j}{z-k_j}\right)^{m_j}, \quad \bar{h}(z) = \sum_{j=1}^{\bar{N}} \ell n \left(\frac{z-K_j}{z-\bar{k}_j}\right)^{\bar{m}_j}, \quad \text{Im}z = 0 \tag{6.1-92}$$

so that

$$h(z) \to 0, \qquad \bar{h}(z) \to 0 \quad \text{as} \quad z \to +\infty$$

then

$$h(z) \to 2\pi i H, \quad \bar{h}(z) \to -2\pi i \bar{H} \quad \text{as} \quad z \to -\infty, \ \text{Im}z = 0 \tag{6.1-93}$$

where

$$H = \sum_{j=1}^{N} m_j, \qquad \bar{H} = \sum_{j=1}^{\bar{N}} \bar{m}_j$$

For real z the functions $\ell n a(z)$ and $\ell n \bar{a}(z)$ are therefore required to have the asymptotic properties

$$\begin{aligned} &\ell n a(z) \to 0, \quad \ell n \bar{a}(z) \to 0 \quad \text{as} \quad z \to \infty \\ &\ell n a(z) \to -2\pi i H, \quad \ell n \bar{a}(z) \to 2\pi i \bar{H} \quad \text{as} \quad z \to -\infty \end{aligned} \tag{6.1-94}$$

From (6.1-91) we obtain

$$a(k) = \prod_{j=1}^{N} \left(\frac{k-k_j}{k-\bar{K}_j}\right)^{m_j} \exp \frac{1}{2\pi i} \left\{\int_{\gamma} \frac{dz}{(z-k)} (\ell n(a(z)) + h(z)) + \int_{\bar{\gamma}} \frac{dz}{(z-k)} (\ell n \bar{a}(z) + \bar{h}(z))\right\} \qquad \text{Im}k > 0 \tag{6.1-95}$$

In order to reconstruct a in terms of $S_{\pm}$ we require to be able to deform the integrals into the real axis. Define functions

$$\begin{aligned} g_j(z) &= \ell n (z-k_j)^{m_j} && |z-k_j| < \delta \\ &\quad 0 && |z-k_j| > \delta \qquad j=N+1,\ldots,M+N \\ \bar{g}_j(z) &= \ell n (z-\bar{k}_j)^{\bar{m}_j} && |z-\bar{k}_j| < \delta \\ &\quad 0 && |z-\bar{k}_j| > \delta \qquad j=\bar{N}+1,\ldots,\bar{M}+\bar{N} \end{aligned}$$

then we rewrite (6.1-95) as

$$a(k) = \prod_{j=1}^{N} \left(\frac{k-k_j}{k-\bar{K}_j}\right)^{m_j} \exp \frac{1}{2\pi i}\left\{ \int_{-\infty}^{\infty} \frac{dz}{(z-k)} (\ell n(a(z)\bar{a}(z)) + h(z)+\bar{h}(z)-\ell(z)-\bar{\ell}(z)) + \int_{\gamma} \frac{dz}{(z-k)}\, \ell(z) + \int_{\bar{\gamma}} \frac{dz}{(z-k)}\, \bar{\ell}(z) \right\} \qquad \text{Im}k > 0$$

where (6.1-96)

$$\ell(z) = \sum_{j=N+1}^{M+N} g_j(z), \quad \bar{\ell}(z) = \sum_{j=\bar{N}+1}^{\bar{M}+\bar{N}} \bar{g}_j(z)$$

However a still depends upon the arbitrary numbers $(K_j, \bar{K}_j)$. An examination of (6.1-96) shows that if $\bar{N} = N$ and $\bar{m}_j = m_j$ then an application of Cauchy's integral formula results in the required representation

$$a(k) = \prod_{j=1}^{N} \left(\frac{k-k_j}{k-\bar{k}_j}\right)^{m_j} \exp \frac{1}{2\pi i} \left\{\int_{-\infty}^{\infty} \frac{dz}{(z-k)} (\ell n(a(z)\bar{a}(z)) - \ell(z) - \bar{\ell}(z)) + \int_{\bar{\gamma}} \frac{dz\, \bar{\ell}(z)}{(z-k)} + \int_{\gamma} \frac{dz \ell(z)}{(z-k)} \right. \qquad \text{Im}k > 0 \tag{6.1-97}$$

where $a(z) \equiv (1-R_+(z)\bar{R}_+(z)) \equiv (1-R_-(z)\bar{R}_-(z))^{-1}$, $\quad \text{Im}z = 0$

The reconstruction of the trace functional from the scattering data is also possible when $\text{Im}k = 0$ on $\text{Im}k < 0$, under the same conditions.

Theorem 6-11: The scattering functions $a(k)$ $\text{Im}k \geqslant 0$, $\bar{a}(k)$ $\text{Im}k \leqslant 0$ are uniquely defined and constructable from the scattering data $S_\pm$ if and only if $\bar{N} = N$ and $\bar{m}_j = m_j$. In this case the relations

$$\text{Ind } a + \sum_{j=1}^{N} m_j = 0$$

$$\text{Ind } \bar{a} - \sum_{j=1}^{N} m_j = 0$$

hold, where Ind a (Ind $\bar{a}$) is the increase divided by 2π of a continuous branch of arg $a(k)$ (arg $\bar{a}(k)$) as k traverses any curve $\gamma(\bar{\gamma})$ from ∞ to $-\infty$ for which the complex zeros of $a(k)$ ($\bar{a}(k)$) are above γ (below $\bar{\gamma}$) and the real zeros are below γ (above $\bar{\gamma}$).

The solvable equations for this problem can be determined in the same way as those for the isospectral Schrödinger operator were determined, Section 3.5. Thus the equation (6.1-13) can be written as

$$\Phi_x = P\Phi \qquad \mathrm{Im}k = 0 \tag{6.1-98}$$

where Φ is the fundamental matrix solution

$$\Phi = \begin{pmatrix} \phi_1 & \bar{\phi}_1 \\ \phi_2 & \bar{\phi}_2 \end{pmatrix}$$

and

$$P = \begin{pmatrix} -ik & Q \\ R & ik \end{pmatrix}$$

Then from a Δ-variation of (6.1-98) we obtain

$$\Phi^{-1}\Delta\Phi \Big|_{x=-\infty}^{x=\infty} = \int_{-\infty}^{\infty} \Phi^{-1}\Delta P\Phi \; dx \tag{6.1-99}$$

Since the fundamental matrix solutions Φ, Ψ are related by the scattering function matrix A (6.1-35), the equation (6.1-99) establishes relationships between the Δ-variation of the scattering functions and the Δ-variation of the scattering potentials. These relationships are the following

$$\begin{aligned} \Delta a &= -I_\Delta(\phi,\psi) \\ \Delta b &= I_\Delta(\phi,\bar{\psi}) \\ \Delta\bar{a} &= I_\Delta(\bar{\phi}\ \bar{\psi}) \\ \Delta\bar{b} &= -I_\Delta(\bar{\phi},\psi) \end{aligned} \tag{6.1-100}$$

where

$$I_\Delta(V,Y) \equiv \int_{-\infty}^{\infty} (\Delta R v_1 y_1 - \Delta Q v_2 y_2)\,dx$$

The representations (6.1-100) have been derived under the conditions that k is real and not a spectral singularity of L and that $\Delta Q \to 0$ "sufficiently fast" as $|x| \to \infty$.

A large class of solvable equations can be derived as a consequence of the requirement that the scattering functions $R_+, \bar{R}_+$ (or alternatively $R_-, \bar{R}_-$) should satisfy linear equations. From (6.1-100) we obtain

$$\begin{aligned} \Delta R_+ &= \frac{1}{a^2} I_\Delta(\phi,\phi) \\ \Delta\bar{R}_+ &= -\frac{1}{\bar{a}^2} I_\Delta(\bar{\phi},\bar{\phi}) \end{aligned} \tag{6.1-101}$$

Thus by imposing the conditions

$$\Omega R_+ = \frac{1}{a^2} I_\Delta(\phi,\phi)$$
$$\bar{\Omega}\bar{R}_+ = -\frac{1}{\bar{a}^2} I_\Delta(\bar{\phi},\bar{\phi}) \tag{6.1-102}$$

where $\Omega, \bar{\Omega}$ are known functions of k, R_+, $\bar{R}_+$ satisfy the linear equations

$$\Delta R_+ = \Omega R_+$$
$$\Delta \bar{R}_+ = \bar{\Omega}\bar{R}_+ \tag{6.1-103}$$

Then just as in the case for the isospectral Schrödinger operator the class of solvable equations we are considering can be obtained from the constraint equations (6.1-102).

For simplicity we shall restrict ourselves to the case when Δ is the total derivative operator with respect to time and we shall assume that $\Omega,\bar{\Omega}$ are functions of k only. Let ${}_kV$, ${}_kY$ be two solutions of (6.1-13), then it is straightforward to obtain the relationships

$$(v_2y_2)_x - R(y_1v_2+v_1y_2) = 2iky_2v_2 \tag{6.1-104}$$

$$(v_1y_1)_x - Q(y_1v_2+v_1y_2) = -2iky_1v_1 \tag{6.1-105}$$

$$(y_1v_2+v_1y_2)(x) = (y_1v_2+v_1y_2)(-\infty) + 2\int_{-\infty}^{x} (R(s)v_1(s)y_1(s) + Q(s)v_2(s)y_2(s))ds \tag{6.1-106}$$

from which we deduce that

$$L_1 V\circ Y(x) = kV\circ Y(x) - \frac{1}{2}(\sigma_2 U(x))(V^T\sigma_1 Y)(-\infty) \tag{6.1-107}$$

where

$$L_1 \equiv \frac{1}{2i}\begin{pmatrix} -\frac{\partial}{\partial x} + 2Q\int_{-\infty}^{x} dsR & 2Q\int_{-\infty}^{x} dsQ \\ -2R\int_{-\infty}^{x} dsR & \frac{\partial}{\partial x} - 2R\int_{-\infty}^{x} dsQ \end{pmatrix}$$

and

$$V\circ Y \equiv \begin{pmatrix} v_1y_1 \\ v_2y_2 \end{pmatrix}, \quad U = \begin{pmatrix} R \\ Q \end{pmatrix}, \quad \sigma_1 = \begin{pmatrix} 0 & 1 \\ 1 & 0 \end{pmatrix} \quad \sigma_2 = \begin{pmatrix} 0 & -i \\ i & 0 \end{pmatrix} \quad \sigma_3 = \begin{pmatrix} 1 & 0 \\ 0 & -1 \end{pmatrix}$$

In particular writing $V^{(2)} \equiv V\circ V$ we have from (6.1-107) and (6.1-106) in the limit $x \to +\infty$

$$L_1\,{}_k\phi^{(2)} = k\,{}_k\phi^{(2)} \qquad ab = \int_{-\infty}^{\infty} U\cdot{}_k\phi^{(2)}(x)\,dx \equiv (U,{}_k\phi^{(2)}) \tag{6.1-108}$$

$$L_1\,{}_k\bar{\phi}^{(2)} = k\,{}_k\bar{\phi}^{(2)} \qquad \bar{a}\bar{b} = \int_{-\infty}^{\infty} U\cdot{}_k\bar{\phi}^{(2)}(x)\,dx \equiv (U,{}_k\bar{\phi}^{(2)}) \tag{6.1-109}$$

Then (6.1-109) with the constraint equations (6.1-101) can be written as

$$\begin{aligned}(\sigma_3 U_t - \Omega(k)U, {}_k\phi^{(2)}) &= 0\\ (\sigma_3 U_t + \bar{\Omega}(k)U, {}_k\bar{\phi}^{(2)}) &= 0\end{aligned} \tag{6.1-110}$$

If $\Omega,\bar{\Omega}$ are either polynomials in k or a ratio of polynomials $\Omega = C_1/C_2$, $\bar{\Omega} = \bar{C}_1/\bar{C}_2$ then (6.1-108) can be used to write (6.1-110) as

$$\begin{aligned}(\sigma_3 U_t - \Omega(L_1^A)U, {}_k\phi^{(2)}) &= 0\\ (\sigma_3 U_t + \bar{\Omega}(L_1^A)U, {}_k\phi^{(2)}) &= 0\end{aligned} \tag{6.1-111}$$

where

$$L_1^A \equiv \frac{1}{2i}\begin{pmatrix} \frac{\partial}{\partial x} + 2R\int_x^{\infty} ds Q & -2R\int_x^{\infty} ds R \\ 2Q\int_x^{\infty} ds Q & -\frac{\partial}{\partial x} - 2Q\int_x^{\infty} ds R \end{pmatrix}$$

In particular if $\bar{\Omega} = -\Omega$ then the constraints (6.1-110) can be satisfied by the nontrivial k-independent constraint

$$C_2(L_1^A)\sigma_3 U_t - C_1(L_1^A)U = 0 \tag{6.1-112}$$

which furnishes a class of equations solvable by the inverse method associated with L. Formally a larger class of solvable equations is obtained by allowing $\Omega,\bar{\Omega}$ to be ratios of entire functions in k. The equations when $\bar{\Omega} \neq -\Omega$ are also solvable and in fact an example within this class is of physical importance. We discuss this case later in the section and again in Section 6.2. For the present we concentrate on the class of equations represented by (6.1-112) where C_1 and C_2 are polynomials in k. Examples of particular equations contained in the class (6.1-112) are the following

Examples in which Ω is entire

$$\Omega(k) \equiv \alpha, \qquad \sigma_3 U_t - \alpha U = 0$$

$$\Omega(k) \equiv \alpha k, \qquad \sigma_3 U_t - \alpha (L_1^A) U = 0$$

$$\begin{pmatrix} R_t \\ -Q_t \end{pmatrix} - \frac{\alpha}{2i} \begin{pmatrix} R_x \\ Q_x \end{pmatrix} = 0$$

$$\Omega(k) \equiv \alpha k^2, \quad \sigma_3 U_t - \alpha (L_1^A)^2 U = 0$$

$$\begin{pmatrix} R_t \\ -Q_t \end{pmatrix} - \frac{\alpha}{(2i)} 2 \begin{pmatrix} R_{xx} - 2R^2 Q \\ Q_{xx} - 2Q^2 R \end{pmatrix} = 0$$

degenerate case: $R = \beta Q^*$, $\alpha = 4i$, $\beta = \pm 1$

$$iQ_t \mp |Q|^2 Q + Q_{xx} = 0$$ the nonlinear Schrödinger equation

$$\Omega(k) \equiv \alpha k^3, \quad \sigma_3 U_t - \alpha (L_1^A)^3 U = 0$$

$$\begin{pmatrix} R_t \\ -Q_t \end{pmatrix} - \frac{\alpha}{(2i)} 3 \begin{pmatrix} R_{xxx} - 6RQR_x \\ -Q_{xxx} + 6RQQ_x \end{pmatrix} = 0$$

degenerate case: $R = \beta Q^*$, $\alpha = 8i$, $\beta = \pm 1$

$$Q_t \pm 6|Q|^2 Q_x + Q_{xxx} = 0$$ a complex modified KdV equation

degenerate case: $R = \beta Q$, $\alpha = 8i$, $\beta = \pm 1$

$$Q_t \pm 6Q^2 Q_x + Q_{xxx} = 0$$ Q real, the modified KdV equation

Examples in which Ω is meromorphic

$$\Omega(k) \equiv \alpha / k, \quad (L_1^A)\sigma_3 U_t - \alpha U = 0$$

$$\frac{1}{2i}\begin{pmatrix} R_{tx} + 2R\int_x^\infty (RQ)_t\,ds \\ Q_{tx} + 2Q\int_x^\infty (RQ)_t\,ds \end{pmatrix} - \alpha\begin{pmatrix} R \\ Q \end{pmatrix} = 0$$

degenerate case: $R = \beta Q$, $\alpha = \pm 1/2i$, $\beta = -1$

$$Q_{tx} - 2Q\int_x^\infty (Q^2)_t\,ds \mp Q = 0$$

The equivalence class of equations determined by Ω also contains the sine-Gordon equation

$$V_{xt} = \pm \sin V$$

upon defining $V_x = -2Q$, provided $4(Q_t)^2 \leqslant 1$.

$$\Omega(k) \equiv -8ik^3 + \frac{i\alpha}{2k}\,, \quad 2(L_1^A)\sigma_3 U_t + i(\alpha - 16(L_1^A)^4)U = 0$$

$$\begin{pmatrix} R_{xt} + 2R\int_x^\infty (RQ)_t ds - \alpha R + R_{4x} - 8RQR_{2x} - 6R_x^2 Q - 4RR_xQ_x + 6R^3Q^2 \\ \qquad - 2R^2Q_{2x} \\ Q_{xt} + 2Q\int_x^\infty (RQ)_t ds - \alpha Q + Q_{4x} - 8RQQ_{2x} - 6Q_x^2 R - 4QR_xQ_x + 6R^2Q^3 \\ \qquad - 2Q^2R_{2x} \end{pmatrix} = 0$$

degenerate case: $R = Q\beta$, $\beta = -1$.

$$Q_{xt} - 2Q\int_x^\infty (Q^2)_t\,ds - \alpha Q + Q_{4x} + 10Q^2Q_{2x} + 10QQ_x^2 + 6Q^5 = 0$$

The equivalence class determined by Ω also contains a combined modified Korteweg de Vries equation and sine-Gordon equation

$$Q_t - 6Q^2Q_x - Q_{xxx} + \frac{\alpha}{2}\sin\left(-2\int_x^\infty Q\,ds\right) = 0$$

Konno et al. (1974) have proposed this equation as modelling weak dislocation effects on wave propagation in an anharmonic crystal.

The function Ω is simply related to the dispersion relations of the linearised solvable equations (6.1-112). Thus we easily see that the linearised equations are

$$C_2\left(\frac{1}{2i}\sigma_3\frac{\partial}{\partial x}\right)\sigma_3 U_t - C_1\left(\frac{1}{2i}\sigma_3\frac{\partial}{\partial x}\right)U = 0 \tag{6.1-113}$$

tting

$$U = (e^{i(2kx-\bar{\omega}t)}, e^{i(-2kx-\omega t)})^T \tag{6.1-114}$$

find that

$$\Omega(k) = -i\bar{\omega}(2k) = i\omega(-2k) \tag{6.1-115}$$

e specification of the linear dispersion relation, that is essentially Ω, termines an equivalence class of solvable nonlinear equations. The members of this ass are related by functions $E(k)$

$$C_1^1(k) = E(k)\, C_1(k), \quad C_2^1(k) = E(k)\, C_2(k) \tag{6.1-116}$$

ere $F(k)$ is entire or meromorphic. In particular the choice $E(k) = C_2^{-1}(k)$ sults in the equations again having the form of evolution equations provided aning can be assigned to $C_2^{-1}(L_1^A)C_1(L_1^A)V$. In fact the degenerate examples given ove, that is the sine-Gordon equation and the equation considered by Konno et al. 974) furnish examples where this is possible. Further discussion of these uations with regard to their initial value problem is given in the next section.

In the above derivation of the solvable equations we have assumed initial nditions on R and Q such that the reconstructed functions satisfy the relevant uation. Thus Ω formally determines an equivalence class of solvable equations. suitable condition of this type is the one given in Section 3.5, namely that the constructed Q and R should belong to the general Schwartz class.

We now return to the determination of the evolution of the scattering data, uation (6.1-112). The techniques of Section 3.5 lead immediately to the following eorem

Theorem 6-12:

Under the conditions of Theorem 6-6 the scattering data $S_+(0)$ is uniquely determined from $R(x,0)$ and $Q(x,0)$. When $\bar{\Omega} = -\Omega$ the function formally determines an equivalence class of solvable equations

$$C_1(L_1^A)_3U_t + C_2(L_1^A)U = 0$$

where $\Omega = C_1/C_2$. Each member of the class has the same $S_+(t)$ defined by

$$(P_{+j}(x,t)e^{-\Omega_j t})_t = \frac{-ie^{-(ik_j x+\Omega_j t)}}{(m_j-1)!} \times$$

$$\times \sum_{s=0}^{m_j-2}\left(\Omega^{(m_j-1-s)}(k)\left[\frac{b}{a}(k,0)(k-k_j)^{m_j}e^{ikx+\Omega(k)t}\right]^{(s)}\right)_{k=k_j}$$

$$(\bar{P}_{+\ell}(x,t)e^{\bar{\Omega}_\ell t})_t = \frac{ie^{(i\bar{k}_\ell x+\bar{\Omega}_\ell t)}}{(\bar{m}_\ell-1)!} \times$$

$$\times \sum_{s=0}^{\bar{m}_\ell-2} \left(\Omega^{(\bar{m}_\ell-1-s)}(k)\left[\frac{\bar{b}}{\bar{a}}(k,0)(k-\bar{k}_\ell)^{\bar{m}_\ell}e^{-(ikx+\Omega(k)t}\right]^{(s)}\right)_{k=\bar{k}_\ell}$$

$$R_{+t}(k,t) = \Omega(k).R_+(k,t), \qquad \bar{R}_{+t}(k,t) = -\Omega(k).\bar{R}_+(k,t)$$

$$\sigma(L(t)) = \sigma(L(0))$$

The numbers Ω_j, $\bar{\Omega}_\ell$ are defined by $\Omega_j = \Omega(k_j)$, $\bar{\Omega}_\ell = \Omega(\bar{k}_\ell)$.
In deriving these equations we assume that if Ω is singular none of the singularities coincide with $\sigma(L)$ or a spectral singularity.

From (6.1-107) it is easy to show that

$$(L_1)^s\phi o\psi = k^s\phi o\psi - \frac{1}{2i}a(k)\left(\sum_{r=0}^{s-1}(L_1)^r\right)(\sigma_2 U) \qquad (6.1\text{-}11$$

The formula

$$\int_{-\infty}^{\infty}(L_1{}^A)^s U.\sigma_2 U\, dx = 0 \qquad (6.1\text{-}11$$

which is established in the exercises then leads to the conclusion that

$$a_t = -I_t(\phi,\psi) = -\int_{\infty}^{\infty} U.\Omega(L_1)\phi o\psi\, dx$$
$$= -\Omega(k)\int_{-\infty}^{\infty} U.\phi o\psi\, dx = 0 \qquad (6.1\text{-}1$$

The last equality is established with the aid of (6.1-106). Alternatively one c use (6.1-97) and the definition of a in terms of S_+ given by (6.1-87) to pro this result. Similarly

$$\bar{a}_t = 0 \qquad (6.1\text{-}1$$

The time evolution of S_- can also easily be written down for this case. T equations defining $\underline{S}_+(t)$ in Theorem 6-12 are exact. The solutions are particula simple if a and $\bar{a}$ have only simple zeros when the sums on the right hand side the expressions for the normalization polynomials vanish and the normalisat polynomials are then just the normalisation "constants". In this case $S_+(t)$ given by

$$\begin{aligned} P_{+j}(t) &= P_{+j}(0)e^{\Omega_j t} \\ \bar{P}_{+\ell}(t) &= \bar{P}_{+\ell}(0)e^{-\bar{\Omega}_\ell t} \\ R_+(k,t) &= R_+(k,0)e^{\Omega(k)t} \\ \bar{R}_+(k,t) &= \bar{R}_+(k,0)e^{-\Omega(k)t} \\ \sigma(L(t)) &= \sigma(L(0)) \end{aligned} \qquad (6.1\text{-}121)$$

For the isospectral Schrödinger operator the scattering function a played a entral role in the Hamiltonian structure of that theory. It served as a generating unction for the infinite family of constants of the motion which were the amiltonians for the solvable equations. For the case $\Omega = -\bar{\Omega}$ when as we have seen $a_t = 0$ and $\bar{a}_t = 0$ a similar interpretation can be given to the functions a and a. A detailed study would take us outside the scope of this book, but we can roceed fairly formally to obtain some of the results.

From the definition of the resolvent operator one can define a regularised trace unctional, $d(k) = \mathrm{Tr}((L - kI)^{-1} - (L_0 - kI)^{-1}$ where $_0 = L(Q = 0, R = 0)$. From (6.1-13) we can obtain the relation

$$-i(\phi_{2k}\psi_1 - \psi_2\phi_{1k})_x = \phi_1\psi_2 + \phi_2\psi_1 \qquad \mathrm{Im}k \geqslant 0 \qquad (6.1\text{-}122)$$

o that we get upon integrating this relationship that

$$(-\ln a)_k = i\int_{-\infty}^{\infty} \left[\frac{\phi_1\psi_2 + \phi_2\psi_1}{a} - 1\right]dx \qquad \mathrm{Im}k \geqslant 0 \qquad (6.1\text{-}123)$$

t is clear therefore from the definition of the trace functional that

$$d(k) = \begin{cases} (-\ln a)_k & \mathrm{Im}k > 0 \\ (-\ln \bar{a})_k & \mathrm{Im}k < 0 \end{cases} \qquad (6.1\text{-}124)$$

he formula for $(-\ln\bar{a})_k$ is obtained upon making the changes $\phi \to \bar{\phi}$, $\psi \to \bar{\psi}$, $a \to \bar{a}$, $i \to -i$ in the right hand side of (6.1-123). From (6.1-106) we get that

$$(\phi_1\psi_2 + \phi_2\psi_1)(x,k) - a(k) = 2\int_{-\infty}^{x} U.(\phi o \psi)dy \qquad (6.1\text{-}125)$$

hen we find using (6.1-123) that formally we have

$$(\ln a)_k = \int_{-\infty}^{\infty}\left(\int_{\infty}^{x}(L_1^A - kI)^{-1}U.\sigma_2 U\, dy\right)dx \qquad \mathrm{Im}k > 0 \qquad (6.1\text{-}126)$$

ormal expansion and integration as $|k| \to \infty$ then gives that

$$\ln a(k) \sim \sum_{j=1}^{\infty} \frac{C_j}{k^j} \quad \text{as} \quad |k| \to \infty \tag{6.1-127}$$

where

$$C_j = \frac{i}{j} \int_{-\infty}^{\infty} \left(\int_{-\infty}^{x} (L_1^A)^j U . \sigma_2 U \, dy \right) \tag{6.1-128}$$

Then since $(\ln a)_t = 0$ it follows that the $\{C_j\}$ are an infinite family of constants of the motion. By direct evaluation of (6.1-128) one can show that the first few members of the family are

$$C_1 = \frac{1}{2i} \int_{-\infty}^{\infty} RQ \, dx \qquad C_2 = \left(\frac{1}{2i}\right)^2 \frac{1}{2} \int_{-\infty}^{\infty} (R_x Q - Q_x R) \, dx$$

$$C_3 = \left(\frac{1}{2i}\right)^3 \frac{1}{3} \int_{-\infty}^{\infty} (R_{2x} Q + RQ_{2x} - R_x Q_x - 3R^2 Q^2) \, dx \tag{6.1-129}$$

These same conserved quantities can be obtained from the analogous expansion of $\ln \bar{a}(k)$, $\mathrm{Im}\, k < 0$, as $|k| \to \infty$. The correctness of these expressions can be established from a direct asymptotic expansion of (6.1-13) for the Jost solutions.

We now establish that the functionals $\{C_i\}$ (or a linear combination of them) are the Hamiltonians of the solvable **evolution** equations for this system (c.f. Section 3.5). The relations established in (6.1-104)-(6.1-106) lead to the expression

$$(L_1^A - kI) \sigma_2 \phi \circ \psi = -\frac{a}{2} U \tag{6.1-130}$$

It follows from (6.1-100) and (6.1-130) that

$$\Delta a = -\int_{-\infty}^{\infty} (\sigma_3 \Delta U . \phi \circ \psi) \, dx = \frac{a}{2i} \int_{-\infty}^{\infty} (\sigma_1 \Delta U . (L_1^A - kI)^{-1} U) \, dx \tag{6.1-131}$$

As we showed in Section 3.5, it is consistent to interpret Δ as an arbitrary variation of the functions Q and R. Thus if we formally expand (6.1-131) and use (6.1-128) we obtain

$$\sigma_1 \frac{\Delta C_j}{\Delta U} = \left(\frac{1}{2i}\right) (L_1^A)^j U$$

where

$$\sigma_1 \frac{\Delta C_j}{\Delta U} \equiv \left(\frac{\Delta C_j}{\Delta Q} , \frac{\Delta C_j}{\Delta R} \right)^T \tag{6.1-132}$$

The entries in $\Delta C_j / \Delta U$ are the functional or Frechet derivatives of the functional C_j with respect to the functions R and Q respectively.

The solvable evolution equations have the form

$$\sigma_3 U_t - C(L_1^A)U = 0 \tag{6.1-133}$$

where

$$C_1(L_1^A) = \sum_{j=0}^{n} a_j (L_1^A)^j \tag{6.1-134}$$

Thus the evolution equations can be written as

$$\sigma_3 U_t = \sigma_1 \frac{\Delta H}{\Delta U}$$

with

$$H = 2i \sum_{j=0}^{n} a_j C_j \tag{6.1-135}$$

which are the required Hamiltonian equations with Hamiltonian H. Notice that besides the asymptotic expansions for large k of the trace functional one can expand it about any regular point. The Hamiltonians generated in this way correspond to nonlinear evolution equations defined by singular dispersion relations (Ω meromorphic). In this case, extra conditions have to be imposed on the scattering functions b, $\bar{b}$ to ensure the validity of this procedure for real k. The expansions we have established for the trace functional have been carried out on the assumption that Q and R have independent variation (are not related by Lemma 6-10). If this is not the case then further modifications are required (c.f. Flaschka and Newell 1975 and Dodd and Bullough 1979).

When $\bar{\Omega} \neq -\Omega$ then we can still find solvable equations but in this case they are not Hamiltonian systems (the trace functional is not a constant of the motion). We include one example of this type here just to emphasize the fact that their are physical solvable equations which fall outside a Hamiltonian framework. Before presenting this example let us reconsider the method we are using to obtain solvable equations. Essentially this hinges upon the fact that we are able to solve the linear equations governing the time evolution of the scattering data. Then we work back from this to see what constraints this imposes upon the potentials and the squared eigen functions. We shall now proceed to obtain a more operator theoretic description for deriving solvable equations.

As we have already seen, $\phi^{(2)}$ and $\bar{\phi}^{(2)}$ are eigenfunctions of L_1 and it is not hard to show from (6.1-104)-(6.1-106) that $\psi^{(2)A} = (\psi_2^2, -\psi_1^2$, $\bar{\psi}^{(2)A} = (\bar{\psi}_2^2, -\bar{\psi}_1^2)$ are eigenfunctions for L_1^A. If we define a bilinear form for a row vector V and a column vector Y by

$$(V,Y) = \int_{-\infty}^{\infty} V(x)Y(x)\,dx \tag{6.1-136}$$

and treat the eigenfunctions for real k as generalised functions then we can evaluate the inner products for the squared eigenfunctions. The result for real k is the following

$$\begin{aligned}
({}_{k_1}\psi^{(2)A}, {}_{k_2}\phi^{(2)}) &= \pi a^2(k_1)\delta(k_1-k_2) \\
({}_{k_1}\bar{\psi}^{(2)A}, {}_{k_2}\bar{\phi}^{(2)}) &= -\pi \bar{a}^2(k_1)\delta(k_1-k_2) \\
({}_{k_1}\psi^{(2)A}, {}_{k_2}\bar{\phi}^{(2)}) &= 0 = ({}_{k_1}\bar{\psi}^{(2)A}, {}_{k_2}\phi^{(2)})
\end{aligned} \tag{6.1-137}$$

In the absence of spectral singularities and eigenvalues it appears that one can deduce the completeness relation from (6.1-137), (Kaup 1976). From now on we proceed formally and for simplicity assume that this is indeed the case. Thus

$$\delta(x-y) = \frac{1}{\pi}\int_{-\infty}^{\infty} dk \, \frac{1}{a^2(k)} \, {}_k\phi^{(2)}(x)\,{}_k\psi^{(2)A}(y) - \frac{1}{\pi}\int_{-\infty}^{\infty} \frac{dk}{\bar{a}^2(k)} \, {}_k\bar{\phi}^{(2)}(x)\,{}_k\bar{\psi}^{(2)A}(y) \qquad (x \geqslant y) \tag{6.1-138}$$

The spectral expansion of $(\sigma_3 U_t(x))^T$ in this basis is therefore

$$(\sigma_3 U_t(x))^T = \frac{1}{\pi}\int_{-\infty}^{\infty} \frac{dk}{a^2(k)} \, (\sigma_3 U_t, {}_k\phi^{(2)})\,{}_k\psi^{(2)A}(x) - \frac{1}{\pi}\int_{-\infty}^{\infty} \frac{dk}{\bar{a}^2(k)} \, (\sigma_2 U_t, {}_k\bar{\phi}^{(2)})\,{}_k\bar{\psi}^{(2)A}(x) \tag{6.1-139}$$

In addition observe that $\theta = (\psi_2\bar{\psi}_2 - \psi_1\bar{\psi}_1)$ satisfies

$$(L_1^A - kI)(\theta^T) = \frac{1}{2i}\binom{R}{Q} \tag{6.1-140}$$

Starting from (6.1-13) we can show that for $V^{(2)}(x,k_1), W \circ Y(x,k_2)$, (V,W,Y are eigenfunctions of L)

$$(W \circ Y^A, V^{(2)}) = -\frac{1}{2i(k_1-k_2)}\left(v_2^2 Y_1 W_1 + v_1^2 Y_2 W_2 - V_1 Y_2 W_1 V_2 - V_1 Y_1 W_2 V_2\right)_{-\infty}^{\infty} \tag{6.1-141}$$

Thus from the completeness relation and (6.1-141) we can derive the spectral expansion for θ . In particular this can be used to show that

$$\binom{R}{Q}^T \equiv (2i(L_1^A - kI)({}_k\theta^T))^T = \frac{1}{\pi}\int_{-\infty}^{\infty} dk \, \frac{b}{a}(k)\,{}_k\psi^{(2)A} + \frac{1}{\pi}\int_{-\infty}^{\infty} dk \, \frac{\bar{b}}{\bar{a}}(k)\,{}_k\bar{\psi}^{(2)A} \tag{6.1-142}$$

If Ω is entire then acting with $\Omega(L_1^A)$ on the transpose of (6.1-142) produces

$$(\Omega(L_1^A)U)^T = \frac{1}{\pi}\int_{-\infty}^{\infty} dk \, \frac{b}{a}(k)\,\Omega(k)\,{}_k\psi^{(2)A} + \frac{1}{\pi}\int_{-\infty}^{\infty} dk \, \frac{\bar{b}}{\bar{a}}(k)\,{}_k\bar{\psi}^{(2)A} \tag{6.1-143}$$

Thus under the conditions (6.1-102) we regain the evolution equations (6.1-112) provided $\bar{\Omega} = -\Omega$. In terms of the nonlinear operator $K = \Omega(L_1^A)$ which defines the solvable equations we have the following statement using the spectral theory developed in Section 3.4. **The operator K is diagonal in the spectral representation of L_1^A** (Kaup and Newell 1979). When Ω and $\bar{\Omega}$ are singular, that is they have poles, then analogous statements can be made about the corresponding nonlinear operators (exercises at end of this chapter). However we shall concentrate on the physical example which we mentioned above. Let Ω and $\bar{\Omega}$ be analytic in the upper and lower half k-planes respectively and vanish as $|k| \to \infty$. Then from (6.1-103) we have in general that

$$
\begin{aligned}
(\sigma_3 U_t)^T &= \frac{1}{\pi}\int_{-\infty}^{\infty} dk\,\Omega(k)\frac{b}{a}(k)\,{}_k\psi^{(2)A} \\
&+ \frac{1}{\pi}\int_{-\infty}^{\infty} dk\,\Omega(k)\frac{\bar{b}}{\bar{a}}(k)\,{}_k\psi^{(2)A}
\end{aligned}
\tag{6.1-144}
$$

whereas from (6.1-138) we get

$$
\begin{aligned}
\frac{1}{\pi}\int_{-\infty}^{\infty} (\Omega(k)-\bar{\Omega}(k))\,{}_k\theta\, dk &= \frac{1}{\pi}\int_{-\infty}^{\infty} dk\,\Omega(k)\frac{b}{a}(k)\,{}_k\psi^{2(A)} \\
&- \frac{1}{\pi}\int_{-\infty}^{\infty} dk\,\bar{\Omega}(k)\frac{\bar{b}}{\bar{a}}(k)\,{}_k\bar{\psi}^{(2)A}
\end{aligned}
\tag{6.1-145}
$$

The system of equations resulting from identifying the right hand sides of these two equations together with equations (6.1-104) and (6.1-105) can be written

$$
\begin{aligned}
U_t &= \frac{1}{\pi}\int_{-\infty} dk\; g(k)\begin{pmatrix} k^{\bar{\lambda}} \\ k^{\lambda} \end{pmatrix} \\
\bar{\lambda}_x &- RN = 2ik\bar{\lambda} \\
\lambda_x &- QN = -2ik\lambda \\
N_x &= 2(R\bar{\lambda}+Q\lambda)
\end{aligned}
\tag{6.1-146}
$$

where $g(k) = \Omega(k) - \bar{\Omega}(k)$, $N = (\psi_1\bar{\psi}_2 + \psi_1\bar{\psi}_2)$, $\bar{\lambda} = \psi_1\bar{\psi}_1$ and $\lambda = \psi_2\bar{\psi}_2$. When $R = \pm Q^*$ this set of equations becomes the Reduced Maxwell-Bloch equations governing the propagation of a pulse through an inhomogeneously broadened medium consisting of 2-level atoms (see Chapter 9). For this system the only motion invariants are the spectrum since

$$
\begin{aligned}
a_t &= -I_t(\phi,\psi) = -\left((\sigma_3 U_t)^T,\ \phi\circ\psi\right) \\
&= \frac{1}{\pi}\int_{-\infty}^{\infty} (\Omega(k)-\bar{\Omega}(k)\,({}_{k'}\theta\,,{}_k\phi\circ\psi)\,dk \\
&= \frac{i}{2}\,a(k)\;P\!\int_{-\infty}^{\infty} \frac{g(k')\,b\bar{b}(k')}{(k'-k)}\,dk'
\end{aligned}
\tag{6.1-147}
$$

6.2 The Inverse Method for the ZS-AKNS Equation

In this section we present the inverse method for the isospectral operator L of the ZS-AKNS scattering problem (6.1-13). We also investigate the conditions on the initial data so that a solution to the initial value problem for a solvable evolution equation exists. This material is therefore the analogue for L of the results given in Chapter 4 for the isospectral Schrödinger operator. The approach adopted here is influenced by the work of Ljance (1967) on the non-self adjoint Schrödinger operator on the half-line and is taken from Dodd (1982).

We begin by deriving the Marchenko equations for L and determining conditions under which Q and R can be uniquely constructed from them. Various methods are available for obtaining the Marchenko equations. We shall begin with the simplest derivation and in addition assume that Q and R have compact support. In this case by Corollary 6-1-2 the Jost solutions and the scattering functions are analytic throughout the k-plane. Then by evaluating the integral

$$I_1(x,k) = \int_C \frac{dz\,\phi(x,z)e^{izx}}{a(z)(z-k)}$$

we obtain

$$I_1(x,k) = i\pi\begin{pmatrix}1\\0\end{pmatrix}$$

$$I_1(x,k) = \int_{C+\bar{C}} \frac{\bar{\psi}(x,z)e^{izx}}{z-k}\,dz + \int_C \frac{b(z)\psi(x,z)e^{izx}}{a(z)(z-k)}\,dz$$

$$- \int_{\bar{C}} \frac{\bar{\psi}(x,z)e^{izx}}{(z-k)}\,dz = 2\pi i\bar{\psi}(x,k)e^{ikx} \tag{6.2-1}$$

$$+ \int_C \frac{b(z)\psi(x,z)e^{izx}}{a(z)(z-k)}\,dz - i\pi\begin{pmatrix}1\\0\end{pmatrix}$$

The expressions on the right hand side of (6.2-1) are found by respectively evaluating the integral on the curve C and by using (6.1-31)-(6.1-33) and Cauchy's theorem. The curves C, $\bar{C}$ are defined in Section 6.1. The result of this procedure repeated for the other Jost solutions is the following

$$\begin{aligned}
\psi(x,k)e^{-ikx} &= \begin{pmatrix}0\\1\end{pmatrix} - \frac{1}{2\pi i}\int_{\bar{C}} \frac{dz}{(z-k)}\frac{\bar{b}}{\bar{a}(z)}\bar{\psi}(x,z)e^{-izx} \\
\bar{\psi}(x,k)e^{ikx} &= \begin{pmatrix}1\\0\end{pmatrix} - \frac{1}{2\pi i}\int_C \frac{dz}{(z-k)}\frac{b(z)}{a(z)}\psi(x,z)e^{izx} \\
\phi(x,k)e^{ikx} &= \begin{pmatrix}1\\0\end{pmatrix} + \frac{1}{2\pi i}\int_{\bar{C}} \frac{dz}{(z-k)}\frac{b(z)}{\bar{a}(z)}\bar{\phi}(x,z)e^{izx} \\
\bar{\phi}(x,k)e^{-ikx} &= \begin{pmatrix}0\\1\end{pmatrix} + \frac{1}{2\pi i}\int_C \frac{dz}{(z-k)}\frac{\bar{b}(z)}{a(z)}\phi(x,z)e^{-izx}
\end{aligned} \tag{6.2-2}$$

If we substitute for the Jost solutions their representation in terms of the transformation operators Theorem 6-7, and then make a Fourier transformation, we obtain the time dependent Marchenko equations

$$\bar{K}_+(x,y,t) + \begin{pmatrix}0\\1\end{pmatrix} F_+(x+y,t) + \int_x^\infty K_+(x,s,t)F_+(s+y,t)\,ds = 0 \qquad y \geq x$$

$$K_+(x,y,t) + \begin{pmatrix}1\\0\end{pmatrix} \bar{F}_+(x+y,t) + \int_x^\infty \bar{K}_+(x,s,t)\bar{F}_+(s+y,t)\,ds = 0 \qquad y \geq x$$

where

$$F_+(x,t) = -\frac{1}{2\pi}\int_C dk\, \frac{b(k,t)}{a(k,t)}\, e^{ikx}$$

$$\bar{F}_+(x,t) = \frac{1}{2\pi}\int_{\bar{C}} dk\, \frac{\bar{b}(k,t)}{\bar{a}(k,t)}\, e^{-ikx} \qquad (6.2\text{-}3)$$

and

$$\bar{K}_-(x,y,t) + \begin{pmatrix}1\\0\end{pmatrix} F_-(x+y,t) + \int_{-\infty}^x K_-(x,s,t)F_-(s+y,t)\,ds = 0 \qquad y \leq x$$

$$K_-(x,y,t) + \begin{pmatrix}0\\1\end{pmatrix} \bar{F}_-(x+y,t) + \int_{-\infty}^x \bar{K}_-(x,s,t)\bar{F}_-(s+y,t)\,ds = 0 \qquad y \leq x$$

where

$$F_-(x,t) = \frac{1}{2\pi}\int_C dk\, \frac{\bar{b}(k,t)}{a(k,t)}\, e^{-ikx}$$

$$\bar{F}_-(x,t) = -\frac{1}{2\pi}\int_{\bar{C}} dk\, \frac{b(k,t)}{\bar{a}(k,t)}\, e^{ikx} \qquad (6.2\text{-}4)$$

When the potentials Q and R do not have compact support but satisfy the conditions of Theorem 6-6 then the time dependent Marchenko equations (6.2-3) and (6.2-4) are still valid but with the contour integrals replaced by

$$F_+(x,t) = \sum_{j=1}^{N} P_{+j}(x,t)e^{ik_j x} + \frac{1}{2\pi}\int_\gamma dk\, R_+(k,t)e^{ikx}$$

$$\bar{F}_+(x,t) = \sum_{\ell=1}^{\bar{N}} \bar{P}_{+\ell}(x,t)e^{-i\bar{k}_\ell x} + \frac{1}{2\pi}\int_{\bar{\gamma}} dk\, \bar{R}_+(k,t)e^{-ikx}$$

$$F_-(x,t) = \sum_{j=1}^{N} P_{-j}(x,t)\bar{e}^{ik_j x} - \frac{1}{2\pi}\int_\gamma dk\, R_-(k,t)e^{-ikx} \qquad (6.2\text{-}5)$$

$$\bar{F}_-(x,t) = \sum_{\ell=1}^{\bar{N}} \bar{P}_{-\ell}(x,t)e^{i\bar{k}_\ell x} - \frac{1}{2\pi}\int_{\bar{\gamma}} dk\, \bar{R}_-(k,t)e^{ikx}$$

Essentially the regularisation involved in the integrals in (6.2-5) replaces rational fractions of the type $(k-k_{\pm})^{-m_{\pm}}$ by the generalised functions $(k-k_{\pm}\pm i0)^{-m_{\pm}}$, where $k_{\pm}$ is a spectral singularity $a(k_+) = 0$, $\bar{a}(k_-) = 0$ of multiplicity $m_{\pm}$ respectively. The functions $R_+ = b/a$, $\bar{R}_+ = \bar{b}/\bar{a}$, $R_- = \bar{b}/a$ and $\bar{R}_- = b/\bar{a}$ are generalised functions on the space of slowly increasing functions. The inverse Fourier transforms of these functions defined by the integrals in (6.2-5) we shall denote by $\hat{R}_+$, $\hat{\bar{R}}_+$, $\hat{R}_-$ and $\hat{\bar{R}}_-$. They are generalised functions on C_0^{∞}, the space of C^{∞} functions having compact support.

<u>**Lemma 6-13:**</u> The Fourier transforms $\hat{R}_+$, $\hat{\bar{R}}_+$, $\hat{R}_-$ and $\hat{\bar{R}}_-$ are defined by

$$\hat{R}_+(x) = \frac{1}{2\pi}\int_{\gamma} dk\, R_+(k)e^{ikx}, \quad \hat{\bar{R}}_+(x) = \frac{1}{2\pi}\int_{\bar{\gamma}} dk\, \bar{R}_+(k)e^{-ikx}$$

$$\hat{R}_-(x) = \frac{1}{2\pi}\int_{\gamma} dk\, R_-(k)e^{-ikx}, \quad \hat{\bar{R}}_-(x) = \frac{1}{2\pi}\int_{\bar{\gamma}} dk\, \bar{R}_-(k)e^{ikx}$$

where the functions R_+, $\bar{R}_+$, R_- and $\bar{R}_-$ are meromorphic in $|\mathrm{Im}\, k| < \varepsilon$. For any δ, $0 < \delta < \varepsilon_0/2$ there exists constants C_δ^+, $\bar{C}_\delta^+$, C_δ^- and $\bar{C}_\delta^-$ such that

$$\int_{-\infty}^{\infty} |e^{\pm\eta x}\hat{R}_{\pm}(x)|^2\, dx < C_\delta^{\pm}$$

$$\delta \leqslant \eta \leqslant \varepsilon_0 - \delta$$

$$\int_{-\infty}^{\infty} |e^{\pm\eta x}\hat{\bar{R}}_{\pm}(x)|^2\, dx < \bar{C}_\delta^{\pm}$$

<u>**Proof:**</u> We prove the result for $\hat{R}_+$. The result for the other transforms is proved in a similar fashion. From the integral representations (6.1-38)-(6.1-41) and Lemma 6-3 follows the result that

$$|R_+(k)| < \frac{C_\delta^{+'}}{|k|}, \quad \delta \leqslant |\mathrm{Im} k| \leqslant \varepsilon_0 - \delta .$$

From the definition of R_+ and Cauchy's theorem we have that

$$\hat{R}_+(x) = \frac{1}{2\pi}\int_{-\infty+i\eta}^{\infty+i\eta} R_+(k)\exp(ikx)\, dk$$

so that

$$e^{\eta x}\hat{R}_+(x) = \frac{1}{2\pi}\int_{-\infty}^{\infty} R_+(\xi + i\eta)\exp i\xi x\, d\xi$$

Finally using the inequality we obtain

$$\int_{-\infty}^{\infty} |e^{\eta x}\hat{R}_+(x)|^2 dx \leqslant \left(\frac{C_\delta^{+'}}{2\pi}\right)^2 \int_{-\infty}^{\infty} \frac{d\xi}{|\xi + i\eta|^2}$$

The Marchenko equations (6.2-2),(6.2-3) can be derived directly from the completeness relation (6.1-86). However, here we shall use the completeness relation to derive a generalised Marchenko equation for the ZS-AKNS system which is the analogue of the one given in Section 4.1 for the Schrödinger operator. It is convenient to use the compact support formulae (6.2-4) for (6.2-5). As we indicated in the previous section under the conditions of Theorem 6-6 this can be properly justified.

Let $\psi_{(i)}, \overline{\psi}_{(i)}$, $i = 1,2$ be the Jost solutions for the potential functions (Q_i, R_i) $i = 1,2$ and let $K_{(i)}$ $i = 1,2$ be the kernels of the corresponding transformation operators. We can also establish the existence and uniqueness of the inverse operators with kernels $M_{(i)}$ $i = 1,2$ such that

$$\tilde{\Psi}_{(i)\infty}(x,k) = \tilde{\Psi}_{(i)}(x,k) - \int_x^\infty M_{(i)}(x,y)\tilde{\Psi}_{(i)}(y,k)dy \quad \text{Im}k > 0$$

$$\tilde{\Psi}_{(i)\infty}(x,k) = \begin{pmatrix} e^{-ikx} & 0 \\ 0 & e^{ikx} \end{pmatrix}, \quad \Psi_{(i)} = (\overline{\psi}_{(i)}, \psi_{(i)}), K_{(i)} = (\overline{K}_{(i)}, K_{(i)}),$$

$$M_{(i)} = (\overline{M}_{(i)}, M_{(i)})$$

where

$$Q_i(x) = -2K_{(i)1}(x,x) = -2M_{(i)1}(x,x)$$

$$R_i(x) = -2\overline{K}_{(i)2}(x,x) = -2\overline{M}_{(i)2}(x,x)$$

$$\lim_{x,y\to\infty} K_{(i)}(x,y) = 0 \qquad \lim_{x,y\to\infty} M_{(i)}(x,y) = 0$$

$$L(x)K_{(i)}(x,y) - i\frac{\partial}{\partial y}K_{(i)}(x,y) = 0, \qquad x < y < \infty \tag{6.2-6}$$

$$M_{(i)}(x,y)\left[L(y) - i\sigma_3\frac{\partial}{\partial y}\right] - i\frac{\partial}{\partial y}M_{(i)}(x,y)\sigma_3 - \frac{\partial}{\partial x}M_{(i)}(x,y) = 0$$

$$x < y < \infty$$

$$\sigma_3 = \begin{pmatrix} 1 & 0 \\ 0 & -1 \end{pmatrix}$$

From the transformation operators associated with the Jost solutions given above one can show that

$$\Psi_{(1)}(x,k) = \Psi_{(2)}(x,k) + \int_x^\infty K(x,y)\Psi_{(2)}(y,k)dy$$

where $\Psi_{(i)} = (\psi_{(i)}, \overline{\psi}_{(i)})$

and

$$K(x,y) = K_{(1)}(x,y) - M_{(2)}(x,y) - \int_x^y K_{(1)}(x,s)M_{(2)}(s,y)ds$$

$$Q_1(x) - Q_2(x) = -2K_{12}(x,x)$$

$$R_1(x) - R_2(x) = -2K_{21}(x,x) \tag{6.2-7}$$

Similarly one can prove the existence of an adjoint operator

$$\psi^A_{(2)}(x,k) = \psi^A_{(1)}(x,k) + \int^\infty \psi^A_{(1)}(y,k)K^A(y,x)dy \tag{6.2-8}$$

The completeness relations for the Jost solutions $\psi_{(i)}, \bar{\psi}_{(i)}$ can be written using the notation of Section 6.1 and equation (6.1-86) as

$$\begin{aligned}\delta(x-y)I = &-\frac{1}{2\pi}\int_C dk\left[R_+^{(i)}(k)\psi_{(i)}(x,k)\psi^A_{(i)}(y,k) + \bar{\psi}_{(i)}(x,k)\psi^A_{(i)}(y,k)\right]\\ &+\frac{1}{2\pi}\int_{\bar{C}} dk\left[\bar{R}_+^{(i)}(k)\bar{\psi}_{(i)}(x,k)\bar{\psi}^A_{(i)}(y,k)\right.\\ &\left.+\psi_{(i)}(x,k)\bar{\psi}^A_{(i)}(y,k)\right]\end{aligned} \tag{6.2-9}$$

The use of (6.2-8) and (6.2-9) for $i = 1$ establishes that

$$\begin{aligned}&\frac{1}{2\pi}\int_C \left[R_+^{(1)}(k)\psi_{(1)}(x,k)\psi^A_{(2)}(y,k) + \bar{\psi}_{(1)}(x,k)\psi^A_{(2)}(y,k)\right]dk\\ &-\frac{1}{2\pi}\int_{\bar{C}}\left[\bar{R}_+^{(1)}(k)\bar{\psi}_{(1)}(x,k)\bar{\psi}^A_{(2)}(y,k) + \psi_{(1)}(x,k)\bar{\psi}^A_{(2)}(y,k)\right]dk\\ &= -\delta(x-y) - \int_y^\infty \delta(x-s)K^A(s,y)ds\ .\end{aligned} \tag{6.2-10}$$

The use of (6.2-9) for $i = 2$ and (6.2-7) then yields the generalised Marchenko equation for this method (Dodd and Bullough, 1979)

$$K(x,y,t) + F(x+y,t) + \int_x^\infty K(x,s,t)F(s+y,t)ds = 0 \qquad x < y$$

where

$$\begin{aligned}F(x,y,t) = &-\frac{1}{2\pi}\int_C \left[R_+^{(1)}(k,t) - R_+^{(2)}(k,t)\right]\psi_{(2)}(x,k)\psi^A_{(2)}(y,k)dk\\ &+\frac{1}{2\pi}\int_{\bar{C}}\left[\bar{R}_+^{(1)}(k,t) - \bar{R}_+^{(2)}(k,t)\right]\bar{\psi}_{(2)}(x,k)\bar{\psi}^A_{(2)}(y,k)dk\end{aligned} \tag{6.2-11}$$

Thus (6.2-11) with (6.2-7) establishes a relationship between pairs of scattering functions. If we put $Q_2 = R_2 = 0$ then (6.2-11) reduces to the time dependent Marchenko equation (6.2-3). This equation can also be used to investigate the Hamiltonian structure associated with the ZS-AKNS system.

We now investigate the conditions which the scattering functions must satisfy in order that the reconstructed Q and R should satisfy the conditions of Theorem 6-6. Of course further conditions must be added if Q and R are to satisfy an arbitrary evolution equation. We investigate these conditions later in this section. We shall work solely with the Marchenko equation (6.2-3) and omit the "+" suffix and suppress the t-dependency of all functions for clarity. The techniques can be similarly applied to (6.2-4). We begin by investigating the conditions that F and $\overline{F}$ satisfy if $|Q(x)| \leqslant C\exp(-2\varepsilon|x|)$ and $|R(x)| \leqslant C\exp(-2\varepsilon|x|)$. Put $x = y$ in (6.2-3) and interpret the resulting equation as an equation for F and $\overline{F}$

$$F(x) = -K\left(\frac{x}{2},\frac{x}{2}\right) - \frac{1}{2}\int_x^{\infty} K\left(\frac{x}{2},\frac{s}{2}\right) F\left(\frac{s+x}{2}\right) ds$$

where

$$F = \begin{pmatrix} 0 & \overline{F} \\ F & 0 \end{pmatrix} \quad \text{and} \quad K = (\overline{K}, K) \tag{6.2-12}$$

Iterating this integral equation we obtain

$$|F(x)| \leqslant \left|K\left(\frac{x}{2},\frac{x}{2}\right)\right| + \frac{1}{2}\int_x^{\infty}\left|K\left(\frac{x}{2},\frac{x_1}{2}\right)\right|\left|K\left(\frac{x+x_1}{4},\frac{x+x_1}{4}\right)\right| dx_1$$

$$+ \dots + \frac{1}{2^{\frac{1}{2}j(j-1)}}\int_x^{\infty}\left|K\left(\frac{x}{2},\frac{x_1}{2}\right)\right|\int_{(x+x_1)}^{\infty}\left|K\left(\frac{x+x_1}{4},\frac{x_2}{4}\right)\right|\dots$$

$$\dots\int_{(x+x_1+\dots x_{j-2})}^{\infty}\left|K\left(\frac{1}{2^{j-1}}(x+x_1+\dots+x_{j-2}),\frac{1}{2^{j-1}}x_{j-1}\right)\right|\left|K\left(\frac{1}{2^j}(x+x_1+\dots+x_{j-1}),\frac{1}{2^j}(x+x_1+\dots+x_{j-1})\right)\right| \times dx_1\dots dx_{j-1} \tag{6.2-13}$$

$$+ \frac{1}{2^{\frac{1}{2}j(j+1)}}\int_x^{\infty}\left|K\left(\frac{x}{2},\frac{x_1}{2}\right)\right|\int_{(x+x_1)}^{\infty}\left|K\left(\frac{x+x_1}{4},\frac{x_2}{4}\right)\right|\dots\int_{(x+x_1+\dots x_{j-1})}^{\infty}\left|K\left(\frac{1}{2^j}(x+x_1+\dots+x_{j-1}),\frac{1}{2^j}x_j\right)\right|$$

$$\times\left|F\left(\frac{1}{2^j}(x+x_1+\dots x_j)\right)\right| dx_1\dots dx_j = I_1 + I_2$$

It is not hard to show by induction using the Corollary 6-7-1 that

$$I_1 \leqslant C_\varepsilon \exp-\varepsilon x \sum_{\ell=1}^{j} \frac{1}{(\ell-1)!}\left(\frac{C_\varepsilon e^{-\varepsilon x}}{\varepsilon}\right)^{\ell-1} \tag{6.2-14}$$

where C_ε is some positive constant which depends upon ε.

$$\int_{(x+x_1+\dots+x_{j-1})}\left|K\left(\frac{1}{2^j}(x+x_1+\dots+x_{j-1}),\frac{1}{2^j}x_j\right)\right|\left|F\left(\frac{1}{2^j}(x+x_1+\dots+x_j)\right)\right| dx_j$$

$$\leqslant C_\varepsilon e^{-2^{-j}\varepsilon(x+x_1+\dots+x_{j-1})}\int_{(x+x_1+\dots+x_{j-1})}^{\infty} e^{-2^{-j}\varepsilon x_j}|F((x+x_1+\dots+x_j)|dx_j \tag{6.2-15}$$

From Lemma 6-13 and (6.2-5) we get

$$\int_y^\infty e^{-2^{-j}\varepsilon x}|F(x+y)|dx \leqslant \left\{\|e^{\eta x}\hat{R}(x)\| + \|e^{\eta x}\hat{\bar{R}}(x)\|\right\}\left(\int_y^\infty e^{-2(\eta+2^{-j}\varepsilon)x}dx\right)^{\frac{1}{2}}$$
$$+\sum_{j=1}^{N}\int_y^\infty e^{-2^{-j}\varepsilon x}|P_j(x)e^{ik_jx}|dx + \sum_{\ell=1}^{\bar{N}}\int_y^\infty e^{-2^{-j}\varepsilon x}|\bar{P}_\ell(x)e^{-i\bar{k}_\ell x}|dx \qquad (6.2\text{-}16)$$
$$\leqslant D_\eta\, \bar{e}^{(\eta+2^{-j}\varepsilon)y} + \int_y^\infty \bar{e}^{(2^{-j}\varepsilon+\gamma)x}P(|x|)dx\ .$$

In the last expression $D_\eta = (\|e^{\eta x}\hat{R}(x)\| + \|e^{\eta x}\hat{\bar{R}}(x)\|)/[2(\eta + 2^{-j}\varepsilon)]^{\frac{1}{2}}$ $\gamma = \min\{\mathrm{Im}k_j, |\mathrm{Im}\bar{k}_\ell|\}$, $0 < \eta < \varepsilon_0 - \delta$ where $0 < \delta < \varepsilon_0/2$ and $P(|x|) = \sum_{j=1}^{N}|P_j(x)| + \sum_{\ell=1}^{\bar{N}}|\bar{P}_\ell(x)|$. Suppose $\deg P(|x|) = m$ and consider the integral in (6.2-16) for an arbitrary monomial $|x|^\ell, 1 \leqslant \ell \leqslant m$ (the term $D_\eta \bar{e}^{(\eta+2^{-j}\varepsilon)y}/(\eta+2^{-j}\varepsilon)$ can be evaluated in the same way as the constant term). Observe that if $\lambda > 0$

$$\int_y^\infty \bar{e}^{\lambda x}|x|^\ell dx = (\sigma)^\ell\, \bar{e}^{\lambda x}\left\{\sum_{k=0}^{\ell-1}k!C_k^\ell\left(\frac{1}{\lambda}\right)^{k+1}y^{\ell-k} + \left(\frac{1}{\lambda}\right)^{\ell+1}(1-(-1)^\ell)\ell!\right\}$$
$$= \frac{\bar{e}^{\lambda x}}{\lambda}P_\lambda(y)\ , \qquad \sigma = \mathrm{sign}(y). \qquad (6.2\text{-}17)$$

The last term on the right hand side of (6.2-17) is absent if $y > 0$, $P_\lambda(y)$ is a polynomial of degree ℓ . Thus a sequence of integrations gives

$$\int_x^\infty \bar{e}^{\varepsilon y_{n-1}}\int_{y_{n-1}}^\infty \bar{e}^{\varepsilon y_{n-2}} \cdots \int_{y_1}^\infty \bar{e}^{(\varepsilon+\frac{1}{2}\gamma)y}|y|^\ell dy\, dy_1 \ldots dy_{n-1}$$
$$\leqslant \frac{e^{-\frac{1}{2}\gamma x}(\bar{e}^{\varepsilon x})^n P_{\gamma,\varepsilon}(x)}{(n\varepsilon+\gamma/2)((n-1)\varepsilon+\gamma/2)\ldots(\varepsilon+\gamma/2)} \leqslant \bar{e}^{\frac{1}{2}\gamma x}P_{\gamma,\varepsilon}(x)\frac{(\bar{e}^{\varepsilon x})^n}{\varepsilon^n n!} \qquad (6.2\text{-}18)$$

$P_{\gamma,\varepsilon}(x)$ is a polynomial of degree ℓ . From this we conclude that $I_2 \to 0$ uniformly as $n \to \infty$ for $-\infty < a \leqslant x$. The estimate for F then follows from (6.2-13) and (6.2-14)

$$|F(x)| \leqslant C_\varepsilon \exp[-\varepsilon x\{\exp(C_\varepsilon\, \varepsilon^{-1}\exp - \varepsilon x)\}] \quad x \geqslant a > -\infty \qquad (6.2\text{-}19)$$

If in addition Q and R are differentiable and obey the estimates in Corollary 2 to Theorem 6-7 then $F_x(x)$ exists and satisfies the inequality

$$|F_x(x)| \leqslant C_\varepsilon(a)\exp(-\varepsilon x) \qquad x \geqslant a > -\infty . \tag{6.2-20}$$

where $C_\varepsilon(x)$ is a monotonic function which decreases as $x \to +\infty$ and increases as $x \to -\infty$. To establish this notice that

$$F(x) = -K\left(\frac{x}{2},\frac{x}{2}\right) - \int_x^\infty K\left(\frac{x}{2},2s-x\right)F(s)\,ds$$

so that

$$F_x(x) = -\frac{1}{2}\left[K^{(1,0)}(\tfrac{1}{2}x,\tfrac{1}{2}x) + K^{(0,1)}(\tfrac{1}{2}x,\tfrac{1}{2}x)\right] + K(\tfrac{1}{2}x,x)F(x)$$

$$-\frac{1}{2}\int_x^\infty K^{(1,0)}(\tfrac{x}{2},2s-x)F(s)\,ds + \int_x^\infty K^{(0,1)}(\tfrac{x}{2},2s-x)F(s)\,ds \tag{6.2-21}$$

where we have used the notation of Section 4.2

$$f^{(i,j,k)}(x,y,t) \equiv \frac{\partial^{i+j+k}}{\partial x^i \partial y^j \partial t^k} f(x,y,t).$$

The estimate (6.2-20) then follows from the Corollaries 6-7-1 and 6-7-2 and the estimate (6.2-19).

The following theorem characterises the scattering data for potentials which satisfy Theorem 6-6.

Theorem 6-14: The scattering data $S_+ = \{R_+, \bar{R}_+, k_j, P_{+j}(x), \bar{k}_\ell, \bar{P}_{+\ell}(k), \text{Im}\, k_j > 0, \text{Im}\bar{k}_\ell < 0, j = 1,\dots,N, \ell = 1,\dots,\bar{N}\}$ for functions Q and R which obey Theorem 6-6 satisfies the following conditions.

(i) The functions $R_+(k)$, $\bar{R}_+(k)$ are meromorphic in the strip $|\text{Im}k| < \varepsilon$ and $R_+(k) = o(1)$, $\bar{R}_+(k) = o(1)$ as $|k| \to \infty$ uniformly within the strip.

(ii) The Fourier transforms $\int R_+(k)\exp(ikx)\,dk$, $\int_\gamma \bar{R}_+(k)\exp(-ikx)\,dk$ define regular functionals on the space of slowly increasing functions. The functions $\hat{F}_+$, $\hat{\bar{F}}_+$ defined by (6.2-5) satisfy the conditions

$$|\hat{F}_+(x)| \leqslant C_\varepsilon^+(a)\exp(-\varepsilon x)$$

$$|\hat{\bar{F}}_+(x)| \leqslant C_\varepsilon^+(a)\exp(-\varepsilon x)$$

for all $x \geqslant a > -\infty$.

If in addition Q and R are differentiable and obey the estimates in Corollary 2, Theorem 6-7 then

$$|F_{+x}(x)| \leqslant C_\varepsilon^+(a)\exp(-\varepsilon x),\ |\bar{F}_{+x}(x)| \leqslant C_\varepsilon^+(a)\exp(-\varepsilon x)$$

and $R(k) = o(|k|^{-1})$, $R(k) = o(|k|^{-1})$ uniformly within the strip $|\text{Im}\, k| < \varepsilon$ as $|k| \to \infty$.

(iii) The degrees of the polynomials $P_{+j}, \bar{P}_{+\ell}$ are m_j-1, $\bar{m}_\ell-1$ respectively where m_j, $\bar{m}_\ell$ are the multiplicities of the non-real numbers k_j, $\bar{k}_\ell$.

(iv) The scattering data S_+ is consistent, that is it defines and is defined by the asymptotic ($|x| \to \infty$) properties of the principal functions, provided Theorem 6-11 is satisfied. In this case the scattering matrices, $\tilde{S}$ and $\tilde{\bar{S}}$ can be defined where

$$\tilde{S}(k) = \begin{pmatrix} T_+(k) & -R_-(k) \\ R_+(k) & T_-(k) \end{pmatrix} \quad \tilde{\bar{S}}(k) = \begin{pmatrix} \bar{T}_-(k) & \bar{R}_+(k) \\ -\bar{R}_-(k) & \bar{T}_+(k) \end{pmatrix}$$

and $T_+ = T_- \equiv a^{-1}, \bar{T}_+ = \bar{T}_- = \bar{a}^{-1}$. The scattering matrices are meromorphic matrix valued functions defined in the strip $|\mathrm{Im}\, k| < \varepsilon$ which satisfy the condition $\tilde{S}\tilde{\bar{S}} = I$.

The scattering data S_- satisfies conditions analogous to (i) - (iii). The estimates for F_-, $\bar{F}_-$ are given by

$$|F_-(x)| \leqslant C_\varepsilon^-(a)\exp \varepsilon x \;, \; |F_{-x}(x)| \leqslant C_\varepsilon^-(a)\exp \varepsilon x$$

$$|\bar{F}_-(x)| \leqslant C_\varepsilon^-(a)\exp \varepsilon x \;, \; |\bar{F}_{-x}(x)| \leqslant C_\varepsilon^-(a)\exp \varepsilon x$$

for all $x \leqslant a < \infty$.

(v) The coefficients of the normalisation polynomials P_{+j}, P_{-j} and $\bar{P}_{+j}$, $\bar{P}_{-j}$ are related. If $p_{+j\ell}$, $p_{-j\ell}$, $\bar{p}_{+j\ell}$ and $\bar{p}_{-j\ell}$ are respectively the coefficients of x^ℓ in these polynomials then

$$p_{-j\ell} = -(i)^{\ell+1} C_\ell^{m_j-1} \sum_{s=0}^{m_j-1-\ell} C_s^{m_j-1-\ell-s} [(f(k))^{-1}]_{k=k_j}^{m_j-1-\ell-s} \left(\left[\frac{(k-k_j)^{m_j}}{a(k)\,(m_j-1)!} \right]^2 \right)^{(s)}_{k=k_j}$$

$$\bar{p}_{-j\ell} = -(-i)^{\ell+1} C_\ell^{\bar{m}_j-1} \sum_{s=0}^{\bar{m}_j-1-\ell} C_s^{\bar{m}_j-1-\ell-s} [(\bar{f}(k))^{-1}]_{k=\bar{k}_j}^{\bar{m}_j-1-\ell-s} \left(\left[\frac{(k-\bar{k}_j)^{\bar{m}_j}}{\bar{a}(k)\,(\bar{m}_j-1)!} \right]^2 \right)^{(s)}_{k=\bar{k}_j}$$

where f, $\bar{f}$ are any analytic functions such that

$$[f(k)]_{k=k_j}^{m_j-1-\ell} = -(-i)^{\ell+1} p_{+j\ell} [C_\ell^{m_j-1}]^{-1} \quad ; \quad [\bar{f}(k)]_{k=\bar{k}_j}^{\bar{m}_j-1-\ell} = -(i)^{\ell+1} \bar{p}_{+j\ell} [C_\ell^{\bar{m}_j-1}]^{-1}$$

Part (v) of this theorem is established from equations (6.1-87) and (6.1-88). The functions $C_\varepsilon^\pm$ are defined by $C_\varepsilon^+(x) = C_\varepsilon(-x) \equiv C_\varepsilon(x)$.

We shall now investigate further properties of $\bar{F}$ which prove essential in solving the inverse problem.

Definition

A matrix valued function $G_{\pm}(x)$ is of type $(G_{\pm},\varepsilon)$ if G satisfies

(a) for $x \geqslant a - \infty(G_+)$ or $x \leqslant a < \infty(G_-)$, $|G_{\pm}(x)| \leqslant C_{\varepsilon}^{\pm}(a)\exp \mp \varepsilon x$

(b) the integral equation

$$ {}_xY_{\pm}(u) = \pm\int_x^{\pm\infty} {}_xY_{\pm}(s)\, G_{\pm}(s+u)\,ds $$

has only the trivial solution ${}_xY_+(u) = 0$, $u \geqslant x$ (${}_xY_-(x) = 0$, $u \leqslant x$) in the class of functions for which

$$ \pm\int_x^{\pm\infty} |{}_xY_{\pm}(u)|\, \exp(\pm\varepsilon u)du < \infty . $$

We now show that F derived from a given L is of type (G_+,ε) . Equation (6.2-19) shows that F satisfies condition (a) of the definition.

Observe that the integral equation in (b) gives the estimate

$$ |{}_xY(u)| \leqslant \int_x^{\infty} |{}_xY(s)||F(s+u)|ds \leqslant C_{\varepsilon}'(a)\exp(-\varepsilon u). \tag{6.2-22} $$

Consequently the Volterra integral equation

$$ {}_xY(u) = {}_xZ(u) + \int_x^u {}_xZ(s)\ K(s,u)ds \quad u \geqslant x \tag{6.2-23} $$

has a unique solution. The kernel K, (6.2-12) satisfies the estimate for K_+ in Corollary 6-7-1. If we substitute (6.2-23) into the Fredholm equation satisfied by ${}_xY$ and use the Marchenko equation

$$ K(x,u) + F(x+u) + \int_x^{\infty} K(x,s)F(s+u)ds = 0 \quad u \geqslant x \tag{6.2-24} $$

then we obtain a Volterra equation satisfied by ${}_xZ$

$$ \begin{aligned} {}_xZ(u) &= \int_x^u {}_xZ(s)K(s,u)ds + \int_x^{\infty}\{{}_xZ(s) - \int_x^s Z(v)K(v,s)dv\}F(s+u)ds \\ &= \int_u^{\infty} {}_xZ(s)F(s+u)ds - \int_x^u {}_xZ(s)\int_s^{\infty} K(s,v)F(v+u)dvds \\ &\quad - \int_x^{\infty}\left[\int_x^s {}_xZ(v)K(v,s)dv\right]F(s+u)ds \\ &= \int_x^{\infty} {}_xZ(s)\left[F(s+u) + \int_s^{\infty} K(s,v)F(v+u)dv\right]ds \end{aligned} \tag{6.2-25} $$

From the estimate for K we see that this is a homogeneous Volterra equation with a rapidly decreasing kernel $K(s,v)$ as $v \to \infty$ which therefore has a unique solution, the trivial solution, ${}_xZ(u) = 0$ for $u \geq x \geq a > -\infty$. It immediately follows from (6.2-23) that ${}_xY(u) = 0$ for $u \geq x \geq a > -\infty$. Thus the function F obtained from a particular L is of type (G_+,ε).

Theorem 6-15:
The functions $F_\pm$ constructed from the scattering data $S_\pm$ of the operator L are of type $(G_\pm,\varepsilon)$.

Corollary 6-15-1:
The scattering data $S_\pm$ of an operator L uniquely determines L.

In the inverse problem we are given the scattering data and we are required to construct Q and R from it. The procedure is similar to that for the Schrödinger inverse problem which is given by the Steps 1 - 5 in Section 4.1. However once F has been constructed from the scattering data the critical part of the analysis is in proving the existence and uniqueness of the solution K to the Marchenko equation. Here existence and uniqueness is for a class of functions to which F and the reconstructed $P = \begin{pmatrix} 0 & Q \\ R & 0 \end{pmatrix}$ belong.

For self-adjoint operators the uniqueness of the solution to the Marchenko equation can be established by making use of the essential fact that the scattering matrix is unitary (see Step 1, Section 4.1). Useful papers on the self-adjoint Dirac problem on the half-line and whole line are by Gasymov (1968) and Frolov (1972). However it is clear from the definition and equation (6.2-24) that if the F constructed from the scattering data is of type (G_+,ε) then the solution K to (6.2-24) exists and is unique in the class of functions $E_\varepsilon(x,\infty)$ such that $Y \in E_\varepsilon(x,\infty)$ if $\int_x^\infty |Y(u)|\, e^{-\varepsilon u} < \infty$ $x \geq a > -\infty$. Similar results are obtained for the Marchenko equation (6.2-4) in the class of functions $E_\varepsilon(-\infty,x)$ } for which $\int_{-\infty}^x |Y(u)|\, e^{\varepsilon u} du < \infty$, $x \leq a$

Theorem 6-16:
If $F_\pm$ are differentiable functions of type $(G_\pm,\varepsilon)$ and $|F_{\pm x}(x)| \leq C_\varepsilon^\pm(a)\, e^{\mp\varepsilon x}$ where (+) $-\infty < a \leq x$ or (-) $x \leq a < \infty$, then the Marchenko equations

$$K_+(x,u) + F_+(x+u) + \int_x^\infty K_+(x,s)F_+(s+u)ds = 0 \qquad u \geq x$$

$$K_-(x,u) + F_-(x+u) + \int_{-\infty}^x K_-(x,s)F_-(s+u)ds = 0 \qquad u \leq x$$

have unique solutions K_+,K_- in the class of functions $E_\varepsilon(x,\infty)$, $E_\varepsilon(-\infty,x)$ respectively. The kernels satisfy the estimates

$$|K_\pm(x,u)| \leq C_\varepsilon^\pm(a)\exp\mp\varepsilon(x+u),$$

$$|K_{\pm x}(x,u)| \leq C_\varepsilon^\pm(a)\exp\mp\varepsilon(x+u),$$

$$|K_{\pm u}(x,u)| \leq C_\varepsilon^\pm(a)\exp\mp\varepsilon(x+u),$$

where ε is a positive number. Suppose that

$$\tilde{\Psi}(x,k) = \tilde{\Psi}_\infty(x,k) + \int_x^\infty K_+(x,u)\tilde{\Psi}_\infty(u,k)\,du$$

where $\tilde{\Psi}_\infty(x,k) = \Phi_\infty(x,k) = \mathrm{diag}(e^{-ikx}, e^{ikx})$.

Let $\sigma = \begin{pmatrix} -1 & 0 \\ 0 & 1 \end{pmatrix}$ and $P_\pm = \begin{pmatrix} 0 & Q_\pm \\ R_\pm & 0 \end{pmatrix}$ where

$$Q_+(x) = -2K_{+1}(x,x), \quad Q_-(x) = 2\bar{K}_{-1}(x,x)$$

$$R_+(x) = -2\bar{K}_{+2}(x,x), \quad R_-(x) = 2K_{-2}(x,x)$$

Then for $-\infty < a \leqslant x < \infty$, $|\mathrm{Im}\, k| < \varepsilon$, the function $\tilde{\Psi}$ satisfies the differential equation

$$Y_x(x,k) = P_+(x)Y(x,k) + ik\sigma Y(x,k)$$

Similarly the function Φ satisfies this equation for $|\mathrm{Im}\, k| \leqslant \varepsilon$ and $-\infty < x \leqslant a < \infty$ with the function P_- replacing P_+.

The components of the functions $P_\pm$ obey the estimates

$$|Q_\pm(x)| \leqslant C_\varepsilon^\pm(a)\exp\mp 2\varepsilon x, \; |Q_{\pm x}(x)| \leqslant C_\varepsilon^\pm(a)\exp\mp 2\varepsilon x$$

$$|R_\pm(x)| \leqslant C_\varepsilon^\pm(a)\exp\mp 2\varepsilon x, \; |R_{\pm x}(x)| \leqslant C_\varepsilon^\pm(a)\exp\mp 2\varepsilon x .$$

Proof: The uniqueness of the solutions $K_\pm$ in $E_\varepsilon(x,\infty)$, $E_\varepsilon(-\infty,x)$ respectively follows from the fact that $F_\pm$ are of type $(G_\pm,\varepsilon)$. Next we derive the estimates for the kernels $K_\pm$. The final part of the proof is similar to Step 4 of Section 4.1.

The Marchenko equations can be written

$$\left(I + {}_xN^\pm\right)K_\pm + F_\pm = 0 \tag{6.2-26}$$

where

$${}_xN^\pm Y(u) = \pm\int_x^{\pm\infty} Y(s)F_\pm(u+s)\,ds$$

Furthermore

$$\pm\int^{\pm\infty} |{}_xN^\pm Y(u)|e^{\mp\varepsilon u}du \leqslant C_\varepsilon^\pm \int^{\pm\infty} e^{\mp 2\varepsilon u}du\int^{\pm\infty} |Y(s)|e^{\mp\varepsilon s}ds$$

so that $\|{}_xN^\pm\|_{E_\varepsilon^\pm}$ exists. By using the method of successive approximations (see Gasymov, 1968) one can show that $(I + {}_xN^\pm)^{-1}$ is uniformly bounded on (+) $-\infty < a \leqslant x < \infty$; (−) $-\infty < x \leqslant a < \infty$. Thus we have

$$\int^{\pm\infty} |K_\pm(x,s)|e^{\mp\varepsilon s}ds \leqslant C_\varepsilon^{\pm'}(a)e^{\mp\varepsilon x}\left\|\left(I + {}_xN^\pm\right)^{-1}\right\|_{E_\varepsilon^\pm}\int_x^{\pm\infty} e^{\mp\varepsilon s}ds \tag{6.2-27}$$

Then from the Marchenko equations and (6.2-27) we obtain

$$|K_{\pm}(x,u)| \leqslant C_{\varepsilon}^{\pm'}(a)e^{\mp\varepsilon u}\left\{e^{\mp\varepsilon x} \pm\int_x^{\pm\infty}|K_{\pm}(x,s)|e^{\mp\varepsilon s}ds\right\}$$
$$\leqslant C_{\varepsilon}^{\pm}(a)\exp\mp\varepsilon(x+u) \quad . \tag{6.2-28}$$

If $F_{\pm}$ are differentiable and obey the estimates in the theorem then one can show from the Marchenko equations that $K_{\pm x}(x,u)$ and $K_{\pm u}(x,u)$ exist and satisfy the estimates

$$|K_{\pm x}(x,u)| < C_{\varepsilon}^{\pm}(a)\exp\mp\varepsilon x, \quad |K_{\pm u}(x,u)| < C_{\varepsilon}^{\pm}(a)\exp\mp\varepsilon x \tag{6.2-29}$$

The functions $K_{\pm x}, K_{\pm u}$ satisfy the equations

$$K_{\pm x}(x,u) + F_{\pm x}(x+u) \pm \int_x^{\pm\infty} K_{\pm x}(x,s)F_{\pm}(s+u)ds \pm K_{\pm}(x,x)F_{\pm}(x+u)=0$$
$$K_{\pm u}(x,u) + F_{\pm u}(x+u) \pm \int_x^{\pm\infty} K_{\pm}(x,s)F_{\pm u}(s+u)ds = 0 \quad . \tag{6.2-30}$$

Since $\sigma F_{\pm} + F_{\pm}\sigma = 0$ it follows that

$$(\sigma K_{\pm x}(x,u) + K_{\pm u}(x,u)\sigma) \pm \sigma K_{\pm}(x,x)F_{\pm}(x+u)$$
$$\pm\int_x^{\pm\infty}(\sigma K_{\pm x}(x,s)F_{\pm}(s+u) + K_{\pm}(x,s)F_{\pm u}\sigma)ds = 0$$

If we now use the Marchenko equations and substitute for $F_{\pm}$ and interchange the ordering of $F_{\pm u}\sigma = -\sigma F_{\pm u}$ we get after an integration by parts that

$$(\sigma K_{\pm x}(x,u) + K_{\pm u}(x,u)\sigma \pm (\sigma K_{\pm}(x,x)K_{\pm}(x,u) - K_{\pm}(x,x)\sigma K_{\pm}(x,u))$$
$$\pm\int_x^{\pm\infty}(\sigma K_{\pm x}(x,s) + K_{\pm s}(x,s)\sigma \pm (\sigma K_{\pm}(x,x)K_{\pm}(x,s)$$
$$- K_{\pm}(x,x)\sigma K_{\pm}(x,s))F_{\pm}(s+u)ds = 0 \tag{6.2-31}$$

Then since $F_{\pm}$ is of type $(G_{\pm},\varepsilon)$ these equations only have the trivial solution. Consequently

$$\sigma K_{\pm x}(x,u) + K_{\pm u}(x,u)\sigma - \sigma P_{\pm}(x)K_{\pm}(x,u) = 0$$

where

$$P_{\pm}(x) = \pm\, \sigma(K_{\pm}(x,x)\sigma - \sigma K_{\pm}(x,x)) \tag{6.2-32}$$

It is now straightforward to show from the definition of $\tilde{\Psi}$ and Φ that

$$\tilde{\Psi}_x = P_+(x)\tilde{\Psi} + ik\sigma\tilde{\Psi}$$

and

$$\Phi_x = P_-(x)\Phi + ik\sigma\Phi \tag{6.2-33}$$

Define $P_{\pm}(x) = \begin{pmatrix} 0 & Q_{\pm} \\ R_{\pm} & 0 \end{pmatrix}$ then we get from (6.2-28) and (6.2-29) that

$$|Q_{\pm}(x)| \leqslant C_{\varepsilon}^{\pm}(a)e^{\mp 2\varepsilon x}, \qquad |Q_{\pm x}(x)| \leqslant C_{\varepsilon}^{\pm}(a)e^{\mp 2\varepsilon x}$$

and

$$|R_{\pm}(x)| \leqslant C_{\varepsilon}^{\pm}(a)e^{\mp 2\varepsilon x}, \qquad |R_{\pm x}(x)| \leqslant C_{\varepsilon}^{\pm}(a)e^{\mp 2\varepsilon x}. \tag{6.2-34}$$

There is in fact no absolute necessity to assume that $F_{\pm}$ is differentiable. The same result can be achieved by replacing $F_{\pm}$ by a sequence of functions (see for example Marchenko and Agranovich (1963) or Gasymov (1968)). It is assumed throughout the proof that a + index indicates the range $-\infty < a \leqslant x$, whereas an - index denotes the range $a \leqslant x < \infty$.

We now establish the conditions a set of data is to satisfy if it is to be the scattering data of an operator L which belongs to the class defined by Theorem 6-6. We assume without loss of generality that the given data is to define S_+.

Theorem 6-17:

Let
$S_+ = \{R_+(k), \bar{R}_+(k), k_j, \bar{k}_\ell, P_j(x), \bar{P}_\ell(x), m_j, \bar{m}_\ell;\ j = 1,\dots,N, \ell = 1,\dots,\bar{N}\}$ be a set of data such that

(i) The functions $R_+(k), \bar{R}_+(k)$ are meromorphic in the strip $|\mathrm{Im}\, k| < \varepsilon_0$ where $\varepsilon_0 < \min\{|\mathrm{Im}\, k_j|, |\mathrm{Im}\, \bar{k}_j|\}$, and $\varepsilon_0 > 0$, with only real poles. Asymptotically $R_+(k) = o(1/|k|)$, $\bar{R}_+(k) = o(1/|k|)$, as $|k| \to \infty$.}

(ii) the degrees of the polynomials P_{+j}, $\bar{P}_{+\ell}$ are m_j-1, $\bar{m}_\ell-1$ where m_j, $\bar{m}_\ell\}$ are respectively the multiplicity of the non-real numbers k_j, $\bar{k}_\ell$, $j = 1,\dots,N$, $\ell = 1,\dots,\bar{N}$.

(iii) $\hat{S}_{\pm}$ satisfies the Theorem 6-11.

(iv) The functions $F_{\pm}$ reconstructed from $\hat{S}_{\pm}$ are of type $(G_{\pm},\varepsilon)$ for some $\varepsilon > \varepsilon_0$ and moreover are differentiable and obey the estimates

$$|F_{\pm x}(x)| < C_\varepsilon^\pm e^{\mp\varepsilon x}$$

Then $\hat{S}_+$ defines an operator L which satisfies (6.1-13) and whose coefficients Q and R satisfy the estimates

$$|Q(x)| \leqslant C_\varepsilon e^{-2\varepsilon|x|}, \qquad |Q_x(x)| \leqslant C_\varepsilon e^{-2\varepsilon|x|},$$

$$|R(x)| \leqslant C_\varepsilon e^{-2\varepsilon|x|}, \qquad |R_x(x)| \leqslant C_\varepsilon e^{-2\varepsilon|x|}.$$

Moreover the scattering data S_+ for L is precisely $\hat{S}_+$.

Proof: Augment $\hat{S}_+$ by including the real poles $k_j(\bar{k}_j)$, $j=N+1,..,N+M$ of R_+ and $\bar{R}_+$ with multiplicities $m_j, \bar{m}_j$ respectively. Because $\hat{S}_+$ satisfies Theorem 6-11 we can construct the functions a, $\bar{a}$ and thus the functions $R_- = \bar{R}_+\bar{a}\,a^{-1}$ and $\bar{R}_- = R_+ a\,\bar{a}^{-1}$ which also satisfy the properties (i). The normalisation polynomials P_{-j}, $\bar{P}_{-j}$ are then constructed from the P_{+j}, $\bar{P}_{+j}$ by using relations (v) given in Theorem 6-14. We then form the functions $F_\pm$ which are required through part (iv) of the theorem to be of type $(G_\pm,\varepsilon)$. It then follows from Theorem 6-15 that the kernels $K_\pm$ of the Marchenko equations are unique and enable the functions

$$\tilde{\Psi}(x,k) = \tilde{\Psi}_\infty(x,k) + \int_x^\infty K_+(x,u)\tilde{\Psi}_\infty(u,k)\,du$$

$$\Phi(x,k) = \Phi_{-\infty}(x,k) + \int_{-\infty}^x K_-(x,u)\Phi_{-\infty}(u,k)\,du \tag{6.2-35}$$

to be uniquely defined for $|\text{Im } k| < \varepsilon$. These functions satisfy (6.1-13) with potential functions $P_\pm$ respectively where

$$|Q_\pm(x)| \leqslant C_\varepsilon^\pm(a)e^{\mp 2\varepsilon x}$$

$$|R_\pm(x)| \leqslant C_\varepsilon^\pm(a)e^{\mp 2\varepsilon x} \tag{6.2-36}$$

We establish the theorem through proving a series of lemmas.

Introduce the functions $H = (H_1, H_2)$, $J = (J_1, J_2)$ by the relations

$$H = \tilde{\Psi}\begin{pmatrix} 1 & \bar{R}_+ \\ R_+ & 1 \end{pmatrix}, \quad J = \Phi\begin{pmatrix} 1 & -R_- \\ -\bar{R}_- & 1 \end{pmatrix} \tag{6.2-37}$$

which are meromorphic functions in the strip $|\text{Im } k| < \varepsilon_0$. Consider the function H. From the Marchenko equation for K_+ we obtain

$$K_+(x,u) + \hat{R}_+(x+u) + \int_x^\infty K_+(x,s)\hat{R}_+(s+u)\,ds = \tilde{K}_+(x,-u) \qquad u \geqslant x$$

where

$$\hat{R}_+(x) = \frac{1}{2\pi}\begin{pmatrix} 0 & \int_{\bar\gamma} \bar{R}_+(k)e^{-ikx}dk \\ \int_\gamma R_+(k)e^{ikx}dk & 0 \end{pmatrix}$$

$$\tilde{K}_+(x,u) = -\sum_{j=1}^{N}\left[M_j(x-u) + \int_x^\infty K_+(x,s)M_j(s-u)ds\right]$$

and

$$M_j(x) = \begin{pmatrix} 0 & \bar{P}_j(x)e^{-i\bar{k}_j x} \\ P_j(x)e^{ik_j x} & 0 \end{pmatrix} \qquad (6.2\text{-}38)$$

As in the previous theorem a + index will denote the range $-\infty < a \leqslant x < \infty$ and a − index the range $-\infty < x \leqslant a < \infty$. If we define $K_+(x,u) = 0$, $u < x$ then taking the Fourier transform of (6.2-38) with respect to the matrix $\hat{\Phi}_\infty(u,k)$ we obtain

$$\tilde{\Psi}(x,k) - \tilde{\Psi}_\infty(x,k) + \tilde{\Psi}(x,k)\begin{pmatrix} 0 & \bar{R}_+(k) \\ R_+(k) & 0 \end{pmatrix} = \int_{-\infty}^{x} \tilde{K}_+(x,u)\,\tilde{\Psi}_\infty(u,-k)du\ . \qquad (6.2\text{-}39)$$

If we compare this with (6.2-37) then we see that

$$H(x,k) = \tilde{\Psi}_\infty(x,k) + \int_{-\infty}^{x}\tilde{K}_+(x,u)\tilde{\Psi}_\infty(u,-k)du \qquad |\mathrm{Im}k| < \varepsilon_0 \qquad (6.2\text{-}40)$$

From the properties of $K_+(x,u)K_{+u}(x,u)$ we have that $\bar{\psi}(x,k)$ is analytic in $\mathrm{Im}k < \varepsilon$, $\psi(x,k)$ is analytic in $\mathrm{Im}\,k > -\varepsilon$ where $\tilde{\Phi} = (\bar{\phi},\psi)$. Furthermore $\tilde{\Phi}$ has the uniform asymptotic behaviour

$$\tilde{\Psi}_\infty^{-1}(x,k)\tilde{\Psi}(x,k) = I + o(1/|k|) \text{ as } |k| \to \infty,\ |\mathrm{Im}k| < \varepsilon \qquad (6.2\text{-}41)$$

The column vectors of $\tilde{\Phi}_\infty^{-1}\tilde{\Phi}$ have this same asymptotic behaviour in their respective half planes of definition.

We now investigate the analytic properties of $_xH(k)$. From the definition of $\tilde{K}_+$ we have that

$$\tilde{K}_+(x,u) = -\sum_{j=1}^{N}\left[P_{+j}\left(-i\frac{\partial}{\partial k}\right)\left(e^{-iku}\psi(x,k)\right)\Big|_{k=k_j},\ \bar{P}_{+j}\left(i\frac{\partial}{\partial k}\right)\left(e^{iku}\bar{\psi}(x,k)\right)\Big|_{k=\bar{k}_j}\right] \qquad (6.2\text{-}42)$$

Put

$$_xD(k)\tilde{\Psi}_\infty(x,k) = H(x,k) - \tilde{\Phi}_\infty(x,k) - \int_\infty^x K_+(x,k)\tilde{\Psi}_\infty(u,-k)du$$

then because of the asymptotic behaviour of R_+ and $\bar{R}_+$ we have from (6.2-37) that $_xD$ has the uniform asymptotic behaviour

$$ {}_xD(k) = o(1/|k|) \text{ as } |k| \to \infty, \quad |\mathrm{Im}k| < \varepsilon_0 \tag{6.2-43}$$

The appropriate Fourier transform of ${}_xD = ({}_xD_1, {}_xD_2)$ follows from its definition in terms of H

$$ {}_x\hat{D}(u) = \frac{1}{2\pi}\left(\int_\gamma {}_xD_1(k)e^{iku}dk, \int_\gamma {}_xD_2(k)e^{-iku}dk\right) \quad |\mathrm{Im}k| < \varepsilon_0. \tag{6.2-44}$$

From its definition or alternatively by re-writing the Marchenko equation in terms of ${}_x\hat{D}$ we get that ${}_x\hat{D}(u) = 0$ $u \geqslant x$ so that the Fourier transform can be written as

$$ {}_xD(k) = \int_{-\infty}^{x} {}_x\hat{D}(u)\Psi_\infty(u,k)\,du . \tag{6.2-45}$$

Choose $0 < \lambda < \varepsilon_0$ and put $\eta = \mathrm{Im}k$. Then from the Schwarz inequality and Parseval's equality we derive the bounds

$$\begin{aligned} \left|\int_{-\infty}^{x} {}_x\hat{D}_1(u)e^{-iku}du\right| &\leqslant \int_{\infty}^{x} |{}_x\hat{D}_1(u)|e^{\eta u}du \\ &\leqslant \left\{\int_{-\infty}^{x} |{}_x\hat{D}_1(u)e^{\lambda u}|_E^2 du\right\}^{\frac{1}{2}}\left\{\int_{-\infty}^{x} e^{2(n-\lambda)u}du\right\}^{\frac{1}{2}} \\ &= \frac{1}{2\pi}\left\{\int_{-\infty}^{\infty} |D_1(\xi+i\lambda)|_E^2 d\xi\right\}^{\frac{1}{2}} \frac{1}{(2(\eta-\lambda))^{\frac{1}{2}}} \end{aligned} \tag{6.2-46}$$

and

$$ \left|\int_{-\infty}^{x} {}_x\hat{D}_2(u)e^{iku}du\right| \leqslant \left\{\frac{1}{2\pi}\int_{-\infty}^{\infty} |{}_xD_2(\xi-i\lambda)|_E^2 d\xi\right\}^{\frac{1}{2}} \frac{1}{(2(\lambda-\eta))^{\frac{1}{2}}} \tag{6.2-47}$$

If we now use the estimate which is implied by (6.2-43) we get from (6.2-46), (6.2-47) that ${}_xD_1$ is analytic in $\mathrm{Im}\,k > \lambda$ and ${}_xD_2$ is analytic in $\mathrm{Im}\,k < \lambda$. If we glue this together with the known analytic behaviour of ${}_xD$ then we deduce that ${}_xD_1$ is meromorphic in $\mathrm{Im}\,k > -\varepsilon_0$ and ${}_xD_2$ is meromorphic in $\mathrm{Im}\,k < \varepsilon_0$. Notice in particular that it is the evaluation of the integral in the definition of ${}_xD$ which is analytically continued. Thus since

$$ \int_{-\infty}^{x}\left(i\frac{\partial}{\partial k}\right)^{\ell} e^{-i(k_j-k)u}du = \left(i\frac{\partial}{\partial k}\right)^{\ell}\left\{\frac{e^{-i(k_j-k)x}}{-i(k_j-k)}\right\} \quad |\mathrm{Im}k| < \varepsilon_0 \tag{6.2-48}$$

Fourier transforming this relation we find that

$$P_{+j}\left(-i\frac{\partial}{\partial k}\right)\left(e^{-iku}\psi(x,k)\right)_{k=k_j} = \frac{1}{2\pi}\sum_{\ell=1}^{m_j}\int_\gamma \left[\frac{F_{j\ell}(x)e^{-i(k_j-k)x}}{(k_j-k)^\ell}\right]e^{-iku}dk \tag{6.2-49}$$

The $F_{j\ell}$ are complex valued functions which are bounded on any finite interval $[a,\infty[$, $a > -\infty$. It is clear that the square bracketed function in (6.2-49) can be analytically continued into $\text{Im } k > 0$ from which the result follows.

As an immediate consequence of this result we have that

$$\text{Res}_{k=k_j}\left\{e^{ikx}H_1(x,k)\right\} = -i\left\{P_{+j}\left(\frac{-i\partial}{\partial k}\right)\left(e^{ikx}\psi(x,k)\right)\right\}_{k=k_j}$$
$$\text{Res}_{k=\bar{k}_j}\left\{e^{-ikx}H_2(x,k)\right\} = i\left\{\bar{P}_{+j}\left(i\frac{\partial}{\partial k}\right)\left(e^{-ikx}\bar{\psi}(x,k)\right)\right\}_{k=\bar{k}_j} \tag{6.2-50}$$

Furthermore H has the uniform asymptotic behaviour

$$e^{ikx}H_1(x,k) = o(1/|k|) \text{ as } |k| \to \infty \quad \text{Im}k > -\varepsilon_0$$
$$\bar{e}^{ikx}H_2(x,k) = o(1/|k|) \text{ as } |k| \to \infty \quad \text{Im}k < \varepsilon_0. \tag{6.2-51}$$

Since $H(x,k)$ is meromorphic in $|\text{Im } k| < \varepsilon_0$ we can apply Cauchy's theorem to (6.2-37). First notice that the Fourier transform of (6.2-37) can be written as

$$\frac{1}{2\pi}\int_\gamma [H(x,k) - \Psi_\infty(x,k)]\Psi_\infty(u,-k)dk$$
$$= \frac{1}{2\pi}\int_\gamma (\tilde{\Psi}(x,k) - \Psi_\infty(x,k))\Psi_\infty(u,-k)dk$$
$$+ \frac{1}{2\pi}\int_\gamma (\tilde{\Psi}(x,k) - \Psi_\infty(x,k))R_+(k)\Psi_\infty(u,-k)dk \tag{6.2-52}$$
$$+ \frac{1}{2\pi}\int_\gamma \Psi_\infty(x,k)R_+(k)\Psi_\infty(u,-k)dk$$

This follows from the analytic properties of the functions. If we complete the contours using $\bar{\gamma}$ and take into account the asymptotic behaviour of the functions as $|k| \to \infty$ we obtain

$$\text{Res}_{k=k_j}\left\{e^{ikx}H_1(x,k)\right\} = \text{Res}_{k=k_j}\left\{R_+(k)\left(e^{ikx}\psi(x,k)\right)\right\} \quad \text{Im}k_j = 0$$
$$\text{Res}_{k=\bar{k}_j}\left\{\bar{e}^{ikx}H_2(x,k)\right\} = \text{Res}_{k=\bar{k}_j}\left\{\bar{R}_+(k)\left(\bar{e}^{ikx}\bar{\psi}(x,k)\right)\right\} \quad \text{Im}\bar{k}_j = 0 \tag{6.2-53}$$

We can establish similar relationships between the functions J and Φ.

Lemma 6-18:

Under the conditions of Theorem 6-15 there exist unique functions $\tilde{\Psi}(x,k)$, $\Phi(x,k)$, $H(x,k)$ and $J(x,k)$ such that

(i) the functions $\psi(x,k)$, $\phi(x,k)$, $(\overline{\psi}(x,k)$, $\overline{\phi}(x,k))$ can be continued analytically into the upper (lower) half-plane $\operatorname{Im} k > -\varepsilon$ $(\operatorname{Im} k < \varepsilon)$, whereas $H_1(x,k)$, $J_2(x,k)$ $(H_2(x,k), J_1(x,k))$ are meromorphic in $\operatorname{Im} k > -\varepsilon_0$ $(\operatorname{Im} k < \varepsilon_0)$. The functions $H_1(x,k)$, $J_2(x,k)$ $(H_2(x,k)$, $J_1(x,k))$ have poles of order m_j at the points k_j $(\overline{k}_j)$, $j = 1,\ldots,N+M$.

(ii) the residues of $H(x,k)$ $(J(x,k))$ are related to the values of the functions $\hat{\Psi}(x,k)$ $(\Phi(x,k))$ by the relations

$$\operatorname{Res}_{k=k_j}\left\{e^{ikx}H_1(x,k)\right\} = -i\left\{P_{+j}\left(-i\frac{\partial}{\partial k}\right)\left(e^{ikx}\psi(x,k)\right)\right\}_{k=k_j} \qquad \operatorname{Im}k_j > 0$$

$$\operatorname{Res}_{k=\overline{k}_j}\left\{\overline{e}^{ikx}H_2(x,k) = i\left\{\overline{P}_{+j}\left(i\frac{\partial}{\partial k}\right)\left(\overline{e}^{ikx}\overline{\psi}(x,k)\right)\right\}_{k=k_j} \qquad \operatorname{Im}\overline{k}_j < 0$$

$$j = 1,\ldots,N$$

$$\operatorname{Res}_{k=k_j}\left\{e^{ikx}H_1(x,k)\right\} = \operatorname{Res}_{k=k_j}\left\{R_+(k)\left(e^{ikx}\psi(x,k)\right)\right\} \qquad \operatorname{Im}k_j = 0$$

$$\operatorname{Res}_{k=\overline{k}_j}\left\{\overline{e}^{ikx}H_2(x,k)\right\} = \operatorname{Res}_{k=\overline{k}_j}\left\{\overline{R}_+(k)\left(\overline{e}^{ikx}\overline{\psi}(x,k)\right)\right\} \qquad \operatorname{Im}\overline{k}_j = 0$$

$$j = N+1,\ldots,N+M$$

Similar relations to these are obtained for $J(x,k)$ and $\Phi(x,k)$. In this case J_2, J_1 replace H_1, H_2; P_-, $\overline{P}_-$ replace P_+, $\overline{P}_+$; R_-, $\overline{R}_-$ replace R_+, $\overline{R}_+$; $\phi,\overline{\phi}$ replace $\psi,\overline{\psi}$ and $-i$ replaces i in the above relations.

(c) for large $|k|$ the functions have the uniform asymptotic behaviour

$$H(x,k)\Psi_\infty^{-1}(x,k) = I + o(1/|k|)$$

$$J(x,k)\Psi_\infty^{-1}(x,k) = I + o(1/|k|) \quad \text{as } |k| \to \infty$$

$$\tilde{\Psi}(x,k)\Psi_\infty^{-1}(x,k) = I + o(1|k|)$$

$$\Phi(x,k)\Psi_\infty^{-1}(x,k) = I + o(1|k|)$$

in their respective domains of analyticity.

(d) for $|\mathrm{Im}\, k| < \varepsilon_0$

$$H(x,k) = \tilde{\Psi}(x,k)\begin{pmatrix} 1 & \bar{R}_+(k) \\ R_+(k) & 1 \end{pmatrix}$$

$$J(x,k) = \Phi(x,k)\begin{pmatrix} 1 & -R_-(k) \\ -\bar{R}_-(k) & 1 \end{pmatrix}$$

We now prove a lemma which effectively solves the inverse problem.

Lemma 6-19:
The functions $\Phi(x,k), \tilde{\Psi}(x,k)$ satisfy the relation

$$\Phi(x,k) = \tilde{\Psi}(x,k)\tilde{A}(k) \qquad |\mathrm{Im}k| < \varepsilon_0$$

where

$$\tilde{A}(k) = \begin{pmatrix} a & \bar{b} \\ b & \bar{a} \end{pmatrix}$$

Proof: Define the functions

$$H' = \Phi\begin{pmatrix} a & 0 \\ 0 & \bar{a} \end{pmatrix}^{-1} \qquad \tilde{\Psi}' = J\begin{pmatrix} \bar{a} & 0 \\ 0 & a \end{pmatrix} \tag{6.2-54}$$

The column vectors of these functions $H' = (H_1', H_2')$, $\tilde{\Psi}' = (\bar{\psi}', \psi')$ have the properties that $\bar{\psi}'$ is analytic in $\mathrm{Im}\, k < \varepsilon_0$ and ψ' is analytic in $\mathrm{Im}\, k > -\varepsilon_0$, whereas H_1' is meromorphic in $\mathrm{Im}\, k > -\varepsilon_0$ and H_2' is meromorphic in $\mathrm{Im}\, k < \varepsilon_0$. From the definition of J we can easily show that

$$H'(x,k) = \tilde{\Psi}'(x,k)\begin{pmatrix} 1 & \bar{R}_+(k) \\ R_+(k) & 1 \end{pmatrix} \qquad |\mathrm{Im}k| < \varepsilon_0 \tag{6.2-55}$$

In the same way as we established (6.2-53) we can show that

$$\mathrm{Res}_{k=k_j}\left\{e^{ikx}H_1'(x,k)\right\} = \mathrm{Res}_{k=k_j}\left\{R_+(k)\left[e^{ikx}\psi'(x,k)\right]\right\}$$

$$\mathrm{Res}_{k=\bar{k}_j}\left\{e^{-ikx}H_2'(x,k)\right\} = \mathrm{Res}_{k=\bar{k}_j}\left\{\bar{R}_+(k)\left[e^{-ikx}\bar{\psi}'(x,k)\right]\right\} \tag{6.2-56}$$

$$j = N+1, \ldots, N+M$$

The analyticity of the functions Φ and $\tilde{\Psi}'$ and their representations (6.2-54) lead to the results that

$$\operatorname{Res}_{k=k_j}\left\{e^{ikx}H_1'(x,k)\right\} = \operatorname{Res}_{k=k_j}\left[e^{ikx}\frac{\phi(x,k)}{a(k)}\right]$$

$$\operatorname{Res}_{k=k_j}\left\{e^{-ikx}J_2(x,k)\right\} = \operatorname{Res}_{k=k_j}\left[e^{-ikx}\frac{\psi'(x,k)}{a(k)}\right] \tag{6.2-57}$$

with similar expressions for the other component functions. Thus from Lemma 6-18 and the second expression in (6.2-57) we obtain

$$\sum_{\ell=0}^{m_j-1} C_\ell^{m_j-1}\left[e^{-ikx}\psi'(x,k)\right]^{(\ell)}_{k=k_j}\left[\frac{(k-k_j)^{m_j}}{(m_j-1)!\,a(k)}\right]^{(m_j-1-\ell)}_{k=k_j}$$

$$= \sum_{\ell=0}^{m_j-1}(i)^{\ell+1}p_{-j\ell}\frac{\partial^\ell}{\partial k^\ell}\left[e^{-ikx}\phi(x,k)\right]_{k=k_j} \tag{6.2-58}$$

The first expression in (6.2-57) and part (v) of Theorem 6-14 now gives the result that

$$\operatorname{Res}_{k=k_j}\left\{e^{ikx}H_1'(x,k)\right\} = -i\left\{P_{+j}\left(-i\frac{\partial}{\partial k}\right)\left[e^{ikx}\psi'(x,k)\right]\right\}_{k=k_j} \tag{6.2-59}$$

This is the first relation in (6.2-50). The second relation in (6.2-50) is satisfied by H_2' and $\overline{\psi}'$.

Furthermore one easily obtains

$$H'(x,k)\Psi_\infty(x,-k) = I + o(1/|k|)$$

$$\text{as } |k| \to \infty \tag{6.2-60}$$

$$\tilde{\Psi}'(x,k)\Psi_\infty(x,-k) = I + o(1/|k|)$$

This asymptotic behaviour is uniform within the domain of analyticity of the component functions of H' and $\tilde{\Psi}'$. Thus by Lemma 6-18 we have that $H' \equiv H$ and $\tilde{\Psi} \equiv \Psi$. The statement of the lemma follows immediately from this.

Lemma 6-19 requires ϕ and $\tilde{\Psi}$ to satisfy the same differential equation. It follows from Theorem 6-16 that $P_+ = P_- \equiv P$ and that $\hat{S}_+$ is the scattering data S_+ of the operator L. The conditions on the coefficients of L also follow from Theorem 6-16.

We have solved the inverse problem under very strong conditions on the scattering data. A study of the solutions for arbitrary initial data (see the next section) indicates that the conditions are too restrictive. However the principal difficulty with this method is the specification of conditions on Q and R which ensure that there are a finite number of spectral singularities. The structure of the spectral singularities when Q and R satisfy a moment equation have yet to be investigated. We remark that any weaker conditions will still lead to a study of the same problems as encountered here (it is clear that many of the results can be immediately adapted to weaker conditions on the potential functions). Finally we mention that the use of the exponential function for the condition on the functions results in formulae which are concise and so lend themselves to a detailed treatment.

We now turn to the initial value problem for the solvable evolution equations of the ZS-AKNS system (see (6.1-112)

$$\sigma_3 U_t - \Omega(L_1^A)U = 0 \qquad x \in \mathbb{R}, \quad t \geqslant 0$$

$$U(x,0) = F(x) \tag{6.2-61}$$

where $U = (R,Q)^T$. The evolution equations have the form $U_t = K(U)$, where K is a local nonlinear operator which only involves the components of U and their x-derivatives. One can easily check that $\Omega(L_1^A)$ is of this type when $\Omega(k)$ is a polynominal in k.

As we showed in Chapter 4 for the solvable evolution equations associated with the isospectral Schrodinger operator, for initial data which is suitably smooth and decays sufficiently fast as $|x| \to \infty$ the initial value problem can be solved by the inverse spectral (scattering) transform method. The same is also true for solvable evolution equations associated with the iso-spectral ZS-AKNS operator. For simplicity we shall assume that Q and R are not linearly related for most of the remaining section. The steps of the method are depicted in Fig. 6-3.

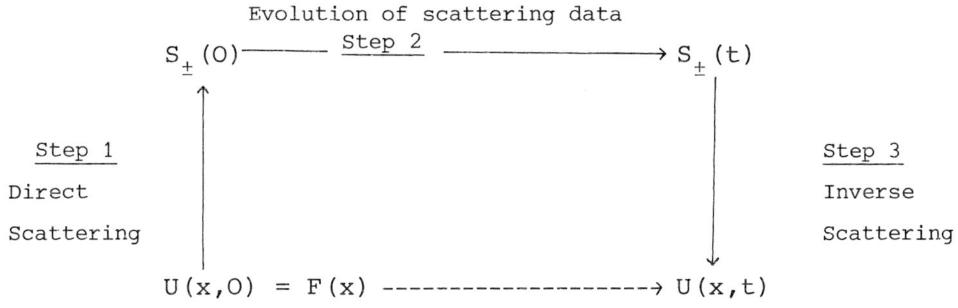

Figure 6-3: The inverse spectral transform method

The reader is referred to Section 4.2 for a full discussion of this method which is explained there quite generally. One merely replaces Q by U. In particular one can show that in the linear limit the method reduces to the method of Fourier analysis for solving linear evolution equations (see exercises at the end of this Chapter).

Essentially then if we wish to apply the inverse method to the solution of (6.2-61) we need to establish the additional conditions on the initial data to ensure a unique construction of $U(x,t)$ in step 3 which is sufficiently differentiability so that it satisfies (6.2-61). This will clearly depend upon the degree of the polynomial $\Omega(k)$. Our arguments here are very similar to those employed in Section 4.2. We now investigate the existence of a solution to (6.2-61) for initial data which belongs to the class of functions given in Theorem 6-6. Uniqueness has to be established separately for each specific equation. The problem is analysed through a series of lemmas and the conditions on $F(x)$ sufficient for the existence of a solution are given in Theorem 6-25.

Lemma 6-20:
If $F(x)$ is C^n and

$$|F(x)^{(j)}| \leqslant C_\varepsilon^j \, e^{-2\varepsilon|x|} \qquad j \leqslant n$$

then $K_\pm^{(j,\ell)}(x,y)$ exist for $j + \ell \leqslant n$ and the following estimates hold

$$|K_\pm^{(j,\ell)}(x,y)| \leqslant C_\varepsilon^{(j+\ell)\pm} \, e^{\mp\varepsilon(x+y)}$$

Proof: We only consider K_+, the result for $\overline{K}_+$ and K_-, $\overline{K}_-$ are proved similarly. Differentiating the equations (6.1-59) we obtain as in the proof of Corollary 6-7-2 that

$$K_{+1}^{(0,1)}(x,y) = -G_{01}(\tfrac{1}{2}(x+y)) - \int_x^{\frac{1}{2}(x+y)} Q(s)K_{+2}^{(0,1)}(s,y+x-s)ds$$

$$K_{+2}^{(0,1)}(x,y) = -\int_x^\infty K_{+1}^{(0,1)}(s,y-x+s)R(s)ds$$

$$K_{+1}^{(1,0)}(x,y) = -\tfrac{1}{4}\dot{Q}(\tfrac{1}{2}(x+y)) + Q(x)K_{+2}(x,y)$$

$$-\tfrac{1}{4}Q(\tfrac{1}{2}(x+y))\int_{\frac{1}{2}(x+y)}^\infty Q(s)R(s)ds - \int_x^{\frac{1}{2}(x+y)} Q(s)K_{+2}^{(0,1)}(s,y+x-s)ds$$

$$K_{+2}^{(1,0)}(x,y) = \int_x^\infty K_{+1}^{(0,1)}(s,y-x+s)R(s)ds + K_{+1}(x,y)R(x)$$

where

$$G_{01}(x) = \tfrac{1}{4}\dot{Q}(x) + \tfrac{1}{4}Q(x)\int_x^\infty Q(s)R(s)ds \qquad (6.2\text{-}62)$$

Then as we showed in establishing the corollary it is easy to deduce the result for $K_+^{(0,1)}$ and $K_+^{(1,0)}$ by using a calculation analogous to the $K_+^{(0,0)}$ case. It is now clear, without going into the details, that induction can be used to establish the result. In all cases when going to the next order once the result for $K_+^{(i,j)}$ has been obtained the result for $K_+^{(i,j+1)}$ is established first.

Now we consider the properties of the scattering functions which are associated with the operator $L_0 \equiv L\,(t=0)$ defined by Lemma 6-13. We can obtain integral representations of these functions in terms of the kernels $K_\pm$ by using the definitions (6.1-31)-(6.1-33) and the transformation operators with kernels $K_\pm$ defined in Theorem 6-7. One finds that

$$e^{2ikx}b(k) = \int_{-\infty}^{\infty} {}_x\pi_1(y)\bar{e}^{2iky}dy$$

$$\bar{e}^{2ikx}\bar{b}(k) = \int_{-\infty}^{\infty} {}_x\bar{\pi}_1(y)e^{2iky}dy$$

$$a(k) = 1 + \int_0^{\infty} \pi_2(y)e^{2iky}dy$$

$$\bar{a}(k) = 1 + \int_0^{\infty} \bar{\pi}_2(y)\bar{e}^{2iky}dy$$

where
$$ {}_x\pi_1(y) = B_{-2}(x,y) - \bar{B}_{+2}(x,y) + \int_{-\infty}^{\min(y,0)} B_{-2}(x,v)\bar{B}_{+1}(x,y-v)\,dv - \int_{\max(y,0)}^{\infty} \bar{B}_{+2}(x,v)B_{-1}(x,y-v)\,dv$$

$$ {}_x\bar{\pi}_1(y) = \bar{B}_{-1}(x,y) - B_{+1}(x,y) + \int_{-\infty}^{\min(0,y)} \bar{B}_{-1}(x,v)B_{+2}(x,y-v)\,dv - \int_{\max(0,y)}^{\infty} \bar{B}_{-2}(x,y-v)B_{+1}(x,v)\,dv \tag{6.2-63}$$

$$\pi_2(y) = B_{-1}(x,-y) + B_{+2}(x,y) + \int_0^y B_{-1}(x,v-y)B_{+2}(x,v)\,dv - \int_0^y B_{-2}(x,v-y)B_{+1}(x,v)\,dv$$

$$\bar{\pi}_2(y) = \bar{B}_{+1}(x,y) + \bar{B}_{-2}(x,-y) + \int_0^y \bar{B}_{+1}(x,v)\bar{B}_{-2}(x,v-y)\,dv - \int_0^y \bar{B}_{+2}(v,x)\bar{B}_{-1}(x,v-y)\,dv$$

$$B_{\pm}(x,y) = 2K_{\pm}(x, 2y + x).$$

The differential properties of the scattering functions are already known. Namely that under the conditions of Theorem 6-6 they are analytic in $|\text{Im } k| < \varepsilon$ and so infinitely differentiable for $|\text{Im } k| < \varepsilon$. However we require to know the asymptotic properties of the scattering functions which are induced by the differentiability and asymptotic properties of the initial data. In the same way that Lemma 4-5 was established one can show that,

<u>**Lemma 6-21:**</u>
If $F(x)$ is C^n then $\left\{{}_x\pi_1^{(m)}(y),\ {}_x\bar{\pi}_1^{(m)}(y),\ m \leqslant n-1\right\}$ are continuous functions.

From Lemmas 6-20, 6-21 and (6.2-63) we can prove the following result.

<u>**Lemma 6-22:**</u> If $F(x)$ is C^n and $|F^{(j)}(x)| \leqslant C_\varepsilon^j\, e^{-2\varepsilon|x|}$ then $\{{}_x\pi_1, {}_x\bar{\pi}_1, \pi_2, \bar{\pi}_2\}$ are class C^{n-1} and furthermore

$$\lim_{y \to \pm\infty} |{}_0\pi_1^{(j)}(y)\exp(\pm 2\varepsilon y)| \leqslant C_\varepsilon^{\pm j} \; ; \quad \lim_{y \to \pm\infty} |{}_0\bar{\pi}_1^{(j)}(y)\exp(\pm 2\varepsilon y)| \leqslant \bar{C}_\varepsilon^{\pm j}$$

$$\lim_{y \to \infty} |\pi_2^{(j)}(y)\exp(2\varepsilon y)| \leqslant C_\varepsilon^j \qquad \lim_{y \to \infty} |\bar{\pi}_2^{(j)}(y)\exp(2\varepsilon y)| \leqslant \bar{C}_\varepsilon^j$$

Lemma 6-22 enables us to deduce the asymptotic properties of the scattering functions.

<u>**Lemma 6-23:**</u>
If $F(x)$ is C^n and $|F^{(j)}(x)| \leqslant C_\varepsilon^j\, e^{-2\varepsilon|x|}$ then $R_+(k)$, $\bar{R}_+(k)$, $R_-(k)$, $\bar{R}_-(k)$ are meromorphic functions in $|\text{Im } k| < \varepsilon$ and they all have the uniform asymptotic behaviour $R^{(m)}(k) = o(|k|^{-(n+1)})$ as $|k| \to \infty$, for $|\text{Im } k| < \varepsilon$, and all m.

<u>**Proof:**</u> We establish the result for $R_+(k)$, the result for the other functions is proved in the same way.

$$a^{(m)}(k) = \int_0^\infty (2iy)^m\, \pi_2(y)e^{2iky}dy$$

$$b^{(m)}(k) = \int_{-\infty}^\infty (-2iy)^m\, {}_0\pi_1(y)e^{-2iky}dy$$

and

$$(2ik)^p b^{(m)}(k) = \int_{-\infty}^\infty \frac{\partial^p}{\partial y^p}(-2iy)^m\, {}_0\pi_1(y)e^{-2iky}dy$$

It is clear from the properties of $\pi_2(y)$ and ${}_o\pi_1(y)$ that $a^{(m)}(k)$ and $b^{(m)}(k)$ exist and are bounded from all m. From the last equation we deduce that $b^{(m)}(k) = o(1/|k|^{n-1})$ for all m. The result then follows from the definition of $R_+ = ba^{-1}$ and the known properties of R_+.

The time evolution of S_+ is defined in Theorem 6-11. In particular, we have the formulae,

$$R_+(k,t) = R_+(k,0)e^{\Omega(k)t}, \quad \bar{R}_+(k,t) = \bar{R}_+(k,0)e^{-\Omega(k)t} \tag{6.2-64}$$

Formally we can write,

$$\begin{aligned} \hat{R}_+(x,t) &= \frac{1}{2\pi}\int_\gamma dk R_+(k,t)e^{ikx} \\ \hat{\bar{R}}_+(x,t) &= \frac{1}{2\pi}\int_{\bar{\gamma}} dk\bar{R}_+(k,t)e^{-ikx} \end{aligned} \tag{6.2-65}$$

and then after deriving expressions for $P_{+j}(x,t)$, $\bar{P}_{+j}(x,t)$ we can form the function $F_+(x,t)$. Provided the conditions of Theorem 6-17 are met it is possible to reconstruct from the Marchenko equations the unique functions $Q(x,t)$, and $R(x,t)$ which satisfy (6.1-13) with Jost solutions $\Psi(x,t)$, $\Phi(x,t)$ constructed according to Theorem 6-16. For the moment we shall assume that $S_+(t)$ satisfies the conditions of Theorem 6-17. Then we can prove the following lemma.

Lemma 6-24:
Provided $S_+(t)$ obeys the conditions of Theorem 6-17 and as $|k| \to \infty$ $R^{(m,o)}(k,t) = o(|k|^{-(n-1)})$ for $|\text{Im } k| < \varepsilon$ and all m, then $U^{(j,\ell)}(x,t), j + \ell \deg \Omega(k) \leqslant n - 2$, exist and $|U^{(j,\ell)}(x,t)| \leqslant C_\varepsilon^{j\ell}(t)\, e^{-2\varepsilon|x|}$.

Proof: Since $R_+(k,t)$ exists and obeys the conditions given in the statement the lemma, we have

$$\hat{R}_+^{(j,\ell)}(k,t) = \frac{1}{2\pi}\int_\gamma dk(ik)^{(j)}(\Omega(k))^{(\ell)}R_+(k,t) \tag{6.2-66}$$

Let $\deg\Omega(k) = s$ then $\hat{R}_+^{(j,\ell)}(x,t)$ exists for $j + \ell s \leqslant n - 2$ since $R_+(k,t)$ has the same asymptotic $|k| \to \infty$, behaviour as $R_+(k,0)$. If we differentiate the Marchenko equation for K_+ we obtain

$$\begin{aligned} &K_+^{(o,j.o)}(x,u,t) + \int_x^\infty K_+^{(o,j,o)}(x,s,t)F_+(s+u,t)ds \\ &= -F_+^{(j,o)}(x+u,t) - \sum_{m=o}^{j} C_m^j \int K_+^{(o,j-m,o)}(x,s,t)F_+^{(m,o)}(s+u,t)ds \end{aligned} \tag{6.2-67}$$

from which we deduce the existence of $K_+^{(o,j,o)}(x,u,t)$ for $j \leqslant n-2$. One can now establish the existence of $K_+(x,u,t)$, $i+j \leqslant n-2$. Then using induction on (6.2-67) we find that

$$|K_+^{(i,j,o)}(x,u,t)| \leqslant C_\varepsilon^{ij+}(a,t)\ \exp[-\varepsilon(x+u)] \qquad -\infty < a \leqslant x \qquad (6.2\text{-}68)$$

where the properties of the functions $C_\varepsilon^{ij+}(a,t)$ have yet to be investigated. The t-derivative of K_+ satisfies the equation

$$K_{+t}(x,u,t) + \int_x^\infty K_{+t}(x,s)F_+(s+u)ds$$
$$= -\sum_{i=o}^{s} a_i\{F_+^{(i,o)}(x+u,t) + \int_x^\infty K_+(x,s)F_+^{(i,o)}(s+u,t)ds\} \qquad (6.2\text{-}69)$$

The same arguments apply as for the previous cases and we obtain the result that

$$|K_+^{(i,j,l)}(x,u,t)| \leqslant C_\varepsilon^{ijl+}(a,t)\ \exp[-\varepsilon(x+u)] \qquad -\infty < a \leqslant x. \qquad (6.2\text{-}70)$$

A similar calculation for K_- and Theorem 6-17 leads to the conclusion that the jth x and ℓth t derivative of $U(x,t)$ exist and are continuous $(j+\ell s \leqslant n-3)$ and that the Jost functions $J(x,t,k)$ exist along with the derivatives $J^{(i,\ell,o)}(x,t,k)$ where $(j+\ell s \leqslant n-2)$. The estimates for for $U^{(j,\ell)}(x,t)$ follow from (6.2-70) and the estimates of the derivatives of the K_- kernel.

Theorem 6-25:
If $F(x)$ is C^n and $|F^{(j)}(x)| \leqslant C_\varepsilon^j e^{-2\varepsilon|x|}$, $j \leqslant n$ where $n \geqslant \deg \Omega(k)+3$, then provided $S_+(t)$ satisfies the conditions of Lemma 6-24 the function $U(x,t)$ satisfies the equation (6.2-61).

Proof: From the definition of L_1^A (equation (6.1-112)) we see that if $\deg \Omega(k) = s$ then we require the existence of $U(x,t)(j,0)$ $(j \leq s+1)$ and $U(x,t)(0,1)$. Thus put $n = s+3$. This ensures that $\Omega(L_1)\phi^{(2)}$ exists. Consequently we can write

$$R_{+t}(k,t) - \Omega(k)R_+(k,t) = \int_{-\infty}^{\infty}\left[\sigma_3 U_t - U\Omega(L_1)\right]\phi^{(2)}(x,k)dx \qquad (6.2\text{-}71)$$

Then by definition the left hand side of (6.2-71) is identically zero and we obtain

$$\sigma_3 U_t(x,t) - \Omega(L_1^A)U(x,t) = 0\ . \qquad (6.2\text{-}72)$$

So far we have not investigated the properties of the functions $C_\varepsilon^{j\ell}(t)$ nor have we studied any constraints on the functions $\Omega(k)$, $F(x)$ which arise from Lemma 6-24.

If $\Omega(k) = \sum_{j=0}^{s} a_j k^j$ then we see from (6.2-65) that for real k, $\hat{R}_+, \hat{\bar{R}}_+$ are only defined if (a) a_j is real and negative and k^j is an even power of k, or (b) a_j is imaginary. Without loss of generality we can consider polynomials $\Omega(k)$ which consist of terms entirely of type (a) or type (b). The functions of type (b) are particularly interesting because solvable equations for the "reduced problem" that is those for which $R = \alpha Q$, $R = \alpha Q^*$ or $R =$ constant are precisely of this type (see exercise 1 in 6.3). This class of solvable equations contains all the important physical equations. For the conditions on the initial data which we are considering $R_+(k,t)$ is analytic in the strip $|\text{Im}\, k| < \varepsilon$. It follows that we also require the integral

$$\hat{R}_+(x,t) = \frac{1}{2\pi}\int_{-\infty+i\eta}^{\infty+i\eta} d\xi R_+(k,t)e^{ikx} \tag{6.2-73}$$

to be defined where $\delta \leqslant \eta \leqslant \varepsilon_0 - \delta$ and $0 < \delta < \varepsilon_0/2$. Let $a_{j\ell} > \varepsilon_0$ then the integral (6.2-73) exists for a large class of polynomials. In particular we see that it is defined when the function $\Omega(k)$ is, $-k^2$ $-k^4 - 6a_{10}k^2$, ik and ik^3. It is in fact possible to write down formulae for large classes of polynomials for which (6.2-73) is defined. For example

$$\Omega_1(k) = \sum_{\ell=0}^{j} ik^{(2j-1)-2\ell} C^{2j-1}_{2\ell+1} a_j^{2(2\ell-1)}$$

with $a_{jo} = (C_+^{2j-1})^{\frac{1}{2}}$ defines one such class of polynomials. For simplicity let us consider the example $\Omega(k) \equiv ik^3$. In this case we obtain

$$|\hat{R}_+(x,t)| \leqslant \frac{1}{2\pi}\int_{-\infty+i\eta}^{\infty+i\eta} d\xi |R_+(k,0)| e^{-\eta x+(-3\eta\xi^2+\eta^3)t}$$

$$\leqslant \frac{1}{2\pi} e^{\eta^3 t}\int_{-\infty+i\eta}^{\infty+i\eta} d\xi |R_+(k,0)| e^{-\eta x} \qquad t \geqslant 0 \tag{6.2-74}$$

From which we deduce that

$$\left[\int_{-\infty}^{\infty} |(\hat{R}_+(x,t)e^{\eta x}|^2 dx\right]^{\frac{1}{2}} \leqslant e^{\eta^3 t} C_\delta^+ \qquad t \geqslant 0 \tag{6.2-75}$$

Since $S_+(t)$ is assumed to satisfy Lemma 6-24 it follows that

$$|F_+(x,t)| \leqslant C(a,t)e^{-\varepsilon x} \quad -\infty < a \leqslant x\,. \tag{6.2-76}$$

Then since $K_+(x,t)$ is unique arguing as we did to establish (6.2-20) from the Marchenko equation and Lemma 6-13 we see that for the case under consideration $\Omega(k) = ik^3$, $C(a,t)$ is a monotonically increasing function of t. It follows that the solution only belongs to the class of functions $|U(x,t)| \leqslant C\, e^{-2\varepsilon|x|}$ for which we have solved the inverse problem for a finite time. Eventually any such initial solution for which R_+ is not identically zero will evolve outside this class. (c.f., the long time asymptotic behaviour of solutions to solvable equations considered in 6.3). Problems of this type do not occur if $R_+(k,0)$, $\bar{R}_+(k,0)$ are identically zero in $S_+(0)$. The class of solvable equations defined earlier $\Omega_1(k)$, also have solutions to their initial value problems as given by Theorem 6-25 which remain within the required class of functions for a finite time.

It follows that although the MKdV can be solved for finite time provided Theorem 6-25 is satisfied for the nonlinear Schrodinger equation we cannot treat the case of a continuous spectrum; our restrictions on the class of solutions precludes this. What is needed is a treatment of spectral singularities when the initial data satisfies a moment equation (see Theorem 6-1).

In the exercise at the end of this chapter we establish existence of solutions to problems of the type (6.2-61) for the reduced class of equations $R = -Q^*$.

Finally we mention that an investigation of the solvable equations defined by singular dispersion relations is needed. If we adopt the procedure of 6.4 then we have formally from (6.1-130) that

$$(L_1^A - kI)^{-1}U = -2a^{-1}\phi \circ \psi \qquad (6.2\text{-}77)$$

Thus if the dispersion relation is $\Omega(k) = (k-k_1)^{-1}$ we obtain for the solvable equation from (6.1-112)

$$\sigma_3 U_t = (L_1^A - k_1 I)^{-1}U = -2\left(\frac{\phi \circ \psi}{a}\right)_{k=k_1} \qquad (6.2\text{-}78)$$

Similarly if $\Omega(k) = (k-k_1)^{-n}$ the solvable equation has the form

$$\sigma_3 U_t = \frac{-2}{(n-1)!}\frac{\partial^{n-1}}{\partial k^{n-1}}\left(\frac{\phi \circ \psi}{a}\right)\Bigg|_{k=k_1} \qquad (6.2\text{-}79)$$

These equations together with the "Bloch equations" and their derivatives (6.1-104)-(6.1-106). If $\mathrm{Im}k_1 = 0$ then from the asymptotic form of (6.2-79) as $|x| \to \infty$ we have additional conditions on the scattering functions R_+ and $\bar{R}_+$. These are easily seen to be that

$$\frac{\partial^m}{\partial k^m} R_+(k)\Bigg|_{k=k_1} = 0 \quad \frac{\partial^m}{\partial k^m}\bar{R}_+(k)\Bigg|_{k=k_1} = 0 \quad m = 0,\ldots,(n-1) \qquad (6.2\text{-}80)$$

A thorough study of solvable equations like (6.2-79) has yet to be made.

6.3 Solutions for Solvable Equations and their Bäcklund Transformations

The richer structure of the spectrum of L compared with the Schrödinger operator (the existence of spectral singularities and repeated eigenvalues) leads to the associated solvable equations possessing a larger variety of exact solutions. Thus besides the soliton solutions for this method we shall also present some solutions which arise from the repeated eigenvalues as well as from the spectral singularities. Further complications arise if $\sigma(L)$ contains a singularity of the dispersion relations or if $\Omega \neq -\Omega^*$, though we shall not consider these cases. We present an extension of the Zakharov-Manakov asymptotic technique (1976). This enables part of the asymptotic form of the solution of an arbitrary solvable equation in the absence of solitons to be determined. The range of validity of the solution is also obtained using this method. Again as in the Schrödinger case Bäcklund transformations can be found which transform between solutions of solvable equations. This material is covered at the end of this section.

We begin be constructing the soliton solutions for this method. These are obtained when $\sigma(L)$ has only simple eigenvalues, spectral singularities are absent and the reflection coefficients are zero. One can use either S_+ or S_- to construct them. We shall use S_+ and the Marchenko equation (6.2-3). The one soliton solutions for Q and R are obtained from taking

$$F_+(x,t) = D_+(t)\, e^{ikx + \Omega(k)t}$$
$$\bar{F}_+(x,t) = \bar{D}_+(t)\, e^{-(i\bar{k}x + \Omega(\bar{k})t)} \tag{6.3-1}$$

where

$$D_+(t) = -\frac{ib(k,t)}{a_k(k,t)}, \qquad \bar{D}_+(t) = \frac{i\bar{b}(\bar{k},t)}{\bar{a}_k(\bar{k},t)} \tag{6.3-2}$$

are the normalisation constants for the bound states, $\sigma(L) = \{k, \bar{k}, \text{Im}k > 0, \text{Im}\,\bar{k} < 0\}$. Introduce the bilinear form

$$\{u,v\} = \int_x^\infty u(y)\; v(y)\,dy \tag{6.3-3}$$

and put

$$\{U,V\} = \{\{(u_1,v_1\},\{u_2,v_2\}\}^T \tag{6.3-4}$$

then the Marchenko equation (6.2-3) can be written after taking appropriate inner products as

$$\{\bar{K}_+(x,s,t),e^{-i\bar{k}s}\} + \binom{0}{1}D_+(t)e^{ikx}\{1,e^{i(k-\bar{k})s}\} + D_+(t)\{1,e^{i(k-\bar{k})s}\} \times \{K_+(x,s,t)e^{iks}\} = 0 \tag{6.3-5}$$

$$\{K_+(x,s,t),e^{iks}\} + \binom{1}{0}\bar{D}_+(t)e^{-ikx}\{1,e^{i(k-\bar{k})s}\} + \bar{D}\;(t)\{1,e^{i(k-\bar{k})s}\} \times \{K_+(x,s,t),e^{-i\bar{k}s}\} = 0 \tag{6.3-6}$$

where s is the integration variable. Solving (6.3-6) for $\{K_+(x,s)\}, \exp(iks)\}$ and $\{\bar{K}_+(x,s),\exp(-i\bar{k}s)\}$ and substituting back into the Marchenko equation gives the solutions

$$Q(x,t) = -2K_{+1}(x,x,t) = 2\bar{D}_+(t)e^{-2i\bar{k}x} E^{-1}(x,t) \qquad (6.3\text{-}7)$$

$$R(x,t) = -2K_{+2}(x,x,t) = 2D_+(t)e^{2ikx} E^{-1}(x,t)$$

where

$$E(x,t) = 1 + \frac{D_+(t)\bar{D}_+(t)e^{2i(k-\bar{k})x}}{(k-\bar{k})^2} \qquad (6.3\text{-}8)$$

and

$$D_+(t) = D_+(0)\ e^{\Omega(k)t}, \qquad \bar{D}_+(t) = \bar{D}_+(0)\ e^{-\Omega(\bar{k})t} \qquad (6.3\text{-}9)$$

provided that $E(x,t) \neq 0$. If, however, $E(x,t) = 0$ then there are infinitely many solutions for $\{\bar{K}_+, \exp(-i\bar{k}s)\}$, $\{K_+(x,s), \exp(iks)\}$ so that the corresponding Marchenko equations do not have unique solutions.

From (6.3-7) we also obtain

$$\int_x^\infty Q(y,t)R(y,t)\,dy = \frac{2i\ e^{2i(k-\bar{k})x} D_+(t)\bar{D}_+(t)}{(k-\bar{k})E}$$

from which we see that $\int_{-\infty}^{\infty} {}_tQ(y)_tR(y)dy$ is not defined for arbitrary t even if the initial functions are bounded. Conditions on Q and R so that the initial value problem has a unique solution are given in Section 6.2.

If we assume that $E(x,t) \neq 0$ then we can write the solutions (6.3-7) in a more soliton-like form as

$$\begin{aligned} Q(x,t) &= A\ \mathrm{sech}\theta\ e^{-i\phi} \\ R(x,t) &= \bar{A}\ \mathrm{sech}\theta\ e^{i\phi} \end{aligned} \qquad (6.3\text{-}10)$$

where

$$\theta(x,t) = \gamma + \tfrac{1}{2}(\Omega(k) - \Omega(\bar{k}))t + i(k-\bar{k})x$$

$$\phi(x,t) = (k+\bar{k})x - \tfrac{1}{2}i(\Omega(k) + \Omega(\bar{k}))t$$

$$e^{2\gamma} = \frac{D_+(0)\bar{D}_+(0)}{(k-\bar{k})^2}$$

$$A = \bar{D}_+^{\frac{1}{2}}(0)D_+^{-\frac{1}{2}}(0)(k-\bar{k})$$

$$\bar{A} = D_+^{\frac{1}{2}}(0)\bar{D}_+^{-\frac{1}{2}}(0)(k-\bar{k})$$

Thus the one-soliton solutions for solvable equations, provided θ and ϕ are real, are single wave packets with distortion-less (sech-shaped) envelopes. We list the form the soliton takes according to the degeneracies given in Lemma 6-10.

(i) $R = +Q$, Q complex

In this case the solution becomes singular after a finite time and does not have the form given by (6.3-10). As an example put $\Omega(k) = -8ik$ and the solvable equation is a complex modified Korteweg-de Vries equation

$$Q_t \pm 6Q^2Q_x - Q_{xxx} = 0$$

The "soliton" solution is

$$Q(x,t) = e^{i\lambda(x,t)} f(x,t) + e^{-i\lambda(x,t)} g(x,t)$$

where

$$f = \xi(\cosh\omega - \sinh\omega)(\cosh^2\omega - \sin^2\lambda)^{-1}$$

$$g = \xi(\cosh\omega - \sinh\omega)(\cosh^2\omega - \sin^2\lambda)^{-1}$$

and

$$\omega(x,t) = 2\eta x + 8\eta(3\xi^2 - \eta^2)t, \quad \lambda = 2\xi x - 8\xi^2(\xi - 3\eta)t + \tau$$

$$\gamma = \frac{\ell n|D_+|}{4\xi} + i\tau, \quad k = \xi + i\eta$$

It is clear from the solution that in general it becomes singular after a finite time. Usually models which give rise to such an equation are non-physical.

(ii) $R = -Q^*$

Notice that soliton solutions cannot occur if $R = Q^*$ because then the eigenvalue problem (6.1-13) is self-adjoint. For (ii) $A = 2\eta\exp(-i\eta)$ and $\gamma = \ell n(|D_+|/2\eta)$ where $k = \xi + i\eta$. As a particular example we have the non-linear Schrödinger equation for which $\Omega = 4ik^2$

$$iQ_t + |Q|^2Q + Q_{xx} = 0$$

The one soliton solution is given by the expression

$$Q(x,t) = 2\eta \exp\{-i(2\xi x + 4(\xi^2 - \eta^2)t + \tau)\}\operatorname{sech}(2\eta(x - x_0) + 8\eta\xi t)$$

where $x_0 = \gamma/2\eta$ is the position of the maximum of the envelope at $t = 0$. The velocity v of the soliton is -4ξ and the amplitude $|A|$, is 2η.

(iii) $R = \alpha Q = \alpha Q^*$, α real

Then necessarily $k = i\eta = -\bar{k}$, $\eta > 0$ and $\bar{D}_+ = -D_+/\alpha = D_+^*/\alpha$ so that $A = \pm 2\eta\alpha^{-\frac{1}{2}}$. As examples we have

The Modified KdV Equation $\Omega = 8ik^3, R = \pm Q$

$$Q_t \pm 6Q^2Q_x + Q_{xxx} = 0$$

$$Q(x,t) = 2\eta\alpha_\pm^{-\frac{1}{2}} \operatorname{sech}(2\eta(x-4\eta^2 t)-\gamma) \qquad \text{where } \alpha_\pm = \pm 1$$

The Sine-Gordon Equation $\Omega = \pm 1/2ik$, $R = -Q = U_x/2$

$$U_{xt} = \pm \sin U$$

$$Q(x,t) = 2\eta \operatorname{sech}(\frac{1}{2\eta}(4\eta^2 x \pm t)-\gamma)$$

$$U(x,t) = 4\tan^{-1}(\exp(\gamma - \frac{1}{2\eta}(4\eta^2 x \pm t)))$$

In this case as we mentioned in Chapter 1, the solution for U is referred to as a kink. The velocity of the MKdV soliton is $4\eta^2$ and that of the sine-Gordon kink $1/4\eta^2$ corresponding to the negative choice of sign in the argument.

An important point to note is that when the dispersion relation (Ω) contains a singular term then, although we are excluding the case when the eigenvalue coincides with the singularity, it is possible that the eigenvalue is very close to the singularity. Thus, in the case of the sine-Gordon considered above and the equation of Konno et al. (1974) the dispersion relation has a singularitiy at $k = 0$. The velocities of the corresponding solitons are $1/4\eta^2$ and $4\eta^2 + \alpha/4\eta^2$ respectively. In the physical coordinate system, these models have solitons which can travel at the limiting velocity of the model (for example the speed of light (unity) in the sine-Gordon equation).

The N-soliton solutions are obtained from the Marchenko equation by assuming the spectrum of L consists of simple eigenvalues only. The Marchenko equation then has a degenerate kernel and can be solved by the techniques of linear algebra. For this case the Marchenko equation (6.2-3) can be written using the transformation operator of Theorem 6-7 as

$$\bar{K}_+(x,y) + \sum_{j=1}^{N} D_{+j}\,\psi_j(x)e^{ik_j y} = 0 \tag{6.3-11}$$

$$K_+(x,y) + \sum_{j=1}^{N} \bar{D}_{+j}\,\bar{\psi}_j(x)e^{-i\bar{k}_j y} = 0 \tag{6.3-12}$$

where

$$\psi_j(x) \equiv \psi(x,k_j) \text{ and } \bar{\psi}_j(x) \equiv \bar{\psi}(x,\bar{k}_j)$$

If we define

$$f_j = \sqrt{iD_{+j}}\,\psi_j, \qquad \bar{f}_j = \sqrt{i\bar{D}_{+j}}\,\bar{\psi}_j \tag{6.3-13}$$

and

$$\lambda_j = (iD_{+j})^{\frac{1}{2}} e^{ik_j x}, \qquad \bar{\lambda}_j = (i\bar{D}_{+j})^{\frac{1}{2}} e^{-i\bar{k}_j x} \tag{6.3-14}$$

then equations (6.3-11),(6.3-12) assume the form

$$\begin{aligned} f_s(x) + \sum_{j=1}^{N} \frac{\lambda_s \bar{\lambda}_j \bar{f}_j(x)}{(k_s - \bar{k}_j)} &= \binom{0}{1}\lambda_s \\ \bar{f}_s(x) + \sum_{j=1}^{N} \frac{\bar{\lambda}_s \lambda_j f_j(x)}{(k_j - \bar{k}_s)} &= \binom{1}{0}\bar{\lambda}_s \end{aligned} \tag{6.3-15}$$

Introducing the 2N column vectors $F = (f_s^{(1)}, \bar{f}_s^{(1)})^T$, $\bar{F} = (f_s^{(2)}, \bar{f}_s^{(2)})^T$ where the indices in parenthesis refer to components and the suffix runs over 1 to N, equations (6.3-15) can be written as the system

$$AF = \Lambda$$

$$A\bar{F} = \bar{\Lambda}$$

where

$$\Lambda = (0,\ldots,0,\ \bar{\lambda}_1,\ldots,\bar{\lambda}_N)^T$$

and

$$\bar{\Lambda} = (\lambda_1,\ldots,\lambda_N,\ 0,\ldots,0)^T \tag{6.3-16}$$

are 2N column vectors. Then by Cramer's rule (assuming $\det A \neq 0$) we have, if $F^{(k)}$ denotes the kth entry of F

$$F^{(k)} = \frac{\det A_{(k)}}{\det A} \tag{6.3-17}$$

where $A_{(k)}$ is obtained from A by replacing the kth column by Λ. In particular we notice that if $1 \leqslant k \leqslant N$ then

$$i\lambda_k F^{(k)} = \frac{\det A_k}{\det A} \tag{6.3-18}$$

where A_k is obtained from A by differentiating the kth column with respect to x. Similarly we find from (6.3-16) that for $N + 1 \leqslant k \leqslant 2N$

$$i\bar{\lambda}_{k-N}\bar{F}^{(k)} = \frac{\det A_k}{\det A} \tag{6.3-19}$$

Summing (6.3-18) and (6.3-19) and then adding results in the expression

$$i\sum_{j=1}^{N}(\lambda_j f_j^{(1)} + \bar{\lambda}_j \bar{f}_j^{(2)}) = \sum_{J=1}^{2N}\frac{\det A_k}{\det A} = \frac{\partial}{\partial x}\ln(\det A) \tag{6.3-20}$$

This result can be used to obtain an explicit formulae for the product of Q and R. This is achieved by equating the asymptotic expansions of ψ and $\bar{\psi}$ resulting from Lemma 6-10 and the corresponding expansions in the N-soliton case derived from equation (6.2-2). The resulting formulae are

$$\begin{pmatrix} Q \\ -\int_x^{\infty} R(y)Q(y)\,dy \end{pmatrix} = 2\sum_{j=1}^{N}\bar{D}_{+j}\bar{\psi}_j\, e^{-ik_j x}$$

$$\begin{pmatrix} \int_x^{\infty} R(y)Q(y)\,dy \\ -R \end{pmatrix} = -2\sum_{j=1}^{N} D_{+j}\psi_j e^{ik_j x} \tag{6.3-21}$$

from which we get the relationship

$$\int_x^{\infty} R(y)Q(y)\,dy = i\sum_{j=1}^{N}(\lambda_j f_j^{(1)} + \bar{\lambda}_j\bar{f}_j^{(2)})) = \frac{\partial}{\partial x}\ln(\det A) \tag{6.3-22}$$

If we differentiate this expression we get a formula for the product RQ which is similar to the N-soliton formula of Section 4.3

$$RQ = -\frac{\partial^2}{\partial x^2}\ln(\det A) \tag{6.3-23}$$

where

$$A = 1 + B\bar{B}$$

and

$$B_{js} = i\frac{\sqrt{D_{+j}\bar{D}_{+s}}}{(k_j - \bar{k}_s)}\exp i(k_j - \bar{k}_s)x$$

$$\bar{B}_{js} = i\frac{\sqrt{\bar{D}_{+j}D_{+s}}}{(\bar{k}_j - k_s)}\exp -i(\bar{k}_j - k_s)x$$

The non-degeneracy of the system of equation (6.3-16) follows from the uniqueness of the solution to the Marchenko equation as shown in Section 6.2. In the case when $R = \alpha Q^*$ or $R = \alpha Q$, Q real, the formula (6.3-23) gives respectively the modulus of the N-soliton solution or the square of the N-soliton solution.

In a similar fashion to the case of the N-soliton solution for the solvable equations associated with the isospectral Schrödinger equation, one can show that the N-soliton solution decomposes under certain conditions into N-soliton solutions as $|t| \to \infty$. Additional restrictions are needed in this case because the richer structure of the solvable equations means that it is possible that the solitons do not separate as $|t| \to \infty$. From (6.3-10) we see that in general the velocity of a soliton is

$$v = \mathrm{Re}\left\{\frac{\Omega(\bar{k}) - \Omega(k)}{2i(k-\bar{k})}\right\} \tag{6.3-24}$$

Thus if $k = \xi + i\eta$, $\bar{k} = \bar{\xi} + i\bar{\eta}$, this leads to an equation of the form

$$h(\xi, \bar{\xi}, \eta, \bar{\eta}; v) = 0 \tag{6.3-25}$$

which defines families of solitons with the same velocity v. If we exclude this case for the moment then we can order the solitons in the N-soliton solution by their velocities

$$v_1 < v_2 \ldots . < v_N \tag{6.3-26}$$

We shall only consider the case $R = \pm Q$, which includes $R = \pm Q^* = \pm Q$ as a specialisation. We easily obtain from (6.3-13), (6.3-14) that

$$|\lambda_j(x,t)| = |\lambda_j(0)| \exp -\eta_j y_j \tag{6.3-27}$$

$$y_j = x - v_j t \tag{6.3-28}$$

It follows that

$$\begin{array}{lll} \text{as} \quad y_j \to +\infty & |\lambda_j| \to 0 & \text{for } j < m \\ \text{and} \quad y_j \to \infty & |\lambda_j| \to \infty & \text{for } j > m \end{array} \tag{6.3-29}$$

along the line $y_m = \text{const}$ as $|t| \to \infty$. Thus in this limit the system (6.3-16) for the N-soliton solution gives the reduced system of equations

$$\begin{aligned} f_m^{(1)} + \frac{|\lambda_m|}{2i\eta_m} f_m^{(2)*} &= -\lambda_m \sum_{j=m+1}^{N} \frac{1}{(k_m - k_j^*)} g_j^{(2)*} \\ \frac{|\lambda_m|^2}{2i\eta_m} f_m^{(1)} + f_m^{(2)*} &= \lambda_m^* + \lambda_m^* \sum_{j=m+1}^{N} \frac{g_j^{(1)}}{(k_m^* - k_j)} \end{aligned} \tag{6.3-30}$$

and

$$\sum_{j=m+1}^{N} \frac{1}{(k_\ell - k_j^*)} g_j^{(2)*} = -\frac{\lambda_m^*}{(k_\ell - k_m)} f_m^{(2)*}$$
$$-\sum_{j=m+1}^{N} \frac{g_j^{(1)}}{(k_\ell^* - k_j)} = 1 + \frac{\lambda_m f_m^{(1)}}{(k_\ell^* - k_m)} \tag{6.3-31}$$

where we have put

$$g_\ell = \lambda_\ell f_\ell \text{ and } f_\ell = (f_\ell^{(1)}, f_\ell^{(2)})^T \tag{6.3-32}$$

We can solve (6.3-31) in the same manner as Zakharov and Shabat (1972) solved the analogous set of equations for the nonlinear Schrödinger equation ($v_j = -4\xi_j$). The solution is

$$g_j^{(1)} = a_j + \frac{2i\eta_m}{a_m} \frac{a_j f_m^{(1)} \lambda_m}{(k_j - k_m)}$$
$$g_j^{(2)} = \frac{2i\eta_m}{a_m} \frac{a_j f_m^{(2)} \lambda_m}{(k_j - k_m)} \tag{6.3-33}$$

where

$$a_j = \prod_{p=m+1}^{N} (k_j - k_p^*) / \prod_{m<p\neq j} (k_j - k_p)$$

$$a_m = 2i\eta_m \prod_{p=m+1}^{N} \frac{k_m - k_p^*}{k_m - k_p}$$

Finally we obtain from (6.3-30) the equations

$$f_{m1} + \frac{|\lambda_m^+|^2}{2i\eta_m} f_{m2}^* = 0$$
$$\frac{|\lambda_m^+|^2}{2i\eta_m} f_{m1} + f_{m2*} = (\lambda_m^+)^* \tag{6.3-34}$$

where

$$\lambda_m^+ = \lambda_m \prod_{p=m+1}^{N} \frac{(k_m - k_p)}{(k_m - k_p^*)}$$

It is easy to see from a comparison of this formula with the expression (6.3-15) for the one soliton case that (6.3-34) represents a soliton with a displaced maximum x_{0m}^+ ($x_0 = \mathrm{Re}\,\gamma/2\eta$) and phase τ^+ ($\tau = \arg(A/2\eta)$) from those at $t = 0$ given by the formulae

$$x_{om}^{+} - x_{om} = \frac{1}{\eta_m} \sum_{p=m+2}^{N} \ln \left| \frac{k_m - k_p}{k_m - k_p^*} \right| < 0 \tag{6.3-35}$$

$$\tau_m^{+} - \tau_m = -2 \sum_{p=m+1}^{N} \arg \left(\frac{k_m - k_p}{k_m - k_p^*} \right) \tag{6.3-36}$$

For $t \to -\infty$ the equivalent formulae are obtained from

$$\lambda_m^{-} = \lambda_m \prod_{p=1}^{m-1} \frac{k_m - k_p}{k_m - k_p^*} \tag{6.3-37}$$

These formulae enable the following interpretation to be given to the N-soliton solution. As $t \to -\infty$ the solution breaks up into N soliton solutions with the slowest soliton in front and the fastest at the back. As $t \to -\infty$ the arrangement of the solitons is reversed. The total change in position of the centre and phase of the mth soliton in the solution as t varies over the range $]-\infty, \infty[$ is then given by

$$\Delta x_{om} = x_{om}^{+} - x_{om}^{-} = \frac{1}{\eta_m} \left(\sum_{j=m+1} \ln \left| \frac{k_m - k_j}{k_m - k_j^*} \right| - \sum_{j=1}^{m-1} \ln \left| \frac{k_m - k_j}{k_m - k_j^*} \right| \right) \tag{6.3-38}$$

$$\Delta \tau_m = \tau_m^{+} - \tau_m^{-} = 2 \sum_{j=1}^{m-1} \arg \frac{k_m - k_j}{k_m - k_j^*} - 2 \sum_{j=m+1}^{N} \arg \left(\frac{k_m - k_j}{k_m - k_j^*} \right) \tag{6.3-39}$$

Thus the N-soliton solution can be considered as representing the pairwise collision of N solitons, the collisions being ordered by the speeds, so that each soliton interacts once with every other soliton. Multi-particle effects do not occur.

For the case mentioned earlier when we have an N-soliton solution, whose soliton components move with the same speed, clearly the above analysis is not applicable and the solution represents a bound state. In the case of the nonlinear Schrödinger equation a bound state results from multi-soliton solutions whose component solitons are characterised by distinct eigenvalues satisfying $\text{Re}\, k = k_0$. For the MkdV and the sine-Gordon equations the constraints on the eigenvalues are $4(\text{Re}\, k)^2 - |k|^2 = k_0$ and $|k| = k_0$ respectively. A specific example for the sine-Gordon equation is given by the pair of eigenvalues $(k, -k^*)$, $\text{Im}\, k > 0$

$$U(x,t) = 4\tan^{-1}\left[\frac{\eta \cos \xi(2x \mp t/2k_0^2 + \alpha)}{\xi \cosh \eta(2x \pm t/2k_0^2 + \beta)}\right] \qquad \begin{array}{l}\alpha = (\gamma + \gamma^*)/2\xi \\ \beta = -i(\gamma - \gamma^*)/2\eta\end{array} \qquad (6.3\text{-}40)$$

In particular in the physical system of coordinates $X = x + t$, $T = x - t$, the value $|k| = 1/2$ gives an oscillatory solution with a fixed location. The solutions can be obtained in a straightforward manner either by directly solving the Marchenko equation or by using Hirota's method discussed in Chapter 1. At the end of the chapter we show how these solutions can also be generated by Bäcklund transformations. This is often the easiest method, at least for the two soliton case.

Another useful application of Bäcklund transformation techniques is in the derivation of solutions which correspond to repeated eigenvalues. The general method requires one to solve the Marchenko equation with the appropriate normalisation polynomial. Alternatively one can view this type of solution as resulting from the coalescence of two or more simple eigenvalues, although the direct application of this idea to the N-soliton solution has not been implemented. In the case of a repeated eigenvalue the normalisation polynomial (6.1-87) for the nonlinear Schrödinger equation can be written as

$$P_+(x,t) = (a_1(t) + a_2(t)x)e^{ikx} \qquad (6.3\text{-}41)$$

where

$$a_1(t) = a_1(0)\ (1+8\gamma\ kt)e^{4ik^2t}$$

$$a_2(t) = a_2(0)\ e^{4ik^2t}$$

Zakharov and Shabat (1972) obtained the solution

$$Q(x,t) = \left(\frac{-4\eta\ f^{*2}\ a_2^*}{1+|a_2|^2|f|^4}\right) \times \left(\frac{-2+|f|^4 g\ a_2^*}{1+|g|^2|f|^4}\right) \qquad (6.3\text{-}42)$$

where

$$g = \frac{2\eta(a_1+2(x + \frac{1}{2\eta})a_2)}{1+|a_2|^2|f|^4} \quad , \qquad f = \frac{e^{ikx}}{2\eta}$$

by directly solving the Marchenko equation. In fact for this case they showed that the bound state corresponding to two distinct eigenvalues ($k = i\eta_1$, $k_2 = i\eta_2$) is periodic with a frequency characterised by the difference $(\eta_1^2 - \eta_2^2)$. In the limit $k_2 \to 0$ the oscillations die out and the solution becomes a soliton. The limit $k_2 \to k_1$ yields the aperiodic solution given by equation (6.3-42). An asymptotic analysis as $|t| \to \infty$ shows that the solution constitutes a superposition of two solitons with amplitude η, the distance between which increases with time like $\ell n(4\eta^2 t)$. We indicate how one can obtain some of these solutions using Bäcklund transformations at the end of this section.

The remaining type of special solution to consider is that which arises from spectral singularities in the initial operator $L_0 = L(Q(x,0))$. This type of solution appears not to have been studied in any great detail except in one particular circumstance, the case when $a(0) = 0$, $\bar{a}(0) = 0$, (Ablowitz et al., 1974).

At $k = 0$ the Jost solutions of (6.1-13) can be explicitly solved if $R = \alpha Q$.

$$ {}_0\phi(x) = \begin{pmatrix} \cosh(\alpha^{\frac{1}{2}} \int_{-\infty}^{x} Q(y)dy) \\ \alpha^{\frac{1}{2}} \sinh(\alpha^{\frac{1}{2}} \int_{-\infty}^{x} Q(y)dy) \end{pmatrix} \tag{6.3-43} $$

$$ {}_0\psi(x) = \begin{pmatrix} \alpha^{-\frac{1}{2}} \sinh(-\alpha^{\frac{1}{2}} \int_{x}^{\infty} Q(y)dy) \\ \cosh(-\alpha^{\frac{1}{2}} \int_{x}^{\infty} Q(y)dy) \end{pmatrix} \tag{6.3-44} $$

For this case

$$ \bar{a}(0) = a(0) = \cosh(\alpha^{\frac{1}{2}} \int_{-\infty}^{\infty} Q(y)dy) \tag{6.3-45} $$

so that $a(0) = 0$ requires that

$$ \alpha^{\frac{1}{2}} \int_{-\infty}^{\infty} Q(y)dy = i(\frac{\pi}{2} + m\pi), \qquad m \in Z \tag{6.3-46} $$

In particular for the sine-Gordon equation $U_x/2 = Q = R$ and U satisfies the boundary condition of a $(\pi/2 + m\pi)$ pulse (this is the area under Q)

$$ \begin{aligned} U &\to 0 \quad \text{as} \quad x \to -\infty \\ U &\to (\frac{\pi}{2} + m\pi) \quad \text{as} \quad x \to +\infty \end{aligned} \tag{6.3-47} $$

As Lamb (1977) has shown these solutions are obtained from the self-similar form of the sine-Gordon equation. Since the sine-Gordon has the scale symmetry $x \to \ell x$, $t \to \ell^{-1} t$ the appropriate similarity variable is $z = xt$ and the sine-Gordon transforms into the equation

$$ zV_{zz} + V_z - \sin V = 0 \qquad V(z) = U(xt) \tag{6.3-48} $$

This equation is in fact transformable into the canonical form for the equation defining the third Painlevé transcendent (Ince 1956). The required solutions are clearly solutions of (6.3-48) with the boundary conditions (6.3-47).

The possibility of spectral singularities makes the stability analysis of solitons technically difficult. In their absence Zakharov and Shabat (1972) showed that a soliton is stable against a perturbation by a continuous spectrum in the (self-focusing) nonlinear Schrödinger equation. Work by Ablowitz and Segur (1977) indicates that spectral singularities lead to collisionless shocks in the asymptotic longtime development of a solution to the nonlinear Schrödinger equation in the absence of solitons. This result presumably applies also to other solvable equations. It corresponds to the anomalous behaviour of the reflection coefficient in the Marchenko equation for the isospectral Schrödinger operator (Section 4.3).

If the scattering function a satisfies the condition

$$|a(k)-1| < 1 \tag{6.3-49}$$

then it is clear that a has no zeros in $\mathrm{Im}k \geqslant 0$. In this case we get from the integral representation (6.1-38) of a that

$$|a(k)-1| \leqslant \int_{-\infty}^{\infty} |Q(y)||e^{iky}\phi_2(y,k)|dy \leqslant \int_{-\infty}^{\infty} |P(y)||e^{iky}\phi(y,k)|dy \tag{6.3-50}$$
$$\leqslant \exp(S_0(\infty))-1$$

Thus a condition for the absence of solitons or solutions corresponding to spectral singularities in the initial operator is that

$$S_0(\infty,0) < \ell n\, 2 \text{ or } \int_{-\infty}^{\infty} |Q(x,0)| + |R(x,0)|)dx < 0.35 \tag{6.3-51}$$

A refinement of this analysis yields the bound

$$(\int_{-\infty}^{\infty} |R(x,0)|dx) \ (\int_{-\infty}^{\infty} |Q(x,0)|dx) < 0.817 \tag{6.3-52}$$

for an initial L_0 which has only a continuous spectrum. An estimate for the eigenvalue of largest magnitude k_0, can be obtained from the asymptotic expansions of Lemma 6-3. Thus provided $\mathrm{Re}\{\int_{-\infty}^{\infty} R(x,0)Q(x,0)dx\} < 0$

$$2i\,k_0 \approx \int_{-\infty}^{\infty} R(x,0)Q(x,0)dx \tag{6.3-53}$$

Better estimates for k_0 can be obtained by including further terms in the asymptotic expansion of $a(k)$ and solving the polynomial resulting from equating the right hand of the expansion to zero after putting $k = k_0$. It appears that if

$$\mathrm{Re}\{\int_{-\infty}^{\infty} R(x,0)Q(x,0)dx\} > 0 \tag{6.3-54}$$

the initial operator does not have a discrete spectrum.

So far we have not considered the general solution of a solvable equation. When the initial data is such that the initial operator $L_0 = (Q(x,0))$ has only a continuous spectrum then it is possible to give a description of the solution in the asymptotic (long time) region using fairly standard asymptotic techniques. As mentioned in Section 4.3, a number of alternative methods are available, and in general one has to use a combination of techniques to obtain a uniformly valid solution for the whole real line. One of the interesting aspects of this analysis which has led to a study of solvable nonlinear **ordinary** differential equations is the existence of a region governed by solutions of the self-similar form of the evolution equation. For the AKNS system some of these solutions cannot be expressed in terms of the familiar transcendental functions of analysis and are therefore called Painlevé transcendents. We shall not give a complete asymptotic description of the long time behaviour of solutions to the AKNS system except in the case of the nonlinear Schrödinger equation.

The reader is referred again to Section 4.3 where the results of the asymptotic analysis of the KdV are given and which illustrate the comments given above.

Here we shall describe and apply the technique due to Zakharov and Manakov (1976) for obtaining solutions to the solvable equations in one of the near-linear asymptotic regions. This method is interesting for a number of reasons. Firstly since the method works from the scattering problem, results can be obtained for the whole class of solvable equations. Specific information about a particular equation (actually the dispersion relation Ω) is only added at the final stage. Secondly the method reconstructs the solution in terms of the initial data. It is in fact an "asymptotic" inverse scattering method.

The method can be applied to two types of equation belonging to the solvable equations which satisfy $\bar{\Omega} = -\Omega$

$$\text{(i)} \quad R = \pm Q^* \qquad \text{(ii)} \quad R = \pm Q^* = \pm Q \tag{6.3-55}$$

If we introduce the change of variables

$$u_1 = -iy_2\, e^{-ikx}\, , \quad u_2 = y_1 e^{ikx} \tag{6.3-56}$$

into (6.1-13) then we get the system used by Zakharov and Manakov for the nonlinear Schrödinger equation (ii)

$$iu_{1x} - (Re^{i\lambda x})u_2 = 0 \tag{6.3-57}$$

$$\lambda = -2k$$

$$iu_{2x} + (Qe^{-i\lambda x})u_1 = 0 \tag{6.3-58}$$

Now we assume that we are in an asymptotic region characterised by near-linear behaviour of WKB type. That is we assume that R and Q can be represented as slowly varying amplitudes modulating a rapidly varying phase. Thus we assume that Q has the representations

$$\text{(i)} \quad Q = \varepsilon^{\frac{1}{2}} A(\varepsilon x)e^{i\Phi}, \quad \Phi = \varepsilon^{-1}p(\varepsilon x) \tag{6.3-59}$$

$$\text{(ii)} \quad Q = \varepsilon^{\frac{1}{2}} A(\varepsilon x)\cos\Phi, \quad \Phi = \varepsilon^{-1}p(\varepsilon x) \tag{6.3-60}$$

in the two cases. The parameter ε has yet to be identified but clearly will be expected to be some function of t . The form assumed for the amplitude, $\varepsilon^{\frac{1}{2}}A(\varepsilon x)$ will be justified later. For the case (i) we can use a multiple scales method which involves only the two scales $X = \varepsilon x$, $\theta = \Phi - \lambda$ whereas case (ii) requires three scales $X = \varepsilon x$, $\theta_1 = \Phi - \lambda x$, $\theta_2 = \Phi + \lambda$ Here we shall treat case (ii) and quote the results for case (i). The equations we require to analyse are

$$\begin{aligned} iu_{1x} &\mp \varepsilon^{\frac{1}{2}} A(e^{i\theta_2}+e^{-i\theta_1})u_2 = 0 \\ iu_{2x} &+ \varepsilon^{\frac{1}{2}} A(e^{i\theta_1}+e^{-i\theta_2})u_1 = 0 \end{aligned} \tag{6.3-61}$$

The assumptions we have made imply that

$$\frac{\partial}{\partial x} = \varepsilon \frac{\partial}{\partial X} + \theta_{1x}\frac{\partial}{\partial \theta_1} + \theta_{2x}\frac{\partial}{\partial \theta_2} \tag{6.3-62}$$

$$u_1(x) \equiv U(X,\theta_j,\varepsilon^{\frac{1}{2}}) = \sum_{i=0} U_i(X,\theta_j)(\varepsilon^{\frac{1}{2}})^i \qquad \text{as } \varepsilon \to 0 \tag{6.3-63}$$

$$u_2(x) \equiv V(X,\theta_j,\varepsilon^{\frac{1}{2}}) = \sum_{i=0} V_i(X,\theta_j)(\varepsilon^{\frac{1}{2}})^i \qquad \text{as } \varepsilon \to 0 \tag{6.3-64}$$

The equations which result from equating the coefficients of powers of $\varepsilon^{\frac{1}{2}}$ are

$$\begin{aligned} i(\theta_{1x}U_{1\theta_1}+\theta_{2x}U_{2\theta_2}) &\mp A(e^{i\theta_2}+e^{-i\theta_1})V_o = 0 \\ i(\theta_{1x}V_{2\theta_1}+\theta_{2x}V_{2\theta_2}) &+ A(e^{i\theta_1}+e^{-i\theta_2})U_o = 0 \qquad (\varepsilon^{\frac{1}{2}}) \end{aligned} \tag{6.3-65}$$

$$\begin{aligned} i(\theta_{1x}U_{2\theta_1}+\theta_{2x}U_{2\theta_2}) + iU_{OX} &\mp A(e^{i\theta_2}+e^{-i\theta_1})V_1 = 0 \\ i(\theta_{1x}V_{2\theta_1}+\theta_{2x}V_{2\theta_2}) + iV_{OX} &+ A(e^{i\theta_1}+e^{-i\theta_1})U_1 = 0 \qquad (\varepsilon) \end{aligned} \tag{6.3-66}$$

The ε^0 equation yields $U_0 = U_0(X)$, $V_0 = V_0(X)$. Since $\theta_{ix} = \theta_{ix}(X)$ we can easily integrate (6.3-65) by a change of independent variable

$$U_1 = \mp A\left(\frac{e^{i\theta_2}}{\theta_{2x}} - \frac{e^{-i\theta_1}}{\theta_{1x}}\right)V_o \tag{6.3-67}$$

$$V_1 = A\left(\frac{e^{i\theta_1}}{\theta_{1x}} - \frac{e^{-i\theta_2}}{\theta_{2x}}\right)U_o \tag{6.3-68}$$

which is valid provided we are not in the resonance region where

$$\theta_{1x} \approx 0 \quad \text{or} \quad \theta_{2x} \approx 0 \tag{6.3-69}$$

If this is the case then the asymptotic expansions (6.3-62)-(6.3-64) are incorrect. In order to prevent secularities occurring at order ε when (6.3-67),(6.3-68) is substituted into (6.3-66) we require

$$iU_{0X} \mp A^2 \left(\frac{1}{\theta_{1X}} - \frac{1}{\theta_{2X}} \right) U_0 = 0$$
$$iV_{0X} \pm A^2 \left(\frac{1}{\theta_{1X}} - \frac{1}{\theta_{2X}} \right) V_0 = 0 \qquad (6.3\text{-}70)$$

If we identify the expansions above with the solution $\underset{x\to-\infty}{L^t}(u_1, u_2) = (0,1)$ then we can integrate (6.3-70) to obtain

$$U_0 = 0$$
$$V_0 = \exp \pm i \int_{-\infty}^{X} A^2 \left(\frac{1}{\theta_{1X}} - \frac{1}{\theta_{2X}} \right) dX \qquad (6.3\text{-}71)$$

We shall assume that there is, for a given λ, a unique point x_0 defined by

$$\Phi_x(x_0) \pm \lambda = 0 \qquad (6.3\text{-}72)$$

for which the above expansions are no longer valid. This assumption is compatible with the class of equations which we examine by this method. Since x_0 varies we write $x_0 = x(\lambda)$. In addition we shall assume that $\lambda > 0$. In the resonance region defined by (6.3-72) we introduce the variable

$$Z = (x-x_0)\Phi^{\frac{1}{2}}_{xx}(x_0) \qquad (6.3\text{-}73)$$

and assume that $\Phi_{xx}(x_0) > 0$. Then since

$$\Phi(x) = \Phi(x_0) + \lambda(x-x_0) + \tfrac{1}{2}(x-x_0)\Phi_{xx}(x_0) + O((x-x_0)^3) \qquad (6.3\text{-}74)$$
$$\text{as } x \to x_0$$

the equations (6.3-61) can be written as

$$i(\Phi_{xx}^{\frac{1}{2}}(x_0)u_{1Z}+\theta_{1x}u_{1\theta_1}+\theta_{2x}u_{1\theta_2})$$
$$\mp \varepsilon^{\frac{1}{2}}A(X_0)(e^{2i\Phi}+1)\exp-i(\Phi(x_0)-\lambda x+\tfrac{1}{2}Z^2)u_2 = 0 \qquad (6.3\text{-}75)$$

$$i(\Phi_{xx}^{\frac{1}{2}}(x_0)u_{2Z}+\theta_{1x}u_{2\theta_1}+\theta_{2x}u_{2\theta_2})$$
$$+ \varepsilon^{\frac{1}{2}}A(X_0)(e^{-2i\Phi}+1)\exp i(\Phi(x_0)-\lambda x_0+\tfrac{1}{2}Z^2)u_1 = 0 \qquad (6.3\text{-}76)$$

From the assumptions on the form of Φ, $\Phi = \varepsilon^{-1}p(\varepsilon x)$ we get that $\Phi_{xx} = \varepsilon\ddot{p}$ so that we can write

$$\Phi_{xx}(x_0) = \varepsilon p''(X_0) \equiv \varepsilon g^2(X_0).$$

Appropriate asymptotic expansions for u_1, u_2 are

$$u_1(x) \equiv \bar{U}(Z,\Phi,\varepsilon^{\frac{1}{2}}) = \sum_{i=0} \bar{U}_i(Z,\Phi)(\varepsilon^{\frac{1}{2}})^i$$
$$u_2(x) \equiv \bar{V}(Z,\Phi,\varepsilon^{\frac{1}{2}}) = \sum_{i=0} V_i(Z,\Phi)(\varepsilon^{\frac{1}{2}})^i \tag{6.3-77}$$

In this case, equating powers of $\varepsilon^{\frac{1}{2}}$ in (6.3-75)-(6.3-76) we obtain for the $O(1)$ equations $\bar{U}_0 = \bar{U}_0(Z)$, $\bar{V}_0 = \bar{V}_0(Z)$.

To prevent the occurrence of singularities at $O(\varepsilon^{\frac{1}{2}})$ we require

$$\bar{V}_{ozz} - iZ\bar{V}_{oZ} \pm \alpha^2\bar{V}_o = 0$$
$$\bar{U}_o = -i\exp[-i(\Phi(x_o)-\lambda x_o+\tfrac{1}{2}Z^2)]\bar{V}_{oz}/\alpha \tag{6.3-78}$$

where

$$\alpha^2(\lambda) = A^2(X_o)g^{-2}(X_o), \quad x_o = x_o(\lambda)$$

It is clear from (6.3-75),(6.3-76) that non-trivial results at this order necessitate the form of amplitude function $(\varepsilon^{\frac{1}{2}}A(X))$ which we have chosen thus justifying our original assumption. The solutions to (6.3-78) can be given in terms of parabolic cylindrical functions (A. Erdelyi et al. 1953)

$$\bar{V}_o = e^{iZ^2/4}(c_1D_{-1\mp i\alpha^2}(Zi^{-\frac{1}{2}}) + c_2D_{-1\mp i\alpha^2}(-Zi^{-\frac{1}{2}})) \tag{6.3-79}$$

For matching between the resonance and asymptotic regions we require the limit $Z \to -\infty$ of (6.3-79) and the limit $X \uparrow X_0(\lambda)$ of (6.3-71). The asymptotic behaviours of the parabolic cylinder functions which we require are

$$D_{-1\mp i\alpha^2}(Ze^{-i\pi/4}) = e^{i\pi/4}e^{iZ^2/4}e^{\mp\pi\alpha^2/4}Z^{\mp i\alpha^2-1} + O(|Z|^{-2})$$
$$\text{as } Z \to +\infty \tag{6.3-80}$$

$$D_{-1\mp i\alpha^2}(-Ze^{-i\pi/4}) = \frac{\sqrt{2\pi}}{\Gamma(1\pm i\alpha^2)}e^{\pm\pi\alpha^2/4}Z^{\pm i\alpha^2}e^{-iZ^2/4}$$
$$+ e^{-3\pi i/4}e^{\pm 3\pi\alpha^2/4}Z^{-1\pm i\alpha^2}e^{iZ^2/4} + O(|Z|^{-2}) \tag{6.3-81}$$
$$\text{as } Z \to +\infty$$

On the overlap region integrating (6.3-71) by parts we obtain

$$U_o = 0$$

$$V_o = \left|\frac{(X-X_o)g^2}{2\lambda}\right|^{\pm i\alpha^2} \exp[\mp i\int_{-\infty}^{X_o} \ell n \left|\frac{\theta_{1x}}{\theta_{2x}}\right| \frac{d}{dX}\left(\frac{A^2}{\Phi_{xX}}\right) dX] \tag{6.3-82}$$

From the equations (6.3-79) and (6.3-82) we find that

$$c_2 = c_1 e^{\pm\pi\alpha^2}$$

$$c_1 = \left(\frac{g\varepsilon^{\frac{1}{2}}}{2\lambda}\right)^{\pm i\alpha^2} \frac{\Gamma(1\pm i\alpha^2)}{\sqrt{2\pi}} e^{\mp\pi\alpha^2/4} \exp[\mp i\int_{-\infty}^{X_o} \ell n \left|\frac{\theta_{1x}}{\theta_{2x}}\right| \times \frac{d}{dX}\left(\frac{A^2}{\Phi_{xX}}\right) dX] \tag{6.3-83}$$

As $X \to +\infty, (u_1,u_2) = (-ib,a)$ and (6.3-70) in this region gives

$$U_o = -i\, b \exp \pm i \int_X^{\infty} A^2 \left(\frac{1}{\theta_{1x}} - \frac{1}{\theta_{2x}}\right) dX$$

$$V_o = a \exp \mp i\int_X^{\infty} A^2 \left(\frac{1}{\theta_{1x}} - \frac{1}{\theta_{2x}}\right) dX \tag{6.3-84}$$

As $X \downarrow X_0(\lambda)$ these functions match onto $\overline{U}_0$, $\overline{V}_0$ in the limit $Z \to +\infty$.

$$U_o = -ib\left[\frac{(X-X_o)g^2}{2\lambda}\right]^{\mp i\alpha^2} \exp[\mp i \int_{X_o}^{\infty} \ell n \left(\frac{\theta_{1x}}{\theta_{2x}}\right) \frac{d}{dX}\left(\frac{A^2}{\Phi_{xX}}\right) dX]$$

$$V_o = a \left[\frac{(X-X_o)g^2}{2\lambda}\right]^{\pm i\alpha^2} \exp[\pm i \int_{X_o}^{\infty} \ell n \left(\frac{\theta_{1x}}{\theta_{2x}}\right) \frac{d}{dX}\left(\frac{A^2}{\Phi_{xX}}\right) dX] \tag{6.3-85}$$

Thus we obtain from equating these two limits

$$a(\lambda,t) = \exp(\pm\pi\alpha^2 \mp i(L_1+L_2))$$

$$b(\lambda,t) = \mp \frac{i\sqrt{2\pi}}{\alpha\Gamma(\mp i\alpha^2)} e^{\pm\pi\alpha^2/2} \exp i[\lambda x_o - \Phi(x_o) - \frac{3\pi}{4} \pm (L_1-L_2) \pm \alpha^2(\ell n\Phi_{xx} - 2\ell n 2\lambda)] \tag{6.3-86}$$

where

$$L_1(\lambda,t) = \int_{X_o(\lambda)}^{\infty} \ell n \left(\frac{\theta_{1x}}{\theta_{2x}}\right) \frac{d}{dX}\left(\frac{A^2}{\Phi_{xX}}(X,t)\right) dX$$

$$= \int_{\lambda}^{\infty} \ell n\left(\frac{\xi-\lambda}{\xi+\lambda}\right)\frac{d}{d\xi}(\alpha^2(\xi,t))\,d\xi \tag{6.3-87}$$

$$L_2(\lambda,t) = \int_{-\infty}^{X_o(\lambda)} \ln\left|\frac{\theta_{1x}}{\theta_{2x}}\right| \frac{d}{dX}\left(\frac{A^2}{\Phi_{xX}}(X,t)\right) dX$$
$$= \int_{-\infty}^{\lambda} \ln\left|\frac{\xi-\lambda}{\xi+\lambda}\right| \frac{d}{d\xi}(\alpha^2(\xi,t))d\xi \tag{6.3-88}$$

If we use the formula

$$|\Gamma(\pm i\alpha^2)|^2 = \pi(\alpha^2 \sinh\pi\alpha^2)^{-1} \tag{6.3-89}$$

then it is straightforward to check directly that

$$|b(\lambda)|^2 \mp |b(\lambda)|^2 = 1 \quad \text{for} \quad \lambda > 0 \tag{6.3-90}$$

for the two cases $R = \pm Q = \pm Q^*$. It is also clear from (6.3-86) that $a(\lambda)$ is analytic in $\text{Im}\,\lambda > 0$ and has no zeros in this half plane.

The problem now, after investigating the forward problem, is to solve the inverse problem with initial data defined by (6.3-86), t large. From the theory of Section 6.1 we have

$$b_t(\lambda,t) = \omega(\lambda)b(\lambda,0)$$
$$a(\lambda,t) = a(\lambda,0), \quad \omega(\lambda) \equiv \Omega(-\tfrac{1}{2}\lambda) \tag{6.3-91}$$

for the class of equations under consideration.

We can therefore relate our problem directly to the given initial data $a(\lambda,0)$, $b(\lambda,0)$ which is obtainable from the initial function $Q(x,0)$. We see immediately that L_1, L_2 are t independent and that

$$\alpha^2(\lambda) = \frac{A^2}{\Phi_{xX}} = \pm\frac{1}{\pi}\ln|a(\lambda)| \tag{6.3-92}$$

In order to obtain Φ define

$$\tilde{\Theta}_{\pm}(\lambda,t) = \mu_{\pm} + \Theta(\lambda,t) \mp (L_1(\lambda) - L_2(\lambda))$$
$$+ \arg\Gamma(\mp i\alpha^2) - \pi/2$$

where

$$\Theta(\lambda,t) = \arg b(\lambda,t), \quad (\mu_+,\mu_-) = \left(-\frac{\pi}{4}, \frac{3\pi}{4}\right) \tag{6.3-93}$$

Then from (6.3-91) we obtain

$$\tilde{\Theta}_{\pm}(\lambda,t) = \text{Im}(\omega(\lambda))t + \tilde{\Theta}_{o\pm}(\lambda)$$
$$\tilde{\Theta}_{o\pm}(\lambda) = \tilde{\Theta}_{\pm}(\lambda,0) \tag{6.3-94}$$

The expression (6.3-86) then yields the relation

$$\Delta_\pm(\lambda,t) = \lambda x - \Phi(x,t) \pm \alpha^2 \ln\Phi_{xx}(x,t) \qquad (6.3\text{-}95)$$
$$\Phi_x = \lambda$$

and

$$\Delta_\pm(\lambda,t) \equiv \tilde{\Theta}_\pm(\lambda,t) \pm 2\alpha^2 \ln 2\lambda = (\mathrm{Im}\omega(\lambda))_t + \Delta_{o\pm}(\lambda) \qquad (6.3\text{-}96)$$
$$\Delta_{o\pm}(\lambda) = \Delta_\pm(\lambda,0)$$

Next we differentiate the **known** function $\Delta\pm(\lambda,t)$ in (6.3-96) with respect to λ which yields as $t \to \infty$.

$$\frac{\partial}{\partial\lambda}(\mathrm{Im}\omega(\lambda)) = \frac{x}{t} \equiv X \qquad (6.3\text{-}97)$$

where we have made the identification $\varepsilon^{-1} = t$. This gives a unique solution λ so that (6.3-95) becomes the identity

$$\varepsilon^{-1}H(X) + K_\pm(X) - \varepsilon^{-1}X\lambda(X) \equiv \Phi(x,\varepsilon) \pm \alpha_\pm^2(X)\ln[\Phi_{xx}(x,\varepsilon)] \qquad (6.3\text{-}98)$$

where $H(X) = \omega(\lambda(X))$ and $K_\pm(X) = \Delta_{o\pm}(\lambda(X))$

A standard asyptotic expansion for Φ then gives (Dodd and Morris 1982)

$$\phi(x,\varepsilon) = \varepsilon^{-1}(X\lambda(X) - \mathrm{Im}\omega(\lambda(X))) \pm \alpha^2\ln\varepsilon - \Delta_{o\pm}(\lambda(X)) \pm \alpha_\pm^2 \ln\frac{\partial\lambda(X)}{\partial X} \qquad (6.3\text{-}99)$$

Thus by solving the algebraic equation for (6.3-97) we can obtain the required asymptotic form of the solution to the solvable nonlinear equation. The formulae simplify considerably if we assume $\omega(\lambda) = -i\lambda^{2n+1}$, $n \geqslant 0$, or $\omega(\lambda) = i\lambda^{-(2n+1)}$, $n \leqslant 0$. Put $X = [-x/(2n+1)t]$, then the solutions in the two cases are the following

$\underline{\omega = -i\lambda^{2n+1}; \quad n \geqslant 0}$

$$(\varepsilon A^2) = \frac{1}{(2n+1)2n\pi t}(X)^{\frac{1}{2n}-1} \ln(|a(X^{\frac{1}{2n}})|^{\mp 1})$$

$$\Phi(x,t) = -t[2n\, X^{\frac{1}{2n}+1}] \mp \alpha^2[\ln(2n+1)t + (1+\frac{1}{2n})\ln X] + \phi_o \qquad (6.3\text{-}100)$$

$$\phi_o = \mp i\pi\alpha^2 + \tilde{\Theta}_{o\pm}(X^{\frac{1}{2n}}) \mp \alpha^2 \ln 8n$$

$\underline{\omega = i\lambda^{-(2n+1)}; \quad n \geqslant 0}$

$$(\varepsilon A^2) = \frac{1}{(2n+2)(2n+1)\pi t}(X)^{-(\frac{2n+3}{2n+2})} \ln(|a(X^{-\frac{1}{(2n+2)}})|^{\pm 1})$$

$$\Phi(x,t) = -t[(2n+2)X^{-(\frac{2n+1}{2n+2})}]\alpha^2[\ln(2n+1)t + \ln X] + \phi_o \qquad (6.3\text{-}101)$$

$$\phi_o = \tilde{\Theta}_{o\pm}(X^{-\frac{1}{(2n+2)}}) \mp \alpha^2 \ln 8(n+1)$$

Notice in the above that $\varepsilon = 1/(2n+1)t$ whilst the region of validity of the expansions is $(-x) >> [(2n+1)t]$ for $\omega = i\lambda^{-(2n+1)}$, $n > 0$, and $(-x) >> [(2n+1)t]^{1+(2n+2)/(2n+3)}$ for $\omega = i\lambda^{-(2n+1)}$, $n \geqslant 0$. In particular, $\omega = i\lambda^3$ and $\omega = i\lambda^{-1}$ give the results for the MKdV and the sine-Gordon equation respectively. For the sine-Gordon equation $\pm U = \pm \sin U$ the results are given in terms of the function $Q = -U_x/2$, but the results for U can easily be found by integration. If $\lambda < 0$ one needs to repeat the above calculation to check that the formula for A and Φ are also valid in this case.

The result for case (i) can be formally obtained from case (ii) by simply replacing $\Delta_\pm$ by $\Theta_\pm$ in the above formulae. The nonlinear Schrödinger equation is a particulary interesting example of this case and was the original equation investigated by Zakharov and Manakov (1976). Ablowitz and Segur (1977) and Segur and Ablowitz (1976) have investigated the effect of respectively spectral singularities and discrete eigenvalues (solitons) on the asymptotic form of the solutions of the nonlinear Schrödinger equation.

The Bäcklund transformations for solvable equations which satisfy $\bar{\Omega} = -\Omega$ can be determined in exactly the same way as was done for the isospectral Schrödinger operation in Section 3.5. Thus we obtain first a relationship between the scattering functions A, A' and the corresponding sets of potentials (Q, R), (Q', R') which satisfy (6.1-13) with fundamental solutions Φ, Φ' respectively

$$(A')^{-1}\Delta A = \int_{-\infty}^{\infty} (\Phi')^{-1}\Delta P\Phi \, dx \tag{6.3-102}$$

where

$$\Delta F \equiv F' - F$$

$$P = \begin{pmatrix} -ik & Q \\ R & +ik \end{pmatrix}, \quad A = \begin{pmatrix} b & \bar{a} \\ a & \bar{b} \end{pmatrix}$$

In this case eigenfunctions ${}_kV$, ${}_kY'$ of L, L' which have the same continuous spectrum but different discrete spectra can be shown from (6.1-13) to satisfy the relationships

$$-(y_1'v_1)_x + (Qy_1'v_2+Q'y_2'v_1) = 2iky_1'v_1 \tag{6.3-103}$$

$$(y_2'v_2)_x - (Ry_2'v_1+R'y_1'v_2) = 2iky_2'v_2 \tag{6.3-104}$$

$$y_2'v_1(x) = y_2'v_1(-\infty) + \int_{-\infty}^{X} (Q(s)y_2'(s)v_2(s) + R'(s)y_1'(s)v_1(s)ds \tag{6.3-105}$$

$$y_1'v_2(x) = y_1'v_2(-\infty) + \int_{-\infty}^{X} (Q'(s)y_2'(s)v_2(s) + R(s)y_1'(s)v_1(s))ds \tag{6.3-106}$$

These relationships can be combined into the formula

$$B_1 \ V \circ Y' = kV \circ Y' - \frac{1}{2i}\begin{pmatrix} Q(x)y_1'v_2(-\infty) & + & Q'(x)y_2'v_1(-\infty) \\ -R(x)y_2'v_1(-\infty) & - & R'(x)y_1'v_2(-\infty)\end{pmatrix} \tag{6.3-107}$$

where

$$B_1 \equiv \frac{1}{2i}\begin{pmatrix} -\frac{\partial}{\partial x} + Q\int_{-\infty}^{x} ds\ R + Q'\int_{-\infty}^{x} dsR', & Q\int_{-\infty}^{x} dsQ' + Q'\int_{-\infty}^{x} ds\ Q \\ -R\int_{-\infty}^{x} dsR' - R'\int_{-\infty}^{x} dsR, & -R\int_{-\infty}^{x} dsQ - R'\int_{-\infty}^{x} dsQ' + \frac{\partial}{\partial x}\end{pmatrix}$$

Then taking the limit $x \to +\infty$ in (6.3-105), (6.3-106) and (6.3-107) for the solutions ϕ, ϕ' yields

$$B_1 \ \phi \circ \phi' = k \ \phi \circ \phi' \tag{6.3-108}$$

$$a'b + b'a = \int_{\infty}^{\infty} (U+U')\cdot_k (\phi\circ\phi')dx \equiv (U+U',_k(\phi\circ\phi')) \tag{6.3-109}$$

Equations (6.3-102) can be written as

$$1 + \bar{b}'b - \bar{a}'a = -I_\Delta(\bar{\phi}',\phi) \tag{6.3-110}$$

$$\Delta\bar{R}_+ = -\frac{1}{\bar{a}\bar{a}'} I_\Delta(\bar{\phi}',\bar{\phi}) \tag{6.3-111}$$

$$\Delta R_+ = \frac{1}{aa'} I_\Delta(\phi',\phi) \tag{6.3-112}$$

$$1 + b'\bar{b} - a'\bar{a} = I_\Delta(\phi',\bar{\phi}) \tag{6.3-113}$$

where $I_\Delta(\cdot,\cdot)$ is the obvious adaption of the definition (6.1-100) to the present situation. If we now combine equations (6.3-111), (6.3-112) and (6.3-109) we obtain

$$\begin{aligned} \Delta R_+ - \Lambda(k)(R_+' + R_+) &= \frac{1}{aa'}(\sigma_2\Delta U - \Lambda(k)(U+U'),\ _k(\phi\circ\phi')) \\ \Delta\bar{R}_+ - \bar{\Lambda}(k)(\bar{R}_+' + \bar{R}_+) &= -\frac{1}{\bar{a}\bar{a}'}(\sigma_2\Delta U + \bar{\Lambda}(k)(U+U),_k(\phi\circ\phi')) \end{aligned} \tag{6.3-114}$$

The functions $\Lambda(k)$ and $\bar{\Lambda}(k)$ are arbitrary polynomials of k or a ratio of polynomials in k. If we form the adjoint of B, with respect to the bilinear form (,) defined in (6.3-109), then we can define a generalised Bäcklund transformation for the solvable equations provided $\bar{\Lambda}(k) = -\Lambda(k)$. It is given by

$$\sigma_2 \Delta U - \Lambda(B_1^A)(U'+U) = 0$$

$$B_1^A \equiv \frac{1}{2i}\begin{pmatrix} \frac{\partial}{\partial x} + R\int_x^\infty ds Q + R'\int_x^\infty ds Q', & -R\int_x^\infty ds R' - R'\int_x^\infty ds R \\ Q\int_x^\infty ds\, Q' + Q'\int_x^\infty ds\, Q, & -\frac{\partial}{\partial x} - Q'\int_x^\infty ds R - Q\int_x^\infty ds R \end{pmatrix} \qquad (6.3\text{-}115)$$

and $$U = (R,Q,)^T$$

The corresponding relationships between the reflection coefficients are

$$R'_+ = \frac{(1+\Lambda(k))}{(1-\Lambda(k))} R_+ \qquad (6.3\text{-}116)$$

$$\bar{R}'_+ = \frac{(1-\Lambda(k))}{(1+\Lambda(k))} \bar{R}_+ \qquad (6.3\text{-}117)$$

These results were first obtained by Calogero and Degasperis (1976). The method used for deriving them here is due to Dodd and Bullough (1977).

The simplest transformation in this set, Λ = constant $=\alpha$, merely gives a scale change of the variables, $R' = \lambda R$, $Q' = \lambda^{-1}Q$, $\lambda = (1+\alpha)(1-\alpha)^{-1}$. The next simplest transformation arises from the choice $\Lambda(k) = (ak + b)^{-1}$ where the numbers $k_1 = (1-b)/a$, $\bar{k}_1 = -(1+b)/a$ satisfy the conditions $\text{Im } k_1 > 0$, $\text{Im } \bar{k}_1 < 0$. In this case the reflection coefficients are related by the formulae

$$R_{1+} = \frac{(k-\bar{k}_1)}{(k-k_1)} R_{0+}, \quad \bar{R}_{1+} = \frac{(k-k_1)}{(k-\bar{k}_1)} \bar{R}_{0+} \qquad (6.3\text{-}118)$$

Thus one can interpret this transformation by saying that $\sigma(L_1)$ contains the additional eigenvalues $k_1, \bar{k}_1$ over $\sigma(L_0)$. The corresponding Bäcklund transformation for the solvable nonlinear equations is given by

$$(R_1-R_0)_x + (R_1+R_0)J_{10} + p_1R_1 - \bar{p}_1R_0 = 0$$

$$(Q_1-Q_0)_x + (Q_1+Q_0)J_{10} - \bar{p}_1Q_1 + p_1Q_0 = 0 \qquad (6.3\text{-}119)$$

$$J_{ij}(x) = \int_x^\infty (R_i(y)Q_i(y) - R_j(y)Q_j(y))dy$$

$$\bar{p}_j = -2i\bar{k}_j, \quad p_j = -2ik_j$$

This corresponds to "one-half" of the classical Bäcklund transformation, the other half is obtained by substituting in the particular solvable equation satisfied by (R_1, Q_1), $(R_0\ Q_0)$ and rearranging so that the equations are first order in t-derivatives of the variable (R_1, Q_1) and (R_0, Q_0).

In this regard notice that if

$$R_{+0t} = \Omega R_{+0} \tag{6.3-120}$$

$$R_{+1t} = \left(\frac{k-\bar{k}_1}{k-k_1}\right) R_{+0t} = \Omega R_{+1} \tag{6.3-121}$$

so that the transformation relates solutions of the same equation. This observation is clearly true also in the case of an arbitrary Bäcklund transformation for which Λ is t-independent. If we allow Λ to be t-dependent then, as in the case of the Bäcklund transformations related to the isospectral Schrödinger equation, we obtain formal transformations between different members of the solvable family. The transformation is not 1 - 1 in either direction. This can be seen by taking either (R_1, Q_1) or (R_0, Q_0) to be zero and integrating (6.3-119). For fixed $k_1, \bar{k}_1$ and Ω a family of solitons (6.3-7) is obtained. It is clear that for Bäcklund related solutions a theorem of commutability exists. Thus for the four solutions (R_i, Q_i), $i = 0,1,2,3$ which are Bäcklund related by transformation (6.3-119) we have

$$\begin{aligned} R_{3+} &= \frac{(k-\bar{k}_2)}{(k-k_2)} R_{1+} = \frac{(k-\bar{k}_2)}{(k-k_2)} \times \frac{(k-\bar{k}_1)}{(k-k_1)} R_{0+} \\ &= \frac{(k-\bar{k}_1)}{(k-k_1)} \times \frac{(k-\bar{k}_2)}{(k-k_2)} R_{0+} = \frac{(k-\bar{k}_1)}{(k-k_1)} R_{1+} \end{aligned} \tag{6.3-122}$$

and of course a similar relationship holds between the $\bar{R}_{i+}$. This imposes relationships between the Bäcklund related solutions (R_i, Q_i), $i=0,\ldots,3$. In the the particular case that R_i is related to Q_i through Lemma 6-10 this relationship is algebraic and furnishes a superposition principle for Bäcklund related solutions. Thus a new solution to a solvable equation (Q_3) can in this case be determined by algebraic means from a given solution (Q_0) and two Bäcklund transforms (6.3-119) (Q_1, Q_2) of it. The general superposition principle is given by the following relationships

$$\begin{aligned} R_3(J_{21}+(p_2-p_1)) + R_2((\bar{p}_1-p_2)-J_{30}) + R_1((p_1-\bar{p}_2)+J_{30}) \\ + R_0((\bar{p}_2-\bar{p}_1)+J_{12}) = 0 \end{aligned} \tag{6.3-123}$$

$$\begin{aligned} Q_3(J_{21}-(\bar{p}_2-\bar{p}_1)) + Q_2((\bar{p}_2-p_1)-J_{30}) + Q_1((p_2-\bar{p}_1)+J_{30}) \\ + Q_0((p_1-p_2)+J_{12}) = 0 \end{aligned} \tag{6.3-124}$$

then multiplying (6.3-123) by $(Q_1 - Q_2)$ and (6.3-124) by $(R_1 - R_2)$, and subtracting we obtain an algebraic relationship between Q_3 and R_3 and the solutions Q_i, $i=0,1,2$ but Q_3 or R_3 has still to be determined from an integral equation.

For the case $R = + Q$ the Bäcklund transformation (6.3-119) is

$$(Q_1-Q_0)_x \pm (Q_1+Q_0) \int_x^\infty (Q_1^{\ 2}(y)-Q_0^{\ 2}(y))dy = p(Q_1+Q_0) \tag{6.3-125}$$

If we multiply this by $(Q_1 - Q_0)$, integrate and then solve the resulting quadratic we obtain

$$\int_x^\infty (Q_1^{\ 2}(y)-Q_0^{\ 2}(y))dy = \pm(p-(p^2 \pm (Q_1-Q_0)^2)^{\frac{1}{2}}) \tag{6.3-126}$$

From which we deduce by a differentation, cancellation and then an integration that for $R = -Q$

$$\begin{aligned} Q_1 &= Q_0 + p\sin(W_1+W_0) \\ W(x) &= \int_x^\infty Q(y)dy \end{aligned} \tag{6.3-127}$$

In the $R = +Q$ case the sine is replaced by a sinh. The superposition principle becomes for distinct eigenvalues k_1, k_2, $\text{Im}\, k_i > 0$

$$\begin{aligned} \sin W_3 &= [\sin W_0((p_1^{\ 2}+p_2^{\ 2})\cos(W_1-W_2)-2p_1p_2) \\ &+ (p_1^{\ 2}-p_2^{\ 2})\cos W_0 \sin(W_1-W_2)][p_1^{\ 2}+p_2^{\ 2}-2p_1p_2\cos(W_1-W_2)]^{-1} \end{aligned} \tag{6.3-128}$$

The superposition principle for a repeated eigenvalue $k_2 = k_1$ can be easily obtained from this formula by putting $Q_2 = Q_1 + \delta Q$, $k_2 = k_1 + \delta k$ and then taking the limit $\delta k \to 0$, or by using L'Hôpital's rule. This yields

$$\begin{aligned} \sin W_3 &= \sin W_0 + 2p_1\dot{W}_1(\cos W_0 - p_1 W_1 \sin W_0)(1+p_1^{\ 2}\dot{W}_1^{\ 2})^{-1} \\ \dot{W}_1 &= \frac{i}{2} W_{1k}\Big|_{k=k_1} \end{aligned} \tag{6.3-129}$$

We emphasize that both of these formulae apply to the whole solvable class of equations for which $R = -Q$. In particular, if we consider the sine-Gordon equation, $U = 2W$, $\Omega(k) = \pm 1/2ik$ then we can obtain directly from these formulae the two soliton solutions considered earlier in this section as well as the repeated eigenvalue solution. In both cases we assume that we start from $Q_0 = 0$ so that the Bäcklund transformed solutions (6.3-118) are solitons. In the first case we obtain for two eigenvalues $k_1 = i\eta_1$, $k_2 = i\eta$ (see Lemma 6-10) the 2-soliton solution. First notice that

$$\tan\tfrac{1}{2}W_3 = \left(\frac{p_1+p_2}{p_1-p_2}\right)\tan\tfrac{1}{2}(W_1-W_2) \tag{6.3-130}$$

so that putting $W_i = 2\tan^{-1}(\exp(\theta_i))$, $\theta_i = \gamma_i - 1/2\eta_i(\eta_i^2 x - t)$ gives the solution

$$U = 4\tan^{-1}\left[\frac{(\eta_1+\eta_2)}{(\eta_1-\eta_2)}\cdot\frac{\sinh\frac{1}{2}(\theta_1-\theta_2)}{\cosh\frac{1}{2}(\theta_1+\theta_2)}\right] \tag{6.3-131}$$

In a similar fashion one can also calculate the bound state solution (6.3-40) for this case (see the exercises for this chapter). Finally we calculate the solution of the sine-Gordon equation corresponding to the repeated eigenvalue $k_1 = i\eta_1$. From (6.3-129) we get

$$\tan W = \frac{4\eta_1\dot{W}_1}{(1-4\eta_1^2\dot{W}_1^2)} \tag{6.3-132}$$

$$\dot{W} = -(2x + \frac{1}{2\eta_1^2}t)\ \mathrm{sech}\theta_1 \tag{6.3-133}$$

so that

$$u = 2\tan^{-1}\left[\frac{2(2\eta_1 x + \frac{1}{2\eta_1}t)\cosh\theta_1}{2(2\eta_1 x + \frac{1}{2\eta_1}t)\cosh^2\theta_1}\right] \tag{6.3-134}$$

For the case $R = \pm Q^*$ the superposition principle is given by equations (6.3-125)-(6.3-126). For a repeated eigenvalue $p = -2ik$, the variations δp and δp^* are independent and then (6.3-125)-(6.3-126) yield if $Q_0 = 0$

$$Q_3 = \frac{-[Q_1 + Q_{1p}((\bar{p} - p) \mp \int_x^\infty |Q_3(y)|^2 dy]}{\pm \int_x^\infty |Q_1(y)|_p^2\ dy} \tag{6.3-135}$$

$$\int_x^\infty |Q_3(y)|^2 dy = \frac{(1 \mp \int_x^\infty |Q_1|_{\bar{p}}^2 dy)[Q_1 + Q_{1p}(\bar{p}-p)] \pm \int_x^\infty |Q_1|_p^2 dy[Q_1 + Q_{1p}(\bar{p}-p)]}{Q_{\bar{p}}\int_x^\infty |Q_1|_p^2 dy \pm (1 \mp \int_x^\infty |Q_1|_{\bar{p}}^2 dy) Q_{1p}}$$

If one uses the one soliton solution for the nonlinear Schrödinger equation then this formula will produce the solution (6.3-42).

Finally we note that the KdV equation (solvable by the Schrödinger inverse method) is related to the MKdV (solvable by the AKNS inverse method) by a particular type of Bäcklund transformation called a Miura transformation (see Section 3.2).

In fact one can relate in this fashion the whole solvable family of the Schrödinger inverse method to the subsystem of solvable equations of the AKNS ionverse method defined by $R = +Q$. This material is developed in the exercises.

6.4 Solvable Nonlinear Equations and Inverse Methods

In this book we have concentrated upon the inverse spectral (scattering) method for solvable nonlinear equations associated with the two most studied linear operators; the Schrodinger operator of Chapters 3 and 4 and the AKNS-ZS operator of Chapter 6, defined on the real line. Even with these inverse methods many questions remain to be answered. For example, a rigorous mathematical treatment of solvable nonlinear integro-differential equations; the initial value problem for nonlinear evolution equations solvable by the AKNS-ZS inverse method for initial data which satisfies less stringent conditions than those demanded by Theorem 6-25; the interpretation of spectral singularities in physical problems. Moreover several extensions are of physical interest; for example the inverse methods for solvable equations with initial data which does not decay to zero as $|x| \to \infty$ (Kawata and Inoue (1978) and Kawata et al. (1980)) and the perturbation theory for nearly solvable equations. The last extension is particularly important. Although solvable equations can serve as models, inevitably we are forced realistically to deal with nonlinear equations which are not solvable, but which in some cases may be construed as perturbed solvable equations. The methods so far developed suffer in many cases from the difficulty of interpreting results (in some cases one can end up with an equation which is as hard to solve numerically as the original equation).

For solvable equations in which the regularised trace functional of the resolvent operator is a constant of the motion a Hamiltonian formulation of the inverse method is possible (Zakharov and Faddeev (1971), Faddeev and Takhtadzhyan (1974), Zakharov and Manakov (1974), Zakharov et al. (1974), Flaschka and Newell (1975), Dodd and Bullough (1979)). The inverse spectral transformation is then seen to correspond to a symplectic map which transforms between manifolds for which the scattering data (S) and the functions appearing in $L(P)$ are the local coordinates. A particular choice of coordinates on S renders the pulled back symplectic form on P canonical. In this system of coordinates the evolution equations of the scattering data become the transformed Hamilton equations on S of the original nonlinear solvable equations written as Hamiltonian systems on P. The canonical coordinates on S then turn out to be of action-angle type.

This formulation of the inverse spectral method, besides its inherent beauty, has enabled several of the physically interesting nonlinear solvable equations to be quantised. For example the nonlinear Schrödinger equation (Kaup 1975) and the sine-Gordon equation (Faddeev and Takhtajan 1974) have been quantised in this way. This is a semi-classical procedure and agrees with results to first order in $\hbar$ obtained by using functional integral or other methods (Dashen et al 1974, 1975, Korepkin and Faddeev 1975 and, for example, Luther 1976).

In the rest of this section we shall exclude from consideration periodic problems and discrete problems and consider generalisations of the methods used in this book to generate other solvable nonlinear equations.

As was well known by mathematicians such as Painlevé, Garnier and Schlesinger who worked on the theory of differential equations at the beginning of this century, solvable nonlinear differential equations arise as the integrability conditions when a linear ordinary differential equation is deformed in such a way as to preserve some "characteristic" of the equation.

Thus solvable nonlinear equations in two independent variables which arise from iso-spectral deformations are obtained from equations of the form

$$Y_x(k) = P(k)\ Y(k) \tag{6.4-1}$$

where $Y(k) = Y(x,t,k)$ is an n-column vector in some function space subject to deformations

$$Y_t(k) = Q(k)\ Y(k) \tag{6.4-2}$$

which preserve the spectrum ($k_t = 0$ if $Y(k)$ is a solution). One easily checks that the integrability conditions $Y_{xt} = Y_{tx}$ give

$$P_t(k) - Q_x(k) - [P(k),\ Q(k)] = 0 \tag{6.4-3}$$

The derivatives in (6.4-3) are total derivatives for the matrix functions P and Q which in general will depend upon a set of functions $\{p_i(x,t),\ i = 1,..,m\}$ and their x-derivatives. The equations (6.4-3) are required to be k-independent. It is easy to see that the two examples of Chapters 3 and 6 fit into this scheme. Other types of deformation are possible which lead to solvable equations. In all cases the preserved quantities enable the deformation of the characteristic data associated with the problem to be simply calculated. For example in the isospectral problems the evolution of the scattering data $S_\pm$ can be easily obtained for the solvable equations. The solution for the problem at arbitrary values of the independent variables is then reconstructed by an inverse transform from this characteristic data. Thus the method adopted for the solution of a solvable nonlinear equation obtained by deformations which preserve the characteristic A is similar to the inverse spectral transform method (see Sections 4.2 and 6.2) and which we shall call an inverse A method. Similarily equation (6.4-2) is then called an iso-A deformation. We shall now briefly review some of the developments in this area which have occurred in the last decade.

6.4.1 Inverse Spectral Methods

The work on these methods has consisted principally in deriving the solvable equations associated with special types of $P(k)$ and $Q(k)$ for equations (6.4-1) and (6.4-2). The inverse method for the solvable equations for the most part is only treated in a cursory fashion although the soliton solutions and their analogues have been obtained in many cases.

Methods for Deriving Solvable Equations

(a) Direct substitution

Assume that P and Q are rational in k, substitute into (6.4-3) and require the resulting matrix equations to be consistent and k-independent.

Example
$$P = \frac{u_x}{k - k_1} \qquad Q = \frac{u_t}{k + k_1}$$

Then a solvable nonlinear equation from (6.4-3) is

$$2k_1 u_{xt} - [u_x, u_t] = 0$$

(Zakharov 1980, Zakharov and Shabat 1975, 1980, Zakharov and Mikhailov 1979).

(b) Techniques developed for specific forms of P

A large class of solvable equations are obtained when the P equation is resolvable into an eigenvalue problem

$$LY = kY$$

(b1) L, A pairs

For L assume the form

$$L = \ell_0 \frac{\partial^n}{\partial x^n} + \sum_{i=1}^{n} u_i \frac{\partial^{n-1}}{\partial x^{n-1}}$$

$u_i = u_i(x,t)$ and ℓ_0 is a constant matrix. Similarly assume that $A(\equiv -Q)$ has the expansion

$$A = a_0 \frac{\partial^m}{\partial x^m} + \sum_{i=1}^{m} v_i \frac{\partial^{m-1}}{\partial x^{m-1}}$$

$v_i = v_i(x,t)$ and a_0 is a constant matrix. The solvable equations are given by

$$L_t = [L, A] \tag{6.4-4}$$

which is the well known Lax formulation. If one imposes the conditions that the matrices $u_i \to \ell_i$, $v_i \to a_i$ "sufficiently fast" as $|x| \to \infty$ where ℓ_i and a_i are constant matrices we see that the corresponding operators $L_0 = L(x = \pm\infty)$, $A_0 = A(x = \pm\infty)$ satisfy

$$[L_0, A_0] = 0 \tag{6.4-5}$$

In fact A is uniquely determined by L and A_0.

Example $L = \ell_0\, \partial/\partial x + u_1$, $A_0 = a_0\, \partial/\partial x$ and assume $u_1 = [\ell_0, u]$ where $u = u(x,t)$. Then from (6.4-5) we have $[\ell_0, a_0] = 0$. The representations for L and A_0 then give that

$$A = a_0 \frac{\partial}{\partial x} + a_1 , \quad a_1 = [a_0, u]$$

$$u_{1_t} = [u_1, a_1] + \ell_0\, a_{1_x} - a_0 u_{1_x}$$

This equation contains the equations for the free movement of an n-dimensional rigid body (Manakov, 1977).

In the case of scalar Lax equations the u_i are real valued functions and

ℓ_0 can be scaled to unity. This case has been extensively studied by Gelfand and Dikii in a series of papers (1975 - 1978). Besides investigating the Hamiltonian structure of the associated evolution equations they show that A is determined completely by the resolvent of the operator $L, R_k = (L - kI)^{-1}$

(b2) Generalised Wronskian techniques

Calogero (1975) and Calogero and Degasperis (1976) are the originators of this technique which has been applied by them to the matrix form of the Scrődinger equation

$$LY = k^2 Y, \quad L = -\frac{\partial}{\partial x^2} + u \tag{6.4-6}$$

This method has also recently been extended to other systems. The method is similar to that developed by AKNS (1974) in that the explicit form of the evolution operator (Q) is not required. The matrix function $u \to 0$ "sufficiently fast" as $|x| \to \infty$. The asymptotic properties of the system are characterised by the scattering matrix (see Chapters 3 and 4) which can be constructed from the scattering data. For simplicity assume that the discrete spectrum is absent, then the reflection coefficient $R(k)$ and transmission coefficient $T(k)$ are defined by the linear relations which exist between the Jost solutions to the system

$$T(k)\phi(x,k) = \psi(x,-k) + R(k)\psi(x,k) \tag{6.4-7}$$

Calogero and Degasperis show by the inclusion of additional entire functions f and g and constant matrices M , N that it is possible to obtain from (6.4-6) two expressions which like the usual Wronskian relation, are independent of x. It is possible to deduce from these relations that if the reflection coefficient is to satisfy a linear equation

$$\begin{aligned} f_0(-4k^2)R_t &= f_n(-4k^2)[M_n, N] \\ &\quad + 2ik\, g_n(-4k^2)\{N_n, R\} + 2ik\, g_0(-4k^2)R \end{aligned} \tag{6.4-8}$$

then the solvable evolution equation is

$$\begin{aligned} f_0(L)u_t &= f_n(L)[M_n, u] \\ &\quad + g_n(L)G\, N_n + 2g_0(L) \end{aligned} \tag{6.4-9}$$

where f_i, g_i are arbitrary entire functions, M_n, N_n arbitrary constant matrices and

$$\begin{aligned} LY &= Y_{xx} - 2\{u, Y\} + G\int_x^\infty dy\, Y(y) \\ GY &= \{u_x, Y\} + [u, \int_x^\infty dy[u(y), Y(y)]] \end{aligned} \tag{6.4-10}$$

The brackets { , } [,] are respectively the anti-commutator and commutator.

Amongst the class of equations given by (6.4-10) the "boomeron equation"

$$\underline{V}_t(x,t) = \underline{b}.\underline{V}_x(x,t)$$

$$\underline{V}_{x_t}(x,t) = V_{xx}(x,t)\underline{b} + \underline{a}x\ \underline{V}_x(x,t) - 2\underline{V}_x(x,t)\ x\ (\underline{V}(x,t)x\underline{b}) \tag{6.4-11}$$

where $\underline{a}$, $\underline{b}$, $\underline{c}$ are constant vectors, has been particularly studied. This has soliton solutions which travel in from $x = +\infty$ and then "boomerang" back out again to $x = +\infty$.

(b3) Generalised AKNS techniques

Newell (1979) considered systems for which $P = kR + U$ where R is a constant trace free diagonal matrix and $U \to 0$ sufficiently fast as $|x| \to \infty$. It is easy to obtain by the procedure of Chapter 6 the relations

$$A^{-1}A_t = \int_{-\infty}^{\infty} \Phi^{-1}\ U_t\Phi\ dx$$
$$A^{-1}[C,A] = \int_{-\infty}^{\infty} \Phi^{-1}[C,U]\ \Phi\ dx \tag{6.4-12}$$

where C is a constant diagonal matrix, A is the scattering matrix connecting the Jost solutions $\Phi = \Psi A$ defined by their asymptotic properties at $x = \pm\infty$. If the entries in R are distinct then one can find from the P equation that there exists an operator D defined by

$$DH_0 = H_{0x} + [H_0,U] + [\int_x^{\infty}[H_0,U]dy,U]$$

where $H_0 = H - \text{diag}\,H$

$$D^A S_0^{ij} = k[R,S_0^{ij}]\quad , \quad \left(S_0^{ij}\right)_{\ell m} = \hat{\phi}_{i\ell}\phi_{mj}$$

and $\hat{\phi}_{il}$, ϕ_{mj} are respectively entries in Φ^{-1} and Φ. Define $\tilde{H}_{jk} = H_{jk}(\alpha_j - \alpha_k)^{-1}$ where diag $R = (\alpha_1,\ldots,\alpha_n)$ so that $DM = DH$. Then it is easy to see from (6.4-12) that if for an arbitrary entire function $\Omega(k)$, A satisfies the linear equation

$$(A^{-1}A_t)_0 = \Omega(k)(A^{-1}[C,A])_0 \tag{6.4-13}$$

that the corresponding solvable equations are given by

$$U_t = \Omega(\tilde{D})[C,U] \tag{6.4-14}$$

It is clear that the generalised Wronskian approach and the generalised AKNS method are very similar. The generalised Wronskian method gives the recursion operator L which occurs in the defining relation for the solvable nonlinear equations immediately but this has to be calculated in the AKNS method. Whether the work (and ingenuity!) in this is comparable to the work in solving the iterations of the generalised Wronskian method would seem to be a matter of personal preference. Note that the Bäcklund transformations which are given by the generalised Wronskian method can also be obtained by a modification of the AKNS approach (see Sections 4.3 and 6.3).

Many further generalisations of the methods are possible. Thus for example some of the α_i in R may be equal. If we take $R = i\,\mathrm{diag}\,(-2,1,1)$ and

$$U = \begin{pmatrix} 0 & Q & R \\ \alpha Q^* & 0 & 0 \\ \beta R^* & 0 & 0 \end{pmatrix} \qquad C = 9i\begin{pmatrix} -1 & 0 & 0 \\ 0 & 0 & 0 \\ 0 & 0 & 0 \end{pmatrix} \tag{6.4-15}$$

then we obtain the equations governing the propagation of the envelope of a polarised wave in a nonlinear medium (Manakov, 1974)

$$\begin{aligned} Q_t &= i(Q_{xx} - 2\alpha Q^2 Q^* - 2\beta\, RR^* Q) \\ R_t &= i(R_{xx} - 2\alpha QQ^* R - 2\beta\, R^2 R^*) \end{aligned} \tag{6.4-16}$$

The solvable nonlinear equations can be extended to higher dimensions by the inclusion of further variables into the formulism. One way of doing this is to add a linear differential operator, which acts on the additional variables, to the A operator of the L,A pair. Thus for example

$$A = \sum_{i=1}^{N} f_i(L)\frac{\partial}{\partial y_i} + A_1 \tag{6.4-17}$$

so that the solvable equation assumes the form,

$$L_t + [L,A_1] + \sum_{i=1}^{N} f_i(L)\frac{\partial L}{\partial y_i} = 0 \tag{6.4-18}$$

The analogous technique for the generalised Wronskian/generalised AKNS methods would be to replace the ∂_t operator in the evolution of the scattering data by an operator

$$\partial = \partial_t + \sum_{i=1}^{N} f_i(k,y)\frac{\partial}{\partial y_i}$$

where the f_i are entire functions of k. The equations generated in this function so far do not seem to have any physical significance. The other alternative is to include the additional operator in L. For example define

$$L_1 = a\frac{\partial}{\partial y} + L$$

then the equation defining the solvable can be written as $L_t - aA_y = [L,A]$. The case of a scalar operator with form

$$L = \partial_x^3 - \frac{3}{4}(u_x + 2u\,\partial_x) + \frac{3}{2}w + \nu^2\partial_x, \quad A = ib(\partial_x^2 - u) \tag{6.4-19}$$

after the change of variables $x \to ix$, $y \to t$, $t \to y$ leads to the equation of Kadomtsev and Petviashvili (1971)

$$u_{xt} = \frac{\partial^2}{\partial x^2}\left(-\nu^2 u + \frac{1}{4}u_{xx} + \frac{3}{4}u^2\right) + \frac{3}{4}\beta^2 u_{yy} \tag{6.4-20}$$

if we choose constants

$$\beta = -i\alpha \quad \text{and} \quad \alpha = i\beta^{-1}\left(\frac{2}{3}\right)^{\frac{1}{2}}$$

Further systems of this type arise from (L,A,B) triads for which the defining equations have the form

$$LY = 0, \; Y_t + AY = 0 \tag{6.4-21}$$

Then since $(L_t - LA)Y = 0$ and $ZLY = 0$ for certain operators Z it follows that the decomposition $Z = A + B$ results in the solvable equations having the form

$$L_t - [L,A] = BL \tag{6.4-22}$$

(Zakharov 1980, Manakov 1976).

The inclusion of the independent variable into the scattering problem leads to a further class of solvable equations (Newell 1979, Calogero and Degasperis 1978). The most interesting example of this kind is the Ernst equation of general relativity. Solutions to this equation determine empty axially symmetric space-times. The Ernst equations are

$$(\mathrm{Re}\,\mathcal{E})\nabla^2\mathcal{E} = \nabla^2\mathcal{E}.\nabla\mathcal{E} \tag{6.4-23}$$

where

$$\nabla^2\mathcal{E} = x_1^{-1}\frac{\partial}{\partial x_1}\left(x_1\frac{\partial\mathcal{E}}{\partial x_1}\right) + \frac{\partial^2\mathcal{E}}{\partial x^2} \quad \text{and} \quad \mathcal{E} = f + ig$$

The function g is the "twist potential", which is defined by $(-g_{x_2}, g_{x_1}) = -x_1^{-1}f\,(\omega_{x_1}, \omega_{x_2})$, where f and ω are coefficients which occur in the Lewis form of the canonical form for the metric

$$ds^2 = f(dt + \omega d\phi)^2 - f^{-1}\left[e^{2\gamma}\left(dx_1^2 + dx_2^2\right) + x_1^2 d\phi^2\right] \tag{6.4-24}$$

A linear deformation problem for the Ernst equation is given by Neugebauer and Kramer (1980), Harrison (1980), and Dodd and Morris (1982).

$$\begin{aligned} Y_z &= -\tfrac{1}{2}f^{-1}\begin{pmatrix} -i\gamma^{\frac{1}{2}}f_z & g_z(1 - i\gamma^{\frac{1}{2}}) \\ -g_z(1 + i\gamma^{\frac{1}{2}}) & i\gamma^{\frac{1}{2}}f_z \end{pmatrix} Y \\ Y_{\bar{z}} &= -\tfrac{1}{2}f^{-1}\begin{pmatrix} i\gamma^{-\frac{1}{2}}f_{\bar{z}} & g_{\bar{z}}(1 + i\gamma^{-\frac{1}{2}}) \\ -g_{\bar{z}}(1 - i\gamma^{-\frac{1}{2}}) & -i\gamma^{-\frac{1}{2}}f_{\bar{z}} \end{pmatrix} Y \end{aligned} \tag{6.4-25}$$

where $\gamma = (\bar{z}+ik)/(z-ik)$, $z = x_1 + ix_2$. Solutions to (6.4-24) have been determined by Zakharov and Belinskii (1978) by an analysis of an operator bundle which is related to (6.4-25) (Dodd and Morris, 1982)

$$Y_z - \frac{2\eta}{(2i\eta - (z+\bar{z}))} \cdot Y_\eta = - \frac{2(z+\bar{z})CY}{(2i\eta - (z+\bar{z}))}$$

$$Y_{\bar{z}} + \frac{2\eta}{(2i\eta + (z+\bar{z}))} \cdot Y_\eta = \frac{2(z+\bar{z})CY}{(2i\eta + (z+\bar{z}))} \qquad (6.4\text{-}26)$$

where $C = \frac{1}{2}h,_z \omega h^{-1}$, $C = \frac{1}{2}h,_z \omega h^{-1}$ and h is the matrix

$$\begin{pmatrix} f & \omega f \\ f & \omega^2 f - x_1^2 f^{-1} \end{pmatrix}$$

As a final example to indicate the richness and diversity of solvable nonlinear equations we present the intermediate equation describing a stratified fluid with finite depth (intermediate equation in the sense that the shallow water limit is the KdV and the deep water limit is the Benjamin-Ono equation),

$$Q_t + 2QQ_x + T(Q_{xx}) = 0$$

where

$$T(f) \equiv P \int_{-\infty}^{\infty} \left[-\frac{1}{2\delta} \coth\left(\frac{\pi(x-y)}{2\delta} \right) + \frac{1}{2\delta} \operatorname{sgn}(x-y) \right] f(y)dy \qquad (6.4\text{-}27)$$

Bäcklund transformations, conserved quantities and the inverse method for this equation has been developed by Kodama et al. (1982). One final comment; the recursion operator (see above) does not necessarily define all the solvable equations associated with a given operator L in an (L,A) pair. Thus for example the equations of simple harmonic generation (Kaup 1978) have the P,Q operators (see also Chapter 8).

$$P = \begin{pmatrix} -ik & Q \\ Q^* & ik \end{pmatrix} \qquad Q = \frac{1}{ik} \begin{pmatrix} R^*R & R^2 \\ -R^{*2} & -RR^* \end{pmatrix} \qquad (6.4\text{-}28)$$

and the equations are $R_x = QR^*$, $Q_t = -2R^2$. The equations are dispersionless and so clearly lie outside the AKNS scheme.

A very important topic which we have omitted from this brief survey is the reduction problem. This occurs whenever the entries in P and Q are related in some way. In general this requires the matrix functions P and Q to admit a symmetry or involution. We refer the reader to the article by Mikhailov (1981) and Calogero and Degasperis (1981) for a study of this problem.

Solution Techniques

(a) The Inverse Method

This is the method used in the book. It is applicable whenever the P and Q equations are resolvable into an L,A pair. For most systems the method has only been applied formally. That is necessary and sufficient conditions on the scattering data for the existence of a unique solution (that is for a unique determination of the coefficient functions u_i in L) have not yet been derived. Similar comments of course also apply to the initial value problems for the solvable equations of these systems. A scalar third order problem has been investigated by Kaup (1980) A general third order scalar problem and an nth order first order system have been investigated by Caudrey (1980, 1982).

There is a variety of methods used to reconstruct the functions u_i from the scattering data.

(a1) Solving a Marchenko equation

The solutions of the nonlinear solvable equation are given in terms of the solution to a Fredholm equation called the Marchenko equation (or fundamental equation in the Russian literature), see Chapters 4 and 6.

(a2) Solving a Riemann-Hilbert problem

Zakharov and Shabat (1972) solved the inverse spectral problem for the nonlinear Schrödinger equation by reconstructing a piecewise analytic function $H(k)$ from its discontinuity on the real axis

$$\phi(\xi) = H(\xi + i0) - H(\xi - i0), \quad \phi(\xi) = R_+(\xi)e^{i\xi x}\psi(x,\xi) \tag{6.4-29}$$

and the residues at the poles of $H(k)$. This results in a set of singular integral equations for the Jost solutions (both proper and improper eigenfunctions) the solution to which determine the solution of the nonlinear Schrödinger equation, (a Fourier transform applied to these equations results in the Marchenko equations for this method, see Section 6.2). Recently this approach has been applied by Zakharov and Shabat (1980), Zakharov and Belinskii (1978) (see also Zakharov 1980) as a solution technique for operator bundles (that is the general equations satisfied by the matrix functions P and Q introduced earlier). Further comments on this method are given under (b) and (c) below.

(b) Dressing Operator Techniques

Zakharov and Shabat (1974) have developed a formal operator technique for obtaining solutions to equations arising from L,A pairs. This includes the case when the operators involve more than two independent variables, briefly outlined earlier. Essentially the method can be viewed as a formal inverse spectral transform. Introduce transformation operators

where

$$K_{\pm}Y = \pm \int_x^{\pm\infty} K_{\pm}(x,y)Y(y)dy, \ \Psi = (I + K_+)\Psi_0, \ \Phi = (I + K_-)\Psi_0 \tag{6.4-30}$$

and Ψ_0 is a fundamental matrix solution of the eigenvalue problem for L_0, the "undressed" or "bare" operator. Clearly the dressed operator L for the eigenfunctions Ψ, Φ is given by $L = (I+K_{\pm})^{-1} L_0(I+K_{\pm})$ assuming the inverse operators exist. Since Ψ and Φ are fundamental solutions there exists a data matrix A which we assume has the form $A = I + R$ which relates the solutions $\Phi = \Psi(I+R)$. It follows from this that we can hypothesise the existence of a Fredholm operator F such that

$$I + F = (I + K_+)^{-1}(I + K_-)$$

where

$$F\Psi_0 = \Psi_0 R \quad \text{and} \quad FY = \int_{-\infty}^{\infty} F(x,y)Y(y)dy \tag{6.4-31}$$

From amongst the class of operators F defined by (6.4-31) those operators are selected which commute with bare differential operators M_0. This can easily be shown to imply that the dressed operator M is uniquely defined and is also a differential operator.

The differential operators M_0 define the "evolution" of the Fourier transform of the scattering data in our previous language. Thus choosing $M_0 = \partial/\partial t + A_0$ results in a linear partial differential equation defining the evolution of F with respect to t.

The dressed operators M_i and L (there may be more than one differential operator which commutes with F) are uniquely defined by their bare operators via the commutation relation with F and the kernels of the transformation operator $M_i(I+K_{\pm})=(I+K_{\pm})M_{0i}$. Furthermore for $y > x$, (6.4-31) gives the equation

$$F(x,y) + K_+(x,y) + \int_x^{\infty} K_+(x,s)F(s,y)ds = 0 \tag{6.4-32}$$

which can be viewed as a Volterra equation defining F given K_+ or a Fredholm equation for K_+ given F. The Cauchy problem for a solvable equation of the form $L_t - aA_u = [L,A]$ for which $M_1 = \partial/\partial t + A$ and $M_2 = a\,\partial/\partial t + L$ may be solved by the following scheme

$$u_i(0) \xrightarrow{I} K_+(0) \xrightarrow{II} F(0) \xrightarrow{III} F(t) \xrightarrow{IV} K_+(t) \xrightarrow{V} u_i(t)$$

Figure 6-4: The Cauchy problem for $L_t - aA_u = [L,A]$

I Determine $K_+(0)$ by solving the Goursat problem

$$aK_{+u}(0) + L(0)K_+(0) - L_0^A K_+(0) = 0$$

II Solve (6.4-32) for $F(0)$

III Solve the linear evolution equation for $F(t)$

IV Solve (6.4-32) with $F(t)$ for $K_+(t)$

V Determine $L(t)$ from

$$aK_{+u}(t) + L(t)K_+(t) - L_0^A K_+(t) = 0$$

Note: Steps IV and V are the inverse transform. L_0^A acts on the second argument (y) of K_+.

Zakharov and Shabat (1980) and Zakharov (1980) have also shown how bare operators $P_0(k)$, $Q_0(k)$ can be dressed to obtain the operator bundles $P(k,x,t)$, $Q(k,x,t)$. Similar ideas can be used to obtain Bäcklund transformations.

(c) Bäcklund Transformations

In Chapters 4 and 6 we have shown how new solutions can be obtained from a given solution of the Schrödinger or AKNS-ZS system. This works well when the solvable equation is generated by the L operator, that is A is determined by L. In general this is not the case and a method due to Zakharov and Shabat (1980) and Zakharov and Mikhailov (1979) can be used. Suppose that we have a given solution (P_0, Q_0) to (6.4-3) and wish to determine a new solution (P,Q).

If P and Q are to satisfy the same equation then we will assume (for simplicity) that we are dealing with a Riemann problem with zeros. In this case we will further assume that we are dealing with a class of special solutions for which the Riemann problem has the solution

$$\Psi(k) = \left(I + \frac{R}{(k-\mu)}\right)\Psi_0(k) \tag{6.4-33}$$

where R is a singular matrix function and Ψ is a solution of (6.4-3) for P and Q. This assumes a particular normalisation which is allowable for the type of equations with which we are dealing. The Bäcklund transformation can now be written

$$\begin{aligned} P &= H_x H^{-1} + HP_0H^{-1} \\ Q &= H_t H^{-1} + HQ_0H^{-1} \end{aligned} \tag{6.4-34}$$

where $H = I + R/(k-\mu)$.

A beautiful example of this technique is given in Zakharov and Belinskii (1978).

(d) The Hirota Technique

Examples of this method have been given in the introductory chapter of this book (Hirota 1976, 1980). Recently (Jimbo and Miwa (1981) and references therein), Jimbo and Miwa have shown that the Hirota technique has a deep theoretical significance. Hirota's dependent variables turn out to be the τ-functions introduced by these authors for the analysis of monodromy preserving deformations (see the next section).

6.4.2 Other Inverse Methods

There is a close connection between solvable nonlinear evolution equations of the class we have been studying and nonlinear ordinary differential equations of Painlevé type. These equations are characterised by the absence of movable critical points (branch points and essential singularities) (Ince, 1956). Lamb (1977), for instance, showed that the π-pulse in the amplifier (Chapter 7) was essentially given by solutions to the similarity form of the sine-Gordon equation which is a special case of the Painlevé type III equation. Ablowitz and Segur (1977) showed that the similarity solution which is governed by the Painlevé II equation played a role in the asymptotic solution of the KdV equation. Indeed it appears that the slowly varying similarity solution governs the main behaviour of the asymptotic solution (see Sections 4.3 and 6.3). In a series of papers Ablowitz et el. (Ablowitz, Ramani and Segur 1978, 1980) showed that special cases of the five Painlevé transcendents arose from the similarity form of well known solvable evolution equations and so by a change of variables could be investigated by the inverse spectral method. However this method did not allow the most general solution to the equations because they imposed boundary conditions of rapid decay on the solutions. Curiously it was found that the problem had been investigated by Garnier (1912) and Schlesinger (1912). As an example consider the MKdV equation, $Q_t - 6Q^2Q_x + Q_{xxx} = 0$. This can be solved by the inverse spectral method with the P and Q operators

$$P = \begin{pmatrix} -i\zeta & Q \\ Q & i\zeta \end{pmatrix}, \quad Q = \begin{pmatrix} -4\zeta^3 - 2i\zeta Q^2 & 4^2Q + 2i\zeta Q_x - Q_{xx} + 2Q^3 \\ 4\zeta^2 Q - 2i\zeta Q_x - Q_{xx} + 2Q^3 & 4\zeta^3 + 2i\zeta Q^2 \end{pmatrix} \tag{6.4-35}$$

The similarity variable for this equation is $z = x/(3t)^{1/3}$. Introduce a new variable $k = \zeta(3t)^{1/3}$, then with a slight modification (the inclusion of the ν/k terms in Q) the new deformation problem can be written

$$Y_z = PY \quad , \quad Y_k = QY \tag{6.4-36}$$

where $P = \begin{pmatrix} -ik & U \\ U & ik \end{pmatrix}$, $Q = \begin{pmatrix} -4ik^2 - i(z + 2U^2) & 4kU + \frac{\nu}{k} + 2ikU_z \\ 4kU + \frac{\nu}{k} - 2iU_z & 4ik^2 + i(z + 2U^2) \end{pmatrix}$

and the solvable equation which results from the integrability conditions is the Painlevé II equation

$$U_{zz} - zU - 2U^3 + \nu = 0 \tag{6.4-37}$$

The question arises what is being preserved by the deformation problem (6.4-36). For this problem the Q equation is treated as the fundamental equation and the P equation defines the deformation. The Q equation has a regular singularity at $k = 0$ and an irregular singularity at $k = \infty$. Connected with the irregular singularity is the Stoke's phenomenon. Fundamental solutions to the Q equation having the same asymptotic expansions at $k = \infty$ are defined on continuous regions of a Riemann surface (in this case the Riemann sphere). Two such solutions defined on adjacent regions are related by a matrix function called a Stoke's multiplier. The fundamental matrix solution to the Q equation at $k = 0$ is multi-valued. On the

Riemann surface ($k = 0$ is a branch point) the relation between two consecutive branches is given by $Y(k\exp(2\pi i)) = Y(k)\,M$ where M is the monodromy matrix connected with $k = 0$. These fundamental solutions defined by asymptotic expansions at $k = 0$ and $k = \infty$ respectively are related by a connection matrix, since each is a linearly independent matrix solution of the Q equation. One can show that the monodromy matrix can be defined in terms of the Stoke's multipliers and the connection matrices and it is these which constitute the monodromy data for the equation. The problem of reconstructing the Q equation from the monodromy data constitutes a generalisation of Riemann's problem from the regular to the irregular singularity case due to Birkhoff (1913). Flaschka and Newell (1980) have shown that the monodromy data is a constant when the Q equation is deformed by the P equation defined in (6.4-36) (that is it is independent of z). In this paper they solve the Painlevé II equation by the inverse monodromy method outlined above. The inverse problem is solved by reconstructing $Y(k)$ at arbitrary values of z from the monodromy data. The solution is a singular integral equation for $Y(k)$ from which U can be reconstructed.

In three very interesting papers Jimbo, Miwa and Ueno (1981) have investigated the general problem when Q has arbitrary rational coefficients and singularities. In addition they show that the single P equation can be replaced by an arbitrary number of equations with a corresponding number of deformation parameters. In this way they are able to incorporate the isospectral deformations within the isomonodromy framework. The papers by Flaschka and Newell (1981) and Jimbo and Miwa (1981) contain further extensions of these ideas.

6.5 Notes

Section 6.1

1. We use a different boundary condition to define   than the one used by AKNS (1974). Their boundary condition arose from their method of generalising the work of Zakharov and Shabat (1972). It is easy to translate the formulae for the two different conditions. In most cases one only needs to remember that $\bar{a}_{AKNS} = -\bar{a}$ and $\bar{\phi}_{AKNS} = -\bar{\phi}$. The definition we have used means that most formulae are converted into their "dual" form under the " - " operation $(\overline{(f)} = \bar{f},\ \overline{(\bar{f})} = f)$. By the dual form of a formula we mean the expression for $(\phi, \bar{\phi}, \psi, \bar{\psi})$ which has the same functional form as the one for $(\bar{\phi}, \phi, \bar{\psi}, \psi)$. This includes formulae involving differentiation with respect to x, t and k, (e.g. (6.1-34)).

2. A generalised spectral measure in a Hilbert space H is a function P which has the properties

(1a) P is specified on some class $D(P)$ of Borel subsets in the complex k-plane.

(1b) The class $D(P)$ contains any Borel subset of each of its elements.

(1c) The class $D(P)$ contains the union of any pair of its elements.

(2a) The values of P are linear operators $P(\Delta)$, $\Delta \in D(P)$ specified on the whole space H and mapping continuously into itself

(2b) $P(\Delta_1)\ P(\Delta_2) = P(\Delta_1 \cap \Delta_2\ ), \Delta_1, \Delta_2 \in D(P).$

(2c) For any decomposition of a set $\Delta \in D(P)$ into pairwise disjoint Borel sets $\Delta_1 \Delta_2, \ldots,$ the series $\Sigma P(\Delta_n)$ converges strongly to $P(\Delta)$.

(2d) If $f \in H$ and $P(\Delta)f = 0$ for all $\Delta \in DP$, then $f = 0$.

(2e) If $f \in H$ and $[P(\Delta)]^* f = 0$ for all $\Delta \in D(P)$ then $f = 0$.

For the operator L the Borel subsets $D(P)$ of the complex k-plane are those whose colsure do not contain the spectral singularities k_j, $\bar{k}_m$, $j=1,\ldots,N$, $m=1,\bar{N}$. For every $\Delta \in D(P)$ and for every $F \in L^2_{(2)}(\mathbb{R})$ define

$$P(\Delta)F(x) = \frac{1}{2\pi}\int_{\Delta\cap]-\infty,\infty[} dk\,(\frac{b}{a}(k)\psi(x,k)\psi(F,k)$$

$$- \frac{\bar{b}}{\bar{a}}(k)\psi(x,k)\psi(F,k) + \psi(x,k)\psi(F,k) - \psi(x,k)\psi(F,k))$$

$$+ \sum_{\substack{k_j \varepsilon \Delta \\ j=1,\ldots,M}} \{(\frac{d}{dk})^{m_j-1} p_j(k)\psi(x,k)\psi(F,k)\}_{k=k_j} + \sum_{\substack{k_j \varepsilon \Delta \\ j=1,\ldots,M}} \{(\frac{d}{dk})^{\bar{m}_j-1} \bar{p}_j(k)\psi(x,k)\psi(F,k)\}_{k=k_j}$$

The function P given by this formula is the generalised spectral measure in $L^2_{(2)}(\mathbb{R})$.

For every finite function $F \in L^2_{(2)}(\mathbb{R})$ we can define the L-Fourier transforms $\psi(F,k)$ and $\bar{\psi}(F,k)$. For any pair of finite functions $F, G \in L^2_{(2)}(\mathbb{R})$ the function $\Psi_{FG}(k) = (\psi(F,k)\,\psi(\tilde{G},k),\ \ \psi(F,k)\,\bar{\psi}(\tilde{G},k), \bar{\psi}(F,k)\,\bar{\psi}(\tilde{G},k), \psi(\bar{F},k)\,\psi(\tilde{G},k))$ where $\tilde{G} = (G_1, G_2)$ belongs to a linear space Z. Marchenko's theory establishes for regular eigen functions (those defined by regular boundary conditions at $x = 0$) the existence of a continuous linear functional R on Z which generalises the Parseval relationship. For our case we assume that this is true also

$$\int_{-\infty}^{\infty} G(x).F(x)dx = R(\Psi_{FG})$$

where the functional R is the generalised spectral function. By analogy with the self-adjoint case we see from the genralised spectral measure defined above that

$$S_+ = \{R_+, \bar{R}_+, k_j, P_{+j}(x), \bar{k}_\ell, \bar{P}_{+\ell}(x),\ j=1,\ldots,M, \ell=1,\ldots,\bar{M}\}$$

is the scattering data for L. The normalisation polynomial $P_j(x)$ has degree m_j-1. It is easy to see from the spectral measure that the coefficient of x^m in $P_{+j}(x)$, $0 \leqslant m \leqslant m_j-1$ is the coefficient of $(\Psi(x,k)\ \Psi(F,k))_{(m)k}$ in the expansion.

<u>Section 6.3</u>

1. As mentioned in 6.1 when $\Omega \neq -\bar{\Omega}$ the inverse problem is still solvable but in this case a and $\bar{a}$ are not constants of the motion. In this case then the system of equations cannot be written as a Hamiltonian system. A particular example of physical significance is afforded by the SIT (self induced transparency) equations of nonlinear optics (Ablowitz, Kaup and Newell, 1974). A more general equation of this

type is

$$\begin{pmatrix} R \\ -Q \end{pmatrix}_t + 2M\begin{pmatrix} L_1^A \end{pmatrix}\begin{pmatrix} R \\ Q \end{pmatrix} = \int_{-\infty}^{\infty} dk\, g(k)\, (\theta^A)^T$$

where

$$\Omega(k) = H(k) + M(k), \quad \bar{\Omega}(k) = \bar{H}(k) + M(k)$$

$$g(k) = \frac{2}{\pi}(\bar{\Omega}(k) - \Omega(k)) \quad \text{and } \theta \text{ is defined in 6.1.}$$

In the cases we considered in the text for which $\Omega = -\bar{\Omega}$ it was automatic that the discrete spectrum was a constant of the motion. However as shown by Kaup and Newell (1979) it is possible for solvable equations to allow moving eigenvalues. In this case the dispersion relations are singular. The novel feature of such equations is that some of the singularities in the dispersion relations coincide with $\sigma(\underline{L})$. A specific example is afforded by the system for which $\Omega(k) = -\bar{\Omega}(k) = \dfrac{M_j}{k-k_j} + \dfrac{M_j}{k-\bar{k}_j}$. Then the equations of motion are

$$\begin{pmatrix} R \\ -Q \end{pmatrix}_t^T = -4iM_j\theta^A(x,k_j) - 4i\bar{M}_j\theta^A(x,\bar{k})$$

together with

$$\begin{pmatrix} L_1^A - k_j \end{pmatrix}\begin{pmatrix} \theta^A(x,k_j) \end{pmatrix}^T = \frac{1}{2i}\begin{pmatrix} R \\ Q \end{pmatrix}$$

$$\begin{pmatrix} L_1^A - \bar{k} \end{pmatrix}\begin{pmatrix} \theta^A(x,\bar{k}) \end{pmatrix}^T = \frac{1}{2i}\begin{pmatrix} R \\ Q \end{pmatrix}$$

The evolution of the scattering data is given by

$$R_{-t} = \Omega(k)R_-, \quad \bar{R}_{-t} = -\Omega(k)\bar{R}_-$$

$$D_{it}^- = \Omega_i D_i^- \qquad k_{it} = 0 \, (k_i \neq k_j)$$

$$D_{it}^- = \bar{\Omega}_i \bar{D}_i^- \qquad \bar{k}_{it} = 0 \, (\bar{k}_i \neq \bar{k}_j)$$

$$D_{jt}^- = \left(\frac{\bar{M}_j}{k_j - \bar{k}_j}\right) D_j^- + M_j D_j{}', \qquad k_{jt} = M_j$$

$$\bar{D}_{jt}^- = \left(\frac{M_j}{\bar{k}_j - k_j}\right)\bar{D}_{j^-} + \bar{M}_j\bar{D}_{j^-}, \qquad \bar{k}_{jt} = \bar{M}_j$$

2. There has been an extensive investigation of the Painleve transcendents in the last few years. Much of it has repeated the studies carried out by mathematicians at the turn of the century but quite forgotten until recently. There are in fact deep connections between the work of these classical workers on solvable ordinary differential equations and solvable nonlinear partial differential equations. This material is briefly reviewed in Section 6.4.

3. The Bäcklund transformation arose as a generalisation of contact transformations. The general theory of contact transformations was developed by Lie (see e.g. Forsyth). These are transformations which preserve the contact module associated with an equation. For equations of first order in two independent variables the contact module is generated (locally) by the one form

$$\theta = dz - pdx - qdt$$

on the manifold with local coordinated (x,t,p,q). When the contact module is restricted to a solution of the equation

$$F(z,z_x,z_t) = 0$$

the contact module is annihilated (in local coordinates $z = z(x,t)$, $p = z\ (x,t)$, $q = z\ (x,t)$). A contact transformation is defined by functions

$$\begin{aligned} x^1 &= X(x,t,z,p,q) \\ t^1 &= T(x,t,z,p,q) \\ z^1 &= Z(x,t,z,p,q) \\ p^1 &= P(x,t,z,p,q) \\ q^1 &= Q(x,t,z,p,q) \end{aligned} \qquad (6.5\text{-}1)$$

which are a symmetry of the contact module. Thus

$$(dZ\text{-}PdX\text{-}Qdt) = \theta(x,t,z,p,q)\ (dz\text{-}pdx\text{-}qdt)$$

The solution of this problem is ultimately given by solving a Mayer system. This theory which arose from surface theory (first order contact of surfaces) provided a general transformation theory for first order partial differential equations. Unfortunately extensions to higher order equations produced only trivial results.

The generalisation due to Bäcklund was to allow transformations of the form

$$\begin{aligned} x^1 &= X(x,t,z,p,q,z^1) \\ t^1 &= T(x,t,z,p,q,z^1) \\ p^1 &= P(x,t,z,p,q,z^1) \\ q^1 &= Q(x,t,z,p,q,z^1) \end{aligned} \qquad (6.5\text{-}2)$$

and require that the transformed contact form be completely integrable. This condition results in a Monge-Ampere equation (essentially a second order partial differential equation linear in its highest order derivatives). Thus (6.5-2) effects a transformation between the equation satisfied by z^1 and if this Monge-Ampere equation does not involve z, and its derivatives an equation satisfied by z^1. If (6.5-2) is invertible then we also can define a transformation from z^1 to the z variables. Generalisations to higher dimensions and more than one variable are easy to obtain. Usually $x^1 = x$ and $t^1 = t$ so that (6.5-2) consists of two equations.

Thus the auto-Bäcklund transformation for the sine-Gordon equation (the equation satisfies by z and z^1 is the same) $z_{xt} = \sin z$ is

$$z^1_x = z_x + 2a \sin\tfrac{1}{2}(z^1 + z)$$

$$z^1_x = -z_t + 2a^{-1}\sin\tfrac{1}{2}(z^1 - z)$$

Further details and extensions can be found in "Bäcklund transformations" (Ed R Muira 1974) and the article by Dodd and Morris (1979).

6.6 Problems

Section 6.1

1.

Show that the Lax operator A for the sine-Gordon equation $U_{xt} = \pm\sin U$ is given by

$$A = \pm\tfrac{1}{4}\begin{pmatrix} \int^x dy\, \cos\tfrac{1}{2}(U(x)+U(y)), & -\int^x dy\, \sin\tfrac{1}{2}(U(x)+U(y)) \\ \int^x dy\, \sin\tfrac{1}{2}(U(x)+U(y)), & \int^x dy\, \cos\tfrac{1}{2}(U(x)+U(y)) \end{pmatrix}$$

in the ZS-AKNS formulation.

2.

When Q and R have compact support then so does $P = (|Q| + |R|)$. Following closely the method of succesive approximations used to establish Theorem 6-1 prove Corollary 6-1-2.

3.

Show that $\phi(x,k)e^{ikx} = \binom{1}{0}$ and R obey Theorem 6-1, but are not necessarily differentiable.

4.

Prove Theorem 6-6 by adapting the proof of Theorem 6-1.

5.

By iterating the equations

$$K_{+1}(x,y) = -\tfrac{1}{2}Q(\tfrac{1}{2}(x+y)) - \int_x^{\frac{1}{2}(x+y)} Q(s)\, K_{+2}(x,y+x-s)\,ds$$

$$K_{+2}(x,y) = 1 - \int_x^{\infty} K_{+1}(s,y-x+s)R(s)\,ds$$

obtain the estimates in equation (6.1-60)

6.

The results of Lemma 6-10 can be established directly from the equation (6.1-13) after substituting in one of the particular functional relationships given in the lemma.

7.

For the self-adjoint case $R = Q^*$ the operator $L \equiv L^S = i\begin{pmatrix} \partial/\partial x & -Q \\ Q^* & -\partial/\partial x \end{pmatrix}$. The standard theory of self-adjoint linear differential operators then applies (see Section 3.4). In particular, because L has no discrete spectrum Parseval's equality assumes a particularly simple form

$$\int_{-\infty}^{\infty} (|v_1(x)|^2 + |v_2(x)|^2)dx = \int_{-\infty}^{\infty} (|u_1(k)|^2 + |u_2(k)|^2) d(\frac{k}{2\pi})$$

where $V \in L_2^{(2)}(\mathbb{R})$ $(\equiv L_x)$ and

$$u_1(k) = \langle V, {}_k m^- \rangle_x \qquad u_2(k) = \langle V, {}_k P^- \rangle_x$$

$\langle .,. \rangle_x$ is the L inner product and

$${}_k m-(x) = \frac{\phi(x,k)}{a(k)}, \qquad {}_k p-(x) = \frac{\psi(x,k)}{a(k)}$$

The vectors U(k) belong to the Hilbert space L_σ of C^2 valued functions on $\mathbb{R}$ which are square integrable with respect to the measure $d\sigma = \frac{1}{2\pi}\, dk$.

8.

For $V \in L_x$ let $U(k) \in L_{\sigma O}$ be the $L_O^S \equiv L^S$ $(Q = 0,\ R = 0)$ representative. Define the Møller operator U_- by

$$U_-V(x) = \frac{1}{2\pi} \int_{\infty}^{\infty} (u_1(k)\,{}_x m^-(k) + u_2(k)\,{}_x p^-(k))dk$$

where ${}_x p(\bar{k}) = {}_k p^-(x)$, ${}_x m^-(k) = {}_k \bar{m}(x)$ are defined in (6.1-10). Since $U_- = U_+ S$ where S is the unitary scattering operator show using the relationship between the fundamental solutions

$$\Phi = \Psi A$$

that the L_O^S representative of S is

$$\tilde{S} = \begin{pmatrix} \frac{1}{a} & \frac{-b^*}{a} \\ \frac{b}{a} & \frac{1}{a} \end{pmatrix}$$

The eigenfunctions are $m^+ = \frac{\Psi}{a}$, $p^- = \frac{\Phi}{a}$.

9.

Define h by the principal branch of the expression $e^h = \phi_1 e^{ikx}$. Then the boundary conditions satisfied by h are $h \to 0$ as $x \to -\infty$ and $h \to \ell na$ as $x \to +\infty$. From (6.1-13) show that h satisfies the equation

$$2ik\, h_x = -QR + Q \frac{\partial}{\partial x}(Q^{-1} h_x) + (h_x)^2$$

Show that this equation admits the asymptotic expansion

$$h_x = \sum_{n=1}^{\infty} \frac{p_n}{(2ik)^n} \quad \text{as} \quad |k| \to \infty$$

and that

$$p_1 = -QR$$

$$p_{n+1} = Q \frac{\partial}{\partial x}(Q^{-1} p_n) + \sum_{j+k=n} p_j p_k \qquad n = 1,2,\ldots,$$

Integrating the asymptotic expansion (which preserves the uniformity of the expansion) show that

$$\ell na = \sum_{n=1}^{\infty} \frac{1}{k^n} \frac{1}{(2i)^n} \int_{-\infty}^{\infty} p_n dx = -\sum_{n=1}^{\infty} \frac{1}{k^n} C_n \qquad \text{as } |k| \to \infty$$

Calculate the first few terms in the expansion

$$C_1 = \frac{1}{2i} \int_{-\infty}^{\infty} Q(x) R(x) dx \qquad C_2 = \frac{1}{(2i)^2} \int_{-\infty}^{\infty} Q(x) R_x(x) dx$$

$$C_3 = ?, \quad C_4 = ?$$

10.

Show that the equation

$$Q_{tx} - 2Q \int_x^{\infty} (Q^2)_+ ds \pm Q = 0$$

is equivalent to the sine-Gordon equation

$$V_{xt} = \pm \sin V$$

$V_x = -2Q$ provide $4(Q_t)^2 \leqslant 1$. Start by showing that the equation can be integrated once using Q_t as an integrating factor

$$Q_t^2 + \left(\int_x^{\infty} (Q^2)_t\, ds\right) \pm \left(\int_x^{\infty} (Q^2)_t\, ds\right) = 0$$

Solve for $y = \int_x^{\infty} (Q^2)_t ds)^2$ and then deduce the result. Obtain the equation

$$Q_t - 6Q^2Q_x - Q_{xxx} + \tfrac{1}{2}\alpha\sin(-2\int_x^\infty Q ds) = 0$$

from an analogous treatment of the equation

$$Q_{xt} - 2Q \int_x^\infty (Q^2)_t ds - \alpha Q + Q_{4x}$$
$$+ 10\, Q^2 Q_{2x} + 10Q\, Q_x^{\,2} + 6Q^5 = 0$$

<u>Section 6.3</u>

<u>1.</u>

When $R = -Q$ show that the scattering problem (6.1-13) can be written as

$$-z_{xx} = Pz = k^2 z$$

where $z = y_1 - iy_2$ and $P = iQ_x - Q^2$. This is just the isospectral Schrödinger equation of Chapters 3 and 4. Thus if P satisfies one of the equations solvable by this method $P = iQ_x - Q^2$ is a Miura transformation (Bäcklund tranformation) from the solvable equation satisfied by Q to this equation. Miura transformations are Bäcklund transformations with the property that $Q \to P$ defines a map (it is one to one) whereas the inverse transformation $P \to Q$ is one to many. Show that when $R = -Q$ that the solvable equations of the ZS-AKNS scattering problem are given by a scalar operator equation. First notice that

$$L_1^A(R = -Q) = \frac{1}{2i}\begin{pmatrix} \frac{\partial}{\partial x} - 2Q\int_x^\infty dsQ & \\ 2Q\int_x^\infty dsQ & -\frac{\partial}{\partial x} + 2Q\int_x^\infty dsQ \end{pmatrix} \equiv \begin{pmatrix} \alpha & \beta \\ -\beta & -\alpha \end{pmatrix}.$$

Thus

$$L_1^A = \sigma_3\gamma_1(\alpha+\beta) + \sigma_3\gamma_2(\alpha-\beta)$$

where

$$\gamma_1 = \frac{1}{2}\begin{pmatrix} 1 & 1 \\ 1 & 1 \end{pmatrix} \qquad \gamma_2 = \frac{1}{2}\begin{pmatrix} 1 & -1 \\ -1 & 1 \end{pmatrix}$$

Then we have

$$\left(L_1^A\right)^{2n} = \gamma_2[(\alpha+\beta)(\alpha-\beta)]^n + \gamma_1[(\alpha-\beta)(\alpha+\beta)]^n$$

$$\left(L_1^A\right)^{2n+1} = \sigma_3\gamma_1(\alpha+\beta)[(\alpha-\beta)(\alpha+\beta)]^n + \sigma_3\gamma_2(\alpha-\beta)[(\alpha+\beta)(\alpha-\beta)]^n$$

Thus provided Ω is odd $\Omega(k) = ikD(k^2)$ we get from (6.1-112) that

$$Q_t + D(G)Q_x = 0$$

$$\text{where} \quad G[Q] = \frac{1}{4}\frac{\partial^2}{\partial x^2} - Q^2 + Q_x \int_x^{\infty} ds\, Q$$

which defines the class of solvable equations in the case $R = -Q$. Show that

$$F[Q^2 - iQ_x]\left(\left(2Q - i\frac{\partial}{\partial x}\right)y(x,t)\right)$$

$$= \left(2Q - i\frac{\partial}{\partial x}\right)\hat{G}[Q]\,y(x,t)$$

$$\text{where} \qquad \hat{G}[Q] \equiv G[Q] + Q_x \int_{-\infty}^{\infty} ds Q$$

$$\text{and} \quad F[Q] \equiv -\frac{1}{4}\left(\frac{\partial}{\partial x^2} - 4Q + 2Q_x \int_x^{\infty} ds\right)$$

F is the recursion operator or generator of the solvable equations associated with the Schrödinger equation (see 3.5). In particular we have

$$(F[Q^2 - iQ_x])^n\left(2Q - i\frac{\partial}{\partial x}\right)y$$

$$\equiv \left(2Q - i\frac{\partial}{\partial x}\right)(\hat{G}[Q])^n y$$

for any integer n. It follows that for any solvable equation associated with the ZS-AKNS system for which $R = Q$ there is a Miura transformation onto a solvable equation associated with the Schrödinger equation. The inter-relation between the equations is given by

$$P_t + D(F)P_x = \left(2Q - i\frac{\partial}{\partial x}\right)(Q_t + D(G)Q_x)$$

(In proving that last formula notice that $\hat{G}Q_x = GQ_x$)

2.

Establish an analogous result to that obtained in (6.3) for the case $R = +Q$.

3.

Use the formula (6.3-135) to obtain the repeated eigenvalue solution (6.3-42) for the nonlinear Schrödinger equation.

4.

For the sine-Gordon equation

$$U_{xt} = \sin U$$

the bound state solution (6.3-40) is obtained from the constraint $2 + 1/2|k|^{-2} = \nu$ on the pair of eigenvalues $(k, -k^*)$. We can quickly show in this case that since $R = -Q$ $\bar{k} = -k$ and $\bar{D} = -D$. Show using (6.3-7) that

$$W_1(x,t) = \int_x^\infty Q(y,t)\,dy = 2\tan^{-1}\left\{\frac{iD(0)}{2k}\exp[2ikx+\Omega(k)t]\right\}$$

Then from (6.3-23) we have for a bound state that

$$U = 4\tan^{-1}\left[\frac{i\eta}{\xi}\tan\frac{1}{2}(W_1 - W_2)\right] \quad \text{where} \quad k = \xi + i\eta$$

from which we obtain the bound state solution

$$U = 4\tan^{-1}\left\{\frac{\eta\cos\xi\left(\alpha + 2x \mp \frac{1}{2k_0^2}2t\right)}{\xi\cosh\eta\left[\beta + 2x \pm \frac{1}{2k_0}2t\right]}\right\}$$

where $\alpha = (2\xi)^{-1}(\gamma + \gamma^*)$, $\beta = -i(2\eta)^{-1}(\gamma - \gamma^*)$, $\exp(i\gamma) = iD(0)/2k$. The locus is given by $|k| = k_0$ where k_0 is a real number.

5.

The following diagram represents a scheme whereby the 4-soliton solution of a solvable equation in the class $R = -Q$ may be obtained by Bäcklund transformations.

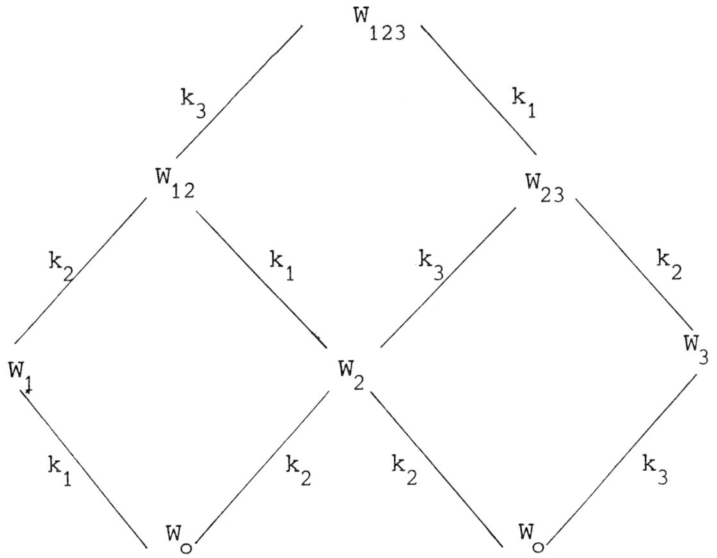

Use the algebraic superposition principle to obtain the 4-soliton for the sine-Gordon equation.

6.

Work through the Zakharov and Manakov asymptotic inverse method for the cases $R = \pm Q^*$.

7.

The Zakharov and Manakov asymptotic inverse method also applies to the hierarchy of solvable equations associated with the isospectral schrödinger equation. The Schrödinger equation is

$$-y_{xx} + Qy = k^2 y \tag{6.6-1}$$

Introduce the functions u_1, u_2

$$y = u_1 e^{ikx} + u_2 e^{-ikx}$$

Then Y is a solution of (1) if u_1 and u_2 satisfy

$$\begin{aligned} &-(u_{1xx} + 2iku_{1x}) + Qu_2 e^{-2ikx} \\ &-(u_{2xx} - 2iku_{2x}) + Qu_1 e^{2ikx} \end{aligned} \tag{6.6-2}$$

Assume

$$Q = 2\varepsilon^{\frac{1}{2}}\tilde{A}(X)\cos\Phi$$

$$\Phi = r(X)\ln\varepsilon \rightarrow p(X)\varepsilon^{-1} + q(X) + O(1)$$

$$X = \varepsilon x .$$

Put $\lambda = -2k$, $\theta_1 = \Phi + \lambda x$, $\theta_2 = \Phi - \lambda x$ and perform a multiple scales expansion of (2). The analysis is very similar to that given in (6.3) Essentially $A = \tilde{A}k^{-1}$ is the main difference in the analysis to $O(1)$.

7. KINKS AND THE SINE-GORDON EQUATION

7.1 Topological Considerations and a Mechanical Model

The world of subnuclear particles displays remarkable regularities. Some of those regularities may be placed in a coherent framework by postulating the existence of approximate symmetry groups. The explanation is only approximate in the sense that certain results prohibited on symmetry grounds are seen experimentally but with a low probability of occurrence. With such symmetries we may associate conserved quantities. An example is the conservation of electric charge. Even in a reaction such as

$$p + \bar{p} \rightarrow \pi^+ + \pi^- + \pi^o$$

in which the collision of a proton with an anti-proton gives rise to three π-mesons, we still have electric charge conserved by the formation of a charged particle pair. Similarly a positively charged positron e^+ in collision with a negatively charged electron e^- can undergo the pair annihilation process

$$e^+ + e^- \rightarrow \gamma + \gamma$$

into a pair of γ-ray photons the quanta of the electromagnetic field. Two photons are produced so that momentum is conserved.

The indestructible nature of such conserved quantities as electrical charge would seem to suggest that they must occupy a very fundamental position in the nature of the real world. How can we build into a mathematical model an entity like electrical charge which cannot be destroyed? There is a very simple conceptual model which illustrates one way in which it can be done.

Consider a rectangular elastic band PQRS as shown in Fig. 7-1a. Suppose that the end PS is fixed in the vertical position shown, but before fixing the other end QR we make a single twist in the band through an angle of 2π. The result is depicted in Fig. 7-1b. The twist between T and T' is now 'trapped' in the band and cannot be removed. The twist has been sealed in by the boundary conditions at the ends of the band. Fig. 7-1c shows the same band but with the unit normal vector at points along the central line C indicated. We assume that the normal at each point on the band lies in a vertical plane. As we move along C from one end to the other the normal vector rotates through the twist angle of 2π. For a band which is very narrow compared to its length, the normal rotates very slowly at long distances from the twist region TT' and only makes a rapid change as we pass through the region itself.

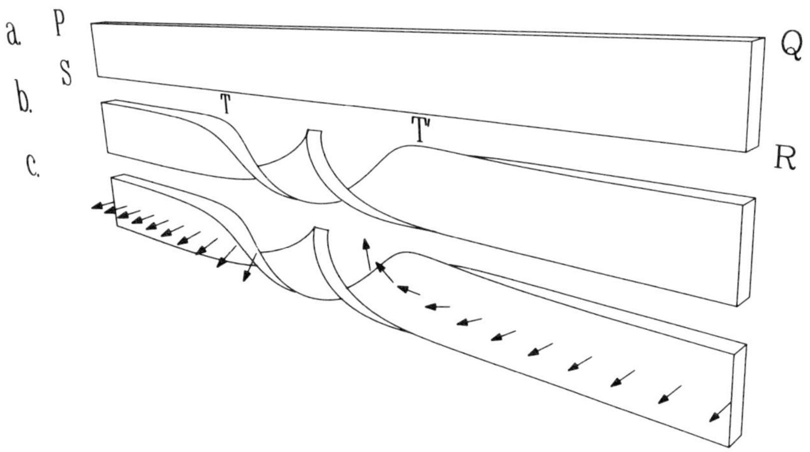

Figure 7-1: Elastic band model

There are several experiments we can carry out on such a band. Suppose that, with our fingers, we push the twist region away from its equilibrium position towards the end PS and hold it there. If we then release our grip, the elastic restoring forces which we overcame with our fingers are free to operate and the twist will spring back to its original equilibrium position. The twist will propagate along the band. Such a localised disturbance which is propagating is very similar to a particle. Now suppose that before we release the twist at PS we untie the band at QR, create a twist in the opposite direction, fix the band again at QR and force the new twist to the end QR and hold it fixed. We have thereby created a configuration which will be equivalent, when released, to the untwisted band. Therefore when we release the system of two constrained twists they will travel along the band towards each other like particles about to collide. When they meet, each annihilates the other leaving an agitated, but untwisted band. The two twists have behaved exactly like a positron and an electron, annihilating and releasing, in this case, their energy into other vibrational modes of the band rather than vibrational modes of the electromagnetic field. A twist in the clockwise direction, say, behaves like a positively charged particle and a twist in the anticlockwise direction like a negatively charged particle.

Any realistic equation describing the dynamics of such a band must admit solutions which describe such localised disturbances. Such solutions are known as **kinks** and one of our aims must be to isolate, in a less intuitive way, the properties we expect such solutions to have. Classical theories which have kink solutions are of considerable interest as their quantised versions provide models which may help us to understand the behaviour of physical elementary particles.

Let us consider in more precise terms the topology of this situation. Each configuration of the band defines a smooth mapping between the centre line of the band and the unit circle. This is established by relating to each point of C the point on the unit circle defined by the tip of the normal vector to the band at that point. When a twist propagates along the band we have defined, at each moment of time, a different function defining the instantaneous state of the band. As a result a consideration of the dynamics of such a system requires us to examine a family of such maps. Consequently, we are inevitably led to topological considerations.

If our band is of length L we may choose our coordinate to be given by $z = x/L$ where x is the distance measured along C from the end PS. Each configuration then defines a mapping $\phi : I \to S^1$ where I denotes the unit interval $[0,1]$ and S^1 the unit circle. In general we will use the notation I^n to denote the Cartesian product of I with itself n times and S^n to denote the sphere in R^{n+1}. The fixed boundary conditions are expressed by the requirement

$$\phi(0) = \phi_o = \phi(1)$$

where ϕ_0 is an arbitrary point on the circle and determined by the exact way we choose to define ϕ. The function ϕ defines a **closed curve** or **loop** in S^1 at the point ϕ_0. Such maps do not distinguish between the boundary points 0 and 1, and these points are effectively identified. Consequently, we are in fact considering smooth mappings from S^1 to S^1. The change in ϕ corresponding to a propagating kink or to manual distortion is, provided we do not crease or tear the band, intuitively a continuous and reversible process. To make that intuitive notion precise we introduce the mathematical concept of **homotopy**.

We say two smooth mappings $\phi_i : I \quad S^1$ $(i = 1,2,)$ are homotopic at the point ϕ_0 if we can find a single mapping $H : I^2 \to S^1$ given in coordinates by $(s,z) \to H(s,z) = H^s(z)$ and having the properties

(i) H^s is a loop at ϕ_0 for all values of $s \in [0,1]$.

(ii) $H^0 = \phi_1$ and $H^1 = \phi_2$.

This clarifies the intuitive observation that when we take the band in configuration ϕ_2 and distort it into another configuration ϕ_2 we pass through an infinite number of physically allowed intermediate configurations. The function H^s labels those interpolating states. A diagrammatic representation of this situation is given in Fig. 7-2.

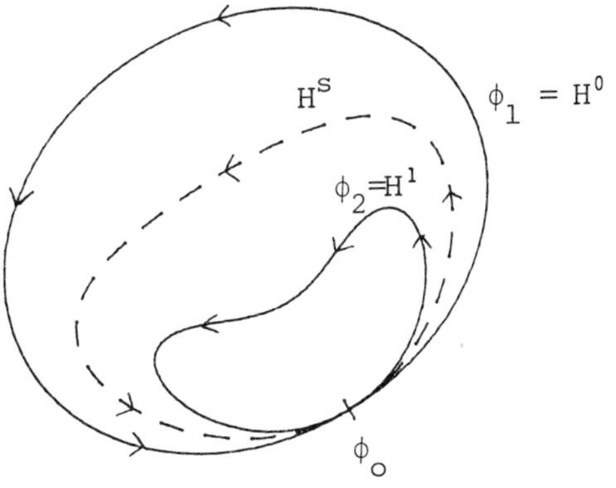

Figure 7-2:

In general S^1 can be replaced by any topological space X. Two functions $f_i : I \to X$ $(i = 1,2)$ are then said to be homotopic to each other at a point $x \in X$ if a continuous function $H : I^2 \to X$ can be found so that

(i) H^s is a loop at $x \in X$ for all values of $s \in [0,1]$.

(ii) $H^0 = f_1$ and $H^1 = f_2$.

The set of all such loops has a natural product operation defined as follows. We define the product $f_1 \circ f_2$ to be the loop obtained by first traversing the loop defined by f_1 and then that defined by f_2. This is expressed in coordinates by

$$f_1 \circ f_2 : z \rightarrow \begin{cases} f_1(2z) & 0 \le z \le \frac{1}{2} \\ f_2(2z-1) & \frac{1}{2} \le z \le 1 \end{cases}$$

and schematically represented in Fig. 7-3.

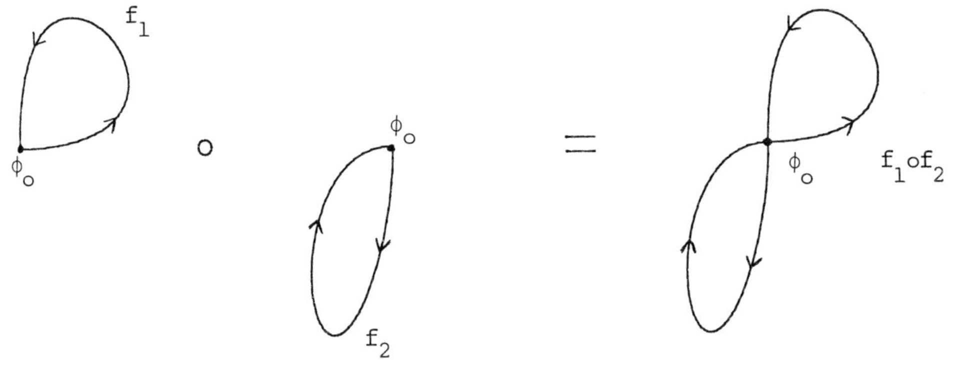

Figure 7-3:

The product we have just defined does not endow the set of all loops at a given point with a group structure. The problem lies in the fact that we are still relating specific maps with the loops. As a result the associative law needed for a group fails to hold. This is because although the loop corresponding to $f_1 \circ (f_2 \circ f_3)$ is the same as that associated with $(f_1 \circ f_2) \circ f_3$ for any three loops f_i $(i = 1,2,3)$ the two mappings have distinct functional forms. To eliminate the dependency upon parameterisation, and thereby concentrate attention upon the loops themselves, we must unite all the different maps that correspond to a single loop. This is done by associating with each loop f the set $\bar{f}$ of all loops homotopic to f at x. This divides the set of loops into classes of mutually homotopic loops which are termed **homotopy classes.**

We can define a product on the set of classes by

$$\bar{f} \circ \bar{g} = \overline{f \circ g}$$

which is easily shown to be well defined. Equipped with that product the set of homotopy classes of loops about a point x of a topological space X form a group called the **fundamental group of** X **at** x and denoted by $\pi_1(X,x)$. If each point of X can be joined to each other point of X by a path lying entirely within X it may be shown that the fundamental groups defined at any two points are isomorphic. The single abstract group to which all the fundamental groups at arbitrary points of such a pathwise connected space are isomorphic is denoted by $\pi_1(X)$ and termed the **fundamental group of X.**

In this case of the band the object of interest is the group $\pi_1(S^1)$. The essential point that we want to stress is that given an initial configuration of the band belonging to some definite homotopy class then, whatever the dynamics may be, it can only deform that initial configuration continuously. As a result the time development of the initial state can only lead to configurations in the same homotopy class. That is an important constraint on the system.

For the band, maps which correspond to different numbers of twists are not deformable into one another and the twist number of the map allocates it to a definite homotopy class. A simple example of a class n map is given by $S_n : z \mapsto (\cos 2\pi nz, \sin 2\pi nz)$. As z covers the interval I the image point of S_n covers the unit circle n times. The integer n counts how many times the image space S^1 is covered by the map S_n. In this particular case the integer n is often referred to as the **winding number** and the group $\pi_1(S^1) = Z$ the additive group of integers.

There is a simple mechanical analogue of the elastic band which will help us to understand how these topological considerations relate to the dynamics of a particular situation.

7.1.1 The Mechanical Pendulum

If we suppose that the mass of our elastic band is concentrated along one edge we can discretise the mass on that edge and replace the continuous distribution by discrete mass units. The capacity to twist is clearly the most important aspect that we need to model. We are led therefore to a model constructed from N penduli linked by torque springs that respond in proportion to the amount of rotation they receive. The restoring torque Γ exerted by such a spring when rotated through an angle θ is assumed to be given by the linear relation $\Gamma = \kappa\theta$ where κ is a constant called the **torque constant** of the spring. Figure 7-4 is a diagrammatic representation of such a mechanical analogue.

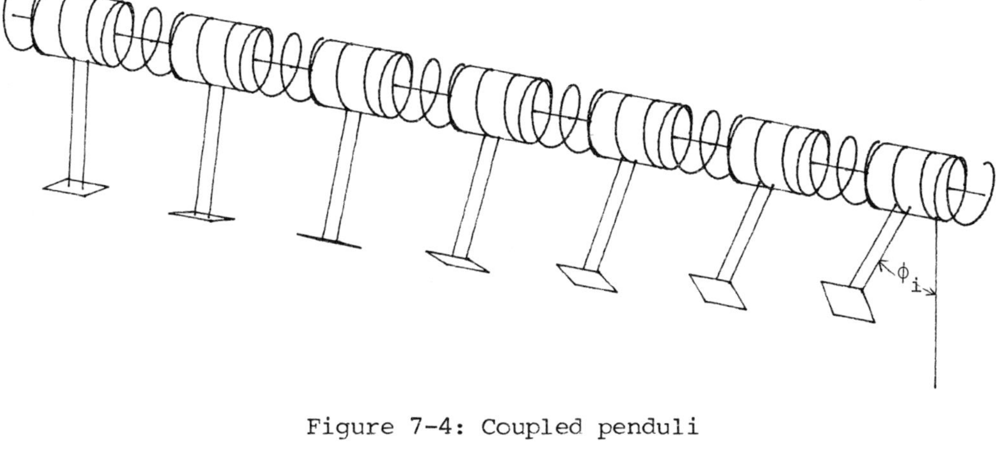

Figure 7-4: Coupled penduli

If the angle made with the downward vertical by the ith pendulum is denoted by ϕ_i then the angular velocity ω_i of the ith pendulum is given by $\dot{\phi}_i$. For an array of identical penduli each with the same moment of inertia J the Newtonian equations of motion become for the ith pendulum

$$J\dot{\omega}_i = \Gamma_i \text{ (due to torque springs) } + \Gamma_i \text{(due to gravity)} \tag{7.1-1}$$

If each pendulum has the same mass M, the torque due to gravity is given by

$$\Gamma_i \text{(due to gravity)} = -Mdg\sin\phi_i \tag{7.1-2}$$

where d is the distance of the centre of mass from the central axis. The torque on the ith pendulum due to the springs comes from the two springs on either side of it. The torques due to the springs joining pendulum i to the one before it and the one after it are given by $\kappa(\phi_{i-1} - \phi_i)$ and $\kappa(\phi_{i+1} - \phi_i)$ respectively. This leads us to the total torque exerted by the springs as

$$\Gamma_i \text{(due to torque springs)} = \kappa(\phi_{i+1} - 2\phi_i - \phi_{i-1}) \tag{7.1-3}$$

and equation (7.1-1) takes the form

$$J\ddot{\phi}_i = \kappa(\phi_{i+1} - 2\phi_i + \phi_{i-1}) - K_G\sin\phi_i \tag{7.1-4}$$

where $K_G = Mdg$.

The equations (7.1-4) may be derived from the Hamiltonian

$$H(\underline{P},\underline{\phi}) = \sum_{i=1}^{N} (\tfrac{1}{2}J(P_i)^2 + K_G(1-\cos\phi_i) + \tfrac{1}{2}\kappa(\phi_i - \phi_{i+1})^2) \tag{7.1-5}$$

by means of the Hamiltonian equations of motion

$$\dot{\phi}_i = \frac{\partial H}{\partial P_i}, \quad \dot{P}_i = -\frac{\partial H}{\partial \phi_i} \tag{7.1-6}$$

The kink solutions we are interested in are the nonlinear normal modes of this discrete dynamical system. As ϕ_i is a mapping from the set $\{1,2,\ldots,N\} \to S^1$ our topological considerations do not yet apply.

Figure 7-5 shows a simulated strobe photograph of a kink excitation propagating from right to left on the type of mechanical analogue we have just described. Note that the kink is slowing down due to friction. One aspect that this figure illustrates is that this system is, in a certain

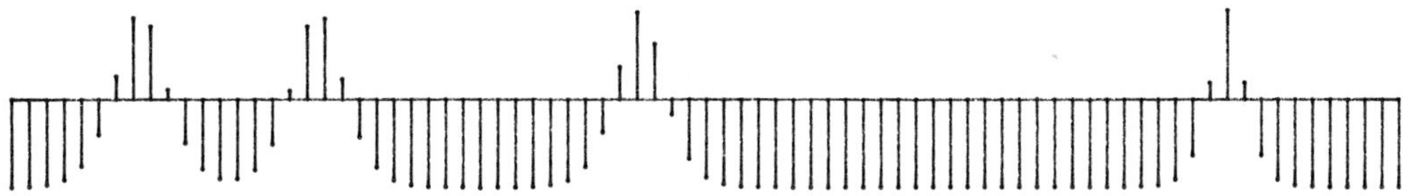

Figure 7-5: Single kink

sense, a 'two-state system'. Unlike a single pendulum, for which the angle ϕ remains small, the penduli in the above chain can and do make complete revolutions. There are two equilibrium configurations corresponding to $\phi = 0 \bmod(2\pi)$ and $\phi = \pi \bmod(2\pi)$ corresponding to the pendulum hanging directly down and standing straight up respectively. The latter is clearly unstable whilst the former is stable. The motion shown consists of a continual transition between these two configurations. It will transpire that the sine-Gordon equation is frequently associated with such quasi-two-state systems.

To obtain a continuum model for which our topological consideration will be pertinent we carry out a limiting process on equation (7.1-4). If the distance between the penduli is h and we set $\kappa = Kh^{-2}$ then allowing $h \to 0$ but retaining K finite yields the partial differential equation

$$J\phi,_{tt} = K\phi,_{xx} - K_G \sin\phi \tag{7.1-7}$$

where the subscripts denote partial derivatives.

If we introduce new space and time variables X and T defined by

$$T = (K_G/J)^{\frac{1}{2}}t \quad \text{and} \quad X = (K_G/K)^{\frac{1}{2}}x \tag{7.1-8}$$

equation (7.1-7) reduces to the standard sine-Gordon form

$$\phi,_{TT} - \phi,_{XX} + \sin\phi = 0 \tag{7.1-9}$$

Appropriate boundary conditions for this situation could be those of the band

$$\phi(t,0) = \phi(t,L) = 0 \mod(2\pi) \tag{7.1-10}$$

or perhaps other alternatives such as

$$\phi,_x(t,0) = \phi,_x(t,L) = 0 \qquad (7.1\text{-}11)$$

might be relevant depending on the specific situation being modelled. The boundary conditions (7.1-10) are the most common for the finite band. For the infinite band we will generally assume the boundary conditions

$$\phi(t,-\infty) = \phi(t,+\infty) = 0 \bmod(2\pi) \qquad (7.1\text{-}12)$$

From the previous chapters we already know that the function

$$\phi_v^t(x) = 4\tan^{-1}(\exp \pm\gamma(x-x_o+vt)) \qquad (7.1\text{-}13)$$

where $\gamma = (1 - v^2)^{-\frac{1}{2}}$ is a kink solution corresponding to the boundary conditions (7.1-12) on the infinite line. This functional form suggests a slightly more general representation that can be used to construct solutions to the sine-Gordon equation with other boundary conditions.

If we seek solutions to (7.1-9) having the form

$$\phi^t(x) = 4\tan^{-1}(f(x)g(t)) \qquad (7.1\text{-}14)$$

we find that the functions f and g must satisfy the equation

$$(f''g-g''f)(1+f^2g^2) + 2((g')^2f^3g-(f')^2g^3f) = fg(1-f^2g^2) \qquad (7.1\text{-}15)$$

In order to find explicit solutions we need to break up this equation into simpler equations for f and g separately. We observe that only powers of f and g are occurring and this suggests that we seek a solution for which f and g satisfy equations of the form

$$\begin{aligned} (f')^2 &= Af^4 + Bf^2 + C \\ (g')^2 &= Dg^4 + Eg^2 + F \end{aligned} \qquad (7.1\text{-}16)$$

From these equations we obtain

$$\begin{aligned} f'' &= 2Af^3 + Bf \\ g'' &= 2Dg^3 + Eg \end{aligned} \qquad (7.1\text{-}17)$$

The substitution of these forms (7.1-16,7.1-17) into equation (7.1-15) gives

$$2f^3g(A+F) - 2g^3f(C+D) + fg(1-f^2g^2)(B-E-1) = 0 \tag{7.1-18}$$

If we choose $A = -F = \alpha$, $B = E + 1 = \beta$ and $C = -D = \gamma$ where α, β, γ are constants, we have an acceptable solution given by any pair of functions f and g which satisfy the equations

$$\begin{aligned}(f')^2 &= \alpha f^4 + \beta f^2 - \gamma \\ (g')^2 &= \gamma g^4 + (\beta-1)g^2 - \alpha\end{aligned} \tag{7.1-19}$$

Before we determine explicit solutions to (7.1-19) we must decide upon the boundary conditions that we wish to impose. Let us select the finite boundary conditions (7.1-11) as these have not been previously considered in this book. The boundary conditions on ϕ require that the function f satisfies the conditions

$$f'(0) = 0 = f'(L) \tag{7.1-20}$$

The Jacobi elliptic function $sn(x,\lambda)$ is defined to be the solution of the equation

$$(f')^2 = (\lambda^2 f^4 - (\lambda^2+1)f^2 + 1) = (1-f^2)(1-\lambda^2 f^2) \tag{7.1-21}$$

with the boundary conditions

$$f(0) = 0 \text{ and } f'(0) = 1 \tag{7.1-22}$$

The constant λ is called the **modulus** of the elliptic function and $0 < \lambda < 1$.

Associated with the elliptic function $sn(x,\lambda)$ are the two functions $cn(x,\lambda)$ and $dn(x,\lambda)$ defined by the algebraic relationships

$$cn^2(x,\lambda) = 1 - sn^2(x,\lambda) \quad\text{and}\quad dn^2(x,\lambda) = 1 - \lambda^2 sn^2(x,\lambda) \tag{7.1-23}$$

The function $cn(x,\lambda)$ is easily shown to satisfy the equation

$$(f')^2 = (1-\lambda^2) + (2\lambda^2-1)f^2 - \lambda^2 f^4 \tag{7.1-24}$$

with the boundary conditions

$$f(0) = 1 \text{ and } f'(0) = 0 \tag{7.1-25}$$

In view of our chosen boundary conditions (7.1-20) the elliptic function $cn(x, \lambda)$ is the relevant one to choose and this leads to the solution

$$f = A\,cn(\kappa x, \lambda_f), \quad g = cn(\Omega t, \lambda_g) \tag{7.1-26}$$

of equations (7.1-19) provided that $A, \kappa, \Omega, \lambda_f$ and λ_g satisfy the equations

$$\begin{aligned} A^2\kappa^2(1-\lambda_f^2) &= \alpha = \lambda_g^2\Omega^2 \\ \Omega^2(1-\lambda_g^2) &= \gamma = A^{-2}\kappa^2\lambda_f^2 \\ 1 + \Omega^2(2\lambda_g^2-1) &= \beta = \kappa^2(2\lambda_f^2-1) \end{aligned} \tag{7.1-27}$$

These equations can be solved for the moduli λ_f and λ_g in terms of A, κ and Ω

$$\begin{aligned} \lambda_f^2 &= A^2(1+A^2)^{-1}\kappa^{-2}(\kappa^2(1+A^2)+1) \\ \lambda_g^2 &= A^2(1+A^2)^{-1}\Omega^{-2}(\Omega^2(1+A^2)+1) \end{aligned} \tag{7.1-28}$$

We also obtain the relationship

$$(\Omega^2-\kappa^2) = (1-A^2)(1+A^2)^{-1} \tag{7.1-29}$$

which is a form of **dispersion relation** similar to those discussed in Chapter 1.

The boundary conditions (7.1-20) must now be imposed on this solution. The function $cn(x, \lambda)$ is periodic with period $4K(\lambda)$ where $K(\lambda)$ is given by

$$K(\lambda) = \int_0^1 dy(1-y^2)^{-\frac{1}{2}}(1-\lambda^2y^2)^{-\frac{1}{2}} \tag{7.1-30}$$

As a result the function $(cn(x,\lambda))'$ is also periodic and has zeros at the points $2nK(\lambda)$ for any integer n. The boundary conditions on the function f of equation (7.1-26) therefore require that the wave number κ takes only the values κ_n given by

$$\kappa_n = 2nL^{-1}K(\lambda_f) \tag{7.1-31}$$

The final solution that we obtain has the form

$$\phi_n = 4\tan^{-1}(A\,\mathrm{cn}(2nK(\lambda_f)L^{-1}x,\lambda_f)\,\mathrm{cn}(\Omega_n t,\lambda_g)) \tag{7.1-32}$$

with $\Omega_n^2 = 4K(\lambda_f)L^{-2}n^2 + (1 - A^2)(1 + A^2)^{-1}$. Figure 7-6 depicts a simulation of this periodic solution.

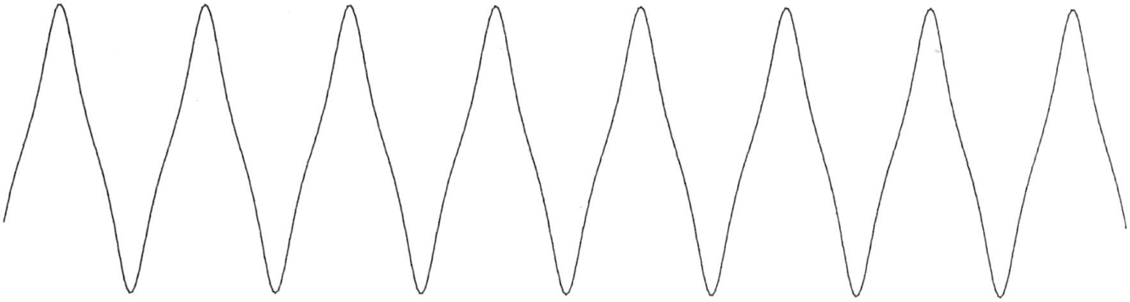

Figure 7-6:

An interesting feature of this solution is that in the small amplitude limit $A \to 0$ both λ_f and $\lambda_g \to 0$ and the solution takes the limiting form

$$\phi_n \sim \cos(n\pi x/L)\cos\Omega t \tag{7.1-33}$$

where $\Omega^2 = n^2\pi^2L^{-2} + 1$. This equation is simply the dispersion relation of the Klein-Gordon equations which results from linearising the sine-Gordon equation. We point this out to emphasise the view of such solutions as nonlinear normal modes of the system.

7.2 Particle Properties

Figure 7-7 shows a series of plots of the function $\psi^{\pm}(x) = 4\tan^{-1}(\pm m(x - q_0))$. In each case we see that the region of variation of the function is centred on the point $x = q_0$. Furthermore, as we increase m the 'length' of the region decreases. Therefore, if such a function represents a localised disturbance, we may interpret the quantity q_0 as its 'position' and the quantity m^{-1} as its 'width'. The function $\psi^{\pm}(x)$ is the static solution of the mass m sine-Gordon equation.

The sine-Gordon equation is invariant under the two dimensional group of Lorentz transformations

$$L(v) : (x,t) \to (\gamma(x-vt),\ \gamma(t-vx)) \tag{7.2-1}$$

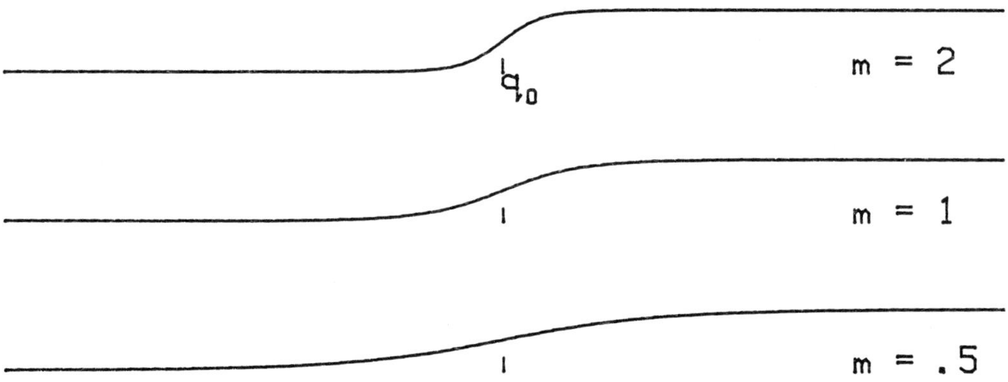

Figure 7-7: Sine-Gordon kink as a function of m

The single soliton (anti-soliton) solution

$$\psi_v^{\pm}(x,t) = 4\tan^{-1}(\pm\gamma m(x-q(t))) \tag{7.2-2}$$

with $q(t) = vt + q_0$ may be obtained from $\psi^{\pm}(x)$ by applying the Lorentz transformation $L(v)$. We see that the localised disturbance described by (7.2-2) has a 'position' which is moving with a linear velocity v and a width $\gamma^{-1} = (1 - v^2)$. Consequently, we have a 'Lorentz contraction' of the pulse when it moves with velocity v. This is exactly analogous to the motion of a relativistic particle.

The analogy between solitons, anti-solitons and positively and negatively charged particles is also readily shown. Figure 7-8 shows the time development of an initial state consisting of two static kinks localised at $x = +d$ and $x = -d$ respectively. The boundary conditions (7.1-12) are applied. We see that the kinks move away from each other and we have a situation of 'repulsion'. Similarly Fig. 7-9 shows the time development of an initial state consisting of a kink at $x = +d$ and an anti-kink situated at $x = -d$. In this case we see that the two disturbances move towards each other and we have a situation of 'attraction'. The diagrams make it intuitively clear why this must happen. It arises as a direct consequence of the boundary conditions. For two like kinks the solution of the sine-Gordon equation which evolves out of the initial configuration has to change by 2π over a short distance. This results in high field gradients and consequently large contributions to the energy. The system moves so as to reduce its energy and as a result the kinks separate. For the solution which develops from the initial condition consisting of two opposite helicity kinks no such rapid change is required and the nearer the two kinks become the lower is the energy. Consequently, two dissimilar kinks approach each other.

The sine-Gordon equation may be derived from the Hamiltonian density

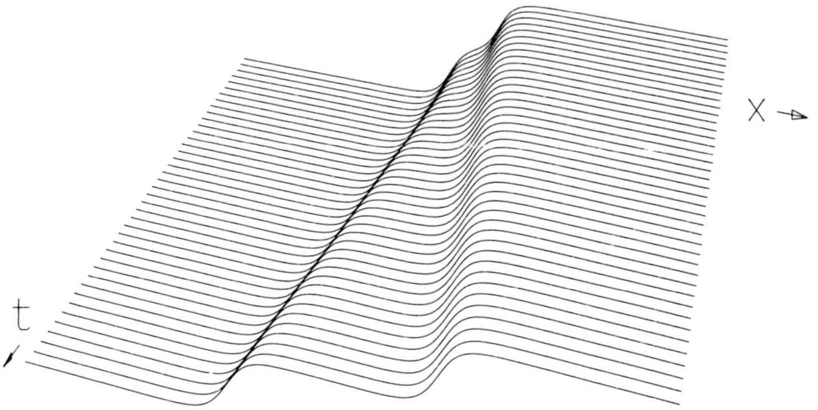

Figure 7-8: kink-kink repulsion

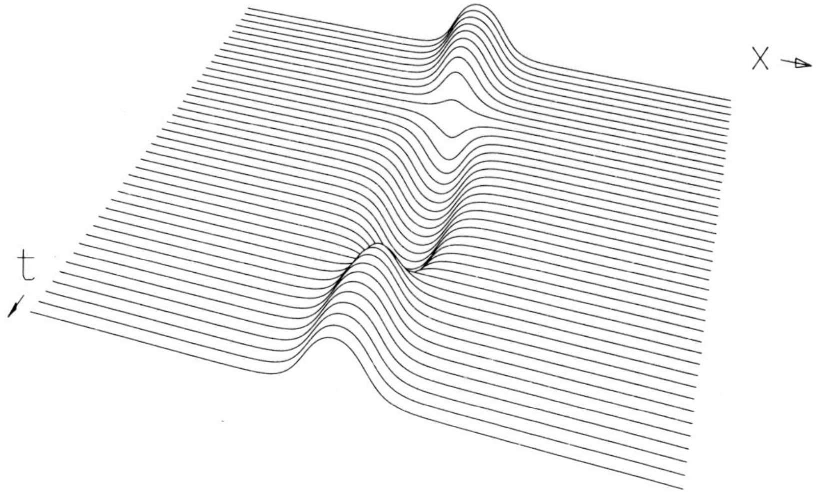

Figure 7-9: kink-antikink attraction

$$H(\pi,\phi) = \tfrac{1}{2}(\pi^2 + \phi_{,x}^2 + 4\sin^2\tfrac{1}{2}\phi) \qquad (7.2\text{-}3)$$

by means of the Hamiltonian field equations

$$\phi,_t = \frac{\delta H}{\delta \pi}$$
$$\pi,_t = \frac{\delta H}{\delta \phi} = -(-\phi,_{xx} + 2\sin\tfrac{1}{2}\phi\cos\tfrac{1}{2}\phi) \qquad (7.2\text{-}4)$$

These equations are the continuum analogue of equations (7.1-6). For any function Ψ, not necessarily a solution of the sine-Gordon equation, we define

$$E_\Psi(x,t) = H(\Psi,_t(x,t),\Psi(x,t)) \qquad (7.2\text{-}5)$$

If Ψ is a solution of the sine-Gordon equation then the energy of that solution is given by

$$\mathrm{E}_\Psi = \int_{-\infty}^{\infty} dx E_\Psi(x,t) \qquad (7.2\text{-}6)$$

For the single kink solution (7.2-2) having unit mass we obtain

$$\mathrm{E}_{\Psi_v^\pm} = 8\gamma \qquad (7.2\text{-}7)$$

which has the standard special relativistic dependency upon the velocity v. The kink solution is therefore an example of a finite energy solution. Such solutions represent time-independent packets of energy which retain their integrity by virtue of strong self- interactions.

Consider a system for which the possible energy values are bounded from below. The lowest energy states of such a system are called its **ground states.** There may be one or many such lowest energy solutions of the same energy which may be chosen to be zero. Let H be the Hamiltonian density of such a system. If the quantity E_Ψ corresponding to H has the property that for all non-singular solutions Ψ it is positive, and zero only in the ground states of the system, then we say that a solution Φ is **dissipative** if

$$\lim_{t\to\infty} \max_x E_\Phi(x,t) = 0 \qquad (7.2\text{-}8)$$

The solutions of most classical linear field theories, such as the Maxwell equations, are dissipative. The kink solution $\psi_v^\pm$ is an example of a non-dissipative solution and such solutions are a feature of nonlinear field theories. We note that the condition that a solution be non-dissipative is a different requirement from that of finite energy. Non-dissipative solutions are not solitons of the type we have concentrated upon in this book. In Chapter 1 we considered the ϕ^4 equations of particle physics. For a real field ϕ they take the form

$$\phi_{,xx} - \phi_{,tt} = \lambda\phi^3 - m^2\phi \tag{7.2-9}$$

The Hamiltonian density H may be taken to be

$$H(\pi,\phi) = \tfrac{1}{2}(\pi^2 + \phi_{,x}^2 + \tfrac{1}{2}\lambda(\phi^2 - m^2/\lambda)^2) \tag{7.2-10}$$

and equation (7.2-9) results from the Hamiltonian field equations

$$\begin{aligned} \phi_{,t} &= \frac{\delta H}{\delta \pi} \\ \pi_{,t} &= -\frac{\delta H}{\delta \phi} = -(-\phi_{,xx} + \lambda\phi^3 - m^2\phi) \end{aligned} \tag{7.2-11}$$

The energy $E_\psi(x,t)$ of a solution is positive and only zero in the two ground states

$$\phi^\pm = \pm(m^2/\lambda)^{\frac{1}{2}} \tag{7.2-12}$$

A solution of (7.2-9) is given by

$$\phi_v^\pm = \phi^\pm \tanh(\frac{m}{\sqrt{2}}\gamma(x-vt) + \delta) \tag{7.2-13}$$

and is shown in Fig. 7-10.

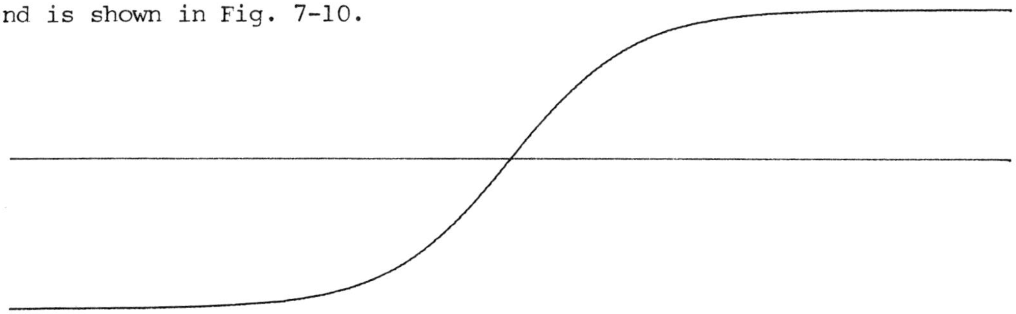

Figure 7-10: single kink solution of ϕ^4 equation

This solution has local energy density

$$E_{\phi_v^\pm}(x,t) = \tfrac{1}{2}\gamma^2(m^4/\lambda)\operatorname{sech}^4(\frac{m}{\sqrt{2}}\gamma(x-vt) + \delta) \tag{7.2-14}$$

and total energy

$$E_{\phi_v^+} = (2\sqrt{2}/3)(m^3/\lambda)\gamma \qquad (7.2\text{-}15)$$

Therefore the kink solution of the ϕ^4 model is also a non-dissipative, finite energy solution. However, we saw in the numerical simulations of Chapter 1 that these kink solutions do not have the collision properties of the solitons we have concentrated upon. We will reserve the term **soliton** for kink solutions which have the property of retaining their integrity even after collison processes have taken place. The term kink will be reserved for the more general non-dissipative solutions such as the ϕ^4 kink considered above.

The result (7.2-7) means that to produce even a static kink a minimum energy of 8 units is required. Consequently, energy must be fed into a system in order to excite its kink modes. For that reason there is another solution which may be physically important.

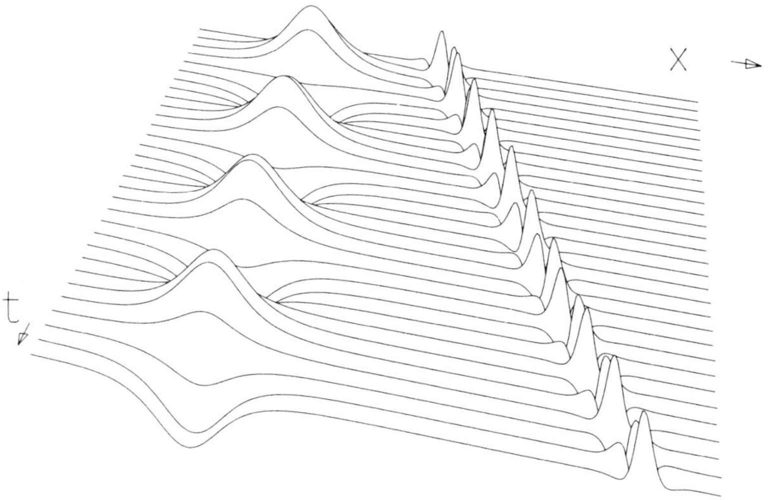

Figure 7-11: Two sine-Gordon breathers, one stationary & one moving.

Figure 7-11 shows the sine-Gordon breather solution

$$\Phi_{Bv} = 4\tan^{-1}((1-\omega^2)^{\frac{1}{2}}\omega^{-1}\sin\omega\gamma(t-vx)\,\mathrm{sech}(\gamma(1-\omega^2)^{\frac{1}{2}}(x-vt)) \qquad (7.2\text{-}16)$$

The importance of this solution may lie in the fact that its rest energy $E_{\Phi_{B_0}}$ is given by

$$E_{\Phi_{Bo}} = 16(1-\omega^2)^{\frac{1}{2}} \qquad (7.2\text{-}17)$$

and varies from 16, the rest mass of two solitons, down to zero as ω tends to one. Consequently, even small amounts of energy are sufficient to excite the breather modes.

If we impose the boundary conditions (7.1-12) the conserved quantity corresponding to the number of twists is given by

$$\mathcal{Q}^t(\phi) = (2\pi)^{-1}\int_{\infty}^{\infty} dx\ \phi^t_{,x}(x) = (2\pi)^{-1}(\phi^t(\infty) - \phi^t(-\infty))$$
$$= (\text{number of kinks}) - (\text{number of anti-kinks}) \tag{7.2-18}$$

and is clearly integer valued. As $\mathcal{Q}^t(\phi)$ depends continuously on t and is integer valued, the only way those two properties can be compatible is by $\mathcal{Q}^t(\phi)$ being conserved regardless of dynamical considerations. This observation corresponds to the fact that from the dynamical viewpoint this is a trivial conservation law following from the continuity equation

$$\rho_{,t} + j_{,x} = 0 \tag{7.2-19}$$

where the kink charge density ρ and kink current density j are given by

$$\rho = \phi_{,x} \qquad j = -\phi_{,t} \tag{7.2-20}$$

These equations bear no relationship to the dynamics at all and simply require ϕ to be a smooth function. The important aspects are the boundary conditions and the fact that the field takes values on the unit circle S^1. The integer valued quantity $\mathcal{Q} = (2\pi)^{-1}\int dx\,\rho(x)$ is known as a **topological charge**.

Apart from its specific role in many physical models, we hope to learn from such a nonlinear field theory the necessary ingredients for the construction of models in more than two dimensions which will have particle-like solutions. Topological charge is an important feature of many field theories that admit kink solutions. As a result it is worthwhile generalising that concept to a more general higher dimensional framework. The associated model equations may be regarded as higher dimensional analogues of the sine-Gordon equation. In a later section we will encounter a physical occurrence of such a model in connection with ferromagnetism.

7.3 Topological Charge

To obtain a more general framework let us recast the sine-Gordon equation into a slightly different form. The function ϕ defines a point on the unit circle having Cartesian coordinates ϕ_1 and ϕ_2 defined by

$$\phi_1 = \cos\phi \qquad \phi_2 = \sin\phi \quad \text{with} \quad \phi_1^2 + \phi_2^2 = 1 \tag{7.3-1}$$

Consequently each function ϕ defines a smooth mapping $\Phi : R^2 \to S^1$ given in coordinates by $(t,x) \mapsto (\phi_1(t,x), \phi_2(t,x))$.

We can generalise this by considering smooth maps $\Phi : R^n \to S^{n-1}$ given in coordinates by $(x_0, x_1, \ldots, x_{n-1}) \to (\phi_1(x_0,\ldots,x_{n-1}),\ldots,\phi_n(x_0,\ldots,x_{n-1}))$ with $\phi_1^2 + \phi_2^2 + \ldots + \phi_n^2 = 1$. It is the simple constraint that the vector $(\phi_1,\ldots,\phi_n)$ have unit length which leads to the generalisation of the continuity equation (7.2-19). As $\langle\underline{\phi},\underline{\phi}\rangle = 1$, differentiation with respect to

x_a yields the equation

$$<\underline{\phi},\partial_a\underline{\phi}> = 0 \tag{7.3-2}$$

This means that the $(n\times n)$ matrix $\partial\phi$ with elements $\partial_a\phi^b$ must be singular. From linear algebra we know that this requires the determinant of $\partial\phi$ to be zero. In terms of the components of $\partial\phi$ this may be written in the form

$$\varepsilon^{a_1\ldots a_n}\varepsilon_{b_1\ldots b_n}\partial_{a_1}\phi^{b_1}\ldots\partial_{a_n}\phi^{b_n} = 0 \tag{7.3-3}$$

This may be re-expressed in the form

$$\partial_{a_1}(\varepsilon^{a_1\ldots a_n}\varepsilon_{b_1\ldots b_n}\phi^{b_1}\partial_{a_2}\phi^{b_2}\ldots\partial_{a_n}\phi^{b_n}) = 0 \tag{7.3-4}$$

As a result we see that the n-component vector current J^a defined by

$$J^a = \varepsilon^{ac_1\ldots c_{n-1}}\varepsilon_{b_1\ldots b_n}\phi^{b_1}\partial_{c_1}\phi^{b_2}\ldots\partial_{c_{n-1}}\phi^{b_n} \tag{7.3-5}$$

is conserved

$$\partial_a J^a = 0 \tag{7.3-6}$$

In the case of the sine-Gordon equation for which $n = 2$ we have

$$J^a = \varepsilon^{ac}\varepsilon_{bd}\phi^b\partial_c\phi^d \tag{7.3-7}$$

with the explicit coordinate functions

$$J^o = (\phi^1\partial_x\phi^2 - \phi^2\partial_x\phi^1), \quad J^1 = (\phi^2\partial_t\phi^1 - \phi^1\partial_t\phi^2) \tag{7.3-8}$$

Substitution of the parameterisation (7.3-1) gives

$$J^o = \phi,_x \qquad J^1 = -\phi,_t \tag{7.3-9}$$

in agreement with equation (7.2-20).

The generalisation of the topological charge (7.2-18) is given by

$$\mathcal{Q}^x{}_o(\phi) = \frac{1}{(n-1)!\Omega_{n-1}} \int_{R^{n-1}} J^o(\phi)dx^1 \ldots dx^{n-1} \qquad (7.3\text{-}10)$$

where $\Omega_{n-1} = 2\pi^{n/2}/\Gamma(n/2)$ is the surface area of the unit sphere S^{n-1} and generalises the factor 2π of the $n = 2$ case.

To evaluate the integral and substantiate the claim that $\mathcal{Q}^x{}_o(\phi)$ is integer valued we need one further mathematical concept, the **Brouwer degree** of a smooth mapping. Let M and N be two compact surfaces of dimension m and $f : M \to N$ a smooth mapping from M to N. For each $y \in N$ define the set $f^{-1}(y) \in M$ by

$$f^{-1}(y) = \{x \in M : f(x) = y\} \qquad (7.3\text{-}11)$$

A point $y \in N$ is said to be **a regular value** of f if the Jacobian ∂f is non-zero at each point of $f^{-1}(y)$. It may then be shown that, provided N is connected, the quantity deg(f) defined by

$$\deg(f) = \sum_{f^{-1}(y)} \text{sgn}(\det(\partial f)) \qquad (7.3\text{-}12)$$

where y is a regular value of f is independent of any specific choice of regular value y. The quantity deg(f) is clearly an integer valued quantity and is called the **Brouwer degree** of the smooth mapping f.

If we consider a real valued function $g : N \to R$ then $(g \circ f) : M \to R$ and we have the integral relationship

$$\int_M dx^1..dx^m (g\circ f)(x^1,..,x^m)\det\partial y = \deg(f) \int_N dy^1..dy^m g(y^1,..,y^m) \qquad (7.3\text{-}13)$$

If we now compactify R^{n-1} by a map $\pi : R^{n-1} \cup \{\infty\} \to S^{n-1}$ the formula (7.3-13) solves our problem. This follows from the result

$$\mathcal{Q}^x{}_o(\phi) = \frac{1}{\Omega_{n-1}} \int dx^1 \ldots dx^{n-1} \; (\det g_{ab})^{\frac{1}{2}} \det\partial y \qquad (7.3\text{-}14)$$

where $(y^1,\ldots,y^{n-1})$ are intrinsic coordinates on the sphere and

$$g_{ab} = \left\langle \frac{\partial \underline{\phi}}{\partial y^a}, \frac{\partial \underline{\phi}}{\partial y^b} \right\rangle \qquad (7.3\text{-}15)$$

is the metric tensor on S^{n-1}. We see immediately that

$$Q^x{}_o(\phi) = \deg(\phi o \pi^{-1})\ \Omega^{-1}_{n-1} \int dy^1..dy^{n-1} (\det g_{ab})^{\frac{1}{2}} = \deg(\phi o\pi^{-1}) \quad (7.3\text{-}16)$$

which is integer valued.

The generalisation of homotopy for maps from S^n onto a topological space X is constructed by defining two maps $f,g : I^n \to X$ to be homotopic at a point $x \in X$ if they have the property

$$f(\partial I^n) = x = g(\partial I^n)$$

so that the mappings f and g are really defined on S^n and there exists a continuous mapping $H : I^{n+1} \to X$ such that $H^s : I^n \to X$ given in coordinates by $H^s : (z_1,\dots,z_n) \to H(s,z_1,\dots,z_n)$ has the properties

(i) $H^s(\partial I^n) = x$ for all $s \in I$

(ii) $H^0 = f$ and $H^1 = g$

The set of equivalence classes of mutually homotopic maps has a natural group structure induced by the product on curves given by

$$(f o g)(z_1,z_2,\dots,z_n) = \begin{cases} f(2z_1,z_2,\dots,z_n) & 0 \le z_1 \le \frac{1}{2} \\ g(2z_1-1,z_2,\dots,z_n) & \frac{1}{2} \le z_1 \le 1 \end{cases}$$

The resulting group is called the **n-th Homotopy group** at the point $x \in X$ and is denoted by $\pi_n(X,x)$. If X is pathwise connected all of the groups at different points $x \in X$ are isomorphic and the single abstract group to which they are all isomorphic is denoted by $\pi_n(X)$ the **n-th Homotopy group of X.**

The topological content of our previous analysis may be summed up in the result $\pi_n(S^n) = Z$ the additive group of integers.

We have only considered models that are a direct generalisation of the sine-Gordon equation as it is difficult to make general statements in other cases. The origin of topological charge as manifest in conservation of 'soliton number' for such equations as the KdV or Nonlinear Schrödinger equation is less easy to get at. To use the results that we have just established we must establish an association between a solution of the relevant equation and some corresponding mapping between spheres. The inverse scattering transform leads to such an association.

7.4 Nonlinear Klein-Gordon Equations

The sine-Gordon and ϕ^4 equations are both examples of the general nonlinear Klein-Gordon equation

$$\phi,_{xx} - \phi,_{tt} = U'(\phi) \tag{7.4-1}$$

corresponding to the Hamiltonian density

$$H(\pi,\phi) = \tfrac{1}{2}(\pi^2 + \phi,_x^2 + 2U(\phi)) \tag{7.4-2}$$

For the sine-Gordon case the potential function is $U_1(\phi)$ given by

$$U_1(\phi) = (1-\cos\phi) \tag{7.4-3}$$

and for the ϕ^4 model we have the potential $U_2(\phi)$ given by

$$U_2(\phi) = \frac{\lambda}{4}\left(\phi^2 - \frac{m^2}{\lambda}\right)^2 \tag{7.4-4}$$

Sketches of these potentialsare shown in Fig. 7-12.

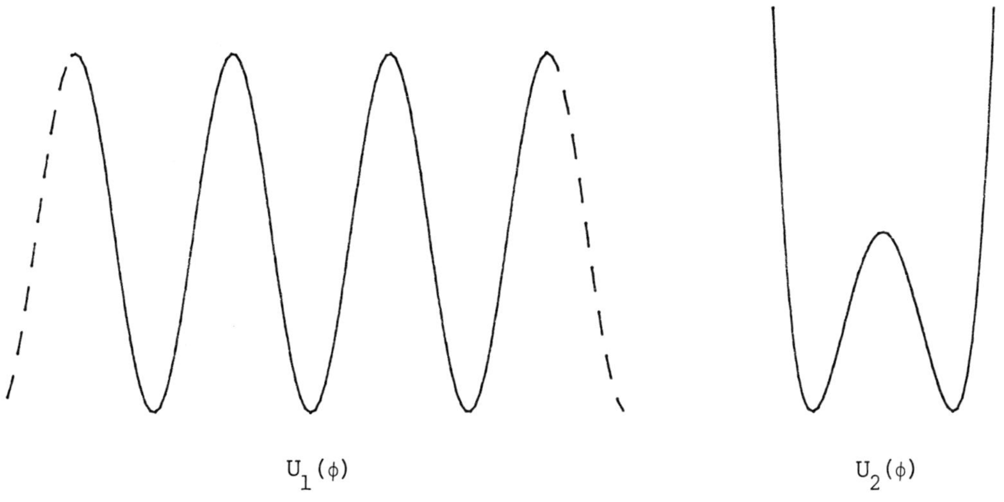

Figure 7-12: (a) Sine-Gordon and (b) ϕ^4 potentials

We can always add an arbitrary constant to $U(\phi)$ and that freedom has been utilised in each of the above cases to ensure that $U(\phi) = 0$ for the ground states of the system. If we assume that this has always been done we see that the ground state values are the zeros of $U(\phi)$.

If a solution is to be of finite energy and has the asymptotic behaviour

$$\phi \to \phi_{\pm}(t) \text{ as } x \to \pm\infty \tag{7.4-5}$$

then we must have

$$U(\phi_{\pm}) = 0 \tag{7.4-6}$$

Consequently, the possible asymptotic values of a finite energy solution are restricted to be the zeros of the potential function $U(\phi)$ and which correspond to the ground states of the system. For the sine-Gordon equation the zeros of $U_1(\phi)$ are $\{2n\pi : n \in \mathbb{Z}\}$ and therefore the only acceptible boundary conditions for a finite energy solution are given by

$$\phi \to 0 \quad \mathrm{mod}(2\pi) \qquad \text{as} \quad |x| \to \infty \tag{7.4-7}$$

For the ϕ^4 model the zeros are $\{\phi_+, \phi_-\}$ and we have alternative asymptotic limiting values. If, as in this case, the zeros of $U(\phi)$ are discrete we must have

$$\partial_t \phi_{\pm\infty}(t) = 0 \tag{7.4-8}$$

and so $\phi_{\pm}$ is a conserved quantity. This gives us an alternative way of looking at topological charge. The space F of non-singular finite energy solutions of the nonlinear Klein-Gordon equation (7.4-1) may be divided into a number of subspaces labelled by the asymptotic values of the fields in that subspace. For example, in the case of the ϕ^4 equation we have four subspaces

$$A^{\alpha\beta} = \{\phi \varepsilon F: \phi \to \{ \begin{matrix} \phi^{\alpha} & x \to +\infty \\ \phi^{\beta} & x \to -\infty \end{matrix} \} \tag{7.4-9}$$

each characterised by a pair of numbers which are ± 1. The topological conserved quantity is not an arbitrary integer in this case but can be defined despite that.

The integral results of the previous section are not directly applicable to nonlinear Klein-Gordon equations in general. However, in the case of a real field ϕ we can consider the direction field $\hat{\phi}$ defined by

$$\hat{\phi} = \phi/|\phi| \tag{7.4-10}$$

The field $\hat{\phi}$ is singular at the zeros of ϕ which we suppose continuous. The kink density $\rho = \hat{\phi}_{,x}$ may still be calculated but has delta function singularities at the zeros of ϕ. If $z_1, \ldots, z_m$ are the zeros of ϕ we obtain

$$\rho = \sum_{j=1}^{m} d(\phi,z_j)\,\delta(x-z_j) \qquad d(\phi,z) = \lim_{x\to z} \operatorname{sgn}(\phi,_x(x)) \tag{7.4-11}$$

Figure 7-13 depicts the situation for a function ϕ with five zeros.

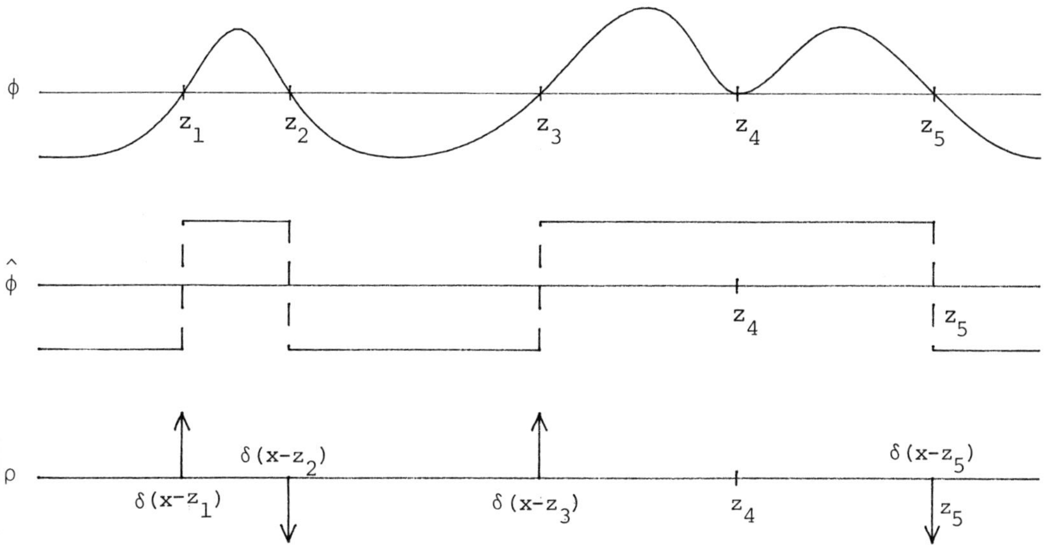

Figure 7-13:

The total topological charge may be defined by

$$\mathcal{Q}^t(\phi) = \int_{-\infty}^{\infty} dx\ \rho(x,t) = \sum_{j=1}^{m} d(\phi,z_j) \tag{7.4-12}$$

As the function ϕ is assumed continuous the possible values of $\mathcal{Q}^t(\phi)$ are ±1 and 0. The single kink solutions $\phi_v^{\pm}$ of the ϕ^4 equations given by (7.2-13) have charge $\mathcal{Q}^t(\phi_v^{\pm}) = 1$ and for the subspaces $A^{\alpha\beta}$ defined in (7.4-9) we have

$$\mathcal{Q}(A^{++}) = 0 = \mathcal{Q}(A^{--}) \quad \text{and} \quad \mathcal{Q}(A^{+-}) = +1 = -\mathcal{Q}(A^{-+})$$

An alternative example is given by the potential

$$U_3(\phi) = (\phi^2 + a^2)(1-\phi^2)^2 \left(\frac{\lambda^2}{8(1+a^2)}\right) \tag{7.4-13}$$

which has two ground states $\phi = \pm 1$. The model corresponding to this potential has the static kink solution

$$\phi(x) = [1+a^{-2}+ \sinh^2(\tfrac{1}{2}\lambda x)]^{-\frac{1}{2}} \sinh(\tfrac{1}{2}\lambda x) \qquad (7.4\text{-}14)$$

which is shown below in Fig. 7-14. For this solution $\hat{\phi} = \text{sgn}\ x$ and it has topological charge +1 as a member of the subspace A^{+-} corresponding to this model.

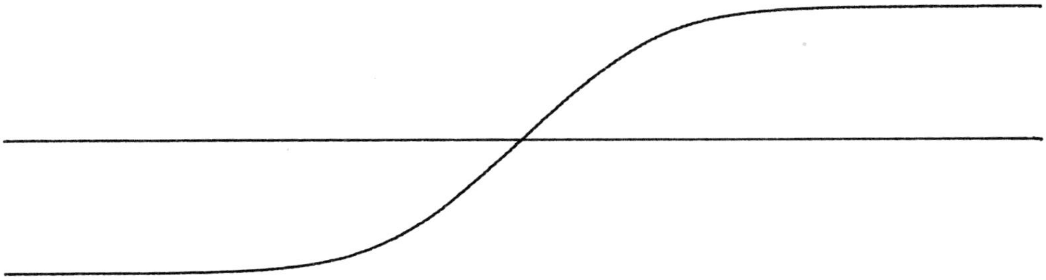

Figure 7-14: The kink solution (7.4-14)

A model which is sometimes used as an alternative to the ϕ^4 model in physical models (for displacive ferroelectrics) is provided by the double-quadratic (DQ) potential given by

$$U_4(\phi) = \tfrac{1}{2}\omega^2 (|\phi|-1)^2 \qquad (7.4\text{-}15)$$

A sketch of the potential $U_4(\phi)$ is shown in Fig. 7-15.

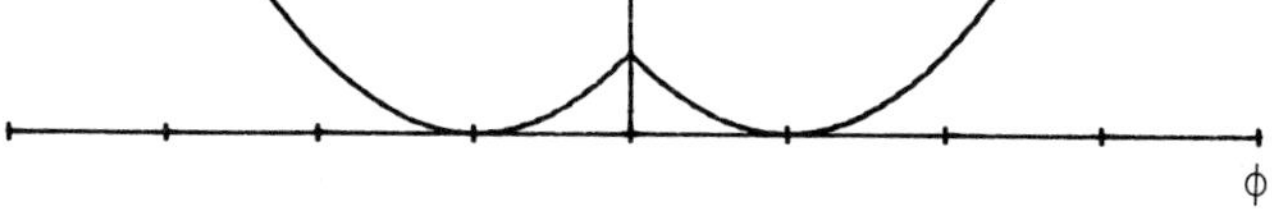

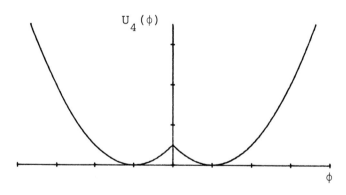

Figure 7-15: The DQ potential

The associated nonlinear Klein-Gordon equation takes the form

$$\phi_{,tt} - \phi_{,xx} + \omega^2(|\phi|-1)\,\text{sgn}\phi = 0 \qquad (7.4\text{-}16)$$

which has the kink solutions

$$\phi_{\pm} = \pm \operatorname{sgn}(\gamma(x-vt))[1-\exp(-\omega\gamma|x-vt|)] \tag{7.4-17}$$

A plot of the static kink wave form is shown in Fig. 7-16

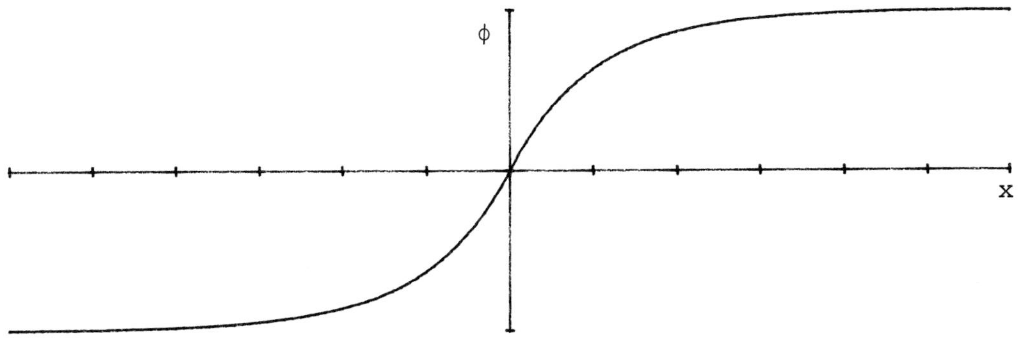

Figure 7-16: The kink solution (7.4-17)

The similarity of this to the ϕ^4 kink is clear. Once again we have a charge +1 non-dissipative solution.

If, as in the cases considered above, there is more than one discrete zero of $U(\phi)$ then not only do we have a nontrivial topological charge but also a non-dissipative solution. If α and β are two zeros of the potential and there exists a non-singular finite energy solution with the asymptotic behaviour

$$\phi \to \begin{cases} \alpha & x \to -\infty \\ \beta & x \to +\infty \end{cases}$$

then

$$E_{\phi}(x,t) > \max_{\psi \in [\alpha,\beta]} U(\psi) > 0 \tag{7.4-18}$$

and consequently (7.2-8) cannot hold.

The situation becomes more interesting if the potential function $U(\phi)$ has a continuum of zeros and especially so if we are in more than one spatial dimension. The Hamiltonian

$$H(\pi,\phi) = \tfrac{1}{2}(\pi^2 + \phi_{,x}^2 + \phi_{,y}^2 + \tfrac{1}{2}\lambda(|\phi|^2-1)^2) \tag{7.4-19}$$

is the complex ϕ^4 model in two spatial dimensions. All points on the circle S^1 are possible ground states of the system. For a model with two spatial dimensions such as this the analogue of allowing the spatial variable x of the one dimensional model to go to $\pm\infty$ is to allow the radial variable $r = (x^2 + y^2)^{\frac{1}{2}}$ to become infinite along rays out from the origin. The result obtained will depend in general upon the ray chosen. As a result each non-singular finite energy solution ϕ defines a mapping $\tilde{\phi} : S^1 \to S^1$ given in coordinates by

$$\tilde{\phi} : \hat{n} \rightarrow \lim_{r\to\infty} \phi(r\hat{n},t) = \tilde{\phi}(\theta,t) \tag{7.4-20}$$

As $\pi_1(S^1) = Z$ it would seem that we can have kink type solutions is this model. There are however, problems. The energy of a solution ϕ is given by

$$E_\phi = \int_0^{2\pi} d\theta \int_0^{\infty} rdr\, \tfrac{1}{2}(\phi_{,t}^2 + \phi_{,r}^2 + r^{-2}\phi_{,\theta}^2 + \tfrac{1}{2}\lambda(|\phi|^2-1)^2) \tag{7.4-21}$$

For a non-singular finite energy solution we see from (7.4-21) that we require $\phi_{,\theta} = 0$ to avoid a logarithmic divergence. As a result the mapping $\tilde{\phi}$ can only be trivial. In the next section we will see that this situation can be salvaged by the introduction of new types of field called gauge fields which are a generalisation of the normal electromagnetic fields. However, there is another more general result which is less easily disposed of.

Let $\underline{\phi}$ be a static solution of a general nonlinear Klein-Gordon equation in N spatial dimensions involving a set of scalar fields ϕ_a. The energy of such a solution derives from two terms

$$E_\phi = E_\phi^1 + E_\phi^2 \quad \text{with} \quad E_\phi^{1,2} \geq 0 \tag{7.4-22}$$

where

$$E_\phi^1 = \tfrac{1}{2} \sum_{a=1}^{m} \int dx^1 \ldots dx^N (\nabla\phi_a)^2$$

$$E_\phi^2 = \int dx^1 \ldots dx^N\, U(\phi_1, \phi_2, \ldots, \phi_m)$$

Consider the one parameter family of field configurations $_\lambda\phi$ defined by

$$_\lambda\phi : \underline{x} \rightarrow \phi(\lambda\underline{x}) \tag{7.4-23}$$

the energy of a member of the family is given by

$$E_{_\lambda\phi} = \lambda^{2-N} E_\phi^1 + \lambda^{-N} E_\phi^2 \tag{7.4-24}$$

As $_1\phi$ is a static solution we must have

$$\left(\frac{\partial}{\partial\lambda} E_{_\lambda\phi}\right)_{\lambda=1} = 0 \tag{7.4-25}$$

which yields the result

$$(N-2)E^1_\phi + N\ E^2_\phi = 0 \tag{7.4-26}$$

If $N = 1$ we find that $E^1_\phi = E^2_\phi$ and that the total energy of a static solution is given by $2E^1_\phi$ in that case. This frequently simplifies its calculation. However, if $N > 2$ the only possible solution consistent with the positivity of energy components is given by $E^1_\phi = E^2_\phi = 0$. This means that there are no finite energy static solutions for $N \geqslant 2$. This negative result is know as **Derricks Theorem**. It is important to realise that we are allowed, and there do exist, models with time-independent non-dissipative solutions.

Another aspect that we must note is that relativistic field theories do not have to be of the non-linear Klein-Gordon type. For example the complex field theory corresponding to the Lagrangian density

$$L = \frac{|\psi,_t|^2 - |\psi,_x|^2}{(1-\lambda^2|\psi|^2)} - m^2|\psi|^2 \tag{7.4-27}$$

has genuine soliton solutions but is nonlinear. There are problems quantising such models but at a classical level they are perfectly acceptable. Other equations such as

$$\psi,_{tt} - \psi,_{xx} = \exp(-\psi) - \exp(+2\psi) \tag{7.4-28}$$

for a complex field ψ do not arise from a potential function $U(\psi,\overline{\psi})$. However, equation (7.4-28) has soliton solutions and may also be given a Hamiltonian formulation but not in the standard form.

7.5 Vortices, Monopoles, and Instantons

In the previous section we experienced some difficulty in constructing a nontrivial charge for finite energy solutions of the complex ϕ^4 model in two spatial dimensions. The existence of such a topological charge is, as we have seen, useful in proving the existence of nondissipative kink solutions in such models. One way of circumventing the problem is to alter the model by the introduction of additional fields known as gauge fields. As well as solving the problem on hand the equations satisfied by the new fields turn out to possess remarkable properties and kink like solutions known variously as vortices, monopoles and instantons. In this section we will attempt to give a brief outline of the nature of these 'soliton type' solutions and their associated topological charges. For convenience we will primarily restrict our attention to the static, time independent situation.

7.5.1 Abelian Gauge Fields

Consider the nonlinear Klein-Gordon equation in two spatial dimensions defined by the equations

$$\Phi,_{xx} + \Phi,_{yy} = 2\frac{\partial U}{\partial \bar{\Phi}}(\Phi,\bar{\Phi}) \tag{7.5-1}$$

These equations may be derived by minimising the energy functional or action defined by

$$E_{\Phi} = \tfrac{1}{2}[\,|\Phi,_x|^2+|\Phi,_y|^2+ 2U(\Phi,\bar{\Phi})\,] \tag{7.5-2}$$

If the potential function $U(\Phi,\bar{\Phi})$ is of the special form

$$U(\Phi,\bar{\Phi}) = V(|\Phi|) \tag{7.5-3}$$

the field equations (7.5-1) are invariant under the symmetry transformation

$$\Phi(x) \to ({}_{\alpha}\Phi)(x) = e^{i\alpha}\Phi(x) = \tilde{\alpha}\Phi \tag{7.5-4}$$

where α is a real constant. Such a transformation is called a **gauge transformation.** The phase α may be chosen to make the classical field $\Phi(x)$ real at any given point x. However, once such a choice has been made the phase of Φ is fixed at all other points.

Consider now the more general transformation

$$\Phi(x) \to ({}_{\beta}\Phi)(x) = e^{i\beta(x)}\Phi(x) = \tilde{\beta}(x)\Phi(x) \tag{7.5-5}$$

in which the phase β is now coordinate dependent $\beta = \beta(x)$ and the function $\tilde{\beta} : R^2 \to S^1$ is defined by $\tilde{\beta} : x \mapsto e^{i\beta(x)}$. We call such a transformation a **coordinate dependent gauge transformation.** The set of all such coordinate dependent phase transformations corresponding to smooth maps $\tilde{\beta} : R^2 \to S^1$ clearly has the structure of a group. We speak of a **continuous group** or **Lie group** of transformations. At any given point the group in this case is $U(1)$ the group of unimodular complex numbers.

Let us pose the question of how we might modify (7.5-1) so that it would be invariant under the action of this **gauge group.** The quantity $\Phi\bar{\Phi}$ is an invariant quantity for these gauge groups

$$\Phi(x)\bar{\Phi}(x) \to ({}_{\beta}\Phi(x))({}_{\beta}\bar{\Phi}(x)) = \Phi(x)\bar{\Phi}(x) \tag{7.5-6}$$

and so the potential term will behave in the same way as in the constant phase case. The problem centres on the derivative terms. The single derivative Φ_a, $(a = x$ or $y)$ transforms according to the rule

$$\Phi_{,a}(x) \rightarrow e^{i\beta(x)}(\Phi_{,a} + \tilde{\beta}^{-1}(x)\tilde{\beta}_{,a}(x)) \tag{7.5-7}$$

As a result we see that (7.5-1) will not remain invariant as it stands but can be made invariant if we can remove the extra term $\tilde{\beta}(x)^{-1}\tilde{\beta}(x)_{,a}$ which occurs. This can be done by the introduction of a new real valued field $A_a(x)$ called an **Abelian gauge field.** If we replace each of the derivatives $\Phi_{,a}$ in (7.5-1) by the combination $D_a\Phi$ defined by

$$D_a\Phi = (\Phi_{,a} - iA_a\Phi) \tag{7.5-8}$$

then this quantity will transform under the action of the gauge group in the simple manner

$$D_a\Phi(x) \rightarrow ({}_\beta D_a\Phi)(x) = e^{i\beta(x)}(D_a\Phi)(x) \tag{7.5-9}$$

just as Φ and $\Phi_{,a}$ did in the constant phase case, provided we define the action of the gauge group on the field $A_a(x)$ to be given by

$$A_a(x) \mapsto ({}_\beta A_a)(x) = A_a(x) - i\bar{\beta}^{-1}(x)\beta(x)_{,a} \tag{7.5-10}$$

The object $D_a\Phi$ is called the **covariant derivative** of Φ and A_a is called an abelian gauge field as $U(1)$ is an abelian group.

The modified field equations now take the form

$$(\partial_x - iA_x)^2\Phi + (\partial_y - iA_y)^2\Phi = \frac{\partial V}{\partial \bar{\Phi}}(|\Phi|) \tag{7.5-11}$$

and are invariant under the continuous group of transformations

$$\Phi \rightarrow {}_\beta\Phi \qquad A_a \rightarrow {}_\beta A_a \tag{7.5-12}$$

defined in (7.5-5) and (7.5-10) with $\beta : R^2 \rightarrow S^1$ a smooth map. We now observe that the new field A_a appears in the equations in exactly the same way that the vector potential of an electromagnetic field does in classical mechanics. This leads us to interpret the new gauge field as the electromagnetic field and to propose that, in the absence of the Φ field, it should satisfy the Maxwell equations in free space. These are usually expressed in the form

$$F_{ab,a} = 0 \tag{7.5-13}$$

(where we are using the summation convention of an implied sum over a repeated index) and the electromagnetic field tensor is defined by

$$F_{ab} = (A_{b,a} - A_{a,b}) \tag{7.5-14}$$

The field tensor F_{ab}, like $|\Phi|^2$, is invariant under the gauge group (7.5-12)

$$F_{ab}(x) \to ({}_\beta F_{ab})(x) = F_{ab}(x) \tag{7.5-15}$$

Coupled field equations (7.5-13) may be obtained by minimising the energy functional

$$E_F = \tfrac{1}{2}\|F\|^2 \tag{7.5-16}$$

where

$$\|F\|^2 = \tfrac{1}{2}F_{ab}F_{ab} \tag{7.5-17}$$

The energy density of the combined system of scalar and electromagnetic fields is given by

$$E_{\Phi,F} = E_\Phi + E_F = \tfrac{1}{2}[|D_x\Phi|^2 + |D_y\Phi|^2 + \|F\|^2 + 2V(|\Phi|)] \tag{7.5-18}$$

The field equations are obtained by minimising this action and take the form

$$F_{ab,a} = J_b \tag{7.5-19}$$

$$(D_x^2 + D_y^2)\Phi = 2\frac{\partial}{\partial\bar{\Phi}} V(|\Phi|) \tag{7.5-20}$$

where

$$J_b = \mathrm{Im}(\Phi\overline{D_b\Phi}) \tag{7.5-21}$$

is the electric current due to the Φ field and acts as a source for the Maxwell equations. We note that this involves a modification of these equations as the current depends explicitly on the gauge field.

The requirement of finite energy requires that the fields Φ and A_a satisfy the boundary conditions

$$|\Phi| \to c \tag{7.5-22}$$

$$|D_x\Phi|^2 + |D_y\Phi|^2 \to 0 \tag{7.5-23}$$

$$\|F\|^2 \to 0 \tag{7.5-24}$$

uniformly as $|x| \to \infty$ with c a zero of V.

$$V(c) = 0 \tag{7.5-25}$$

There are two different ways in which the relevant topological charge may be constructed. From (7.5-24) we see that as $|x| \to \infty$ we must have

$$F_{ab} \to 0 \tag{7.5-26}$$

However, this does not require the gauge fields A_a to become zero. The field F_{ab} must go to zero faster than $|x|^{-2}$ and so we must have

$$A_a = -i\tilde{\chi}^{-1}\tilde{\chi}_{,a} + O(|x|^{-1-\epsilon}) \text{ as } |x| \to \infty \quad (\epsilon>0) \tag{7.5-27}$$

where $\tilde{\chi} = \exp i\chi$ is a function of the polar angle ϕ and may be thought of as a function from $S^1 \to S^1$. A gauge transformation with a smooth gauge function $\omega(x)$ sends A_a to ${}_\omega A_a$ with the asymptotic behaviour

$${}_\omega A_a = -i(e^{-i\chi}\partial_a e^{i\chi}) - i(e^{-i\omega}\partial_a e^{i\omega}) + O(|x|^{-1-\epsilon}) \tag{7.5-28}$$

$$= -i(e^{-i(\chi+\omega)}\partial_a e^{i(\chi+\omega)}) + O(|x|^{-1-\epsilon}) \tag{7.5-29}$$

It might therefore be supposed that we can always choose $\exp i\omega = \exp -i\chi$ to eliminate $\tilde{\chi}$. However, unlike $\tilde{\chi}$, $\tilde{\omega}$ must be defined everywhere. In particular we must have $\tilde{\omega}$ independent of ϕ when $r = 0$ and $\tilde{\omega}$ is homotopic to the identity element. This means that a gauge transformation such as (7.5-28) cannot alter the homotopy class of $\tilde{\chi} = \exp i\chi$. The mapping $\tilde{\chi}$ therefore endows this model with a topological charge, the winding number of $\tilde{\chi}$. A simple expression for the charge can be obtained by considering the vector V_a defined by

$$V_a = \frac{1}{2\pi}\epsilon_{ab}A_b \tag{7.5-30}$$

As $|x| \to \infty$ the radial component $V_r = \hat{\underline{r}}\cdot\underline{V}$ has the asymptotic behaviour

$$V_r = \frac{1}{2\pi|x|}\chi_{,\theta} + O(|x|^{-1-\epsilon}) \tag{7.5-31}$$

and so integrating the function $|x|V_r$ around a circle of radius R and then allowing $R \to \infty$ we obtain

$$\lim_{R\to\infty}\int_0^{2\pi} d\theta|x|V_r(x) = \frac{1}{2\pi}\int_0^{2\pi} d\theta\, \chi_{,\theta} = \mathcal{Q}[\tilde{\chi}] \tag{7.5-32}$$

where $\mathcal{Q}[\tilde{\chi}]$ is the topological degree of $\tilde{\chi}: S^1 \to S^1$.

Alternatively, we can appeal to the divergence theorem, provided the vector field V_a is sufficiently well behaved, and obtain

$$\lim_{R\to\infty} \int_0^{2\pi} d\theta\, RV_r = \lim_{R\to\infty} \int_{|x|=R} \underline{V}.\underline{ds} = \int_{R^2} \text{div}\, \underline{V}\, d^2x \tag{7.5-33}$$

but

$$\text{div}\, \underline{V} = \frac{1}{2\pi}\, \epsilon_{ab} A_{b,a} = \frac{1}{2\pi}\, (A_{y,x} - A_{x,y}) = \frac{1}{2\pi}\, F_{xy} \tag{7.5-34}$$

Our final result is the integral formula

$$\frac{1}{2\pi} \int_{R^2} F_{xy} d^2x = Q[\tilde{\chi}] = N \tag{7.5-35}$$

where N is an integer. Physically F_{xy} is the magnetic field and $\int F_{xy} d^2x$ is the total magnetic flux through the x-y plane. Equation (7.5-35) states that this total flux is quantised into integral multiples of 2π. We shall return to this in Section 7.6 when it will occur in the context of superconductivity.

A second approach consists in considering the Φ field. In the particular example of Section 7.4 we attempted to construct the topological charge by considering the mapping $\tilde{\Phi}:S^1 \to X$ defined by

$$\tilde{\Phi} : \hat{n} \to \lim_{r\to\infty} \Phi(r\hat{n}) \tag{7.5-36}$$

where X is the set of zeros of the potential $V(|\Phi|)$. The particular example considered was the complex ϕ^4 potential

$$V(|\Phi|) = \frac{\lambda}{4} (|\Phi|^2 - 1)^2 \tag{7.5-37}$$

The set of zeros in that case are the points on the unit circle and consequently $X \cong S^1$. As $\pi_1(S^1) = Z$ this would have given us an integer valued charge had it been nontrivial. The construction failed because the mapping $\tilde{\Phi}$ was forced to be trivial by virtue of the requirement of finite energy which led to

$$\Phi_{,\theta} \to 0 \quad \text{as } |x| \to \infty \tag{7.5-38}$$

Once gauge fields are introduced the probem represented by (7.5-38) is immediately solved. For a finite energy solution we now require

$$\Phi_{,\theta} - i|x|A_\theta\Phi \to 0 \quad \text{as } |x| \to \infty \tag{7.5-39}$$

This means that for any finite energy solution we must have

$$A_\theta = O(|x|^{-1}) \qquad |x| \to \infty \tag{7.5-40}$$

From (7.5-27) and (7.5-39) we see that modulo a constant gauge transformation we have

$$\tilde{\Phi} = \tilde{\chi} \tag{7.5-41}$$

and the two different routes lead to the same topological charge.

7.5.2 Vortices

The model which consists of the ϕ^4 equations with the potential (7.5-37) modified by the inclusion of a U(1) gauge field is called the **Abelian Higgs Model.** The action functional is given by

$$E_{\Phi,F} = \tfrac{1}{2}[\,|D_x\Phi|^2 + |D_y\Phi|^2 + \|F\|^2 + \tfrac{\lambda}{2}(|\Phi|^2-1)^2\,] \tag{7.5-42}$$

For the special parameter value $\lambda = 1/2$ this can be recast into the form

$$E_{\Phi,F} = \tfrac{1}{2}[(\Phi_{1,x}+A_1\Phi_2) \mp (\Phi_{2,y}-A_2\Phi_1)]^2 \tag{7.5-43}$$

$$= \tfrac{1}{2}[(\Phi_{1,y}+A_2\Phi_2) \pm (\Phi_{2,x}-A_1\Phi_1)]^2 + \tfrac{1}{2}[F_{xy} \pm \tfrac{1}{2}(\Phi_1^2+\Phi_2^2-1)]^2 + \tfrac{1}{2}F_{xy}$$

where

$$\Phi = \Phi_1 + i\Phi_2 \tag{7.5-44}$$

We see that the total energy $\int E_{\Phi,F}\,d^2x$ satisfies, as a result of (7.5-35), the bound

$$\int_{R_2} E_{\Phi,F}\,d^2x \geq \pi N \tag{7.5-45}$$

The energy will be minimised and the bound saturated if and only if

$$(\Phi_{1,x}+A_1\Phi_2) - \epsilon(\Phi_{x,y}-A_2\Phi_1) = 0 \tag{7.5-46}$$

$$(\Phi_{1,y}+A_2\Phi_2) + \epsilon(\Phi_{2,x}-A_1\Phi) = 0 \tag{7.5-47}$$

$$F_{xy} + \tfrac{1}{2}\epsilon(\Phi_1^2+\Phi_2^2-1) = 0 \tag{7.5-48}$$

when

$$\epsilon N > 0 \quad , \quad \epsilon = \pm 1 \tag{7.5-49}$$

No closed form solutions to these equations are known but existence proofs can be obtained. It can be shown that for a given $N(N > 0)$ and a set of N points $\{z_i\}$ $(i = 1,\ldots,N)$ in C, there exists a finite action solution of the equations (7.5-46)-(7.5-49), unique to within gauge transformations, with the properties

(i) The solution is globally C^∞.

(ii) The zeros of $\Phi(z,\bar{z})$, $z = x + iy$, are the set of points $\{z_i\}$ $(i = 1,\ldots,N)$ and $\Phi(z,\bar{z}) \sim (z - z_j)^{n_j}$, $z \to z_j$ where n_j is the multiplicity of z_j in the set of zeros.

Such solutions with $N > 0$ are called **N-vortex solutions** and the analogous solutions with $N < 0$ are called **N-anti-vortex solutions.**

Although a closed form expression for an N-vortex solution is not known the system of equation (7.5-46)-(7.5-49) can be reduced to a simpler form. We can use our topological considerations to help us guess a form that the solutions might take. The map $\tilde{\chi}_n : S^1 \to S^1$ defined by

$$\tilde{\chi}_n : z \to z^n \qquad |z| = 1 \tag{7.5-50}$$

has charge n and thus suggests that for Φ we seek a solution of the form

$$\Phi_1 + i\Phi_2 = e^{in\theta} f(r) \tag{7.5-51}$$

where

$$f(\infty) = 1 \tag{7.5-52}$$

For this mapping $\chi = n\theta$ and so asymptotically the gauge field takes the form

$$A_a = (n\theta)_{,a} = -\frac{n\epsilon_{ab}\hat{x}_b}{r} \quad , \qquad \hat{x}_b = x_b/r \tag{7.5-53}$$

We may therefore hope to find a solution in the form

$$A_a = -\frac{n\epsilon_{ab}\hat{x}_b}{r} a(r) \tag{7.5-54}$$

where

$$a(\infty) = 1 \tag{7.5-55}$$

Substituting the forms (7.5-51) and (7.5-54) into the equations (7.5-46)-(7.5-49) for $N > 0$ reduces those equations to the following pair of ordinary differential equations

$$r\frac{df}{dr} - n(1-a)f = 0 \tag{7.5-56}$$

$$\frac{2n}{r}\frac{da}{dr} + (f^2 - 1) = 0 \tag{7.5-57}$$

Near to $r = 0$ the solutions have the asymptotic forms

$$f \sim c\, r^n \qquad (7.5\text{-}58)$$

$$a \sim \frac{1}{4n}\, r^2 \qquad (7.5\text{-}59)$$

and so Φ has a zero of order n at zero. Figure 7-17 shows the single vortex solution for Φ and for the gauge field near the singular point $r = 0$ where it takes the asymptotic form

$$\underline{A} = \pm \frac{r}{4}\, \hat{e}_\theta + O(r^2) \qquad (7.5\text{-}60)$$

It is clear from this diagram why we refer to these solutions as **vortices**. A similar analysis can be made when $\lambda \neq \frac{1}{2}$ and it can be shown that finite energy solutions of charge N exist for all N when $0 < \lambda \leqslant \frac{1}{2}$. Solutions with positive N we call **vortices** and those with negative N we call **anti-vortices.**

The time dependent equations take the form

$$F_{ab} = (A_{b,a} - A_{a,b}) \qquad (7.5\text{-}61)$$

$$F_{tb} = (A_{b,t} - A_{t,b}) = E_b \qquad (7.5\text{-}62)$$

(which define the fields F_{ab} and E_b) together with the dynamical equations

$$E_{b,t} = (F_{ab,a} - J_b) \qquad (7.5\text{-}63)$$

$$E_{b,b} = -\rho \qquad (7.5\text{-}64)$$

$$(D_t^{\,2} - D_x^{\,2} - D_y^{\,2})\,\Phi = -\lambda\Phi(|\Phi|^2 - 1) \qquad (7.5\text{-}65)$$

where J_b is defined in (7.5-21) and we define the charge density ρ by

$$\rho = -\operatorname{Im}(\Phi\overline{D_t\Phi}) \qquad (7.5\text{-}66)$$

Roman subscripts take the usual values. From (7.5-61),(7.5-62) we have the continuity equation

$$F_{xy,t} = E_{y,x} - E_{x,y} = \operatorname{div}(E_y, -E_x) \qquad (7.5\text{-}67)$$

From this equation we see, by means of the divergence theorem, that the charge

$$\mathcal{Q} = \frac{1}{2\pi}\int_{R^2} F_{xy}\, d^2x \qquad (7.5\text{-}68)$$

is independent of time provided $|\underline{E}| \to 0$ as $|x| \to \infty$ sufficiently fast. This means that if we start the system in say the 2-vortex state it can only evolve in

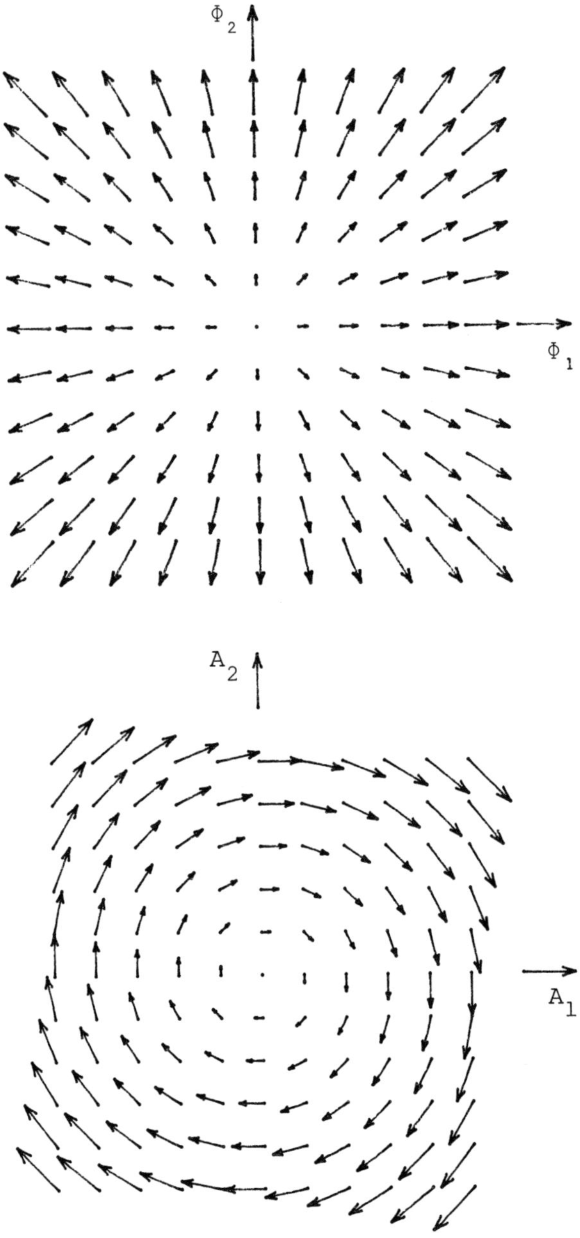

Figure 7-17: Single vortex solution

:ime to a state in the same homotopy class. In Section 7.2 we considered the force)etween the solitons of the sine-Gordon equation. A similar analysis can be made of .he forces between vortices. Jacobs and Rabbi have made a computer study of the two- 'ortex potential of the abelian Higgs model (7.5-42) for various values of λ. 'igure 7-18 shows the qualitative structure of their results. For $\lambda = 1/2$, the special value which led to the reduced equations (7.5-46)-(7.5-48), we see that there .s a zero interparticle force and the vortices do not interact. In this case -vortex solutions, such as those found by solving (7.5-56)-(7.5-57), can exist This 'alue of λ separates the λ-values into two distinct regions. For $\lambda < 1/2$ the .wo vortex potential is negative and the force is attractive. However, for $\lambda > 1/2$, .he potential is positive and leads to the repulsive force.

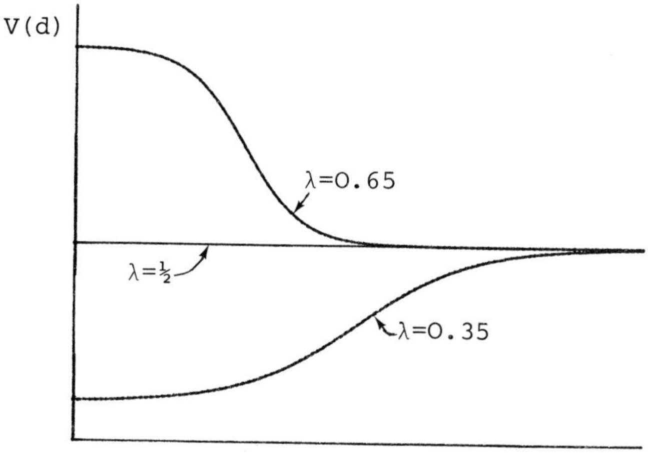

Figure 7-18: Intervortex potential as a function of the separation distance

s may readily be seen this system (and gauge field modified nonlinear Klein-Gordon quations in general) has some of the features of the exactly integrable soliton ystems concentrated upon in this book. However there does not seem to be either an nfinite sequence of conserved densities nor a Bäcklund transformation. We return to related system which does have both integrally charged kinks and a Bäcklund ransformation in Section 7.9.

.6 Crystal Dislocations and Order Parameters

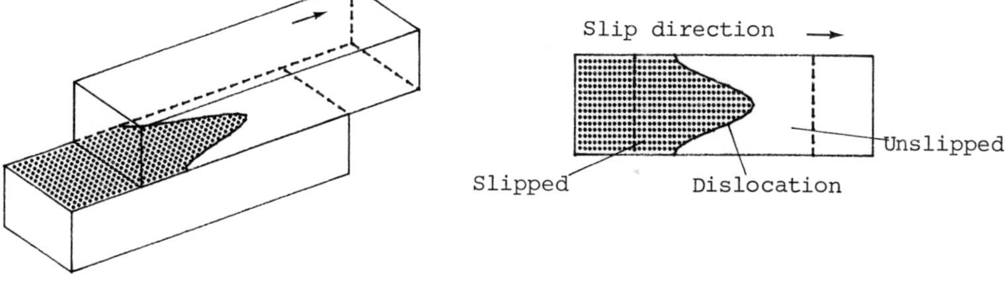

Figure 7-19: The slip process

When a metal crystal is continuously deformed by a shearing force it responds by the sliding of one plane of atoms in the crystal over another. Figure 7-19 schematically represents two regions of such a metal crystal. The upper region is slipping over the lower region under the action of a shearing force in the direction shown. The leading edge of the top region and an area behind it have all slipped but there is a region further back which has yet to slip. This delay leads to a distortion of the crystal lattice in a localised region of the crystal, in this case a planar region perpendicular to the direction of slip, which propagates through the crystal. This type of dislocation is known as an **edge dislocation.**

A very simple model of this situation may be constructed by regarding the lower region as remaining static and providing a periodic potential in which the atoms of the upper region move. Figure 7-20 shows the two layers immediately adjacent to the slip plane.

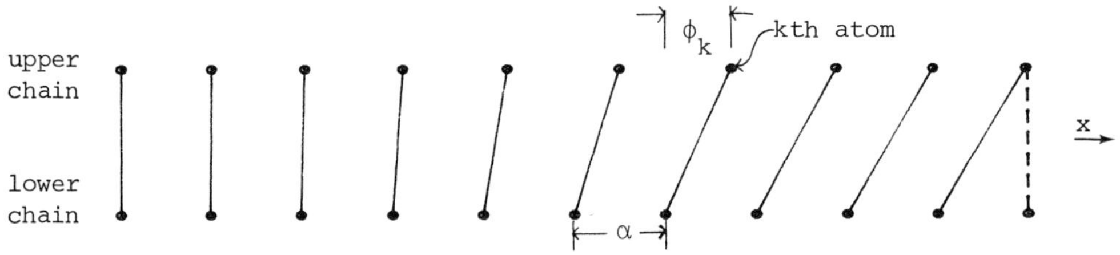

Figure 7-20:

If the lattice spacing in the direction of the slip plane is α then a possible form for the periodic potential resulting from the static lower layers as experienced by the kth atom of the upper layer is given by

$$U(\phi_k) = A(1-\cos(2\pi\phi_k/\alpha)) \tag{7.6-1}$$

where ϕ_k is the displacement of the kth atom from its equilibrium position. If the forces between the atoms in the crystal layer are assumed to be like those due to linear springs with a force constant of β, as considered in Chapter 1 concerning the FPU problem, the equation of motion of the kth atom is given by

$$m\ddot{\phi}_k = -2\pi A\alpha^{-1}\sin(2\pi\phi_k/\alpha) + \beta(\phi_{k+1}-2\phi_k + \phi_{k-1}) \tag{7.6-2}$$

Equation (7.6-2) is the same as that of the mechanical pendulum (7.1-4) after trivial scaling. If we define $\Phi_k = (2\pi\phi_k/\alpha)$, $A = \alpha^2\overline{A}$, $\beta = \overline{\beta}\alpha^{-2}$ and allow $\alpha \to 0$ we obtain in the continuum limit the sine-Gordon equation

$$m\Phi_{,tt} = \overline{\alpha}\Phi_{,xx} - (2\pi)^2\overline{A}\sin\Phi \tag{7.6-3}$$

The two equilibrium states $\Phi = 0$ and $\Phi = \pi$ of this system correspond to the situation of lattice allignment and of being a half plane out of step as shown in Fig. 7-19. The dislocation moves by a rapid sequence of 'half-slips'.

In this example we see clearly that elastic energy is stored up in such a dislocation. The energy required to create the static one soliton solution

$$\Psi = 4\tan^{-1}(\exp(2\pi x(\bar{A}/\bar{\beta})^{\frac{1}{2}})) \qquad (7.6\text{-}4)$$

is, from equation (7.2-6) using the Hamiltonian appropriate to (7.6-3)

$$E_{min} = 4\pi^{-1}(\overline{A\alpha})^{\frac{1}{2}} \qquad (7.6\text{-}5)$$

It is interesting to note that it is found from experiments that metals with a low value of E_{min} are more plastic. This may be explained by their increased ease of production of soliton-like dislocations. In the above we have only considered edge dislocations. We note for completeness that in general the boundaries between the slipped and non-slipped regions in a metal crystal are curves and are called **line defects.** The energy stored in such a defect is proportional to the length of the defect and we have a situation similar to that which occurs with surface tension. When the system evolves in time it does so in a manner which reduces its energy and this can lead to line defects which are in fact closed curves in a way analogous to the formation of soap bubbles. If we ask how two such line defects interact in a crystal we have a problem in which topological notions of the type we have considered will be important.

An ordered system such as a crystal allows us to specify, at each point of the region of space occupied by the system, a quantity Ψ called an **order parameter.** An order parameter characterises the extent to which the system is ordered in the region of a given point of the volume occupied by the system. In terms of cartesian coordinates $x \in R^n$ we assume the function Ψ, given in these coordinates by $\Psi : x \to \Psi(x)$, is continuous in some sense and its possible values constitute a space known as the **'order parameter space'** S_Ψ of the system.

There may be many different ways of defining an order parameter and a particular choice is usually associated with the study of a particular physical property of the system. A medium is said to be in an **ordered state** if the function Ψ allocates the same value to all points of the system.

The function Φ used above to describe the continuum analogue of the discrete crystal is an order parameter for that system. Similarly, the rotation angle ϕ of the continuum model of the mechanical pendulum considered in section two is also an order parameter. A single twist in the elastic band, or a one soliton dislocation in the crystal, separates two regions in which the system is in an orderd state. The order parameter varies rapidly only in the vicinity of the twist or line defect. In each of these two cases the order parameter space was the unit circle S^1. In the case of the band this was because the order parameter ϕ was an angle. In the case of the edge dislocation it arose as a result of the periodic nature of the crystal lattice. This is most easily seen by looking at a physical quantity such as the density. If $\rho_0(x)$ is the density of the perfect, undistorted, one-dimensional crystal layer then if the lattice spacing is α we have

$$\rho_o(x + \alpha) = \rho_o(x) \tag{7.6-6}$$

If the density $\bar{\rho}$ of the distorted crystal can be represented in the form

$$\bar{\rho}(x) = \rho_o(x+\phi(x)) \tag{7.6-7}$$

then ϕ is an order parameter for the system taking the value 0 in the ordered state. The function ϕ is not periodic but a change in the value of ϕ by a multiple of the lattice spacing α produces a configuration indistinguishable, by density measurements alone, from one in which it had not changed at all. The transformation (7.6-7) is a member of the group T of translations

$$T_L : x \to x + L \quad , \qquad (\bar{T}_L f)(x) = f(x+L) \tag{7.6-8}$$

If $H_\alpha = \{\bar{T}_L \in T : L = n\alpha,\ n \in Z\}$ the subgroup of translations by multiples of the lattice spacing α we see that the order parameter space $S_\Phi = T/H_\alpha \cong S^1$.

The advantage of considering the spaces that arise as quotient spaces is that it greatly facilitates the determination of the relevant homotopy groups. These in turn help us to decide on the existence of kink-like solutions. Most of the order parameters that commonly occur in condensed matter physics have associated with them a group of transformations G which acts transitvely on the order parameter space. If ψ is a particular value of an order parameter we define H_ψ the **little-group** or **isotropy group** of ψ, to be the subgroup of G defined by

$$H_\psi = \{g \in G : g\psi = \psi\}$$

In general H_ψ is not a normal subgroup of G. As G acts transitively on S_ψ any two little groups are isomorphic via an inner automorphism. The structure we have outlined enables us to identify S_ψ with the quotient space G/H_ψ of left cosets of H_ψ. The existence of kink solutions in condensed matter physics is therefore closely related to the structure of the homotopy groups $\pi_n(G/H)$.

There are two physical phenomena which are related to the existence of an ordered state of matter and for which the sine-Gordon equations are of particular relevance. These are Ferromagnetism and Superconductivity. Ferromagnetism will be considered in Section 7.7. In this section and also Section 7.8 we will consider Superconductivity. For the present we will describe the physical phenomena and introduce the classical theory of Landau and Ginzberg which is expressed in terms of a complex valued order parameter. When we return to a further discussion of superconductivity in Section 7.8 we will be concerned with quantum mechanical tunneling and the Josephson effect. This will provide us with the opportunity to develop some of the necessary quantum mechanical ideas in the earlier part of that section.

Superconductivity

The superconducting state of matter is an ordered phase characterised by a number of distinct and remarkable properties. The most important of these phenomena are infinite conductivity and the Meissner effect.

Infinite Conductivity

If certain metals such as lead, mercury or niobium are cooled to within a few degrees of absolute zero they suddenly lose, at a well defined temperature $T = T_c$, all traces of electrical resistance. An electric current once established within a ring of such material will circulate indefinitely and experimentally a lower bound of 10^5 years has been determined for the decay time of such a current loop.

Theoretically the current should persist for at least $10^{10^{10}}$ years being destroyed by thermodynamic fluctuations which lead to an effective electrical resistance at temperatures below $T = T_c$. It is interesting to note that those same thermodynamic fluctuations can also give rise to superconducting effects at temperatures above $T = T_c$. This overlapping of the two regimes due to fluctuation effects is typical of quantum systems. Such quantum mechanical behaviour is to be expected as the critical temperature of a superconducting metal such as mercury is 4.2°K and at such low temperatures we are in a realm of physics far removed from the world of our experience. In such uncharted territory, in the domain of sub-atomic particles, quantum mechanical considerations are clearly of paramount significance and the normal laws of physics may not hold.

The Meissner effect

If the metal is cooled in the presence of a constant magnetic field H then the critical temperature is dependent upon H, $T_c = T_c(H)$. Figure 7-21 shows the experimentally determined shape of the curve $T_c = T_c(H)$.

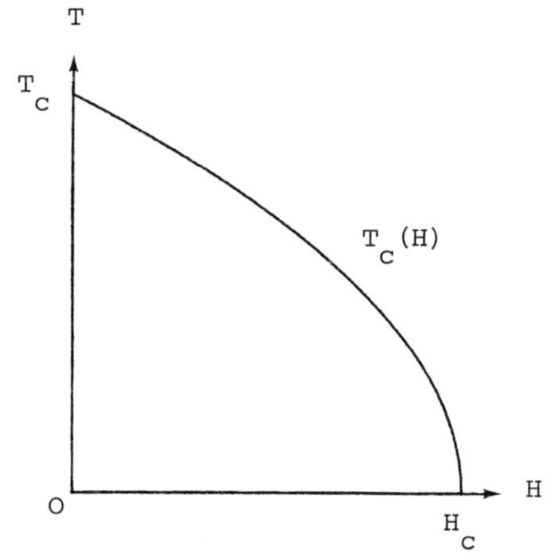

Figure 7-21:

We note that if $H > H_c$ then cooling even to $T = 0$ will not induce a transition to the superconducting phase and such fields destroy the superconducting state.

As the temperature is decreased through the critical temperature the magnetic field, which in the normal phase penetrates into the metal, is forcefully excluded from the bulk of the superconductor. This expulsion of the field from the sample in the normal phase as it makes its transference into the superconducting phase is called the Meissner effect. The field is not totally excluded but penetrates to a finite distance λ into the surface of the material. The distance λ is called the **penetration depth.** A typical value of the penetration depth is about 10^{-5} cm.

The superconducting state represents a definite thermodynamic phase marked by a discontinuous change in thermodynamic quantities such as the free energy F as the temperature makes the transition through the critical temperature.

An essential element in most of the theories that have been developed to explain the special properties of the superconducting state is the idea of Cooper pairs or superconducting electrons.

Cooper pairs

A fundamental notion in the theory developed by Bardeen, Cooper and Schreiffer is that the free conduction electrons in a superconducting metal can bind together in pairs as a result of the interaction with the lattice of ion cores. The idea is simple. If we consider two isolated electrons but concentrate initially on just one of the pair we can see that is will attract positive ions towards it. The cloud of positive ions that surround the first electron will then attract the second electron of the pair. Although intuition suggests that this should be a small effect it turns out that what appears to be a minor screening effect can lead to an attractive force. This binding force leads to the formation of a bound state, a composite object having a lower energy than its two components separately, and called either a Cooper pair or sometimes a superconducting electron. This bound state has twice the electronic charge, approximately twice the electron mass and most important of all it has no intrinsic spin. A normal electron has an internal quantum number called spin which can take the values $\pm 1/2$. Such particles with spin are called **fermions** and a fundamental property of such particles is that in an aggregate consisting of many fermions each fermion must exist in a different quantum state from all the other fermions in the system. Particles with zero intrinsic spin are called **bosons** and in an aggregate of bosons all the constituents may occupy the same quantum configuration. Fermions are said to obey Fermi-Statistics and bosons Bose-Statistics. The bound pair acts as a boson because if both electrons in a pair are exchanged, the sign of the wave function changes twice.

The Cooper pairs have a spatial extent of the order of a second characteristic length ξ known as the **coherence length.** For a typical superconductor ξ is approximately 10^{-4} and the ratio of the penetration depth to the coherence length, known as the Landau-Ginzberg parameter, is usually less than one. The two electrons in a pair are quite widely spread and the average distance between pairs is much smaller than the size of the pair.

As a result of the attractive inter-electron forces, resulting from the interaction with the ion lattice, the gas of electrons is unstable against the formation of Cooper pairs and pairing continues until some equilibrium point is reached. Because of the Boson nature of the pairs a system consisting of a large condensate of Cooper pairs, most of which are in their lowest energy state, is therby formed. One can now see why superconductors are so different from normal metals.

One of the earliest and most successful phenomenological theories formulated in order to codify the experimental facts known at the time was devised by Landau and Ginzberg. It was a product of remarkable physical intuition and the bare outline that we will give below involves many interpretations made long after their theory was originally proposed. Indeed, it was not until 1959 that Gorkov was able to link their highly successful theory directly to microscopic theory.

Landau-Ginzberg Theory

We recall from Chapter 2 that we interpret the wave function of a quantum particle as giving the probability density. This was not the original interpretation of Schrödinger but a later re-interpretation due to Born. Schrödinger originally interpreted the wave function in terms of the charge density and current density given by $\rho = e\psi\psi^*$ and $\underline{j} = ie\hbar/2m\ (\psi^*\nabla\psi - \psi\nabla\psi^*)$ which he thought of as the physical electric charge density and electric current of the quantum mechanical electron described by the wave-function ψ. If this interpretation were correct the electric currents should generate electromagnetic fields and a system such as a hydrogen atom should emit light and consequently be unstable. As this was not the case the re-interpretation of Born was essential to the success of the theory. However, there is a situation in which the original view of Schrödinger is tenable. Consider the situation in which a large number of particles, necessarily bosons, are in the same state and let Φ be the wave function describing that configuration. In that situation $|\Phi|^2$ can be interpreted as the density of particles and if each has charge q then $q|\Phi|^2$ is the electric charge density. This is precisely the situation within a superconductor. Landau and Ginzberg introduce a complex pseudowave-function Φ as an order parameter for the Bose-condensate of Cooper pairs. The quantity n defined by

$$n(x) = |\Phi(x)|^2 \tag{7.6-9}$$

is interpreted, in the original Schrödinger way, as giving the local density of superconducting electrons. If a material consists of regions of normal and superconducting phases n is zero in the normal phase regions and approximately constant within the superconducting regions. The essential idea of Landau and Ginzberg is that Φ varies slowly in space except at regions of interface between normal and superconducting zones. This idea is implemented by supposing that the free-energy density of the superconducting phase can be expanded in a series of the form

$$f = f_o + f_2|\Phi|^2 + f_4|\Phi|^4 + \frac{1}{2m^*}|(-i\hbar\underline{\nabla}\Phi - e^*\underline{A}\Phi)|^2 \tag{7.6-10}$$

where f_0 is the free-energy density of the normal phase, the quantities f_i $(i = 2,4)$ are temperature dependent constants, m^* is the effective mass of a Cooper pair and $e^* = 2e$ is its charge. The wave-function of the superconducting electrons has been coupled to the electromagnetic field described by the vector potential A in the gauge invariant manner described in Section 7.5.

If we make a change of scale and add a physically irrelevant constant term (7.6-10) can be recast into the form

$$\tilde{f} = \tfrac{1}{2}\sum_{a=1}^{3}|D_a\Phi|^2 + \frac{\lambda}{4}(|\Phi|^2-1)^2 \tag{7.6-11}$$

which we immediately recognise as the static, $d = 3$, abelian Higgs model of Section 7.5. The Landau-Ginzberg equations are obtained by minimising the free-energy just as the action of the abelian Higgs model was minimised in the previous section. The equations are completely equivalent. In the previous section we discovered that the Higgs model had vortex solutions. The forces between two vortices depend upon whether λ is greater than or less than $1/2$. The properties of superconductors described so far correspond to systems described by Higgs models with $\lambda < 1/2$ and are called type I superconductors. Systems which correspond to Higgs models with $\lambda > 1/2$ are called type II superconductors.

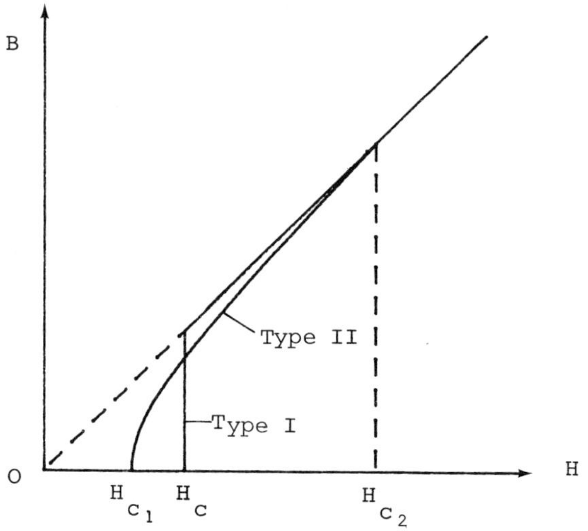

Figure 7-22: Type I and II Superconductors

Figure 7-22 shows the difference in flux penetration between type I and type II superconductors. In each case the flux is eventually excluded but in a type II material the expulsion does not occur at a sharp value of the external field. For a type II superconductor there are two critical values of the external field H_{c_1} and H_{c_2}. As the field is increased past H_{c_1} the flux begins to penetrate into the superconductor. However, unlike the penetration in type I materials which is laminar, the superconductor is pierced by a number of vortex tubes which are regularly spaced and form a lattice. The individual vortex tubes are held apart by their mutual repulsion and the total number of tubes is defined by the total flux which, by (7.5-35), is quantised into N units of flux. Each tube carries a single quantum of flux into the body of the superconductor. This filamentation process continues until $H = H_{c_2}$ when the internal magnetic field matches the external field and the system is in its normal phase. For values of H between H_{c_1} and H_{c_2} the system is said to be in the intermediate or Schubnikov phase. The lattice of vortices is an experimentally observable structure as shown by Essmann & Träuble.

As vortices are associated with topological aspects of the equations that describe the superconducting state it will not be surprising if they occur again. In our next example, Ferromagnetism, vortices will again occur and many of the remarks made here will be relevant there also.

7.7 Ferromagnetism and Solitons

A magnetic crystal may be viewed as a lattice of magnetic dipoles. In a paramagnetic material the magnetic ions forming the lattice must be aligned by an external magnetic field before the crystal displays a magnetic moment. In the case of ferromagnetic materials no such applied magnetic field is required. Ferromagnetic crystals possess a magnetic dipole moment even in the absence of an external field. The interactions between the magnetic ions in the ferromagnetic case are sufficient to induce the long range orderings of the elementary dipoles which are needed to produce a macroscopic affect.

A ferromagnet may not always seem to possess a magnetic moment and may need to be 'magnetized' by an applied field. Even a field of very low intensity will suffice. When such an external field is removed a permanent magnetic moment is retained. This resultant magnetization can be destroyed by mechanical shock or heat.

At high temperatures ferromagnetic substances behave as paramagnets which spontaneously develop a permanent magnetic moment at a critical temperature known as the Curie temperature T_c. Below the Curie temperature the mutual alignments which are produced consist of large regions in which all of the dipoles are parallel to one another. These regions are known as **ferromagnetic domains.** The direction of magnetization changes from one domain to the next. More than one domain is formed as this allows the magnetic energy of the crystal to be substantially reduced.

The boundaries separating these domains are known as **Bloch walls.** We will see that for certain configurations there can be motion of these boundaries which are described by the sine-Gordon equation. The Bloch walls represent surfaces within a ferromagnetic system upon which the magnetization cannot be defined. They are singular surfaces akin to line defects.

In a continuum model we consider the lattice distributed over space according to some density function. In place of $\underline{m}_i$ the magnetic moment of the dipole at lattice site i we have a magnetic moment field $\mathbf{m}(\mathbf{x})$ where $\mathbf{m}(\mathbf{x})$ is the magnetic moment density at the point with coordinates $\mathbf{x}$. The function $\mathbf{m}(\mathbf{x})$ is a typical order parameter. The magnetic moment vector has constant length M

$$m_x^2 + m_y^2 + m_x^2 = M^2 \tag{7.7-1}$$

for all $\mathbf{x}$ and so the order parameter space for this system is S^2.

If each of the magnetic moments in a crystal were free and independent of the other magnetic moments its variation in time would be determined by the external magnetic field. In the intuitive Landau-Lifshitz theory of such magnetic materials the influence of the interactions between the different dipoles is characterised by an 'effective magnetic field'.

Let us denote the magnetic energy density in the absence of an external field by W. In the presence of an external field it is given by

$$\bar{W} = W - \underline{H}.\underline{m} \tag{7.7-2}$$

where $\underline{H}$ is the magnetic field strength within the crystal.

The distribution of moments within the crystal is determined by the requirement that the total energy $\underline{E}$ given by

$$\bar{E} = \int d^3x\ \bar{W} \tag{7.7-3}$$

is a minimum. Varying $\underline{m}$ but keeping $\mathbf{H}$ fixed we obtain

$$\delta\bar{E} = \int d^3x \{ \left(\frac{\delta W}{\delta \underline{m}}\right) - \underline{H} \}.\delta\underline{m} = 0 \tag{7.7-4}$$

From this equation we see that the quantity $(\delta W/\delta \underline{m} - \mathbf{H})$ is always parallel to $\mathbf{m}$. This follows from the fact that $\delta \mathbf{m}.\ \mathbf{m} = 0$ as $\mathbf{m}$ is a vector of constant length. The magnetic dipoles are aligning themselves with the field $(\mathbf{H} - \delta W/\delta \underline{m})$ and this led Landau and Lifshitz to interpret the quantity $(\mathbf{H} - \delta W/\delta \underline{m})$ as the 'effective magnetic field' denoted by $\underline{\mathbf{H}}_{eff}$.

$$\underline{H}_{eff} = \underline{H} - \left(\frac{\delta W}{\delta \underline{m}}\right) \tag{7.7-5}$$

If the torque $\underline{\Gamma}$ on the elementary magnetic moments is taken to be proportional to $\mathbf{H}_{eff}$

$$\underline{\Gamma} = \gamma\, \underline{H}_{eff} \tag{7.7-6}$$

we have the equation of motion

$$\underline{m}_{,t} = -\gamma\, \underline{H}_{eff} \wedge \underline{m} \tag{7.7-7}$$

The macroscopic field $\mathbf{H}$ is determined by the Maxwell equations of magneto-statics within the crystal. These take the form,

$$\begin{aligned} &\text{curl}\ \underline{H} = 0 \\ &\text{div}(\underline{H} + 4\pi \underline{m}) = 0 \end{aligned} \tag{7.7-8}$$

together with the boundary conditions. At the boundary between two different media 1 and 2 we must have

$$\begin{aligned} &\underline{H}_{1t} = \underline{H}_{2t} \\ &(\underline{H} + 4\pi \underline{m})_{1n} = (\underline{H} + 4\pi \underline{m})_{2n} \end{aligned} \tag{7.7-9}$$

where the subscripts t and n correspond to the tangential and normal components of the vectors at the boundary.

It is often found experimentally that ferromagnetic crystals are easier to magnetise along one axis than any other. Such an axis is called the **axis of easiest magnetization.** If we restrict our attention to the situation of magnetized layers stacked up along the axis of easiest magnetization and chosen to be the z-axis, we may seek solutions that depend only upon the horizontal x-coordinate in the layers. The analogy between this case and that of edge dislocations considered in Section 7.9 is readily seen.

For this one-dimensional case the Maxwell equations (7.7-8) reduce to

$$H_{y,x} = 0 \qquad H_{z,x} = 0 \qquad (H_x + 4\pi m_x)_{,x} = 0 \tag{7.7-10}$$

If there is no external field the solution of these equations subject to the boundary conditions (7.7-9) at the surface of the crystal, which is assumed to be horizontal, is given by

$$H_x = -4\pi m_x \qquad H_y = 0 \qquad H_z = 0 \tag{7.7-11}$$

This solution may be recast into the form

$$\underline{H} = -\frac{\delta}{\delta \underline{m}} (2\pi m_x^2) \tag{7.7-12}$$

Using this result the equations of motion can be written as

$$\underline{m}_{,t} = -\gamma \frac{\delta \overline{W}}{\delta \underline{m}} \wedge \underline{m} \tag{7.7-13}$$

where

$$\overline{W} = W + 2\pi m_x^2$$

is the total magnetic energy. We will refer to equation (7.7-13) as the **Landau-Lifshitz equation.**

Consider a ferromagnetic crystal with an axis of easiest magnetization. In the Landau-Lifshitz theory of such crystals the total magnetic energy density W is assumed to consist of two contributions W_I and W_A with

$$W = W_I + W_A \tag{7.7-14}$$

The contribution W_I is due to the nonhomogeneous distribution of the magnetic moments within the crystal. The energy per unit volume that results from this inhomogeneity is modelled by

$$W_I = A\left[(\nabla \underline{m}_x)^2 + (\nabla \underline{m}_y)^2 + (\nabla \underline{m}_z)^2\right] \tag{7.7-15}$$

In the one-dimensional case this reduces to

$$\tilde{W}_I = A(\underline{m},_x)^2 \tag{7.7-16}$$

The contribution W_A is due to existence of an axis easiest magnetization. If we suppose the coordinate system chosen so that the z-axis lies along the axis of easiest magnetization then this **magnetic anisotropy energy** per unit volume is modelled by

$$W_A = K(m_x^2 + m_y^2) \tag{7.7-17}$$

which is zero when m points along the axis of easiest magnetization. As the vector m is of constant modulus we may replace W_A by

$$\tilde{W}_A = -Km_z^2 \tag{7.7-18}$$

This adds only a constant to the total energy of the system and does not affect the Landau-Lifshitz equation (7.7-13).

For the one-dimensional case the total energy density is given by

$$\tilde{W} = \left[2\pi m_x^2 + A(m,_x)^2 - Km_z^2)\right] \tag{7.7-19}$$

Substitution of this into the Landau-Lifshitz equations yields the equations of motion for the one-dimensional model.

Let us introduce the polar representation of m given by

$$\underline{m} = M(\sin\theta\cos\psi, \sin\theta\sin\psi, \cos\theta) \tag{7.7-20}$$

This is shown diagrammatically in Fig. 7-23.

The Landau-Lifshitz equations (7.7-13) may then be written in the form

$$\begin{aligned} M\sin\theta\ \psi,_t &= \gamma\frac{\delta\tilde{W}}{\delta\theta} \\ M\sin\theta\ \theta,_t &= -\gamma\frac{\delta\tilde{W}}{\delta\psi} \end{aligned} \tag{7.7-21}$$

where W takes the form

$$\tilde{W} = 2\pi M^2\sin^2\theta\cos^2\psi - K\cos^2\theta + A\left[\sin^2\theta(\psi,_x)^2 + (\theta,_x)^2\right] \tag{7.7-22}$$

The resulting equations are

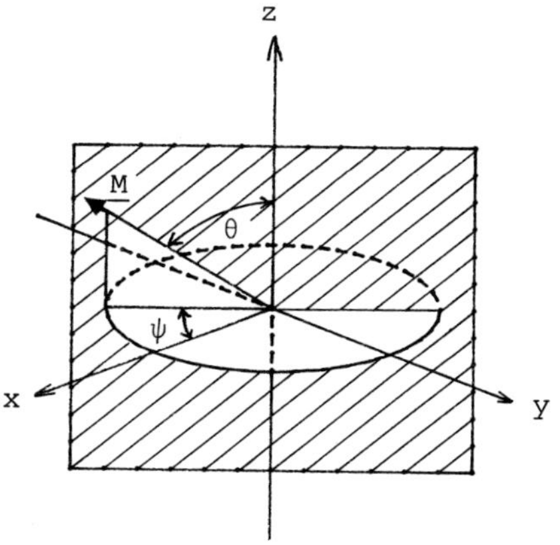

Plane containing the magnetization vector

Figure 7-23:

$$\gamma^{-1} M\psi_{,t} \sin\theta = 4\pi M^2\cos^2\psi\cos\theta\sin\theta+2K\sin\theta\cos\theta-2A\theta_{,xx}$$
$$\gamma^{-1} M\theta_{,t} \sin\theta = 4\pi M^2\sin^2\theta\sin\psi\cos\psi-2A(\sin^2\theta\psi_{,x})_{,x} \tag{7.7-23}$$

Now suppose that

$$\psi = \pi/2 + \epsilon \tag{7.7-24}$$

with $\epsilon <<1$ and a sufficiently slowly varying function of x that the final term in the second equation of (7.7-23) may be neglected. The equations (7.7-23) then reduce to the approximate form

$$M\gamma^{-1}\epsilon_{,t} \sin\theta \sim K\sin 2\theta-2A\theta_{,xx}$$
$$M\gamma^{-1}\theta_{,t} \sim 4\pi M^2\sin\theta\ \epsilon \tag{7.7-25}$$

Eliminating ϵ we obtain

$$\theta_{,tt} = -\ 4\pi\gamma^2(K\sin 2\theta-2A\theta_{,xx}) \tag{7.7-26}$$

If we define

$$\Phi = 2\theta, \quad X = (\frac{A}{K})^{-\frac{1}{2}} x, \quad T = (8\pi\gamma^2 K)^{-\frac{1}{2}} t \tag{7.7-27}$$

equation (7.7-26) takes the standard form

$$\Phi_{,tt} - \Phi_{,xx} + \sin\Phi = 0 \tag{7.7-28}$$

of the sine-Gordon equation. The soliton solutions of this sine-Gordon equation are 180° ferromagnetic domain walls. The solitons interpolate between two equilibrium states which differ by a multiple of 2π determined by the topological charge of the solution. Consequently, they describe how the magnetization changes between up and down domains. The single soliton corresponds to a moving wall between an up and a down zone. The soliton-antisoliton solution and soliton-soliton solution correspond to two walls separating three domains, two up separated by a single down region. The breather solution is time periodic and represents a localized deviation of the magnetization density from a direction parallel or antiparallel from the axis of easiest magnetization.

7.7.1 The Isotropic Heisenberg Ferromagnet

The Landau-Lifshitz equations define a general class of models defined by a single function W. The isotropic Heisenberg ferromagnet is defined by the magnetic density

$$W = A(\nabla \underline{m})^2 \tag{7.7-29}$$

with corresponding Landau-Lifshitz equation

$$\underline{m}_{,t} = -\gamma A\, \underline{m} \wedge \Delta \underline{m} \qquad \underline{m}^2 = M^2 \tag{7.7-30}$$

By appropriately scaling the variables this equation may be reduced to the canonical form

$$\underline{S}_{,t} = \underline{S} \wedge \Delta \underline{S} \qquad |\underline{S}|^2 = 1 \tag{7.7-31}$$

Equation (7.7-31) is of the type considered in Section 7.4. Each solution defines a mapping $\tilde{S}^t : S^2 \to S^2$ given in coordinates by

$$\tilde{S}^t : \hat{n} \to \lim_{r\to\infty} \underline{S}(r\hat{n},t) \qquad (\hat{n})^2 = 1 \tag{7.7-32}$$

similar to the mapping defined on S^1 by equation (7.4-20). For such mappings we have the topologically conserved charge,

$$Q = \frac{1}{4\pi} \int_{S^2} d\theta d\phi \;\; \underline{\tilde{S}}^t \cdot [\tilde{S}^t_{,\theta} \wedge \tilde{S}^t_{,\phi}] \tag{7.7-33}$$

which is independent of t.

Solutions which conform with the boundary condition

$$\underline{S}(x,t) \to \underline{S}_0 \qquad \text{a constant for} \quad |x| \to \infty \tag{7.7-34}$$

will have the charge (7.7-33) zero. However, we should note that this boundary condition means that each solution $\underline{S}$ actually defines a mapping from S^3 (compactified R^3) to S^2. Viewed in this way each solution of (7.7-31) with boundary condition (7.7-34) may be associated with an element of $\pi_3(S^2)$. As $\pi_3(S^2) = Z$ it is possible to define an integral valued topological charge for such solutions but it is nonlocal in character. We wish only to point out that there may be more than one topological charge associated with a problem.

A particularly interesting set of solutions to (7.7-31) are the static, time independent solutions. These solutions satisfy the equations,

$$\underline{S} \wedge \Delta\underline{S} \tag{7.7-35}$$

As $\underline{S}$ is of unit length it follows that

$$\Delta\underline{S} = \lambda(\underline{x})\underline{S} \tag{7.7-36}$$

for some scalar function $\lambda(\mathbf{x})$. It is easily shown that

$$\lambda(x) = - \nabla S \,.\, \nabla\underline{S} \tag{7.7-37}$$

and equation (7.7-36) becomes

$$\Delta\underline{S} + (\nabla\underline{S}.\nabla\underline{S})\underline{S} = 0 \tag{7.7-38}$$

The solutions of this equation may be classified according to their topological charge defined by (7.7-33). The elements of the homotopy group $\pi_2(S^2)$ are characterising the possible **singular points** in the ferromagnet where the magnetic moment vector is not defined.

In the case of planar spins in the x-y plane, equation (7.7-30) takes the form

$$\underline{S}_{,xx} + \underline{S}_{,yy} + (|\underline{S}_{,x}|^2 + |\underline{S}_{,y}|^2)\underline{S} = 0 \qquad \underline{S} \in S^2 \tag{7.7-39}$$

For this planar case the solution function $\underline{S}$ maps R^2 into S^2. From the theory of Section 7.3 we may therefore define a topological charge for such solutions by

$$Q[\underline{S}] = \frac{1}{8\pi} \iint \epsilon_{\mu\nu}\epsilon^{abc} S,_{\mu}{}^{a} S,_{\nu}{}^{b} S^{c} d^2x \tag{7.7-40}$$

The equations (7.7-39) result from the minimisation of the functional

$$W[\underline{S}] = \iint d^2x \; (|S,_x|^2 + |S,_y|^2) \tag{7.7-41}$$

subject to the constraint $|\underline{S}|^2 = 1$.

Following the analysis of vortices in Section 7.5 we can utilise the formula (7.7-40) to write $W[\underline{S}]$ in the following form

$$W[\underline{S}] = \iint d^2x \; \tfrac{1}{2} \sum_{\mu=1}^{2} |K_{\pm}^{\mu}|^2) \quad \mp 8\pi \; Q[S] \tag{7.7-42}$$

where

$$K_{\pm}^{\mu a} = \left(S^a,_\mu \pm \epsilon_{\mu\nu} \epsilon^{abc} S,_\nu^b \, S^c \right) \tag{7.7-43}$$

This means that

$$W[\underline{S}] \geq 8\pi \; |Q[\underline{S}]| \tag{7.7-44}$$

and that the functional will be minimised within the class of charge N solutions when

$$K_{\epsilon}^{\mu a} = 0 \qquad \epsilon N < 0 \qquad (\epsilon = \pm 1) \tag{7.7-45}$$

These are directly analogous to equations (7.5-46)-(7.5-49). The solutions of (7.7-39) which satisfy (7.7-45) are also called vortices.

If we introduce the paramerisation

$$\underline{S} = (\sin\theta\cos\Phi, \sin\theta\sin\Phi, \cos\theta) \tag{7.7-46}$$

equation (7.7-45) takes the form

$$\Phi,_\mu = - \epsilon \; \operatorname{cosec}\theta \; \epsilon_{\mu\nu} \theta,_\nu \tag{7.7-47}$$

This equation can be reduced to its simplest form by mapping the image sphere of the mapping $\underline{S}$ into the complex plane. If $\underline{S}$ satisfies the boundary condition

$$\underline{S} \to (0,0,-1) \qquad |x| \to \infty \tag{7.7-48}$$

a convenient choice of complex variable is given by

$$\omega = \cot(\theta/2)e^{-i\Phi} = \omega_1 + i\omega_2 \tag{7.7-49}$$

and equation (7.7-47) reduces to the linear form

$$\omega_{1,\mu} = - \epsilon\epsilon_{\mu\nu}\omega_{2,\nu} \tag{7.7-50}$$

These are simply the **Cauchy-Riemann** equations.

We may now write down a general N-vortex solution. If we define

$$\xi = x + i\epsilon y \tag{7.7-51}$$

and assume the absence of essential singularities, then the general solution of (7.7-50) is given by

$$\omega(\xi) = \omega_0 \prod_i (\xi-\xi_i)^{n_i} \prod_j (\xi-\tilde{\xi}_j)^{-m_j} \quad (n_i, m_j > 0) \tag{7.7-52}$$

As in Section 7.5, the charge corresponding to $\omega(\xi)$ is given by the zeros of $\omega(\xi)$ and no mixed vortex-antivortex states exist. If the spin fields corresponding to $\omega(\xi)$ is denoted by $\underline{S}[\omega]$, then

$$Q\,[S[\omega]] = \sum_i n_i \tag{7.7-53}$$

To obtain a 1-vortex solution we may choose $\omega_1 = i\xi$ and obtain

$$\Phi = \phi - \pi/2 \qquad \theta = 2\cot^{-1}\rho \tag{7.7-54}$$

with the corresponding spin vector

$$\underline{S} = \left(\frac{2\rho}{1+\rho^2}\sin\phi, \ -\frac{2\rho}{1+\rho^2}\cos\phi \ , \ \frac{1-\rho^2}{1+\rho^2}\right) \tag{7.7-55}$$

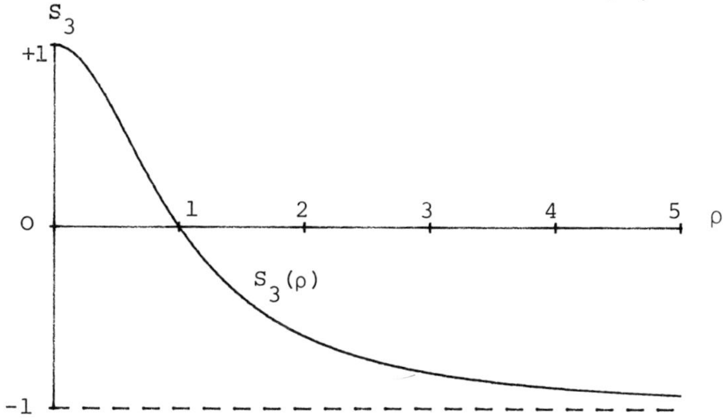

(a) The spin component S_3

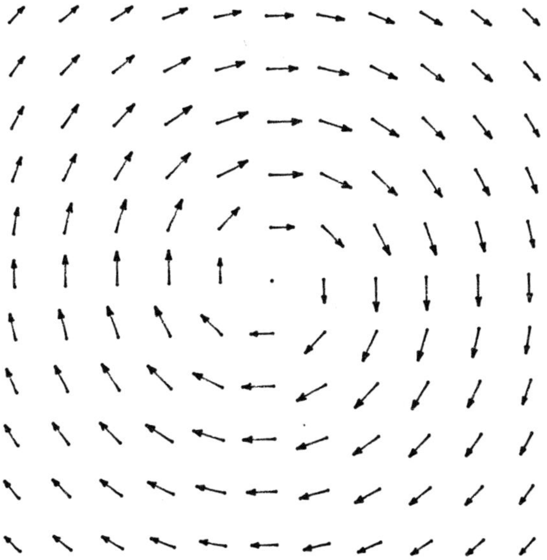

(b) The projected vector field $\underline{S}$ in the x-y plane
Figure 7-24:

Figure 7-24a shows the spin component S_3. We see that the spin field is an ordered state perpendicular to the spin plane except in a circular region about the origin. Figure 7-24b illustrates the direction field in the x-y plane and we see it has the typical vortex structure familiar from Section 7.5.

Although (7.7-39) is not an evolution equation, it may be associated with an inverse scattering problem of a nonstandard type.

If we introduce the complex variable

$$\xi = x + iy \tag{7.7-56}$$

then equation (7.7-39) results from the integrability conditions of the linear scattering problem

$$\psi,_{\xi} = (1-\gamma)\ (\underline{S} \wedge \underline{S},_{\xi}) \cdot \frac{\underline{\sigma}}{2i}\ \psi \tag{7.7-57}$$

$$\psi,_{\bar{\xi}} = (1-\gamma^{-1})\ (\underline{S} \wedge \underline{S},_{\bar{\xi}}) \cdot \frac{\underline{\sigma}}{2i}\ \psi \tag{7.7-58}$$

where the $\underline{\sigma}$ are the Pauli matrices defined by

$$\sigma_x = \begin{pmatrix} 0 & 1 \\ 1 & 0 \end{pmatrix} \qquad \sigma_y = \begin{pmatrix} 0 & -i \\ i & 0 \end{pmatrix} \qquad \sigma_z = \begin{pmatrix} 1 & 0 \\ 0 & -1 \end{pmatrix} \tag{7.7-59}$$

and γ plays the role of the spectral parameter. The solution techniques used to analyse problems of this type are beyond the scope of this book, and are the subject of much current research. A brief survey of such techniques can be found in Section 6.4. We will return to equations (7.7-39) at the end of this section where it will be shown that they are equivalent, in a special way, to a euclidian form of the sine-Gordon equation.

7.7.2 The Continuous Heisenberg Chain Model

In one space dimension equation (7.7-31) takes the form

$$\underline{S},_t = \underline{S} \wedge \underline{S},_{xx} \qquad |\underline{S}^2| = 1 \tag{7.7-60}$$

These equations together with the boundary condition

$$\underline{S}(x,t) \to \underline{S}_0 \quad \text{a constant for} \quad |x| \to \infty \tag{7.7-61}$$

serve to define the **continuous Heisenberg spin chain model.** As $\pi_1(S^2) = 0$ we might suspect that there are no non-dissipative kink solutions to this equation. As we shall see that is very far from the truth. It is important to remember that the appearance of an obvious topologically conserved quantity generally implies the existence of kink solutions but the apparent lack of such a quantity allows us to draw no conclusion about their absence.

The unit vector $\underline{S}$ may be parameterised as in (7.7-46) as

$$\underline{S} = (\sin\theta(x,t)\cos\phi(x,t), \sin\theta(x,t)\sin\phi(x,t), \cos\theta(x,t)) \tag{7.7-62}$$

To determine a soliton type of solution we put

$$\theta(x,t) = \theta(x-vt), \qquad \phi(x,t) = \Omega t + \hat{\phi}(x-vt) \tag{7.7-63}$$

Substitution of these forms into equations (7.7-60,7.7-61) yields equations which may be integrated once to give

$$\hat{\phi},_{\chi} = v(1+\sin\theta)^{-1} \tag{7.7-64}$$

$$(\theta,_{\chi})^2 = 4\Omega\left(\frac{1-\sin\theta}{1+\sin\theta}\right)\left(\frac{1+\sin\theta}{2} - \frac{v^2}{4\Omega}\right) \tag{7.7-65}$$

where $\chi = (x - vt)$.

If we use the standard half angle formulae from trigonometry, equation (7.7-65) may be simplified to

$$(\beta,_{\chi})^2 = \Omega\,\frac{\sin^2\beta}{\cos^2\beta}\,(\cos^2\beta - \cos^2\beta_0) \tag{7.7-66}$$

with β and β_0 defined by

$$\beta = \tfrac{1}{2}\theta, \quad \frac{v^2}{4\Omega} = \cos^2\beta_0 = 1-b^2 \tag{7.7-67}$$

Equation (7.7-66) may be easily integrated to give

$$\sin\beta = b\,\mathrm{sech}(b\sqrt{\Omega}(\chi - x_0)) \tag{7.7-68}$$

This gives the solution

$$S_3 = \cos\theta = 1-2b^2\mathrm{sech}^2(b\sqrt{\Omega}(x-x_0-vt) \tag{7.7-69}$$

using this equation (7.7-64) may now be integrated to give

$$\hat{\phi}(x,t) = \phi_0 + \tfrac{1}{2}v(x-x_0-vt) + \tan^{-1}\left[\left(\frac{b^2}{1-b^2}\right)^{\frac{1}{2}}\tanh(b\sqrt{\Omega}(x-x_0-vt)\right] \tag{7.7-70}$$

Figure 7-25 shows a plot of $S_3(\chi)$. This is a non-dissipative solution with finite energy

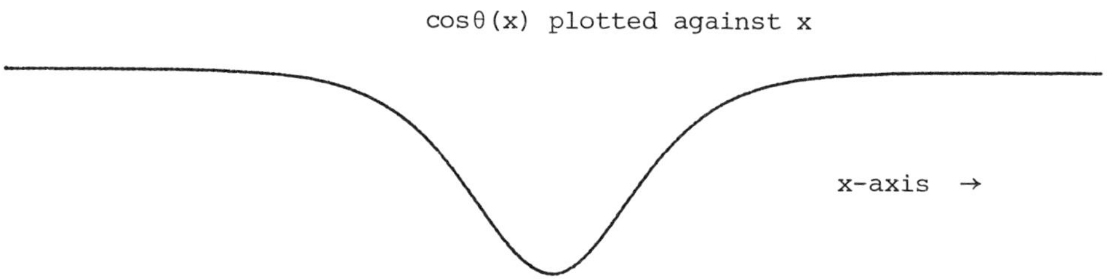

Figure 7-25: S_3 from equation (7.7-69)

$$E = 4b\sqrt{\Omega} \tag{7.7-71}$$

and is therefore a kink solution for equation (7.7-60). However, we note that it is similar to the breather solution of the sine-Gordon in that its energy can be arbitarily small depending on the magnitude of Ω. This means that only a small amount of energy is needed to excite such a solution.

The question of whether this equation has solutions of the soliton type considered in this book is harder to answer.

Tjon and Wright have carried out numerical studies of the collision of two such kink solutions. The collision appears to be elastic with the two kinks interpenetrating and then reemerging as single kinks again. There is no radiation apparent but there is a clear phase shift.

The equations (7.7-60) do indeed comprise an exactly integrable system. If we define the matrix S by

$$S = \underline{S}.\underline{\sigma} \tag{7.7-72}$$

where the $\underline{\sigma}$ are the Pauli matrices

$$\sigma_x = \begin{pmatrix} 0 & 1 \\ 1 & 0 \end{pmatrix} \qquad \sigma_y = \begin{pmatrix} 0 & -i \\ i & 0 \end{pmatrix} \qquad \sigma_2 = \begin{pmatrix} 1 & 0 \\ 0 & -1 \end{pmatrix} \tag{7.7-73}$$

then equations (7.7-60) may be recast in the form

$$S,_t = \frac{1}{2i}[S,S,_{xx}] \tag{7.7-74}$$

$$S^2 = I,\ S^+ = S,\ \text{tr}\ S = 0 \tag{7.7-75}$$

If we take $\underline{S}_o = (0,0,1)$ equation (7.7-61) becomes

$$S(x,t) \to \sigma_z \qquad \text{as} \quad |x| \to \infty \tag{7.7-76}$$

Equations (7.7-74) are the integrability conditions of the inverse scattering problem

$$\psi,_x = i\lambda S\psi \tag{7.7-77}$$

$$\psi,_t = (\lambda SS,_x + 2i\lambda^2 S)\psi \tag{7.7-78}$$

The inverse scattering problem defined by (7.7-77) may be treated in a way similar to the AKNS system considered in Chapter 6. Gelfand-Levitan-Marchenko equations can be constructed and soliton solutions including (7.7-69,7.7-70) determined.

The inverse scattering problem for the nonlinear Schrödinger equation

$$i\phi,_t + \phi,_{xx} + 2|\phi|^2\phi = 0 \tag{7.7-79}$$

is given by

$$\begin{aligned} \tilde{\psi},_x &= [i\lambda\sigma_2 + \begin{pmatrix} 0 & \phi \\ -\phi & 0 \end{pmatrix}]\tilde{\psi} \\ \tilde{\psi},_t &= [-i\begin{pmatrix} \phi\bar{\phi} & \bar{\phi},_x \\ \phi,_x & -\phi\bar{\phi} \end{pmatrix} + \lambda\begin{pmatrix} 0 & 2\bar{\phi} \\ -2\phi & 0 \end{pmatrix} + 2i\lambda^2\sigma_z]\tilde{\psi} \end{aligned} \tag{7.7-80}$$

It may be shown that by making a transformation of the form

$$\psi = g(x,t)\tilde{\psi} \tag{7.7-81}$$

where $g(x,t)$ is a matrix independent of λ the system (7.7-77,7.7-78) may be converted into one having the form (7.7-80). As a result the nonlinear Schrödinger equation and the equations of the Heisenberg linear chain model are, in a well defined sense, equivalent. A transformation of the type (7.7-81) on the wave function of the scattering problem is known as a **gauge transformation.** Two systems linked in this way are said to be **gauge equivalent.**

Another relevant example of this type of relationship concerns the Heisenberg Ferromagnet equations (7.7-39). The scattering problem

$$\psi_\xi = [\zeta\begin{pmatrix} 0 & 1 \\ -1 & 0 \end{pmatrix} + \frac{i}{2}\begin{pmatrix} \phi,_\xi & 0 \\ 0 & -\phi_\xi \end{pmatrix}]\psi \tag{7.7-82}$$

$$\psi,_{\bar{\xi}} = \zeta^{-1}\begin{pmatrix} 0 & e^{-1\phi} \\ -e^{i\phi} & 0 \end{pmatrix} \tag{7.7-83}$$

where $\xi = x + iy$ may be associated with the **elliptic sine-Gordon** equation

$$\phi,_{xx} + \phi,_{yy} = 4\sin\phi \tag{7.7-84}$$

Let Ψ be a fundamental matrix solution of (7.7-82)-(7.7-83). It is not difficult to show that Ψ may be chosen to belong to SU(2). If we define the matrices g and Φ by

$$g = \Psi(\zeta=1) \qquad \Psi = g\Phi \tag{7.7-85}$$

then a simple calculation shows that Φ satisfies the equations

$$\Phi,_{\xi} = (\zeta-1)g^{-1}\begin{pmatrix} 0 & 1 \\ -1 & 0 \end{pmatrix} g\Phi \tag{7.7-86}$$

$$\Phi,_{\xi} = (\zeta^{-1}-1)g^{-1}\begin{pmatrix} 0 & e^{i\phi} \\ -e^{-i\phi} & 0 \end{pmatrix} g\Phi \tag{7.7-87}$$

The matrix S defined by

$$S = g^{-1}\begin{pmatrix} 0 & 1 \\ -1 & 0 \end{pmatrix} g \tag{7.7-88}$$

has the properties

$$SS,_{\xi} = 2g^{-1}\begin{pmatrix} 0 & 1 \\ -1 & 0 \end{pmatrix} g \tag{7.7-89}$$

$$SS,_{\bar{\xi}} = 2g^{-1}\begin{pmatrix} 0 & e^{i\phi} \\ -e^{-i\phi} & 0 \end{pmatrix} g\Phi \tag{7.7-90}$$

If S is expressed in terms of the Pauli matrices by $S = \underline{S}.\underline{\sigma}$ then $SS,_{a} = i\underline{S}\wedge\underline{S},_{a}.\underline{\sigma}(a = \xi,\bar{\xi})$ and by means of (7.7-89)-(7.7-90) may be expressed as

$$\Phi,_{\xi} = \frac{1}{2i}(1-\zeta)\underline{S}\wedge\underline{S},_{\xi}\cdot\underline{\sigma}\Phi \tag{7.7-91}$$

$$\Phi,_{\bar{\xi}} = \frac{1}{2i}(1-\zeta^{-1})\underline{S}\wedge\underline{S},_{\bar{\xi}}\cdot\underline{\sigma}\Phi \tag{7.7-92}$$

These are none other than the inverse scattering equations of the Heisenberg Ferromagnet given in equations (7.7-57)-(7.7-58). Consequently, we have a means of constructing from solutions of the elliptic sine-Gordon equation (7.7-84) solutions of the Ferromagnet equations (7.7-39). The process is very similar to that involved with Bäcklund transformation. Given an initial solution of (7.7-84) equations (7.7-82)-(7.7-83) must be solved for $\zeta = 1$ and a fundamental matrix constructed. If a full inverse scattering transform were available for (7.7-82)-(7.7-83) solving the Gelfand-Levitan-Marchenko equations would yield both the soliton and the matrix g. Unfortunately, in the elliptic case the necessary equations are not available. Once g has been found constructing $\underline{S}$ is easy. If we write

$$g = \begin{pmatrix} a & b \\ -b^* & a^* \end{pmatrix} \tag{7.7-93}$$

then from (7.7-88)

$$S_1 = i(a^*b-b^*a) \tag{7.7-94}$$

$$S_2 = (a^*b+b^*a) \tag{7.7-95}$$

$$S_3 = |a|^2-|b|^2 \tag{7.7-96}$$

7.8 Quantum Mechanics and the Sine-Gordon Equation in Quantum Optics

In Chapter 2 we explained that we could associate with each state of a quantum system a complex square integrable wave function Ψ^t. That wave function develops in time according to the time dependent Schrödinger equation

$$i\hbar\, \Psi^t_{,t} = \hat{H}\, \Psi^t \tag{7.8-1}$$

where $\hat{H}$ is an operator representing the classical Hamiltonian or energy operator. For notational convenience we will suppress the t index on Ψ^t unless essential for clarity. As Ψ is square integrable we are able to interpret the real number $|\Psi|^2$ as the probability density corresponding to the quantum configuration described by the wave function Ψ. Several examples of such normalisable functions were explicitly constructed in Chapter 2 where the operator $\hat{H}$ was given by

$$\hat{H} = \left(-\frac{\hbar^2}{2m}\cdot\frac{d^2}{dx^2} + Q(x)\right) \tag{7.8-2}$$

Any operator representing a measurable quantity, such as m_z the z-component of the magnetic moment of a charged paricle, is required to have the property of being self-adjoint $m_z = m_z^\dagger$. For example, the z-component of the magnetic moment of a particle of charge e governed by the non relativistic Hamiltonian (7.8-2) may be taken to be

$$\hat{m}_z = \frac{ie\hbar}{2m}\left(x\frac{\partial}{\partial y} - y\frac{\partial}{\partial x}\right) = \frac{e}{2m}\hat{J}_z \tag{7.8-3}$$

This ensures that the eigenvalues of such an operator, which are interpreted as giving the allowable measured values of the quantity represented, are real. Any such self-adjoint operator is called an **observable.** A quantum state represented by an eigenfunction of an observable $\hat{U}$ is called an eigenstate of $\hat{U}$ and we will use the same symbol to represent both the quantum state and its associated wave function.

One of the basic assumptions of quantum mechanics is that the eigenfunctions associated with any observable $\hat{U}$ are **complete.** This means that any quantum state, that is any wave function, may be expanded as a linear combination of those eigenfunctions. For example, suppose that the eigenvalues λ_n are discrete and that there is only one normalisable wave function u_n corresponding to each eigenvalue λ_n.

$$\hat{U} u_n = \lambda_n u_n \tag{7.8-4}$$

The wave functions u_n may be normalised so that they form an orthonormal sequence of functions relative to the inner product $\langle u,v\rangle = \int u^*(x)v(x)dx$. We may then write,

$$\langle u_n, u_m \rangle = \delta_{nm} \tag{7.8-5}$$

Completeness of the u_n means that any wave function Ψ can be expanded as a linear combination of the u_n,

$$\Psi = \sum_{n=1}^{\infty} a_n u_n \tag{7.8-6}$$

From (7.8-5) we see that the coefficients a_n in (7.8-6) are given by

$$a_n = \langle u_n, \Psi \rangle \tag{7.8-7}$$

If Ψ is normalised we also see from (7.8-6) that

$$\langle \Psi, \Psi \rangle = 1 = \sum_{n=1}^{\infty} |a_n|^2 \tag{7.8-8}$$

As a result we are led to interpret the a_n defined by (7.8-7) as probability densities. The quantity $p_n = |a_n|^2$ is regarded as giving the probability that the state Ψ will be found in the eigenstate u_n, that is the probability that an experiment to measure U will yield the result λ_n. The **expectation value** of an observable $\hat{U}$ in a state Ψ is defined by

$$\langle \hat{U} \rangle_{\Psi} = \langle \Psi, \hat{U}\Psi \rangle = \sum_{n=1}^{\infty} |a_n|^2 \lambda_n \tag{7.8-9}$$

Given the interpretation of $|a_n|^2$ as the probability of obtaining the U-value λ_n in a measurement of U on a system represented by wave function Ψ we see that the right hand side of (7.8-9) is just the statistical expectation value of the experimental quantity U having possible values λ_n with corresponding probabilities p_n of occurence. The real number $\langle U \rangle_{\Psi}$ is the nearest that we can come to ascribing a definite value of the observable $\hat{U}$ to the quantum state with wave function Ψ.

An arbitrary wave function Ψ may be expanded in terms of a time independent set of eigenfunctions ϕ_n,

$$\Psi = \sum_{n=1}^{\infty} a_n \phi_n \tag{7.8-10}$$

ıbstituting this representation into the time-dependent Schrödinger equation yields ıe following equations for the time evolution of the coefficients a_n

$$i\hbar\, a_{n,t} = \sum_{m=1}^{\infty} \tilde{H}_{nm}\, a_m \tag{7.8-11}$$

ıere the matrix $\tilde{H}_{nm}$ is given by

$$\tilde{H}_{nm} = \langle u_n, \hat{H}\, u_m \rangle \tag{7.8-12}$$

ıis linear set of equations for the a_n is simply a reformulation of the ime-dependent Schrödinger equation.

The expectation value of an observable $\tilde{A}$ in the state Ψ is given by

$$\langle \hat{A} \rangle_\Psi = \sum_{nm} a_n^* a_m \langle u_n, \hat{A}\, u_m \rangle \tag{7.8-13}$$

ıich may be expressed in the matrix form

$$\langle \hat{A} \rangle_\Psi = \mathrm{Tr}(\rho \tilde{A}) \tag{7.8-14}$$

ıere ρ and $\tilde{A}$ are the two self-adjoint matrices defined by

$$\tilde{A}_{nm} = \langle u_n, \hat{A}\, u_m \rangle = \tilde{A}^{\dagger}_{nm} \tag{7.8-15}$$

$$\rho_{nm} = a_n a_m^* \tag{7.8-16}$$

ıe matrix ρ is called the **density matrix** corresponding to the state Ψ. As Ψ is ırmalised we must have

$$\mathrm{Tr}\ \rho = 1 \tag{7.8-17}$$

ıe time development of the a_n is known from equation (7.8-11) and so we may easily ɛtermine that the time development of the density matrix is governed by the equation

$$i\hbar\, \rho_{,t} = [\tilde{H}, \rho] \tag{7.8-18}$$

which is known as **Liouville's equation** by analogy with its counterpart in classica mechanics. For an arbitrary observable $\tilde{A}$ the Schrödinger equation takes the form

$$i\hbar \frac{d\tilde{A}}{dt} = i\hbar \frac{\partial \tilde{A}}{\partial t} + [\tilde{H},\tilde{A}] \qquad (7.8\text{-}19)$$

of which (7.8-18) is a special case.

As an example of a quantum system that may be reduced to the sine-Gordon equatic we are going to consider the interaction of a light wave with a gas. The atoms whic comprise the gas may be regarded, for the purposes of the analysis, as two stat quantum systems with a ground state and single excited state. Therefore as a preluc to our consideration of that situation we will specialise the general results jus obtained to the particular case of the two state system.

The two-state quantum system:

Consider a two-state quantum system described by a Hamiltonian $\hat{H}_0$ with two ener eigenvalues E_0 and E_1 with corresponding eigenfunctions ϕ_0 and ϕ_1.

$$\hat{H}_o\phi_o = E_o\phi_o \,, \qquad \hat{H}_o\phi_1 = E_1\phi_1 \qquad (7.8\text{-}20)$$

with $E_0 < E_1 = E_0 + \hbar\omega$. When placed in interaction with some external environme the Hamiltonian is changed to

$$\hat{H} = \hat{H}_o + \hat{H}_I^t \qquad (7.8\text{-}2$$

where the H_I^t is an additional energy due to the interaction and is called t **interaction Hamiltonian.** The interaction, which may be explicitly time dependen induces transitions between the two states of the unperturbed system.

For this two-dimensional situation a convenient vector representation for t Liouville equation exists. This is introduced as follows. The Pauli matrice introduced in Section 7.7 equation (7.7-73), have the commutation relations

$$[\sigma_a,\sigma_b] = 2i\epsilon_{abc}\sigma_c \qquad (7.8\text{-}2$$

Together with the unit matrix I they constitute a basis for the vector space 2 × 2 matrices. As a result we may write ρ and $\tilde{H}$ in the form

$$\rho = \tfrac{1}{2}(1+\underline{\rho}\cdot\underline{\sigma}) \qquad \tilde{H} = \tfrac{1}{2}\hbar(\omega_o I+\underline{\omega}\cdot\underline{\sigma}) \qquad (7.8\text{-}2$$

where

$$\omega_0 = \mathrm{Tr}\,\tilde{H}, \quad \underline{\rho} = \mathrm{Tr}(\rho\underline{\sigma}), \quad \underline{\omega} = \frac{1}{\hbar}\,\mathrm{tr}(\tilde{H}\underline{\sigma}) \tag{7.8-24}$$

Equation (7.8-23) has taken into account equation (7.8-17). With density matrix (7.8-18) becomes

$$\tfrac{1}{2}i\hbar\rho,_t\cdot\underline{\sigma} = \frac{\hbar}{4}[\underline{\sigma}\cdot\underline{\rho},\underline{\sigma}\cdot\underline{\omega}] = \frac{\hbar i}{2}\,(\underline{\rho}\wedge\underline{\omega})\cdot\underline{\sigma} \tag{7.8-25}$$

and this equation takes the compact form

$$\underline{\rho},_t = \underline{\rho}\wedge\underline{\omega} \tag{7.8-26}$$

which are known as the **Bloch equations** and are similar to the Landau-Lifshitz equations of Section 7.7.

For a two state atom in interaction with an external electric field the interaction Hamiltonian operator takes the form

$$\hat{H}^t_I = -\,\underline{E}^t\cdot\hat{\underline{P}} \tag{7.8-27}$$

where $\underline{P}$ is the dipole operator given by

$$\hat{\underline{P}} = -e\,\hat{\underline{x}}$$

If ϕ_0 and ϕ_1 are chosen to be time independent eigenfunctions of $\hat{H}_0$ and in addition we choose their phases so that they are real then the matrix representation $\underline{P}$ takes the form

$$\underline{P} = \mathcal{P}\sigma_x \tag{7.8-28}$$

where

$$\underline{\mathcal{P}} = -e\int\phi_0\,\hat{x}\phi_1\,d^3x \tag{7.8-29}$$

The macroscopic polarisation $\underline{p}$ of this single two-state system is given by

$$\underline{p} = \mathrm{Tr}(\rho\tilde{P}) = \underline{\mathcal{P}}\,\mathrm{Tr}(\rho\sigma_x) = \underline{\mathcal{P}}\,\rho_1 \tag{7.8-30}$$

The matrix representation of the total Hamiltonian is given by

$$\tilde{H} = \begin{pmatrix} E_0 & 0 \\ 0 & E_0+\hbar\omega \end{pmatrix} - \underline{E}^t \cdot \underline{P}\ \sigma_x \tag{7.8-31}$$

and the corresponding ω of the Bloch equations takes the form

$$\underline{\omega} = -\frac{2}{\hbar}(\underline{E}^t \cdot \underline{P})\hat{i} - \omega\hat{k} \tag{7.8-32}$$

The Bloch equations which result are given by

$$\begin{aligned} \rho_{1,t} &= \omega\rho_2 \\ \rho_{2,t} &= -\omega\rho_1 + \frac{2}{\hbar}\underline{E}^t \cdot \underline{P}\rho_3 \\ \rho_{3,t} &= -\frac{2}{\hbar}\underline{E}^t \cdot \underline{P}\rho_2 \end{aligned} \tag{7.8-33}$$

We must add to these the Maxwell equations obeyed by the electric field **E(x,t)**. In such a quasi-classical model the electric field remains un-quantized and the appropriate equations for an electric field incident upon a gas of two-state atoms of density n_0 are,

$$\Delta\underline{E} - \underline{E}_{,tt} = 4\pi n_o \underline{p}_{tt} = 4\pi n_o \underline{P}\rho_{1,tt} \tag{7.8-34}$$

Equations (7.8-33,7.8-34) are the classical **Maxwell-Bloch equations**. A scaled form of these equations in two dimensions were already considered in section nine of Chapter one where numerical experiments indicated that they were not an exactly integrable system. Clearly some form of limiting operation will have to be applied to them if we hope to obtain an exactly integrable system like the sine-Gordon equation.

Let E^* be a measure of the electric field strength, $\underline{E}(x,t)$ and $\underline{P}$ be vectors in the same direction $\hat{e}$, and the whole problem reduced to one space and one time dimension. Define the dimensionless quantities ϵ and $\overline{E}(x,t)$ by

$$\epsilon = \frac{E^*|\underline{P}|}{\hbar\omega} \qquad \overline{E}(x,t) = \frac{\underline{E}.\hat{e}}{E^*} \tag{7.8-35}$$

We can now write equations (7.8-33,7.8-34) in the form

$$\begin{aligned} \overline{E}_{,xx} - \overline{E}_{,tt} &= -2\gamma\omega^2\epsilon(\rho_1 - 2\epsilon\overline{E}\rho_3) \\ \rho_{1,tt} + \omega^2\rho_1 &= 2\epsilon\overline{E}\rho_3 \\ \rho_{3,t} &= -2\epsilon\overline{E}\rho_2\omega = -2\epsilon\overline{E}\rho_{1,t} \end{aligned} \tag{7.8-36}$$

where we have assumed that for low density gases

$$\frac{2\pi}{E^*}|\underline{P}|n_o = \epsilon\gamma \, , \qquad \gamma = O(1)$$

For the situation in which $\epsilon \ll 1$ we can solve these equations by the method of multiple scales. Define the slow space and time variables X and T by

$$X = \epsilon\omega x \qquad T = \epsilon\omega x \tag{7.8-37}$$

and assume the following asymptotic forms for $\bar{E}(x,t)$, ρ_1 and ρ_3

$$\begin{aligned}\bar{E}(x,t) &= E_o(x,t,X,T) + \epsilon E_1(x,t,X,T) + O(\epsilon^2) \\ \rho_1(x,t) &= \rho_{10}(x,t,X,T) + \epsilon\rho_{11}(x,t,X,T) + O(\epsilon^2) \\ \rho_3(x,t) &= \rho_{30}(x,t,X,T) + \epsilon\rho_{31}(x,t,X,T) + O(\epsilon^2)\end{aligned} \tag{7.8-38}$$

Substituting these into equations (7.8-33) we obtain at $O(\epsilon^0)$

$$E_{o,xx} - E_{o,tt} = 0 \tag{7.8-39}$$

$$\rho_{10,tt} + \omega^2\rho_{10} = 0 \tag{7.8-40}$$

$$\rho_{30,t} = 0 \tag{7.8-41}$$

and at $O(\epsilon)$

$$\begin{aligned}E_{1,xx} - E_{1,tt} + 2\omega(E_{o,xX} - E_{o,tT}) &= -2\omega^2\gamma\rho_{10} \\ \rho_{11,tt} + \omega^2\rho_{11} + 2\omega\rho_{10,tT} &= 2\omega^2 E_o\rho_{30} \\ \rho_{31,t} + \omega\rho_{30,T} &= -2E_o\rho_{10,t}\end{aligned} \tag{7.8-42}$$

As solution to (7.8-39) we take the travelling wave solution

$$E_o = \mathrm{Re}\left[e^{i\omega(x-t)}\varepsilon_o(X,T)\right] \tag{7.8-43}$$

and for equation (7.8-40) the related choice

$$\rho_{10} = \mathrm{Re}\left[e^{i\omega(x-t)}U_0(X,T)\right] \tag{7.8-44}$$

Substituting these forms into (7.8-42) and eliminating secularities we obtain

$$\begin{aligned} i(\varepsilon_{0,X}+\varepsilon_{0,T}) &= -\gamma U_0 \\ -iU_{0,T} &= \varepsilon_0\rho_{30} \\ \rho_{30,T} &= -\frac{i}{2}(\varepsilon_0\bar{U}_0-\bar{\varepsilon}_0 U_0) \end{aligned} \tag{7.8-45}$$

If we assume that ε_0 is real and write

$$U_0 = -iP \qquad \rho_{30} = -N \tag{7.8-46}$$

where P and N are real then equations (7.8-45) further reduce to

$$\varepsilon_{0,X} + \varepsilon_{0,T} = \gamma P \tag{7.8-47}$$

$$P_T = \varepsilon_0 N \tag{7.8-48}$$

$$N_T = -\varepsilon_0 P \tag{7.8-49}$$

From (7.8-48,7.8-49) we see that $(P^2 + N^2)_{,T} = 0$ and we set

$$P^2 + N^2 = 1 \tag{7.8-50}$$

This allows us to parameterise P and N by

$$P = \pm\sin\Phi \qquad N = \pm\cos\Phi \tag{7.8-51}$$

From (7.8-49) it follows that

$$\varepsilon_0(x,t) = \Phi_{,T} \tag{7.8-52}$$

and substitution of (7.8-51,7.8-52) into (7.8-47) gives us

$$\Phi_{,XT} + \Phi_{,TT} = \pm\ \gamma\ \sin\Phi \tag{7.8-53}$$

If we introduce the new space and time variables $\bar{x}$, $\bar{t}$ defined by

$$\bar{x} = \sqrt{\gamma}(T-2x) \qquad \bar{t} = \sqrt{\gamma T} \tag{7.8-54}$$

this equations takes the standard form

$$\Phi_{,xx} - \Phi_{tt} \pm \sin\Phi = 0 \tag{7.8-55}$$

of the sine-Gordon equation in laboratory coordinates. Further details concerning the physics of this example can be found in Chapter 9. For our second example we return to the phenomena of superconductivity. We first develop the theory of Section 7.6 somewhat further and then we will be ready to show that the sine-Gordon equation appears once again as a model of flux propagation in a Josephson transmission line.

7.8.1 Time Dependent Landau-Ginzberg Theory

In the time dependent form of the Landau-Ginzberg theory of superconductivity the Cooper pairs are all in the same state and are described by a single wave-function Φ which in this case satisfies a phenomenological Schrödinger equation in the time-dependent form

$$i\hbar\,\frac{\partial\Phi}{\partial t} = \frac{1}{2m^*}\,(-i\hbar\underline{\nabla}-e^*\underline{A})^2\Phi + V(x)\Phi + \lambda\Phi|\Phi|^2 \tag{7.8-56}$$

The function $V(x)$ is a scalar potential and e^* and m^* are the charge and mass of the Cooper pair. The quantity $\underline{A}(x)$ is the vector potential of an external electromagnetic field.

By analogy with the results for the one-dimensional Schrödinger equation found in Chapter 2, it is easily shown that (7.8-56) can be written in the conservation form

$$\rho_{,t} = \underline{\nabla}\cdot\underline{j} \tag{7.8-57}$$

where

$$\rho = e^*|\Phi|^2 \tag{7.8-58}$$

$$\underline{j} = \frac{i\hbar}{2m^*}\,e^*(\Phi^*\underline{\nabla}\Phi-\Phi\underline{\nabla}\Phi^*) + \frac{(e^*)^2}{2m^*}\,|\Phi|^2\underline{A} \tag{7.8-59}$$

For a normalisable wave function Φ, the scalar quantity $|\Phi|^2$ is still interpreted as giving the time dependent distribution of Cooper pairs within the superconducting material. The vector quantity $\underline{j}$ is interpreted as the electric current vector which is the source of the electromagnetic field $\underline{H}$ given by the Maxwell equation

$$\text{curl}\ \underline{H} = \underline{j} \tag{7.8-60}$$

We note that this involves an effective modification of the Maxwell equations as the vector potential $\underline{A}$ appears explicitly in the current $\underline{j}$.

Consider the experimental configuration in which two superconductors are separated from each other by a barrier of insulating material.

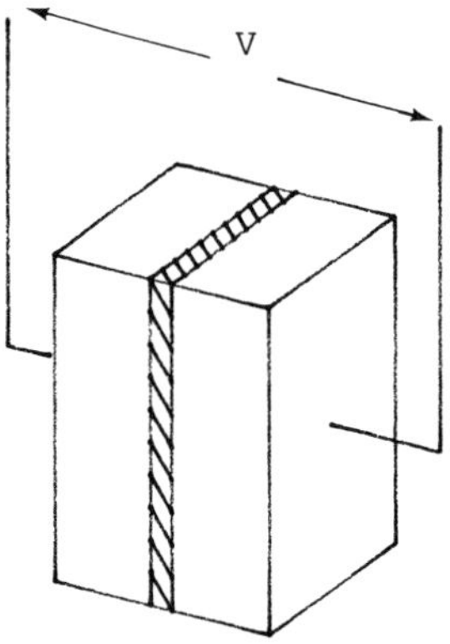

A Josephson junction consisting of two layers of superconducting material separated by a non--superconducting barrier.

Figure 7-26: A Josephson Junction

Figure 7-26 depicts such a situation. A typical arrangement is a layer of lead and a layer of niobium separated by a layer of niobium oxide. In Chapter 2 we saw that a quantum particle had a non zero probability of being able to penetrate to the other side of a potential barrier which would be inpenetrable to the corresponding classical particle. That phenomenon of barrier penetration is usually referred to as **quantum tunnelling.** In an exactly analogous way it is possible for a Cooper pair to tunnel through an intervening layer separating two superconductors. It is this superconducting tunnelling phenomena which was considered by Josephson for his Ph.D thesis at Cambridge in 1962. We shall now consider that phenomenon within the context of the Landau-Ginzberg theory.

Josephson Junctions

Before considering the quantum aspects of the problem we will first consider the details which relate to the classical electromagnetic aspects of the problem.

If we write

$$\Psi = \rho^{\frac{1}{2}} e^{i\Phi} \tag{7.8-61}$$

then substitution into (7.8-59) shows that the superconducting electric current is given by

$$\underline{j} = -\frac{e^{*2}\rho}{m^*}\left(\underline{A} - \frac{\hbar}{e^*}\underline{\nabla}\Phi\right) \tag{7.8-62}$$

which may be rearranged into the form

$$\underline{\nabla}\Phi = \frac{e^*}{\hbar}\left(\underline{A} + \frac{m^*}{e^{*2}\rho}\,\underline{j}\right) \tag{7.8-63}$$

This expression is only valid within the superconductors. If we define ϕ to be the change in phase of the wave function across the barrier

$$\phi(x,y,t) = \Phi(x,y,0+,t) - \Phi(x,y,0-,t) \tag{7.8-64}$$

the following simple argument shows that ϕ is nontrivial. Figure 7-27 shows the interface taken as the x-y plane P and Q two arbitrary points in the barrier.

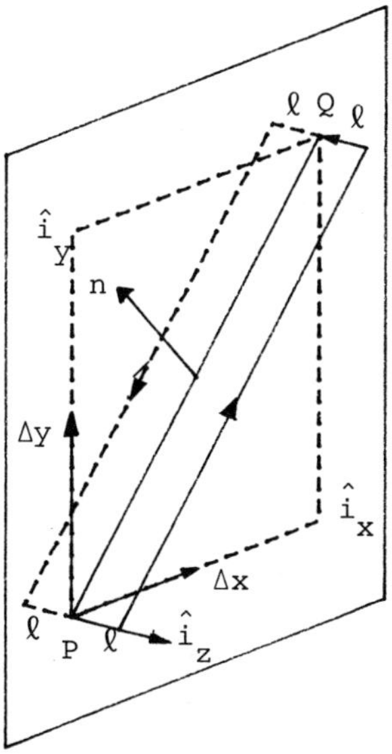

Figure 7-27:

Integrating over the closed curve C shown in Fig. 7-27 we obtain, provided ℓ is greater than the penetration depth

$$\phi(Q) - \phi(P) = \frac{e^*}{\hbar} \oint_C \left(\underline{A} + \frac{m^*}{e^{*2}\rho} \underline{j} \right) d\underline{r} \tag{7.8-65}$$

where P and Q are two points in the insulating barrier with coordinates (x,y,0) and (x+Δx,y+Δy,0) respectively. It follows that

$$\phi(Q) - \phi(P) = \phi_{,x}\Delta x + \phi_{,y}\Delta y \tag{7.8-66}$$

By Stokes' theorem

$$\oint_C \underline{A}.d\underline{r} = \int_S \underline{B}.d\underline{S} \tag{7.8-67}$$

where S is any surface having boundary curve C. Taking S to be the rectangular plate shown in Fig. 7-27 we easily discover that

$$\underline{S} = 2\ell(\hat{\underline{j}}\Delta x - \hat{\underline{i}}\Delta y) \tag{7.8-68}$$

and if we assume that $\underline{B}$ is constant across the plate

$$\int_S \underline{B}.d\underline{S} \approx \Delta\underline{S}.\underline{B} = (B_y\Delta x - B_x\Delta y)2\ell \tag{7.8-69}$$

Allowing Δx and Δy to tend to zero we obtain the equations

$$\phi_{,x} = \frac{e^*}{\hbar} B_y . 2\ell = \alpha B_y \tag{7.8-70}$$

$$\phi_{,y} = -\frac{e^*}{\hbar} B_x . 2\ell = -\alpha B_x \tag{7.8-71}$$

where

$$\alpha = \frac{2e^*\ell}{\hbar} \tag{7.8-72}$$

These equations give us information about the spatial variations of ϕ but we still need to know how it varies with time and how it is related to the superconducting current through the barrier.

To obtain the relationship that we need we will follow the analysis of Feynman which models the situation as a two state quantum system. Figure 7-28 below is a schematic representation of the situation and such a configuration is known as a **Josephson junction**.

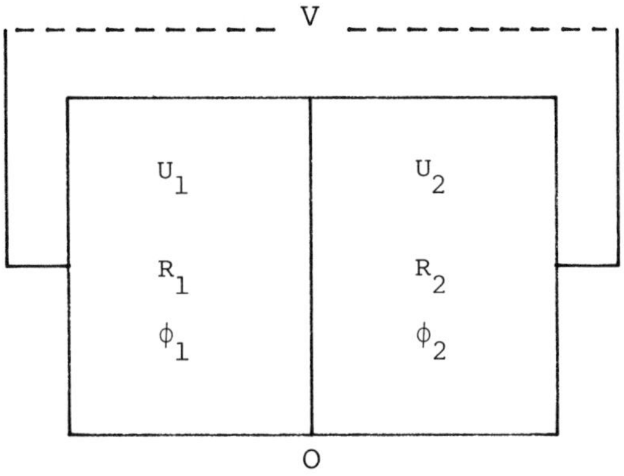

A schematic Josephson junction consisting of two regions of identical superconductor separated by a thin layer of insulation at O.

Figure 7-28: Schematic Josephson Junction

In each of the superconducting regions the system is described by a pseudo-wave-function. In region 1 let it be ϕ_1 and in region 2 let it be ϕ_2 . If we knew something about the potential barrier corresponding to the insulator layer we could try to solve the scattering problem for the nonlinear Schrödinger equation (7.8-56). A simple model due to Jacobson considers a static situation in which the wave function obeys a time-independent form of nonlinear Schrödinger equation in each of the superconducting regions with the potential V(x). Within the insulating barrier the wave function satisfies a normal Schrödinger equation with a repulsive potential. The relevant Schrödinger equations are solved in each region and matched together at the interfaces just as we solved the elementary scattering problems in Chapter 2. However, these calculations are involved and similar results can be obtained in a simpler, if rather intuitive, manner by exploiting an analogy between this arrangement and a two state quantum system.

Let us suppose, for simplicity, that the two superconductors are of the same material. If the two superconducting regions were not connected the wave function in each region would satisfy a time-dependent Schrödinger equation of the form

$$i\hbar\phi_{i,t} = \hat{H}_O\phi_i \qquad (i = 1,2) \tag{7.8-73}$$

Suppose further that in each of the regions 1 and 2 the system is in an energy eigensate of energy U_1 and U_2 respectively.

$$\hat{H}_O\phi_i = U_i\phi_i \qquad (i = 1,2) \tag{7.8-74}$$

Each superconducting electron is confined to its particular region and so the wave function ϕ_1 is zero in region 2 and the wave function ϕ_2 is zero in region 1. The U_i are the self energies of the superconducting electrons in separate domains and are not related. Consider now the situation of a Josephson junction in which the insulator layer is thin and there is a possibility of quantum tunnelling. The wave function of the interacting system is now nonzero in both regions. The self-energies are no longer independent. If there is a potential difference of V across the junction the self-energies must be related by

$$U_2 - U_1 = e^*V \tag{7.8-75}$$

We assume that the presence of the insulating layer can be modelled by an interaction Hamiltonian $\hat{H}_T$ known as the **tunnelling Hamiltonian** with the total quantum system decribed by the Hamiltonian

$$\hat{H} = \hat{H}_O + \hat{H}_T \tag{7.8-76}$$

A particularly simple model results if we suppose that a single Cooper pair is tunnelling through the barrier. This is the model we obtain if we consider the system as a two-state system. We think of the junctions as being in essentially two states. One state (described by wave-function ϕ_1) in which the Cooper pair is on the left of the barrier and a second state (described by wave-function ϕ_2) in which the Cooper pair is to the right of the barrier. The wave function describing the interacting system will be a linear combination of these two basic states following the general theory developed earlier. Equations (7.8-11) become

$$\begin{aligned} i\hbar\, a_{1,t} &= U_1a_1 + Ka_2 \\ i\hbar\, a_{2,t} &= U_2a_2 + Ka_1 \end{aligned} \tag{7.8-77}$$

where we have assumed that

$$(\phi_i, \hat{H}_T\phi_j) = \delta_{ij}K \tag{7.8-78}$$

Choosing the zero of energy halfway between U_1 and U_2 equation (7.8-77)can be written in the symmetrical form

$$\begin{aligned} i\hbar\, b_{1,t} &= \tfrac{1}{2}e^*Vb_1 + Kb_2 \\ i\hbar\, b_{2,t} &= -\tfrac{1}{2}e^*Vb_2 + Kb_1 \end{aligned} \tag{7.8-79}$$

The quantities $|b_1|^2$ and $|b_2|^2$ are the probabilities of finding the Cooper pair on the left or the right of the insulator layer. If we choose the phases of the basic states ϕ_1 and ϕ_2 so that they are real we may write

$$b_i = \sqrt{\rho_i}\, e^{i\theta_i} \tag{7.8-80}$$

and identify the phase change ϕ to be

$$\phi = (\theta_2 - \theta_1) \tag{7.8-81}$$

substituting (7.8-79) and separating real and imaginary parts we obtain

$$\rho_{1,t} = \frac{1}{\hbar} K\sqrt{\rho_1\rho_2}\,\sin\phi \tag{7.8-82}$$

$$\rho_{2,t} = \frac{1}{\hbar} K\sqrt{\rho_2\rho_1}\,\sin\phi \tag{7.8-83}$$

$$\theta_{1,t} = \frac{K}{\hbar}\sqrt{\frac{\rho_2}{\rho_1}}\,\cos\phi - \frac{e^*V}{2\hbar} \tag{7.8-84}$$

$$\theta_{2,t} = \frac{K}{\hbar}\sqrt{\frac{\rho_1}{\rho_2}}\,\cos\phi + \frac{e^*V}{2\hbar} \tag{7.8-85}$$

The wave function in region R_i is given by

$$\psi_i = \sqrt{\rho_i}\, e^{i\theta_i}\, \phi_i \tag{7.8-86}$$

and each of the basic wave functions ϕ_i is normalised to one on R_i. For each region we have the conservation law (7.8-57) and an integration over a region R_i gives us the result

$$j_{iz} = \rho_{i,t} \tag{7.8-87}$$

In an actual Josephson junction ρ_1 and ρ_2 are both approximately equal to a common value ρ_0 and also very nearly constant in time. At first sight this latter property would seem to be contradicted by equations (7.8-82)-(7.8-85). However, this is not the case as not all the features of the problem have been included in the equations. From (7.8-87) we see that $\rho_{1,t}$ is the superconducting current from region 1. That current would soon charge up region 2 were it not for the fact that there is an external battery providing the potential difference across the barrier. The currents that flow in the battery circuit have not been included and their effect is to allow ρ_1 and ρ_2 to attain a constant value ρ_0. The omission of the auxilliary circuit means that equations (7.8-74) only express how the densities would start to change and what current would initially start to flow. Combining (7.8-82)-(7.8-85) and (7.8-87) we obtain our final result for the superconducting current passing through the barrier

$$j_{iz} = \bar{J}\,\sin\phi \tag{7.8-88}$$

where

$$\bar{J} = \frac{2K\rho_0}{\hbar} \tag{7.8-89}$$

Within the insulator layer the normal Maxwell equations hold and there is an additional displacement current density

$$j_{dz} = c_s V_{,t} \tag{7.8-90}$$

where c_s is the capacitance per unit area of the junction. The Maxwell equation (7.8-60) now becomes

$$B_{y,x} - B_{x,y} = \mu_0 (j_{1z} + j_{dz}) \tag{7.8-91}$$

From (7.8-84),(7.8-85) we obtain by subtraction our final basic equation

$$\phi_{,t} = \frac{e^*}{\hbar} V \tag{7.8-92}$$

If we now incorporate equations (7.8-70)-(7.8-72),(7.8-88),(7.8-92) together we arrive at the familiar equation

$$\phi_{,xx} + \phi_{,yy} - \frac{1}{c^2}\phi_{,tt} = \frac{1}{\beta^2}\sin\phi \tag{7.8-93}$$

where

$$\beta = (\hbar/\mu_0 e^* \bar{j}\ell)^{\frac{1}{2}}, \quad c = (\mu_0 c_s \ell)^{-\frac{1}{2}} \tag{7.8-94}$$

If we have a very narrow junction, in the sense that variations in the y-direction can be neglected, then the equation reduces to the standard sine-Gordon equation in one space and one time dimension.

The single soliton solution is given by

$$\phi(x,t) = 4\tan^{-1}\exp[\beta^{-1}\gamma(x-cvt)] \tag{7.8-95}$$

and the related magnetic field is in the y-direction

$$B_y = 2\gamma\ \alpha^{-1}\beta^{-1}\mathrm{sech}[\beta^{-1}\gamma(x-cvt)] \tag{7.8-96}$$

A typical sech shaped pulse of magnetic field propagates along the junction. The current density j_{1z} is easily found to be

$$j_{1z} = -2\bar{j}\ \mathrm{sech}[\beta^{-1}\gamma(x-cvt)]\tanh[\beta^{-1}\gamma(x-cvt)] \tag{7.8-97}$$

and a graph of this function for $t = 0$ is shown in Fig. 7-29.

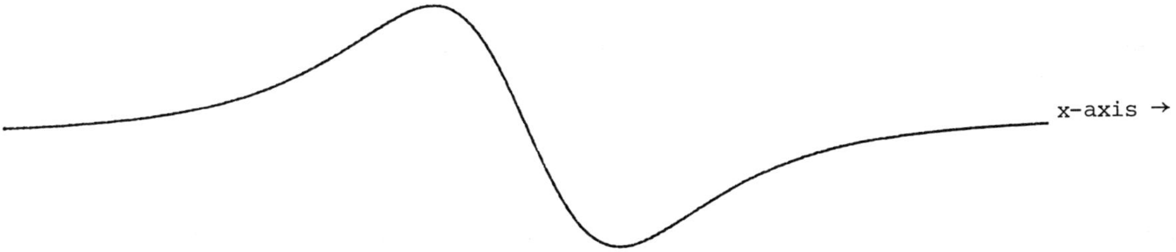

Figure 7-29:

From (7.8-69) we see that the total flux of the magnetic field passing through the barrier is given by

$$\frac{\hbar}{e^*}\int_{-\infty}^{\infty}\phi_{,x}dx = \frac{2\pi\hbar}{e^*}\,Q(\phi) \qquad (7.8\text{-}98)$$

As a result we see that the flux is 'quantised' into units of size $2\pi\hbar/e^*$. This effect was originally predicted by London but he assumed that the charge e^* was the electronic charge. However, when the quantization of flux was experimentally demonstrated by Deaver and Fairbank the quantum unit was found to be half the expected value. We now understand this as e^* is the charge on a Cooper pair which is 2e. Thus, the single soliton solution (7.8-95) represents a single quantum of flux propagating along the junction. Similarly the N-soliton solutions carry N units of flux and are often referred to as the **N-fluxon solution.** The solution (7.8-95) is called the **fluxon.** Alternatively these solutions are sometimes called **vortex solutions** by analogy with the corresponding solutions of the type II superconductors mentioned in Section 7.6.

The soliton solutions above refer to the standard infinite boundary conditions. These are less relevant for Josephson junctions where finite boundary conditions of the type

$$\phi_{,x}(0) = A, \qquad \phi_x(L) = B \qquad (7.8\text{-}99)$$

more commonly describe experimental configurations. This leads to the more general elliptic solutions of the type constructed in Section 7.2.

7.9 Nonabelian Gauge Fields, Monopoles, and Instantons

In Section 7.5 we saw that the introduction of a gauge field enabled us to construct nonlinear Klein-Gordon equations with nontrivial topological charges. Kink-like solutions such as vortices were found, but these were not generally solitons in ths sense of this book. The essential features of (i) an infinite number of conservation laws, and (ii) the existence of a Bäcklund transformation, did not seem to be present. To arrive at related systems which do have both integrally charged kinks and Bäcklund transformations, we must move up one dimension and generalise our notion of a gauge field.

7.9.1 Non Abelian Gauge Fields

In Section 7.5 we introduced the abelian gauge field associated with U(1) the group of one-dimensional complex unitary matrices. To generalise we may consider a more general group of unitary matrices G contained within the general linear group of all non-singular $n \times n$ complex matrices GL(n,C). We will restrict our attention to groups of unitary matrices parameterised by $p < n^2$ real parameters. It can be shown that elements g of such p-parameter continuous matrix groups may be represented in the form

$$g = \exp\, iL_A\alpha^A \qquad (A=1,\ldots,p),\ \alpha^A \in R \qquad (7.9\text{-}1)$$

where $L_A (A = 1,2,\ldots,p)$ are a set of p, $n \times n$ hermitian matrices which satisfy a set of commutation relations of the form

$$[L_A, L_B] = iC_{ABC}L_C \qquad (7.9\text{-}2)$$

The C_{ABC} are constants characteristic of the p-parameter group and known as the

structure constants of the group. The matrices L_A are called the generators and span a vector space $\underline{L}$. As a result of (7.9-2) the commutator product [,] defines a mapping of $\underline{L} \times \underline{L} \to \underline{L}$ and makes $\underline{L}$ into an algebra known as the **Lie algebra** of G.

The natural inner product on GL(n,C) is defined by

$$\langle A,B\rangle = \mathrm{Tr}(A^{\dagger}B) \tag{7.9-3}$$

It is frequently possible to select the generators L_A so that they form an orthogonal set relative to this inner product

$$\langle L_A, L_B\rangle = C_A \delta_{AB} \tag{7.9-4}$$

where it is sometimes convenient to choose $C_A \neq 1$.

When the group is nonabelian there is a natural action of G on its Lie algebra $\underline{L}$, known as the adjoint action, and defined by

$$\Phi \to {}_g\Phi = g\Phi g^{-1} \tag{7.9-5}$$

We choose this to be the generalisation (7.5-5) with the matrix Φ replacing the single scalar field ϕ. Each of the p scalar fields which parameterise Φ are transformed by a $p \times p$ matrix which replaces the 1×1 matrix $\tilde{\beta}$. The advantage of making this choice is that the modulus $\|\Phi\|$ defined by

$$\|\Phi\|^2 = \langle\Phi,\Phi\rangle \tag{7.9-6}$$

is invariant and generalises the invariant $|\phi|^2$ of the single scalar field ϕ. We must note however, that there may be other G-invariant combinations of the field Φ from which a G-invariant potential could be constructed. In d-spatial dimensions our generalisation of (7.5-1) which is invariant under G is defined by the action

$$E_{\Phi} = \tfrac{1}{2}\left[\sum_{a=1}^{d} \|\Phi_{,a}\|^2 + 2V(\|\Phi\|)\right] \tag{7.9-7}$$

The transformation (7.9-5) is now the generalisation of a **gauge transformation** and the group G is known as the **gauge group**.

Consider now coordinate-dependent gauge transformations defined by a smooth map $g{:}R^d \to G$ given in coordinates by

$$g{:}\ x \to \exp\left[iL_A\beta^A(x)\right] \tag{7.9-8}$$

The functions $\beta^A{:}R^d \to R$ are all smooth functions $(A = 1,2,\ldots,p)$.

To make the equations corresponding to (7.9-7) invariant under these coordinate-dependent gauge transformations we introduce d gauge fields A_a belonging to $\underline{L}$. We construct a covariant derivative by

$$D_a\Phi = (\Phi_{,a} - i[A_a, \Phi]) \tag{7.9-9}$$

following the pattern of (7.5-8) but allowing for the choice of the adjoint action on the matrix Φ.

The action of the gauge group on the gauge fields A_a is determined by requiring $D_a\Phi$ to transform in the same way as Φ

$$(D_a\Phi)(x) \to ({}_gD_a\Phi)(x) = g(x)(D_a\Phi)(x)g^{-1}(x) \tag{7.9-10}$$

The result is the transformation

$$A_a(x) \to ({}_gA_a)(x) = g(x)A_a(x)g^{-1}(x) + i\, g(x)\, g^{-1}(x)_{,a} \tag{7.9-11}$$

In the abelian case the field tensor F_{ab} was invariant but in the nonabelian case it is not and must be replaced by the relevant invariant combination. The field tensor G_{ab} defined by

$$G_{ab} = F_{ab} - i[A_a, A_b] \tag{7.9-12}$$

has the same transformation property as Φ and $D_a\Phi$,

$$G_{ab}(x) \to ({}_gG_{ab})(x) = g(x)G_{ab}(x)\, g^{-1}(x) \tag{7.9-13}$$

and the action defining the gauge modified equations is (if we choose $C_A = 1$ $(A = 1,2,\ldots,p)$) given by

$$E_{\Phi,G} = \tfrac{1}{2}\left[\sum_{a=1}^{d} \|D_a\Phi\|^2 + \|G\|^2 + 2V(\|\Phi\|)\right] \tag{7.9-14}$$

where $\|G\|^2$ is the invariant quantity defined by

$$\|G\|^2 = \tfrac{1}{2}\sum_{ab} \langle G_{ab}, G_{ab}\rangle \tag{7.9-15}$$

If the scalar field is absent the equations corresponding to the term $\tfrac{1}{2}\|G\|^2$ are a generalisation of Maxwell's equations and are called the **Yang-Mills equations,**

$$D_aG_{ab} = 0 \tag{7.9-16}$$

If we suppose that the potential function has a finite number of zeros c_i $(i = 1,2,\ldots,M)$ then F, the space of finite energy solutions, divides into M sectors A_i defined by

$$A_i = \{(\Phi, A_a) \in F : \|\Phi\| \to c_i \text{ as } \|x\| \to \infty\} \tag{7.9-17}$$

In each sector the function $\|\Phi\|$ takes its asymptotic values on the sphere in $\underline{L}$ of radius c_i.

As a result a mapping $\tilde{\Phi}: S^{d-1} \to S^{p-1}$ is defined by

$$\tilde{\Phi}: \hat{n} \to \lim_{|x|\to\infty} \Phi(|x|\hat{n})\, c_i^{-1} \tag{7.9-18}$$

for $\Phi \in A_i$. The existence of a topological charge is determined by the nature of $\pi_{d-1}(S^{p-1})$.

7.9.2 SU(2) Invariant Nonlinear Klein-Gordon Equations

To illustrate the above construction let us consider the group of 2×2 unitary matrices with determinant $+1$. This is the group $SU(2)$. Any matrix belonging to $SU(2)$ may be parameterised by a pair of complex number a and b and written

$$g = \begin{pmatrix} a & b \\ -\bar{b} & \bar{a} \end{pmatrix} \qquad a\bar{a} + b\bar{b} = 1 \tag{7.9-19}$$

and we see that $SU(2) \cong S^3$. A matrix of this type can be expressed in the form

$$g = \exp[iS_A \beta^A] \tag{7.9-20}$$

where the matrices S $(A = 1,2,3)$ satisfy the commutation relations

$$[S_A, S_B] = i\epsilon_{ABC} S_C \tag{7.9-21}$$

A set of hermitian matrices which satisfy (7.9-21) is given by $S_A = \tfrac{1}{2}\sigma_A$ where σ_A are the Pauli matrices defined by

$$\sigma_1 = \begin{pmatrix} 0 & 1 \\ 1 & 0 \end{pmatrix} \quad \sigma_2 = \begin{pmatrix} 0 & -i \\ i & 0 \end{pmatrix} \quad \sigma_3 = \begin{pmatrix} 1 & 0 \\ 0 & -1 \end{pmatrix} \tag{7.9-22}$$

If we write

$$A_a = A_a{}^R S_R \qquad \Phi = \phi^R S_R \tag{7.9-23}$$

where the fields Φ and A_a are real then $D\Phi$, G_{ab} and $\|\Phi\|$ may be expressed in terms of the real vectors $\underline{\Phi} = (\Phi^1, \Phi^2, \Phi^3)$ and $\underline{A}_a = (A_a^1, A_a^2, A_a^3)$. Using the commutation relations (7.9-21) it is easily shown that

$$D_a\underline{\Phi} = \underline{\Phi}_{,a} + (\underline{A}_a \wedge \underline{\Phi}) \tag{7.9-24}$$

$$\underline{G}_{ab} = \underline{A}_{b,a} - \underline{A}_{a,b} + \underline{A}_a \wedge \underline{A}_b \tag{7.9-25}$$

$$\|\Phi\|^2 = (\phi^1)^2 + (\phi^2)^2 + (\phi^3)^2 = |\underline{\Phi}|^2 \tag{7.9-26}$$

where the normalised inner product on $\underline{L}$ is $\langle a,b\rangle = 2\mathrm{tr}(a^\dagger b)$ and $\langle S_A, S_B\rangle = \delta_{AB}$.

The dimension of the Lie algebra is $p = 3$ and therefore as $\pi_2(S^2) = \underline{Z}$ we will obtain a nontrivial topological charge if we take $d = 3$. With that choice the action (7.9-14) may be expressed in terms of the vector fields by

$$E_{\Phi,G} = \tfrac{1}{2}\left[\sum_{a=1}^{3} |(\underline{\phi}_{,a}+\underline{A}_a\wedge\underline{\Phi})|^2 + \tfrac{1}{2}\sum_{ab} |(\underline{A}_{b,a}-\underline{A}_{a,b}+\underline{A}_a\wedge\underline{A}_b)|^2+2V(|\Phi|)\right] \tag{7.9-27}$$

with the corresponding field equations

$$D_a^2\underline{\phi} = \frac{\partial}{\partial\phi_i} V(\|\underline{\phi}\|) \tag{7.9-28}$$

$$D_b\underline{G}_{ab} = -\underline{\phi}\wedge D_a\underline{\phi} \tag{7.9-29}$$

If we choose the potential function (7.5-37) these are the equations of the **SU(2) nonabelian Higgs Model.**

The analogue of formula (7.5-35) is the integral formula for the topological charge given by

$$N = \frac{1}{4\pi}\int_{R^3} <\epsilon_{abc}G_{ab},D_c\Phi> d^3x \tag{7.9-30}$$

This gives the topological charge of the map $\tilde{\Phi}:S^2 \to S^2$ defined in (7.9-18).

For the SU(2) Higgs model we may use the result (7.9-30) to recast (7.9-14) in a way analogous to (7.5-43),

$$E_{\Phi,G} = \tfrac{1}{2}\|\tfrac{1}{2}\epsilon_{abc}G_{bc}\mp D_a\Phi\|^2 + \frac{\lambda}{4}(\|\Phi\|^2-1)^2 \pm 4\pi<\epsilon_{abc}G_{ab},D_c\Phi>) \tag{7.9-31}$$

From this we see that

$$\int_{R^3} d^3x E_{\Phi,G} \geq 4\pi N \tag{7.9-32}$$

with equality only if $\lambda = 0$ and the following equations hold,

$$\tfrac{1}{2}\epsilon_{abc}G_{bc} = -\epsilon D_a\Phi, \quad \epsilon N > 0 \qquad \epsilon = \pm 1 \tag{7.9-33}$$

In terms of the vectors $\underline{\Phi}$ and $\underline{A}_a$ these equations may be written

$$\underline{A}_{b,a} - \underline{A}_{a,b} + \underline{A}_a\wedge\underline{A}_b = -\epsilon\ \epsilon_{abc}(\underline{\Phi}_{,a}+\underline{A}_a\wedge\underline{\Phi}) \tag{7.9-34}$$

There are several ways in which particular solutions to these equations may be found. We have chosen to use the existence of a Bäcklund transformation for this system as it shows that its kink solutions are solitons of the type we have concentrated upon in this book. Equations (7.9-33) are known as the **Bogomolny equations.**

7.9.3 The Bäcklund Transformation and Monopole Solutions

Manton has proposed a parameterisation in which the equations (7.9-34) take on a simpler form. This ansatz is given in cylindrical polar coordinates by

$$\underline{\Phi} = (0, \phi_1, \phi_2) \tag{7.9-35}$$

$$\underline{A}_\phi = (0, \eta_1, \eta_2) \tag{7.9-36}$$

$$\underline{A}_z = (w_1, 0, 0) \tag{7.9-37}$$

$$\underline{A}_\rho = (w_2, 0, 0) \tag{7.9-38}$$

where $x = \rho\cos\phi$, $y = \rho\sin\phi$ and η_a, ϕ_a and w_a are functions of z and ρ alone. The polar components of the gauge field are related to the cartesian components by

$$\underline{A}_x = \underline{A}_\rho \cos\phi - \underline{A}_\phi \frac{\sin\phi}{\rho} \tag{7.9-39}$$

$$\underline{A}_y = \underline{A}_\rho \sin\phi + \underline{A}_\phi \frac{\cos\phi}{\rho} \tag{7.9-40}$$

$$\underline{A}_z = \underline{A}_z \tag{7.9-41}$$

and in terms of them equations (7.9-34) become

$$\rho^{-1}(\underline{A}_{\phi,\rho} + \underline{A}_\rho \wedge \underline{A}_\phi) = -(\underline{\phi}_{,z} + \underline{A}_z \wedge \underline{\phi}) \tag{7.9-42}$$

$$\underline{A}_{z,\rho} - \underline{A}_{\rho,z} + \underline{A}_\rho \wedge \underline{A}_z = \rho^{-1}\underline{A}_\phi \wedge \underline{\phi} \tag{7.9-43}$$

$$\rho^{-1}(\underline{A}_{\phi,z} - \underline{A}_\phi \wedge \underline{A}_z) = (\underline{\phi}_{,\rho} + \underline{A}_\rho \wedge \underline{\phi}) \tag{7.9-44}$$

Substituting the expressions (7.9-35)-(7.9-38) into these equations we obtain the system of scalar equations

$$\phi_{1,z} - w_1\phi_2 = \rho^{-1}(\eta_{1,\rho} - w_2\eta_2) \tag{7.9-45}$$

$$\phi_{2,z} + w_1\phi_1 = \rho^{-1}(\eta_{2,\rho} + w_2\eta_1) \tag{7.9-46}$$

$$w_{1,\rho} - w_{2,z} = \rho^{-1}(\phi_1\eta_2 - \phi_2\eta_1) \tag{7.9-47}$$

$$\phi_{1,\rho} - w_2\phi_2 = -\rho^{-1}(\eta_{1,z} - w_1\eta_2) \tag{7.9-48}$$

$$\phi_{2,\rho} + w_2\phi_1 = -\rho^{-1}(\eta_{2,z} + w_1\eta_1) \tag{7.9-49}$$

This enables us to determine an important subclass of solutions to equations (7.9-34) which includes multi-kink solutions. The equations (7.9-45)-(7.9-49) can be further reduced to a single complex equation if we introduce two real functions f and ψ and parameterise the Manton fields in the following way

$$\phi_1 = f^{-1}\psi_{,z} = -w_1 \tag{7.9-50}$$

$$\phi_2 = -f^{-1}f_{,z} \tag{7.9-51}$$

$$\eta_1 = -\rho f^{-1}\psi_{,\rho} = \rho w_2 \tag{7.9-52}$$

$$\eta_2 = \rho f^{-1}f_{,\rho} \tag{7.9-53}$$

The complex field E defined by

$$E = f + i\psi \tag{7.9-54}$$

may then be shown to satisfy the single complex equation

$$\mathrm{Re}E(E_{,\rho\rho}+\rho^{-1}E_{,\rho}+E_{,zz}) = (E_\rho{}^2+E_z{}^2) \tag{7.9-55}$$

This is the celebrated **Ernst equation** which gives an alternative formulation of the static axially symmetric gravitational field problem.

In the same way that Bäcklund transformations were constructed in earlier chapters so may a Bäcklund transformation be determined for the Ernst equation. The situation is more complicated as the linear scattering problem appropriate to the Ernst equation is of a more involved type than the normal AKNS-ZS system.

We will not derive the Bäcklund transformation but simply explain how it is used to derive the one kink solution of the Bogomolny equations.

Given a solution E^0 of the Ernst equation (7.9-55) construct the following quantities,

$$M_1^0 = \tfrac{1}{2}f_0{}^{-1}E^0{}_{,\zeta} \tag{7.9-56}$$

$$M_2^0 = \tfrac{1}{2}f_0{}^{-1}\bar{E}^0{}_{,\zeta} \tag{7.9-57}$$

$$N_1^0 = \tfrac{1}{2}f_0{}^{-1}\bar{E}^0{}_{,\bar{\zeta}} = \overline{M_1^0} \tag{7.9-58}$$

$$N_2^0 = \tfrac{1}{2}f_0{}^{-1}E_{,\bar{\zeta}} = \overline{M_2^0} \tag{7.9-59}$$

where ζ is the complex variable defined by

$$\zeta = \rho + iz \tag{7.9-60}$$

If it is possible to solve the Ricatti equations

$$q_{,\zeta} = -\left[(M_2^0 - M_1^0)q + \gamma(w)(M_2^0 - M_1^0 q^2)\right] \tag{7.9-61}$$

$$q_{,\bar\zeta} = -\left[(N_1^0 - N_2^0)q + \gamma^{-1}(w)(N_1^0 - N_2^0 q^2)\right] \tag{7.9-62}$$

where

$$\gamma(w) = [(w - i\bar\zeta)(w + i\zeta)^{-1}]^{\frac{1}{2}} \tag{7.9-63}$$

and w is a real number which plays the role of a spectral parameter for the inverse problem of the Ernst equation, then a new solution of the Ernst equation is given by

$$M_1 = \left(\frac{\gamma+q}{1+\gamma q}\right)\left(q^{-1}M_2^0 + \frac{\gamma}{4\rho}\right) = \overline{N_1} \tag{7.9-64}$$

$$M_2 = \left(\frac{\gamma+q}{1+\gamma q}\right)\left(qN_2^0 + \frac{1}{4\gamma\rho}\right) = \overline{N_2} \tag{7.9-65}$$

The one kink solution may be constructed by starting from the initial solution

$$f = e^z, \qquad \psi = 0 \tag{7.9-66}$$

which corresponds to the trivial solution of the Bogomolny equations

$$\underline{\Phi} = (0,0,-1), \quad \underline{A}_\phi = \underline{A}_z = A_\rho = 0 \tag{7.9-67}$$

The Ricatti equations (7.9-61),(7.9-62) take the form

$$q_{,\zeta} = i/4(1-q^2)\gamma(w) \tag{7.9-68}$$

$$q_{,\zeta} = -i/4\,(1-q^2)\gamma^{-1}(w) \tag{7.9-69}$$

and may be combined to give

$$\frac{dq}{(1-q^2)} = \tfrac{1}{2}d\left[(w-i\bar\zeta)^{\frac{1}{2}}(w+i\zeta)^{\frac{1}{2}}\right] = \tfrac{1}{2}dR, \qquad R^2 = (w-z)^2 + \rho^2 \tag{7.9-70}$$

and hence easily integrated to provide for q the expression

$$q(\zeta,\bar\zeta) = \tanh(\tfrac{1}{2}R + K) \tag{7.9-71}$$

where K is a constant. If we choose $K = 0$ we obtain for M_1 and N_2 from equations (7.9-64),(7.9-65)

$$M_1 = \left(\frac{\gamma+\tanh R/2}{1+\gamma\tanh R/2}\right)\left(-\frac{i}{4}\coth R/2 + \gamma/4\rho\right) \tag{7.9-72}$$

$$N_2 = \left(\frac{\gamma+\tanh R/2}{1+\gamma\tanh R/2}\right)\left(\frac{i}{4}\tanh R/2 + \frac{1}{4\gamma\rho}\right) \tag{7.9-73}$$

The Manton fields are related to M_1 and N_2 by

$$\phi_1 + i\phi_2 = 2(M_1 - N_2) \tag{7.9-74}$$

$$\eta_2 + i\eta_1 = 2\rho(M_1 + N_2) \tag{7.9-75}$$

and using the fact that

$$\gamma(w) = \frac{1}{R}[(w-z)-i\rho] \tag{7.9-76}$$

and the double angle formulae for hyperbolic functions we obtain

$$\phi_1 + i\phi_2 = -i/2\left[\frac{w-z+R\tanh R/2 - i\rho}{R+(w-z)\tanh R/2 - i\rho\tanh R/2}\right]\left(\coth R - \frac{1}{R}\right) \tag{7.9-77}$$

from which ϕ_1 and ϕ_2 may be explicitly determined.

As a result of the special form of the Bogomolny equations it can be shown that the charge N may be expressed in terms of the fields ϕ_i alone in the form

$$N = \frac{1}{8\pi}\lim_{r\to\infty}\int_{|x|=r} d\underline{S}\cdot\underline{\nabla}(\phi_1^2+\phi_2^2) \tag{7.9-78}$$

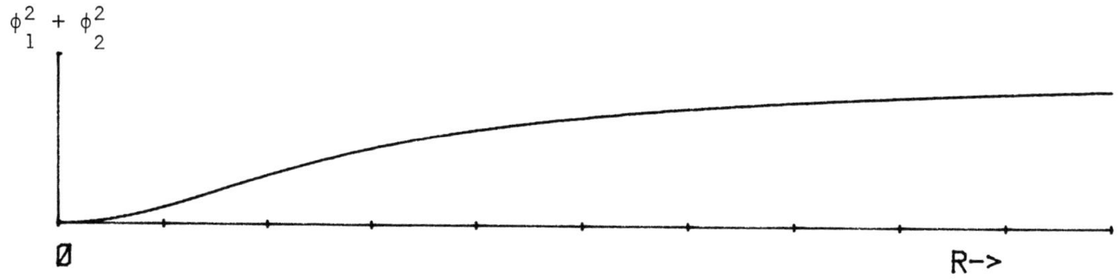

Figure 7-30:

From (7.9-77) we see that

$$\phi_1^2+\phi_2^2 = \left(\coth R - \frac{1}{R}\right)^2 \tag{7.9-79}$$

and Fig. 7-30 shows a plot of this function as a function of R. Substitution of this function into (7.9-78) shows that we have a solution with $N = 1$. In a similar way to the case of the sine-Gordon equation a nonlinear superposition principle can be constructed from the Ricatti equations (7.9-68),(7.9-69) for use in constructing exact solutions of the Ernst equation. By using that superposition principle, in conjunction with the 1-kink solution we have just determined, exact N-kink solutions may be found. They are extremely complicated functions to look at but in principle no more difficult to find than the N-soliton solutions of the sine-Gordon equation. These kink solutions are known as **monopoles** as they are analogous to the singular solutions of the Maxwell equations corresponding to isolated magnetic charges.

7.9.4 The Self-dual Yang-Mills Equations and Instantons

The last model with its monopole solutions had $\lambda = 0$ and so the explicit form of the potential was not involved. We were not, in fact, solving a genuine Klein-Gordon equation at all. The field Φ appears in a very similar way to the gauge fields A_a and this leads us to ask whether there might not be a formulation in which they are exactly the same type of field. This proves to be the case.

Consider a pure SU(2) Yang-Mills theory in R^4 and look for solutions which are independent of the x_4-coordinate. The field tensor $\underline{G}_{\mu\nu}$ $(\mu,\nu = 1,\ldots,4)$ has components G_{ab} $(a,b = 1,2,3)$ in the three-dimensional subspace with coordinates (x_1,x_2,x_3) given by

$$\underline{G}_{ab} = \underline{A}_{b,a} - \underline{A}_{a,b} + \underline{A}_a \wedge \underline{A}_b \tag{7.9-80}$$

and when $\mu = 4$ the components

$$\underline{G}_{4a} = \underline{A}_{a,4} - \underline{A}_{4,a} + \underline{A}_4 \wedge \underline{A}_a = -D_a \underline{A}_4 \tag{7.9-81}$$

as $A_{a,4} = 0$. The free Yang-Mills equations are given by

$$\underline{G}_{\mu\nu,\nu} + \underline{A}_\nu \wedge \underline{G}_{\mu\nu} = 0 \tag{7.9-82}$$

and separate into the two sets of equations

$$\underline{G}_{ab,b} + \underline{A}_b \wedge \underline{G}_{ab} = -\underline{A}_4 \wedge D_a \underline{A}_4 \tag{7.9-83}$$

and

$$\underline{G}_{4a,a} + \underline{A}_a \wedge \underline{G}_{4a} = 0 \tag{7.9-84}$$

If we put

$$\underline{\Phi} = -\underline{A}_4 \tag{7.9-85}$$

we see that these equations are none other than equations (7.9-28),(7.9-29) with $V = 0$. This means that monopoles are, in fact, solutions of the four-dimensional Euclidean Yang-Mills equations. This suggests that there may be more general kink-like solutions to these equations. This is indeed the case and exact N-kink solutions, usually called either **instantons** or **pseudoparticles**, can be constructed. We cannot enter into a discussion of why these instanton solutions are considered so important in particle physics except to say that they relate to phenomena which cannot be analysed with normal perturbation theory techniques. We shall complete this brief encounter with these 'solitons' of particle phsyics by showing how the topological notions we have used so far can help us to determine instanton solutions to the Yang-Mills equations.

For a finite energy solution to the Yang-Mills equations we require

$$\|G\| \to 0 \quad \text{as} \quad |x| \to \infty \tag{7.9-86}$$

As in the case of vortices this does not mean that the gauge field A_μ must be asymptotically trivial but only that it must be asymptotically a pure gauge transformation

$$A_\mu = ig(x)\, g^{-1}(x)_{,\mu} + O(|x|^{-3-\epsilon}), \quad |x| \to \infty \tag{7.9-87}$$

This defines a mapping $g:S^3 \to SU(2)$ ($\cong S^3$ from (7.9-19)) which has an integral valued topological charge. An integral formula can be found for this charge which is the analogue of formulae (7.5-35) and (7.9-30)

$$N = \frac{1}{4\pi^2} \int_{R^4} d^4x\, \underline{G}_{\mu\nu} \cdot \underline{\tilde{G}}_{\mu\nu} \tag{7.9-88}$$

where $\tilde{G}_{\mu\nu}$ is defined by

$$\tilde{G}_{\mu\nu} = \tfrac{1}{2} \epsilon_{\mu\nu\sigma\tau} G_{\sigma\tau} \tag{7.9-89}$$

and is called the **dual field tensor.** This result can be used to rewrite the action functional as

$$E_G = \frac{1}{8} |\underline{G}_{\mu\nu} \pm \underline{\tilde{G}}_{\mu\nu}|^2 \mp \frac{1}{4} \underline{G}_{\mu\nu} \underline{\tilde{G}}_{\mu\nu} \tag{7.9-90}$$

and we obtain the bound

$$\int d^4x\, E_G \geq \mp \pi^2 N \tag{7.9-91}$$

The bound is attained if and only if

$$G_{\mu\nu} = \tilde{G}_{\mu\nu}, \qquad \epsilon N > 0 \quad \epsilon = \pm 1 \tag{7.9-92}$$

These equations are known as the **self-dual Yang-Mills equations.**

If we are seeking a charge one solution it will have the asymptotic behaviour (7.9-87) with $g:S^3 \to SU(2)$ of charge one. The identity mapping $I:S^3 \to SU(2) \cong S^3$ is defined by

$$I(x) = \frac{x_4 + i\underline{\sigma}.\underline{x}}{r}, \quad r^2 = x_4^2 + |\underline{x}|^2 \tag{7.9-93}$$

and has charge one. This suggests that we seek a solution of the form

$$A_\mu = if(r)I(x)(I^{-1}(x))_{,\mu} \tag{7.9-94}$$

involving the single unknown function f(r). The guess (7.9-94) implies the following forms for the vectors $\underline{A}_\mu$,

$$\underline{A}_1 = (x_4, -x_3, x_2)\ a(r) \tag{7.9-95}$$

$$\underline{A}_2 = (x_3, x_4, -x_1)\ a(r) \tag{7.9-96}$$

$$\underline{A}_3 = (-x_2, x_1, x_4)\ a(r) \tag{7.9-97}$$

$$\underline{A}_4 = (-x_1, -x_2, -x_3)\ a(r) \tag{7.9-98}$$

where

$$a(r) = 2f(r)r^{-2} \tag{7.9-99}$$

Substitution into (7.9-92) with $\epsilon = 1$ gives us

$$\frac{1}{4}\, a'(r) = -a^2(r) \tag{7.9-100}$$

which may be integrated to give

$$a(r) = \frac{2}{(r^2+c)} \tag{7.9-101}$$

where C is an arbitrary constant. The final form for A_μ is then

$$A_\mu = \frac{ir^2}{r^2+c}\, I(x)(I^{-1}(x))_{,\mu} \tag{7.9-102}$$

This solution is the famous **Belavin, Polyakov, Tyupkin and Schwartz (BPTS) one-instanton solution.**

More general solutions can be constructed by examining, and generalising, the structure of the vector equations (7.9-95)-(7.9-98). These equations have the form

$$\underline{A}_\nu = \underline{n}_{\mu\nu} x^\mu\ a(r) \tag{7.9-103}$$

where the constant vectors $\eta_{\mu\nu}$ are explicitly defined by (7.9-103) and are known as the **t'Hooft matrices.** If we adopt the parameterisation

$$\underline{A}_\nu = -\ \tilde{\underline{n}}_{\mu\nu} \partial^\mu \log\phi \tag{7.9-104}$$

where $\tilde{\underline{n}}_{\mu\nu}$ is defined by

$$\sim \quad [(-1)^{\delta_{\mu 4}+\delta_{\nu 4}}]\underline{\eta}_{\mu\nu} \tag{7.9-105}$$

then (7.9-95)-(7.9-98) are replaced by

$$\underline{A}_1 = (\phi_{,4}, \phi_{,3}, -\phi_{,2})\phi^{-1} \tag{7.9-106}$$

$$\underline{A}_2 = (-\phi_{,3}, \phi_{,4}, \phi_{,1})\phi^{-1} \tag{7.9-107}$$

$$\underline{A}_3 = (\phi_{,2}, -\phi_{,1}, \phi_{,4})\phi^{-1} \tag{7.9-108}$$

$$\underline{A}_4 = (-\phi_{,1}, -\phi_{,2}, -\phi_{,3})\phi^{-1} \tag{7.9-109}$$

Direct substitution shows that (7.9-106)-(7.9-109) gives a solution if ϕ satisfies the linear wave equation

$$\Box\phi = \phi_{,11} + \phi_{,22} + \phi_{,33} + \phi_{,44} = 0 \tag{7.9-110}$$

If we seek solutions which depend only upon r then this equations is reduced to

$$\phi'' + \frac{3}{r}\phi' = 0 \tag{7.9-111}$$

with the singular solution

$$\phi = c_0 + c_1 r^{-2} \tag{7.9-112}$$

The solution corresponding to this choice may be shown to be a charge one instanton which is gauge equivalent to the BPTS solution.

A more general solution to (7.9-110), which follows from the linear superposition principle for linear equations, is given by

$${}_k\phi = c_0 + \sum_{j=1}^{k} \frac{c_j}{|(x-x_j)|^2} \tag{7.9-113}$$

where the x_i are an arbitrary set of points in R^4. The field corresponding to ${}_k\phi$ may be shown to be of charge k and is a **k-instanton solution.**

In this chapter we have tried to give an idea of the way in which kink solutions have arisen, and continue to arise, in physics and engineering. We have stressed the sine-Gordon equation in particular, owing to its repeated occurance and of course its complete integrability as demonstrated in previous sections. It should be remembered however, that the equation considered then, was the sine-Gordon equation with trivial boundary conditions and in light cone coordinates. In normal laboratory coordinates, on finite intervals with boundary conditions such as (7.8-99), the theory we have developed has much less to say.

Particular emphasis was placed on the existence of topologically conserved quantities as these help in both recognising potentially integrable equations and also in the construction of explicit kink solutions.

Our examples have been chosen from a wide range of physical theory, but we have by no means exhausted all the possible choices. In other chapters some of the examples will be developed further, and also new applications of the sine-Gordon equation in meterology developed.

7.10 Notes

Section 7.1

1. A mechanical model may easily be constructed and does not need to be elaborate to demonstrate kink behaviour. A length of elastic of square cross section, available from any craft or modelling enthusiasts shop, provides a suitable 'torque spring'. A piece about two feet long will suffice. Insert large headed pins such as those used in mapping at approximately 1/2" intervals (on the same side of the elastic) for its whole length. You will have now constructed a very servicable visual aid which can be used for classroom demonstrations or personal experimentation.

2. An elementary account of Homotopy theory suitable for mathematics students can be found, for example, in the books by Hocking and Young (1961) or Croom (1978). There are very few books available which are oriented towards the Physics or Engineering communities and we cite instead the articles by Mermin (1979) and Michel (1980). For a more advanced treatment the books by Spanier (1966) and Hilton (1966) are standard references. Some of the earliest papers which examine the relationship between topological notions and `particle-like' characteristics are due to Skyrme (1958) and Finkelstein and his co-workers (Finkelstein and Misner, 1959, Finkelstein, 1966, Finkelstein and Rubenstein, 1968, Finkelstein and Weil, 1978). Most of these papers can be easily, and profitably, read by undergraduates.

3. For further details of the mechanical pendulum the reader may consult the work of Scott (1969, 1970) or the more recent simulation studies of Fulton (19??).

4. An elementary treatment of elliptic functions suitable for mathematics students can be found in Whittaker and Watson (1962) or from a more modern point of view in Lang (1973). For physicists and engineers the book by Byrd and Freidman (1954) is a useful source. The special solutions of the type $\phi^t=4\tan^{-1}[f(x)g(x)]$ were first considered in detail by Lamb (1971) and have recently been generalised by Costabile et al. (1978).

Section 7.2

1 The first papers to consider in detail the collisional and particle properties of sine-Gordon solitons are due to Perring and Skyrme (1962) and Rubinstein (1970).

Section 7.3

1 Our treatment of topological charge in this section follows the unpublished work of Patani et al. (1976) and the paper by Arafune et al. (1975). For further mathematical information and results the books by Milnor (1965) and Guillemin and Pollack (1974) may be consulted. The Homotopy groups for spheres are tabulated and can be found in the book by Toda (1962).

2. One topic which we did not consider in the main text was that of **spin**. We have seen in this section that the set of all maps $\phi : R^n \to X$ which have the property

$\phi(x_1,\ldots,x_n) \to x$ as $|\underline{x}| \to \infty$ can be divided into a set of equivalence classes Q_μ of homotopically equivalent paths. The nature of the index set μ depends, of course, on X and the collection of Homotopy classes Q_μ form the nth Homotopy group $\pi_n(X,x)$. If, for example, $X = S^n$ it can be shown that $\pi_n(S^n) = Z$ the additive group of integers and the homotopy classes can be labelled by Z and we can denote them by Q_m. Functions which belong to the class Q_n are all of degree m. In the case $n = 1$ they would correspond to m twists through 2π.

Suppose that $n = 3$ and we are considering fields defined on normal Euclidean space. If the equations which the field satisfies are invariant under the action of the rotation group on $\mathbf{R}^3$, we expect the set of solutions of the equation to be mapped into itself under such a transformation. As a rotation through 2π about any axis in R^3, does not change R^3 it is normally supposed that under the induced action on the field corresponding to such a 2π rotation the field will be unchanged also. A field has spin 1/2 if under such a 2π rotation it is not left invariant but changes its sign. Clearly a rotation through 4π will return the field to its original value. If ϕ belongs to the homotopy class Q then if we consider the 1-parameter group of rotations about a single axis we obtain a 1-parameter family of functions in Q corresponding to the action of a member of that group on ϕ. As the angle is increased from 0 to 2π the field traverses a path in Q which starts at ϕ but which returns to $-\phi$. It is now clear that the question of whether spin 1/2 fields exist in a given model is concerned with the first homotopy group $\pi_1(Q)$ of an equivalence class of $\pi_n(X)$. Given a field theory which has kink solutions we say that the theory 'admits spin' if paths beginning and ending at some $\phi \in Q$ and corresponding to a rotation of 2π about some axis can be found which are not in the same homotopy class. This means that $\pi_1(Q)$ must contain an element of order 2 but that condition alone is not enough as the element in question must also correspond to a 2π rotation. For details of how $\pi_1(Q)$ can be calculated we refer to the papers of Williams et al. (Williams, 1970, Williams and Zvengrowski, 1977).

Section 7.4

1 In this section we follow the lectures of Coleman (1977). A slightly more rigorous treatment of some aspects of these lectures can be found in the book by Jaffe and Taubes (1980).

2. For a classical nonlinear Klein-Gordon equation the ground states are given by the solutions of the equation $U'(\phi) = 0$. In the neighbourhood of any given solution η of this equation, we can linearise the Klein-Gordon equation. When the system is quantized this linear field theory describes the elementary excitations about the ground state ϕ determined by $\langle\phi\rangle = \eta$ where the brackets denote the vacuum expectation value. In the classical model we introduce the new field χ defined by $\phi = \eta + \chi$ and the field equations become $(\Box\chi + U''(\eta)\chi) = O(\chi^2)$. The quantity $U''(\eta)$ now replaces m in this local form of the Klein-Gordon equation and is identified with the square of the mass of the elementary excitation involved. If Φ is a multicomponent field we obtain a **mass matrix** $M_{ab} = \partial^2 U/\partial\phi_a\partial\phi_b$ and its eigenvalues determine the squares of the masses of elementary excitations. In this case we obtain a **mass specrum.** If the potential U is invariant under an internal symmetry group G then the application of a transformation $g \in G$ will generally change the ground state η to an alternative ground state $g\eta$. In the quantum case the vaccum state is not invariant under G. The elementary particle states which correspond to the linearisation about η are known as **Higgs bosons** and their mass spectrum will not display the symmetry G. However, if we define the subgroup H_η

of G by $H_\eta=\{h \in G : h\eta=\eta\}$ the mass spectrum of the Higgs bosons will have the symmetry H. The symmetry G is said to be a **broken symmetry** and the mechanism by which it is broken is the failure of the ground state of the system to be invariant under G even though the Hamiltonian describing the fundamental interactions is invariant. This is in distinction to the mechanism by which a symmetry G is broken by the addition of extra interaction terms to the Hamiltonian which are not invariant under G which is a symmetry of the basic Hamiltonian. The former mechanism is known as **spontaneous symmetry breaking.** An elementary account can be found in the paper by Dashen (1969). If dim G = g and dim H = h the mass matrix has a kernel of dimension g - h. This means that there are g - h Higgs bosons of zero mass. These massless particles are known as Goldstone bosons. It may seem that this is an embarassing feature as there are very few massless particles known in the real world. However, it turns out that when gauge fields are introduced these particles acquire a mass. We refer to the papers in Mohapatra and Lai (1981) for further information.

We see that, provided the action of G is transitive, the set of ground states can be identified with the quotient space G/H of left cosets of H in G. This means that the question of whether or not kinks exist in such models is closely related to the homotopy groups of the type $\pi_n(G/H)$. The reader should note the similarity of this situation with that of general order parameters discussed in Section 7.6. An analysis of the homotopy groups of this type which is related to the treatment in the text can be found in the paper by Goddard et al. (1977).

3. The double quadratic potential is slightly different as U is not differentiable at the origin. An anlysis of the kink solution and reference to its use in modelling displacive ferroelectrics can be found in the paper by Trullinger (1979).

4. Derrick's Theorem only applies to static solutions, but is still an important result if only on account of its generality. It can be found in Derrick's paper (1964). For an example of a nonstatic kink see the paper by T.D. Lee (1976).

5. Equation (7.4-27) is the model first considered by Getmanov (1977) and is equivalent to the Regge-Lund model (Lund and Regge, 1976, Lund, 1977) which is a completely integrable generalisation of the sine-Gordon equation.

6. Equation (7.4-28) is the Dodd-Bullough equation (1977). This equation is associated with the same third order scattering problem as the Boussinesq equation (Fordy and Gibbons, 1981).

7. A great many simulation studies of nonlinear Klein-Gordon equations in two and higher dimensions have been carried out. As a general reference the reader may consult the review by Makhankov (1978). The final chapter of this book also contains further details of this work. Numerical simulations of the so-called sine-Gordon chain have been performed by Schneider and Stoll (see references and articles in Bishop and Schneider (1978)) using a molecular dynamics technique with periodic boundary conditions. These simulations have also formed the subject of several computer-produced films by the IBM Zurich group.

Section 7.5

1 Our basic references remain the lectures of Coleman (1977) and the book by Jaffe and Taubes (1980). For more references and information on gauge fields in particle physics the book by Mohapatra and Lai (1981) is useful.

2. For a discussion of electromagnetic fields in the context of classical mechanics see the book by Goldstein (1980).

3. Equations (7.5-46)-(7.5-48) are sometimes called the Bogomolny equations of the abelian Higgs model (Bogomolny, 1976).

4. The numerical simulations of Jacobs and Rebbi can be found in their papers (Jacobs and Rebbi, 1979, Rebbi, 1980). In general if a theory described by an energy density $E(\Phi)$ has single kink solutions ψ then the interparticle potential $V(d)$ is defined by

$$V(d) = E({}_{12}\Phi) - E({}_{1}\psi) - E({}_{2}\psi)$$

where ${}_1\psi$ and ${}_2\psi$ are two single kink solutions located at $t = 0$ a distance d apart and ${}_{12}\psi$ is the solution of the field equations which develops from that initial state. For the sine-Gordon equation (Perring and Skyrme, 1962)

$$V \sim \begin{cases} 32\exp(-d) & d \to \infty \\ 2\pi d^{-1} & d \to 0 \end{cases}$$

The work of Jacobs and Rebbi follows on from earlier work of de Vega and Schapesnik (1976).

5. The vortex solutions corresponding to $\lambda = \frac{1}{2}$ behave very much like the solitons we have considered in earlier chapters. In fact the Bogomolny equations (7.5-46-(7-6-45c)) can be reduced to the single equation

$$\Delta u = \exp(u) - 1 \qquad (*)$$

where $u = 2\mathrm{Re}(\log\Phi)$. This is very similar to the exactly integrable Liouville equation

$$\Delta u = \exp(u)$$

This suggests that (*) may also be associated with some form of inverse problem.

Section 7.6

1. An elementary account of crystal defect theory can be found in the book by Rosenberg (1975).

2. The simple periodic potential model which leads to the sine-Gordon equation was first put forward by Frenkel and Kontorova (1939). This work was followed by numerous authors amongst whom we may cite Frank and van der Merwe (1949, 1950), Seeger et al. (1953) and Seeger and Schiller (1966).

3. The ability to represent order parameter spaces as quotient spaces of groups is convenient because there are algebraic results which simplify the calculation of homotopy groups of the type π (G/H) once those of G and H are known. These results are best expressed in terms of **exact sequences**. If G_1, G_2 and G_3 are three groups and $i: G_1 \to G_2$ and $j: G_2 \to G_3$ are homomorphisms, the diagram

$$G_1 \overset{i}{\rightarrow} G_2 \overset{j}{\rightarrow} G_3$$

is called an exact sequence if

$$\ker j = \operatorname{im} i$$

Special relationships between three groups can be very neatly expressed in this notation. For example the exact sequence

$$(1) \qquad 0 \rightarrow G_1 \rightarrow G_2 \rightarrow 0$$

is equivalent to the statement that G_1 and G_2 are isomorphic. Similarly, if the G_i are abelian the exact sequence

$$(2) \qquad 0 \rightarrow G_1 \rightarrow G_2 \rightarrow G_3 \rightarrow 0$$

is equivalent to the statement that G_3 is isomorphic to G_2/G_1.

A basic result of homotopy theory (Hilton, 1966) is that the sequence

$$(3) \qquad \rightarrow \pi_n(H) \rightarrow \pi_n(G) \rightarrow \pi_n(G/H) \rightarrow \pi_{n-1}(H) \rightarrow \ldots\ldots$$

is exact at every stage. The sequence terminates at the zero order homotopy groups defined by

$$\pi_0(H) = \begin{cases} H & \text{if } H \text{ is discrete} \\ 0 & \text{if } H \text{ is connected} \end{cases}$$

If we know some of the groups in the long exact sequence (3) we may be able to find segments in forms such as (1) and (2) which will enable us to identify isomorphic groups. Some useful results are

(i) $\pi_n(H) = 0 \quad n \neq 0$ if H is discrete

(ii) $\pi_2(G) = 0$ for all compact groups G.

(iii) if $G = G_1 \times G_2$ then $\pi_n(G_1 \times G_2) = \pi_n(G_1) \quad {}_n(G_2)$

For example, we saw that the order parameter of a crystal was S^1 and could be expressed in the form T/H_α. From the long exact sequence

$$\rightarrow \pi_1(H_\alpha) \rightarrow \pi_1(T) \rightarrow \pi_1(S^1) \rightarrow \pi_0(H_\alpha) \rightarrow \pi_0(T)$$

(which follows from (3) by choosing $G = T$ and $H = H_\alpha$) we obtain

$$0 \rightarrow \pi_1(S^1) \rightarrow H_\alpha \rightarrow 0$$

because $T \cong R$ which has $\pi_1(R) = 0$ and $\pi_0(R) = 0$. From (1) we deduce that

$$\pi_1(S^1) \cong H \cong Z$$

This example is trivial but serves to illustrate this approach to the determination of homotopy groups. A more interesting example concerns multiple topological charges. In section nine we consider nonabelian Higgs models. If the basic theory is invariant under a simply connected compact gauge group G but the ground states are invariant under a subgroup J the existence of kink solutions depends upon the homotopy groups of G/J. If $d = 3$ we need to be able to calculate the homotopy group $\pi_2(G/J)$. In such models it turns out (Goddard et al., 1977) that J is necessarily the direct product of a torus T_1 and a simply connected subgroup G_1. The exact sequence above make the calculation of $\pi_2(G/J)$ trivial. The long exact sequence (3) gives us

$$\rightarrow \pi_2(G) \rightarrow \pi_2(G/J) \rightarrow \pi_1(J) \rightarrow \pi_1(G) \rightarrow \cdots\cdots$$

As G is compact and simply connected

$$\pi_2(G) = 0 \qquad \pi_1(G) = 0$$

and we obtain a sequence of the form (1). Thus

$$\pi_2(G/J) = \pi_1(J)$$

As $J = G_1 X T_1$ the result (iii) gives

$$\pi_1(J) = \pi_1(G_1) \times \pi_1(T_1)$$

As G_1 is simply connected $\pi_1(G_1) = 0$. Therefore if the rank of the torus T_1 is m we must have

$$\pi_1(J) = Z^m = \pi_2(G/J)$$

This is very interesting as it means that instead of a single topological charge we have a set of m integral valued topological charges.

4. The original work of Landau and Ginzberg (1950) was followed by the phenomenological theory of Bardeen, Cooper and Schrieffer (1957). There are many basic texts on superconductivity. Amongst those available we suggest the

comprehensive selection of articles edited by Parks (1969) and the text books by Rickayzen (1965) and Tinkham (1975). The final link between the early theory of Landau and Ginzberg and the microscopic BCS theory was made by Gorkov (1959, 1960). The experimental observation of the vortex lattice is described by Essman and Trauble(1967).

Section 7.7

1. An elementary introduction to Ferromagnetism suitable for our needs can be found in the book by Rosenberg (1975).

2. The original theory of Landau and Lifshitz can be found in the collected works of Landau (1935, 1969) and the background theory in magnetostatics in (1960). The latter book is a useful source of more recent references.

3. The earliest derivations of the sine-Gordon equation as a description of the motion of Bloch Walls are due to Doring (1948) and Enz (1964). An interesting recent paper on the sine-Gordon model of Bloch walls is the paper by Currie (1977). In particular Currie shows how the breather solution can be obtained by analytic continuation of the Hirota formula for the N-soliton solutions.

4. The energy functional $W[S]$ of the isotropic ferromagnet can be expressed in terms of the complex variable w. If we denote the energy by $\tilde{W}(w)$ we find that

$$\tilde{W}(w) = 8 \iint d^2x |w,_{\xi}|^2 (1 + |w|^2)^{-2} = 8\pi \mathcal{Q}(w)$$

where $\mathcal{Q}(w)$ is the charge expressed as a functional of w. As the energy of a 2-vortex solution is determined by its charge it is clear that the intervortex potential will be zero. This is the case for any Bogomolny type equation simply by construction.

5. The Minkowski space analogue of the isotropic Ferromagnet is defined by the equations

$$\partial^\mu \partial_\mu q^a + (\partial^\mu q^b \partial_\mu q^b) q^a = 0$$

where the indices μ and ν take the values 0,1,2,3 with the metric $-g_{11} = -g_{22} = -g_{33} = 1$. This model is called the **nonlinear σ -model.** An analysis of a linear eigenvalue problem associated with this problem in one space and one time dimension was first carried out by Lüscher and Pohlmeyer (1978). This model is gauge equivalent (in the sense of equation (7.7-81) to the normal sine-Gordon equation.

6. The fact, established by Hopf, that $\pi_3(S^2)$ is nontrivial is of historical importance because it established the difference between homotopy and homology. Hopf constructed a specific generator for $\pi_3(S^2)$ known as the **Hopf map.** This map is constructed as follows. The domain S^3 is represented by pairs of complex numbers (z_1, z_2) with the property that $z_1\bar{z}_1 + z_2\bar{z}_2 = 1$. The range S^2 is represented by

the quotient space of this S^3 by the equivalence relation

$$(z_1, z_2) \sim (z'_1, z'_2)$$

if there exists a $\lambda \in C$ such that $z_1 = \lambda z'_1$ and $z_2 = \lambda z'_2$. Denote the equivalence class of (z_1, z_2) by $[z_1 : z_2]$. Notice that this gives a realisation of S^2 as a quotient space,

$$S^2 \equiv SU(2)/U(1) \equiv S^3/S^1$$

The Hopf map H is defined to be the natural map $H: S^3 \rightarrow S^2$ given in coordinates by

$$(z_1, z_2) \rightarrow [z_1, z_2]$$

This mapping can be expressed in a different way by the **Spin map.** Spin $S^3 \rightarrow S^2$ is defined in terms of the Pauli matrices (7.7-73) by

$$\text{Spin: } Z \quad \underline{S} = Z^{\dagger} \underline{\sigma} Z \qquad \text{where } Z = (z_1, z_2) \quad z_1\bar{z}_1 + z_2\bar{z}_2 = 1$$

Notice that this map appears in a later section in equations (7.7-94)-(7.7-96). Each mapping $F: S^3 \rightarrow S^3$ defines a mapping $f = (\text{Spin}_0 F): S^3 \rightarrow S^2$. As $\pi_3(S^3) = Z$ each such mapping F has a definite topological charge N. As a result mappings from $S^3 \rightarrow S^2$ can inherit such a charge by being pulled back to S^3 by the Spin map. From (7.3-5) and (7.3-10) we know that the quantity

$$Q[F] = \frac{1}{3!\Omega_3} \int_{S^3} J^0(F)$$

where

$$J^0 = \varepsilon_{ijk} \varepsilon_{b_1 b_2 b_3 b_4} F^{b_1} \partial_i F^{b_2} \partial_j F^{b_3} \partial_k F^{b_4}$$

is an integral valued topological charge for $F: S^3 \rightarrow S^3$. If we introduce the expressions

$$A_i = -i(Z^\dagger \partial_i Z)$$

$$F_{jk} = -i(\partial_j Z^\dagger \partial_k Z - \partial_k Z^\dagger \partial_j Z)$$

the charge density J^o can be expressed in the form

$$J^o = \frac{3}{2}\varepsilon_{ijk}A_i F_{jk} = 3\underline{A}.\mathrm{curl}\underline{A}$$

From the defining relationship $\underline{S} = Z^\dagger \underline{\sigma} Z$ it can be shown that

$$(\mathrm{curl}\underline{A})_i = \varepsilon_{ijk}\varepsilon_{abc}S_a \partial_j S_b \partial_k S_c$$

Therefore, given $\underline{S}$, this equation defines the vector field $\underline{A}$ as a nonlocal functional of $\underline{S}$ and a topological charge for $\underline{S}$ is given by

$$\Omega_3 Q[S] = \tfrac{1}{2}\int_{S^3} \underline{A}.\mathrm{curl}\underline{A}$$

This invariant is known as the **Hopf invariant** (Whitehead, 1947).

7. The solitons of the Heisenberg spin chain were first examined numerically by Tjon and Wright (1977) and the system was shown to be completely integrable by Takhtajan (1977). The gauge equivalence of this model to the nonlinear Schrodinger equation was shown by Zakharov and Takhtajan (1979).

Section 7.8

1. There are numerous texts on basic quantum mechanics. Representative are the books by Mertzbacher (1961), Gottfried (1966) and Messiah (1961, 1962). A treatment of the quantum theory of magnetism can be found in the book by White (1971).

2. For a nonrelativistic particle the magnetic moment vector is given by $\hat{\underline{m}} = e(2m)^{-1}\hat{\underline{j}}$ and to find the thermal average of $\underline{m}$ we must find the expectation value of this operator in some state and then average over the states through which the system evolves in time. An alternative approach is to consider not the same system at different times but an ensemble of systems at the same time. If there are N systems then each will be described by a density matrix $\rho^{(j)}$ $j = 1,..,N$. The average over time can now be replaced by an average over the ensemble. In this way we obtain an ensemble averaged density matrix defined by

$$\bar{\rho} = N^{-1}\sum_n \rho^{(n)}$$

The density matrix $\bar{\rho}$ satisfies all of the normal density matrix equations and the thermally averaged magnetic moment is given by

$$\underline{M} = \langle\langle\hat{\underline{m}}\rangle\rangle = \mathrm{tr}(\bar{\rho}\tilde{m})$$

The dynamical equation for this mean magnetic moment is given by (7.8-19) to be

$$\hbar\frac{d\underline{M}}{dt} = -i\langle\langle[\tilde{\underline{m}},\tilde{H}]\rangle\rangle \qquad (*)$$

If the Hamiltonian H has the form

$$\tilde{H} = \tilde{H}_o - \tilde{\underline{M}}.\underline{H}$$

where $\underline{H}$ is the magnetic field strength within the crystal due to an externally applied field and $\tilde{H}_o$ is assumed to commute with $\tilde{\underline{M}}$ we have the situation envisaged in the Landau-Lifshitz model of section seven. In general, $\underline{M}$ is proportional to the total angular momentum $\underline{J}$

$$\tilde{\underline{M}} = \beta\tilde{\underline{J}}$$

and the quantum mechanical angular momentum operators satisfy the commutation relations

$$[\tilde{J}_i,\tilde{J}_j] = \varepsilon_{ijk}\tilde{J}_k$$

The equation (*) then reduces to

$$\frac{d\underline{M}}{dt} = -\gamma\underline{M}\wedge\underline{H}$$

where $\beta = \gamma\hbar$. This is the quantum derivation of the Bloch equations.

3. The model consisting of a two-state atom system in interaction with a laser is developed in the paper by Lamb (1964).

4. The Bloch equations receive their name from a similar set of equations which occur in the paper by Bloch (1946).

5. Our derivation of the sine-Gordon equation from the Nonlinear Landau-Ginzberg equation follows that in the book by Solymar (1972).

6. The time dependent Landau-Ginzberg equation was first developed from microscopic theory by Gorkov and Eliashberg (1969). The equation takes the form (Tinkham, 1975)

$$D^{-1}(\partial_t + ie^*\psi\hbar^{-1})\Delta + \xi^{-2}(|\Delta|^2 - 1)\Delta + (i\underline{\nabla} + \frac{e^*}{\hbar c}\underline{A})^2\Delta = 0$$

where D is a diffusion constant and is the electrochemical potential divided by the electronic charge. The wave function Δ is the so called 'gap parameter' which plays an important role in the BCS theory of superconductivity.

7. The model of Jacobson can be found in his paper (1965).

8. The idea of looking at the Josephson junction as a two state system is due to Feynman (1969).

9. The review article by Parmentier (1978) develops some of the periodic solutions of section four in the context of Josephson functions.

Section 7.9

1. An elementary account of gauge fields can be found in the original paper by Yang and Mills (1954) or in Bernstein (1974). There have been several recent reviews on the mathematical aspects of gauge theories for physicists. For example one may consult Eguchi et al (1980) or Madore (1981).

2. The Bogomolny equation (7.9-34) were first derived in Bogomolny,(1976).

3. The ansatz which allows the reduction of the Bogomolny equations to the Ernst (1968) equation was derived by Manton (1977). The actual reduction was carried out by Forgais et al (1980).

4. The inverse problem for the Ernst equation was first discovered by Harrison (1978) and in related form by Neugebaur and Kramer (1980), Maison (1979), Dodd and Morris (1980) and Zakharov and Belinskii (1978).

5. For further information on instantons the lectures of Coleman (1977) or Olive et al. (1979) are a useful source.

7.11 Problems

Show that

$$\phi = 4\tan^{-1}\{A\mathrm{dn}[\beta(x - x_o);\lambda_f]\,\mathrm{sn}[\Omega(t - t_o);\lambda_g]\}$$

where

$$\lambda_f = 1 - \{1 - \beta^2(1 + A^2)\}\{\beta^2 A^2(1 + A^2)\}^{-1}$$

$$\lambda_g = \{A^2[1 - \Omega^2(1 + A^2)]\}\{\Omega^2(1 + A^2)\}^{-1}$$

is a solution to the sine-Gordon equation with the boundary condition

$$\phi_x(0) = 0 = \phi_x(\ell)$$

provided that we have the **dispersion relation**

$$\beta = A\Omega$$

Show that the boundary conditions result in the eigenvalue condition

$$\beta_n = (n/\ell)K(\lambda_f)$$

Choose β so that the modulus $\lambda_f \equiv 1$ and use the identities

$$\mathrm{dn}(x,1) = \mathrm{sech}x \qquad \mathrm{sn}(x,0) = \sin x$$

to show that this solution is a generalisation of the breather solution on the infinite line. What solution results if β is chosen so as to make $\lambda_f \equiv 0$.

2. Consider the sine-Gordon equation

$$\phi_{tt} - \phi_{xx} + \sin\phi = 0 \tag{1}$$

To examine the stability of the 1-soliton solution of this equation we linearise equation (1) by introducing the new field f defined by

$$\phi = 4\tan^{-1}\exp(x) + f(x)\exp(-i\omega t)$$

and retain only terms of first order. Show that this leads to the Schrodinger solution

$$-f'' + (1 - 2\text{sech}^2 x)f = \omega^2 f$$

Use the results of Chapter 2 to show that there is one 'bound state' with energy zero and wave function

$$_b f(x) = 2\text{sech}x \qquad (2)$$

and that the remaining eigenfunctions form a continuum with $\omega_k^2 = k^2+1$ and have the functional form

$$_k f(x) = (2\pi)^{-\frac{1}{2}}(\omega_k)^{-1}(k + i\tanh x)\exp(ikx) \qquad (3)$$

The soliton solution breaks the translational symmetry of the model and so the zero mode may be thought of as a Goldstone boson.

As the functions f_b and f_k are eigenfunctions of a self adjoint operator, they satisfy certain orthogonality and completeness relations. Determine those relations.

The sine-Gordon equation often occurs in the modified form

$$\phi_{tt} - \phi_{xx} + \sin\phi = F$$

where F is a constant driving force. For small F this equation is a perturbed form of the standard sine-Gordon equation. Let us suppose that a single soliton solution of the free sine-Gordon equation enters a region where these perturbing forces are operative. We anticipate that the form of the soliton solution will be retained but that its physical attributes such as velocity and width will be altered. We consider an initial single kink solution and seek a new perturbed solution in the form

$$\phi = 4\tan^{-1}\exp(x - vt) + \psi(x,t)$$

Show that to first order the field ψ will satisfy the equation

$$\psi_{tt} - \psi_{xx} + (1 - 2\text{sech}^2 x)\psi = F \qquad (4)$$

This equation can be solved by using the Fourier time transform

$$\tilde{\psi}(z,\omega) = (2\pi)^{-\frac{1}{2}} \int_{-\infty}^{\infty} \exp(i\omega t)\psi(x,t)dt$$

and expanding $\tilde{\psi}(z,\omega)$ in the complete set (2) and (3)

$$\tilde{\psi}(z,\omega) = \psi_b(\omega)\,{}_bf(x) + \int_{-\infty}^{\infty} dk\psi(k,\omega)\,({}_kf(x))$$

Hence solve equation (4).

3. Prove that the topological change $Q[\phi]$ defined

$$Q[\phi] = \frac{1}{(n-1)!\Omega_{n-1}} \int_{R^{n-1}} \varepsilon^{c_1 \ldots c_n} \varepsilon_{b_1 \ldots b_n} \phi^{b_1} \partial c_1 \phi^{b_2} \partial c_2 \ldots dx^1 \ldots dx^n$$

can be recast into the form

$$Q[\phi] = \frac{1}{\Omega_{n-1}} \int dx^1 \ldots dx^{n-1} (\det g_{ab})^{\frac{1}{2}} \det \partial y$$

where $(y^1, \ldots y^{n-1})$ are intrinsic coordinates on the sphere and

$$g_{ab} = \left\langle \frac{\partial \phi}{\partial y^a}, \frac{\partial \phi}{\partial y^b} \right\rangle$$

is the metric tensor on S^{n-1}.

4. An elementary particle in Euclidean space R^4 with coordinates $(t,\underline{x})$ is described by a mapping $\Phi: R^4 \to R^3$ given in coordinates by $(x_o, \underline{x}) \to (\phi_1, \phi_2, \phi_3)$. By introducing the direction field $\Phi: R^4 \to S^2$ defined by

$$\hat{\phi} = \phi / \|\phi\|$$

we can define an integer valued topological charge for such fields by

$$\bar{Q}[\hat{\phi}] = \frac{1}{8\pi} \int d^3x \varepsilon^{c_1 c_2 c_3} \varepsilon_{b_1 b_2 b_3} \partial_{c_1} \hat{\phi}^{b_1} \partial_{c_2} \hat{\phi}^{b_2} \partial_{c_3} \hat{\phi}^{b_3}, \quad \partial_t \bar{Q}[\hat{\phi}] = 0$$

Calculate the topological change of the t-independent fields

(i) $\phi^1 = x^1\ f(x_1,x_2,x_3)$, $\phi^2 = x^2\ f(x_1,x_2,x_3)$, $\phi^3 = x^3\ f(x_1,x_2,x_3)$

(ii) $\phi^1 = 2x^1\ f(x_1,x_2,x_3)$, $\phi^2 = 2x^2 f(x_1,x_2,x_3)$, $\phi^3 = (x_1^2, x_2^2, x_3^2 - 1)\ f(x_1,x_2,x_3)$

(iii) $\phi^1 = x^1 f(x_1,x_2,x_3)$, $\phi^2 = x^2 x^3 f(x_1,x_2,x_3)$, $\phi^3 = ((x^3)^2 - 1) f(x_1,x_2,x_3)$

The projection of $\hat{\phi}$ onto the plane $x_1 = 0$ as a circuit vector ${}_2\hat{\phi}$. Sketch the vector field ${}_2\hat{\phi}$ around the singular points of each of the above fields by drawing the phase portrait of the system of ordinary differentialequations in each case

$$\dot{x}_2 = {}_2\hat{\phi}^2(x_2,x_3) \qquad \dot{x}_3 \doteq {}_2\hat{\phi}^3(x_2,x_3)$$

Construct a field having change 3.

5. Exercise 4 involved a particular example of a mapping from $R^n \to S^{n-2}$ when $n = 4$. Show that if $\hat{\phi}: R^n \to S^{n-2}$ is a smooth mapping except for a finite number of points the techniques of section three can be generalised to show that

$$\bar{J} = \frac{1}{(n-1)!\,\Omega^{n-1}}\ \varepsilon^{ac_1,\ldots,c_n}\ \varepsilon_{b_1 b_2 - b_n}\ \partial_{c_1}\hat{\phi}^{b_1}\partial_{c_2}\hat{\phi}^{b_2}\ldots\partial_{c_n}\hat{\phi}^{b_n}$$

is a conserved current. In order to obtain finite change solutions this requires that

$$\hat{\phi} \to \hat{\phi}_0 \qquad \text{a constant as } |\underline{x}| \to \infty$$

A conserved charge can then be defined by

$$\bar{Q}[\hat{\phi}] = \int \bar{J}^0 dx'.-dx'', \qquad \partial_t \bar{Q}[\hat{\phi}] = 0$$

Show that if ${}_t\hat{\phi}: S^{n-1} \to S^{n-2}$ is defined by

$${}_t\tilde{\phi}(\hat{n}) = \lim_{r\to\infty} \hat{\phi}(t, r\hat{n})$$

then

$$\bar{Q}[\hat{\phi}] = \deg({}_t\tilde{\phi})$$

6. The Dodd-Bullough equation is the special case of the equations

$$\theta_{xt} = e^{2\theta} - e^{-\theta}\cos 3\phi \tag{1}$$

$$\phi_{xt} = e^{-\theta}\sin 3\phi \tag{2}$$

corresponding to $\phi = 0$ (Fordy and Gibbons, 1981).

By directly determining a travelling wave of the form

$$\theta = \theta(x-vt) \qquad \phi = \phi(x-vt)$$

find the single soliton solution

$$\theta = \tfrac{1}{2}\ln\{\frac{1+e^{3\eta}}{1+e^{\eta})^3}\} \tag{3}$$

$$\phi = \tan^{-1}\{\frac{\sqrt{3}\,e^{\eta}}{2-e^{-\eta}}\} \tag{4}$$

where $= kx + 3k^{-1}t$.

Show that the equation

$$\partial_x(\bar{\theta}^{(n)}-\theta^{(n+1)}) = -k\{\exp(\theta^{(n+1)}-\bar{\theta}^{(n+1)})-\exp(\theta^{(n)}-\bar{\theta}^{(n)})\}$$

$$\partial_t(\theta^{(n)}-\bar{\theta}^{(n)}) = -k^{-1}\{\exp(\bar{\theta}^{(n)}-\bar{\theta}^{(n+1)})-\exp(\bar{\theta}^{(n-1}-\theta^{(n)})\}$$

(where $\theta^{(1)}=\theta+i\phi$, $\theta^{(2)}=-2i\phi$, $\theta^{(3)}=-\theta+i\phi$ and the indices are to be understood mod 3), define a Backlund transformation for the equations (1) and (2) and use them to rederive the solution (3) and (4).

Let a Backlund transformation with parameter k_1 take a solution (θ,ϕ) to a solution $(\bar{\theta},\bar{\phi})$ and another with parameter k_2 take that same solution to the solution $(\hat{\theta},\hat{\phi})$. If we require that those same transformations take $(\hat{\theta},\hat{\phi})$ and $(\bar{\theta},\bar{\phi})$ into the same solution $(\tilde{\theta},\tilde{\phi})$ (permutability) we obtain a generalisation of the superposition formula for the sine-Gordon equation. In this case it takes the form

$$\exp(\theta^{(n+1)}+\tilde{\theta}^{(n)}) = \{\frac{k_1\exp(-\bar{\theta}^{(n)})-k_2\exp(-\hat{\theta}^{(n)})}{k_1\exp(-\bar{\theta}^{(n+1)})-k_2\exp(-\hat{\theta}^{(n+1)})}\} \times \exp(\bar{\theta}^{(n)}+\hat{\theta}^{(n)})$$

Use this result to construct a 2-soliton solution to equations (1) and (2).

7. Consider a model described in terms of three real valued scalar fields ϕ^i (i=1,2,3) with field equations

$$\Box\phi^i + \langle \partial_\mu\phi, \partial^\mu\phi\rangle \phi^i = 0 \tag{1}$$

$$\langle\phi,\phi\rangle = 1 \tag{2}$$

(a) Show that the parameterisation

$$\phi = \begin{pmatrix} \cos\beta & \sin\alpha \\ \sin\beta & \sin\alpha \\ \cos\alpha & \end{pmatrix}$$

reduces the field equations (1) and (2) to the form

$$\Box\alpha = \tfrac{1}{2}\sin 2\alpha\,\partial_\mu\beta\,\partial^\mu\beta \tag{3}$$

$$\partial^\mu(\sin^2\alpha\,\partial_\mu\beta) = 0 \tag{4}$$

(b) By seeking a solution in the form

$$\beta = \lambda\theta \qquad \alpha = \alpha(r)$$

where r and θ are polar coordinates, show that a satisfies the equation

$$\alpha_{,rr} + r^{-1}\alpha_{,r} - \frac{\lambda^2}{2} r^{-2}\sin 2\alpha = 0$$

(c) By introducing the new variable z defined by $r = \exp zn^{-1}$ show that the model has the solution

$$\phi = \operatorname{sgn}(n)\begin{pmatrix} \begin{pmatrix}\cos n\theta \\ \sin n\theta\end{pmatrix} \dfrac{2r^n}{1+r^{2n}} \\ \\ \dfrac{1-r^{2n}}{1+r^n} \end{pmatrix}$$

(d) Show that the topological charge for this model takes the form

$$Q[\phi] = \frac{1}{4\pi} \int d^2x \ \sin\alpha \left| \frac{\partial(\alpha,\beta)}{\partial(x_1,x_2)} \right|$$

(e) Show that $Q[\phi_n] = n$

8. A general class of models of interest in elementary particle theory and general relativity is defined by the field equations

$$\partial^\mu(\partial_\mu g g^{-1}) = 0 \qquad \mu = 0,1,2,3. \tag{1}$$

where g is an element of a matrix group.

In Minkowski space we can introduce the curvilinear coordinates (s,θ,ρ,ψ) defined by

$$x = (s\times\cosh\theta, s\times\sinh\theta, \rho\cos\psi, \rho\sin\psi)$$

which are a generalised form of axial polar coordinates in R .

Show that if one seeks solutions to the field equations (1) which are functions of s and ρ alone then equation (1) reduces to the form

$$(g_s g^{-1})_s + s^{-1} g_s g^{-1} = (g_\rho g^{-1})_\rho + \rho^{-1} g_\rho g^{-1}$$

By means of the introduction of the new variables $r = s\rho$ and $t = (s^2 + \rho^2)$ reduce this equation to the form

$$(g_t g^{-1})_t = (g_r g^{-1})_r + r^{-1} g_r g^{-1}$$

(2)

If g is chosen to belong to the group of 2×2 unitary matrices with the property $g^2 = -I$ show that it can be written in the form $g = i\underline{\sigma}.\underline{n}$ where n is a unit vector in R^3. Using this parameterisation show that the field equations satisfied by $\underline{n}$ are given by

$$\underline{n}_{tt} - r^{-1}(r\underline{n}_r)_r + (\underline{n}_t^2 - \underline{n}_r^2)\underline{n} = \underline{0}$$

This is the same equation that would be obtained by seeking radially symmetric solutions to the Minkowski space form of the Heisenberg ferromagnet equations in two space dimensions. If one allows t to be pure imaginary we obtain the normal Euclidean equations.

Show that the integrability conditions for the linear problem

$$[(\mu-\mu^{-1})\partial_r + (\mu+\mu^{-1})\mu/_r\partial_\mu - g_t g^{-1} + \mu^{-1} g_r g^{-1}]\Psi = 0 \quad (3)$$

$$[\mu(\partial_r - \tfrac{\mu}{r}\partial_\mu) + \partial_t - g_t g^{-1}]\Psi = 0 \quad (4)$$

are precisely equation (2). This is a generalised inverse scattering problem for equation (2) of the type mentioned in section 6.4.

Seek a solution to equations (3) and (4) in the form of a series

$$\Psi = 1 + \sum_{k=1}^{\infty} \psi_k(r,t)\mu^{-k}$$

and hence show that, assuming the existence of any necessary integrals, the quantities Q_k defined by

$$Q_k = \lim_{r\to\infty} r^k \psi_k(r,t)$$

are independent of t. This shows that this model has an infinite set of nonlocal conservation laws like the sine-Gordon.

Show that the first two conserved quantities are given by

$$Q_1 = \int^{\infty} g_t g^{-1}\, r\, dr \qquad Q_2 = \int^{\infty} g_t g^{-1} \int^{r} g_t g^{-1} r' dr'\, r\, dr$$

9. The SU(2) Bogolmony equations take the form

$$\partial_a \underline{A}_b - \partial_b \underline{A}_a + \underline{A}_a \times \underline{A}_b = -\varepsilon_{abc}(\partial_c \underline{\Phi} + \underline{A}_c \times \underline{\Phi})$$

Show that solutions of this equation may be found in the form

$$\Phi_j = f(r)\frac{x_j}{r} \qquad (A_j)^k = -g(r)\varepsilon_{jkt}\frac{x}{r}t$$

provided f and g satisfy the coupled ordinary differential equations

$$(fr^{-1})' + (gr^{-1})' = -r^{-1}g(g+f) \tag{1}$$

$$r(gr^{-1})' + fg = -r^{-1}(2g-f) \tag{2}$$

Show that these equations can be simplified if we introduce the new functions F and G defined by

$$F = f + r^{-1} \qquad G = g - r^{-1}$$

and equations (1) and (2) reduce to the form

$$F' = -G^2 \tag{3}$$

$$G' = -GF \tag{4}$$

Show that (3) and (4) imply that

$$F^2 - G^2 = C^2 \quad \text{a constant}$$

and that as a result (3) may be trivially integrated to yield

$$F = C\{\tanh(\frac{r-r_o}{C})\}^{-1}$$

$$G = C\{\sinh(\frac{r-r_o}{C})\}^{-1}$$

If we impose the boundary condition $F \to 1$ as $r \to \infty$ and $F(0) = 1$ we obtain directly the **Prasad-Sommerfeld monopole solution** obtained in section nine by Backlund transformation techniques. Although this is much easier it gives no information about how to determine multi-monopole solutions.

8. THE NONLINEAR SCHRÖDINGER EQUATION AND WAVE RESONANCE INTERACTIONS

8.1 Introduction

In Chapter 1 and again in Chapter 6 we mentioned the equation

$$\beta \frac{\partial^2 A}{\partial x^2} + \gamma A|A|^2 = i \frac{\partial A}{\partial t} \tag{8.1-1}$$

as an integrable evolution equation although we have not yet discussed its physical significance. The name "Nonlinear Schrödinger" (NLS) has been coined precisely because its structure is that of the Schrödinger equation of quantum mechanics with $|A|^2$ as a potential, although for most of the situations in which it ocurs it has no relationship with the real quantum Schrödinger equation other than in name. In fact, it plays a significant role in the theory of the propagation of the envelopes of wave trains in many stable dispersive physical systems in which no dissipation occurs. Since it has such a broad application we will therefore devote the whole of this section to discussing the perturbation method of multiple scales in order to introduce the unfamiliar reader to the simple technique of finding the evolution equation for the envelope of a wave train propagating through a nonlinear dispersive system. Physical examples are considered specifically in later sections.

Many intrinsically nonlinear systems will admit harmonic wave train solutions

$$\phi = a \exp[i(kx-\omega(k)t)] \tag{8.1-2}$$

provided the amplitude a is small enough. This requires that nonlinear terms are small enough to be neglected and if this is the case then the amplitude will remain constant in time, as shown in Fig. 8-1.

The effect of nonlinearity on these sinusoidal oscillations is to cause a variation in the amplitude in both space and time. This is due to the production of higher harmonics, originating from the nonlinear terms, which react back on the original wave. We want to consider a situation in which the basic state of a system is a linear harmonic solution which, although small in amplitude, is nevertheless large enough that the effect of nonlinearity cannot be neglected. We will restrict ourselves to situations where the change of the wave envelope is slow in both space and time in comparison to the carrier wave.

Such examples are common in everyday life. For instance AM (amplitude modulated) radio waves are a specific example of a fast oscillating carrier wave with a relatively slowly varying envelope. A radio receiver picks up

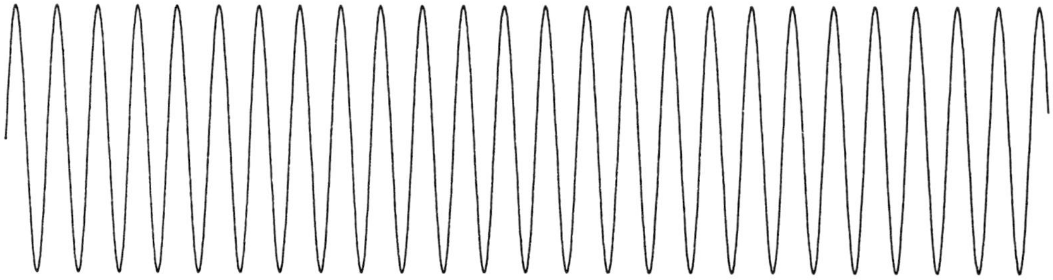

Figure 8-1: A sinusoidal oscillation with constant amplitude

the magnitude and frequency of the wave envelope and converts this into sound waves: the listener hears the wave envelope, not the carrier wave. Figure 8-2 depicts a sinusoidal oscillation with a slowly varying envelope.

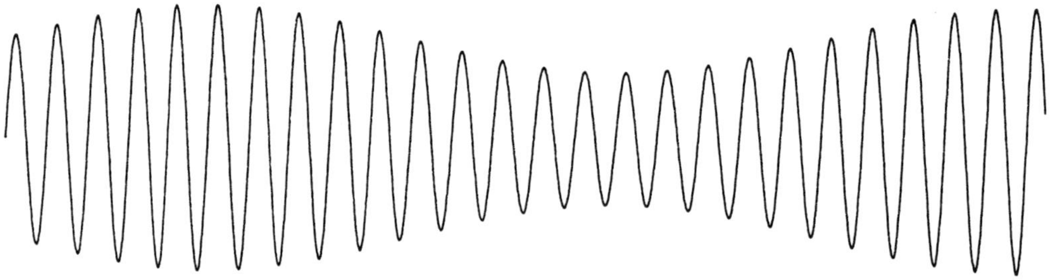

Figure 8-2: A sinusoidal oscillation with a slowly varying amplitude

A further important area in which a slowly varying envelope modulating a carrier wave is relevant is in coherent laser optics. For instance, a coherent optical pulse of a nanosecond in duration in the blue light region (say) would have about 10^6 cycles of the carrier within the envelope pulse. In nonlinear optics there are ways and means of dealing with the original equations of motion from which the evolution equation for a slowly varying envelope (and phase) is desired. This usually involves neglecting higher derivatives of the envelope and higher harmonics of the carrier in the original equations of motion. This is most commonly used when the input carrier wave frequency is resonant with the natural frequency of the host medium (e.g. with the atomic frequency of a two level atomic system). These ideas are applicable to a great number of areas such as plasma physics and fluids as well as optics but the essential idea is to find the weakly nonlinear development of oscillations in the system in question. For resonant systems an instability is created in the oscillations which gives different results to off-resonant situations. In this chapter we shall deal with the weakly nonlinear off-resonant case only, and show that the NLS equation is the typical amplitude equation in this case. The resonant case (unstable) will be dealt with in the next chapter.

In an informal fashion it is possible to show how the NLS equation arises as the evolution equation for a carrier wave envelope (Hasegawa, 1975). A truly linear system has a dispersion relation which is independent of the amplitude. However, let us assume that the development of a harmonic wave in a weakly nonlinear system can be represented by a dispersion relation which is amplitude dependent. Such a situation actually occurs in nonlinear optics and plasmas where the refractive index or dielectric constant of a medium can be dependent on the electric field.

$$\omega = \omega(k; |A|^2) \tag{8.1-3}$$

A Taylor expansion around some suitable wave number k_o and frequency ω_o gives

$$\omega - \omega_o = \left[\frac{\partial \omega}{\partial k}\right]_o (k-k_o) + \tfrac{1}{2}\left[\frac{\partial^2 \omega}{\partial k^2}\right]_o \cdot (k-k_o)^2 + \left[\frac{\partial \omega}{\partial(|A|^2)}\right]_o \cdot |A|^2 \ \ldots \tag{8.1-4}$$

Equation (8.1-4) is the Fourier space equivalent of an operator equation which when operating on A, yields

$$i\left(\frac{\partial}{\partial t} + \left[\frac{\partial \omega}{\partial k}\right]_o \frac{\partial}{\partial x}\right) A + \tfrac{1}{2}\left[\frac{\partial^2 \omega}{\partial k^2}\right]_o \cdot \frac{\partial^2 A}{\partial x^2} - \left[\frac{\partial \omega}{\partial(|A|^2)}\right]_o \cdot A|A|^2 = 0 \tag{8.1-5}$$

where higher terms have been neglected. Equation (8.1-5) is the NLS equation and this rather heuristic derivation shows how the effect of the nonlinear term can be crudely modeled by thinking of the system as having an amplitude dependent dispersion relation. This quick method tells us how the NLS equation arises but unfortunately, for a specific set of model equations, it does not tell us the values of the coefficients in the final equation; in particular the $\partial\omega/\partial(|A|^2)$ term. As we shall see in the next section, the sign on this term is rather important. Instead it is more desirable at this point to introduce a more formal mathematical method which can be applied in general to a large range of nonlinear equations where we want to know the development of a slowly varying envelope modulating a fast carrier wave.

This latter property means that many wavelengths of the carrier wave are contained in just one wavelength of the envelope. Consequently $\lambda_c \lambda_e^{-1} \ll 1$ where λ_c, λ_e are typical wavelengths of the carrier and envelope respectively. Since x and t are normal space and time variables for the carrier we can define a set of "slow" space and time variables $X_n = \epsilon^n x$; $T_n = \epsilon^n t$ ($\epsilon \ll 1$). These slow space and time variables are the variables for the envelope motion and from now on will be considered as independent variables. In approaching the method of finding the evolution equation for the envelope of oscillations for a given nonlinear equation, it is better to proceed by example than in generality. The method is often given the name of "multiple scales" or "two timing" for obvious reasons.

Let us choose a cubically nonlinear Klein Gordon equation

$$\phi_{xx} - \phi_{tt} = \alpha\phi - \beta\phi^3 \tag{8.1-6}$$

which has a Hamiltonian density

$$H = \tfrac{1}{2}(\phi_x{}^2 + \phi_t{}^2) + \tfrac{1}{2}\alpha\phi^2 - \tfrac{1}{4}\beta\phi^4 \tag{8.1-7}$$

and consequently a potential

$$V(\phi) = \tfrac{1}{2}\alpha\phi^2 - \tfrac{1}{4}\beta\phi^4 \tag{8.1-8}$$

A plot of the potential of this equation shows that $\phi = 0$ is a minimu[m] while $\phi = \pm(\alpha/\beta)^{\frac{1}{2}}$ are maxima. $\phi = 0$ is therefore a stable equilibriu[m] solution of (8.1-6) and so we shall expand about this solution in a[n] asymptotic series in ϵ^p where p is, as yet, some unknown positiv[e] number.

$$\phi = \epsilon^p\phi^{(1)} + \epsilon^{2p}\phi^{(2)} + \ldots \tag{8.1-9}$$

The differential operators $\partial/\partial x$ and $\partial/\partial t$ in (8.1-6) must also b[e] transformed in order to take account of the x,t dependence of the slo[w] scales X_n and T_n. On these slow space, time and amplitude scales, th[e] system is said to be weakly nonlinear. $\partial/\partial x$ and $\partial/\partial t$ now transform into

$$\frac{\partial}{\partial x} \to \frac{\partial}{\partial x} + \sum_{n=1} \epsilon^n \frac{\partial}{\partial X_n}$$
$$\frac{\partial}{\partial t} \to \frac{\partial}{\partial t} + \sum_{n=1} \epsilon^n \frac{\partial}{\partial T_n} \tag{8.1-10}$$

From now on we treat the variables x, X_n, t and T_n as independent Substituting (8.1-9) and (8.1-10) into (8.1-6) we obtain

$$\left\{\left(\frac{\partial}{\partial x} + \epsilon\frac{\partial}{\partial X_1} + \epsilon^2\frac{\partial}{\partial X_2}\ldots\right)^2 - \left(\frac{\partial}{\partial t} + \epsilon\frac{\partial}{\partial T_1} + \epsilon^2\frac{\partial}{\partial T_2}\ldots\right)^2 - \alpha\right\}\times$$
$$\times\left(\epsilon^p\phi^{(1)} + \epsilon^{2p}\phi^{(2)}\ldots\right) + \beta\{\epsilon^p\phi^{(1)} + \epsilon^{2p}\phi^{(2)}\ldots\}^3 = 0 \tag{8.1-11}$$

We then find that the terms at different orders of ϵ are

$$O(\epsilon^p): \quad \left(\frac{\partial^2}{\partial x^2} - \frac{\partial^2}{\partial t^2} - \alpha\right)\phi^{(1)} \tag{8.1-12}$$

$$O(\epsilon^{p+1}): \quad 2\left(\frac{\partial^2}{\partial x\partial X_1} - \frac{\partial^2}{\partial t\partial T_1}\right)\phi^{(1)} \tag{8.1-13}$$

$$O(\epsilon^{2p}): \quad \left(\frac{\partial^2}{\partial x^2} - \frac{\partial^2}{\partial t^2} - \alpha\right)\phi^{(2)} \tag{8.1-14}$$

$$O(\epsilon^{2p+1}): \qquad 2\left(\frac{\partial^2}{\partial x \partial X_1} - \frac{\partial^2}{\partial t \partial T_1}\right)\phi^{(2)} \tag{8.1-15}$$

$$O(\epsilon^{p+2}): \qquad 2\left(\frac{\partial^2}{\partial x \partial X_2} - \frac{\partial^2}{\partial t \partial T_2}\right)\phi^{(1)} + \left(\frac{\partial^2}{\partial X_1^2} - \frac{\partial^2}{\partial T_1^2}\right)\phi^{(1)} \tag{8.1-16}$$

$$O(\epsilon^{3p}): \qquad \beta(\phi^{(1)})^3 + \left(\frac{\partial^2}{\partial x^2} - \frac{\partial^2}{\partial t^2} - \alpha\right)\phi^{(3)} \tag{8.1-17}$$

We now need to use a plausibility argument as in Chapter 5 on how to find a sensible value for p. Obviously, equation (8.1-12) is just the linear problem and gives no information about p. We have only two choices for nontrivial p at these lower orders: (a) $p + 1 = 2p$, (b) $p + 1 = 3p$. Choice (b) corresponds to $p = \frac{1}{2}$ but this situation leads to no sensible conclusions in solving for $\phi^{(1)}$. The other alternative is choice (a) which corresponds to $p = 1$. Setting the various terms to zero at appropriate orders of ϵ now gives:

$$O(\epsilon): \quad \left(\frac{\partial^2}{\partial x^2} - \frac{\partial^2}{\partial t^2} - \alpha\right)\phi^{(1)} = 0 \tag{8.1-18}$$

$$O(\epsilon^2): \quad \left(\frac{\partial^2}{\partial x^2} - \frac{\partial^2}{\partial t^2} - \alpha\right)\phi^{(2)} = -2\left(\frac{\partial^2}{\partial x \partial X_1} - \frac{\partial^2}{\partial t \partial T_1}\right)\phi^{(1)} \tag{8.1-19}$$

$$O(\epsilon^3): \quad \left(\frac{\partial^2}{\partial x^2} - \frac{\partial^2}{\partial t^2} - \alpha\right)\phi^{(3)} = -2\left(\frac{\partial^2}{\partial x \partial X_1} - \frac{\partial^2}{\partial t \partial T_1}\right)\phi^{(2)} \tag{8.1-20}$$

$$- 2\left(\frac{\partial^2}{\partial x \partial X_2} - \frac{\partial^2}{\partial t \partial T_2}\right)\phi^{(1)} - \beta(\phi^{(1)})^3 - \left(\frac{\partial^2}{\partial X_1^2} - \frac{\partial^2}{\partial T_1^2}\right)\phi^{(1)}$$

We can now take a harmonic solution of (8.1-18) but since the differential operators are only in the fast scales x and t we may write

$$\phi^{(1)} = A(X_1, X_2, \ldots; T_1, T_2, \ldots)\ \exp(i\theta) + \text{c.c.} \tag{8.1-21}$$

$$\theta = kx - \omega t + \delta \tag{8.1-22}$$

$$\omega^2 = k^2 + \alpha \qquad\qquad \alpha > 0 \tag{8.1-23}$$

The function A in (8.1-21) is an arbitrary (for the moment) complex amplitude function of the slow scales. We have now achieved one stage of the calculation in that we have managed to obtain, at lowest order, a linear oscillatory wave which has an envelope which is a function of the slow scales only. We have therefore achieved a separation of the fast and slow scales. Using (8.1-21) in the $O(\epsilon^2)$ equation we obtain

$$\left(\frac{\partial^2}{\partial x^2} - \frac{\partial^2}{\partial t^2} - \alpha\right)\phi^{(2)} = -2i\left(k\frac{\partial A}{\partial X_1} + \omega\frac{\partial A}{\partial T_1}\right)\exp(i\theta) + \text{c.c.} \tag{8.1-24}$$

In solving for $\phi^{(2)}$ from this equation we immediately run into a problem:

the homogeneous part of the solution is like (8.1-21), precisely because the linear operator in each is identical. Consequently the right hand side "resonates" with this solution as it also contains an $\exp(i\theta)$ term. Consequently the particular integral for $\phi^{(2)}$ will contain terms like $\theta\exp(i\theta)$. As $t \to \infty$ this sort of term will blow up and the perturbation theory will become invalid for times $t > \epsilon^{-1}$. This type of term is often called a **secular term.** To prevent such an occurrence we are left with only two recourses. The first is to have A a constant so the right hand side of (8.1-24) would be zero. The second and more general one is to demand that A satisfies the equation

$$k\frac{\partial A}{\partial X_1} + \omega\frac{\partial A}{\partial T_1} = 0 \tag{8.1-25}$$

This requirement forces us to take a new variable $\bar{X}$

$$\bar{X}_1 = X - c_g T_1 \tag{8.1-26}$$

where $c_g = d\omega/dk$ - the group velocity of the wave train. Hence the amplitude function must be written as $A = A(\bar{X}, X_2, T_2)$. Consequently, in the frame travelling with the group velocity, the secular terms vanish. We still retain the X_2 and T_2 variables since the retention of these enables us to have two or more independent variables in A.

Moving on to the $O(\epsilon^3)$ terms we obtain

$$\left(\frac{\partial^2}{\partial x^2} - \frac{\partial^2}{\partial t^2} - \alpha\right)\phi^{(3)} = -2i\left(k\frac{\partial A}{\partial X_2} + \omega\frac{\partial A}{\partial T_2}\right)\exp(i\theta)$$

$$-\left[(1-c_g^2)\frac{\partial^2 A}{\partial \bar{X}^2} + 3\beta A^2 A^*\right]\exp(i\theta) - \beta A^3\exp(3i\theta) + c.c. \tag{8.1-27}$$

We have missed out the term which includes $\phi^{(2)}$ because this term is automatically zero because of (8.1-26). We again see that all the $\exp(i\theta)$ terms in (8.1-27) are secular. However the $\exp(3i\theta)$ term is not secular because it does not resonate with the homogeneous solution. Therefore in order to prevent the perturbation theory becoming invalid for times $t > \epsilon^{-1}$ we must demand that the coefficient of $\exp(i\theta)$ be zero. This then forces A to evolve according to the equation

$$(1-c_g^2)\frac{\partial^2 A}{\partial \bar{X}^2} + 3\beta A^2 A^* + 2i\left(k\frac{\partial A}{\partial X_2} + \omega\frac{\partial A}{\partial T_2}\right) = 0 \tag{8.1-28}$$

We can now see that it is not entirely necessary to have both X_2 and T_2 since we can always go into a frame of reference to exclude one of them. For the moment we shall drop the X_2 and re-write equation (8.1-28) as

$$\frac{\partial^2 A}{\partial \bar{X}^2} + 3\beta\omega^2\alpha^{-1}A|A|^2 + 2ik\omega^2\alpha^{-1}\frac{\partial A}{\partial T_2} = 0 \tag{8.1-29}$$

This is the Nonlinear Schrödinger equation of Chapters 1 and 6 [see equations (1.6-1)) and (6.1-1)]. The "space" variable is $\bar{X}$ and the "time" variable is T_2 and therefore the automatic time scale on which the envelope equation operates is quite long, since one unit of time on the T_2 scale is ϵ^{-2}

units of real time. $A(\bar{X}, T_2)$ is a complex function and therefore contains information about the phase of the wave. If we split A into its real and imaginary parts:

$$A(\bar{X},T_2) = a(\bar{X},T_2)\ \exp[i\Phi(\bar{X},T_2)] \tag{8.1-30}$$

where a and Φ are real, we have

$$\phi^{(1)} = 2a(\bar{X},T_2)\ \cos[kx-\omega t + \Phi(\bar{X},T_2)] \tag{8.1-31}$$

$2a(\bar{X},T_2)$ is the real slowly varying amplitude function and Φ is the slowly varying phase. It is obviously necessary that initial data is specified for both $a(\bar{X},T_2)$ and $\Phi(\bar{X},T_2)$ in solving the initial value problem for (8.1-29). The single envelope soliton solution for (8.1-29) is given by [see equation (1.6-2)]

$$A = a\sqrt{2/\beta}\ \exp[i\Phi]\ \text{sech}[a(x-bt)] \tag{8.1-32}$$

$$\Phi = \frac{1}{2}bx-(\frac{1}{4}\ b^2-a^2)t \tag{8.1-33}$$

where the exponential is the phase term as we showed above, and a and b are arbitrary constants. The possibility of a slowly varying phase means that the frequency and wave number of the wave train may by pulled off its central value (ω,k) and this difference will vary slowly in space and time. In nonlinear optics this is known as a "chirp".

The result (8.1-29) holds equally well when considering the evolution of small oscillations for the sine-Gordon equation

$$\phi_{xx} - \phi_{tt} = \sin\phi \tag{8.1-34}$$

Since we have been expanding about the stable state $\phi = 0$, we need only consider the $\phi - \phi^3/6$ terms on the right hand side as quintic terms and higher will not contribute until $O(\epsilon^5)$. We can therefore take $\alpha = 1$ and $\beta = 1/6$ in (8.1-29) to obtain the result for the sine-Gordon equation.

8.2 A Class of Equations Which Yield the NLS Equation

The multiple scales calculation of Section 8.1 makes it clear that a certain degree of structure occurs in the calculation which indicates that a generalisation may be possible. For instance, the operator $\partial^2/\partial x^2 - \partial^2/\partial t^2 - \alpha$ always occurred on the left hand side and other operators such as $2(\partial^2/\partial x\partial X_1-\partial^2/\partial t\partial T_1)$are obviously the next term in a Taylor expansion of that particular operator using ϵ as a small expansion parameter.

We therefore consider a class of partial differential equations of the form

$$L(\frac{\partial}{\partial t};\frac{\partial}{\partial x})\phi = \sum_i (M^{(i)}\phi)(N^{(i)}\phi) + \sum_i (P^{(i)}\phi)(R^{(i)}\phi)(Q^{(i)}\phi) \tag{8.2-1}$$

where L, M, N, P, Q and R are scalar differential operators in $\partial/\partial x$ and $\partial/\partial t$. We will assume that the original equations of motion have already been expanded about some stable equilibrium point ϕ: $\phi \to \phi_o + \phi$. We then expand ϕ as an asymptotic series in ϵ:

$$\phi = \epsilon\phi^{(1)} + \epsilon^2\phi^{(2)} + \ldots \tag{8.2-2}$$

where

$$\frac{\partial}{\partial t} \to \frac{\partial}{\partial t} + \epsilon\frac{\partial}{\partial T_1} + \epsilon^2\frac{\partial}{\partial T_2} + \ldots$$

$$\frac{\partial}{\partial x} \to \frac{\partial}{\partial x} + \epsilon\frac{\partial}{\partial X_1}$$

and expand the operators as a Taylor series about $\partial/\partial t$ and $\partial/\partial x$:

$$L\left(\frac{\partial}{\partial t} + \epsilon\frac{\partial}{\partial T_1} + \epsilon^2\frac{\partial}{\partial T_2}\,;\,\frac{\partial}{\partial x} + \epsilon\frac{\partial}{\partial X_1} + \ldots\right) \to L\left(\frac{\partial}{\partial t}\,;\,\frac{\partial}{\partial x}\right)$$
$$+ \epsilon\left(L_1\frac{\partial}{\partial X_1} + L_2\frac{\partial}{\partial T_1}\right) + \tfrac{1}{2}\epsilon^2\left(L_{11}\frac{\partial^2}{\partial X_1^2} + 2L_{12}\frac{\partial^2}{\partial X_1\partial T_1} + L_{22}\frac{\partial^2}{\partial T_1^2} + 2L_1\frac{\partial}{\partial T_1}\right) + \ldots \tag{8.2-3}$$

The suffixes 1 and 2 refer to partial differentiation with respect to $\partial/\partial x$ and $\partial/\partial t$ respectively. Substitution of these expansions into (8.2-1) gives

$$O(\epsilon):\quad L\phi^{(1)} = 0 \tag{8.2-4}$$

$$O(\epsilon^2):\ L\phi^{(2)} = -\left(L_1\frac{\partial}{\partial X_1} + L_2\frac{\partial}{\partial T_1}\right)\phi^{(1)} + \sum_i \left(M^{(i)}\phi^{(1)}\right)\left(N^{(i)}\phi^{(1)}\right) \tag{8.2-5}$$

$$O(\epsilon^3):\ L\phi^{(3)} = -\left(L_1\frac{\partial}{\partial X_1} + L_2\frac{\partial}{\partial T_1}\right)\phi^{(2)} - \frac{1}{2}\Big[L_{11}\frac{\partial^2}{\partial X_1^2}$$
$$+ 2L_{12}\frac{\partial^2}{\partial X_1\partial T_1} + L_{22}\frac{\partial^2}{\partial T_1^2} + 2L_2\frac{\partial}{\partial T_2}\Big] \tag{8.2-6}$$
$$+ \text{cubic nonlinear terms}$$

Equation (8.2-4) is no more than the linearised version of (8.2-1). As explained in Section 8.2, we shall take as the basic state the harmonic wave with an amplitude function which is a function of the slow scales

$$\phi^{(1)} = A(X_1, T_1, T_2)\,\exp(i\theta) + \text{c.c.} \tag{8.2-7}$$

$$\theta = k\omega - \omega t$$

and

$$\ell(-i\omega, ik) = 0 \tag{8.2-8}$$

We are using the notation that

$$L\left(\frac{\partial}{\partial t};\frac{\partial}{\partial x}\right)\exp(i\theta) = (\exp(i\theta))\,\ell(-i\omega;ik)$$

where capitals are operator functions of $\partial/\partial x$ and $\partial/\partial t$ and small letters are functions of k and ω.

Equation (8.2-8) is the dispersion relation. At this point we should point out that we are considering only those systems which yield real values of ω for all k, that is, purely dispersive systems. Those which give imaginary or complex values for ω for certain values of k will be unstable because an exponential growth term will arise in (8.2-7). This type of system behaves rather differently to completely stable ones and we shall be considering these in the next chapter.

The substitution of (8.2-7) into (8.2-5) gives

$$L\phi^{(2)} = -i\left(\ell_\omega \frac{\partial A}{\partial T_1} - \ell_k \frac{\partial A}{\partial X_1}\right) \exp(i\theta) + \sum_i m^{(i)} n^{(i)} A^2 \exp(2i\theta) + \sum_i m^{(i)} n^{(i)*} |A|^2 + \text{c.c.} \tag{8.2-9}$$

As we explained in the last section, the $\exp(i\theta)$ term is a secular term and to prevent the perturbation theory becoming invalid we therefore require that A evolves according to the equation

$$\ell_\omega \frac{\partial A}{\partial T_1} - \ell_k \frac{\partial A}{\partial X_1} = 0 \tag{8.2-10}$$

Total differentiation of the dispersion relation (8.2-8) with respect to k gives

$$\begin{aligned} &\ell_k + \left(\frac{d\omega}{dk}\right)\ell_\omega = 0 \\ &\ell_{kk} + 2\ell_{k\omega}\left(\frac{d\omega}{dk}\right) + \left(\frac{d\omega}{dk}\right)^2 \ell_{\omega\omega} + \left(\frac{d^2\omega}{dk^2}\right)\ell_\omega = 0 \end{aligned} \tag{8.2-11}$$

and consequently we have from equation (8.2-10)

$$A = A(\bar{X}, T_2) \tag{8.2-12}$$

$$\bar{X} = X_1 - \left(\frac{d\omega}{dk}\right)T_1 \tag{8.2-13}$$

Hence, on the first slow space and time scales, the waves travel at the group velocity.

The $mn^* AA^*$ term in (8.2-9) can also cause secularities for certain types of L. For instance, if L has a constant in it, as in equation (8.1-2) of the last section, then this term will cause no problems since no secular terms like $x \exp(i\theta)$ or $t \exp(i\theta)$ will occur in $\phi^{(2)}$. However if L contains no constant term and every term is an operator then $\phi^{(2)}$ will contain secular terms. If L is of this type then M and N must satisfy the condition

$$\sum_i m^{(i)} n^{(i)} + c.c. = 0 \tag{8.2-14}$$

if no secular terms are to occur.

The particular integral of $\phi^{(2)}$ is of the form

$$\phi^{(2)} = \alpha_1 A^2 \exp(2i\theta) + \alpha_2 |A|^2 + B(\bar{X}, T_2) + c.c. \tag{8.2-15}$$

where B is admitted as an integration constant, but is allowed to be a function of the slow scales. α_2 may or may not be zero depending on the form of L. We use (8.2-15) in (8.2-6) and look for secular terms. There may be of two sorts. The first will be the usual ones in $\exp(i\theta)$ terms which will always occur. The second type which are functions of the slow scales only (e.g. B and $|A|^2$ and their slow derivatives) will be secular if L contains no constant. Since (8.2-13) forces all the slow derivatives to be in $\partial/\partial\bar{X}$, we will always find B to be either zero or proportional to $|A|^2$. Removal of the $\exp(i\theta)$ secular terms then yields the NLS equation in the form

$$2i\ell_\omega \frac{\partial A}{\partial T_2} + \left(\frac{\partial^2 \omega}{\partial k^2}\right) \frac{\partial^2 A}{\partial \bar{X}^2} + \gamma A|A|^2 = 0 \tag{8.2-16}$$

The $\partial\omega/\partial k$ coefficient can be obtained from (8.2-10) by use of equation (8.2-11) and may turn out to be a complicated function of both k and ω. Note that equation (8.2-16) is the NLS equation in the same form as (8.1-5) except we have ignored the X_2 spatial scale. The sign will depend on L and the form of the nonlinearity and the sign of $\partial^2\omega/\partial k^2$ for some k will depend on the form of the dispersion relation

Before we turn to some physical examples, it is desirable at this point to return to the inverse scattering operator of the NLS equation to consider the criteria for the production of solitons. The relative signs of β and γ in (8.1-1) are important as they determine whether the isospectral operator of the NLS equation is self or skew adjoint. Repeating the calculation of Chapter 6, we find that the relevant Zakharov-Shabat (1972, 1972) form of the operator which has a spectrum constant for all time is

$$\begin{aligned} \frac{\partial \psi_1}{\partial x} + i\lambda\psi_1 &= q\psi_2 \\ \frac{\partial \psi_2}{\partial x} - i\lambda\psi_2 &= r\psi_1 \end{aligned} \tag{8.2-17}$$

where in this case

$$\begin{aligned} q &= (\gamma/2\beta)^{\frac{1}{2}} A \\ r &= -(\gamma/2\beta)^{\frac{1}{2}} A^* \end{aligned} \tag{8.2-18}$$

Obviously, if $\beta\gamma > 0$ then $r = -q^*$ and so the operator is skew adjoint giving imaginary eigenvalues. The results of Chapter 2,3,4 and 6 showed that solitons arise from a discrete spectrum which in turn arises from negative (bound) energy states. The negative energy states are associated with the imaginary eigenvalues. If $\beta\gamma < 0$ then $r = q^*$ and the operator is self-adjoint giving real eigenvalues. No solitons are possible in this case.

The sign of $\beta\gamma$ varies from problem to problem since it will be dependent on the form of the nonlinearity and the various parameters of the example in question. One could consider $\beta\gamma > 0$ as being the case where focusing or bunching of the wave envelope occurs whereas $\beta\gamma < 0$ would be the case when defocusing occurs. In Section 8.4 we will consider optical focusing and defocusing in a medium where the refractive index varies nonlinearly with the field. A similar example is also mentioned in the section notes concerning the possible propagation of solitons down optical fibres. The general principle of the sign of $\beta\gamma$ corresponding to focusing or defocusing of waves applies to any off-resonant, stable, weakly nonlinear system whether in a fluid, a plasma or a laser optical device.

As we have seen above, the criterion $\beta\gamma \gtrless 0$ is important in determining how some given initial data will evolve. The two possibilities of either bunching into solitons or evolving in a self-similar form without bunching is reminiscent of the Benjamin-Feir criterion (Benjamin and Feir 1967; Benjamin 1967) for the stability of instability of water waves due to the effect of side-bands on the fundamental frequency. One needs to determine whether the underlying carrier wave is destablised by the effect of the nonlinearity in the original equations of motion. Although the NLS equation is the final amplitude equation in either case, the sign of $\beta\gamma$ is, in fact, the determining factor in deciding whether the carrier is unstable under the influence of side bands. Using a very simple and elegant method of Stuart and DiPrima (1978), it is quite easy to show that $\beta\gamma > 0$ is the criterion for instability, as we would expect. Firstly, we linearise around an x-independent solution. Such a solution is easily found:

$$A_o(t) = a_o \exp(i\gamma|a_o|^2 t) \qquad (8.2\text{-}19)$$

where a_o is some arbitrary complex number. Let us now take

$$A_o(x,t) = A_o(t)(1+B(x,t)) \qquad (8.2\text{-}20)$$

and then using (8.2-20) linearise the NLS equation. We easily find that the equation for B is

$$i\frac{\partial B}{\partial t} + \beta\frac{\partial^2 B}{\partial x^2} + \gamma|a_o|^2(B+B^*) = 0 \qquad (8.2\text{-}21)$$

Since (8.2-21) is linear we take

$$B = B_1 \exp[i(\ell x+\Omega t)] + B_2 \exp[-i(\ell x+\Omega^* t)] \qquad (8.2\text{-}22)$$

Taking Ω complex allows for the fact that $\mathrm{Im}(\Omega) \neq 0$ will indicate that the waves are unstable. We obtain two equations

$$\gamma|a_o|^2 B_1^* + B_2(\Omega^* - \beta\ell^2 + \gamma|a_o|^2) = 0 \qquad (8.2\text{-}23)$$

$$(-\Omega - \beta\ell^2 + \gamma|a_o|^2)B_1 + \gamma|a_o|^2 B_2^* = 0 \qquad (8.2\text{-}24)$$

and the condition that this pair of equations has a nontrivial solution is simply

$$\Omega^2 = \beta\ell^2(\beta\ell^2 - 2\gamma|a_o|^2) \qquad (8.2\text{-}25)$$

Regardless of the sign of β, it is easy to show that $\Omega^2 < 0$ if

$$\beta\gamma > \frac{\beta^2\ell^2}{2|a_o|^2} > 0 \qquad (8.2\text{-}26)$$

If $\beta\gamma > 0$ then wave numbers $\ell < 2|\gamma\beta^{-1}|^{\frac{1}{2}} a_0$ will be unstable. This criterion now agrees with that obtained above from the inverse scattering transform.

Finally, in this section we will give one last brief example of how the NLS equation arises as an envelope equation by using a generalised KdV equation as an underlying equation which falls into the general category (8.2-1). This KdV equation we will take in the form

$$u_t + u_x + \beta_1(u^3)_x + \beta_2(u^2)_x + u_{xxx} = 0 \qquad (8.2\text{-}27)$$

In this case, the operator L is

$$L = \frac{\partial}{\partial t} + \frac{\partial}{\partial x} + \frac{\partial^3}{\partial x^3} \qquad (8.2\text{-}28)$$

The dispersion relation is $\omega = k - k^3$ in this case. Expanding u as

$$u = \sum_{n=1} \epsilon^n u^{(n)} \qquad (8.2\text{-}29)$$

we find as expected at $O(\epsilon)$, that

$$u^{(1)} = A(X_1,T_1,T_2)\exp(i\theta) + \text{c.c.} \tag{8.2-30}$$

and at $O(\epsilon^2)$ we also find

$$\left(\frac{\partial}{\partial t}+\frac{\partial}{\partial x}+\frac{\partial^3}{\partial x^3}\right)u^{(2)} = -\left[\frac{\partial A}{\partial T_1}+(1-3k^2)\frac{\partial A}{\partial X_1}\right]\exp(i\theta)$$
$$-2\beta_1 ikA^2\exp(2i\theta) + \text{c.c.} \tag{8.2-31}$$

The $\exp(i\theta)$ term is secular and so we take

$$\bar{X} = X_1 - (1-3k^2)T_1 \tag{8.2-32}$$

Integrating equation (8.2-31) to find $u^{(2)}$, we obtain

$$u^{(2)} = \frac{\beta_1}{3k^2}\left[A^2\exp(2i\theta) + A^{*2}\exp(-2i\theta)\right] + B(\bar{X},T_2) \tag{8.2-33}$$

where $B(\bar{X},T_2)$ is an integration constant with respect to the fast scales x and t but can be made a function of the slow scales. At $O(\epsilon^3)$ we now find

$$\left(\frac{\partial}{\partial t}+\frac{\partial}{\partial x}+\frac{\partial^3}{\partial x^3}\right)u^{(3)} + \left(3\frac{\partial^3}{\partial x^2\partial X_1}+\frac{\partial}{\partial X_1}+\frac{\partial}{\partial T_1}\right)\left[\frac{\beta_1}{3k^2}A^2\exp(2i\theta)\right.$$
$$\left.+\frac{\beta_1}{3k^2}A^{*2}\exp(-2i\theta)+B\right] \tag{8.2-34}$$
$$+\left(3\frac{\partial^3}{\partial x\partial X_1^2}+\frac{\partial}{\partial T_1}\right)\left[A\exp(i\theta)+A^*\exp(-i\theta)\right] = -\beta_2\frac{\partial}{\partial x}\left[A|A|^2\exp(i\theta)+\ldots\right]$$
$$-2\beta_1\frac{\partial}{\partial x}\left[\left(AB+\frac{\beta_1}{3k^2}A|A|^2\right)\exp(i\theta)+\ldots\right]$$
$$-\beta_1\frac{\partial}{\partial X_1}\left[A^2\exp(2i\theta)+2|A|^2+A^{*2}\exp(-2i\theta)\right]$$

There are two types of secular terms in (8.2-34). The first are ones which are functions of the slow scales only and which will give rise to terms explicitly in x and t in $u^{(3)}$. Removal of these gives

$$\left(\frac{\partial}{\partial X_1}+\frac{\partial}{\partial T_1}\right)B = -2\beta_1\frac{\partial}{\partial X_1}\left(|A|^2\right) \tag{8.2-35}$$

Use of (8.2-32) shows that

$$B = -\frac{2\beta_1}{3k^2}|A|^2 \qquad (8.2\text{-}36)$$

Removal of the $\exp(i\theta)$ secular terms gives

$$3ik\frac{\partial^2 A}{\partial \bar{X}^2} + \frac{\partial A}{\partial T_2} = -i\beta_2 kA|A|^2 + \frac{2i\beta_1^2}{3k}A|A|^2 \qquad (8.2\text{-}37)$$

which finally gives the NLS equation

$$i\frac{\partial A}{\partial T_2} = 3k\frac{\partial^2 A}{\partial \bar{X}^2} + k\left(\beta_2 - \frac{2\beta_1^2}{3k^2}\right)A|A|^2 \qquad (8.2\text{-}38)$$

The criterion $\beta\gamma > 0$ for solitons in this case is given by

$$3k^2\beta_2 > 2\beta_1^2 \qquad (8.2\text{-}39)$$

On its own ($\beta_2 = 0$) the KdV equation will not allow envelope solitons on top of the oscillatory waves of the continuous spectrum, whereas the MKdV equation on its own ($\beta_1 = 0$) will allow these if $\beta_2 > 0$. Recalling the work of Chapter 6, we showed there that breather solutions (complex conjugate eigenvalues) can exist for the MKdV equation when $\beta_2 > 0$ but cannot exist when $\beta_2 < 0$ nor for the KdV equation. This fits with the result of this section on the NLS equation, since a breather solution in the high frequency limit will become an envelope soliton modulating an oscillating carrier wave.

Finally, we have shown in this section that stable conservative systems yield the NLS equation as a generic envelope equation. In the next two sections we will concentrate on just two worked examples, the first of which is a slight exception to the general case in that a 2-dimensional NLS equation is derived for optical self-focusing in which the spatial variables are transverse to the direction of carrier wave propagation.

8.3 Optical Self-Focusing

When a neutral dielectric of n atoms undergoes interaction with intense laser light which is off-resonant, then it is possible that the polarisation of the dielectric P may depend nonlinearly on the field E. For weak fields, we normally write

$$\underset{\sim}{P} = \underline{\underline{\alpha}}\underset{\sim}{E} \qquad (8.3\text{-}1)$$

where $\underline{\underline{\alpha}}$ is the polarisability tensor, the elements of which are determined by the crystal structure. For simplicity in this case we will consider a smooth distribution of n atoms/c.c. which are centro-symmetric and which form an isotropic medium. We will write the polarisation P as

$$P = \alpha_1 E + \alpha_3 E^3 + \ldots \qquad (8.3\text{-}2)$$

where α_1 is the linear and α_3 the nonlinear polarisability. Maxwell's equation governs the evolution of the field E and in this case is

$$\nabla^2 E - \frac{1}{c^2} E_{tt} = \frac{4\pi n}{c^2} P_{tt} \tag{8.3-3}$$

which can be re-written as

$$\nabla^2 E - \beta E_{tt} = \gamma (E^3)_{tt} \tag{8.3-4}$$

where

$$\beta = c^{-2}(1+4\pi n\alpha_1) \tag{8.3-5}$$

$$\gamma = 4\pi n\alpha_3 c^{-2} \tag{8.3-6}$$

We assume a linearly polarised wave propagating down the z-axis only, and so we want to consider slow amplitude variations of $\exp[i(kz - \omega t) + \delta]$. Introducing slow space and time scales

$$X_n = \epsilon^n x \; ; \quad Y_n = \epsilon^n y \quad ; \quad Z_n = \epsilon^n z \tag{8.3-7}$$

and expanding E as

$$E = \epsilon E^{(1)} + \epsilon^2 E^{(2)} + \ldots \tag{8.3-8}$$

equation (8.3-4) becomes

$$\begin{aligned}
&[(\nabla^2-\beta\, \partial^2/\partial t^2) + 2\epsilon(\partial^2/\partial x\partial X_1 + \partial^2/\partial y\partial Y_1 + \partial^2/\partial z\partial Z_1 - \beta\, \partial^2/\partial t\partial T_1) \\
&+ \epsilon^2(\partial^2/\partial X_1^2 + \partial^2/\partial Y_1^2 + \partial^2/\partial Z_1^2 - \beta\, \partial^2/\partial T_1^2 + 2\partial^2/\partial x\partial X_2 + 2\partial^2/\partial y\partial Y_2 \\
&+ 2\partial^2/\partial z\partial Z_2 - 2\beta\, \partial^2/\partial t\partial T_2)\ldots](\epsilon E^{(1)}+\epsilon^2 E^{(2)}+ \ldots) \\
&= \gamma(\frac{\partial^2}{\partial t^2} + \ldots)(\epsilon E^{(1)}+\epsilon^2 E^{(2)} \ldots)
\end{aligned} \tag{8.3-9}$$

At successive orders of ϵ we find

$$O(\epsilon): \quad (\nabla^2-\beta\, \partial^2/\partial t^2)E^{(1)} = 0 \tag{8.3-10}$$

$$\begin{aligned}
O(\epsilon^2): \quad (\nabla^2-\beta\, \partial^2/\partial t^2)E^{(2)} = &-2(\partial^2/\partial x\partial X_1 + \partial^2/\partial y\partial Y_1 + \partial^2/\partial z\partial Z_1 \\
&- \beta\, \partial^2/\partial t\partial T_1)E^{(1)}
\end{aligned} \tag{8.3-11}$$

$$\begin{aligned}
O(\epsilon^3): \quad (\nabla^2-\beta\, \partial^2/\partial t^2)E^{(3)} = &-2(\partial^2/\partial x\partial X_1 + \partial^2/\partial y\partial Y_1 + \partial^2/\partial z\partial Z_1 \\
&- \beta\, \partial^2/\partial t\partial T_1)E^{(2)} + \gamma\, \partial^2/\partial t^2[(E^{(1)})^3] \\
&-(\partial^2/\partial X_1^2 + \partial^2/\partial Y_1^2 + \partial^2/\partial Z_1^2 - \beta\, \partial^2/\partial T_1^2 + 2\partial^2/\partial x\partial X_2 + 2\partial^2/\partial y\partial Y_2 \\
&+ 2\partial^2/\partial z\partial Z_2 - 2\beta\, \partial^2/\partial t\partial T_2)E^{(1)}
\end{aligned} \tag{8.3-12}$$

Equation (8.3-10) is the linear problem and we can write as a solution

$$E^{(1)} = E(X_n, Y_n, Z_n, T_n) \exp(i\theta) + \text{c.c.} \tag{8.3-13}$$

where

$$\theta = kz - \omega t + \delta \tag{8.3-14}$$

$$\omega^2 = k^2/\beta \qquad d\omega/dk = \beta^{-\frac{1}{2}} \tag{8.3-15}$$

This solution represents a wave polarised along the z-axis with a slowly varying envelope E. Since the exponential $\exp(i\theta)$ contains only z as a space variable, the first two terms on the right hand side of (8.3-11) drop out, leaving

$$(\nabla^2 - \beta\, \partial^2/\partial t^2)E^{(1)} = -2\beta i\omega\left(\frac{\partial E}{\partial T_1} + \beta^{-\frac{1}{2}}\frac{\partial E}{\partial Z_1}\right)\exp(i\theta) + \text{c.c.} \tag{8.3-16}$$

In common with most of these problems, the term on the right hand side of (8.3-16) is secular and must be removed. We therefore define

$$\xi = Z_1 - \left(\frac{d\omega}{dk}\right)T_1 \tag{8.3-17}$$

The $O(\epsilon^3)$ problem now becomes

$$\begin{aligned}(\nabla^2 - \beta\partial^2/\partial t^2)E^{(3)} &= -\,[\partial^2 E/\partial X_1^2 + \partial^2 E/\partial Y_1^2 + 2ik\,\partial E/\partial Z_1 \\ &+ 2i\beta\omega\,\partial E/\partial T_2]\exp(i\theta) \\ &+ \gamma\omega^2[9E^3\exp(3i\theta) - 3E^2E^*\exp(i\theta)] + \text{c.c.}\end{aligned} \tag{8.3-18}$$

No derivatives in ξ appear because of exact cancellation between the $\partial^2 E/\partial z_1^2$ and $\partial^2 E/\partial T_1^2$ terms. Removal of secular terms from equation (8.3-18) finally yield the amplitude equation for E:

$$\frac{\partial^2 E}{\partial X_1^2} + \frac{\partial^2 E}{\partial Y_1^2} + 3\gamma\omega^2 E|E|^2 + i\,\partial E/\partial \bar{Z} = 0 \tag{8.3-19}$$

where we have transformed into a frame of reference moving with the group velocity along the z-axis.

Equation (8.3-19) is the NLS equation in two space dimensions. In one transverse space dimension, if $\gamma > 0$, then equation (8.3-19) admits envelope soliton solutions and since it is integrable by inverse scattering (see Chapter 6, Section 3 for full solutions) an initial profile will produce "solitons". Asymptotically $(\bar{Z} \to \infty)$, these solutions will look like

$$|E|^2 = 4\sum_{i=1} a_i^2 \operatorname{sech}^2\{2a_i X_1 - 8a_i b_i \bar{Z} + \delta_i\} \tag{8.3-20}$$

where it is understood that $\bar{Z}$ now plays the role of "time".

Each "soliton" represents a channel or filament diverging from the z-axis at an angle $\tan^{-1}(4b_i)$. This filamentation of an initial profile is called self-focusing. If however, γ is negative, then the inverse scattering eigenvalue problem is self-adjoint and no solitons are possible and only the continuous spectrum appears which disperses as Z increases. This would be equivalent to the de-focusing case. Because X_1 and Y_1 are transverse variables, focusing in this case refers to bunching into filaments in a transverse fashion. The longitudinal variable ξ, defined in (8.3-17), does not appear because there is no dispersion in this direction. In other examples where soliton propagation is longitudinal, focusing would refer to the bunching of the carrier wave into pulses (solitons) along the wave propagation direction as explained in Section 8.2. This would occur provided longitudinal dispersion was present.

8.4 Langmuir Waves in a Plasma

We now turn to a problem in plasma physics which is very similar to the one considered in Chapter 5. In that problem we considered the motion of long waves in the density of charged ions in a plasma but neglected the electron mass. We now consider a situation where the ions are perfectly cold and therefore stationary. We no longer consider the electrons to be just a charged gas, but now include the electron inertia and seek an amplitude equation which describes the development of electron plasma oscillations (Langmuir waves). The only contribution from the ions is in Poisson's equation where the effect of their charge must be taken into account. We shall write down the continuity and momentum equations and Poisson's equation for the electrons in dimensionless form as the procedure for scaling such equations was given in Section 5.2. We use the notation that n and v are the electron density and velocity respectively and ϕ is the electrostatic potential. Boundary conditions are: $n \to 1;\ \phi,\ v \to 0$ as $|x| \to \infty$.

$$n_t + (nv)_x = 0 \tag{8.4-1}$$

$$v_t + vv_x = \phi_x - n_x/n \tag{8.4-2}$$

$$\phi_{xx} = n-1 \tag{8.4-3}$$

The factor of -1 in (8.4-3) is the scaled to unity constant ion charge and the term n_x/n in (8.4-2) is the pressure of the electrons, a term which we derived in Chapter 5. In that chapter, because the electron mass was neglected, the terms $v_t + vv_x$ were not taken into account which enabled us to integrate the equation once to obtain $n = \exp(\phi)$. We are including electron inertia here so we cannot take this latter relationship between the density and the potential to be true.

In the spirit of this chapter we now want to analyse the evolution of the amplitude of oscillations in the electron density. These are known as Langmuir waves. The scaling procedure is just the same as before except we now have a set of coupled nonlinear p.d.e's with three dependent variables and not one. Taking slow space and time scales $X_1 = \epsilon x$, $T_1 = \epsilon t$, $T_2 = \epsilon^2 t$ and expanding n, v and ϕ about their equilibrium values, we have

$$n = 1 + \sum_{i=1} \epsilon^i n^{(i)} \tag{8.4-4}$$

$$v = \sum_{i=1} \epsilon^i v^{(i)} \tag{8.4-5}$$

$$\phi = \sum_{i=1} \epsilon^i \phi^{(i)} \tag{8.4-6}$$

We will miss out the X_2 scale because this can be tranformed away by a change in the frame of reference. Instead of expanding the differential operators and the dependent variables in full, which takes a lot of space, we will instead state the ordered equations at $O(\epsilon)$, $O(\epsilon^2)$ and $O(\epsilon^3)$ for equations (8.4-1 - 8.4-3). For (8.4-1)

$$O(\epsilon): \quad n_t^{(1)} + v_x^{(1)} = 0 \tag{8.4-7}$$

$$O(\epsilon^2): \quad n_t^{(2)} + v_x^{(2)} = -\left[\frac{\partial n^{(1)}}{\partial T_1} + \frac{\partial n^{(1)}}{\partial X_1} + (n^{(1)}v^{(1)})_x\right] \tag{8.4-8}$$

$$O(\epsilon^3): \quad n_t^{(3)} + v_x^{(3)} = -\left[\frac{\partial n^{(2)}}{\partial X_1} + \frac{\partial n^{(2)}}{\partial T_1} + \frac{\partial n^{(1)}}{\partial T_2} + (v^{(2)}n^{(1)} + v^{(1)}n^{(2)})_x\right] \tag{8.4-9}$$

For (8.4-2)

$$O(\epsilon): \quad v_t^{(1)} - \phi_x^{(1)} + n_x^{(1)} = 0 \tag{8.4-10}$$

$$O(\epsilon^2): \quad v_t^{(2)} - \phi_x^{(2)} + n_x^{(2)} + \frac{\partial v^{(1)}}{\partial T_1} + \tfrac{1}{2}(v^{(1)2} - n^{(1)2})_x + \frac{\partial n^{(1)}}{\partial X_1} - \frac{\partial \phi^{(1)}}{\partial X_1} = 0 \tag{8.4-11}$$

$$O(\epsilon^3): \quad v_t^{(3)} - \phi_x^{(3)} + n_x^{(3)} + \frac{\partial v^{(2)}}{\partial T_1} - \frac{\partial \phi^{(2)}}{\partial X_1} + \frac{\partial n^{(2)}}{\partial X_1} + \frac{\partial v^{(1)}}{\partial T_2}$$
$$+ \frac{1}{2}\frac{\partial}{\partial X_1}(v^{(1)^2} - n^{(1)^2}) + (v^{(1)}v^{(2)} - n^{(1)}n^{(2)})_x + n^{(1)^2}n_x^{(1)} = 0 \tag{8.4-12}$$

For (8.4-3)

$$O(\epsilon): \quad \phi_{xx}^{(1)} = n^{(1)} \tag{8.4-13}$$

$$O(\epsilon^2): \quad \phi_{xx}^{(2)} + 2\frac{\partial^2 \phi^{(1)}}{\partial x \partial X_1} = n^{(2)} \tag{8.4-14}$$

$$O(\epsilon^3): \quad \phi_{xx}^{(3)} + 2\frac{\partial^2 \phi^{(2)}}{\partial x \partial X_1} + \frac{\partial^2 \phi^{(1)}}{\partial X_1^2} = n^{(3)} \tag{8.4-15}$$

At $O(\epsilon^3)$ in (8.4-1) and $O(\epsilon^2)$ and $O(\epsilon^3)$ in (8.4-2) we have re-arranged the formulae and appealed to others in turn in order to simplify them. The linear problem can be cast in matrix form:

$$\begin{pmatrix} \frac{\partial}{\partial t} & \frac{\partial}{\partial x} & 0 \\ \frac{\partial}{\partial x} & \partial/\partial t & -\partial/x \\ -1 & 0 & \frac{\partial^2}{\partial x^2} \end{pmatrix} \cdot \begin{pmatrix} n^{(1)} \\ v^{(1)} \\ \phi^{(1)} \end{pmatrix} = 0 \tag{8.4-16}$$

Remembering that we are looking for oscillating solutions $\exp(i\theta)$, $\theta = kx - \omega t + \delta$; the dispersion relation is found by putting the determinant of the matrix in (8.4-16) to zero which is no other than the condition that nontrivial solutions exist for $n^{(1)}$, $\phi^{(1)}$ and $v^{(1)}$. Hence we find that

$$\det \begin{pmatrix} -i\omega & ik & 0 \\ ik & -i\omega & -ik \\ -1 & 0 & -k^2 \end{pmatrix} = 0 \qquad (8.4\text{-}17)$$

which gives

$$\omega^2 = k^2 + 1 \qquad d\omega/dk = k/\omega \qquad (8.4\text{-}18)$$

and we have a solution of the linear problem in the form

$$\begin{pmatrix} n^{(1)} \\ v^{(1)} \\ \phi^{(1)} \end{pmatrix} = A(X_1, T_1, T_2) \begin{pmatrix} 1 \\ \omega/k \\ -1/k^2 \end{pmatrix} \exp(i\theta) + \text{c.c.} \qquad (8.4\text{-}19)$$

where $A(X_1, T_1, T_2)$ is the complex amplitude function.

Alternatively we could have eliminated $v^{(1)}$ and $\phi^{(1)}$ to obtain

$$n_{tt}^{(1)} - n_{xx}^{(1)} + n^{(1)} = 0 \qquad (8.4\text{-}20)$$

At this point we should note that as in all these types of problems, as we go to the next order of ϵ, the form of the linear part of the problem remains the same, a result made plain by Section 8.2. At each order therefore we will obtain equations in the form

$$n_t^{(i)} + v_x^{(i)} = \alpha^{(i)} \qquad (8.4\text{-}21)$$

$$v_t^{(i)} - \phi_x^{(i)} + n_x^{(i)} + \beta^{(i)} = 0 \qquad (8.4\text{-}22)$$

$$\phi_{xx}^{(i)} + \gamma^{(i)} = n^{(i)} \qquad (8.4\text{-}23)$$

where the $\alpha^{(i)}$, $\beta^{(i)}$ and $\gamma^{(i)}$ are the remainders in the expressions in the $O(\epsilon^2)$ and $O(\epsilon^3)$ equations.

These yield

$$n^{(i)}_{tt} - n^{(i)}_{xx} + n^{(i)} = \alpha^{(i)}_{t} + \beta^{(i)}_{x} + \gamma^{(i)} \tag{8.4-24}$$

Obviously, if the right hand side of (8.4-24) contains terms in $\exp(i\theta)$, these will be secular and must be removed in order to keep the perturbation theory valid. For $i = 2$ one finds that

$$\begin{aligned}\alpha^{(2)}_{t} + \beta^{(2)}_{x} + \gamma^{(2)} &= 2i\omega\left[\frac{\partial A}{\partial T_1} + \frac{k}{\omega}\frac{\partial A}{\partial X_1}\right] \exp(i\theta) \\ &\quad - (2+4\omega^2)\, A^2 \exp(2i\theta) + \text{c.c.}\end{aligned} \tag{8.4-25}$$

Since k/ω is the group velocity we have the usual behaviour and so we define

$$\bar{X} = X_1 - \left(\frac{k}{\omega}\right) T_1 \tag{8.4-26}$$

in order to remove the secular term in (8.4-25). Without going into all the working we can now integrate the equations at $O(\epsilon^2)$ to obtain

$$n^{(2)} = \frac{2}{3}(2\omega^2+1) A^2 \exp(2i\theta) + \text{c.c.} + B_1 \tag{8.4-27}$$

$$v^{(2)} = \frac{\omega}{3k}(4\omega^2-1) A^2 \exp(2i\theta) + \frac{i}{\omega k^2}\cdot\frac{\partial A}{\partial \bar{X}} \exp(i\theta) + \text{c.c.} + B_2 \tag{8.4-28}$$

$$\phi^{(2)} = -\frac{1}{6k^2}(2\omega^2+1) A^2 \exp(2i\theta) - \frac{2i}{k^3}\cdot\frac{\partial A}{\partial \bar{X}} \exp(i\theta) + \text{c.c.} + B_3 \tag{8.4-29}$$

The secular terms at $O(\epsilon^3)$ in the slow scales only give $B_1(\bar{X},T_2) = B_2(\bar{X},T_2) = 0$ and

$$B_3 = \frac{1}{2} k^{-2} |A|^2 \tag{8.4-30}$$

and finally the $\exp(i\theta)$ secular terms give, after some algebra

$$2i\frac{\partial A}{\partial T_2} + \frac{1}{\omega^3}\cdot\frac{\partial^2 A}{\partial \bar{X}^2} - \frac{1}{3\omega}(8k^4+21k^2+12) A|A|^2 = 0 \tag{8.4-31}$$

We note, as expected, that the coefficient of the $\partial^2 A/\partial \bar{X}^2$ term is exactly $d\omega^2/dk^2$.

8.5 Quadratic Wave Resonance Interaction

In the derivation of the NLS equation given in Section 2, the equation was always produced as a result of the removal of secular terms at $O(\epsilon^3)$ in order to prevent the perturbation theory becoming invalid. We could refer to this as a "cubic resonance". If however, the underlying system has quadratic nonlinearities, it is possible to obtain a quadratic resonance at $O(\epsilon^2)$ if we include more than one wave number and frequency into the problem; a situation which can physically occur when different modes interact in the same system. In order to study the effect of several interacting modes, let us return to our calculation of Section 8.2 and take the linear solution at $O(\epsilon)$ to be

$$\phi^{(1)} = A_1\exp(i\theta_1) + A_2\exp(i\theta_2) + A_3\exp(i\theta_3) + c.c. \qquad (8.5\text{-}1)$$

Let us also suppose that the waves form a "triad" such that

$$\begin{aligned} k_3 &= k_1 + k_2 \qquad \text{(conservation of momentum)} \\ \omega_3 &= \omega_1 + \omega_2 \qquad \text{(conservation of energy)} \end{aligned} \qquad (8.5\text{-}2)$$

and so

$$\theta_3 = \theta_1 + \theta_2 \qquad (8.5\text{-}3)$$

At $O(\epsilon^2)$ the quadratic term $(M\phi)(N\phi)$ in (8.2-1) gives various harmonics, most of which will not resonate with L. However, because $\theta_3 = \theta_1 + \theta_2$, three particular terms will resonate with L. Writing out the $O(\epsilon^2)$ terms, we have

$$\begin{aligned} L\phi^{(2)} = &- \sum_{j=1}^{3} \left(\ell_{\omega_j} \cdot \frac{\partial A}{\partial T_1} - \ell_{k_j} \cdot \frac{\partial A}{\partial X_1}\right) \exp(i\theta_j) \\ &+ (m_1n_2+m_2n_1)A_1A_2\exp[i(\theta_1+\theta_2)] + (m_3n_1^*+m_1^*n_3)A_3A_1^*\exp[i(\theta_3-\theta_1)] \\ &+ (m_3n_2^*+m_2^*n_3)A_3A_2^*\exp[i(\theta_3-\theta_2)] + c.c. \\ &+ \text{nonsecular quadratic terms} \end{aligned} \qquad (8.5\text{-}4)$$

ℓ_{k_j} means the derivative of ℓ with respect to k, evaluated at $k = k_j$ $(j = 1,2,3)$. Using the fact that $\theta_3 = \theta_1 + \theta_2$ we can remove the three separate secular terms to obtain finally

$$\left(\frac{\partial}{\partial T_1} + c_1 \frac{\partial}{\partial X_1}\right)A_1 = \mu_1 A_2^* A_3 \qquad (8.5\text{-}5)$$

$$\left(\frac{\partial}{\partial T_1} + c_2 \frac{\partial}{\partial X_1}\right)A_2 = \mu_2 A_3 A_1^* \qquad (8.5\text{-}6)$$

$$\left(\frac{\partial}{\partial T_1} + c_3 \frac{\partial}{\partial X_1}\right)A_3 = \mu_3 A_1 A_2 \qquad (8.5\text{-}7)$$

These are generally known as the three wave resonance equations. The quadratic terms only produce a resonance because of the relation between the frequencies and wave numbers. Although it might appear difficult physically to satisfy (8.5-2), many systems operate in such a fashion that each of these conditions is satisfied automatically by creation of a third wave. For instance, in nonlinear optics, a laser of suitable frequency acting as a pump, can induce a Stokes wave through a Raman process and the third wave is produced as an acoustic (phonon) wave in the material. Equations (8.5-2) are much easier to satisfy in higher space dimensions and are also more physically relevant.

$$\underline{k}_3 = \underline{k}_1 + \underline{k}_2$$
$$\omega_3 = \omega_1 + \omega_2 \tag{8.5-8}$$

For a given dispersion relation we now have a much greater degree of freedom such that (8.5-8) is satisfied. In two space dimensions for instance we must include the further slow scale $Y_1 = \epsilon y$ to obtain equations with six group velocities c_{ij} which can be calculated from the three frequencies.

$$\left(\frac{\partial}{\partial T_1} + c_{11}\frac{\partial}{\partial X_1} + c_{12}\frac{\partial}{\partial Y_1}\right)A_1 = \mu_1 A_2^* A_3 \tag{8.5-9}$$

$$\left(\frac{\partial}{\partial T_1} + c_{21}\frac{\partial}{\partial X_1} + c_{22}\frac{\partial}{\partial Y_1}\right)A_2 = \mu_2 A_3 A_1^* \tag{8.5-10}$$

$$\left(\frac{\partial}{\partial T_1} + c_{31}\frac{\partial}{\partial X_1} + c_{32}\frac{\partial}{\partial Y_1}\right)A_3 = \mu_3 A_1 A_2 \tag{8.5-11}$$

The 3-wave resonance or "mixing" phenomenon is a very important effect in all nonlinear systems since it allows an exchange of energy between modes. It also allows the system to be pumped with energy in one mode and have this energy transferred into another. Zakharov and Manakov (1973; 1976; 1976) followed by Kaup (1976) have shown that the one-space dimensional version of the 3-wave equations is integrable by the inverse scattering transform and Zakharov (1976) has shown the same for the two dimensional version. Obviously, we can see from the way the equations arise that this resonance interaction will occur naturally in a whole batch of physical systems including nonlinear optics, electronics, plasmas and hydrodynamics. A large and excellent review has been written by Kaup, Rieman and Bers (1979) followed by Reiman (1979) on almost every aspect of the 3-wave interaction and a complete list of references can be found in these including a discussion of the decaying and explosive instabilities. Kaup (1981) has also discussed the 3-wave resonance in three spatial dimensions.

The problem of dealing with the integrability of equations (8.5-5)-(8.5-7) by inverse scattering is rather a difficult problem and is out of the scope of this book. However, in simpler terms, we can consider a 2-wave version of equations (8.5-5)-(8.5-7). A repeat of the above calculation with the condition $k_2 = 2k_1$ yields the 2-wave resonance equations

$$\left(\frac{\partial}{\partial T_1} + c_1\frac{\partial}{\partial X_1}\right)A_1 = \mu_1 A_1^* A_2$$
$$\left(\frac{\partial}{\partial T_1} + c_2\frac{\partial}{\partial X_1}\right)A_2 = \mu_2 A_1^2 \tag{8.5-12}$$

Making the transformation

$$\xi = \mu_1 (X - c_2 T_1)/(c_1 - c_2)$$
$$\tau = -\mu_2 (X - c_1 T_1)/(c_1 - c_2) \qquad (8.5\text{-}13)$$

equations (8.5-12) become

$$\frac{\partial A_1}{\partial \xi} = A_1^* A_2$$
$$\frac{\partial A_2}{\partial \tau} = A_1^2 \qquad (8.5\text{-}14)$$

The integrability conditions of the 2 x 2 scattering problem of Chapter 6 (see equations (6.1-12)) can be satisfied by choosing

$$A = -\frac{i}{2i\lambda}|A_1|^2 \qquad B = \frac{1}{2i\lambda}A_1^2 \qquad C = -\frac{1}{2i\lambda}A_1^{*2} \qquad (8.5\text{-}15)$$

$$q = r^* = A_2 \qquad (8.5\text{-}16)$$

See Kaup et al (1979) for the various solutions obtainable through the inverse scattering transform including the decaying and explosive instabilities.

8.6 Long wave short wave resonances

In 1977, Benney suggested a new form of triad resonance between three wave numbers k_1, k_2 and k_3 and three frequencies ω_1, ω_2 and ω_3. We take

$$k_1 = k_2 + k_3 \qquad (8.6\text{-}1)$$

$$\omega_1 = \omega_2 + \omega_3 \qquad (8.6\text{-}2)$$

but consider k_1 and k_2 to be very close: $k_1 = k + \epsilon\kappa$, $k_2 = k - \epsilon\kappa$ and $k_3 = 2\epsilon\kappa$ ($k, \kappa \sim O(1)$ and $\epsilon \ll 1$). The wave number triad given in (8.6-1) is automatically achieved by this choice but the frequency triad can only be achieved if

$$\omega(k+\epsilon\kappa) - \omega(k-\epsilon\kappa) = \omega_3 \qquad (8.6\text{-}3)$$

which, to $O(\epsilon)$ becomes

$$2\epsilon\kappa\frac{d\omega}{dk} = \omega_3 \qquad (8.6\text{-}4)$$

From (8.6-4) we see that (8.6-2) is satisfied to $O(\epsilon)$ provided the group velocity of the short wave (wave number k) equals the phase velocity of the long wave (wave number $2\epsilon\kappa$). Benney called this the long wave short wave resonance. At first sight this resonance would appear hard to achieve but a geometrical picture of how this might be done is given in Fig. 8-3. It

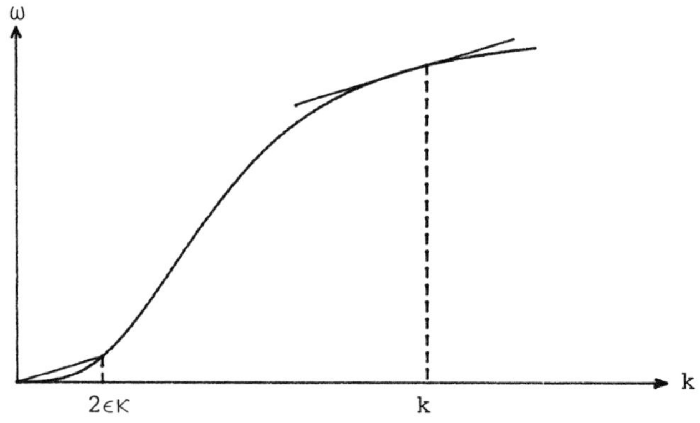

Figure 8-3:

requires a special form of the dispersion relation to achieve this resonance condition. A much more likely and more general type of system for which $c_p(2\epsilon\kappa) = c_g(k)$ is one which has a double branch. Two types of double branches are given in Figs. 8-4 and 8-5.

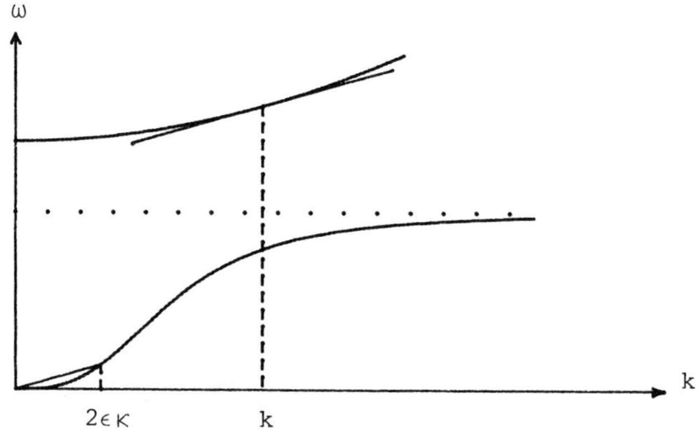

Figure 8-4:

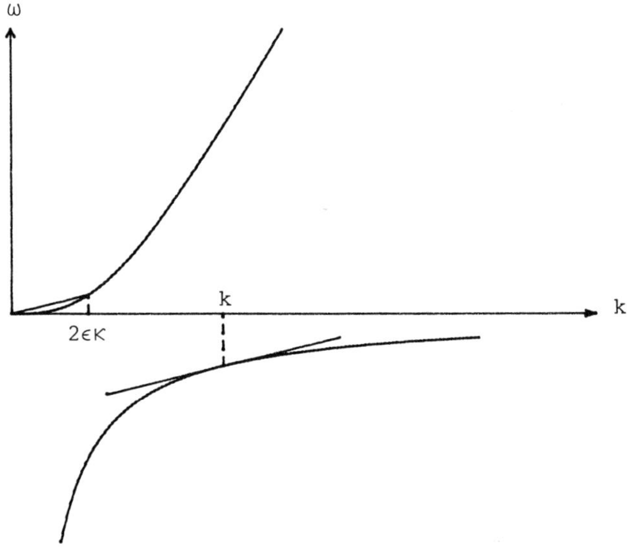

Figure 8-5:

Provided the gradients of the various branches are of the same order of magnitude then $c_p = c_g$ is achieved. Double or multiple branch dispersion relations are very common in physics and it is obviously these which provide the best set of examples to study this type of resonance. Indeed, double branched systems provide a configuration in ω, k space which allows this special type of triad resonance when the ordinary 3-wave resonance cannot be achieved in one space dimension. Figure 8.4 in particular is a typical double branch dispersion relation for a whole category of systems which have an upper (optical) branch and a lower (acoustic) branch. The long wave short wave resonance mechanism works very simply. We can think of the two waves $k + \epsilon\kappa$; $k - \epsilon\kappa$ as two side-bands of the main wave k which beat together to form a slow variation. This can be mathematically expressed as

$$\begin{aligned}\phi &= a\exp(i\theta_1) + a\exp(i\theta_2) + c.c. \\ &= 4a\cos(kx - \omega t)\cos\left[\epsilon(\kappa x - \Omega t)\right]\end{aligned} \tag{8.6-5}$$

The fast variation in the upper branch is far too fast to be felt by the lower branch on the k^{-1} and ω^{-1} space and time scales, but the slow variation caused by the beat from the upper branch can drive or excite the lower branch. The two halves of the system, which would normally be considered to be independent, are now coupled together. Whether this coupling can be achieved can be determined geometrically, as in Figs. 8-4 - 8-5, from a properly drawn set of dispersion curves.

Instead of trying to obtain general amplitude equations, by the method described in the last section, for the three wave resonance by matching θ_1 terms with $\theta_2 + \theta_3$ terms, it is simpler to use a mixture of the multiple scales method to find the slow amplitude variation of the upper branch and the 'stretched co-ordinate' method of Chapter 5 to determine the long wave associated with the lower branch.

We split the equations of motion into the two constituent parts corresponding to the upper and lower branches. Dealing with the lower branch first we write this part of the equations of motion in the general form

$$L^{(u)}\{\frac{\partial}{\partial x};\frac{\partial}{\partial t}\} = f(\phi, N) \tag{8.6-6}$$

where ϕ and N are the dependent variables corresponding to the upper and lower branches. We write

$$\begin{aligned}\phi &= \epsilon\phi^{(1)} + \epsilon^2\phi^{(2)} + \ldots \\ N &= \epsilon^2 n + \ldots.\end{aligned} \tag{8.6-7}$$

and take slow scales $X = \epsilon x$, $T = \epsilon t$, $\tau = \epsilon^2 t$.
From (8.6-6) we expand in a Taylor series and find

$$L^{(u)}\phi^{(1)} = 0 \tag{8.6-8}$$

$$L^{(u)}\phi^{(2)} = -(L_1^{(u)}\frac{\partial}{\partial T} + L_2^{(u)}\frac{\partial}{\partial X})\phi^{(1)} \tag{8.6-9}$$

where the subscripts 1 and 2 refer as usual to differentiation with respect to the first and second positions within $L^{(u)}$. Taking

$$\phi^{(1)} = A(X,T,\tau)\exp(i\theta) + c.c. \tag{8.6-10}$$

we must, as usual, connect X and T in the group velocity travelling frame to remove secular terms at $O(\epsilon^2)$. We therefore define a new variable $\xi = \epsilon(x - c_g t)$. Using this variable at $O(\epsilon^3)$ we find that

$$L^{(u)}\phi^{(3)} = -\tfrac{1}{2}\{\beta\frac{\partial^2}{\partial\xi^2} + i\frac{\partial}{\partial\tau}\}\phi^{(1)} + \gamma n\phi^{(1)} \tag{8.6-11}$$

where the final term in (8.6-11) arises from quadratic terms which couple ϕ and N together in (8.6-6). The final amplitude equation turns out to be

$$i\frac{\partial A}{\partial\tau} + \tilde{\beta}\frac{\partial^2 A}{\partial\xi^2} = \tilde{\gamma}An \tag{8.6-12}$$

where $\tilde{\beta}$ and $\tilde{\gamma}$ are constants. The second equation, deriving from the lower branch is trickier to obtain and requires care. Models with lower branches depicted as in Fig. 8-4, such as plasmas, usually behave as wave equations coupled to the upper branch by a nonlinear driving term

$$L^{(\ell)}\{\frac{\partial}{\partial t};\frac{\partial}{\partial x}\} = g(\phi,N)$$
$$L^{(\ell)} \equiv \frac{\partial^2}{\partial t^2} - c_p^2\frac{\partial^2}{\partial x^2} \tag{8.6-13}$$

where c_p is the phase speed of waves in the lower branch. Rescaling the x and t into ξ and τ and using these as stretched co-ordinates as in Chapter 5, the wave operator in (8.6-13) now becomes

$$L^{(\ell)} = \epsilon^2(c_g^2 - c_p^2)\frac{\partial^2}{\partial\xi^2} - 2\epsilon^3 c_g\frac{\partial^2}{\partial\xi\partial\tau} + \epsilon^4\frac{\partial^2}{\partial\tau^2} \tag{8.6-14}$$

Since N is of order ϵ^2, the only terms on the rhs of (8.6-13) which can match at $O(\epsilon^4)$ are terms such as $\partial^2|A|^2/\partial\xi^2$ which would immediately imply that n is proportional to $|A|^2$. This result implies that the lower branch would be inertialess as it would just follow the upper branch motions. The lowest order terms on the left hand side of (8.6-14) can be removed by using Benney's resonance condition $c_p = c_g$ and the right hand side can only couple to the left at $O(\epsilon^5)$. This would in effect require a weak coupling constant of $O(\epsilon)$ in the terms on the right hand side of (8.6-13) on this scaling. Other scalings are discussed in the chapter notes. We finally obtain

$$\frac{\partial n}{\partial\tau} = -\Delta\frac{\partial}{\partial\xi}\{|A|^2\} \tag{8.6-15}$$

The amplitude equations (8.6-15) and (8.6-12) together constitute an integrable system, a result which is discussed and referenced in the chapter notes. Since n must be real, its solutions are very like KdV solitons, whereas the amplitude function $A(\xi,\tau)$ is complex and is a wave envelope with solutions very like NLS envelope solitons. These equations form a bridge between long wave solitons associated with the lower branch, and envelope solitons associated with the upper branch, and allow an exchange of energy between them, even though they operate on different scales. In many problems, long and short waves are considered to be independent and this energy exchange between long and short waves has important physical consequences in some physical models, particularly molecular chains. We would like to emphasise, however, that the derivation of the second equation is model dependent and the sketch given above of the derivation of (8.6-15) may need to be modified depending on the model.

A good example of double branching occurs in a plasma of electrons and ions. We have already studied ion acoustic effects in Chapter 5 where we studied long ion waves with a background of hot electrons which was modelled as a charged gas. Conversely, in Section 8.4, we studied the evolution of Langmuir (electron density) waves with a static ion background. Each of the two constituent parts has a separate dispersion relation. In dimensionless form the Langmuir branch was found to be

$$\omega^2 = 1 + k^2 \qquad \text{upper branch} \tag{8.6-16}$$

and the ion acoustic branch was

$$\omega^2 = k^2(1 + k^2)^{-1} \qquad \text{lower branch} \tag{8.6-17}$$

Coupling between the Langmuir oscillations and the ion sound can be achieved through Benney's resonance mechanism. Zakharov (1972) studied this problem and obtained equation (8.6-12) and a two way version of (8.6-15) which can be rescaled into (8.6-15) itself if the ξ,τ scaling is used. These equations are often known as Zakharov's equations. They occur because of the long wave short wave coupling and so from now on we shall call this type of resonance the Zakharov-Benney long wave short wave resonance. The type of plasma coupling problem mentioned here has been discussed extensively by Gibbons et al. (1977) and Gibbons (1978).

Benney (1977) orginally formulated the idea of this type of resonance to deal with the problem of coupling the gravity and capillary modes of surface water waves; a problem discussed also by Grimshaw (1978) and Djordjevic and Redekopp (1977).

8.6.1 Davydov's alpha-helix model

Certain types of protein molecules consists of peptide groups (H-N-C=O) bonded in periodic arrays in three spiralling chains forming a helical molecular chain. The so-called α-helix, in which neighbouring peptide groups in each chain are linked by hydrogen bonds, is one such example. Cross hydrogen bonding between the peptide groups gives a rigid structure to the helix. This type of molecular chain is a classic example of a double branched system. Within each peptide group, or sub-molecule, quantum transitions occur due to the vibrational structure of the C=O double bond in the region of the infra-red frequency spectrum. Photons will propagate through the system from one group to another, thereby introducing dispersive effects. On top of that, on much longer space and time scales, purely classical elastic longitudinal waves will propagate down the chain with the helix behaving like a spring. The typical wavelengths of these mechanical vibrations are much longer than the light wavelength, so we would expect the long wave short wave resonance to be applicable. Davydov and Kislukha (1976) have considered a mathematical model designed to couple these to types motions. A general discussion of molecular chain models can be found in Davydov (1971, 1979). The Hamiltonian for the system is taken to be

$$H = E + \sum_n \{(E_{int} - D_n)B_n^\dagger B_n - J(B_{n+1}^\dagger B_n + B_{n+1}B_n^\dagger)\} \tag{8.6-18}$$

where E is the kinetic plus potential energy, E_{int} is the constant intermolecular excitation energy (between the sub-molecules) and $B_n^\dagger$ and B_n are the intermolecular boson creation and annihalation operators satisfying the commutation relations

$$[B_n, B_m^\dagger] = \delta_{nm}$$

$$[B_n, B_m] = 0$$

The label n stands for the nth molecule which has mass M and position r_n. The term $B_{n+1}^\dagger B_n$ destroys a photon on the nth molecule and creates one on the (n+1)th. The number J depends on the electric dipole moment along the chain axis of the vibration transition within the double bond. Coupling between the mechanical motions and quantum effects is modelled by introducing the function D_n in the Hamiltonian. An elastic wave passing down the chain will alter the intermolecular bonding energy and hence alter E_{int}. Since $B_n^\dagger B_n$ stands for the total number of photons in the system (being the number operator), the total energy on the nth molecule is modified by a passing elastic wave.

We define ρ_n to be the disturbance from equilibrium of the nth molecule and assume that the D can be expressed in terms of nearest neighbour interactions

$$D_n = G(r_{n+1} - r_n) + G(r_n - r_{n-1})$$

which, to first order in ρ_n becomes

$$D_n \simeq (1 + \gamma\rho_n)D \tag{8.6-19}$$

We expect the departure from equilibrium to be small so the positive constant γ will be small. D is also a constant and depends on the particular chain. The quantum mechanical part of the problem can be characterised by a wave function $|\psi\rangle$ defined as

$$|\psi> = \sum_n a_n(t)B_n^\dagger\,|0> \tag{8.6-20}$$

where the $a_n(t)$ are time dependent amplitudes, which, if $|\psi>$ is normalised must satisfy

$$\sum_n a_n^2 = 1$$

Using the Schrodinger equation

$$ih\,\frac{\partial|\psi>}{\partial t} = H\,|\psi> \tag{8.6-21}$$

we obtain for the amplitudes

$$ih\,\frac{\partial|\psi>}{\partial t} = [(E_{int} + E) - (1 + \gamma\rho_n)D]a_n - J(a_{n+1} + a_{n-1}) \tag{8.6-22}$$

The mechanical vibrations will obey an equation of motion which will be the equivalent classical Hamiltonian to (8.6-18).

$$\begin{aligned} H &= \sum_n \{[(E_{int}+E) - (1+\gamma\rho_n)D]a_n^* a_n - Ja_n^*(a_{n+1}+a_{n-1})\} \\ E &= T + U \end{aligned} \tag{8.6-23}$$

Without the $\rho_n D$ coupling term, usual Hamiltonian mechanics for a set of coupled linear oscillators would give the discrete wave equation

$$\ddot{\rho}_n = \mu M^{-1}(\rho_{n+1} + \rho_{n-1} - 2\rho_n) \tag{8.6-24}$$

where μ is the coupling coefficient from the potential energy U. The extra coupling term in the Hamiltonian now gives a modified form of (8.6-24)

$$\ddot{\rho}_n = \mu M^{-1}(\rho_{n+1} + \rho_{n-1} - 2\rho_n) + \gamma D\{2|a_n|^2 - |a_{n+1}|^2 - |a_{n-1}|^2\} \tag{8.6-25}$$

Equations (8.6-22) and (8.6-25) are differential equations which can be converted into partial differential equations by introducing a continuous space variable $x = Rn$ where R is the lattice spacing

$$\rho_n(t) \to \rho(x,t) \qquad\qquad a_n(t) \to a(x,t)$$

We obtain

$$ih\frac{\partial a}{\partial t} - (\lambda - \gamma D\rho)a + J\frac{\partial^2 a}{\partial x^2} = 0 \tag{8.6-26}$$

$$\frac{\partial^2\rho}{\partial t^2} - c_p^2\frac{\partial^2\rho}{\partial x^2} = -\gamma DRM^{-1}\frac{\partial^2}{\partial x^2}(|a|^2) \tag{8.6-27}$$

where $c_p = (\mu R^2/M)^{\frac{1}{2}}$, which is the longitudinal sound speed in the chain and $\lambda = E_{int} + E - D - 2J$. Equations (8.6-26) and (8.6-27) are Davydov's chain equations connecting the probability wave amplitude $a(x,t)$ and the longitudinal displacement ρ, and take the same form as Zakharov's equations of plasma physics. These on their own are not integrable by inverse scattering, but in the Chapter Notes we give an outline of how they will reduce either to the NLS limit or to the long wave-short wave resonance equations when the stretched co-ordinates $\xi = \epsilon(x - c_g t)$; $\tau = \epsilon^2 t$ are used. The expansion for a

$$a = \epsilon^q a^{(1)} + \epsilon^{q+1} a^{(2)} + \ldots. \tag{8.6-28}$$

is chosen by determining q in terms of the coupling constant γDRM^{-1} which we shall call G. If we write $G = \epsilon^g \bar{G}$ then as a summary from the Chapter Notes, if $c_p \neq c_g$, then we choose $q = (2-g)/2$ and obtain the NLS limit from (8.6-26), (8.6-27). If, however, we are at the value of k such that $c_p = c_g$, then we need to choose $q = (3-g)/2$ and the final equations are not the NLS but are the long wave-short wave equations given in (8.6-12)-(8.6-15). Hyman et al. (1981) discuss the NLS limit for Davydov's equations and Scott (1982) generalises the equations to include the effects of a 3 chain helical spiral and other interactions. In a real protein the value of G will depend very much on the constituents of the protein. Whether $c_p = c_g$ can be achieved again depends on the protein characteristics and the frequency range used in the experiment.

8.7. Notes

Section 8.1

A full description of the method of multiple scales used in these sections and first developed by Cole and Kevorkian can be found in Nayfeh's book (1973) on perturbation methods, Cole (1968), Nayfeh and Mook (1979) and Jordan and Smith (1977). There are many perturbation techniques for dealing with nonlinear equations some of which are more appropriate than others depending on the situation; for example, the Krylov-Bogoliubov-Mitropolsky (1961) method of averaging, the Poincare-Lindstedt method, Haken's method (1978) of slaved modes, Eckhaus's method (1965) and Whitham's slowly varying Lagrangian method (1974). The method of multiple scales is ideal for finding amplitude equations associated with systems in which we are considering weak nonlinearity and has been further developed by Stuart (1960) for dealing with unstable systems (see next chapter). As one might expect for a generic equation, the NLS equation has arisen in many unconnected places in the literature over a span of many years. On the applied mathematics side the paper by Benney and Newell (1966) appears to have been the first formal derivation using multiple scales although the equation appeared in a variety of subjects through the use of various methods. The use of perturbation methods by Taniuti and co-workers around the same period has had a major influence on many of the subsequent derivations of the NLS equation in various regions of physics and hydrodynamics. We mention some but by no means a complete list of references: Taniuti and Yajima (1969, 1973), Asano, Taniuti and Yajima (1969), Taniuti and Wei (1968), Montgomery and Tidman (1964), Davidson (1972), Fried and Ichikawa (1972), Kakutani and Sugimoto (1974). See also references in Karpman (1975), Whitham (1974) and Newell (1974), Davey (1972), Stuart and DiPrima (1978), Chu and Mei (1970, 1971).

Section 8.2

(a) Hasimoto and Ono (1972) have shown that the NLS equation arises for the amplitude of gravity wave packets on water of depth h. In this case the dispersion relation for the carrier wave is nonpolynomial in ω (see Chapter 5)

$$\omega^2 = (gkh)\tanh(kh)$$

Davey and Stewartson (1974) and in a slightly different form, Benney and Roskes (1969) have shown that in two dimensions with slow scales

$$\bar{X} = \epsilon(x-c_g t) \qquad Y = \epsilon y \qquad T_2 = \epsilon^2 t \tag{8.7-1}$$

the following equations arise

$$2i\frac{\partial\zeta}{\partial t} - \frac{\partial^2\zeta}{\partial\bar{X}^2} + \frac{\partial^2\zeta}{\partial Y^2} = \frac{9}{2}\zeta|\zeta|^2 + 3\zeta\frac{\partial\phi}{\partial\bar{X}} \tag{8.7-2}$$

$$\frac{\partial^2\phi}{\partial\bar{X}^2} + \frac{\partial^2\phi}{\partial Y^2} = -3\frac{\partial}{\partial\bar{X}}|\zeta|^2 \tag{8.7-3}$$

which are a two-dimensional generalisation of the NLS equation. Freeman and Davey (1975) have looked at the limiting procedure through which equations are derived. Rather remarkably, Anker and Freeman (1977) have used the two space dimensional inverse scattering procedure of Zakharov and Shabat (1974) to show that the equations (8.7-2),(8.7-3) are integrable. These equations, together with the Kadomsev-Petviashvili equation (2-dimensional KdV equation - see Chapter 5) are one of the few physically relevant 2-space dimensional equations known to be solvable by the inverse scattering transform. If the Y-dimension is dropped, equations (8.7-2),(8.7-3) reduce to the NLS equation.

(b) The Benjamin-Feir instability (Benjamin and Feir 1967; Benjamin 1967) is a mechanism associated originally with the instability of a Stokes' wave on deep water due to the development of sidebands and their interaction with harmonics. Stuart and DiPrima (1978) have shown that the Eckhaus (1965) resonance mechanism is essentially equivalent to the Benjamin-Feir mechanism. They have shown that the instability mechanism arises because a mutual reinforcement of resonance occurs between the first harmonic of a wave and its upper and lower sidebands. Their analysis for the NLS equation shown in this section, which showed that an exponential growth in time occurs, can be explained in terms of harmonic-sideband interaction in the following way. A wave $a\exp[i(kx - \omega t)]$ will have a first harmonic, forced by the nonlinearity of the form $a \exp[2i(kx - \omega t)]$. Suppose now that two perturbations arise in the form of the upper and lower sidebands of k; $a_1\exp[i(k_1 x - \omega_1 t)]$ and $a_2\exp[i(k_2 x - \omega_2 t)]$. Terms of the form

$$a^2 a_1 \exp\{i[(2k-k_1)x - (2\omega-\omega_1)t]\} \tag{8.7-4}$$

and

$$a^2 a_2 \exp\{i[(2k-k_2)x - (2\omega-\omega_2)t]\} \tag{8.7-5}$$

will appear. If $k_1 + k_2 = 2k$ and $\omega_1 + \omega_2 = 2\omega$ then a "resonance" appears since each of these terms will be secular in any weakly nonlinear perturbation theory. Taking $k_1 = k(1 + \Delta)$; $k_2 = k(1 - \Delta)$; $\omega_1 = \omega(1 + \delta)$ and $\omega_2 = \omega(1 - \delta)$ then the resonance conditions are satisfied. Benjamin and Feir's original calculation was for waves on deep water for which the dispersion relation is $\omega^2 = kg$. To first order, this implies that δ and Δ must be related by $\delta = \frac{1}{2}\Delta$.

The criterion $\beta\gamma \gtrless 0$ has been discussed in several papers: Karpman and Krushkal (1968), Newell (1974), Lange and Newell (1974). See also references in Drazin and Reid (1981).

For experimental and numerical work see the papers by Yuen and Lake (1975) and Lake et al. (1977). Yuen and Lake's paper (1975) also contains a nice derivation of the NLS equation in hydrodynamics using Whitham's slowly varying Lagrangian method.

(c) The function $B(X_1, T_1)$ which occurs as an integration constant in (8.2-15) sometimes is not determined at $O(\epsilon^3)$. For certain equations no secular terms purely in the slow scales occur and so at $O(\epsilon^3)$ terms in AB are left in the final amplitude equation. In this case, B is determined in terms of $|A|^2$ at $O(\epsilon^4)$ and in performing any calculation of this sort, one should always check at least at $O(\epsilon^4)$ anyway. Whether or not B is annihalated by the fast scale operators, or whether it is determined in terms of A at $O(\epsilon^3)$ or $O(\epsilon^4)$ depends on the problem concerned and care needs to be taken with each example.

Section 8.3

The problem of considering the full equation (8.3-19) is very interesting and has been looked at by Zakharov and Synakh (1976). If we consider (8.3-19) in circularly symmetric form:

$$2i \frac{\partial E}{\partial z} + \frac{1}{r}\frac{\partial}{\partial r}\left(r \frac{\partial E}{\partial r}\right) + \beta E|E|^2 = 0$$

then it is possible to construct two integrals of the motion

$$I_1 = \int_0^\infty r|E|^2 dr \qquad I_2 = \int_0^\infty [|E_r|^2 - \tfrac{1}{2}\beta|E|^4]r\,dr \tag{8.7-6}$$

These can be checked by direct calculation. It is also possible to show, by direct calculation that

$$I_2 = \beta \frac{\partial^2}{\partial z^2} \int_0^\infty |E|^2 r^3 dr \tag{8.7-7}$$

which can be directly integrated to

$$\int_0^\infty |E|^2 r^3 dr = \frac{1}{2}\beta^{-1} I_2 z^2 + c_1 z + c_2 \tag{8.7-8}$$

Now, if $\beta > 0$, then quite a large class of data on E makes $I_2 < 0$. We only need an envelope which has sufficiently gentle slope compared to its amplitude such that the integrand in I is negative. Since $\beta > 0$, we have the right hand side of (8.7-8) turning from positive to negative at some point $z = z_0$ while the left hand side of (8.7-8) is an integral with positive definite integrand. This therefore indicates a singularity in the amplitude function E for some finite value of z and the solution ceases to exist. There is an analogy with classical mechanics here. The integral in (8.7-8) can be likened to a moment of inertia of the system which collapses in on itself after a finite value of z. Berkshire and Gibbon (1982) have considered this problem further and shown that the analogy with classical mechanics is not accidental. The solution ceases to exist when blow-up occurs, which is equivalent to collapse. The results of Sundman on collapse in the N-body problem are directly applicable (Siegel and Moser, 1971) and using his methods the nature of the singularity can be found. Zakhanov and Synakh (1976) and subsequently Konno and Suzuki (1979) have confirmed this result by numerically integrating (8.7-6) with a Gaussian initial profile at $z = 0$. The amplitude reached 2000 times its initial value. Of course such an enormous growth invalidates the approximations which went into deriving the NLS equation in Section 8.3 to begin with, but such behaviour nevertheless indicates this type of "self-focusing" of energy is an important mechanism in optics. There is of course a large literature on the subject and the reader is referred to Chiao et al. (1964), Kelley (1965), Askar'yan (1974), Lugovoi and Prokhorov (1974), and particularly the paper by Zakharov and Synakh (1976) and the references therein. It has been shown experimentally that when the nonlinearity is cubic, then the field amplitude focuses at a particular point, a result which confirms to some degree the singular behaviour of the two-dimensional NLS equation. The focusing and filamentation of energy is important experimentally to the laser physicist because it enables him to obtain considerable information on the medium in which the focusing is taking place.

The rather remarkable occurrence of the self-focusing singularity has consequences not just in nonlinear optics because, as we have seen, the procedure for obtaining the NLS equation is a general one and applies to quite a large class of systems. For instance, let us consider our example from Section 8.2 in two space dimensions

$$\phi_{xx} + \phi_{yy} - \phi_{tt} = \alpha\phi - \beta\phi^3 \qquad \alpha > 0 \tag{8.7-9}$$

which would also describe small amplitude perturbations of the two-dimensional sine-Gordon equation. Introducing slow scales $X = \epsilon x$, $Y = \epsilon y$, $T_1 = \epsilon t$, $T_2 = \epsilon^2 t$ and proceeding as usual we find that at $O(\epsilon)$

$$\phi^{(1)} = A(X,Y,T_1,T_2)\ \exp(i\theta) + \text{c.c.} \tag{8.7-10}$$

$$\theta = kx + \ell y - \omega t + \delta$$

$$\omega^2 = k^2 + \ell^2 + \alpha \qquad \alpha > 0 \tag{8.7-11}$$

At $O(\epsilon^2)$, secular terms need to be removed as usual and these are

$$\frac{\partial A}{\partial T_1} + \left(\frac{d\omega}{dk}\right)\frac{\partial A}{\partial X} + \left(\frac{d\omega}{d\ell}\right)\frac{\partial A}{\partial Y} = 0 \tag{8.7-12}$$

while at $O(\epsilon^3)$ the final amplitude equation for A turns out to be

$$2i\omega\frac{\partial A}{\partial T_2} + \beta A^2A^* + \left(\frac{\partial^2}{\partial X^2} + \frac{\partial^2}{\partial Y^2} - \frac{\partial^2}{\partial T_1^2}\right)A = 0 \tag{8.7-13}$$

We can, of, course eliminate the derivative terms in T_1 from (8.7-13) to obtain

$$\left[1-\left(\frac{d\omega}{dk}\right)^2\right]\frac{\partial^2 A}{\partial X^2} + \left[1-\left(\frac{d\omega}{d\ell}\right)^2\right]\frac{\partial^2 A}{\partial Y^2} - 2\left(\frac{d\omega}{dk}\right)\left(\frac{d\omega}{d\ell}\right)\frac{\partial^2 A}{\partial X\partial Y}$$
$$+ \beta A^2A^* + 2i\omega\frac{\partial A}{\partial T_2} = 0 \tag{8.7-14}$$

The second order operator on A can be simplified by a co-ordinate transformation

$$\xi = X \quad ; \quad \eta = \omega\alpha^{-\frac{1}{2}}\{[1-(\frac{d\omega}{dk})^2]Y + (\frac{d\omega}{dk})(\frac{d\omega}{d\ell})X\} \tag{8.7-15}$$

to give finally

$$2i\omega \frac{\partial A}{\partial T_2} + \beta A^2A^* + [1-(\frac{d\omega}{dk})^2]\cdot\left(\frac{\partial^2 A}{\partial \xi^2} + \frac{\partial^2 A}{\partial \eta^2}\right) = 0 \tag{8.7-16}$$

Since

$$\frac{d\omega}{dk} = k/\omega \qquad\qquad \frac{d\omega}{d\ell} = \ell/\omega$$

then

$$1-(\frac{d\omega}{dk})^2 = (\alpha+\ell^2)\omega^{-2} > 0$$

Consequently, in (8.7-16), with $\beta > 0$, we have the same situation as (8.7-6) and blow-up will occur at a finite value of T_2 (the slow time). The fact that the two dimensional NLS equation blows up in finite time simply indicates that $A >> \epsilon^{-1}$ and consequently the perturbation procedure breaks down. Therefore, unlike the one-space dimensional case, the amplitude of oscillations will grow and fully nonlinear solutions of the system must be taken into account at this point.

The 2-dimensional NLS equation arises in other areas too. It occurs as the amplitude equation below criticality for the two-layer Kelvin-Helmholtz instability (inviscid) and under certain circumstances, focusing can occur (Gibbon and McGuinness 1980). This indicates that while a system can be **linearly** stable, it can be **nonlinearly** unstable. For a general discussion of nonlinear self-focusing, see Newell (1978,1979).

A rather interesting recent experiment has been reported by Mollenauer et al. (1980), who sent a 7 picosecond pulse down a 700 metre long single mode silica glass fibre at a wavelength of 1.55 micrometres. They chose a particular fibre and light frequency such that $\gamma > 0$ and $\partial^2\omega/\partial k^2 > 0$ for equation (8.2-16), the NLS equation being the appropriate equation for the propagation of an optical pulse down the fibre. In very much the same vein as the derivation of the NLS equation in (8.1-5), Hasegawa (1975) and Hasegawa and Tappert (1973) have shown that a refractive index which is quadratically dependent on the electric field envelope will yield the NLS equation for the evolution of that field envelope. This will occur when there is longitudinal dispersion, i.e. when $d^2\omega/dk^2 \neq 0$. The experiment of Mollenauer et al. (1980) has confirmed this very nicely. The stability of a soliton over a length of 700 metres shows how robust these solutions are.

Section 8.4

The derivation of the NLS equation in plasma physics for various situations has an enormous literature associated with it. We have given a long list of references in Section 1 of these notes. Further references can be obtained from the articles by Ichikawa (1979), Weiland, Ichikawa and Wilhelmson (1979) and Weiland and Wilhelmsson (1977), which include methods and references with regard to perturbation approaches to the NLS equation which take into account physical effects such as Landau damping (see also Karpman and Maslov 1978; Keener and McLaughlin 1977).

Section 8.6

(a) As far as equations (8.6-12) and (8.6-15) are concerned, it is simpler to scale out the constants. When $\tilde{\beta}$ and $\tilde{\gamma}$ are positive they can be cleaned up and scaled into the form

$$i\frac{\partial S}{\partial t} + \frac{\partial^2 S}{\partial x^2} = LS \tag{8.7-17}$$

$$\frac{\partial L}{\partial t} = -\frac{\partial}{\partial x}(|S|^2) \tag{8.7-18}$$

where $\xi = x\tilde{\beta}^{\frac{1}{2}}$, $\tau = t$, $\tilde{\gamma}n = L$ and $S = (\Delta\gamma\beta^{-\frac{1}{2}})^{\frac{1}{2}}A$.

Ma (1978) and Yajima and Oikawa (1976) have shown that equations (8.7-17), (8.7-18) are integrable by the inverse scattering transform. For simplicity we shall make the transformations

$$S = \sqrt{2}\, A(-x;-\tfrac{1}{2}t)$$

$$L = B(-x;-\tfrac{1}{2}t) \tag{8.7-19}$$

which transform (8.7-17),(8.7-18) into

$$\frac{1}{2}\, i\, \frac{\partial A}{\partial t} = \frac{\partial^2 A}{\partial x^2} + AB$$

$$\frac{\partial B}{\partial t} = -4\frac{\partial}{\partial x}(|A|)^2 \tag{8.7-20}$$

The Lax pair of operators $\hat{L}$ and $\hat{P}$ for the eigenvalue problem

$$\hat{L}\tilde{\psi} = i\lambda\tilde{\psi} \tag{8.7-21}$$

with the time dependence of the eigenfunctions as

$$\tilde{\psi}_t = \hat{P}\tilde{\psi} \tag{8.7-22}$$

are given by

$$\hat{L} = \begin{pmatrix} \frac{1}{3}\cdot\frac{\partial}{\partial x} & -\frac{1}{3}A & -iB/3 \\ 0 & \frac{1}{2}\frac{\partial}{\partial x} & -\frac{1}{2}A^* \\ i & 0 & \partial/\partial x \end{pmatrix} \tag{8.7-23}$$

and

$$\hat{P} = \begin{pmatrix} i\lambda^2 & 2(\lambda A - iA\ ) & 2i|A|^2 \\ -2A^* & -i\lambda^2 & 2(iA^* - \lambda A^*) \\ 0 & -2A & i\lambda^2 \end{pmatrix} \qquad (8.7\text{-}24)$$

The integrability condition for constant eigenvalues of $\hat{L}$ is the Lax condition: $\hat{L}_t = [\hat{P},\hat{L}]$. This gives equations (8.7-20). Various soliton solutions are given in Yajima and Oikawa (1976) and Ma (1978). A single soliton solution is given by

$$\begin{aligned} B &= -2b^2 \text{sech}^2\theta \\ A &= ib\sqrt{2a}\exp(\eta)\text{sech}\theta \end{aligned} \qquad (8.7\text{-}25)$$

where

$$\begin{aligned} \theta &= bx + 4abt + \delta \\ \eta &= 2ax + 2i(a^2 - b^2)t \end{aligned} \qquad (8.7\text{-}26)$$

Because (8.7-21) is a third order eigenvalue problem the solution is rather difficult, as in the 3-wave resonance interaction, and the reader is referred to the given references for details. Newell (1978) has shown that a slightly modified set of equations to (8.7-17),(8.7-18) can also be integrated by the inverse scattering transform.

(b) Zakharov's equations (Zakharov 1972) have been used by plasma physicists as model equations which describe the interaction between ion sound (density N) and Langmuir waves (envelope amplitude E), a coupling which was not taken into account in Section 8.4. In the calculation in that section, the ions were taken to be just a static background charge. Zakharov's equations take the form

$$\nabla^2 E + iE_t = EN \qquad (8.7\text{-}27)$$

$$N_{tt} - c_p^2 \nabla^2 N = \nabla^2(|E|^2) \qquad (8.7\text{-}28)$$

where the second equation takes the form of a wave equation which is driven by the electric field. In the static limit for the ions, $N \propto |E|^2$ and equation (8.7-27) becomes the NLS equation. In the chapter notes for Section 8.3 we showed that collapse is possible ($E \to \infty$ in finite time) for negative energy, a process which has been interpreted as a possible mechanism for Langmuir turbulence (Zakharov, 1972, Nicholson and Goldman 1978).

Davydov's chain equations (8.6-26)-(8.6-27) take the same form as Zakharov's equations. On their own, in one dimension, they are not a completely integrable system. In the long wave-short wave resonance limit or the NLS limit they then become one. In Section 8.6 we gave an argument which outlined how the integrable system (8.6-12)-(8.6-15) could be obtained on one particular scaling. The argument can be generalised and the necessary scalings depend on the coupling constant" on the right hand side of (8.6-13). In the Davydov case this would be γDRM^{-1}. Let us call any coupling constant G which we shall write as $G = \epsilon^g \bar{G}$ and expand ϕ ($\equiv a$ in (8.6-26))

$$\phi = \epsilon^q \phi^{(1)} + \epsilon^{q+1} \phi^{(2)} \dots\dots \tag{8.7-29}$$

which generalises (8.6-7). The number q is to be determined in terms of g. Equation (8.6-13) (or in the Davydov case (8.6-27)) now becomes

$$\begin{aligned}&[\epsilon^2 (c_g^2 - c_p^2)\frac{\partial^2}{\partial \xi^2} - 2\epsilon^3 c_g \frac{\partial^2}{\partial \tau \partial \xi} + \epsilon^4 \frac{\partial^2}{\partial \tau^2}] (\epsilon^2 n) \\ &= \epsilon^{2+g+2q} \bar{G} \frac{\partial^2}{\partial \xi^2}[(|A|^2) + \dots.]\end{aligned} \tag{8.7-30}$$

We can now immediately identify the two cases (i) $c_g \neq c_p$, (ii) $c_g = c_p$.
(i) $c_g \neq c_p$. In this case choose $q = (2-g)/2$ and we find that $n \propto |A|^2$ giving the NLS limit for the final equation in A.
(ii) $c_g = c_p$. Choose $q = (3-g)/2$ and we have

$$n_\tau = - \bar{G}(|A|^2)_\xi /2c_g \tag{8.7-31}$$

which is the long wave-short wave limit.

Obviously the value of k for which $c_g = c_p$ is a special resonance and is known as the k^* limit in plasma physics. For the other values of k for which $c_g \neq c_p$, the acoustic brance appears to be "slaved" to the optical branch. This is the inertialess case referred to in Section 8.6. However when $c_g = c_p$ then the acoustic branch is allowed some freedom. On the scalings used in Section 8.6 in which $q = 1$ it is necessary to have $g = 1$ to achieve $c_g = c_p$. In this case the acoustic branch is not so severely tied to the optical branch to prevent any acoustic inertia. The more general formula $q = (3-g)/2$ allows a generalization of the expansion for ϕ in the special limit when $c_g = c_p$.

9. AMPLITUDE EQUATIONS IN UNSTABLE SYSTEMS

9.1 Introduction

The examples of the previous chapter, in which the NLS equation arose as the governing amplitude equation, were specifically chosen as cases where the physical systems concerned were stable for all parameter values in the whole of the k-space, and consequently no imaginary parts occured in the dispersion relation. Such systems are closed or conservative in that no dissipation of energy occurs through frictional or other effects and also no pumping of the systems occur from external energy sources. The NLS equation, which governs the evolution of the wave envelopes of conservative systems in the weakly nonlinear limit, describes the competition between nonlinearity and dispersion. The nonlinear terms produce harmonics, (provided the initial amplitude is sufficient for these to be taken into account), and the harmonics and the fundamental compete along with dispersive effects to produce a final balance.

However, a second category of problems occur in wave propagation which are significantly different to the situation described above. If a system is driven or pumped with energy through some mechanism, for example a rotation, a background flow or a heat gradient, then potential energy is made available to the waves. There may be a "control parameter" within the mathematical model, whose role can be important in that the system in question may become unstable under the influence of the background energy flow when this parameter passes through a critical value. In the post-critical region, the wave train can draw energy from the background of available potential energy. This is a further factor governing the evolution of the wave in addition to those given in the first paragraph above. It also means that infinitesimally small disturbances will grow if the control parameter is in the post-critical region, whereas in the wholely stable systms of the last chapter such waves will always remain small.

There are many physical examples of this type of behaviour. One well known example is the case of a fluid heated from below. For small temperature gradients the fluid is able to conduct the heat away but as the gradient is increased conduction is not sufficient, and the fluid starts to convect. At the critical value of the heat gradient, at which the initial stationary state of the fluid becomes unstable, and which therefore marks the onset of convection, we say that a bifurcation has occured since one state of the system has become unstable and it moves to another stable state. This example is treated in more detail in the chapter notes. Another famous example of hydrodynamic instability is the onset of turbulence in laminar fluid flow between two infinite horizontal plates as the Reynolds number is increased through a critical value. The problem of hydrodynamic stability is extremely complicated and has been an active area of research for hydrodynamicists for over half a century. The interested reader should consult the chapter notes of this section for references to reading matter. Possibly the most well known example outside fluid mechanics is the laser, where atoms in an optical cavity will radiate individually, and out of phase with one another, if the occupation number of the sample is below a critical value. Above this critical value they can radiate in phase and produce coherent light. Bifurcations of the type described above can be thought of very simply as a phase transition, a concept which is usually used in ferromagnetics when a ferromagnet will show different magnetic properties either side of a critical temperature.

In order to illustrate how this instability comes about we need to consider how the dispersion relation depends on k and ω. We shall confine ourselves to one space dimension for simplicity, and consider model partial differential equations written in the general matrix form

$$\underline{\underline{L}}\left(\frac{\partial}{\partial t};\frac{\partial}{\partial x};\mu\right)\underline{\phi} = \{\ldots,(\underline{\underline{M}}\underline{\phi}).(\underline{\underline{N}}\underline{\phi}),\ldots\}^{T} \qquad (9.1\text{-}1)$$

The quadratically nonlinear term on the right hand side of (9.1-1) can be suitably generalised to higher orders if necessary. The parameter μ is a control parameter. It is understood that $\underline{\phi}$, the vector of dependent variables, is a variation around some stable equilibrium state of the system, and that μ enters the problem as a result of having expanded around this state. Other than the introduction of μ, the features of (9.1-1) are essentially the same as in Chapter 8. In that chapter we deliberately restricted ourselves to forms of $\underline{\underline{L}}$ which gave purely real roots ω for all k. In this chapter we will now discuss how two possible types of instability can arise. They can be categorised by considering the dispersion relation for small amplitude harmonic wave solutions

$$\underline{\phi} = \underline{b}\exp(i\theta) + c.c \qquad (9.1\text{-}2)$$

$$\theta = kx-\omega t + \delta \qquad (9.1\text{-}3)$$

We therefore have

$$\underline{\underline{\ell}}(-i\omega;ik;\mu)\underline{b} = 0 \qquad (9.1\text{-}4)$$

From now on we shall use the notation that small letters are functions of k and ω and large letters are functions of $\partial/\partial x$ and $\partial/\partial t$. Nontrivial solutions of (9.1-4) exist iff

$$\det \underline{\underline{\ell}} = 0 \qquad (9.1\text{-}5)$$

Equation (9.1-5) is the dispersion relation. Whether the underlying model contains damping terms or not now shows in the dispersion relation. This enables us to split the analysis into two categories:

Category I:

If damping is present, then the dispersion relation (9.1-5) will be a complex function of ω and k, because odd and even powers of i will occur. Complex roots can therefore occur.

Let us consider a complex root

$$\omega = \omega_R(k,\mu) + i\omega_I(k,\mu) \tag{9.1-6}$$

Equation (9.1-2) is now

$$\underline{\phi} = \underline{b} \exp[i(kx-\omega_R t)] \exp(\omega_I t) + c.c \tag{9.1-7}$$

As μ is varied it is possible that ω_I may change sign. If $\omega_I > 0$, this solution will grow exponentially in time and is unstable. If $\omega_I < 0$, it decays and the solution is asymptotically stable.

When $\omega_I(k,\mu) = 0$ the solution is said to be neutrally or marginally stable. Inversion of this equation into

$$\mu = R(k) \tag{9.1-8}$$

gives a boundary curve in (ω,k) space between asymptotic stability and instability. There are many forms of such curves, but a typical one of parabolic shape is given in Fig. 9-1.

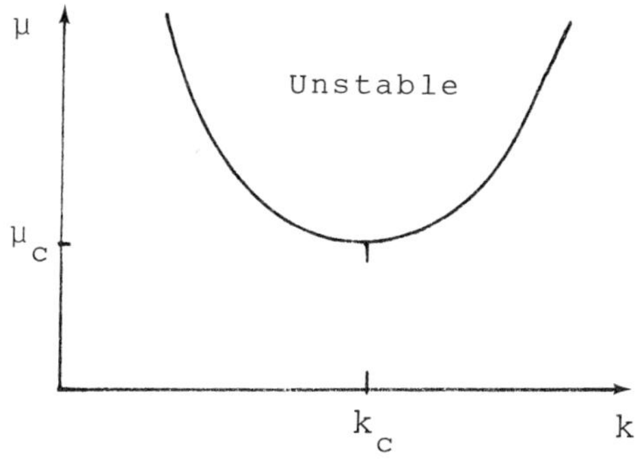

Figure 9-1:

The minimum point at $k = k_c$, $\mu = \mu_c$, is called the bifurcation or critical point because for $\mu < \mu_c$, $\underline{\phi}$ always decays for every value of k. However, for $\mu > \mu_c$, a finite band width of modes become unstable and will cause a growth in the wave amplitude. The critical wave number k is a preferred mode of the system because it is the first mode to become unstable as μ is increased through its critical value. The system is said to undergo a bifurcation as the qualitative nature of harmonic wave solutions change around $\mu = \mu_c$.

The very simple analysis above has been for systems in which dissipation plays a major role. It is possible however, for certain systems to become unstable with dispersion and not dissipation playing a major role.

Category II:

The second type of instability occurs when no dissipation is present. Although the system is purely dispersive, with a real dispersion relation, it is possible for roots of the dispersion relation to occur in complex conjugate pairs

$$\omega = \omega_R \pm i\omega_I \tag{9.1-9}$$

Equation (9.1-2) is now

$$\underline{\phi} = \underline{b} \exp[i(kx-\omega_R t)] \cdot \exp(\pm\omega_I t) + c.c \tag{9.1-10}$$

Because of the pair of roots, a growth occurs if $\omega_I \neq 0$ and asymptotic stability is not possible. However, it is possible to achieve neutral stability if $\omega_I = 0$. That such a case can occur can easily be seen by considering a dispersion relation which is quadratic in ω : $a\omega^2 + b\omega + c = 0$ where a, b and c are functions of k and ω. For values of μ such that $b^2 > 4ac$ we obviously have $\omega_I = 0$ but as the value of μ is changed such that $b^2 < 4ac$, we have $\omega_I \neq 0$. $\omega_I(k,\mu) = 0$ now gives the boundary in (μ,k) space between neutral stability in the whole of the region below the neutral curve $\mu = R(k)$ and instability above the curve. For some point (μ,k) which lies below the curve, a harmonic wave of this wavenumber neither grows nor decays, as in the first case for dissipative systems, but propagates dispersively.

For both categories, the main interest in the behaviour of the system lies in the region $\mu \simeq \mu_c$, when the wave amplitude starts to grow by drawing available potential energy from the background source. As the amplitude grows, the original assumption that the nonlinear terms can be neglected becomes invalid. At this stage the nonlinearity, the instability and the natural dispersive effects in the system all compete resulting in a restriction of the amplitude growth. The interesting point about unstable systems is that the nonlinearity is forced to take effect because of an instability in the linear solution in the region $\mu = \mu_c$. This is in contrast to the stable systems considered in the previous chapter, where nonlinearity only came into play if the initial wave was of sufficient amplitude that weakly nonlinear effects could not be neglected.

Another important effect is the tendency of the system to operate through the preferred mode k_c. In physical terms, this can be thought of as the resonant mode of the system. Before embarking on any scaling analysis of our two categories, it is worth returning to the heuristic derivation of the NLS equation which we pursued in Section 8.1, in which we considered a dispersion relation which was weakly amplitude dependent. Since we were considering only stable cases in that chapter, it was possible to consider only first order ω terms. We must consider all roots here. Furthermore, if we make the dispersion relation $|A|^2$ dependent, then we must also add in this case an extra function which we shall call B, which can be thought of as a potential energy function. This function gives a measure of how much energy is being exchanged between the background of available potential energy and the waves. Let us write our "nonlinear dispersion relation" as

$$D(k,\omega,|A|^2,B,\mu) = 0 \tag{9.1-11}$$

Taking a Taylor expansion about $k = k_c$; $\omega = \omega_{Rc}$; $|A|^2 = B = 0$ in the region of $\mu \simeq \mu_c$ we find

$$(k-k_c)D_k + (\omega-\omega_{Rc})D_\omega + |A|^2 D_{|A|^2} + BD_B + (\mu-\mu_c)D_\mu$$

$$+ \tfrac{1}{2}[(\omega-\omega_{Rc})^2 D_{\omega\omega} + 2(k-k_c)(\omega-\omega_{Rc})D_{k\omega} + (k-k_c)^2 D_{kk}\ldots\ldots] \qquad (9.1\text{-}12)$$

$$\ldots\ldots\ldots = 0$$

We can see immediately that for the Category II type of instability (the purely dispersive type) that a significant difference in behaviour occurs. Since roots occur in conjugate pairs, on the neutral curve $\mu = R(k)$, this conjugate pair tends to a **double root** of the dispersion relation. Consequently D_ω is zero at the critical point and so to is D_k. The consequent amplitude equation for A must therefore be **second order in time** as well as space. No double root occurs for Category I and so the amplitude equation for this type will be first order in time.

Using (9.1-12), an example of the difference between the purely stable dispersive waves of the last chapter and the cases considered here given by considering a laser pulse propagating through an atomic medium. Equation (8.1-3) gives a good physical description of the way the waves behave, as it is well known experimentally that the refractive indices of such media depend on $|A|^2$ **provided** the incoming carrier wave is well away from the atomic resonance frequency of the host medium. In the last chapter we described this as the "off-resonant" case. However, if the incoming carrier wave is on or close to the atomic resonant frequency, and if the atoms of the medium are also pumped initially so that a proportion of them are in the upper atomic state, then we have a situation which is very similar to the one described in this section. The NLS equation is no longer the evolution equation of the wave envelope, because of the available potential energy stored in the atoms ready to be used by the waves oscillating at the preferred or "resonant" frequency. The phenomenon of self-induced transparency (SIT) in nonlinear optics is one physical example of this resonant case, and this will be dealt with in Section 9.4.

The method of finding the evolution equation for the amplitude A is very similar to that of the last chapter. The multiple scales approach has the advantage over other more heuristic methods in that an exact ordering occurs and one is left in no doubt about which should or should not be included. Using Taylor expansions as in (8.1-4) or (9.1-12) is useful as an illustration but not sufficient for obtaining precise answers. For both categories I and II, we will consider the carrier wave to be operating around the critical or resonant wavenumber k_c and for μ to be close to μ_c.

This involves taking a small band width of unstable modes $(k_c-\epsilon;\ k_c+\epsilon)$ about the critical wavenumber. We depict this bandwidth in Fig. 9-2. Formally we can expand $\mu = R(k)$ as a Taylor series about k_c:

$$\mu = R(k_c) \pm \tfrac{1}{2}\epsilon^2 R''(k_c) + \ldots. \qquad (9.1\text{-}13)$$

What is ϵ? From (9.1-13), we see that ϵ is a measure of how far the system is from exact criticality

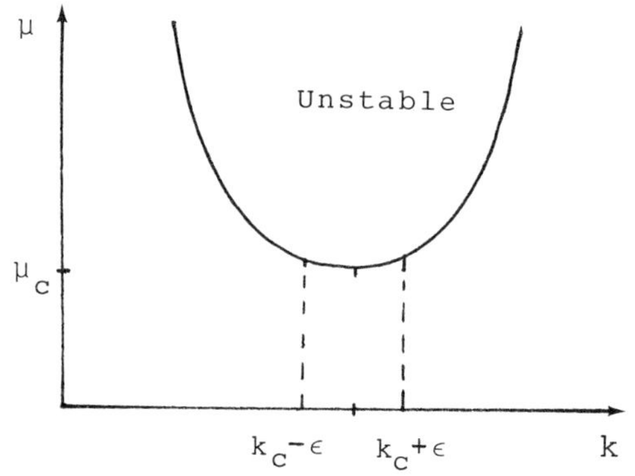

Figure 9-2:

$$\epsilon \sim (\mu-\mu_c)^{\frac{1}{2}} \tag{9.1-14}$$

and is therefore determined by the choice of μ under the necessary restriction that ϵ remains small. The negative sign in (9.1-13) formally allows us to take account of a slightly sub-critical state: $\epsilon \simeq (\mu_c - \mu)^{\frac{1}{2}}$. Now that ϵ has been determined, it can be used as the small parameter which determines slow space and time scales $X_1 = \epsilon x$; $X_2 = \epsilon^2 x; \ldots$ $T_1 = \epsilon t$; $T_2 = \epsilon^2 t$; ... . From this point onwards the same procedure is followed as in Chapter 8. $\underline{\phi}$ is expanded as an asymptotic series in ϵ and L is expanded about $\partial/\partial t$ and $\partial/\partial x$ and k and ω are evaluated at k_c and ω_{Rc} respectively. In Fourier space, this is equivalent to a Taylor expansion about the critical point (k_c, μ_c). The envelope amplitude $A(X_1, X_2, \ldots, T_1, T_2 \ldots)$ arises at $O(\epsilon)$ in the formal multiple scales problem

$$\underline{\phi}^{(1)} = \underline{b}\, A \exp(i\theta_c) + c.c$$
$$\theta_c = k_c x - \omega_{Rc} t + \delta \tag{9.1-15}$$

At this point in this section, instead of taking the reader through a rather long justification for the occurrence of the various amplitude equations which arise for our two categories of instabilities, we shall summarise the results, and leave the formal asymptotics to the next section. Any reader who is not interested in the scaling method can skip the next section.

The simplest equation, in one dimension, which arises for Category I type instabilities, is a cubically nonlinear diffusion equation

$$\frac{\partial A}{\partial T_2} = \pm\alpha_1 A - \beta_1 \frac{\partial^2 A}{\partial \overline{X}^2} + \gamma_1 A|A|^2 \tag{9.1-16}$$
$$\overline{X} = X_1 - c_g T_1$$

The coefficients α_1, β_1 and γ_1 are often complex for this category of problem. Equation (9.1-16) is often known as a "cubically nonlinear Landau-Ginsburg" equation in theoretical physics, because of its similarity to a result in superconductivity. In fluid mechanics, it goes under a variety of names including the Newell and Whitehead (1969) equation. This equation and its generalisations is the "typical" equation one would expect to arise in examples of Category I. It arises particularly in problems in convection and plane parallel flow. Some references and further comments are given in the section notes.

For Category II type instabilities, as predicted, we obtain amplitude equations which are second order in time and space. Equation (9.1-16), being an equation with a dissipative character, is not integrable in the sense described in this book. For Category II type instabilities, two types of "typical" amplitude equations arise. The first **is** integrable and the second is **not** integrable. The first is a coupled pair of equations in which the function B, first introduced in (9.1-11), arises:

$$\left(\frac{\partial}{\partial T_1}+c_1\frac{\partial}{\partial X_1}\right)\left(\frac{\partial}{\partial T_1}+c_2\frac{\partial}{\partial X_1}\right)A = \pm\alpha A - \beta AB \tag{9.1-17}$$

$$\left(\frac{\partial}{\partial T_1}+c_2\frac{\partial}{\partial X_1}\right)B = \left(\frac{\partial}{\partial T_1}+c_1\frac{\partial}{\partial X_1}\right)|A|^2 \tag{9.1-18}$$

The coefficients α and β are real and the c_1 and c_2 are the two values of the group velocity evaluated at (k_c, μ_c). It is fairly obvious that while c_g is single valued in (9.1-16), because the neutral curve arises from a single root of the dispersion relation, the group velocity in Category II will be double valued due to the coalescence of the complex conjugate pair of roots into a double root on the neutral curve. The $\pm$ sign on the αA term in (9.1-17) is chosen according to the choice of a slightly super or sub critical state around μ_c. Equations (9.1-17) and (9.1-18) are the main topic of this chapter and the two examples considered later are both examples in which (9.1-17) and (9.1-18) occur. These equations, which from now on we shall call the AB equations, are integrable by the inverse scattering transform, and can be transformed into a more familiar form already dealt with in Chapter 6. The two examples considered here are

(i) Self-Induced Transparency in ultra-short optical pulse propagation;

(ii) The inviscid two layer model of baroclinic instability.

These two examples will be the subjects of Section9.3 and 9.4. The SIT example will be performed line by line to see how the calculation works out in practice. The two-layer model, which is rather long to do in full, will be given in outline only.

Before reviewing some simple properties of the AB equations, we should mention that the second type of amplitude equation arises for the Category II type instability:

$$\left(\frac{\partial}{\partial T_1}+c_1\frac{\partial}{\partial X_1}\right)\left(\frac{\partial}{\partial T_1}+c_2\frac{\partial}{\partial X_1}\right)A = \pm\alpha A-\beta A|A|^2 \tag{9.1-19}$$

The function B plays no role in this system. The inviscid Kelvin-Helmholtz instability (Weissman 1979) and a buckling problem of elastic shells (Lange and Newell 1973) are two physical examples where (9.1-19) occurs. We note however that a co-ordinate transformation

$\alpha^{\frac{1}{2}}T_1 = \xi+\tau$; $\alpha^{\frac{1}{2}}X_1 = c_1\xi + c_2\tau$, and a rescaling of A: $\phi = (\beta/\alpha)^{\frac{1}{2}}.A$, turn (9.1-19) into the well-known ϕ^4 equation

$$\phi_{\xi\tau} = \pm\phi - \phi|\phi|^2 \tag{9.1-20}$$

which was noted in previous chapters to be a non-integrable system.

Returning to the AB equations, it is worthwhile spending a few lines showing that they are a system already considered in Chapter 6. To do this it is easier if we transform them in the following way:

Define

$$R = \sqrt{2\beta}\ A \tag{9.1-21}$$

$$S = \pm 1 - \beta\alpha^{-1}B \tag{9.1-22}$$

with semi-characteristic co-ordinates

$$\xi = -\ (X_1-c_1T_1)(c_1-c_2)^{-1}\alpha^{-\frac{1}{2}}\ ;\ \tau = (X_1-c_2T_1)(c_1-c_2)^{-1}\alpha^{\frac{1}{2}} \tag{9.1-23}$$

The AB equations now become

$$R_{\xi\tau} = RS \qquad\qquad R\to 0\ ;\quad S\to \pm 1 \tag{9.1-24}$$

$$S_\xi = -\tfrac{1}{2}(|R|^2)_\tau \qquad\qquad |\xi| \to \infty \tag{9.1-25}$$

The Zakharov-Shabat-AKNS method of 2 x 2 inverse scattering dealt with in Chapter 6 shows that (9.1-24),(9.1-25) are the integrability conditions of the eigenvalue problem

$$\begin{aligned}\frac{\partial\psi_1}{\partial\xi} + i\lambda\psi_1 &= \tfrac{1}{2}R\psi_2\\ \frac{\partial\psi_2}{\partial\xi} - i\lambda\psi_2 &= -\tfrac{1}{2}R^*\psi_1\end{aligned} \tag{9.1-26}$$

with the eigenfunctions subject to

$$\begin{aligned}\frac{\partial\psi_1}{\partial\tau} &= (-S\psi_1 + R_\tau\psi_2)/4i\lambda\\ \frac{\partial\psi_2}{\partial\tau} &= (R^*_\tau\psi_1 + S\psi_2)/4i\lambda\end{aligned} \tag{9.1-27}$$

The matrix Schrödinger equation derivable from (9.1-26)

$$\left(\frac{\partial^2}{\partial\xi^2} + \lambda^2\right)\underline{\psi} = \underline{\underline{V}}\underline{\psi} \tag{9.1-28}$$

has the potential matrix $\underline{\underline{V}}$ in the form

$$\underline{\underline{V}} = \begin{pmatrix} \tfrac{1}{4}|R|^2 & \tfrac{1}{2}R \\ -\tfrac{1}{2}R^* & \tfrac{1}{4}|R|^2 \end{pmatrix} \tag{9.1-29}$$

Soliton solutions are only possible if the energy states are negative. These bound states corresponds to purely imaginary eigenvalues of $\underline{\underline{V}}$. (9.1-24) is only skew-adjoint if $\beta > 0$, and so this is the criterion that the AB equations should yield soliton solutions. This is equivalent to the $\beta\gamma > 0$ Benjamin-Feir criterion of the last chapter for the NLS equation.

For $\beta > 0$ and A complex, R is complex and equations (9.1-24), (9.1-25) cannot be simplified any further. If, however, A and R are real, which corresponds to no phase variation in the carrier wave, it is possible to integrate these equations once. Take $R = \phi_\xi$ and assuming that $\phi_{\xi\tau} = F(\phi)$ we find

$$\phi_{\xi\xi\tau} = \phi_\xi S \tag{9.1-30}$$

$$S_\xi = -\phi_\xi \phi_{\xi\tau} \tag{9.1-31}$$

Both these equations give the result

$$F'' + F = 0 \tag{9.1-32}$$

and so we may take $F(\phi) = \pm\sin\phi$ without loss of generality. The equation

$$\phi_{\xi\tau} = \pm\sin\phi \tag{9.1-33}$$

is the famous sine-Gordon equation. Along with this we also have

$$S = \pm\cos\phi \tag{9.1-34}$$

and in each equation the $\pm$ signs refer to super and subcritical states respectively. If $\beta < 0$ then (9.1-25) has a positive sign and (9.1-32) turns out to be $F'' - F = 0$, and the sinh-Gordon equation results.

The various multi-soliton solutions can be found from Chapter 6; it is not necessary to repeat them here. The $\pm$ signs give, of course, different results in these formulae. For the subcritical case (-ve sign), the single soliton solution (real R) is:

$$\begin{aligned} R &= 2a_1 \operatorname{sech}\theta \\ \phi &= 4\tan^{-1}(\exp\theta) \\ S &= -1 + 2\operatorname{sech}^2\theta \\ \theta &= a\xi - \frac{1}{a}\tau + \delta \end{aligned} \tag{9.1-35}$$

In the (X_1, T_1) frame of reference we have

$$\begin{aligned} \theta &= \kappa(X_1 - vT_1) \\ v &= (c_1\beta a^2\alpha^{-1} + c_2)/(\beta^2 a^2 \alpha^{-1} + 1) \\ \kappa &= (\alpha + \beta a^2)\beta^{\frac{1}{2}} a(c_2 - c_1) \end{aligned} \tag{9.1-36}$$

It is not difficult to prove that $c_2 < v < c_1$ and so the soliton velocity must lie **between** the two group velocities c_1 and c_2. This soliton is stable to small perturbations. In Chapter 8 we discussed how the sine-Gordon equation arises as a model equation for a mechanical set of pendula where ϕ plays the role of the angle of twist. This soliton solution (9.1-35) plays the role of a twist from $\phi = 0$ to $\phi = 2\pi$. $\phi = 0$ is the "down" position for the pendulum where $S = -1$ (initially). S can be thought of as the potential energy of the system. As τ increases, ϕ becomes equal to π at its maximum, the "up" position where $S = +1$. Then, as the pendulum falls over, S decreases to -1 and ϕ goes to 2π. A pendulum which starts and ends in the down position is obviously stable to small perturbations around the states $\phi = 0 \pmod{2\pi}$. The fact that the subcritical soliton velocity lies between the two group velocities of the underlying carrier wave, qualitatively agrees with this conclusion, in the sense that the pulse on the envelope is neither running ahead nor behind the energy velocity in the carrier.

The supercritical state (+ve sign) has, for real R

$$R = 2b\,\text{sech}\,\eta$$
$$S = 1-2\,\text{sech}^2\eta$$
$$\eta = b\xi + \frac{1}{b}\tau + \delta \qquad (9.1\text{-}37)$$

with the η in the (X_1, T_1) frame being given by

$$\eta = \kappa(X_1 - vT_1)$$
$$v = c_1 + (c_1-c_2)[\beta b^2\alpha^{-1}-1]^{-1} \qquad (9.1\text{-}38)$$
$$\kappa = (\alpha-\beta b^2)/b\beta^{\frac{1}{2}}(c_1-c_2)$$

In this case, either $v \geqslant c_1$ or $v \leqslant c_2$ and so the soliton velocity is either faster than the highest or slower than the lowest group velocity of the carrier. This shows that this soliton is non causal. For instance, in Section 9.3 we will find that $c_1 = 1$ and $c_2 = 0$, where unity is the velocity of light in dimensionless units. We would expect this soliton to be unstable since it is trying to run too fast (or too slow) for the energy propagation of the carrier. Examination of (9.1-37) shows that the system begins and ends in the "up" position $S = +1$, and is only at $S = -1$, the "down" position, when $\phi = \pi$. This is the same as having a pendulum starting initially in the up position, which is an unstable position. The above formulae have been given for real R and so are really the single soliton formulae for the sine-Gordon equation where $R = \phi_\xi$. The inclusion of a variable phase term, making R a complex function of ξ and τ does not really alter the analogy with the pendulum. Equations (9.1-24) and (9.1-25) have a conserved quantity

$$|R_\tau|^2 + S^2 = 1 \qquad (9.1\text{-}39)$$

provided $R \to 0$; $S \to \pm 1$ as $|\xi| \to \infty$. This conservation law is equivalent to

$$\sin^2\phi + \cos^2\phi = 1 \tag{9.1-40}$$

when R is real.

The next section is a starred section which can be ignored on a first reading. In it we attempt to fill out some of the details of how the various amplitude equations arise. There are some significant mathematical differences between the dissipatively unstable examples which give the cubically nonlinear diffusion equation (9.1-16), and the second order in time amplitude equations (9.1-17)-(9.1-18) and (9.1-19), which arise for dispersively unstable examples.

9.2 *Secular Perturbation Theory for the Derivation of the Amplitude Equations

The aim of this section is to elaborate on some of the differences between the examples of this chapter, which have an imaginary part to their dispersion relations, and the examples of Chapter 8 which were deliberately restricted to examples with real dispersion relations. We will follow the ideas and notation of Chapter 8 and of the last section, and continue with the multiple scales analysis. To remind the reader, we are using slow space and time scales $X_m = \epsilon^m x$; $T_m = \epsilon^m t$ $(m=1,2,\ldots)$. The small parameter ϵ has been defined in (9.1-14) and determines how far the system is from criticality. In Chapter 8 we used a Taylor expansion about $\partial/\partial x$ and $\partial/\partial t$ for the operator L. We will also perform this expansion but we must also perform a third expansion about the point $\mu = \mu_c$. The result is

$$O(\epsilon): \quad \underline{\underline{L}}\underline{\phi}^{(1)} = 0 \tag{9.2-1}$$

$$O(\epsilon^2): \quad \underline{\underline{L}}\underline{\phi}^{(2)} = -\left(\underline{\underline{L}}_1 \frac{\partial}{\partial T_1} + \underline{\underline{L}}_2 \frac{\partial}{\partial X_1}\right)\phi^{(1)} + (\underline{\underline{M}}\phi^{(1)}).(\underline{\underline{N}}\phi^{(1)}) \tag{9.2-2}$$

$$O(\epsilon^3): \quad \underline{\underline{L}}\phi^{(3)} = -\left(\underline{\underline{L}}_1 \frac{\partial}{\partial T_1} + \underline{\underline{L}}_2 \frac{\partial}{\partial X_1}\right)\phi^{(2)} - \tfrac{1}{2}\Big\{\underline{\underline{L}}_{11} \frac{\partial^2}{\partial T_1^{\,2}} + 2\underline{\underline{L}}_{12} \frac{\partial^2}{\partial X_1 \partial T_1}$$
$$+ \underline{\underline{L}}_{22} \frac{\partial^2}{\partial X_1^{\,2}} + 2\underline{\underline{L}}_1 \frac{\partial}{\partial T_2} + 2\underline{\underline{L}}_2 \frac{\partial}{\partial X_2} \pm R''(k_c)\underline{\underline{L}}_\mu\Big\}\phi^{(1)} \tag{9.2-3}$$
$$+ \text{nonlinear terms}$$

Taking $\phi^{(1)}$ as in (9.1-15) with the carrier wave operating at the critical frequency and wave number, we find that (9.2-2) becomes

$$\underline{\underline{L}}\phi^{(2)} = -i\left(\underline{\underline{\ell}}_\omega \frac{\partial A}{\partial T_1} - \underline{\underline{\ell}}_k \frac{\partial}{\partial X_1}\right)\underline{b}\exp(i\theta_c) + \text{c.c.} \tag{9.2-4}$$

where we are using small letters when L, M and N are functions of k and ω.

When the dispersion relation is purely real, as in Chapter 8, we showed that the first terms in (9.2-4) are secular terms and invalidate the perturbation theory after finite time. The difference between the two categories of instabilities lies precisely in these terms. It is now necessary to determine under what conditions the terms on the right hand side of (9.2-4) produce secular terms in $\underline{\phi}^{(2)}$. The simplest case is to consider the scalar version of (9.2-4). The two conditions which prevent secular terms occurring are either to take $\bar{X} = X_1 - (d\omega/dk)T_1$ as in Chapter 8, or to have both ℓ_ω and $\ell_k = 0$. This latter condition can only occur if the scalar dispersion relation $\ell = 0$ has a **double root** in both ω and k at the critical point. In matrix form, the equivalent set of conditions for no secular terms is $\partial(\det \ell)/\partial\omega = 0$ and $\partial(\det \ell)/\partial k = 0$. The relation $\det(\ell) = 0$ is the matrix dispersion relation equivalent to $\ell = 0$ in the scalar case. This latter result is not difficult to prove; it just requires more algebra. The criterion of a double root in the dispersion relation is obviously the crucial point and divides our instabilities into the two categories of Section 1. It is now possible to deal with these separately:

Category I (Dissipatively Unstable)

In Section 9.1 we considered a single complex root only

$$\omega = \omega_R + i\omega_I \tag{9.2-5}$$

which can occur when damping (for example, viscosity) is present. At the critical point and everywhere on the neutral curve, $\omega_I = 0$ and so, in general, only **single roots** of the dispersion relation occurs on the neutral curve. It is mathematically possible to have a double root but physically this would be a highly pathological case. Because no double root exists, the ℓ_ω and ℓ_k terms in (9.2-4) produce secular terms which must be removed. The removal of secular terms can be performed in a similar manner to the last chapter, by introducing a new variable $\bar{X}$

$$\bar{X} = X_1 - \left(\frac{d\bar{\omega}}{dk}\right)T_1 \tag{9.2-6}$$

where $d\omega/dk$ (the group velocity) is obtained from the dispersion relation and is single valued

$$\frac{d}{dk}(\det \underline{\underline{\ell}}) = \left[\frac{\partial}{\partial k} + \left(\frac{d\omega}{dk}\right)\frac{\partial}{\partial\bar{\omega}}\right](\det \underline{\underline{\ell}}) = 0 \tag{9.2-7}$$

The variable ω is defined as $\bar{\omega} = \bar{\omega}(k)$ after we have set $\mu = R(k)$. Using the new variable $\bar{X}$ in (9.2-3) and removing secular terms again we find that the amplitude equation we obtain is (9.1-16):

$$\frac{\partial A}{\partial T_2} = \pm\alpha_1 A + \beta_1 \frac{\partial^2 A}{\partial \bar{X}^2} + \gamma_1 A|A|^2 \tag{9.2-8}$$

where, for convenience, we have changed the frame of reference such that the X_2 scale can be excluded. For a dissipative system, the operator L and its derivatives L_1, L_2 etc., will contain odd and even terms in $\partial/\partial x$ and $\partial/\partial t$, and consequently the coefficients α_1, β_1 and γ_1 will be, in general, complex. Note that the system of Chapter 8, which have single roots of the dispersion relation but which are purely real roots, will give the NLS equation in which α_1, β_1 and γ_1, would be pure imaginary. Equation (9.2-8) is not integrable by inverse scattering **unless** α_1, β_1 and γ_1 are pure imaginary. In that case it becomes a dispersive nonlinear wave equation with no diffusion (see equ. (6.1-1) and also the chapter notes for Sections 9.1 and 9.2).

Category II (Dispersively unstable)

This second category of instabilities, as explained in Section 9.1, is characterised by the occurrence of a complex conjugate pair of roots;

$$\omega = \omega_R \pm i\omega_I \tag{9.2-9}$$

Everywhere on the neutral curve we have $\omega_I = 0$ and so the dispersion relation has a **double root** in ω on this curve, and therefore $\partial(\det \ell)/\partial\omega = 0$. We also have the relationship

$$\left(\frac{\partial}{\partial k} + \frac{d\omega}{dk}\frac{\partial}{\partial \omega} + \frac{dR}{dk} \cdot \frac{\partial}{\partial R}\right) \det \underline{\underline{\ell}} = 0 \tag{9.2-10}$$

where ω, in this instance, is $\omega = \omega(k,\mu)$, and so $\partial(\det\ell)/\partial k = 0$ only if $dR(k)/dk=0$ also. This is only true at minimum when $k = k_c$. Our conclusion is therefore, that the $\partial A/\partial T_1$ term never gives rise to secular terms anywhere on the neutral curve, whereas the $\partial A/\partial X_1$ term is only nonsecular at the critical point. If these terms do not invalidate the perturbation theory, it is not necessary to introduce the $\bar{X}$ variable, and X_1 and T_1 can be left as independent variables. Equation (9.2-4) will now integrate to

$$\underline{\phi}^{(2)} = \left(\underline{a}_1 \frac{\partial A}{\partial T_1} + \underline{a}_2 \frac{\partial A}{\partial X_1}\right)\exp(i\theta_c) + \underline{a}_3 A^2 \exp(2i\theta_c) + \text{c.c.} + \underline{D}(X_1,T_1) \tag{9.2-11}$$

The harmonic terms come from the nonlinearities and the vector $\underline{D}$ is a constant of integration with respect to the "fast" scales x and t. We are assuming, as in Chapter 8, that the forms of L, M and N are such that no secular terms in AA^* are thrown up in (9.2-4).

We will point out only the salient features of the latter part of the calculation, since many of the points are obvious though rather long to prove in detail. Firstly, the role of the function $\underline{D}$ is important. Since $\underline{D}$ is a function of the slow scales only, it will be annihilated by the L_1 and L_2 operator terms in (9.2-3), unless L contains terms which are either first derivatives in time or space (thus producing at least one element of L which contains a constant). If this occurs, then $\underline{D}$ is preserved. The slow derivatives on $\underline{D}$ and slow derivatives on AA^* from the nonlinearities now form a second type of secular form which will produce terms explicitly in x and t in $\underline{\phi}^{(3)}$ but with no exponential attached. The other type of secular term produces an amplitude equation which is second order in T_1 and

X_1. The final result is the combinations of equations which we have called the AB equations

$$\left(\frac{\partial}{\partial T_1} + c_1 \frac{\partial}{\partial X_1}\right)\left(\frac{\partial}{\partial T_1} + c_2 \frac{\partial}{\partial X_1}\right) A = \pm \alpha A - \beta AB \tag{9.2-12}$$

$$\left(\frac{\partial}{\partial T_1} + c_2 \frac{\partial}{\partial X_1}\right) B = \left(\frac{\partial}{\partial T_1} + c_1 \frac{\partial}{\partial X_1}\right) |A|^2 \tag{9.2-13}$$

The first equation arises from the $\exp(i\theta_c)$ secular terms, and the second from the D-terms described above. The function B arises as an appropriate linear combination of the elements of $\underline{D}$. The velocities c_1 and c_2 are the two values of the **group velocity** of the carrier at the critical point. Differentation of (9.2-9) with respect to k, after taking the limit $\mu \to R(k), k \to k_c$, shows that two distinct values of the group velocity occur because of the double root.

A further total derivative of (9.2-7) shows that

$$\ell_{kk} + 2 \frac{d\bar{\omega}}{dk} \ell_{k\bar{\omega}} + \left(\frac{d\bar{\omega}}{dk}\right)^2 \ell_{\bar{\omega}\bar{\omega}} + \left(\frac{d^2\bar{\omega}}{dk^2}\right) \ell_{\bar{\omega}} = 0 \tag{9.2-14}$$

Since $\ell_{\bar{\omega}} = 0$ we have a simple quadratic for the group velocity which yields distinct roots. These are the c_1 and c_2 in (9.2-12). The time and space derivatives in (9.2-3) will factor exactly into this form. Then, if necessary, a linear combination of the elements of $\underline{D}$ can be taken such that the coefficients of the $\partial/\partial X_1$ terms in (9.2-13) turn out to be c_1 and c_2. The term $\pm\alpha A$ comes from the term $\pm R''(k_c)\underline{\underline{L}}_\mu$ in (9.2-3), with the positive/negative signs referring to a super/sub-critical state respectively.

The occurrence or non-occurrence of B is crucial to the structure of the equations. The examples of self induced transparency (SIT) (discussed in Section 9.3) and the two layer model of baroclinic instability (considered in Section 9.4) are cases where the operator $\underline{L}$ contains first derivatives. The AB type of equation arises in these two examples. However, if $\underline{L}$ takes such a form that $\underline{D}$ is annihilated by the fast scale operators, then no AB form of nonlinearity can occur. In this case, no secular terms of the second type that gave rise to equation (9.2-13) occur, and we are left with just one equation

$$\left(\frac{\partial}{\partial T_1} + c_1 \frac{\partial}{\partial X_1}\right)\left(\frac{\partial}{\partial T_1} + c_2 \frac{\partial}{\partial X_1}\right) A = \pm \alpha_2 A - \beta_2 A|A|^2 \tag{9.2-15}$$

The factorisation of the operator containing the two distinct group velocities occurs in the same way as the AB equations. The two layer model of the Kelvin-Helmholtz instability gives rise to equation (9.2-15), a calculation performed by Weissman (1979). However, we shall not study this example in this book. The main purpose of this section has been to identify the essential points of difference between the Category I and Category II type of instabilities. Type I (dissipative) operate on a $T_2 = \epsilon^2 t$ scale whereas Type II (dispersive) operate on a $T_1 = \epsilon t$ scale, and are second order in time also. This makes the initial value problems entirely different, and also means that it is necessary to use different types of finite difference schemes when numerical integration is necessary.

Our next section will be concerned with calculating the AB and hence sine-Gordon equations directly from the Maxwell-Bloch equations. The detail of how the main points emphasised in this section occur will show up as the calculation progresses.

9.3 Ultra-short Optical Pulse Propagation and Self-induced Transparency

In Chapter 8 the interaction of light with an atomic medium far off resonance was considered. In this case, no transitions are induced between the atomic energy levels. The strongest transitions in the optical region are electric dipole transitions. Consequently, in any situation in which electromagnetic radiation interacts with electric dipoles, we must include spatial propagation effects because of the short wavelengths. This is not the case for magnetic resonance phenomena because magnetic dipole radiation has wavelengths of at least a millimetre or more. The intention of this section is to study the propagation of electromagentic radiation which is resonant or near resonant with the energy levels of a 2-level atomic system. In Section 7.8, we studied such a system using the Hamiltonian of a two level system and Heisenberg's equation of motion, and obtained the so-called Bloch equations in (7.8-33) for the interaction of a field E with a 2-level atomic system. A generalisation of these equations is

$$\dot{P} = \frac{2p}{h} EN + \omega_s' Q - P/T_{tr} \tag{9.3-1}$$

$$\dot{Q} = -\omega_s' P - Q/T_{tr} \tag{9.3-2}$$

$$\dot{N} = -\frac{2p}{h} EP - \frac{(N-N_{eq})}{T_L} \tag{9.3-3}$$

where damping has been included phenomenonologically. These are called the optical Bloch equations. Here E is the electric field, P and Q are polarisation functions and N is the occupation number $(-1 < N < 1)$ which gives a measure of the atomic inversion. The constant p is the electric dipole strength, ω_s' is the frequency of the resonant two level atoms, and ω_s the frequency of the carrier of the incoming field. In general the two are exactly equal, but vibrations of each individual atom will Doppler shift each individual atomic frequency, and so ω_s' will take on a continuum of values centred around ω_s. With present day technology, it is not difficult to produce light pulses of either nanosecond (10^{-9}sec) or even a few picoseconds (10^{-12}sec) duration. Since these times are the pulse envelope duration, we are considering a situation in which an initial pulse has the form of a carrier wave which is **resonant** with the two level atoms. For longer pulse durations, it is crucial to include the important damping terms in (9.3-1)-(9.3-3). T_{tr} and T_L are called atomic relaxation times, so named because the upper atomic states may have finite lifetimes and N_{eq} is the equilibrium value of the occupation number when no field is applied. However, for nanosecond pulses or less, particulary if the right medium is chosen such as a dilute gas, it is possible to have T_{tr} and T_L significantly greater than the pulse duration. In consequence the pulse has interacted with the medium before any relaxation effects have taken place and in this case it is then possible to ignore the damping terms in the Bloch equations. However, these will be mentioned again in Section 9.5.

Under this restriction it is now becoming clear that the system with which we are dealing is of the Category II type discused in Section 9.1. The fact that the atoms are able to absorb and re-emit the incoming pulse of light indicates strongly that purely neutrally stable wave propagation no longer occurs and that the NLS equation is unlikely to occur as the amplitude equation for the pulse envelope. Indeed in Section 7.8, it was shown that a simple slowly varying envelope analysis of the Bloch equations gave the sine-Gordon equation and not the NLS equation. Our aim in this section is to confirm this result by showing that the system discussed here is unstable, and gives the AB equations, which reduce to the sine-Gordon equation. In fact, the Bloch equations, coupled with the Maxwell equations for the field E, is a classic example of a Category II type of instability. Our calculation, although longer than that of Section 7.8, is designed to take the inexperienced reader through the steps of the calculation in detail. Before starting on the calculation we should add Maxwell's equation, which couples the polarisation Q to the field E

$$\left(\frac{\partial^2}{\partial x^2} - \frac{1}{c^2}\frac{\partial^2}{\partial t^2}\right)E = \frac{4\pi np}{c^2}\frac{\partial^2}{\partial t^2}\int_{-\infty}^{\infty} Q(x,t;\Delta)g(\Delta)d\Delta \tag{9.3-4}$$

The integral on the right hand side of (9.3-4) is important in optics as it takes into account the Doppler shift of the resonant frequencies of all the individual atoms which are vibrating at random. This effect is called "inhomogeneous broadening" and the function $g(\Delta)$ is a symmetric normalised statistical distribution function; possibly a Gaussian or Lorentzian. $\Delta = \omega_s' - \omega_s$ and so is the amount each atom is off-resonance. Since we have $\geqslant 10^{12}$ atoms, the summation of the effects of all these atoms may be represented by an integral. Inhomogeneous broadening is important mathematically, as well as physically, in that it gives us what is known as the "Area Theorem" which will be dealt with later on in this section. For the moment, however, we lose no real physics by taking $g(\Delta) = \delta(\Delta)$ (sharp-line on-resonance exactly) for the following calculations. The effect of leaving $g(\Delta)$ as a general distribution will be mentioned as we proceed. Equations (9.3-1)-(9.3-4) are easier to handle if they are scaled into dimensionless variables. Defining dimensionless electric field, polarisation, space and time variables as

$$\frac{2p}{h\omega_s}\cdot E \to E; \quad P \to P; \quad Q \to Q; \quad N \to N$$

$$\omega_s t \to t; \quad \omega_s c^{-1} x \to x$$

We find that the Bloch equations and the Maxwell equation become

$$\begin{aligned} E_{xx} - E_{tt} &= \bar{\alpha} Q_{tt} \\ P_t &= EN+Q \\ Q_t &= -P \\ P_t &= -EN \end{aligned} \qquad \bar{\alpha} = \frac{8\pi np^2}{h\omega_s} \tag{9.3-5}$$

where $\bar{\alpha}$ is a dimensionless constant which gives a measure of the coupling between the atoms and the field. Our approach to equations (9.3-5) will be as follows. Firstly we will show that an instability of the dispersive type (Category II) exists. To do this, it is necessary to consider an equilibrium solution of (9.3-5) $E = P = Q = 0$, $N = N_o$. This means that **initially** the atoms may start with some occupation number N_o $(-1 < N_o < 1)$.

Experiments in attenuators start with $N_o = -1$, but we shall see that starting with a general value of N_o makes the calculation easier to handle. The significance of having $N_o = -1$ initially will be seen later. Linearising (9.3-5) around $N = N_o$, and eliminating Q, we find that the linear part can be written in the form

$$\begin{pmatrix} \frac{\partial^2}{\partial x^2} - \frac{\partial^2}{\partial t^2} & \bar{\alpha}\frac{\partial}{\partial t} & 0 \\ -N_o \frac{\partial}{\partial t} & \frac{\partial^2}{\partial t^2} + 1 & 0 \\ 0 & 0 & \partial/\partial t \end{pmatrix} \cdot \begin{pmatrix} E \\ P \\ \bar{N} \end{pmatrix} \tag{9.3-6}$$

where $N = N_o + \bar{N}(x,t)$. The dispersion relation is given by the determinant of the matrix on the left hand side of (9.3-6) in Fourier space:

$$\omega[(\omega^2-k^2)(1-\omega^2)-\bar{\alpha}N_o\omega^2] = 0 \tag{9.3-7}$$

Ignoring the $\omega = 0$ branch we obtain a quadratic equation in ω:

$$\omega^4-\omega^2[k^2+1-\bar{\alpha}N_o] + k^2 = 0 \tag{9.3-8}$$

solutions of which are

$$2\omega^2 = (k^2+1-\bar{\alpha}N_o) \pm [(k^2+1-\bar{\alpha}N_o)^2 - 4k^2]^{\frac{1}{2}} \tag{9.3-9}$$

It is now easy to see that we have a dispersive instability of the type mentioned in Section 9.1. Neutral stability occurs when

$$(k^2+1-\bar{\alpha}N_o)^2 - 4k^2 \geqslant 0 \tag{9.3-10}$$

and instability occurs when the discriminant is negative. The inequality (9.3-10) can now be written

$$(k+1)^2 \le \bar{\alpha}N_o \le (k-1)^2 \qquad (9.3\text{-}11)$$

Ignoring the left hand part which only takes account of the negative wave numbers, we have a perfectly parabolic neutral curve

$$N_o = (\bar{\alpha})^{-1}(k-1)^2 \qquad (9.3\text{-}12)$$

$$(N_o)_c = 0 \quad k_c = 1 \qquad (9.3\text{-}13)$$

with a critical point (0, 1). It is not surprising physically that the most unstable wavenumber (and therefore frequency in dimensionless units) is the resonant frequency . From (9.3-9) it is easily found that since the discriminant is zero at criticality the phase velocity is unity. The calculation of the group velocity is mcre tricky because of the square root. Differentiating (9.3-9) with respect to k before using (9.3-12), and then taking the limit $\bar{\alpha}N \to (k-1)^2$ we find that

$$\left(\frac{d\omega}{dk}\right)_c = \left.\frac{2k \pm 2k}{4\omega}\right|_{k,\omega\to 1} = 1 \text{ and } 0 \qquad (9.3\text{-}14)$$

We now have the two values of the group velocity predicted in Section 9.1 for the carrier wave.

Let us now return to the multiple scales caculation of (9.3-5) performed around $(k_c - \epsilon ; k_c + \epsilon)$. We have

$$\bar{\alpha}N_o = \pm \epsilon^2 + \ldots. \qquad (9.3\text{-}15)$$

$$N = N_o + \epsilon N^{(1)} + \epsilon^2 N^{(2)} + \ldots. \qquad (9.3\text{-}16)$$

It is simpler to eliminate Q from equations (9.3-5) and use only E, P and N.

We must also remember that we are evaluating $\bar{\alpha}N$ at the critical point and so it must be written as in (9.3-15). Eliminating P and Q from (9.3-5) the equations at the first three orders of ϵ become

$$O(\epsilon):\quad \left\{\frac{\partial^2}{\partial x^2} - \frac{\partial^2}{\partial t^2}\right\}\left\{\frac{\partial^2}{\partial t^2} + 1\right\}E^{(1)} = 0 \qquad (9.3\text{-}17)$$

$$O(\epsilon^2):\quad \hat{\mathcal{L}}E^{(2)} = -2\left\{\left(\frac{\partial^2}{\partial x \partial X_1} - \frac{\partial^2}{\partial t \partial T_1}\right)\left(\frac{\partial^2}{\partial t^2} + 1\right) + \frac{\partial^2}{\partial t \partial T_1}\left(\frac{\partial^2}{\partial x^2} - \frac{\partial^2}{\partial t^2}\right)\right\} E^{(1)} \qquad (9.3\text{-}18)$$
$$- \bar{\alpha}\, \partial^2/\partial t^2 (E^{(1)}N^{(1)})$$

$$O(\epsilon^3): \quad \hat{\mathcal{L}}E^{(3)} = \hat{M}E^{(2)} - \{(\frac{\partial^2}{\partial X_1^2} - \frac{\partial^2}{\partial T_1^2} - 2\frac{\partial^2}{\partial t\partial T_2}(\frac{\partial^2}{\partial t^2}+1)$$
$$+ 2(\frac{\partial^2}{\partial t^2}+1)(\frac{\partial^2}{\partial x\partial X_2} - \frac{\partial^2}{\partial t\partial T_2}) + 2(\frac{\partial^2}{\partial x^2} - \frac{\partial^2}{\partial t^2})\frac{\partial^2}{\partial t\partial T_2} +$$
$$+ (\frac{\partial^2}{\partial x^2} - \frac{\partial^2}{\partial t^2})(\frac{\partial^2}{\partial T_1^2} + 2\frac{\partial^2}{\partial t\partial T_2}) + \bar{\alpha}\frac{\partial^2}{\partial t^2}\{\pm E^{(1)} + N^{(1)}E^{(2)} + N^{(2)}E^{(1)}\}$$
$$+ 4\frac{\partial^2}{\partial t\partial T_1}(\frac{\partial^2}{\partial x\partial X_1} - \frac{\partial^2}{\partial t\partial T_1})\}E^{(1)} - 2\bar{\alpha}\frac{\partial^2}{\partial t\partial T_1}(E^{(1)}N^{(1)}) \tag{9.3-19}$$

$\hat{\mathcal{L}}$ is the operator in (9.3-17) and $\hat{M}$ is the operator on $E^{(1)}$ on the right hand side of (9.3-18).

Before we discuss solutions of the above we need to look at $N_t = -EP$ in combination with the Maxwell equation. The latter gives

$$O(\epsilon): \quad -\bar{\alpha}\frac{\partial P^{(1)}}{\partial t} = (\frac{\partial^2}{\partial x^2} - \frac{\partial^2}{\partial t^2})E^{(1)} \tag{9.3-20}$$

$$O(\epsilon^2): \quad -\bar{\alpha}[\frac{\partial P^{(2)}}{\partial t} + \frac{\partial P^{(1)}}{\partial T}] = (\frac{\partial^2}{\partial x^2} - \frac{\partial^2}{\partial t^2})E^{(2)} + 2(\frac{\partial^2}{\partial x\partial X_1} - \frac{\partial^2}{\partial t\partial T_1})E^{(1)} \tag{9.3-21}$$

and the former gives

$$O(\epsilon): \quad N_t^{(1)} = 0 \tag{9.3-22}$$

$$O(\epsilon^2): \quad N_t^{(2)} + \frac{\partial N^{(1)}}{\partial T_1} = -E^{(1)}P^{(1)} \tag{9.3-23}$$

$$O(\epsilon^3): \quad \frac{\partial N^{(1)}}{\partial T_2} + \frac{\partial N^{(2)}}{\partial T_1} + N_t^{(3)} = -E^{(1)}P^{(2)} - E^{(2)}P^{(1)} \tag{9.3-24}$$

We now turn to considering these equations in order to find a set of evolution equations for the envelope. Remembering that we are at the critical point $k_C = \omega_{Rc} = 1$, we have a solution of (9.3-17) in the form

$$E^{(1)} = \mathcal{E}(X,T_1,T_2)\exp(i\theta) + c.c.$$
$$\theta = (x-t) + \delta \tag{9.3-25}$$

where $\mathcal{E}$ is now a slowly varying amplitude function. (9.3-20) immediately tells us that $P_t^{(1)} = 0$ and so in general $P^{(1)} = P^{(1)}(x,X_1,T_1,T_2)$. However, this will produce a constant secular term in (9.3-21) so we must take $P^{(1)} = 0$. Again (9.3-22) gives $N^{(1)} = N^{(1)}(x,X_1,T_1,T_2)$. This will produce a secular term in the $\partial^2(E^{(1)}N^{(1)})/\partial t^2$ term in (9.3-18) and so we take $N^{(1)} = 0$. (9.3-23) then gives $N^{(2)} = D(x,X_1,T_1)$.

We shall ignore the possibility of x-dependence and take $N^{(2)}$ to be

$$N^{(2)} = D(X_1, T_1) \tag{9.3-26}$$

This function D plays the same role as the D-function of equation (9.1-32), although here we need only one component. Returning now to equation (9.3-18) and evaluating at $N_o = 0$, $k_c = \omega_{Rc} = 1$ we find immediately that $\hat{\mathcal{L}} E^{(2)} = 0$. This is the same as the original linear problem (9.3-17) for $E^{(1)}$. $E^{(2)}$ will then contain $\exp(i\theta)$ in the solution which will give zero at $k_c = \omega_c = 1$ in the first term on the right hand side of (9.3-21). Since $P^{(1)} = 0$ this equation can be integrated to give

$$P^{(2)} = \frac{2}{\bar{\alpha}} \left(\frac{\partial E}{\partial X_1} + \frac{\partial E}{\partial T_1}\right) \exp(i\theta) + \text{c.c.} \tag{9.3-27}$$

Using this result for $P^{(2)}$ in (9.3-24), we obtain

$$N_t^{(3)} = -\frac{\partial D}{\partial T_1} - \frac{2}{\bar{\alpha}} \left[\left(\frac{\partial}{\partial X_1} + \frac{\partial}{\partial T_1}\right) |E|^2\right] \tag{9.3-28}$$

The first two terms on the right hand side (9.3-28) are functions of the slow scales only and on integration would yield an explicit t-dependence in $N^{(3)}$. This clearly must be removed to prevent the perturbation theory becoming invalid as $t \gg \epsilon^{-1}$ and so we must have

$$\frac{\partial D}{\partial T_1} = -\frac{2}{\bar{\alpha}} \left(\frac{\partial}{\partial X_1} + \frac{\partial}{\partial T_1}\right) |E|^2 \tag{9.3-29}$$

Finally we turn to the $O(\epsilon^3)$ problem of equation (9.3-19). Using $N^{(1)} = 0$, $N^{(2)} = D$ etc, we remove secular terms in $\exp(i\theta)$ from (9.3-19) to obtain at $k_c = 1$

$$\frac{\partial}{\partial T_1} \left(\frac{\partial}{\partial X_1} + \frac{\partial}{\partial T_1}\right) E = \tfrac{1}{4}\bar{\alpha}[\pm E + ED] \tag{9.3-30}$$

We note that equations (9.3-30) and (9.3-29) are the AB equations of Section 9.1 and Section 9.2 with $\alpha_1 = \bar{\alpha}/4$; $B = -\tfrac{1}{2}\bar{\alpha}D$; $\beta_1 = \tfrac{1}{2}$. $c_2 = 0$; $c_1 = 1$. It is not necessary to take linear combinations to form B or from the operators $\partial/\partial T : \partial/\partial T + \partial/\partial X$ as only one D function actually occurs in the calculation and the form of the pair of operators with $c_1 = 1$, $c_2 = 0$ comes out exactly as predicted from (9.3-14). Note also that all the X_2 and T_2 derivatives vanish at the critical point.

In Section 9.2 we discussed the various behaviours of solutions of the AB equations with regard to the $\pm$ signs i.e. super or sub critical states. The negative (positive) sign corresponds to starting the system either below (above) $N_o = 0$. There appears, at first sight, to be a contradiction with experiment here, as soliton propagation can occur when $N_o = -1$ initially. However, our estimate of ϵ, from equation (9.1-15) with $\mu = N_o = -1$, is

$$\epsilon \sim (\bar{\alpha})^{\frac{1}{2}} \tag{9.3-31}$$

SIT experiments are usually performed at low densities and an estimate for ϵ, which depends on the atomic density, is that $\bar{\alpha} \simeq 0.01$ (see Eilbeck et al., 1973). Since $N_o = -1$, we have $\epsilon \simeq 0.1$ which is sufficiently small to play the role of a small expansion parameter. If the density is too high, then $N_o = -1$ (whole system in the ground state) makes ϵ too large and the AB equations are no longer valid. Hence, low densities are necessary for soliton propagation when $N_o = -1$ initially.

In order to discuss the multisoliton solutions of (9.3-29),(9.3-30), we shall make the following transformations: let

$$\left(\frac{\partial}{\partial X_1} + \frac{\partial}{\partial T_1}\right)E = \pm(\tfrac{1}{4}\bar{\alpha})P \qquad D = \pm(N-1) \tag{9.3-32}$$

Equations (9.3-29),(9.3-30) become

$$\left(\frac{\partial}{\partial X_1} + \frac{\partial}{\partial T_1}\right)E = \pm\alpha' P \qquad \alpha' = \tfrac{1}{4}\bar{\alpha} .$$

$$\frac{\partial P}{\partial T_1} = EN \qquad E,P \to 0 \qquad \text{as } |X_1| \to \infty \tag{9.3-33}$$

$$\frac{\partial N}{\partial T_1} = -\tfrac{1}{2}(E^*P+EP^*) \qquad N \to \pm 1$$

and we note that $N^2 + PP^*$ is a conserved quantity and equal to unity. Equations (9.3-29),(9.3-30) are more familarly known as the SIT (Self Induced Transparency) equations where the positive (negative) sign stands for the amplifier (attenuator). From the direct and inverse scattering problems of Chapter 6, the N-soliton formulae (without phase variation) are:

$$\begin{aligned} E^2 &= 4\frac{\partial^2}{\partial T_1^2}\log\det|M| \\ M_{ij} &= \frac{2}{E_i+E_j}\cosh[\tfrac{1}{2}(\theta_i+\theta_j)] \\ \theta_i &= \tfrac{1}{2}(E_iT_1-\Omega_iX_1)+\tfrac{1}{2}\delta_i \\ v_i &= \frac{\Omega_i}{E_i}\,1 \mp \frac{4\alpha'}{E_i^2} \end{aligned} \tag{9.3-34}$$

The single soliton formula is

$$E = E_1\operatorname{sech}\tfrac{1}{2}[E_1T_1-\Omega_iX_1+\delta_i] \tag{9.3-35}$$

and the two and three soliton formulae can be calculated from the determinant formula. In the experimental and theoretical literature concerning nonlinear optics, the soliton given in (9.3-35) is called a "2π-pulse" because its time area is 2π :

$$\int_{-\infty}^{\infty} E\,dT_1 = 2\pi \tag{9.3-36}$$

It is not difficult to show that an n-soliton solution has time area $2n\pi$. Figure 9-3 shows an initial pulse of time area 4π splitting into two 2π pulses or solitons. As well as pure imaginary eigenvalues (negative energy bound states) it is possible to have the "breather" solutions discussed in Chapter 6 which are complex conjugate pairs of eigenvalues. These oscillating breather solutions have as much area below as above the axis and so have zero area. In optics they are called zero-π pulses. Such a solution is shown in Fig. 9-4.

In the language of the inverse scattering transform, any initial data that has N discrete eigenvalues (N bound states) will yield N 2π pulses, plus possibily an oscillatory part corresponding to the continuous part of the

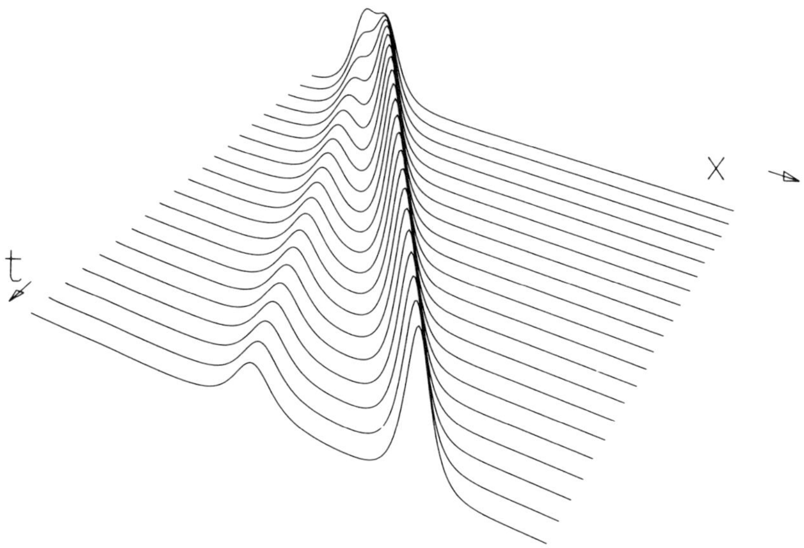

Figure 9-3:

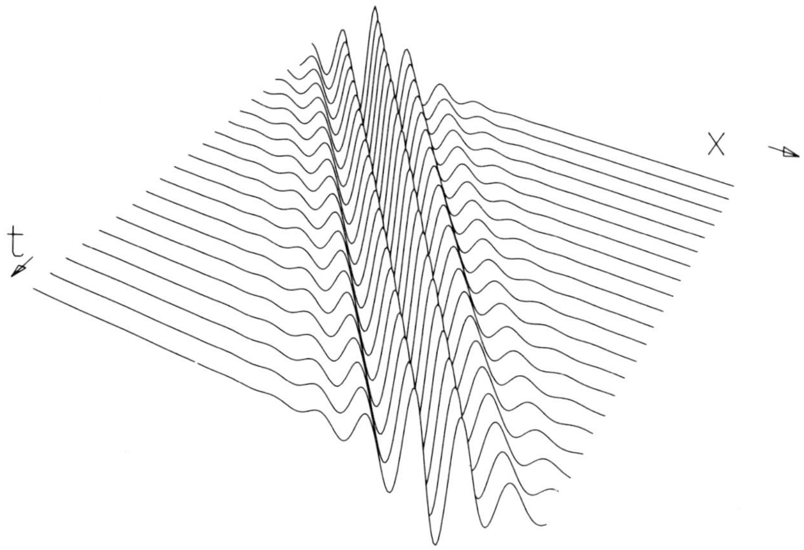

Figure 9-4:

spectrum of the initial data.

We note from (9.3-34) that the soliton velocities v_i are greater than unity for the amplifier (upper sign) but less than unity in the attenuator (lower sign). Consequently **the 2π pulse in the amplifier is unstable** since it travels faster than light and so is non-causal. However the 2π pulse in the attenuator is stable. The amplifier is equivalent to starting the atoms in their upper states (all pendulae up) and so the incoming 2π-pulse will break up as it has a background of potential energy on which to draw. The 2π formulae which we have given are for exact resonance with no phase variation. In general, E is complex and so carries a phase term with it which varies slowly in space and time compared to the carrier frequency and wavenumber. The inclusion of a slowly varying phase slightly alters the carrier frequency and wavenumber, a phenomenon known as "chirping".

Inhomogeneous broadening and the Area Theorem

Up to this point, we have ignored the effect of inhomogeneous broadening on the structure of solutions of the SIT equations. This was done by taking $g(\Delta)$ as a Dirac-delta function centred around the resonant frequency. The effect of including the integral in Maxwell's equation makes the multiple scales calculation marginally more difficult.

The linear stability problem includes the inhomogeneous integral average. It is necessary to make the assumption that the departure from resonance for each atom is small compared to the resonant frequency in order to find the neutral curve. We shall not repeat the calculation here; suffice to say that the choice of

$$E = E_R \exp(i\Phi); \quad P = (u+iv)\exp(i\Phi) \tag{9.3-37}$$

yields:

$$\left(\frac{\partial}{\partial X_1} + \frac{\partial}{\partial T_1}\right) E_R = \pm\alpha' \int_{-\infty}^{\infty} g(\Delta)\, u\, d\Delta$$
$$E_R\left(\frac{\partial}{\partial X_1} + \frac{\partial}{\partial T_1}\right)\Phi = \pm\alpha' \int_{-\infty}^{\infty} g(\Delta)\, v\, d\Delta \tag{9.3-38}$$

$$\frac{\partial u}{\partial T_1} = E_R N + \frac{\partial \Phi}{\partial T_1}\cdot v \tag{9.3-39}$$
$$\frac{\partial v}{\partial T_1} = -\frac{\partial \Phi}{\partial T_1}\cdot u$$
$$\frac{\partial N}{\partial T_1} = -E_R u$$

The choice of $\Phi = \Delta T_1$, with the assumptions (a) that the two differential operators are independent (which they are in the IST calculations of Chapter 6) and (b) that v is anti-symmetric in Δ, mean that equations (9.3-38) and (9.3-39) reduce to

$$
\begin{aligned}
\left(\frac{\partial}{\partial X_1} + \frac{\partial}{\partial T_1}\right) \mathcal{E}_R &= \pm\alpha' \int_{-\infty}^{\infty} u g(\Delta)\, d\Delta \\
\frac{\partial u}{\partial T_1} &= \mathcal{E}_R N + \Delta v \\
\frac{\partial v}{\partial T_1} &= -\Delta u \\
\frac{\partial N}{\partial T_1} &= -\mathcal{E}_R u
\end{aligned}
\tag{9.3-40}
$$

Note that the functions u, v and N are functions of Δ but $\mathcal{E}_R$ is not, since it is a macroscopic variable. In order to show that (9.3-40) is integrable by inverse scattering, it is necessary to show that (9.3-40) is integrable **without** the inhomogeneous broadening integral. The resulting equations

$$
\begin{aligned}
\left(\frac{\partial}{\partial X_1} + \frac{\partial}{\partial T_1}\right) \mathcal{E}_R &= \pm\alpha' u \\
\frac{\partial u}{\partial T_1} &= \mathcal{E}_R N + \Delta v \\
\frac{\partial v}{\partial T_1} &= -\Delta u \\
\frac{\partial N}{\partial T_1} &= -\mathcal{E}_R u
\end{aligned}
\tag{9.3-41}
$$

are known as the Reduced Maxwell Bloch (RMB) equations. We note that this system was shown to be integrable by the ZS-AKNS inverse scattering method in Chapter 6. In that chapter, it was shown that the RMB equations have N-soliton solutions of the same structural form as the sine-Gordon/SIT equations (as does the MKdV equation), since the eigenvalue problem is the same; only the time variation of the eigenfunctions is different. The form of the N-soliton solution of the RMB equations is exactly the same as (9.3-34) with only a change in Ω_i :

$$
v_i = \frac{\Omega_i}{E_i} = 1 \mp \frac{4\alpha'}{E_i^2 + 4\Delta^2}
\tag{9.3-42}
$$

The only effect of the inhomogeneous broadening integral is to take the averaging procedure through into the soliton velocities:

$$
v_i = \frac{\Omega_i}{E_i} = 1 \mp \int_{-\infty}^{\infty} \frac{4\alpha' g(\Delta)\, d\Delta}{E_i^2 + 4\Delta^2}
\tag{9.3-43}
$$

and the rest of the soliton formulae are exactly the same as in (9.3-34). This result can be confirmed easily for the one and two soliton solutions. We refer the reader to the chapter notes for references and further comments on the inverse scattering aspect of solving (9.3-40).

As we can see, inhomogeneous broadening has very little effect on soliton solutions of the SIT equations other than changing the soliton velocities. However, inclusion of inhomogeneous broadening makes one significant difference in the information we can extract from (9.3-40). It is possible, provided the averaging integral is included, to obtain what is known as an "area theorem" for the **time area** of a given pulse. We shall define the total time area of a pulse as $\theta(x)$

$$\theta(x) = \int_{-\infty}^{\infty} E_R(X_1,T_1)dT_1 \tag{9.3-44}$$

First of all we integrate the first equation in (9.3-41) from $T_1 = -\infty$ to $T_1 = \tau$ to obtain

$$E_R(\tau) + E_R(-\infty) + \int_{-\infty}^{\tau} E_R(X_1,T_1)dT_1 = \int_{-\infty}^{\infty} d\Delta g(\Delta) \int_{-\infty}^{\tau} u(X_1,T_1,\Delta)dT_1 \tag{9.3-45}$$

In order to calculate u, we eliminate v, from (9.3-41)

$$\frac{\partial^2 u}{\partial T_1^2} + \Delta^2 u = \frac{d}{dT_1}(E_R N)$$

and then write the solution in Green's function form

$$u(X_1,T_1,\Delta) = \int_{-\infty}^{\tau} dT' \Delta^{-1} \sin[\Delta(T_1 - T')] d(E_R' N')/dT' \tag{9.3-46}$$

Since we will be taking the limit $T_1 = \tau \to \infty$, we can write, in this limit

$$\lim_{t\to\infty} \frac{\sin \Delta t}{\Delta} = 2\pi\delta(\Delta) \tag{9.3-47}$$

and so on the assumption that we can exchange integrals we find that (9.3-45) becomes

$$E_R(\tau) + \frac{d}{dX_1} \int_{-\infty}^{\tau} E_R(X_1,T_1)dT = \pm 2\pi g(0)\alpha' \int_{-\infty}^{\tau} E_R(X_1,T_1,0)N(X_1,T_1,0)dT_1 \tag{9.3-48}$$

We are assuming that $E_R \to 0$ as $T_1 \to \pm\infty$. Since the integral of (9.3-48) contains E_R and N at $\Delta = 0$ only, we can write $P = \sin\phi$; $N = \cos\phi$; $E_R = \partial\phi/\partial T_1$ for the three Bloch equations in (9.3-40) when $\Delta = 0$. Performing the integration in (9.3-48) and taking the limit $\tau \to \infty$, we finally obtain

$$\frac{d\theta}{dX_1} = \pm 2\alpha' g(0) \sin\theta \tag{9.3-49}$$

Solving (9.3-49) we find that

$$\tan[\tfrac{1}{2}\theta(X_1)] = [\tan\tfrac{1}{2}\theta(0)] \exp[\pm 2\pi\alpha' g(0) X_1] \tag{9.3-50}$$

This beautiful result was first obtain by McCall and Hahn (1967, 1969) and gives the form of the change of the time area of an arbitrary pulse at any point in the medium. Obviously, this result is entirely dependent on the inclusion of inhomogeneous broadening; each atom radiates at a slightly off-resonant frequency and the Area Theorem is just a mathematical way of expressing the co-operation of the atoms in forming the final pulse area. It is also an expression of the fact that inhomogeneous broadening, far from being a destructive effect on the soliton behaviour, is actually a co-operative effect and enables us to evaluate the progression of an arbitrary pulse area through the medium. It is not possible to obtain this type of result for other soliton equations.

We have already noted that what we call a soliton in other contexts is called a 2π pulse in nonlinear optical pulse propagation. The 2π pulse is stable in an attenuator (lower sign) but unstable in an amplifier (upper sign). This result fits in very nicely with the results of the area theorem and can be explained diagrammatically. Figure 9-5 below shows a plot of $\theta(x)$ against x where one should follow left to right along the curves for the attenuator but right to left for the amplifier.

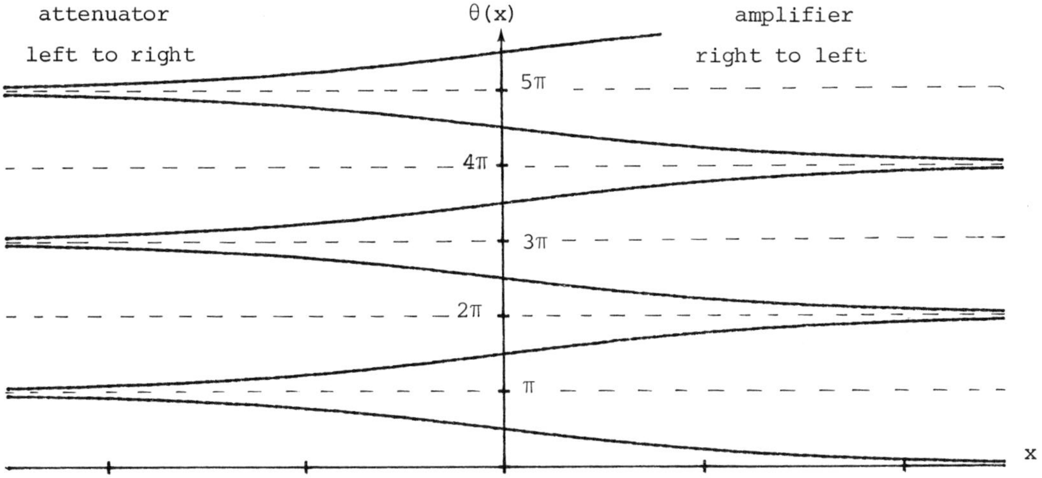

Figure 9-5:

The asymptotes for the attenuator (left to right) in Fig. 9-5 are $2n\pi$ and any pulse of area $\pi < \theta < 2\pi$ will grow to 2π, while any pulse of area $0 < \theta < \pi$ will drop to zero. This growth or decay is not true growth or decay, as in a dissipative system which may absorb or lose energy, but is more a re-shaping of the time area of the pulse. This will, at the same time, cause a re-shaping of the spatial area. We have already indicated that $2n\pi$ pulses in the attenuator are stable since they are causal. It is these stable pulses which are the asymptotic states into which an initial pulse of arbitrary time area will become eventually in space. However, the opposite is true for the amplifier. The 2π pulse in the amplifier is unstable as it needs to travel faster than the speed of light in vacuo and Fig. 9-5 shows that a $2\pi-\varepsilon$ pulse will eventually become a π-pulse. This pulse would have become a 2π pulse in the attenuator. The asymptotic areas are therefore $(2n+1)\pi$ for the amplifier. These are **not** soliton solutions of the SIT equations but are equivalent to self-similar solutions (see notes) of the SIT/sine-Gordon equations.

The Reduced Maxwell Bloch equations: an alternative approach to the SIT equations

Until now we have used a multiple scales approach on the Maxwell-Bloch equations (9.3-1)-(9.3-4), expanded around the critical point of the neutral stability curve and derived the slowly varying envelope equations using secular perturbation theory.

An alternative approach to this method is to return to the original Maxwell-Bloch set of equations (9.3-1)-(9.3-4) and consider waves travelling to the right only. This reduction is equivalent to ignoring backscattering or in mathematical terms is equivalent to considering only one characteristic of the Maxwell equation. Equation (9.3-4) now becomes

$$\frac{\partial E}{\partial x} + \frac{1}{c}\frac{\partial E}{\partial t} = -\frac{2\pi np}{c^2}\frac{\partial}{\partial t}\int_{-\infty}^{\infty} Q(x,t,\Delta)\,d\Delta \tag{9.3-51}$$

This approximation is valid provided the coupling constant $\bar{\alpha}$ between the field and the atoms is small compared to unity. This constant $\bar{\alpha}$ is proportional to the atomic density and has a value ~ 0.01 for gaseous densities : $(n \leqslant 10^{18}$ atoms/cc.). Rescaling the reduced Maxwell equation (9.3-51) and the Bloch equations in the following manner $\omega_a t \to t$; $\omega_a c^{-1}x \to x$; $2p\omega_a^{-1}hE \to E$; $\omega_s'\omega_a^{-1} \to \omega_s'$, we find that they reduce to

$$\begin{aligned}
\frac{\partial E}{\partial x} + \frac{\partial E}{\partial t} &= -\alpha_a \int_{-\infty}^{\infty} g(\Delta)\, Q_t(x,t,\Delta)\,d\Delta \\
Q_t &= -\omega_s' P \\
P_t &= EN + \omega_s' Q \\
N_t &= -EP
\end{aligned} \tag{9.3-52}$$

where ω_a is a typical atomic frequency such that the new ω_s' is of order unity and α_a has the definition

$$\alpha_a = \frac{4\pi np^2}{h\omega_a} \tag{9.3-53}$$

The RMB equations (9.3-52) have exactly the same **mathematical** structure as the SIT equations of (9.3-40) although their physical meaning is different. In (9.3-40), E_R is the real part of the electric field envelope, whereas in (9.3-52), E is still the full electric field. As we have already mentioned, the RMB equations are an integrable system: we gave their soliton solutions in the last subsection. The single and double soliton solution of the RMB equations are

$$E(x,t) = E_1 \operatorname{sech}\{\tfrac{1}{2}E_1[t-\Omega_1 x\ ]\} \tag{9.3-54}$$

$$\Omega_i = 1 + \int_{-\infty}^{\infty} \frac{g(\Delta)\alpha_a \omega_s' d\omega_s'}{E_i^2+4\omega_s'^2} \tag{9.3-55}$$

$$E(x,t) = \left(\frac{E_1^2-E_2^2}{E_1^2+E_2^2}\right) \cdot \frac{E_1\operatorname{sech}\theta_1+E_2\operatorname{sech}\theta_2}{1-B_{12}\tanh\theta_1\tanh\theta_2+B_{12}\operatorname{sech}\theta_1\operatorname{sech}\theta_2} \tag{9.3-56}$$

$$B_{12} = \frac{2E_1E_2}{E_1^2+E_2^2}$$

Although these are mathematically exact solutions of the RMB equations, each soliton is of femto-second duration ($\tfrac{1}{2}E_1 \simeq \omega_s$) and such a pulse is impossible to realise in the laboratory. It would need to be both extremely short and ultra-intense (1000 Tera Watts/cm^2), an intensity which is neither obtainable nor sensible since real dielectrics would disintegrate. These solutions are for the **electric field itself** since we have made no approximations with regard to slowly varying envelopes. However, we recall that a "breather" type of solution is possible if we take E_1 and E_2 as anti-complex conjugate pairs $E_1 = -E_2^* = E_o + 2i\omega_c$. If we take the high frequency limit of the breather solution from (9.3-56) we should recover a solution which has the form of a fast oscillation modulated by a slowly varying envelope. This ought to match with the SIT equation results. We obtain, after a little algebra, the breather solution of the RMB equations

$$E(x,t) = 2E_o \operatorname{sech}\theta_R \left(\frac{\cos\theta_I - \gamma\sin\theta_I \tanh\theta_R}{1+\gamma^2\sin^2\theta_I \operatorname{sech}^2\theta_R}\right) \tag{9.3-57}$$

$$\theta_R = \tfrac{1}{2}E_o(t-\Omega_R x) \tag{9.3-58}$$

$$\theta_I = \omega_c(t-\Omega_I x)$$

$$\gamma = \tfrac{1}{2}E_o/\omega_c$$

$$\Omega_R = 1+\int_{-\infty}^{\infty} g(\omega_s'-\omega_s)\,4D\alpha_a\omega_s'[E_o^2+4\omega_s'^2+4\omega_c^2]\,d\omega_s' \tag{9.3-59}$$

$$\Omega_I = 1+\int_{-\infty}^{\infty} g(\omega_s'-\omega_s)\,4D\alpha_a\omega_s'[4(\omega_s'^2-\omega_c^2)-E_o^2] \tag{9.3-60}$$

$$D^{-1} = E_o^4+8E_o^2(\omega_s'^2+\omega_c^2) + 16(\omega_s'^2-\omega_c^2)^2 \tag{9.3-61}$$

If we now choose $\omega_c \simeq \omega_s$ (the resonant frequency), and take E_o such that $\gamma = E_o/\omega_c \ll 1$, then the expression for the field $E(x,t)$ now becomes

$$E(x,t) = 2E_o \operatorname{sech}[\tfrac{1}{2}E_o(t-\Omega x)]\cos[\omega_s(t-x)]$$
$$\Omega = 1+2\alpha_a\omega_s E_o^{-2} \tag{9.3-62}$$

Since $2\alpha_a\omega_s = \alpha'$, we have reproduced exactly the SIT single soliton pulse modulating a cosine carrier wave moving at the velocity of light.

This calculation therefore confirms the results of the multiple scales expansion of the original Maxwell-Bloch equations and justifies the approximations involved. In summary we can say that high frequency breather solutions of the RMB equations are equivalent to SIT solitons modulating a fast carrier wave provided that in both cases the atomic density is small.

Finally we note that the next order in γ in (9.3-57) yields only an extra phase term $\phi(x,t) = \gamma\tanh\theta_R$ in the carrier wave. The chirping frequency will therefore be $O(\gamma^2)$, which is of order 10^{-6} in the pico second region, and therefore is neglible. Although the RMB approach is more unwieldy than using the SIT equations, it has the advantage that, provided the atomic density is small so the reduction of the Maxwell equation is valid, the equations are valid for all frequencies and intensities and could be used in other frequency requires where the approximations used in deriving the SIT equations may be invalid.

9.4 The Two Layer Baroclinic Instability

We now turn our attention away from optical physics and turn to a problem in rotating fluid mechanics. Our concern will be a study of a very simple model which manifests what have become known as baroclinic waves. These waves occur in both the atmosphere and oceans of the Earth, with wavelengths of the order of 1000km and 10km respectively. They can also occur in the atmospheres of some of the major planets.

Baroclinic instability has long been identified as one of the basic mechanisms supplying kinetic energy to the large-scale weather systems (depressions and their associated frontal structures) of mid-latitudes. Generally, the instability can arise when an equilibrium state exists in which surfaces of constant density are not parallel to surfaces of constant gravitational potential. In this case, particles of fluid moving in trajectories which lie between the potential surfaces and density surfaces can release some of the potential energy of the system and gain kinetic energy.

As one would expect, the problem of mathematically modelling such behaviour beginning from the Navier-Stokes equations is extremely difficult, partly because of the geometry of the problem.

One way theoreticians have of studying this type of phenomenon is to construct a simple fluid model, which although idealised, nevertheless displays baroclinic instability. The state of equilibrium mentioned above, between surfaces of constant gravitational potential and constant pressure, can be achieved in the atmosphere, the ocean and in laboratory experiments from a horizontal gradient of temperature together with a rapid rotation of the entire system. An analogous model, but one simpler to handle theoretically, is to consider what has been called the "two layer model" (Philips 1954), in which two layers of immiscible, inviscid fluid of different densities, the lighter overlaying the heavier, are placed in an infinite channel and maintained in relative horizontal motion. The entire system is rotated about its vertical axis. One cannot pretend that this is an exact model of the atmosphere, but it does display baroclinic instability, which, if viscosity is small and is neglected, will yield a dispersive type of instability (Category II in Section 9.1). For our application we shall suppose that the upper and lower layers of fluid move at velocities U_1 and U_2 respectively in the x-direction in an infinite straight channel with stress-free side-walls at $y = 0,1$. (See notes for comments on these side-wall boundary conditions). As in Section 5.4, where we described the beta-plane approximation, the effects of planetary sphericity are incorporated by using a beta-plane approximation in which the Coriolis parameter is taken to be $2\Omega_o + \beta y$. The model is described pictorially in Fig. 9-6. It is not necessary to start the analysis of the instability from the Navier-Stokes equations since it is possible to considerably simplify the model depicted in Fig. 9-6 by using the geostropic approximation described in Section 5.4. This approximation, which allows a balance between the Coriolis force and the pressure gradient, reduces the Navier-Stokes equations for the two layer model to a coupled pair of vorticity equations. Unfortunately it

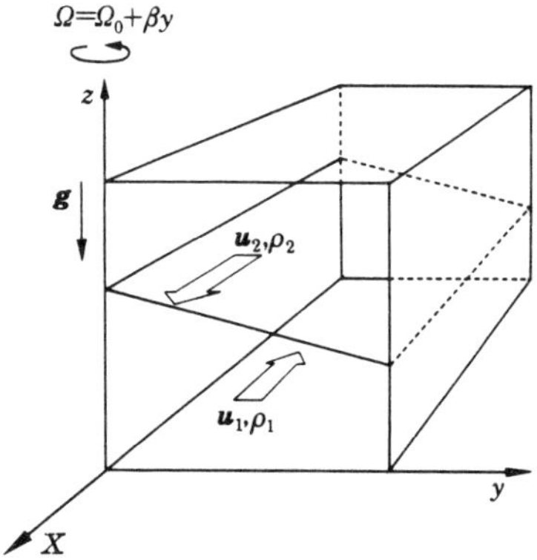

Figure 9-6:

is not possible here to derive these vorticity equations from first principles, because of the sheer length of the calculation. In such models boundary layers occur, called Ekman layers, the effects of which must be included. Unfortunately, we do not have the time to explain the details of how this operates. A clear derivation of the two layer vorticity equations can be found in a paper by Pedlosky (1970). In the geostrophic limit, the potential vorticity equations governing the flow are:

$$\left[\frac{\partial}{\partial t}+\frac{\partial\psi_1}{\partial x}\frac{\partial}{\partial y}-\frac{\partial\psi_1}{\partial y}\frac{\partial}{\partial x}\right]\left[\nabla^2\psi_1+(\psi_2-\psi_1)F+\beta y\right]=0,$$
$$\left[\frac{\partial}{\partial t}+\frac{\partial\psi_2}{\partial x}\frac{\partial}{\partial y}-\frac{\partial\psi_2}{\partial y}\frac{\partial}{\partial x}\right]\left[\nabla^2\psi_2+(\psi_1-\psi_2)F+\beta y\right]=0, \qquad (9.4\text{-}1)$$

where ψ_1 and ψ_2 are the upper and lower layer streamfunction and F is the internal rotational Froude number defined as $F = 4\Omega^2 L^2 p/(\Delta p/\bar{p})gD/2$, where Ω is the rotation rate, L is the channel width, D its depth, $\Delta p/\bar{p}$ the fractional density constrast between the fluids and g the acceleration due to gravity. We take F and β both to be fixed for the purposes of the present discussion. The streamfunctions ψ_i are taken to be the constant background zonal flow plus a perturbation:

$$\psi_i = -U_i y+\phi_i(x,y,t) \qquad (9.4\text{-}2)$$

The shear between the two layers, $\Delta U = U_1 - U_2$ will be treated as a variable parameter and will be seen to play the role of μ in the discussion of the previous sections. Equations (9.4-1) may now be written:

$$\left(\frac{\partial}{\partial t}+U_1\frac{\partial}{\partial x}+\frac{\partial\phi_1}{\partial x}\cdot\frac{\partial}{\partial y}-\frac{\partial\phi_1}{\partial y}\cdot\frac{\partial}{\partial x}\right)\left(\nabla^2\phi_1+(\phi_2-\phi_1)F\right)+(\beta+F\Delta U)\frac{\partial\phi_1}{\partial x}=0$$
$$\left(\frac{\partial}{\partial t}+U_2\frac{\partial}{\partial x}+\frac{\partial\phi_2}{\partial x}\cdot\frac{\partial}{\partial y}-\frac{\partial\phi_2}{\partial y}\cdot\frac{\partial}{\partial x}\right)\left(\nabla^2\phi_2+(\phi_1-\phi_2)F\right)+(\beta-F\Delta U)\frac{\partial\phi_2}{\partial x}=0 \qquad (9.4\text{-}3)$$

which in turn can be linearized to give:

$$\left(\frac{\partial}{\partial t} + U_1 \frac{\partial}{\partial x}\right)(\nabla^2\phi_1 + (\phi_2 - \phi_1)F) + (\beta + F\Delta U)\frac{\partial \phi_1}{\partial x} = 0$$
$$\left(\frac{\partial}{\partial t} + U_2 \frac{\partial}{\partial x}\right)(\nabla^2\phi_2 + (\phi_1 - \phi_2)F) + (\beta - F\Delta U)\frac{\partial \phi_2}{\partial x} = 0 \qquad (9.4\text{-}4)$$

Note the Jacobian form of the nonlinear term in equations (9.4-4); a form which is common to any two-dimensional advection problem.

The pair of equations (9.4-4) can be writen in the form of (9.1-1) where the operator $\underline{\underline{L}}$ is given by

$$L = \begin{pmatrix} \left(\frac{\partial}{\partial t} + U_1 \frac{\partial}{\partial x}\right)(\nabla^2 - F) + (\beta + F\Delta U)\frac{\partial}{\partial x} & F\left(\frac{\partial}{\partial t} + U_1 \frac{\partial}{\partial x}\right) \\ F\left(\frac{\partial}{\partial t} + U_2 \frac{\partial}{\partial x}\right) & \left(\frac{\partial}{\partial t} + U_2 \frac{\partial}{\partial x}\right)(\nabla^2 - F) + (\beta - F\Delta U)\frac{\partial}{\partial x} \end{pmatrix} \qquad (9.4\text{-}5)$$

The Jacobian nonlinearity can easily be set in the quadratic product form of the right hand side of (9.1-1). For instance, the Jacobian of ϕ_1 with ϕ_2 can be written as

$$J_{x,y}(\phi_2;\phi_1) = (M\underline{\phi}).(N\underline{\phi}) \qquad (9.4\text{-}6)$$

where

$$M = \frac{1}{\sqrt{2}}\begin{pmatrix} 1 & 1 \\ -1 & 1 \end{pmatrix}\frac{\partial}{\partial x}, \qquad N = \frac{1}{\sqrt{2}}\begin{pmatrix} 1 & -1 \\ 1 & 1 \end{pmatrix}\frac{\partial}{\partial y} \qquad (9.4\text{-}7)$$

Taking $\det(\ell) = 0$ as in (9.1-5) gives a quadratic dispersion relation, the roots of which are

$$\frac{\omega}{k} = \tfrac{1}{2}(u_1 + u_2) + \frac{\beta(F+a^2)}{a^2(2F+a^2)} \pm \frac{[4F^2\beta^2 - (\Delta u)^2 a^4(4F^2 - a^4)]^{\frac{1}{2}}}{2a^2(2F+a^2)} \qquad (9.4\text{-}8)$$
$$a^2 = k^2 + m^2\pi^2$$

where $a^2 = k^2 + m^2\pi^2$.

Furthermore, we also find that

$$b_1 = 1$$
$$b_2 = F^{-1}(a^2+F) + (F\Delta U + \beta)[F(\omega/k - u_1)]^{-1} \qquad (9.4\text{-}9)$$

For the present discussion we are considering values of k and F such that $a^2 < 2F$. Under this circumstance, as ΔU is increased, the sign within the square root change from positive to negative.

The shear ΔU plays the role of μ, the bifurcation parameter, and the neutral curve is therefore given by

$$\mu = \Delta U(k) = 2\beta F/a^2(4F^2 - a^4)^{\frac{1}{2}} \qquad (9.4\text{-}10)$$

When $\mu < \Delta U(k)$, the system is neutrally stable and hence the instability is purely dispersive, but for $\mu > \Delta U(k)$ baroclinic instability occurs. The neutral stability curve given in (9.4-10) is of the kind illustrated in Fig.

9-1. It has a minimum at

$$a_c^2 = k_c^2 + m^2\pi^2 = \sqrt{2F} \tag{9.4-11}$$

and

$$\mu_c = (\Delta U)_c = \beta/F \tag{9.4-12}$$

In keeping with the results of Section 9.1, we find at the critical point $((\Delta U)_c, k_c)$ that while the phase speed is single valued and equal to U_2, the group velocity has two values, which we denote c_1 and c_2:

$$\begin{aligned} c_1 &= u_2 + \beta k_c^2 (1+b_c)^2 F^{-2} \\ c_2 &= u_2 \end{aligned} \tag{9.4-13}$$

where $b_c = \sqrt{2} - 1$. There is also a linearly independent solution with $b_c = \sqrt{2} + 1$, corresponding to the unphysical total wavenumber $a^2 = -2F$. As explained in Section 9.1, for shears somewhat greater than $(\Delta U)_c$, wavelike solutions will amplify until the nonlinear terms in equations (9.4-4) become important. Laboratory experiments indicate thay they can act so as to limit the amplitude of the waves and the result for a wide range of imposed parameters is that there exists a regime of highly organised "regular baroclinic" waves. The review by Hide and Mason (1975) contains many details of such waves.

Pedlosky (1972) was the first to perform the multiple scales expansion about the critical point. The only difference between his calculation and our general approach is his choice of expansion about β/F. Pedlosky took

$$\Delta U = \beta/F \pm \epsilon^2 \tag{9.4-14}$$

and chose to be $\pm\epsilon^2$ from critical instead of performing a formal Taylor expansion as in Section 9.2. This however, only changes the numerical value of α, the square of the growth rate. Expanding the perturbation stream functions as

$$\phi_i = \epsilon\phi_i^{(1)} + \epsilon^2\phi_i^{(2)} + \cdots \tag{9.4-15}$$

we obtain at $O(\epsilon)$

$$\begin{pmatrix} \phi_1^{(1)} \\ \phi_2^{(1)} \end{pmatrix} = A(X_1,T_1) \begin{pmatrix} 1 \\ b_2 \end{pmatrix} \sin(m\pi y)\exp[i(kx-\omega t)] + c.c. \tag{9.4-16}$$

The perturbation analysis of (9.4-4) proceeds in the same way as that described in Section 9.2. Since $\underline{\underline{L}}$ is a dispersive operator (we are considering here only the inviscid case), the results of Section 9.1 apply and no secular terms occur at the bifurcation point. Furthermore, the linear operator $\underline{\underline{L}}$ has first derivatives in every element and consequently the vector $\underline{D}(X_1,T_1)$ occurs in the analysis. The situation in which it occurs is complicated by the presence of the y-variation. The function $B(X_1,T_1)$

turns out to include an integral over y because an adjoint form of the operator $\underline{\underline{L}}$ must be used to remove secular terms (see Nayfeh, 1973). We find the expression for B to be

$$B(X_1,T_2) = \frac{-4\beta}{m\pi F^2}\int_0^1 \left[\frac{\partial^2 D_2}{\partial y^2} + F(D_1-D_2)\right]\sin(2m\pi y)\,dy \tag{9.4-17}$$

Physically, $B(X_1,T_1)$ is a measure of the mean flow correction resulting from the self-rectification of the nonlinear wave. The final amplitude equations are in the category of (9.1-17,9.1-18) and turn out to be

$$\left(\frac{\partial}{\partial T_1} + c_1\frac{\partial}{\partial X_1}\right)\left(\frac{\partial}{\partial T_1} + c_2\frac{\partial}{\partial X_1}\right) = \pm\alpha A - NAB$$

$$\left(\frac{\partial}{\partial T_1} + c_2\frac{\partial}{\partial X_1}\right) B = \left(\frac{\partial}{\partial T_1} + c_1\frac{\partial}{\partial X_1}\right)|A|^2 \tag{9.4-18}$$

Because of the Jacobian form of the nonlinearity, there are no second harmonic terms at $O(\epsilon^2)$ making any contribution to (9.3-18). The square of the "growth rate" α and the constant N are defined as

$$\alpha = \tfrac{1}{2}k_c^{\,2}b_c^{\,2}\beta F^{-1} \tag{9.4-19}$$

$$N = (\tfrac{1}{2}k_c b_c m\pi)^2 \tag{9.4-20}$$

The application of such analysis to real physical systems such as the atmosphere or laboratory annulus experiments requires care, for two principal reasons. First, the departure of the shear from the critical values, ϵ^2, is not always small. Consequently, the t, T_1 and T_2 timescales are not especially well separated, and the multiple scales analysis breaks down. Secondly, the physical dimensions of the system constrain the long space scale. For example, in the atmosphere and in many laboratory experiments, the system accommodates five or six wavelengths, so that the wavelength of the carrier wave cannot be very short compared to the lengthscale of the envelope variation. Nonetheless, these analyses are useful in the fluid mechanics context since they do lead to results which bear some similarity to laboratory observation.

We now turn our attention to the two layer model when the beta-effect is excluded. Equation (9.4-8) now becomes

$$\omega/k = \tfrac{1}{2}(u_1+u_2) \pm \tfrac{1}{2}\Delta U(a^4-4F^2)^{\frac{1}{2}}(2F+a^2)^{-1} \tag{9.4-21}$$

Obviously, the value of ΔU now plays no role in the instability but if $a^2 < 2F$ then an imaginary part appears in (9.4-21). Since F contains the fluid density values, we use this as the parameter μ and the neutral curve is now

$$\mu = F = \tfrac{1}{2}a^2 = \tfrac{1}{2}(k^2+m^2\pi^2) \tag{9.4-22}$$

The minimum value of k is $k_c = 0$ with $F_c = \tfrac{1}{2}m^2\pi^2$. As we showed in Section 9.1, it is necessary to consider unstable modes up the side of the neutral curve away from $k_c = 0$ which prevents the admission the X_1-scale but allows the use of the X_2-scale. Because the X_1-scale is not present, the equivalent of equation (9.1-34) integrates to $B = |A|^2$ and the final amplitude equation takes the form

$$i\chi \frac{\partial A}{\partial X_2} = \pm\alpha_3 A - \beta_3 A|A|^2 + \gamma_3 \frac{\partial^2 A}{\partial T_1^2} \tag{9.4-23}$$

Since the $\pm\alpha_3 A$ term can be absorbed into the left hand side of (9.4-23), we obtain the NLS equation with T_1 and X_2 exchanging places as the "space" and "time" variables in the usual inverse scattering formulation of Chapter 6. Technically, for this equation we would not be solving an initial value problem. Instead the inverse scattering transform has data specified for A at $X_2 = 0$ (say) which is a form of boundary value problem.

9.5 The Effect of Weak Dissipation

The main aim of this chapter, and indeed this whole book, has been a study of integrable partial differential equations. In this final section we will depart slightly from this area by considering the effect of dissipation on the dispersively unstable examples of this chapter. In all the soliton equations, damping will destroy their exactly integrable character and any consideration of the inverse scattering transform becomes vacuous as the scattering data will have decayed to zero at infinity. Of course every physical system is subject to some energy losses, however small.

It would appear that the inclusion of weak energy losses in the modelling calculations in this book, though physically relevant, will not yield any new mathematically interesting phenomena. For instance, the inclusion of weak viscosity in the fluid and plasma examples which yield the KdV equation in the lossless limit, only produces a u_{xx} term, turning the KdV equation into the KdV-Burgers equation.

The AB equations are one example which gives some interesting results when damping is included. These equations were originally derived, in the weakly nonlinear approximation, in the neighbourhood of the critical points of those systems which have roots of their dispersion relations which occur in conjugate pairs. If it is not to destroy the structure of the AB equations entirely, the inclusion of any damping must firstly be weak ($O(\epsilon)$) and secondly must not fundamentally shift the underlying neutral curve upon which the derivation of the AB equations is based. With present day knowledge it is not known how to deal with the full p.d.e's since damping terms destroy the integrability properties. For this reason we shall consider the AB equations with time variation only

$$\begin{aligned} \frac{d^2 A}{dT_1^2} &= \pm\alpha A - \beta AB \\ \frac{dB}{dT_1} &= \frac{d}{dT_1}(|A|^2) \end{aligned} \tag{9.5-1}$$

Equations (9.5-1) have the structure of a cubically nonlinear oscillator. The inclusion of weak damping (for example, viscosity) will modify (9.5-1) to produce decay terms in the oscillator:

$$\frac{d^2A}{dT_1^2} + \Delta_1 \frac{dA}{dT_1} = \pm\alpha A - \beta AB \tag{9.5-2}$$

$$\frac{dB}{dT_1} + \Delta_2 B = \left(\frac{d}{dT_1} + \Delta_3\right)|A|^2 \tag{9.5-3}$$

The effect of the spatial derivatives on the space-independent solutions of (9.5-2) and (9.5-3) is an open question, but brings into play the Benjamin-Feir instability mechanism in which a time dependent (only) solution may be unstable to sidebands.

The derivation of (9.5-2) and (9.5-3) is intuitively obvious and can be performed by adding in the extra damping terms to the calculation of Section 9.2. The only restriction on this calculation is that the original neutral curve must not be shifted by more than $O(\epsilon)$, otherwise the original critical point, around which all the Taylor expansions occurred, will have moved. If purely weak dissipative effects have been added to the original model, then all the Δ_i and α are real. If extra weak dispersive effects are included, then Δ_1 and α must obviously be complex. Dispersive effects never effect either the damping or growth rate terms in any system, and so this latter type of effect will contribute only to the imaginary part of Δ_1 and α. The term "extra weak dispersive effect" means that, even though we are considering an undamped dispersive system as the fundamental unstable model, it is possible to consider phenomenologically the effect of other weak dispersive terms in the equation of motion in the same way as the weak damping terms. This procedure is valid provided the character of the underlying stability is not disturbed. Two physical examples of such dispersive effects are the beta-plane terms used in rotating fluids to take into account curvature effects and the so-called detuning terms in the laser in which the cavity resonance does not quite match the resonant frequency of the atomic sample. These effects will be mentioned in more detail in the examples to follow.

Equations (9.5-2) and (9.5-3) do not look particularly interesting on their own. However we will transform them into a first order set of three o.d.e's by the following transfers.

$$\tau = \Omega T_1 \qquad \Omega = \mathrm{Re}\Delta_1 - \tfrac{1}{2}\Delta_3 \tag{9.5-4}$$

$$X = (2\beta)^{\frac{1}{2}}\,\Omega^{-1}\,A \tag{9.5-5}$$

$$Z = 2\beta\,\Omega^{-1}\Delta_3^{-1}B \tag{9.5-6}$$

Equations (9.5-2) and (9.5-3) now become

$$\dot{X} = -\sigma X + \sigma Y \tag{9.5-7}$$

$$\dot{Y} = -XZ + rX - aY \tag{9.5-8}$$

$$\dot{Z} = -bZ + \tfrac{1}{2}(X^*Y + XY^*) \tag{9.5-9}$$

where $r = r_1 + ir_2$, $a = 1 - ie$, and the five parameters are defined by

$$\sigma = \tfrac{1}{2}\Delta_3\Omega^{-1} \qquad b = \Delta_2\Omega^{-1}$$
$$r_2 = [\Delta_3 \mathrm{Im}(\Delta_1) + 2\mathrm{Im}(\alpha)]/(\Delta_3\Omega) \ : \ r_1 = 1 + 2\mathrm{Re}(\alpha)/\Delta_3\Omega \qquad (9.5\text{-}10)$$
$$e = -(\mathrm{Im}\Delta_1)\Omega^{-1}$$

The extra variable Y is defined by equation (9.5-7) and the variable X must not be confused with the slow spatial scale X_1. Equations (9.5-7) to (9.5-9) are known as the Lorenz equations (Lorenz 1963) in complex form. They reduce to the real form of the Lorenz equations when only weak dissapative effects and no weak dispersive effects are included. In this case, $\mathrm{Im}\,\Delta_1 = \mathrm{Im}\,\alpha = 0$ and so $a = 1$ and r is real. The complex nature of X and Y in this case can be removed by a phase transformation and the equations reduce to the form originally considered by Lorenz

$$\begin{aligned} \dot{X} &= -\sigma X + \sigma Y \\ \dot{Y} &= (r_a - Z)X - Y \\ \dot{Z} &= -bZ + XY \end{aligned} \qquad (9.5\text{-}11)$$

The Lorenz equations are famous for displaying a sequence of bifurcations, which end up, at higher values of r, with phase space trajectories lying on what is called a strange attractor. It is not our intention to spend any time on this topic since it is not the subject of this book, but the interested reader can turn firstly to Lorenz' orginal paper or to articles contained or referenced in Marsden and McCracken (1976) and Haken (1978). The point about the solutions of these o.d.e's is that while periodic or quasi-periodic solutions represent an ordered state in the sense that the Fourier analysed power spectrum contains one or more independent frequencies, the power spectrum of a strange attractor is broad. This is thought of as a chaotic or turbulent state particularly since a chaotic attractor is highly sensitive to minute changes in initial conditions.

In his original paper Lorenz (1963) was studying a problem in fluid convection in two dimensions. In that problem σ was the Prandtl number and r_a the Rayleigh number. The occurrence of the Lorenz equations in this section is totally different from Lorenz' convection problem (which is described in the chapter notes) and the two must not be confused.

We will now investigate the behaviour of the real version of the Lorenz equations (9.5-11) as far as we can analytically. They have fixed points at the origin $X = Y = Z = 0$ when $0 < r_a < 1$ and also at

$$\left.\begin{aligned} X = Y &= \pm[b(r_a-1)]^{\frac{1}{2}} \\ Z &= (r_a - 1) \end{aligned}\right\} \quad r_a > 1 \qquad (9.5\text{-}12)$$

Linearising around the origin gives

$$\begin{pmatrix} \dot{X} \\ \dot{Y} \\ \dot{Z} \end{pmatrix} = \begin{pmatrix} -\sigma & \sigma & 0 \\ r_a & -1 & 0 \\ 0 & 0 & -b \end{pmatrix} \begin{pmatrix} X \\ Y \\ Z \end{pmatrix} \qquad (9.5\text{-}13)$$

Solutions of the form $\underline{a}\exp(\lambda t)$ are found by obtaining the eigenvalues of the matrix in (9.5-13). The characteristic equation is

$$(\lambda+b)\ \ [\lambda^2+\lambda(\sigma+1)+\sigma(1-r_a)] \ = 0 \tag{9.5-14}$$

The roots of this are $\lambda = -b$ and

$$\lambda = \tfrac{1}{2}\{-\sigma-1\pm[\sigma+1)^2+4\sigma(1-r_a)^{-\frac{1}{2}}\} \tag{9.5-15}$$

when $0 < r_a < 1$, all three roots are negative. The origin is stable and is said to be an attracting fixed point, as trajectories in phase space will spiral into this point. If $r_a > 1$ then one root becomes positive and the origin is now unstable and is said to be a repelling fixed point. When $r_a > 1$, however, the two other fixed points given in (9.5-12) appear. A linearisation around these yields a characteristic equation

$$\lambda^3+\lambda^2(\sigma+b+1) + \lambda b(\sigma+r_a) + 2\sigma b(r_a-1) = 0 \tag{9.5-16}$$

Considering the possibility of a complex conjugate pair of roots $\lambda = \lambda_o \pm i\Omega$, it is obvious that as r_a is varied, the point $r_a = r_{ac}$ at which λ_o changes from negative to positive is the point at which these fixed points become unstable. At this point we can factor a term $\lambda^2 + \Omega^2$ out of (9.5-16) and write the characteristic equation as

$$(\lambda^2+\Omega^2)(\lambda+\delta) = 0 \tag{9.5-17}$$

whence

$$\delta = \sigma+b+1\ ; \qquad \Omega^2 = b(\sigma+r_a) \qquad\qquad (9.5\text{-}18)$$
$$\delta\Omega^2 = 2\sigma b(r_a-1)$$

These three equations give the value of r_a as

$$r_{ac} = \frac{\sigma(\sigma+b+3)}{\sigma-b-1} \tag{9.5-19}$$

For this value of r_a to be positive, it is necessary to have $\sigma > b + 1$. In summary, when $0 < r_a < 1$, the origin is stable and is the only fixed point. When $1 < r_a < r_{ac}$, the origin is unstable, but the other two fixed points are stable; trajectories in phase space will be repelled by the origin but attracted to the other two fixed points. When $r_a > r_{ac}$ (provided $\sigma > b + 1$), these two fixed points are also unstable. The next step which Lorenz took was to perform a numerical integration of these equations when $r_a > r_{ac}$ to see how the trajectories in phase space behaved. He chose $b = 8/3$, $\sigma = 10$ which gives $r_{ac} = 470/19 = 24.73$ and $r_a = 28$. A numerical plot of the phase space trajectories showed that they were aperiodic and moved on an attracting "surface" which is known as the Lorenz attractor. Without resorting to some very difficult topology, it is impossible to describe the Lorenz attractor adequately except to say that it is a type of Riemann surface with infinitely many sheets. The trajectories orbit around each of the two fixed points given in (9.5-12). As the orbits move from one fixed point to another they move onto a different sheet of the attractor. In this way, they avoid crossing.

Furthermore, Lorenz showed that the switching of the orbits from one fixed point to the other is entirely unpredictable since no periodicities appear in the motion. A topological description of the Lorenz attractor and associated properties can be found in the papers referenced in the book by Marsden and McCracken (1976).

Complex Lorenz equations

Before turning to an example of how the Lorenz equations occur through the AB equations, we shall briefly look at the complex version of the Lorenz equations, to see the differences between the real and complex versions. We would expect some differences, as the complex equations are a five dimensional system, as opposed to a three dimensional system in the real case. One immediate difference is in the number of fixed points of each system.

The origin $X = Y = Z = 0$ is still a fixed point. As well as this fixed point we also have $X = Y$, whence $Z = r - a$. The third equation implies that

$$|X|^2 = b(r-a) \tag{9.5-20}$$

Since Z is real, it follows that such points can only exist if $e + r_2 = 0$. It is exactly this latter equation which is the condition such that the imaginary parts of Δ_1 and α can be scaled away by a phase rotation in (9.5-2), thus reducing the complex to the real case. For this reason, we will ignore this pathological condition, and assume that e and r_2 do not satisfy $e + r_2 = 0$. In this case therefore, only one fixed point exists. Examing the stability of the origin, we obtain $\lambda = -b$ or

$$(\sigma+\lambda)(a+\lambda) - \sigma r = 0 \tag{9.5-21}$$

for the characteristic equation. Roots of (9.5-21) are given by

$$\lambda = \tfrac{1}{2}[-\sigma-a\pm(p+iq)] \qquad p > 0 \tag{9.5-22}$$

where

$$p+iq = [(\sigma+a)^2+4\sigma(r-a)]^{\frac{1}{2}} \tag{9.5-23}$$

We therefore have

$$\mathrm{Re}(\lambda) = \tfrac{1}{2}[\pm p-\sigma-1] \tag{9.5-24}$$

One eigenvalue has negative real part and the other yields a critical stability limit when $p = \sigma + 1$. Taking real and imaginary parts of (9.5-23) we find that

$$\begin{aligned} p^2-q^2 &= (\sigma+1)^2 + 4\sigma(r_1-1)-e^2 \\ pq &= 2\sigma(e+r_2) - e(\sigma+1) \end{aligned} \tag{9.5-25}$$

Taking $r_1 = r_{1c}$ when $p = \sigma + 1$, we find that r_{1c} satisfies

$$r_{1c} = 1 + \frac{(e+r_2)(e-\sigma r_2)}{(\sigma+1)^2} \tag{9.5-26}$$

$$\tfrac{1}{2}(e+q) = \frac{\sigma(e+r_2)}{\sigma+1} \tag{9.5-27}$$

Equation (9.5-27) denotes the frequency of the critically stable eigenmode since $\omega = \mathrm{Im}(\lambda) = (e+q)/2$. We may observe that if $e + r_2 \neq 0$ then the origin becomes oscillatorily unstable at $r_1 = r_{1c}$. For most o.d.e's it is not possible to find an exact periodic solution in closed form. The Hopf bifurcation theorem (Marsden and McCracken, 1976) gives the necessary conditions which are needed for a bifurcation to a periodic solution, but gives no indication of the analytic form of the solution. We are fortunate in this case that we can find an exact periodic solution

$$\begin{aligned} X &= A\exp(i\omega t) \\ Y &= (1+i\omega\sigma^{-1})\exp(i\omega t) \\ Z &= r_1 - r_{1c} \\ |A|^2 &= b(r_1 - r_{1c}) \end{aligned} \tag{9.5-28}$$

where r_{1c} and ω are defined in (9.5-26) and (9.5-27).

The difference between the real and complex cases is now obvious. To return to the real case, we either take $e = r_2 = 0$ or $e + r_2 = 0$. In either case, $\omega = 0$ and $r = 1$ and the limit cycle reduces to a circle of fixed points. This is none other than a rotation of the two fixed points of the real Lorenz equations into a continuum, and equations (9.5-28) reduce to (9.5-12). Consequently, in the complex case, when the origin becomes unstable at r_{1c}, the limit cycle takes the place of the two fixed points of the real case. These two points became unstable when $r > r_{ac}$ and the Lorenz attractor appeared. The limit cycle of the complex equations in turn is only stable for a finite range of values of $r_1 : r_{1c} < r_1 < r'_{1c}$. We will not pursue the calculation further here because the details are too long, but in principle it is not difficult to study the stability of the limit cycle by transforming into a rotating frame of frequency ω and then performing a stability analysis. By this method, r can be found analytically. Numerical integration is necessary to find the behaviour when $r_1 > r'_{1c}$. This problem has been studied by Fowler et al. (1981) who found that doubly periodic behaviour occurs when $r_2, \sigma, e \simeq 1$, which represents motion on a 2-torus. Figure 9-7 shows plots of $\mathrm{Re}(X)$ and Z versus time.

A power spectral analysis of this data shows that two frequencies of irrational ratio occur. Aperiodic motion of the Lorenz attractor type only occurs in the limit $r_2 \to 0$. The conclusion is that the complex version of the equations displays a greater degree of order, in the sense that provided r_2 or e are not close to zero, either singly or doubly periodic motion is predominant in parameter space. r_2 and e are themselves the imaginary parts of r and a and stemmed directly from the inclusion of weak dispersive effects in the original problem. Earlier we quoted the beta-effect as an example of this type of effect. As an example we will use the 2-layer model of baroclinic instability discussed in Section 4 and include weak viscosity and a weak beta-effect to show how the complex Lorenz

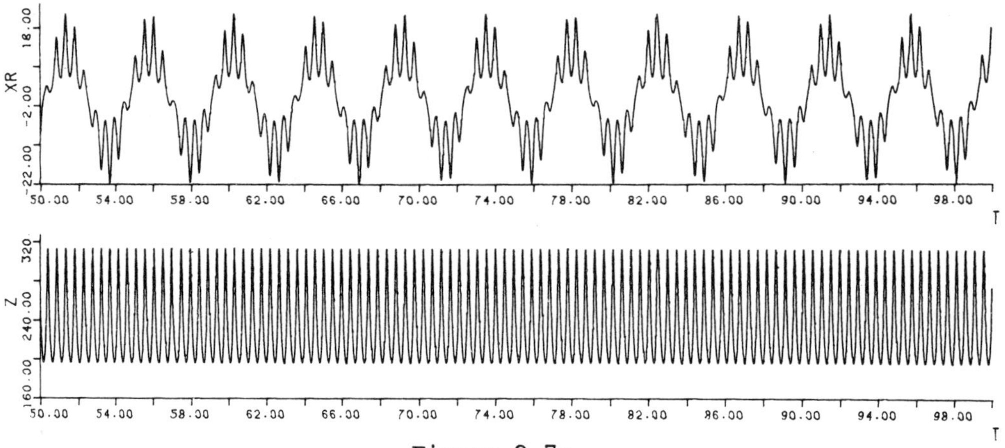

Figure 9-7:

equations arise.

The quasi-geostrophic potential vorticity equations given in (9.4-1) are modified by the inclusion of viscosity to become

$$\left[\frac{\partial}{\partial t} + \frac{\partial \psi_i}{\partial x}\frac{\partial}{\partial y} - \frac{\partial \psi_i}{\partial y}\frac{\partial}{\partial x}\right]\left[\nabla^2\psi_i + F(\psi_j - \psi_i) + \beta y\right] = -\nu\nabla^2\psi_i \qquad i \neq j = 1,2 \tag{9.5-29}$$

where ν is the viscosity coefficient. Expanding around $\psi_i = -U_i y$, as in Section 9.4, we can find a new neutral curve with the viscosity included. This is found by setting $\omega_I = 0$ and is

$$\mu = \Delta u = \frac{4}{(2F-a^2)}\left[\frac{\beta^2F^2}{a^2(a^2+F)^2} + \frac{a^2\nu^2}{k^2}\right] \tag{9.5-30}$$

This is **not** the same curve in the limit $\nu \to 0$ as the inviscid neutral curve found in Section 9.4 which was

$$\mu = \Delta u = \frac{2\beta F}{a^2(4F^2-a^4)^{\frac{1}{2}}} \tag{9.5-31}$$

The two curves do not match in the inviscid limit, and we conclude that even weak viscosity destabilizes the neutral curve (see Romea, 1977). However, in the limit $\beta \to 0$, the two curves coalesce, although the nature of the instability changes. In this limit, the neutral curve is

$$F = \tfrac{1}{2}a^2 = \tfrac{1}{2}(k^2+m^2\pi^2) \tag{9.5-32}$$

It is still possible to take $\nu = \epsilon\bar{\nu}$ and $\beta = \epsilon\bar{\beta}$ which only adds a correction at $O(\epsilon^2)$ to (9.5-32). As we mentioned in Section 9.4, (9.5-32) has a minimum at $k_c = 0$ and so it is necessary to take a wave number k away from the minimum, and confine ourselves in general to an ϵ^2 bandwidth: $F \to F \pm \epsilon^2$ about the whole neutral curve, instead of an ϵ^2 bandwidth above and below the minimum. The calculation here follows exactly as in the outline of Section 9.4. At $O(\epsilon^2)$ we can integrate the equations to obtain

$$\phi_1^{(2)} = D_1(T_1,y)$$
$$\phi_2^{(2)} = D_2(T_1,y) + \frac{4i}{k\Delta u}\left[\frac{dA}{dT_1} + (\bar{\nu} - \frac{ik\bar{\beta}}{a^2})A\right]\sin(m\pi y)\times\exp(i\theta) + \text{c.c.} \tag{9.5-33}$$

We note that the coefficient of A is complex, the imaginary part arising from the dispersive beta-effect. At $O(\epsilon^3)$, the removal of secular terms which give t terms only in $\phi^{(3)}$ gives

$$\frac{\partial}{\partial T_1}\left(\frac{\partial^2 D_1}{\partial y^2} - a^2 D_1\right) + \bar{\nu}\frac{\partial^2 D_1}{\partial y^2} = \frac{2a^2 m\pi}{\Delta u}\sin(2m\pi y)\left[\frac{\partial}{\partial T_1} + 2\bar{\nu}\right]|A|^2 \tag{9.5-34}$$

and an equation in D_2 which immediately shows that $D_2 = -D_1$. Removal of secular terms which give terms like $t\exp(i\theta)$ in $\phi^{(3)}$ yields:

$$\frac{d^2A}{dT_1^2} + \frac{3}{2}\left(\bar{\nu} - \frac{i\beta k}{a^2}\right)\frac{dA}{dT_1} = a^{-2}\left[\pm\frac{1}{4}k^2(\Delta u)^2 + \frac{5}{9}k^2\bar{\beta}^2 a^{-2} + ik\bar{\beta}\bar{\nu}\right]A + \left(\frac{k^2\Delta u}{2a^2}\right)A\int_0^1 \frac{\partial^2 D_1}{\partial y^2}\sin(2m\pi y)\,dy \tag{9.5-35}$$

Next we define

$$\bar{B}(T_1) = \int_0^1 \frac{\partial^2 D_1}{\partial y^2}\sin(2m\pi y)\,dy$$

subject to the boundary conditions $D = 0$ at $y = 0$ and 1. Equation (9.5-34), on multiplication by $\sin(2m\pi y)$ and integration over $y = 0$ to 1 gives

$$\left(1 + \frac{a^2}{4m^2\pi^2}\right)\frac{d\bar{B}}{dT_1} + \bar{\nu}\bar{B} = \frac{m\pi a^2}{\Delta u}\left[\frac{d}{dT_1} + 2\bar{\nu}\right]|A|^2 \tag{9.5-36}$$

Equation (9.5-35) now becomes

$$\frac{d^2A}{dT_1^2} + \frac{3}{2}(\bar{\nu} - ik\bar{\beta}a^{-2})\frac{dA}{dT_1} = a^{-2}\left[\pm\frac{1}{4}k^2(\Delta u)^2 + \frac{5}{9}k^2\bar{\beta}^2 a^{-2} + ik\bar{\beta}\bar{\nu}\right]A - \tfrac{1}{2}k^2(\Delta u)^2 a^{-2}A\bar{B} \tag{9.5-37}$$

By redefining the variable B as

$$\bar{B} = \left(\frac{4m^2\pi^2a^2(\Delta u)^{-1}}{a^2 + 4m^2\pi^2}\right)B \tag{9.5-38}$$

to make the coefficient of $d(|A|^2)/dT$ unity, as in equation (9.5-3), we find that the equivalent values of Δ_1, Δ_2 & Δ_3 in equations (9.5-2) and (9.5-3) are

$$\begin{aligned}\Delta_1 &= \tfrac{3}{2}(\bar{\nu}-ik\bar{\beta}a^{-2})\\ \Delta_2 &= \frac{4m^2\pi^2\bar{\nu}^2}{a^2+4m^2\pi^2}\\ \Delta_3 &= 2\bar{\nu}\end{aligned} \tag{9.5-39}$$

and the values of the coefficients in the complex Lorenz equations are given by

$$\begin{aligned}\sigma &= 2; \quad b = \frac{8m^2\pi^2}{a^2+4m^2\pi^2}\\ r_2 &= (k\bar{\beta})(a^2\bar{\nu})^{-1}; \quad e = 3r_2\\ r_1 &= 1+\frac{2}{\bar{\nu}^2}\left[\pm\frac{1}{4}k^2(\Delta u)^2a^{-2}+\frac{5}{9}\bar{\beta}^2a^{-4}\right]\end{aligned} \tag{9.5-40}$$

We note finally that $e + r_2 = 4r_2$ and so the periodic solution given in (9.5-28) exists.

Finally the laser example of Section 9.3 also yielded the AB equations. The inclusion of the homogeneous broadening terms originally shown in (9.3-1)-(9.3-3) and conductivity leakage does indeed give the Lorenz model under certain circumstances. This has been investigated closely by Haken and co-workers (see references in Haken, 1978).

9.6 Notes

(a) A specific example of a dissipative instability is to consider the onset of convection in a fluid heated from below; a problem originally studied by Rayleigh (1916). Let us consider the flow occurring in a fluid held between two infinite, flat, horizontal plates a distance L apart. The fluid is heated from below uniformly such that the temperature difference between the top and bottom plates is a constant value ΔT. We will take α to be the coefficient of thermal expansion, κ the coefficient of thermal conductivity, ν the coefficient of viscosity and g the acceleration due to gravity. ψ is the stream function and θ the temperature variation at a point in space and time. It is simplest to consider motions which are two dimensional in the vertical plane $(x$ & $z)$ and in this case the equations governing the flow are:

$$\left(\frac{\partial}{\partial t}+\frac{\partial\psi}{\partial x}\frac{\partial}{\partial z}-\frac{\partial\psi}{\partial z}\frac{\partial}{\partial x}\right)\nabla^2\psi = \nu\nabla^4\psi + \alpha g\frac{\partial\theta}{\partial x} \tag{9.6-1}$$

$$\left(\frac{\partial}{\partial t}+\frac{\partial\psi}{\partial x}\frac{\partial}{\partial z}-\frac{\partial\psi}{\partial z}\frac{\partial}{\partial z}\right)\theta = \frac{\Delta T}{L}\cdot\frac{\partial\psi}{\partial x} + \kappa\nabla^2\theta \tag{9.6-2}$$

In this particular problem, the boundary conditions are important at the upper and lower plates and so we shall take "free" boundary conditions i.e. ψ and $\nabla^2\psi$ are zero at $z = 0$ and $z = L$. To satisfy these we take an oscillatory small amplitude solution of the linearised version of (9.6-1,9.6-2) to be

$$\begin{pmatrix} \psi \\ \theta \end{pmatrix} = [\underline{b}\ \exp i\theta + c.c.]\sin\left(\frac{\pi z}{L}\right) \qquad \theta = kx - \omega t \tag{9.6-3}$$

after the form of (9.1-2,9.1-3). The linear part of (9.6-1,9.6-2) can be written as

$$\begin{pmatrix} \frac{\partial}{\partial t}\nabla^2 - \nu\nabla^4 & -\alpha g\partial/\partial x \\ -\frac{\Delta T}{L}\frac{\partial}{\partial x} & \frac{\partial}{\partial t} - \kappa\nabla^2 \end{pmatrix}\begin{pmatrix} \psi \\ \theta \end{pmatrix} = \begin{matrix}\text{Quadratic} \\ \text{nonlinear} \\ \text{terms}\end{matrix} \tag{9.6-4}$$

Taking $k = \pi a/L$ for simplicity, the determinant of the matrix in (k,ω) space in (9.6-4) gives the dispersion relation:

$$(1+a^2)\left(i\omega - \frac{\nu\pi}{L^2}\right)[-i\omega + \kappa(1+a^2)\pi^2/L^2] + \frac{\Delta T\alpha g a^2}{L} = 0 \tag{9.6-5}$$

which gives a quadratic in ω with an imaginary coefficient in the ω term. The immediate solution of this quadratic is easily seen to be

$$2\omega = i\{-(\nu+\kappa)(1+a^2)\pi^2/L^2 \pm [(\nu-\kappa)^2(1+a^2)\frac{\pi^4}{L^4} + \frac{4\Delta T\alpha g a^2}{(1+a^2)L}]^{\frac{1}{2}}\} \tag{9.6-6}$$

Since the terms in the square bracket are positive definite, ω is pure imaginary: $\omega = i\omega_I$ with $\omega_R = 0$ in the notation of Section 9.1. In this case therefore, there is no harmonic wave motion with growth or decay but a switch from stationary growth to stationary decay. We obviously have many parameters in the problem, the adjustable one being ΔT. To find the curve of neutral stability, as in Section 9.1, we set $\omega_I = 0$ which is equivalent in this case to setting $\omega = 0$ in (9.6-6). After squaring, we obtain very easily the condition

$$\frac{\Delta T\alpha g L^3}{\nu\kappa} = \frac{\pi^2(1+a^2)^3}{a^2} \tag{9.6-7}$$

The expression on the left hand side of (9.6-7) is a dimensionless quantity called the Rayleigh Number R_a

$$R_a = \frac{\Delta T\alpha g L^3}{\nu\kappa} = \frac{\pi^4(1+a^2)^3}{a^2} \tag{9.6-8}$$

R_a now plays the role of μ, the bifurcation parameter of Section 9.1 and the expression in (9.6-8) is now the neutral curve $\mu = R(k)$ (remember that $k = \pi a/L$). The critical point can very quickly be determined from (9.6-8) by finding the minimum of the curve. This is easily seen to occur when $a^2 = \frac{1}{2}$ and $R_{ac} = 27\pi^4/4$ and so (k_c,μ_c) is, in this problem, $(\pi/\sqrt{2}L, 27\pi^4/4)$

(b) Much literature is associated with the nonlinear diffusion equation (9.1-16). Its role as a weakly nonlinear equation in dissipative systems is almost universal. Landau (1959) conjectured heuristically that, at a critical Reynolds number for a viscous fluid flow, the amplitude equation with no spatial dependence should be of the form

$$\partial A/\partial T_2 = \alpha A - \beta A|A|^2 \tag{9.6-9}$$

It was Stuart (1960), however, who showed that (9.6-9) is indeed the relevant amplitude equation near the critical Reynolds number for the specific example of plane Poiseuille flow. Obtaining the neutral curve in this case is more complicated than just setting $\omega_I = 0$ because other space dimensions and boundary conditions cannot be ignored. In this case it is necessary to solve numerically the fourth-order Orr-Sommerfeld ordinary differential equation. The Stuart-Landau equation (9.6-9) describes the evolution of discrete waves, but the inclusion of spatial variation for this category of instabilities leads in general near the critical point to (9.1-16) which we shall repeat (see Davey 1972, Newell 1974)

$$\partial A/\partial T_2 = \alpha_1 A - \beta_1 A|A|^2 + \gamma_1 \partial^2 A/\partial \bar{X}^2 \qquad (9.6\text{-}10)$$

$$\bar{X} = X_1 - (d\omega/dk)T_1 \qquad (9.6\text{-}11)$$

The name of (9.6-10) varies from subject to subject. In theoretical physics, biology and chemistry, it tends to be called a generalized Landau-Ginsburg equation (see the review by Kuramoto (1978) and recent work by Lin and Kahn (1980) on delayed population models). This latter equation arises in superconductivity and has the structure of a quantum Schrödinger equation with $|A|$ as a potential. However, in fluid mechanics, Newell and Whitehead (1969) seem to have been the first to have derived the form of (9.6-10) for the particular case of Benard convection in the region of the critical Rayleigh number (β_1 & γ_1 real). In this case again, for real convection problems, other space dimensions need to be taken into account but the structural form of (9.6-10) is the fundamental one. Stewartson and Stuart (1971) followed by Hocking et al. (1972) showed that (1.14) occurs as the generalization of the Stuart-Landau equation (1.13) when spatial variation is included in the plane Poiseuille flow problem (β_1 real, γ_1 complex). Ball (1977) has shown that when $\gamma_1 > 0$, the solution of (9.6-10) breaks down after finite time (β_1, γ_1 real) when $\beta_1 < 0$. Lange and Newell (1974) have shown that a criterion exists $\beta_{1R}\gamma_{1R} + \beta_{1I}\gamma_{1I} \gtrless 0$ which determines whether the system achieves or does not achieve a monochromatic state. For the NLS equation of Chapter 8, a monochromatic state would be equivalent to having $\beta\gamma < 0$ in that problem. The book by Drazin and Reid (1981) on hydrodynamic stability contains most references on problems in which (9.6-10) occurs.

As one would expect, there are other disciplines where (9.6-10) arises. For instance, Pawlik and Rowlands (1975) have shown that it occurs in the nonlinear propagation of waves in piezo-electric semi-conductors (β_1 complex; γ_1 imaginary) near the critical value of a d.c. field. Newell's (1974) calculation of (9.6-10) was for a scalar L but we have generalized this result to a matrix $\underline{\underline{L}}$ as a corollary to our main analysis in Section 9.2. Two final but crucial points about (9.6-10) should be noted. First, it is first-order in time on the T_2-scale. The T_1-timescale has been absorbed into the travelling frame moving with the group velocity. This arises because of the necessary removal of secularities at $O(\epsilon^2)$. Secondly, the group velocity c_g is single-valued on the neutral curve.

(c) With regard to the second type of instability (Category II), although there are many models known to produce complex conjugate pairs of roots, as a general class this type of instability has received less attention in the literature. Often the two types have been mixed in together despite the fact that the behaviour of the two is markedly different. We give a list of seven different dispersive instabilities:

(i) the inviscid two-layer model for baroclinic instability on a beta-plane (Philips 1954, Pedlosky 1970, 1972);

(ii) the Eady model (continuously stratified) for baroclinic instability on a beta-plane (Eady 1949; Drazin 1970, 1972);

(iii) self-induced transparency (SIT) - lossless ultrashort optical pulse propagattion through two-level atomic media (McCall and Hahn 1967, 1969, Lamb 1971);

(iv) the inviscid two-layer Kelvin-Helmholtz instability (Weissman 1979 and references therein);

(v) buckling of elastic shells and beams (Lange and Newell 1971);

(vi) The symmetric two-stream plasma instability (Pawlik 1977 and references therein);

(vii) inviscid two-layer model for baroclinic instability on the F-plane (no beta-plane) (Pedlosky 1970).

The above list is not comprehensive but gives the most well known ones. Examples (i), (iii) and (vii) are dealt with in Sections 9.3 and 9.4 of this chapter. The Eady model of baroclinic instability is similar to the two layer model but consists of a single fluid which is continuously stratified. The dispersion relation is not polynomial in ω but a complex conjugate pair of roots can nevertheless occur. A full discussion of which amplitude equation is appropriate to which model is given in Gibbon and McGuinness (1981).

(d) A technical difficulty occurs for those dispersive instabilities which have a minimum at the origin; that is, $k_c = 0$. This means that no carrier wave exists as its wavelength would be infinite. In this case, side modes away from zero must be considered. As we showed in Section 9.2, the spatial terms at $O(\epsilon^2)$ are secular everywhere except at the minimum. This means that the X_1-scale must be excluded but a further longer scale $X_2 = \epsilon^2 x$ can be used instead. The final amplitude equation now becomes ($\alpha_3, \beta_3, \gamma_3$ real)

$$i \frac{\partial A}{\partial X_2} = \pm\alpha_3 A + \gamma_3 \frac{\partial^2 A}{\partial T_1^2} - \beta_3 A|A|^2 \tag{9.6-12}$$

which is a type of NLS equation with space and time exchanged. This type of behaviour occurs for the lossless symmetric two-stream plasma instability (example (vi)) and the two layer baroclinic instability when no beta-plane is included (example (vii)).

Section 9.3

The original numerical and analytical calculations and experimental observations of SIT were made by McCall and Hahn (1967, 1969). As an atomic sample, they used a ruby rod cooled with liquid helium. The laser was a Q-switched ruby laser controlled to provide the plane polarised $E(2E) \leftrightarrow 4A_2(\pm 3/2)$ output line with typical pulse widths of 10-20 nanoseconds. Further elegant experiments have been performed by Gibbs and Slusher (1970, 1972) using a 5-10 nanosecond coherent optical pulse obtained from a 202 Hg II laser and a Rubidium 85 sample.

(a) An excellent review of the theoretical spects of ultra-short optical pulse progagation, including the Area Theorem, has been written by Lamb 1971. Many of the relevant experimental and theoretical references can be found in this lucidly written paper. This review played a significant role in the early development of the study of solitons in that it pulled together much of the work on the sine-Gordon equation, particularly concerning Bäcklund Transformations. It was Lamb who traced the older and almost forgotten results buried in Eisenhart 1960 and Forsyth 1959 on the sine-Gordon equation and its connection with surfaces of constant negative curvature in differential geometry. Other papers and reviews on SIT, particularly concerned with the soliton aspect, can be found in Bullough et al. (1973) Bullough (1973, 1975, 1976, 1979).

The ideas concerning the RMB equations were first presented by Eilbeck et al. (1973) and Caudrey et al. (1973) as an alternative to the SIT equations and they were shown to be integrable by IST by Gibbon et al. (1973). The authors who have dealt with the inverse scattering aspect of the SIT equations with and without inhomogeneous broadening are Lamb (1970, 1971, 1973) and Ablowitz, Kaup and Newell (1974). The purely soliton aspects, using Hirota's method, has been dealt with by Caudrey et al. (1973a,b,c, 1973, 1974, 1975).

(b) In both the multiple scales calculation and in the approximation which reduced the Maxwell equation to form the RMB equations, it was found to be necessary to have $\bar{\alpha}$ small. For example $\bar{\alpha} \simeq 0.01$ in the optical frequencies when $n \sim 10^{18}$ atoms/cc. In the experiments of Gibbs and Slusher (1972) in Rubidium vapour n takes on the value of about 10^{12} atoms/cc and so $\bar{\alpha} \sim 10^{-8}$.

Section 9.4

(a) A derivation of equations (9.4-18) for the two layer model has been given by Pedlosky (1970), who goes into considerable detail regarding the approximations under which they are valid. In this paper he derived the AB equations with just the T_1-scale only, (no X_1 or X_2 scales) and so obtained an equation of the form

$$\frac{d^2A}{dT_1^{\,2}} = \pm\,\alpha A - \beta A|A|^2 \tag{9.6-13}$$

In his 1972 paper, Pedlosky then included the X_1-scale to obtain the AB equations. The two layer model as such is obviously a convenient construction, but is used because it is essentially the simplest mathematical model which displays baroclinic instability. Another slightly more complex model is the so-called Eady model (Eady 1949) which allows a continuous stratification in a single fluid with a variety of shears (Drazin 1970). The AB equations arise also in the Eady model with linear shear and stratification (Moroz 1981).

(b) The integrability of the AB equations with reference to the two layer model was proved by Gibbon et al. (1979), and for the Eady model with linear shear and stratification by Moroz and Brindley (1981). Laboratory studies of baroclinic instability using the rotating fluid annulus show a rich variety of motions including periodic, multiply periodic and aperiodic flow. Much of this work has been pioneered by Hide (1958, 1969), using the thermally driven rotating fluid annulus. Hart (1972) has considered the two layer system in the laboratory. For references to recent work, see Hide and Mason (1975) and Hide et al. (1977).

(c) Technically, the choice of $\sin(m\pi y)$ for the y-variation of the stream functions is incorrect since both the y-derivatives of the ψ_i should be zero at $y = 0$ and 1. This involves the use of a Fourier series for the y-variation which complicates the calculation considerably (Smith 1974). Pedlosky has pointed out that, for this particular case in which the fluid is inviscid, the approximation of using $\sin(m\pi y)$ is quite good.

(d) The beta-effect (see also Chapter 5) can be included in annulus experiments by taking sloping end-walls at the top and bottom of the annulus. This experimentally takes into account the earth's curvature. Mason (1977) has made a detailed study of wave motion when sloping end walls are included, and there are significant differences in experiments when sloping end walls are included or excluded. Theoretically we saw that, in the simple two layer model, the inclusion or exclusion of the beta-plane term βy made considerable differences in the theory both in the inviscid case and in the weakly damped case. See Section 9.5 in which the Lorenz model can be derived for no beta-effect but cannot when the beta-effect is excluded. For general references on geophysical fluid dynamics, see the notes of Chapter 5.

Section 9.5

(a) The Lorenz equations were derived in 1963 for the problem described in the notes for Section 9.1 of this chapter; that is, the two dimensional study of fluid convection between two horizontal plates with free boundary conditions at the upper and lower boundaries. The stream function ψ and temperature variation θ between the plates are then expanded as a Fourier series at criticality and inserted into the Navier-Stokes equations (9.6-1) and (9.6-2).

$$\psi = \sqrt{2}\kappa a^{-2}(1+a^2)\left[X(\tau)\sin\left(\frac{\pi a x}{L}\right)\sin\frac{\pi z}{L}\right]\ldots$$

$$\theta = \sqrt{2}\pi^{-1}R_a R_{ac}^{-1}\Delta T\left[Y(\tau)\cos\left(\frac{\pi a x}{L}\right)\sin\left(\frac{\pi z}{L}\right)-\tfrac{1}{2}Z(\tau)\sin\left(\frac{2\pi z}{L}\right)\right]\ \ldots \qquad (9.6\text{-}14)$$

The coefficients outside the amplitudes are there as a scaling device. X, Y and Z are Fourier amplitudes where

$$\tau = \pi^2 L^{-2}(1+a^2)\kappa t$$

which is a dimensionless time variable. Insertion of these into the Navier-Stokes equations and truncating at this number of modes (one in ψ

and two in θ yields the Lorenz equations (9.5-11) with

$$r_a = R_a/R_{ac} \qquad \sigma = \nu\kappa^{-1} \qquad b = 4(1+a^2)^{-1}$$

The mode expansion and truncation procedure described here is a typical method in such problems. In theory, the larger the number of modes one includes, the more accurate the model is, although this assumption rests on the uniform convergence of the expansion which does not always occur. However, Howard and Krishnamurti (1980) have developed a 6 variable truncated convection model which contains the Lorenz model on a three dimensional (3 mode) invariant sub-space.

(b) The application of the correct side-wall boundary conditions for the two-layer and Eady models introduces an infinity of Z-functions into the Lorenz model. They are modified to become

$$\begin{aligned} \dot{X} &= -\sigma X + \sigma Y \\ \dot{Y} &= (r-\sum_{n=1}^{\infty} Z_n)X - aY \\ \dot{Z}_n &= -b_n Z_n + \tfrac{1}{2}\gamma_n (X^*Y+XY^*) \end{aligned} \qquad (9.6\text{-}15)$$

In the real case $a = 1$ Pedlosky and Frenzen (1981) have considered the behaviour of (9.6-15) in comparison to the single Z case.

(c) An account of the ideas of how the AB equations plus weak dissipative and dispersive effects can be found in Gibbon and McGuinness (1981).

10. NUMERICAL STUDIES OF SOLITONS

10.1 Introduction

The diligent reader who has reached this chapter may by now have formed the opinion that all nonlinear wave equations with solitary wave solutions have multi-soliton solutions, and that the Inverse Spectral Transform is a universally powerful tool in the study of such problems. Those with more acute memories will have kept in mind the warning given at the end of Chapter 1 and will have realized that the techniques developed in this book apply only to a variety of interesting special cases. There are many other equations of equal interest for which the methods we have described cannot be applied. Despite the enormous progress made in the theory of the IST in the last decade, there has been relatively little development in the analysis of equations for which no IST tools are available. Indeed, it is in this area that many of the unsolved problems related to soliton-like behaviour occur.

When detailed analytic tools are lacking, a useful approach is to study the evolution of solutions numerically. Historically the computer was the tool which prompted the initial research into the subject: Zabusky (1981) has recently surveyed both initial and more recent developments and has argued a strong case for the innovative contribution which numerical studies make to nonlinear problems in many areas.

In this chapter we describe suitable numerical methods for nonlinear wave equations with solitary wave solutions. These methods can be tested on equations for which analytic results and exact solutions are known, and then applied to equations for which analytic results are not available. A major part of this chapter will be devoted to summaries of the results of various numerical investigations of such equations. Often the exact soliton behaviour possessed by equations studied elsewhere in this book no longer hold - instead we shall find a more diffuse sort of "quasi-soliton behaviour". In addition the equations with exact multi-soliton solutions in one space dimension may exhibit some form of "soliton-like" behaviour when extended to two or more space dimensions.

Lack of space will prevent more than a brief overview of the literature, with many gaps and omissions. We begin with a brief introduction to some common numerical techniques to give a framework for the subsequent discussions.

10.2 Basic Numerical Methods

As a tool for the study of partial differential equations, numerical analysis offers a confusing variety of techniques. For specific equations, the question of the 'best' numerical method is an extremely complicated one. Even a partial answer to this question depends on many factors: for example the final accuracy required for a specific range of the independent variables, the limitations of time and storage space, and machine word length. Equally important is the amount of time and effort available for the development of the appropriate software. Most computational physicists or applied mathematicians are content to derive one reasonably efficient method and few detailed comparisons are made with other methods. For this reason no outstanding claims are made for the methods described below except in the cases where meaningful tests have been carried out. Most work is required, not just in the theoretical analysis of the accuracy and efficiency of different methods, but in practical field tests involving near-optimum codes. A further consideration which is becoming increasingly important is the suitability of selected algorithms for the new generation of parallel processing computers now becoming widely available.

In this brief section we can do no more than outline some basic numerical tools. Full descriptions will be found in the books by Richtmyer and Morton (1967), Mitchell and Griffiths (1980), Ames (1977), Gottlieb and Orszag (1977), and Meis and Marcowitz (1978).

Most of the equations considered in this book can be put either in the form

$$u_t = L(u) \tag{10.2-1}$$

or in the form

$$u_{tt} = L(u) \tag{10.2-2}$$

where $L(u)$ is some general nonlinear differential operator in the space variable x. Initially we shall consider the one space-dimensional case, and defer the extensions to higher space dimensions to later sections. In the same spirit we shall normally consider u to be a scalar field, although many of the methods described here generalize to the case of a vector field. Equation (10.2-2) can of course be written as a two component form of (10.2-1), but it is useful to consider methods designed to deal with (10.2-2) directly.

In order to treat the equations numerically we must replace boundary conditions at infinity by conditions at some finite boundary. To minimise the effect of an infinite boundary, we can take the boundary a large distance from regions where u is nonconstant. At these boundaries we take u or its space derivative zero. Another more sophisticated modification to this method is to use so-called outflow boundary conditions: we assume any part of the solution to reach the boundary satisfies some linear or nonlinear uni-directional wave equation which approximates the original wave equation in some way. This reduces the problems caused by reflections at finite boundaries. Of course other boundary conditions may be appropriate to model some physical situations, even though such conditions may invalidate the theoretical treatments described in earlier chapters.

A useful special case of the above equations is obtained by taking $L(u) = au_{xx}$. With this choice and $a = 1$ then (10.2-1) gives the linear heat equation discussed in Chapter 1. With $a = i$ in (10.2-1) we get a linear Schrödinger equation, and with $a = 1$ in (10.2-2) we get the linear wave equation.

Two basic tchniques are used to reduce the problem to one involving only a finite number of parameters: the function approximation approach and the finite difference approach.

10.2.1 Function Approximation Methods

This method, as its name implies, approximates the exact solution $u(x,t)$ by an approximate solution $\tilde{u}$ defined on a finite dimensional subspace (usually in the x variable only):

$$u(x,t) \approx \tilde{u}(x,t) = \sum_{i=0}^{n} c_i(t)\phi_i(x) \tag{10.2-3}$$

The $\phi_i(x)$ are appropriately chosen basis functions. One common choice for these are the trigometric functions, leading to a finite Fourier Transform or spectral method. Another popular choice are piecewise polynomial functions with a local basis, giving the Finite Element Method. In the simplest version of the Finite Element Method the function $\tilde{u}$ is a piecewise linear function, and the basis functions $\phi_i(x)$ are the so-called "hat" functions shown in Fig. 10-1, and described by the equations

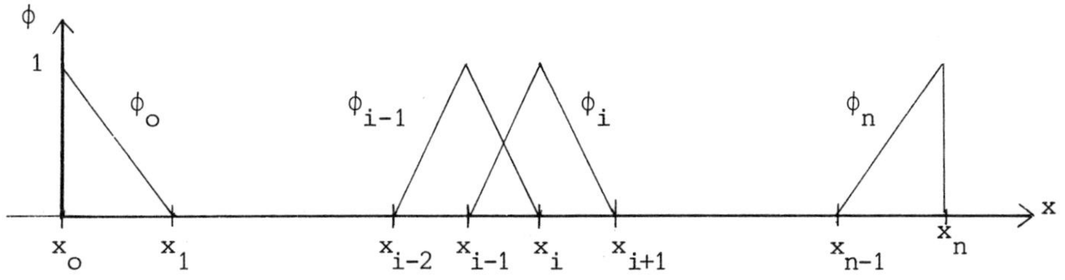

Figure 10-1: Basis functions for piecewise linear $\tilde{u}$

$$\begin{aligned}\phi_i(x) &= (x - x_{i-1})/(x_i - x_{i-1}) & x_{i-1} \leqslant x \leqslant x_i \\ &= (x_{i+1} - x)/(x_{i+1} - x_i) & x_i \leqslant x \leqslant x_{i+1} \\ &= 0 & \text{elsewhere}\end{aligned} \tag{10.2-4}$$

Here the x_i are the "nodes", or the points at which a derivative of $\tilde{u}$ becomes discontinuous: usually they are equally spaced in x. Similar diagrams can be drawn for the basis functions of higher order piecewise polynominals; the essential point about the choice of basis functions is that they are only nonzero over a small range of the x-axis, which leads to efficient numerical handling of the resulting set of simultaneous equations.

Once a specific choice of subspace has been made, and assuming the $\phi_i(x)$ have been chosen to satisfy the boundary conditions, substituting (10.2-3) in (10.2-1) gives

$$\sum_{o}^{n} \dot{c}_i(t)\phi_i(x) - L\{\sum_{o}^{n} c_i(t)\phi_i(x)\} = r(x,t) \tag{10.2-5}$$

where the dots represent differentiation with respect to t. If $\tilde{u}$ was an exact solution, the residual $r(x,t)$ would be zero. To obtain equations for the coefficients $c_i(t)$ we adopt the criteria that the residual is to be made small in some sense. One common choice is the Galerkin approach

$$\int r(x,t)\psi_j(x)\,dx = 0 \qquad j = 0,1,\ldots n \tag{10.2-6}$$

where the functions $\psi_j(x)$ are known as "test" functions: often, but not always, these test functions are chosen to be the same set as the basis functions $\phi_i(x)$. The conditions (10.2-6) lead to a system of ordinary differential equations in the coefficients $c_i(t)$. Here the integral is over the finite region outside which u is assumed to be zero.

As an example, consider the linear heat equation with both basis and test functions given by (10.2-4), and with equally spaced nodes separated by a step length h. In this case the integrals in (10.2-6) are easily evaluated, and the resulting ordinary differential equations are in the tridiagonal form

$$\dot{c}_{i-1} + 4\dot{c}_i + \dot{c}_{i+1} = \frac{6a}{h^2}(c_{i-1} - 2c_i + c_{i+1}) \qquad i = 1,\ldots n \tag{10.2-7}$$

An important departure from the strict definition of the Galerkin method concerns the treatment of nonlinear terms under the integral in (10.2-6). Terms of the form

$$\int F\{\sum_o^n c_i(t)\phi_i(x)\}\psi_j(x)\,dx \tag{10.2-8}$$

where $F(\tilde{u})$ is a nonlinear function, require a numerical quadrature to evaluate. Instead one can use the so-called "product approximation" method (Christie et al., 1981) and replace (10.2-8) by

$$\sum_{i=o}^{n} F(c_i)\int \phi_i(x)\psi_j(x)\,dx$$

The product approximation method, used with local basis functions like the hat functions for obvious reasons, gives much simpler algorithms and even in some cases an unexpected bonus of higher accuracy.

A useful survey of finite element methods for soliton equations is given by Mitchell and Schoombie (1981).

An alternative approach to (10.2-6) is to set the residual to be zero at a given set of points $x_o,\ldots,x_n$.

$$r(x_j,t) = 0, \qquad j = 0,1,..,n. \tag{10.2-9}$$

This is the collocation method: although little used at present in partial differential equations, it is gaining popularity in the field of ordinary differential equations. The collocation method is formally equivalent to the Galerkin method when the test functions are chosen to be Dirac delta functions.

10.2.2 Finite Difference Methods

Function approximation methods are relatively new, and for many applications the more familiar finite difference methods are still the most popular. For this approach we seek approximations u_m^n to the original function $u(x,t)$ at a set of points x_m, t_n on a rectangular grid in the x,t plane, where $x = hm$ and $t = kn$. By expanding function values at grid points in a Taylor series, approximations to the differential equation involving algebraic relations between grid point values can be obtained. Assuming some suitable approximation L_m^n for $L(u)$ at the grid point (m,n) has been derived, various types of scheme can be constructed according to the type of approximation for the time derivative. Two simple choices, together with the appropriate "computational molecule" (Ames 1977) for equation (10.2-1) are shown below

$$u_m^{n+1} - u_m^n = kL_m^n \tag{10.2-10}$$

$$u_m^n - u_m^{n-1} = kL_m^n \tag{10.2-11}$$

In the specific case where $L(u) = au_{xx}$, then the usual finite difference approximation to this term is

$$L_m = (u_{m+1}^n - 2u_m^n + u_{m-1}^n)/h^2 \tag{10.2-12}$$

Equations (10.2-10) and (10.2-11) are known as the simple explicit and simple implicit method respectively. Here 'explicit' means that each $n+1$ time step value is given directly, whereas 'implicit' means that a set of algebraic equations has to be solved at each time level to give the set of grid point values. Although the latter case would seem to be much more costly in computational time, the resulting equations have a simple banded structure which enables very efficient methods to be constructed. Any small overheads involved in the implicit method schemes are usually balanced by the fact that such methods are often much more stable (see below), and a larger time step can be taken in the scheme.

By combining (10.2-10) at level n and (10.2-11) at level $(n+1)$ we get the more accurate Crank-Nicholson scheme, another implicit method.

$$u_m^{n+1} - u_m^n = \tfrac{1}{2}k(L_m^{n+1} + L_m^n) \tag{10.2-13}$$

Alternatively we can combine (10.2-10) and (10.2-11) at the nth level to get a three level scheme, the so-called leapfrog scheme

$$u_m^{n+1} - u_m^{n-1} = 2kL_m^n \tag{10.2-14}$$

One further permutation which provides a useful scheme is the so-called Hopscotch scheme (Gourlay, 1971). This is obtained by applying (10.2-10) at odd values of $(n+m)$ along the nth time level and (10.2-11) at even values of $(n+m)$ along the $(n+1)$th time level. This cunning combination can make (10.2-11) explicit, since two of the three unknowns have been calculated during the pass of the scheme (10.2-10).

The computational molecules shown above are for the case where L is a second order operator and the simplest difference schemes are used. For more complicated cases such as the KdV equation some modification will be necessary. Not all the schemes are independent, for example the leapfrog scheme (10.2-14) is in some circumstances equivalent to two succesive applications of the Hopscotch scheme.

A hybrid method which is related to the finite difference method is the so-called "method of lines". In this scheme, only the spatial variable(s) are discretised, and the resulting (large) system of ordinary differential equations is solved using a standard library package. At first sight this method appears to be an attractive and simple approach to the numerical solution of partial differential equations. However an unwise choice of o.d.e. solver, or an over-optimistic error bound input, can give rise to excessive computing bills. This method is no short cut to efficient numerical approximations of the problem, and should only be used with caution.

The truncation error of a finite difference scheme is defined as the difference between the partial differential equation and its finite difference approximation. If this error is expanded in a Taylor series in h and k, the order of the scheme is the order of the lowest order terms in h and k. For example the Crank-Nicholson scheme is second order in both h and k since its truncation error is $O(h^2) + O(k^2)$.

Another hybrid method which gives much greater accuracy than the method of lines is the so-called pseudo-spectral method. In this scheme, the solution is also calculated on a spatial grid, using some simple finite difference scheme in the time variable, but the approximations to the spatial derivatives are calculated in Fourier transform space. The tranforms in and out of transform space can be calculated efficiently using the Fast Fourier Transform (FFT) algorithm. Fornberg (1981) has shown that this method is in a sense equivalent to a finite difference scheme with **infinite** order accuracy in the space step h.

10.2.3 Convergence, Consistency, and Stability

If the function approximation method, with local basis functions, is chosen, and the ordinary differential equations generated by (10.2-6) or (10.2-9) are solved by some suitable discretisation in time, then the end result is a set of difference equations in the coefficients $c_m^n = c_m(t_n)$ very like finite difference equations. Whatever approach is chosen, one would ideally like to show that the method is convergent, i.e. the global error for some fixed and finite value of t tends to zero as the step lengths (nodal separations) tend to zero, or as the number of basis functions tend to infinity. Unfortunately, the proof of convergence is usually extremely difficult, even when possible, and a satisfactory theory is only available in general for linear equations (Meis and Marcowitz, 1978). Most discussions of nonlinear methods are based on an analysis of the equations linearised about some constant solution. In all the following we shall adopt such a course, and discuss only necessary conditions for convergence, instead of the more tricky and technical problem of sufficiency.

To ensure convergence, methods in general must be consistent and stable. A consistent method is one in which the truncation error tends to zero as the step lengths in the problem tend to zero. Sometimes this limit must be taken in a specific way; for example in the Hopscotch method the limit must be such that $k/h \to 0$ as $k,h \to 0$. Stability means that some norm of the approximate solution must remain bounded as $n \to \infty$, where $nk = t$ for fixed t. Even stability is usually difficult to prove for nonlinear difference equations, and the usual approach is to study the stability of a linearised version of the equations. We present here a simple description of one approach, the method of Fourier stability analysis, and refer the reader to the literature for more details of this and other stability analysis methods.

In the Fourier method, the discrete approximation u_m^n is decomposed into Fourier modes, and the behavior of each mode analysed. A typical mode will behave like $\mu^n \exp(i\beta m)$, where i is the square root of -1, and β is a discrete Fourier variable in the range $[0, \pi]$. Substituting such a term into the difference equation gives an equation for μ as a function of β and the step lengths of the problem, and a necessary condition for stability is that $|\mu| \leqslant 1$ for all β. For example, substituting this term into the simple explicit scheme (10.2-10) for the linear heat equation gives

$$\mu = 1 - 4(k/h^2)\sin^2(\beta/2) \tag{10.2-15}$$

We see that, for this scheme, stability requires $k \leqslant h^2/2$. In this case we say the scheme is conditionally stable, i.e. it is only stable if the time step k is bounded by some function of the space step h. For the heat equation, we leave it as an exercise for the reader to show that the simple implicit, Crank-Nicholson, and Hopscotch scheme are all unconditionally stable, whereas the leapfrog scheme is unstable for all finite k. However the leapfrog scheme is stable for the linear Schrödinger equation when $k \leqslant h^2/4$ and for the linear wave equation when $k \leqslant h$. For three-level schemes such as the leapfrog scheme, the Fourier method leads to a quadratic in the "amplification coefficient" μ. Although in these simple examples the resulting roots can be easily analysed, in more complicated cases the results of Miller (1973) are often useful.

For systems of equations the amplification factor becomes an amplification matrix: it is then necessary to show that the eigenvalues of this matrix are bounded in absolute value by unity.

When stability requirements impose a limit on the time step size, this has important practical consequences, since for smaller k a larger amount of computer time will be required to reach a given value of t. The size of k and h is also restricted by accuracy considerations. These restrictions are especially important in higher space dimensions.

10.3 Nonlinear Klein-Gordon Equations

The equations

$$u_{xx} - u_{tt} = F(u) \tag{10.3-1}$$

for various choices of F, have played an important role in the study of solitons. Some of the most common are

$$F(u) = \sin u \tag{10.3-2}$$

$$F(u) = -u + u^3 \tag{10.3-3}$$

$$F(u) = \pm(\sin u + \tfrac{1}{2}\sin(u/2)) \tag{10.3-4}$$

We have already treated the sine-Gordon (SG) equation (10.3-2) in some detail in Chapters 1 and 7: it is completely integrable and has exact analytic solutions describing the interaction of N solitons (kinks). The phi-four equation (10.3-3) has also been mentioned in earlier chapters: it is named from the Lagrangian from which it is derived (although we have used u instead of ϕ as the independent variable here!) It is an important model in solid state physics (Bishop et al., 1980) and in high energy particle physics (Makhankov, 1978). The Double sine-Gordon (DSG) equation (10.3-4) with the +ve sign has applications in nonlinear optics (Duckworth et al. 1976, Bullough and Caudrey 1978); with the -ve sign it is found in nonlinear optics and in the study of the B phase of liquid helium (Bullough and Caudrey 1978, Kitchenside et al. 1979).

All these equations have kink solutions (different asymptotic values as $x \to \pm\infty$), and are Lorentz invariant. Despite the similarity between these equations, the theoretical tools developed earlier in this book apply only to the sine-Gordon equation (10.3-2). For the other equations a numerical study is necessary.

10.3.1 The Sine-Gordon Equation

Historically, Perring and Skyrme (1962) were the first to investigate the SG equation numerically. They considered two simple schemes. First was a simple leapfrog method, equivalent to

$$u_m^{n+1} = -u_m^{n-1} + r^2(u_{m+1}^n + u_{m-1}^n) + 2(1-r^2)u_m^n - k^2 \sin u_m^n \qquad (10.3\text{-}5)$$

where $r = k/h$. Numerical tests, and a linear stability analysis, showed that this scheme was unstable for $k = h$, but it was found that reducing k to $0.95h$ was sufficient to remove this instability. The second scheme involved re-writing the equation as a pair of first order equations, posibly in the form

$$\begin{aligned} u_x + u_t &= v \\ v_x - v_t &= \sin u \end{aligned} \qquad (10.3\text{-}6)$$

and introducing new variables $\eta, \xi = t \pm x$ so that (10.3-6) become

$$u_\eta = \tfrac{1}{2}v \,, \qquad v_\xi = -\tfrac{1}{2}\sin u \qquad (10.3\text{-}7)$$

the so-called characteristic form. Since the characteristics in this case are straight lines, (10.3-7) can be solved by standard prediction-corrector techniques for ordinary differential equations (Ames, 1977). This method, as used by Perring and Skyrme, was probably based on the familiar trapezium rule approximation which gives a second order accurate method. Although slightly more accurate than (10.3-5) it involves more work since an iteration of the corrector step is required at each time step.

Using both these schemes the kink-kink interaction was investigated and an analytic formula was derived following the numerical results. The bound state kink-antikink pair (meson/bion/breather) was also investigated and a numerical study showed that a kink-bion interaction was also stable. The analytic solutions obtained by Perring and Skyrme on the basis of their numerical study had been derived earlier and independently by Seeger et al. (1953).

For many years numerical interest in the SG equation declined with the developments of the analytic results described elsewhere in this book, but interest in other NLKG equations led to a revival of computational studies in the mid '70's. One of the most useful numerical schemes to emerge was a simple modification to (10.3-5), due to Ablowitz, Kruskal and Ladik (1979). They showed that the leapfrog scheme can be stabilized for $k = h$ by using a space average $\tfrac{1}{2}(u_{m+1}^n + u_{m-1}^n)$ instead of u_m^n in the final term of (10.3-5). This gives, for $k = h$

$$u_m^{n+1} = -u_m^{n-1} + u_{m+1}^n + u_{m-1}^n - h^2F\{\tfrac{1}{2}(u_{m-1}^n + u_{m+1}^n)\} \qquad (10.3\text{-}8)$$

This modification reduces the number of multiplications by one, but more importantly the term in u_m^n is now absent, and calculations can be performed on a diagonal grid involving only even (or odd) values of $(n+m)$. This reduces the running time by a factor of 2. (An important practical point to note is that if the results on both independent diagonal grids are calculated, the solutions on the two different grids can get slightly out of step, producing an apparently oscillating solution with a similar form to that produced by an unstable scheme. This is an another reason why only one of the two grids should be retained.) In practice only two time levels need to be stored and the final code is extremely simple and efficient. Numerical tests carried out by Eilbeck (1978) have confirmed that at least for medium size step lengths ($h \approx 0.1$) the scheme (10.3-8) is more efficient than both second and fourth order predictor-corrector methods based on characteristics. We have used the scheme (10.3-8) to calculate all the figures produced in this section.

Another interesting numerical scheme for nonlinear KG equations has been proposed by Strauss and Vazquez (1978). Their scheme, in the one space dimensional case, is the implicit finite difference scheme

$$u_m^{n+1} = -u_m^n + u_{m+1}^n + u_{m-1}^n - \frac{h^2\,[G(u_m^{n+1}) - G(u_m^{n-1})]}{u_m^{n+1} - u_m^{n-1}} \tag{10.3-9}$$

where

$$G'(u) = F(u)$$

This scheme has the advantage that it has a conserved "energy" given by

$$\begin{aligned} E &= \frac{h}{2}\sum_m (u_m^{n+1} - u_m^n)^2/h^2 + \frac{h}{2}\sum_m (u_{m+1}^{n+1} - u_m^{n+1})(u_{m+1}^n - u_m^n)/h^2 \\ &\quad + \frac{h}{4}\sum_m \{(u_m^{n+1})^2 + (u_m^n)^2\} + \frac{h}{2}\sum_m \{G(u_m^{n+1}) + G(u_m^n)\} \end{aligned} \tag{10.3-10}$$

In practice the scheme (10.3-9) gives very similar results to the simpler scheme (10.3-8), but no detailed comparisons of the two schemes have been made.

Although analytic results for the SG equation provide a full description of solutions on unbounded domains, in real life the equation is used to model physical systems with finite boundaries and fixed boundary conditions. In addition, physical models often have extra terms involving dissipation and energy sources. With these complications, the exact theory breaks down, although perturbation theory can often give useful results in special cases. For more general problems, numerical calculations are a useful tool. Recently much work has been done on modelling Josephson junctions using these techniques. Some results on the finite boundary problem are outlined in papers by the Salerno group, cf. Constabile et al. (1978). Currently there is much interest in a perturbed sine-Gordon equation

$$u_{xx} - u_{tt} = \sin u + \alpha u_t - \beta u_{xxt} - \gamma \tag{10.3-11}$$

which models Josephson junctions. In this equation, the term in α represents a dissipation term, the term in β is another damping term due to surface impedance effects, and γ is a driving term representing a bias current imposed on the system. More details are given in the papers by Christiansen et al. (1981), Lomdahl et al. (1981) and in Lomdahl's Ph.D. thesis (1982), and papers cited therein. This study used an implicit second order finite difference scheme: details are also given in

Lomdahl's thesis.

The effects of dissipation and driving terms are fairly predictable: kinks are slowed down or speeded up respectively, and when both α and γ terms are present, an asymptotic velocity is reached where these two terms balance. The numerical results in this case are in good agreement with simple perturbation schemes based on a Hamiltonian approach.

Numerical studies on the full version of (10.3-11) show another interesting effect: a 'bunched fluxon' mode, in which two or more kinks bunch together and shuttle back and forth across the Josephson junction in a stable configuration. A typical 2-kink solution is plotted in Fig. 10-2.

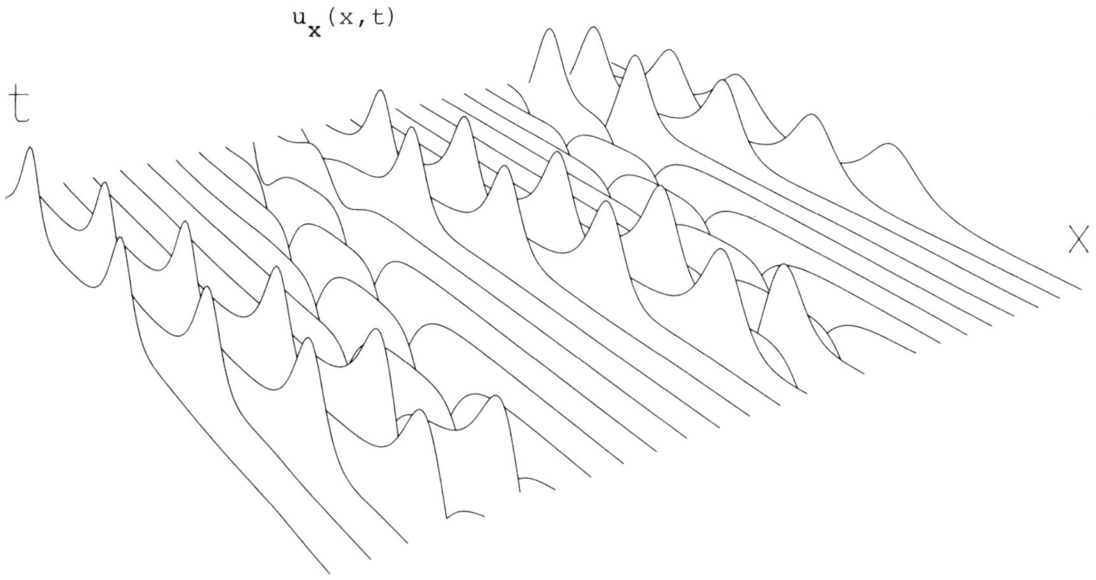

Figure 10-2: Bunched fluxon mode in the Josephson junction

Note that in this figure it is the spatial derivative of $u(x,t)$ that is plotted.

The calculations reported in the references cited above also reveal good agreement with experimental measurements of voltage current characteristics and oscillator power outputs on real Josephson junctions.

Computer film buffs may like to know that many of the results for the unperturbed and perturbed sine-Gordon equation are now illustrated on a 16 mm cine film (Eilbeck and Lomdahl, 1982). This film also shows some of the solutions of the sine-Gordon equation in 2+1 dimensions discussed in Section 10.6.

10.3.2 The Phi-four Equation

The phi-four equation has been studied by a number of groups. This equation has the single kink (antikink) solution

$$u(x,t) = \pm\tanh(\xi/\sqrt{2}), \qquad \xi = (x - vt)(1 - v^2)^{-\frac{1}{2}} \tag{10.3-12}$$

with the kink (antikink) solution taking the +(-) sign. Note that the kink (antikink) takes the $u = -1$ ($u = 1$) solution to the $u = 1$ ($u = -1$) solution. In contrast to the SG equation, the phi-four equation has only the two zero-energy "vacuum states" $u = \pm 1$. The only topologically possible initial conditions involving two kinks are a kink and an anti-kink.

Kudryavtsev (1975) was the first to publish the results of a numerical study of the collision of a kink and an anti-kink for a fixed centre-of-mass velocity of 0.1. He showed that at this energy the two kinks formed a long-lived oscillating bound state, which decayed slowly by radiating energy to infinity. He also reported that for higher collision velocities the kinks repelled each other, and part of their energy was lost to radiation. We have already shown a high energy kink-antikink collision for the phi-four case in Fig. 1-14, and for the low energy case in Fig. 1-15. These results clearly show the lack of exact soliton properties for this equation.

Aubry (1976) was the first to show that the formation of a bound-state kink-antikink pair was not a simple threshold effect, but exhibits a more complicated dependence on the centre of mass velocity v. Subsequent studies have revealed an increasingly complex picture of the kink-antikink interaction. The most recent and detailed study is due to Campbell and co-workers at Los Alamos, and is yet unpublished. The Los Alamos group carried out a series of calculations using a fourth order method, for the complete range of $0 < v < 1$, in steps of 0.001. The results show that for $v > .258$, the kink and antikink bounce off each other, in a similar way to Fig. 1-14, leaving behind energy in a variety of oscillating modes and in low amplitude radiation. For small velocities, $v < .194$, a bound state breather/bion solution was always formed, as shown in Fig 1-15. Figure 10-3 shows the central value of $u(x,t)$ as a function of t in such a collision. Note that the oscillating state is remarkably stable over a large number of oscillations, even though energy is radiating away to the boundaries throughout the interaction.

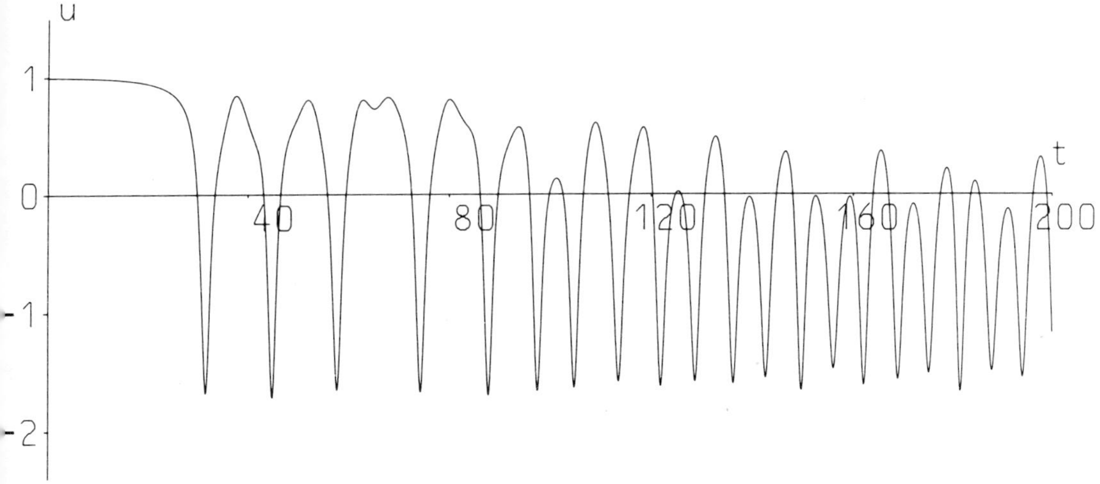

Figure 10-3: u(0,t) for kink-antikink collision, v = 0.1, phi-four model

For values of v between 0.193 and 0.258 there is a sequence of regions in which the two kinks are alternatively reflected or trapped. Figures 10-4 and 10-5 show two calculations on either side of the boundary between two such regions.

Note that for $v = 0.223$, after the initial collision a quasi-stable oscillating state appears. For the case $v = 0.224$, the system passes through **two** oscillations before separating. Campbell et al. report a total of 8 reflection bands, separated

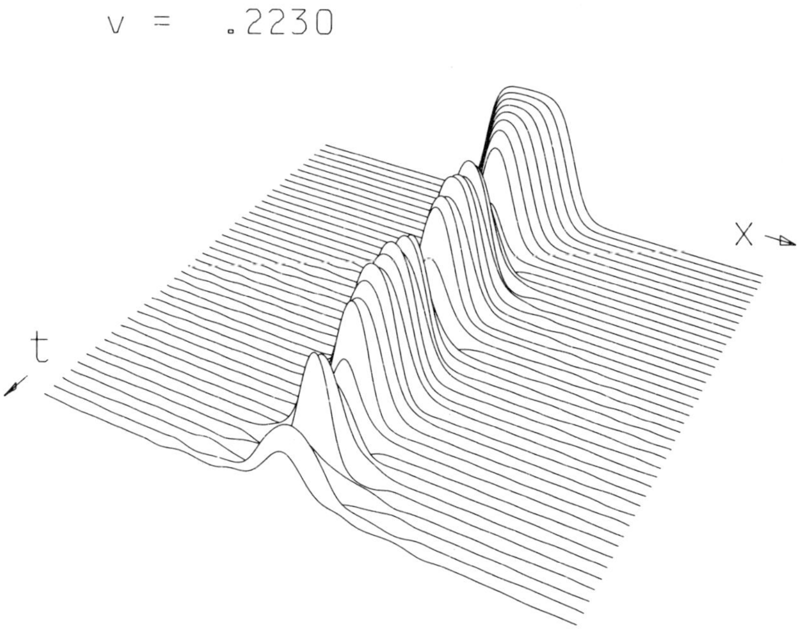

Figure 10-4: Kink-antikink collision, phi-four model, v = 0.223

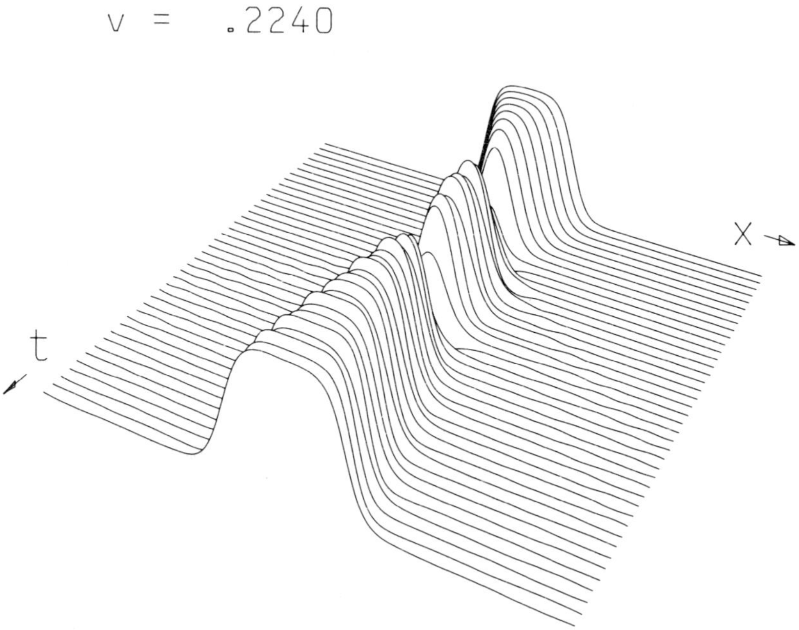

Figure 10-5: Kink-antikink collision, phi-four model, v = 0.224

by capture bands, some bands being only 0.001 thick. More detailed calculation reveals even more detailed band structure. Cambell (unpublished) has put forward a convincing theoretical explanation for these attracting and repelling windows based on a resonance mechanism involving linear oscillation modes excited by the kink-antikink collision. Similar results for a modified sine-Gordon equation have also been reported by Peyrard and Remoissenet (unpublished).

Further work by the Los Alamos group has also shown numerical evidence, based on a Newton-Kantorowitz quasilinearisation, for an exact breather solution similar to the sine-Gordon breather. No analytic form for this solution has yet been found, but numerical tests show strong evidence for stability.

10.3.3 The Double Sine-Gordon Equation

We turn now to the Double sine-Gordon equation, considering the +ve sign first. This has a stable kink solution (Ablowitz et al., 1979)

$$u = -4\ \arctan(\sqrt{5}\operatorname{cosech}\theta)$$
$$\theta = \tfrac{1}{2}\sqrt{5}(x - vt)(1 - v^2)^{-\frac{1}{2}} \qquad (10.3\text{-}13)$$

which can be set into a more interesting "double-kink" form

$$u = 4\arctan[\exp(\theta+\Delta)] + 4\arctan[\exp(\theta-\Delta)] \qquad (10.3\text{-}14)$$

with

$$\Delta = \ell n(\sqrt{5}+2) = 2.1180339\ldots \qquad (10.3\text{-}15)$$

Note that although (10.3-14) looks like a double-kink, it is a single travelling wave solution with the fixed value of Δ given by (10.3-15). A plot of the spatial derivative of u shows its double peaked structure clearly (see Fig. 10-6).

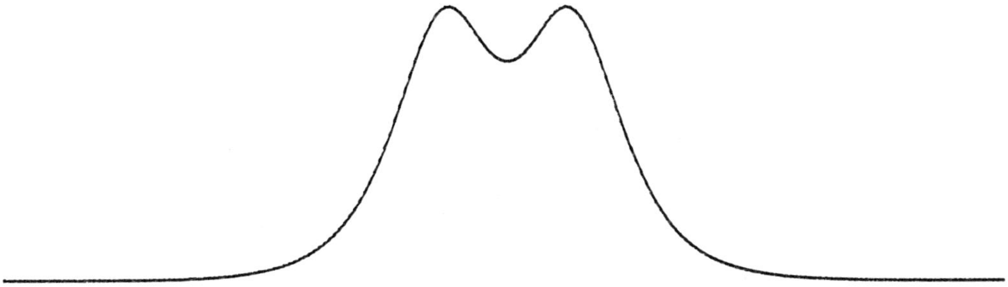

Figure 10-6: Plot of x-derivative of (10.3-14)

The kink (10.3-14) takes u through 4π from any integer multiple of 4π to the next.

With a different choice for Δ than that given by (10.3-15) the solution (10.3-14) is no longer an exact travelling wave. Bullough & Caudrey (1978) have studied the equation numerically using a high order predictor-corrector scheme integrating along characteristics. They studied the DSG equation in a different coordinate system related to the nonlinear optics applications, which explains why their equation for the kink differs from the form given above. With a small perturbation on Δ they found that the peaks oscillated about their exact travelling wave separation, and they called these objects "wobblers". No analytic solution for such behaviour has yet been found. When a large perturbation was applied to Δ , a curious form of leapfrogging behaviour was observed, in which each peak grows, overtakes the other, and then dies away, this cycle repeating indefinitely. They also ran two kinks on a collision course, and found that to within reasonable numerical accuracy the kinks behaved as true solitons and regained their form after collision. However when a kink and antikink solution were collided, there was a clear energy loss by radiation, suggesting that no true 2-kink solution exists. This was confirmed by the studies of Ablowitz et al.

Ablowitz et al. (1979) studied the +ve DSG equation in the more general form

$$F(u) = \sin u + \lambda \sin 2u \tag{10.3-16}$$

This equation, with the choice of $\lambda = 2$ can be rescaled into (10.3-4). They found that for collisions of kink-antikink pairs there was a very small energy loss to radiation at high collision velocities, but the radiation losses increased at lower collision velocites. For some values of λ, bound states were formed at low collision velocities with similar properties to the phi-four bound states discussed earlier.

Kitchenside et al. (1979) also studied the -ve sign version of the DSG equation. This version has two different forms of kink solutions, one with a jump of $4\pi - 2\delta$ and the other with a jump of 2δ respectively, with analytic forms given by:

$$u = 2\pi + 4\arctan(\sqrt{3/5}\tanh\theta/2)$$
$$u = 4\arctan(\sqrt{5/3}\tanh\theta/2) \tag{10.3-17}$$

Here the constant δ is $2\cos^{-1}(-1/4)$, and

$$\theta = \kappa(x - vt), \qquad \kappa = \sqrt{15/16}\,(1 - v^2)^{-\frac{1}{2}}$$

With appropriate boundary conditions only certain kink and antikink initial configurations are topologically possible. However the number of possible combinations is still large and we refer the reader to the original paper for details of the results. In all cases it was found that kink-kink and kink-antikink collisions gave rise to some energy losses due to radiation, so it seems unlikely that any exact multi-kink solutions exist. As in the previous cases considered, at low enough collision velocities a bound-state "breather" solution could be formed.

10.4 'Long Wave' Equations

We have already studied the KdV as an example of a long wave equation in some details in earlier chapters. In fluid mechanics such equations arise when the waves in question are long compared with the depth of fluid in question, or compared with some other typical scale occuring in the problem. However such physical assumptions are often ignored in the subsequent analyses.

10.4.1 The KdV and Related Equations

In laboratory co-ordinates the KdV equation takes the dimensionless form

$$u_t + u_x + uu_x + u_{xxx} = 0 \tag{10.4-1}$$

The linear term in u_x can be removed by a transformation to a frame moving with unit velocity to give the more familiar version

$$u_t + uu_x + u_{xxx} = 0 \qquad (10.4\text{-}2)$$

The KdV equation was first studied numerically by Zabusky and Kruskal (Zabusky 1967, Zabusky and Kruskal 1965). They used the following leapfrog finite difference scheme

$$\begin{aligned} u_m^{n+1} = u_m^{n-1} - \frac{k}{3h}(u_{m+1}^n + u_m^n + u_{m-1}^n)(u_{m+1}^n - u_{m-1}^n) \\ - \frac{k}{h^3}(u_{m+2}^n - 2u_{m+1}^n + 2u_{m-1}^n - u_{m-2}^n) \end{aligned} \qquad (10.4\text{-}3)$$

The space average term for u in the finite difference approximation for uu_x is a device to conserve the energy $\frac{1}{2}\Sigma(u_m^n)^2$ to within terms of order $O(k^2)$. The linear stability requirement for this scheme is that (Greig and Morris 1976)

$$k/h^3 \leqslant (4 + h^2|u_o|)^{-1} \qquad (10.4\text{-}4)$$

which means that a very small time step must be used to preserve stability (here u_o is the maximum value of u in the range of interest). The nonlinear stability properties of this scheme have been extensively studied by Newell (1977). A survey of some finite difference schemes for the KdV equation appeared in the paper by Vliegenthart (1971). In 1976 Greig and Morris proposed a Hopscotch scheme for this equation. This scheme has the less stringent stability requirement

$$k/h^3 \leqslant (2 - h^2 u_o)^{-1} \qquad (10.4\text{-}5)$$

The Hopscotch scheme can be shown to possess only a small phase error in its Fourier modes. However the small time step required even with (10.4-5) is apparent: a two soliton collision required 4800 time steps to achieve an accuracy of 1% in the final amplitudes.

The application of spectral and pseudospectral methods to the KdV equation has been studied by Schamel (1978) and by Abe and Inoue (1980). A number of authors have studied the application of Galerkin methods to this equation, a recent reference being the paper by Sanz-Serna and Christie (1981). These authors use piecewise linear basis functions with cubic test functions to derive a method which is fourth order in space. They found that this scheme with $h = 0.033$ and $k = 0.01$ gave better accuracy than the Zabusky-Kruskal scheme with $h = 0.01$ and $k = 0.0005$. These impressive results suggest that this is one of the best schemes currently available for the KdV equation. Further schemes are surveyed by Schoombie (1982).

A different approach to the numerical study of the KdV equation has been adopted by Osborne and Provenzale (1981). In this study, direct use is made of the Inverse Scattering Transform to solve the initial value problem, with initial data being approximated by a piecewise constant function. This generalisation of the usual spectral methods promises to be an useful technique, with the only drawback being the fact that it is not applicable to equations for which no inverse scattering problem is known.

Fornberg and Whitham (1978) have published details of numerical studies of the KdV equation and related equation, especially of the type

$$u_t + uu_x + \int_{-\infty}^{\infty} K(x-\xi)u_\xi(\xi,t)d\xi = 0 \tag{10.4-6}$$

where $K(x)$ can be chosen to give various dispersive effects. These authors used a pseudospectral method in the x variable together with a leapfrog method in t. This approach is ideally suited to equations with linear integrals terms such as (10.4-6). Their scheme has the stability requirement

$$k/h^3 \leqslant 3/(2\pi^2) \approx 0.152 \tag{10.4-7}$$

The accuracy of this method is claimed to be high, but unfortunately there is no direct comparison with other schemes. Fornberg and Whitham also consider the higher order KdV equations

$$u_t + u^p u_x + u_{xxx} = 0 \tag{10.4-8}$$

and numerical results show that with $p \geqslant 3$ the soliton collision is inelastic. However, results for (10.4-6) with $K(x) = \pi/4 \exp(-|x|\pi/2)$ show some evidence for an exact two-soliton solution. Unfortunately the calculations were carried out with a step length such that only moderate accuracy could be achieved, and further tests are required before any strong claim for exact soliton behavior can be made.

10.4.2 The Regularized Long-wave Equation

As an alternative model to the KdV equation, Peregrine (1966) and Benjamin et al. (1972) have proposed the so-called regularized long-wave (RLW) equation

$$u_t + u_x + uu_x - u_{xxt} = 0 \tag{10.4-9}$$

Numerical schemes for this equation were studied by Eilbeck and McGuire (1975,1977). They derived a three level finite difference scheme which is stable for all practical values of k, and in fact works best if $k = h$. This means that a considerably greater time step can be taken in the numerical study of (10.4-9) as compared to the KdV equation (10.4-1). The scheme can be made conservative by a space-averaged nonlinear term as in the scheme (10.4-3).

The numerical study by Eilbeck and McGuire (1977) showed that after a collision of two solitary waves the pulses re-appeared with amplitudes within 0.3% of their original amplitude. On the basis of this evidence, and on the evidence of a computer produced film showing the two and three solitary wave collision, it was concluded that the RLW equation (10.4-9) had exact multisoliton solutions. A sequence of frames showing the three 'soliton' solution is shown in Fig. 2 in Eilbeck (1978).

Unfortunately the pictorial and numerical evidence in this study was not sufficiently accurate to pick up the inelastic radiation occurring for the two solitary wave collison in this model. Abdullcev et al. (1976) pointed out that when very large solitary waves (~10) collide in the RLW equation, a very small oscillating tail ($\sim 10^{-3}$) appeared, which could not be adequately explained as numerical error. These authors showed that for higher order RLW equations corresponding to (10.4-8) the inelasticity was even more pronounced. A more striking demonstration of this effect has been achieved by Santarelli (1978) who made use of the fact that (10.4-9) has solitary wave solutions travelling in both directions. When two travelling in opposite directions are allowed to collide the inelastic effect is much greater and extra solitary waves appear. This collision of a "soliton" and "antisoliton" has been studied in more detail by Lewis and Tjon (1979), who found evidence for "resonance" effects and a rich structure of multi-soliton production.

The accuracy of the calculations showing these inelastic effects has been further verified by higher-order calculations with a variety of exotic schemes. Khalifa (1979) developed a fourth order collocation method using quartic or quintic splines to study the interaction of two solitary wave solutions of the RLW equation, and found the small oscillatory tail predicted by lower order schemes.

Bona et al. (1980) have derived a numerical scheme based on recasting the equation into integral form. This scheme can be shown to be fourth order in space and time, and the authors have demonstrated rigorously the existence of the small oscillating tail for a range of two solitary wave collisions. More details of the scheme and detailed comparisons with experimental results for wave train experiments are given in Bona et al. (1981).

Although it is now clear that the two-soliton solution for the RLW equation does not exist, there seems to be no explanation for the size of the oscillating tail. In the NLKG cases discussed above, the breakdown of exact soliton behaviour is always much more pronounced, at least for some range of collisional velocities. Why it should be such a small effect in the case of the RLW equation is still a puzzle.

Two more important points arise from the study of the KdV and RLW equation. Firstly, if testing for exact soliton behaviour, a very careful analysis and an accurate numerical scheme are required. Secondly, in deriving an efficient numerical scheme to model a physical situation, it may be necessary to return to the original mathematical model and modify it in a way which will lead to well-behaved numerical schemes. It was precisely for this reason that Peregrine first proposed the RLW equation.

10.5 Other Equations in One Space Dimension

Nonlinear Optics has always provided a fertile area for interaction between experimental results, numerical studies and analytic work. The discovery of the phenomena of Self-Induced Transparency (SIT) by McCall and Hahn (1967) followed from the numerical and analytic study of solitary wave solutions of the appropriate equations for the envelope of an intense coherent light pulse in a low density dielectric. Other numerical work in this field is reported in the reviews by Lamb (1971) and Bullough et al. (1979). Although the envelope and approximate field equations have exact multisoliton solutions, numerical studies of the exact field equations (Caudrey and Eilbeck 1977) based on fourth-order predictor corrector methods have shown that exact soliton behaviour breaks down at high densities.

A rational finite difference scheme for the nonlinear Schrödinger equation, similar to the Strauss-Vazquez scheme for nonlinear KG equations (10.3-9), has been advocated by Delfour et al. (1981). Griffiths et al. (1982) have suggested a finite element scheme for this equation, which compares well with more classical finite difference schemes.

The field of plasma physics has also provided many nonlinear wave equations with soliton-like properties. Some interesting studies on KdV-type equations have been reported by Tappert (1974) and Schamel (1979). A great deal of work on soliton phenomena in plasma physics has been performed in the USSR, and the reviews by Makhankov (1978, 1980) are valuable sources of reference. Part of the 1978 review deals with numerical studies of various modifications of the Boussinesq equation. Also treated are the nonlinear Schrödinger equation with self-consistent potential, with emphasis on soliton-like phenomena in plasma physics. In all cases the modifications to the basic equations introduces some degree of inelasticity into the soliton interaction. Details of the numerical scheme used in the case of the modified Boussinesq equation are given in the paper by Bogolubsky (1977).

The reviews by Makhankov (1978, 1980) also deal with realistic field theory models. In one space dimension, in addition to the SG and NLKG equations described in Section 10.3, he also reports work on various coupled relativistic field equations. However the most interesting results from these reviews lie in in three-space dimensional calculations, which will be discussed in Section 10.6.

Before we move on to higher space dimensions, one further development worth mentioning is work on nonlinear difference equations with exact multisoliton solutions. Essentially, this approach is to seek a difference approximation to the original wave equation which has exact solutions possessing many of the properties of the analytic solutions of the original p.d.e. These equations are of interest in their own right, and may also provide useful numerical schemes of a novel kind. It may be that these equations will also lead to a better understanding of numerical methods for nonlinear wave equations, and point the way to better schemes for equations not having multisoliton properties, though there is little evidence of progress in this direction as yet.

Ablowitz and Ladik (1977) have developed a number of schemes which are solvable by the method of inverse scattering. Although these schemes involve complicated nonlocal terms, more recent unpublished work suggest that modified versions of these methods may be competetive with more conventional methods. More details are given in Ablowitz and Segur (1981). Hirota has devoted a series of papers (cf. Hirota 1977) to some simpler nonlinear difference equations which have exact multisoliton solutions. As an example, the following is one difference version of the SG equation.

$$\sin[(u_{m+1}^{n+1} + u_{m-1}^{n-1} - u_{m+1}^{n-1} - u_{m-1}^{n+1})/4]$$

$$= h^2 \sin[(u_{m+1}^{n+1} + u_{m-1}^{n-1} + u_{m+1}^{n-1} + u_{m-1}^{n+1})/4] \qquad (10.5\text{-}1)$$

10.6 Numerical Studies in Higher Space Dimensions

The study of soliton-like solutions of nonlinear wave equations in two or more space dimensions is currently (1982) an extremely active field. Any review of this topic is necessarily incomplete and will be out of date by the time the reader turns to this page. Over the last few years a mass of numerical data has accumulated concerning solutions of various equations, and recent advances in theory promise to give important advances in this area. However, as in the case of one space dimension, it is likely that advances in theory may often apply to different equations from those with interesting numerical solutions.

On the practical side the problems of computing in two or three space dimensions should not be underated. With reasonably efficient numerical methods it is possible to treat time-dependent problems in two space dimensions on a modern high-speed computer, but the computing budgets required are not trivial. In three space dimensions the problems become more pronounced, and only the biggest instalations are equipped to deal with such computing requirements. In some cases it is possible to reduce the problem by one or more dimensions by assuming cylindrical or spherical symmetry. However the increasing availability of high speed parallel processing computers or array processors will undoubtedly make previously impossible calculations feasible.

Almost as big a problem is that of exhibiting the results of the calculation when it is complete, and here some sophisticated graphical hardware and software is required. Computer produced cine film is an ideal output medium, but such facilities are not available to all users. Computer produced video film will inevitably form a cheaper alternative to cine film when appropriate hardware is developed.

In two or more space dimensions the types of solutions we can expect to find become more varied in comparison with the one dimensional case. An obvious generalization of the solitary wave solution in one space dimension is a plane wave solitary wave, having the same cross section as the 1-D case along its direction of motion, and having translational invariance along directions perpendicular to the velocity vector. This sort of solution is rather unphysical due to its infinite extent (and infinite energy), but it can exist between parallel finite boundaries if it is moving parallel to the boundary and no-flux conditions exist at the boundary.

When two or more nonparallel plane solitary waves occur in an infinite region, there is a complicated interaction in the region where the waves overlap. This situation is not easily amenable to numerical calculation since it is necessary to introduce finite boundaries, and clearly at least one of the plane waves will not be normal to the boundary. Rather surprisingly, it is possible in some special cases to find exact analytic n-soliton solutions describing the interaction of plane wave solitary waves crossing at arbitrary angles. In dealing with localized solutions the nomenclature becomes somewhat loose. Exact localized soliton solutions are still somewhat rare, and the objects receiving the most study are the equivalent of a single solitary wave, i.e. a stable or "almost-stable" finite energy solution which may have some translational velocity but in addition usually has some internal time-dependant structure. Despite the fact that these objects are not completely stable, even without considering interactions with other solutions, this type of solution is still commonly referred to as a "soliton". This usage is especially prevalent in high energy physics. We prefer to keep the word "soliton" for those pulses having exact collisional stability properties, and we shall refer to the more general types of solution as "quasi-solitons" or speak of "soliton-like" behaviour.

Much of the interest in localized solutions has been in the study of solutions of NLKG equations in two or three dimensions. Before sketching some results for these equations we shall first discuss some work on the KdV and NLS equations in two and three space dimensions.

10.6.1 The KdV and NLS Equations in 2- and 3-D

Maxon and Viecelli (1974) have derived cylindrically and spherically symmetric versions of the KdV equation for small amplitude acoustic waves in a collisionless plasma of warm electrons and cold ions. The equations are

$$u_\eta + (k\eta)^{-1}u + uu_\xi + \tfrac{1}{2}u_{\xi\xi\xi} = 0 \qquad (10.6\text{-}1)$$

where η is the time-like coordinate, ξ the radial coordinate, and $k = 1,2$ for spherical and cylindrical symmetry respectively. The authors examined the evolution of a single solitary wave using a leapfrog finite difference scheme based on that used by Zabusky and Kruskal for the KdV equation. It was found that the pulse steepened and narrowed as time increased, a result confirmed by the analytic work of Cumberbatch (1978). More recently Calogero and Degasperis (1978) have extended the spectral transform method to deal with (10.6-1) in the case $k = 2$. This work confirms that a singularity in the solution develops after a finite time.

An two-dimensional version of the KdV equation has received much attention in the last few years. This is the so-called Kadomtsev-Petviashvili equation (Kadomtsev and Petviashvili, 1970)

$$(u_t + 6uu_x \pm u_{xxx})_x + 3u_{yy} = 0 \qquad (10.6\text{-}2)$$

The n-soliton solution of this equation was derived by Satsuma (1976): the elucidation of its structure was carried out by Miles (1977) and Freeman et al. (cf. the useful review paper by Freeman (1980)).

One interesting use of the computer is to plot known analytic solutions: when the solutions are complicated a diagram is often worth pages of analysis. Figure 10-7 shows the 3-soliton solution of equation (10.6-2), taken with the + sign.

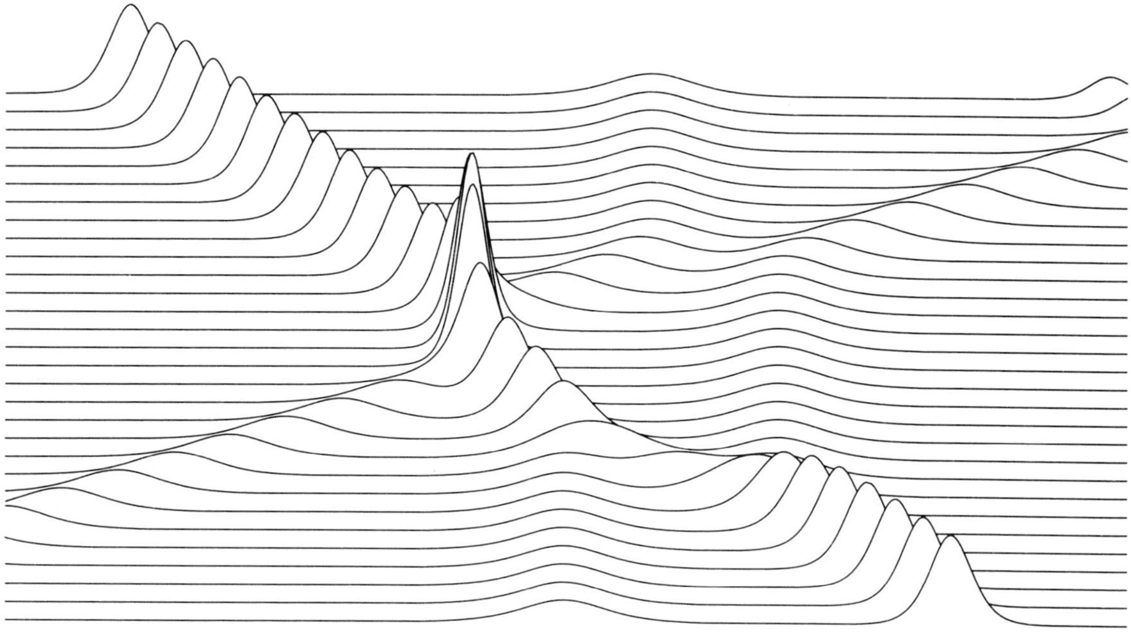

Figure 10-7: 3-soliton solution of the Kadomtsev-Petviashvili equation

It shows the three plane wave solitons with an interaction region in which several short lengths of wave appear. There is a phase shift in each plane soliton caused by the interaction. This diagram corresponds to Fig. 6a in Freeman's review, and the reader is refered to this paper for further details.

Even more interesting are solutions which are localized in space and have finite energy. This type of solution is more amenable to numerical study for since the boundary conditions are simpler. Such localized solutions to (10.6-2) (taken with the - sign) were first found by Manakov et al. (1977) and are described in detail in Freeman's review paper. These were the first exact soliton-like solutions with collisional stability in two space dimensions to be discovered. The study of such solutions is currently an area of active interest, but a detailed treatment of this area is outside the scope of this book and could in any case become quickly dated.

The KP equation has also been studied numerically by Makhankov, Litvinenko, and Shvanka (1981), using a two-dimensional version of the Hopscotch finite difference scheme. However this study involved only work on the unstable line solitons of the KP equation with the - sign.

Some results for the NLS equation in n space dimensions

$$-iu_t + \nabla^2 u + \lambda u|u|^{p-1} = 0 \qquad (10.6\text{-}3)$$

have been reviewed by Strauss (1978). He states that for $\lambda > 0$, $p > 1 + 4/n$, there exists initial conditions $u(x,0)$ such that the solution fails to exist for all t. If $\lambda < 0$ then the solution for $p = 3$ exists for all time and is smooth and unique.

The failure of solutions to exist in these cases is not just a technicallity, but has important applications in plasma physics and nonlinear optics, where the collapse of cylindrically symmetric beams leads to self-trapping or self-focussing effects. Such collapse has been observed in numerical studies of the NLS equation by Konno and Suzuki (1979) and Lomdahl et al. (1980) and Zakharov and Synakh (1976) (see Chap 8.7)

10.6.2 NLKG Equations in 2- and 3-D

The NLKG equations (10.3-1) described in Section 10.3 can easily be generalized to higher space dimensions

$$\nabla^2 u - u_{tt} = F(u) \qquad (10.6\text{-}4)$$

where ∇^2 is the Laplacian in n dimensions. In this form the equations maintain their Lorentz invariance, and have the conserved energy

$$E = \int (\tfrac{1}{2}u_t^2 + \tfrac{1}{2}|\nabla u|^2 + G(u))dx \qquad (10.6\text{-}5)$$

where $G' = F$. The regularity of solutions of (10.6-4) has been reviewed by Strauss (1978). When $F(u) = -u + u^3$, (the phi-four equation), no solutions blow up in finite time. For the SG equation (10.3-2), solutions are well behaved and as smooth as the initial data. For other NLKG equations the behaviour of solutions is sensitive to the nature of the nonlinear term $F(u)$. For example, some solutions of the equation with $F(u) = u - u^3$ may blow up in finite time.

Most numerical studies of (10.6-4) have taken $n = 3$ and assumed spherical or cylindrical symmetry. With spherical symmetry (10.6-4) becomes

$$u_{rr} + \frac{2}{r}u_r - u_{tt} = F(u) \qquad (10.6\text{-}6)$$

It is often convenient for numerical work to make the substitution $v = ru$ to get

$$v_{rr} - v_{tt} = rF(v/r) \qquad (10.6\text{-}7)$$

In this form we can use the same numerical schemes as in Section 10-3 with the modified right-hand side of (10.6-7).

It has been known for some time that no stable steady state solution of the Lorentz invariant scalar field equations exist (Derrick, 1964). However this result does not rule out time dependent oscillating solutions.

The first NLKG equation to be studied under the assuption of spherical symmetry was the phi-four equation. Bogolubsky and Makhankov (1976) took as initial conditions a "stationary" kink solution

$$u(r,0) = \tanh[(r-R)/\sqrt{2}], \qquad R \gg 1 \tag{10.6-8}$$

This is a spherical kink which becomes an exact stationary solution only in the limit $R \to \infty$.

With these initial conditions the spherical kink initially collapses towards the origin. Near the origin, some violent oscillations occur, during which the kink is reflected, and following this the kink returns to a state close to the initial conditions, although some energy in the form of small amplitude oscillations are radiated off to infinity. The whole cycle is then repeated. Since energy is lost during each cycle, the process cannot continue indefinitely, and eventually after a large number of oscillations the kink collapses completely and sends out a final burst of radiation. The number of oscillations depend on the value of R in a complicated way. The existence of these quasi-stable pulsations has lead to the name pulson for pulses displaying this behaviour.

Bogolubsky and Makhankov (1977) also investigated pulson solutions of the SG equation and found similar behaviour. In the SG case there is the added possibility of pulsons which take initially a double or multiple kink form. These pulsons have been investigated by Bogolubsky (1977); he shows that an initial double kink is relatively stable, but eventually cascades into a single kink pulson which has a similar lifetime. The SG equation with rotational symmetry has also been studied independently by Christiansen and Olsen (1978), and further studies of kink-kink and pulson-pulson interactions in two space dimensions have been carried out by Christiansen and Lomdahl (1981). The numerical technique used in this latter study was a simple extension of the finite difference scheme (10.3-8). In two space dimensions, this becomes

$$\begin{aligned} u^{n+1}_{\ell m} &= -u^{n-1}_{\ell m} + \tfrac{1}{2}(u^n_{\ell,m+1} + u^n_{\ell,m-1} + u^n_{\ell-1,m} + u^n_{\ell+1,m}) \\ &\quad - (h^2/2)F[\tfrac{1}{4}(u^n_{\ell,m+1} + u^n_{\ell,m-1} + u^n_{\ell-1,m} + u^n_{\ell+1,m})] \end{aligned} \tag{10.6-9}$$

Here the extra space variable y_ℓ is ℓh, and for this scheme the time step k is $h/\sqrt{2}$ to preserve the linear stability condition. Again only a diagonal grid is needed, saving a factor of two in both storage and time requirements. This scheme would convert easily to a parallel computer with little modification. In use on a normal computer, it has proved reliable and efficient. Figure 10-8 shows some of the initial stages in the evolution of a sine-Gordon pulson, taken from the film by Eilbeck and Lomdahl (1982). In this figure, $\sin(u/2)$ is plotted as a function of x and y for successive t values.

In addition to the phi-four and SG equations, the Russian group at JINR, Dubna, have studied several other NLKG equations, in particular the two equations

$$F(u) = u - u^3 \tag{10.6-10}$$

$$F(u) = u - u|u^2|/(1 + |u|^2) \tag{10.6-11}$$

Their work has been reviewed by Makhankov (1978,1980).

A most interesting development of the JINR group has been the extension of the numerical studies to problems involving the head-on collision of two pulson solutions for various equations in two and three space dimensions (assuming cylindrical symmetry in the latter case). Some details are given in the review by Makhankov

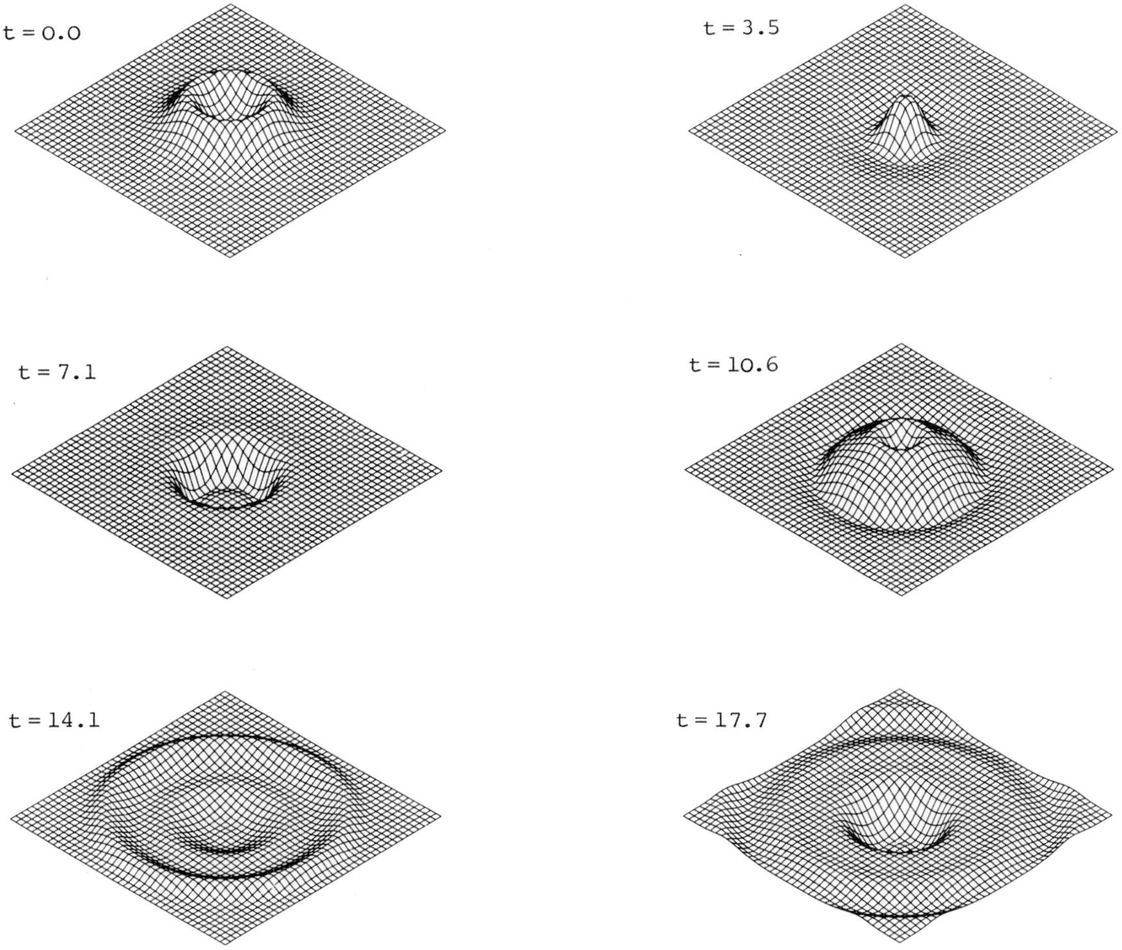

Figure 10-8: Pulson solution of the SG equation

(1980), and the reader is referred to the current literature for recent developments in this interesting field. Makhankov et al. have identified the following qualitatively different interaction possibilities:

- elastic and weakly elastic interactions
- breakup following collision
- decay through a resonant state
- creation of a long-lived bound state

A final intriguing possibility is the formation of **stable** bound states from the collision of two **unstable** pulsons (Makhankov, Kummer and Shvanchka, 1980).

10.7 Notes

Section 10.2

In equation (10.2-5), some care is needed concerning the effect of L on the basis functions $\phi_i(x)$. If the $\phi_i(x)$ are the "hat" functions, any second derivatives in in the space variable in (10.2-5) are only defined in the sense of a distribution. In the Galerkin method $r(x,t)$ appears only inside an integral, and so the resulting equations for the c_i are well defined. To use this approach for the collocation method it is necessary for the basis functions to posses a sufficient number of derivatives, i.e. for L a second order operator we would need to work with piecewise cubic or quadratic functions.

Section 10.3

One of the more outstanding ommissions in this short chapter has been any mention of the role of numerical studies of sine-Gordon and other equations as models in solid state physics. A good introduction to this area is the review by Bishop et al. (1981). and the proceedings of the Oxford Conference on Solions and Condensed Matter Physics (Bishop and Schneider, 1978). Many significant contributions to this field have been made by Schneider and Stoll and co-workers at the IBM Zurich Research Laboratories. As an example we refer the reader to surveys of the static and dynamic properties of the discretized sine-Gordon and phi-four chains (Schneider and Stoll 1980, 1981), in which details of the classical statistical mechanics of such systems are presented.

Recent numerical studies of the harmonically driven SG equation have revealed the presence of chaos in some cases. A study by Eilbeck, Lomdahl and Newell (1981) used a spatially inhomogeneous driving force to achieve this effect, whereas more recent work by Trullinger et al. (unpublished) has demonstrated the onset of chaos with a homogeneous driving force but inhomogeneous initial conditions.

Section 10.4

Pen-Yu and Sanz-Serna (1981) have recently proved convergence for a family of methods for the KdV equation, including some of the finite difference and finite element methods described in this section.

Section 10.5

The topic of Davydov solitons on the alpha-helix protein molecule has already been introduced in Section 9.6. Numerical studies into this discrete system have proved a vital tool in investigating energy transport along these molecules. A recent reference in this area is Scott (1982).

REFERENCES

At the time of writing (October, 1982) there have appeared several other books and collections of edited papers devoted to solitons and related topics (Lamb, 1980, Bullough and Caudrey (Eds.), 1980, Eilenberger, 1980, Ablowitz and Segur, 1981, Eckhaus and Van Harten, 1981, Toda, 1981, Calogero and Degasperis, 1982). Each book gives different emphasis to physical intuition, mathematical rigor, applications, etc., and the choice of other texts depends on the readers interests. Some other recent books on nonlinear waves which include introductory discussions of soliton theory include those by Bhatnagar (1979), Karpman (1975), and Whitham (1974). An introduction to the theory of Bäcklund transformations will be found in Anderson and Ibragimov (1979). In addition there is a growing list of conference proceedings devoted wholly or partly to soliton theory: Bishop and Schneider, 1978, Boiti et al. 1980, Calogero, 1978, Leibovich and Seebass, 1974, Lonngren and Scott, 1978, Miura, 1976, Moser, 1975, Newell, 1974, Ranada, 1979.

The observant reader will note that some of the entries in this bibliography are not referred to directly in the text. This is in order to include references which we thought would prove useful to the reader, but which were not to hand for various reasons when the chapters were completed. Naturally our choice is spasmodic and incomplete, and we apologise to those authors whose works we have inadvertently overlooked.

Bibliography

Abdulloev, Kh.O., Bogolubsky, I.L., and Makhankov, V.G. 1976. One more example of inelastic soliton interaction. Phys. Lett., 56A, 427-428.

Abe, K., and Inoue, O. 1980. Fourier expansion solution of the KdV equation. J. Comp. Phys., 32, 202-210.

Ablowitz, M.J., Kaup, D.J., Newell, A.C., and Segur, H. 1973. Method for solving the sine-Gordon equation. Phys. Rev. Lett., 30, 1262-1264.

Ablowitz, M.J., Kaup, D.J., Newell, A.C., and Segur, H. 1973. Nonlinear evolution equations of physical significance. Phys. Rev. Lett., 31, 125-127.

Ablowitz, M.J., Kaup, D.J., and Newell, A.C. 1974. Coherent pulse propagation, a dispersive, irreversible phenomena. J. Math. Phys., 15, 1852-1858.

Ablowitz, M.J., Kaup, D.J., Newell, A.C., and Segur, H. 1974. The inverse scattering transform-Fourier analysis for nonlinear problems. Studies in Appl. Math., 53, 249-315.

Ablowitz, M.J., and Ladik, J.F. 1977. On the solution of a class of nonlinear partial difference equations. Studies in Appl. Math., 57, 1-12.

Ablowitz, M.J. and Segur, H. 1977. Asymptotic solutions of the KdV equation. Studies in Appl. Math., 57, 13-44.

Ablowitz, M.J., Ramini, A. and Segur, H. 1978. Nonlinear evolution equations and ordinary differential equations of Painleve type. Lett. Nuovo Cim., 23, 333-8.

Ablowitz, M.J., Kruskal, M.D., and Ladik, J.F. 1979. Solitary wave collisions. SIAM J. Appl. Math., 36, 428-437.

Ablowitz, M., Kruskal, M., and Segur, H. 1979. A note on Miura's transformation. J. Math. Phys., 20, 999-1003.

Ablowitz, M.J., Ramini, A. and Segur, H. 1980. A connection between nonlinear evolution equations and ordinary differential equations of P-type . I. J. Math. Phys., 21, 715-721.

Ablowitz, M.J. and Segur, H. 1981. Solitons and the Inverse Spectral Transform. Philadelphia: SIAM.

Agranovich, A.K., and Marchenko, V.A., 1963. The inverse problem of scattering theory. London: Gordon and Breach.

Ames, W.F. 1977. Numerical Methods for Partial Differential Equations. London: Nelson.

Anderson, L. and Ibragimov, N.H. 1979. Lie-Bäcklund transformations in applications. Amsterdam: North-Holland.

Anker, D., and Freeman, N.C. 1978. Interpretation of three-soliton interactions in terms of resonant triads. J.Fluid.Mech., 87, 17-31.

Anker, D., and Freeman, N.C. 1978. On the soliton solutions of the Davey-Stewartson equations for long waves. Proc. Royal Soc. Lond., A360, 529-40.

Arafune, J., Freund, A.G. and Goebel, C.J. 1975. Topology of Higgs fields. J. Math. Phys., 16, 433-7.

Asano, N., Taniuti, T., and Yajima, N. 1969. Perturbation method for non-linear wave modulation II. J. Math. Phys., 10, 2020-24.

Askar'yan, G.A. 1974. The self-focussing effect. Sov. Phys. Uspekhi, **16**, 680-686.

Aubry, S. 1976. A unified approach to the interpretation of displacive and order-disorder systems.II. J. Chem. Phys., **64**, 3392-3402.

Bäcklund, A.V. 1875. Einiges uber Curven und Flachentransformationen. Lund Universitets Arsskrift, 10, 1-12.

Ball, J.M. 1977. Remarks on blow-up and non-existence theorems for nonlinear evolution equations. Quart. J. of Maths (Oxford), 28, 473-486.

Bardeen, J., Cooper, L.N. and Schrieffer, J.R. 1957. Theory of superconductivity. Phys. Rev., 108, 1175-1204.

Bargmann, V. 1949. On the connection between phase shifts and scattering potential. Rev. Mod. Phys., 21, 488-493.

Barone, A., Esposito, F., Magee, C.J., and Scott, A.C. 1971. Theory and applications of the sine-Gordon equation. Riv. Nuovo Cimento, 1, 227-267.

Belinskii, V.A. and Zakharov, V.E. 1978. Integration of the Einstein equations by means of the inverse scattering problem technique and construction of exact soliton solutions. Sov. Phys. JETP, 48, 985-994.

Bender, C. and Orszag, S. 1978. Advanced mathematical methods for scientists and engineers. New York: McGraw Hill.

Benjamin, T.B., and Feir, J.E. 1967. The disintegration of wave trains in deep water. Part 1. J. Fluid Mech., 27, 417-30.

Benjamin, T.B. 1967. Instability of periodic wave-trains in nonlinear dispersive systems. Proc. Royal Soc. Lond., **A299**, 59-75.

Benjamin, T.B. 1972. The stability of solitary waves. Proc. Roy. Soc. Lond. A., 328, 153-183.

Benjamin, T.B., Bona, J.G., and Mahoney, J.J. 1972. Model equations for long waves in nonlinear dispersive systems. Phil. Trans. Roy. Soc. London, **272A**, 47-78.

Benney, D.J., and Newell, A.C. 1967. The propagation of nonlinear wave envelopes. J. of Mathematics and Physics, **46**, 363-93.

Benney, D.J., and Roskes, G.J. 1969. Wave Instabilities. Studies in Appl. Math., 48, 377-85.

Benney, D.J. 1976. Significant interactions between long and short waves. Studies in App. Math., 55, 93-106.

Benney, D.J. 1977. A general theory for interactions between long and short waves. Studies in Appl. Math., 56, 81-94.

Berezin, Y.A., Karpman, V.I. 1964. Theory of non-stationary finite amplitude waves in a low density plasma. Sov. Phys. JETP, 19, 1265-71.

Berezin, Y.A. and Karpman, V.I. 1967. Nonlinear evolution of disturbances in plasmas and other dispersive media. Sov. Phys. JETP, 24, 1049-56.

Berkshire, F.H. and Gibbon, J.D. 1982. Blow-up in the NLS equation - a parallel with the N-body problem. Imperial College Maths. Dept. preprint, 1982, submitted to Studies in Appl. Math.

Bernstein, J. 1974. Spontaneous symmetry breaking, gauge theories, the Higgs mechanism and all that. Rev.Mod. Phys., 46, 7-48.
Bhatnagar, P.L. 1979. Nonlinear waves in one-dimensional dispersive systems. Oxford: Clarendon Press.
Bialynicki-Birula, I., and Mycielski, J. 1979. Gaussons: solitons of the logarithmic Schrödinger equation. Physica Scripta, 20, 539-544.
Bishop, A.R., and Schneider, T. 1978. Solitons and Condensed Matter Physics. Berlin: Springer. Springer Series in Solid State Sciences 8.
Bishop, A.R. 1979. Solitons in condensed matter physics. Physica Scripta, 20, 409-423.
Bishop, A.R., Krumhansl, J.A., and Trullinger, S.E. 1980. Solitons in condensed matter: a paradigm. Physica D, 1, 1-44.
Blij, F., van der. 1978. Some details of the history of the Korteweg-de Vries equation. Nieuw Archief voor Wiskunde (3), 26, 56-64.
Bloch, F. 1946. Nuclear induction. Phys. Rev., 70, 460-474.
Bogoliubov, N.N., and Mitropolsky, Y.A. 1961. Asymptotic methods in the theory of nonlinear oscillations. New York: Gordon and Breach.
Bogolubsky, I.L. 1977. Some examples of inelastic soliton interaction. Comp. Phys. Comm., 13, 149-155.
Bogolubsky, I.L., Makhankov, V.G., and Shvachka, A.B. 1977. Dynamics of the collisions of two space-dimensional pulsons in field theory. Phys. Lett., 63A, 225-227.
Bogolubsky, I.L. and Makhankov, V.G. 1977. Dynamics of heavy spherically-symmetric pulsons. JETP Lett., 25, 107-110.
Bogomolny, E.B. 1976. The stability of classical solutions. Sov. J. Nucl. Phys., 24, 449-454.
Boiti, M., Pempinelli, F. and Soliani, G. (Eds.). 1980. Nonlinear evolution equations and dynamical systems. Berlin: Springer. Lecture notes in physics Vol. 120.
Bona, J.L. and Smith, R. 1975. The initial value problem for the KdV equation. Phil. Trans. Roy. Soc., A278, 255-604.
Bona, J.L., Pritchard, W.G., and Scott, L.R. 1980. Solitary-wave interaction. Phys. Fluids, 23, 438-441.
Bona, J.L., Pritchard, W.G., and Scott, L.R. 1981. An evaluation of a model equation for water waves. Phil. Trans. Roy. Soc., 302, 457-510.
Borg, G. 1949. Uniqueness theorems in the spectral theory of $y'' + (\lambda - q(x))y = 0$. In Johan Grundt (Ed.), Proceedings of the Eleventh Congress of Scandanavian Mathematicians, held at Trondheim, August 22-25, : Tanums Forlag.
Boussinesq, J. 1872. Theorie des ondes et des remous qui se propagent le long d'un canal rectangulaire horizontal, en communiquant au liquide contenu dans ce canal des vitesses sensiblement pareilles de la surface au fond. J. Math. Pures. Appl., ser. 2, 17, 55-108.
Bullough, R.K., Caudrey, P.J., Eilbeck, J.C. and Gibbon, J.D. 1974. A general theory of self-induced transparency. Opto-electronics, 6, 121-140.
Bullough, R. K. 1978. Solitons. Phys. Bull., 29, 78-82.
Bullough, R.K., and Caudrey, P.J. 1978. The multiple sine-Gordon equations in nonlinear optics and in liquid He. In Calogero, F. (Ed.), Nonlinear evolution equations solvable by the spectral transform, London: Pitman.
Bullough, R.K., Jack, P.M., Kitchenside, P.W., and Saunders, R. 1979. Solitons in laser physics. Physica Scripta, 20, 364-381.
Bullough, R.K. and Caudrey, P.J. (Eds.). 1980. Solitons. Berlin: Springer. Topics in Current Physics, Vol. 17.
Burchnall, J.L. and Chaundy, T.W. 1922. Commutative ordinary differential operators. Proc. Lond. Math. Soc. (2), 21, 420-440.
Byrd, P.F., and Friedman, M.D. 1971. Handbook of elliptic integrals. Berlin: Springer. 2nd. ed., revised.
Calogero, F. 1975. A method to generate soluble nonlinear evolution equat-

ions. Lett. Nuovo Cim., 14, 443-7.

Calogero, F. and Degasperis, A. 1976. Nonlinear evolution equations solvable by the inverse spectral transform. - I. Nuovo Cimento, 32B, 201-242.

Calogero, F. and Degasperis, A. 1977. Nonlinear evolution equations solvable by the inverse spectral transform. Nuovo Cim., 39B, 1-54.

Calogero, F. 1978. Nonlinear evolution equations solvable by the spectral transform. London: Pitman.

Calogero, F. and Degasperis, A. 1978. Exact solution via the spectral transform of a nonlinear evolution equation with linearly x-dependent coeeficients. Lett. Nuovo Cim., 22, 138-141.

Calogero, F. and Degasperis, A. 1978. Solution by the spectral transform method of a nonlinear evolution equation including as a special case the cylindrical KdV equation. Lett. Nuovo Cim., 23, 150-154.

Calogero, F. and Degasperis, A. 1981. Reduction techniques for matrix nonlinear evolution equations solvable by the spectral transform. J. Math. Phys., 22, 23-31.

Calogero, F. and Degasperis, A. 1982. Spectral transform and solitons. Amsterdam: North-Holland.

Caudrey, P.J., Gibbon, J.D., Eilbeck, J.C., and Bullough, R.K. 1973. Exact multisoliton solutions of the self-induced transparency and sine-Gordon equations. Phys. Rev. Lett., 30, 237-239.

Caudrey, P.J., Eilbeck, J.C., Gibbon, J.D., and Bullough, R.K. 1973. Exact multisoliton solution of the inhomogeneously broadened self-induced transparency equations. J. Phys. A., 6, L53-L56.

Caudrey, P.J., Eilbeck, J.C., Gibbon, J.D., and Bullough, R.K. 1973. Multiple soliton and bisoliton bound state solutions of the sine-Gordon equation and related equations in nonlinear optics. J. Phys. A., 6, L112-L115.

Caudrey, P.J., Eilbeck, J.C. and Gibbon, J.D. 1974. An N-soliton solution of the Reduced Maxwell-Bloch equations. J.I.M.A., 14, 375-386.

Caudrey, P.J., Eilbeck, J.C., and Gibbon, J.D. 1975. The sine-Gordon equation as a model classical field theory. Nuovo Cim., 25B, 497-512.

Caudrey, P.J., and Eilbeck, J.C. 1977. Numerical evidence for breakdown of soliton behavoir in solutions of the Maxwell-Bloch equations. Phys. Lett., 62A, 65-66.

Caudrey, P.J. 1980. The inverse problem for the third order equation $u_{xxx} + q(x)u_x + r(x)u = -i\xi^3 u$. Phys. Lett., 79A, 264-7.

Caudrey. 1982. The inverse problem for a general NxN spectral equation. To be published in Physica D.

Chadan, K. and Sabatier, P.C. 1977. Inverse problems in quantum scattering theory. New York: Springer.

Charney, J.G. 1947. The dynamics of long waves in a baroclinic westerly current. J. Met., 4, 135-162.

Chiao, R.Y., Garmire, E., and Townes, C.H. 1964. Self trapping of optical beams. Phys. Rev. Lect., 13, 479-82.

Chillingworth, D.W. 1976. Differential Topology with a View to Applications. London: Pitman.

Christiansen, P. L., and Olsen, O.H. 1979. Ring-shaped quasi-soliton solutions to the two- and three- dimensional sine-Gordon equation. Physica Scripta, 20, 531-538.

Christiansen, P.L., and Olsen, O.H. 1980. Fluxons on a Josephson Line with loss and bias. Wave Motion, 2, 185-196.

Christiansen, P.L., Lomdahl, P.S., Scott, A.C., Soerensen, O.H. and Eilbeck, J.C. 1981. Internal dynamics of long Josephson junction oscillators. Appl. Phys. Lett., 39, 108-110.

Christiansen, P.L., and Lomdahl, P.S. 1981. Numerical study of 2+1 dimensional sine-Gordon solitons. Physica D, 2, 482-494.

Christie, I., Griffiths, D.F., Mitchell, A.R. and Sanz-Serna, J.M. 1981. Product approximation for nonlinear problems in the finite element method.

IMA J. Num. Anal., 1, 253-266.
Chu, V.H., and Mei, C.C. 1970. On slowly varying Stokes waves. J. Fluid Mech. 41, 873-87.
Chu, V.H., and Mei, C.C. 1971. The nonlinear evolution of Stokes waves in deep water. J. Fluid Mech., 47, 337-57.
Coddington, E.A. and Levinson, N. 1955. Theory of ordinary differential equations. New York: McGraw-Hill.
Cohen, A. 1979. Existence and regularity of solutions of the KdV equation. Arch. Rat. Mech. Anal., 79, 143-175.
Cole, J.D. 1951. On a quasi-linear parabolic equation occuring in aerodynamics. Quart. Appl. Math., 9, 225-236.
Cole, J.D. 1968. Perturbation Methods in Applied Mathematics. Waltham: Blaisdell.
Coleman, S. 1977. Classical lumps and their quantum descendents. In Zichichi, A. (Ed.), New phenomena in subnuclear physics, . New York: Plenum Press.
Coleman, S. 1979. The uses of instantons. In Zichichi, A. (Ed.), 1977 Erice Lectures, . New York: Plenum Press.
Constabile, G., Parmentier, R., Savo, B., McLaughlin, D.W. and Scott, A.C. 1978. Exact solutions of the sine-Gordon equation describing oscillations in a long (but finite) Josephson junction. Appl. Phys. Lett., 32, 587-9.
Costabile, G., Parmentier, R.D., and Savo, B. 1978. Fluxon-breather-plasma oscillation decay in long Josephson junctions. J. de Physique, 39, C6-567,C6-568. Colloque C6, supp. au #8.
Courtnay Lewis, J. and Tjon, J.A. 1979. Resonant production of solitons in the RLW equation. Phys. Lett., 73A, 275-279.
Croom, F.H. 1978. Basic concepts in algebraic topology. Berlin: Springer.
Cumberpatch, E. 1978. Spike solutions for radially symmetric solitary waves. Phys. Fluids, 21, 374-6.
Currie, J.F. 1977. Aspects of exact dynamics for general solutions of the SG equation with applications to domain walls. Phys. Rev. A, 16, 1692-9.
Dashen, R., Hasslacher, B. and Neveu, A. 1974. Non-perturbative methods and extended hadron models in field theory: I Semiclassical functional methods. Phys. Rev. D, 10, 4114-29.
Dashen, R., Hasslacher, B. and Neveu, A. 1975. Particle spectrum in model field theories from semiclassical functional integral techniques. Phys. Rev. D, 11, 3424-50.
Davey, A. 1972. The propagation of a weak nonlinear wave. J. Fluid Mech., 53, 769-81.
Davey, A., and Stewartson, K. 1974. On three dimensional packets of surface waves. Prov. Royal Soc. Lond., A338, 101-110.
Davidson, R. 1972. Methods in nonlinear plasma theory. New York: Academic Press.
Davydov, A.S. 1971. Theory of molecular excitons. New York: Plenum Press.
Davydov, A.S. and Kislukha, N.I. 1976. Solitons in one-dimensional molecular chains. Sov. Phys. JETP, 44, 571-575.
Davydov, A.S. 1979. Solitons in molecular systems. Physica Scripta., 20, 387-394.
De Vega, H.J. and Schaposnik, F.D. 1976. Classical vortex solutions of the abelian Higgs model. Phys. Rev. D, 14, 1100-6.
Deift, P. 1978. Applications of a commutation formula. Duke Math. J., 45, 267-310.
Deift, P. and Trubowitz, E. 1979. Inverse scattering on the line. Comm. Pure Appl. Math., 32, 121-251.
Delfour, M., Fortin, M., and Payre, G. 1981. Finite differences solutions of nonlinear Schrödinger equation. J. Comp. Phys., 44, 277-288.
Derrick, G. 1964. Comments on nonlinear wave equations as models for elementary particles. J. Math. Phys., 5, 1252-4.
Dillon, J.F. 1963. Domains and domain walls. In Rado, G. and Suhe, S.H.

(Ed.), Magnetism III, , pp 415-461. London: Academic Press.

Djordjevic, V.D., and Redekopp, L.G. 1977. On two dimensional packets of capillary-gravity waves. J. Fluid Mech., 79, 703-14.

Dodd, R.K. and Bullough, R.K. 1977. Polynominal conserved densities for the sine-Gordon equation. Proc. R. Soc. Lond., A352, 481-503.

Dodd, R.K. and Bullough, R.K. 1977. Bäcklund transformations for the AKNS inverse method. Phys. Lett., 62A, 70-74.

Dodd, R.K. 1978. Generalised Bäcklund transformations for some nonlinear partial differential-difference equations. J. Phys. A, 11, 81-92.

Dodd, R.K. and Bullough, R.K. 1979. The generalized Marchenko equation and the canonical structure of the AKNS-ZS inverse method. Phys. Scripta, 20, 514-530.

Dodd, R.K. 1982. On the AKNS-ZS inverse method and the initial value problem for solvable equations. Preprint, Trinity College, Dublin.

Doring, W. 1948. Uber die Trägheit der Wände zunschen wiesschon Bezirkon. Zeits. Naturf., 39, 373-9.

Drazin, P.G. 1970. Nonlinear baroclinic instability of a continuous zonal flow. Quart. J. Roy. Met. Soc., 96, 667-676.

Drazin, P.G., and Reid, W.H. 1981. Hydrodynamic Stability. Cambridge: Cambridge University Press.

Dubrovin, B. and Novikov, S.P. 1975. Periodic and conditionally periodic analogs of the many-soliton solutions of the KdV equation. Sov. Phys. JETP, 40, 1058-1063.

Dubrovin, B.A., Matveev, V.B., and Novikov, S.P. 1976. Nonlinear equations of KdV type, finite-zone linear operators and Abelian varieties. Russ. Math. Surveys, 31, 59-146.

Duckworth, S., Bullough, R.K., Caudrey, P.J., and Gibbon, J.D. 1976. Unusual soliton behaviour in the self-induced transparency of Q(2) vibration-rotation transitions. Phys. Lett., 57A, 19-22.

Eady, E.T. 1949. Long waves and cyclone waves. Tellus, 1, 33-52.

Eckhaus, W. 1965. Studies in nonlinear stability theory. Berlin: Springer.

Eckhaus, W. and Van Harten, A. 1981. The inverse scattering transformation and the theory of solitons. Amsterdam: North-Holland.

Eguchi, T., Gilkey, P.B. and Hanson, A.J. 1980. Gravitation, gauge theory and differential geometry. Phys. Rep., 66, 213-393.

Eilbeck, J.C. and Bullough, R.K. 1972. The method of characteristics in the theory of resonant or nonresonant nonlinear optics. J. Phys. A., 5, 820-829.

Eilbeck, J.C., Gibbon, J.D., Caudrey, P.J. and Bullough, R.K. 1973. Solitons in nonlinear optics I. A more accurate description of the 2 pulse in self-induced transparency. J. Phys. A., 6, 1337-1347.

Eilbeck, J.C., and McGuire, G.R. 1975. Numerical study of the regularized long wave equation I: Numerical Methods. J. Comp. Phys., 19, 43-57.

Eilbeck, J.C., and McGuire, G.R. 1977. Numerical study of the regularized long wave equation II: Interaction of solitary waves. J. Comp. Phys., 23, 63-73.

Eilbeck, J.C. 1978. Numerical studies of solitons. In Bishop,A.R., and Schneider, T. (Ed.), Springer Series in Solid-State Sciences (8): Solitons and Condensed Matter Physics, Berlin: Springer-Verlag.

Eilbeck, J.C., Lomdahl, P.S., and Newell, A.C. 1981. Chaos in the inhomogeneously driven sine-Gordon equation. Phys. Lett., 87A, 1-4.

Eilbeck, J.C. 1981. Kink Collisions in the ϕ^4 Model. 16 mm cine film, white on blue, silent, approx 4 mins. (1981).

Eilbeck, J.C. and Lomdahl, P. S. 1981. Sine-Gordon Solitons. 16 mm cine film, white on blue, with sound track, approx 12 mins. (1981).

Eilenberger, G. 1980. Solitons. Berlin: Springer.

Eisenhart, L.T. 1960. A Treatise on the Differential Geometry of Curves and Surfaces. New York: Dover.

Emmerson, G.S. 1977. John Scott Russell. London: John Murray.
Enz, U. 1964. Die Dynamik der blochschon Wand. Phys. Acta., 37, 245-251.
Ernst, F.J. 1968. New formulation of the axially symmetric gravitational field problem. Phys. Rev., **167**, 1175-8.
Essmann, U. and Träuble, H. 1967. The direct observation of individual fluxons on type II semiconductors. Phys. Lett., **24A**, 526-7.
Faddeev, L.D. 1963. The inverse problem in the quantum theory of scattering. J. Math. Phys., **4**, 72-104.
Faddeev, L.D. 1964. Properties of the S-matrix of the one-dimensional Schrödinger equation. Amer. Math. Soc. (2), **65**, 139-166.
Faddeev, L.D. and Takhtadzhyan, L.A. 1974. . Usp. Mat. Nauk., 29, 249-???
Faddeev, L.D. and Konepkin, V.E. 1978. Quantum theory of solitons. Phys. Rep. **42**, 1-87.
Fermi, E., Pasta, J.R., and Ulam, S.M. 1955. Studies of nonlinear problems. Technical Report LA-1940, Los Alamos Sci. Lab., . also in 'Collected Works of E. Fermi, vol. II., Chicago: Univ. Chicago Press, 1965, 978-988.
Feynman, R.P., Leighton, R.B. and Sands, M. 1969. ? In Feynmann, R.P. (Ed.), The Feyneman Lectures on Physics, Vol. III, , pp (21-1)-(21-18). New York: Addison Wesley.
Finkelstein, D. and Misner, C. 1959. Some new conservation laws. Ann. Phys., 6, 230-243.
Finkelstein, D. 1966. Kinks. J. Math. Phys., 7, 1218-25.
Finkelstein, D. and Rubenstein, J. 1968. Connection between spin, statistics, and kinks. J. Math. Phys., **9**, 1762-79.
Finkelstein, D. and Weil, D. 1978. Magnetohydrodynamic kinks in astrophysics. Int. J. Theo. Phys., 17, 201-217.
Flaschka, H. 1974. On the Toda lattice I. Phys. Rev., **B9**, 1924-1925.
Flaschka, H. 1974. On the Toda lattice II: Inverse scattering solution. Prog. Theor. Phys., 51, 703-716.
Flaschka, H. and Newell, A.C. 1975. Integrable systems of nonlinear evolution equations. In Moser, J. (Ed.), Dynamical systems, theory and applications, , pp 355-440. Berlin: Springer.
Flaschka, H. and Newell, A.C. 1981. Multiphase similarity solutions. Physica D, 3, 203-221.
Foias, C. 1960. An application of vector distributions to spectral theory. Bull. Sci. Math., **87**, 147-158.
Fokas, A.S. 1980. A symmetry approach to exactly solvable evolution equations. J. Math. Phys., 21, 1318-25.
Forgacs, P., Horvath, Z. and Palla, L. 1980. Generating the Bogomolny-Prasad-Sommerfeld one-monopole solution by a Bäcklund transformation. Phys. Rev. Lett., 15, 505-8.
Fornberg, B., and Whitham, G.B. 1978. A numerical and theoretical study of certain nonlinear wave phenomena. Phil. Trans. R. Soc. Lond., 289, 373-404.
Fornberg, B. 1978. Pseudospectral calculations on 2-D turbulence and nonlinear waves. SIAM-AMS Proceedings, 11, 1-17.
Fornberg, B. 1981. Numerical computation of nonlinear waves. In Enns, R.H., Jones, B.L., Miura, R.M. and Rangnekar, S.S. (Ed.), Nonlinear Phenomena in Physics and Biology, , pp 157-184. New York: Plenum Pub.
Forsyth, A.R. 1959. Theory of differential equations. New York: Dover.
Fowler, A.C., Gibbon, J.D., and McGuinness, M.J. 1982. The complex Lorenz equations. Physica D: Nonlinear Phenomena, 4, 139-163.
Fröberg, C. 1948. Calculation of the interaction between two particles from the asymptotic phase. Phys. Rev., 72, 519-520.
Frank, F.C. and van der Merwe, J.H. 1949. One dimensional dislocations I. Proc. R. Soc. Lond., **A198**, 205-216.
Frank, F.C. and van der Merwe, J.H. 1950. One dimensional dislocations II. Proc. R. Soc. Lond., **A201**, 261-268.

Freeman, N.C. and Johnson, R.S. 1970. Shallow water waves on shear flows. J. Fluid Mech., 42, 401-9.
Freeman, N.C., and Davey, A. 1975. On the evolution of packets of long surface waves. Proc. Royal Soc. Lond., A344, 427-33.
Freeman, N.C. 1979. A two dimensional distributed soliton solution of the KdV equation. Proc. Roy. Soc. Lond. A., 366, 185-204.
Freeman, N.C. 1980. Soliton interactions in two dimensions. Adv. Appl. Mech., 20, 1-37.
Frenkel, J. and Kontorova, T. 1939. On the theory of plastic deformation and twinning. J. Phys. USSR, 1, 137-149.
Fried, B., Ichikawa, Y.H. 1973. On the nonlinear Schrödinger equation for Langmuir waves. J. Phys. Soc. Japan, 34, 1073-82.
Fulton, T. 1977. Equivalent circuits and analogs of the Josephson effect. In Schwarz, B.B. and Foner, S. (Ed.), Superconductor applications: SQUIDS and machines, , pp 125-187. : Plenum Press.
Fultz, D. 1951. . In Compendium of Meteorology, Boston: Amer. Met. Soc.
Garabedian, R.R. 1964. Partial differential equations. New York: J. Wiley.
Gardner, C.S., Greene, J.M., Kruskal, M.D., and Miura, R.M. 1967. Method for solving the Korteweg-de Vries equation. Phys. Rev. Lett., 19, 1095-1097.
Gardner, C.S. and Morikawa, G.M. 1969. Similarity in the asymptotic behaviour of collision free hydromagnetic wave and water waves. Report NYO-9082, Courant Inst. of Math. Sciences.
Gardner, C.S. 1971. Korteweg de Vries equation and generalizations IV. The Korteweg de Vries equation as a Hamiltonian system. J. Math. Phys., 12, 1548-1551.
Gardner, C.S., Greene, J.M., Kruskal, M.D., and Miura, R.M. 1974. Korteweg-de Vries equation and generalizations. VI. Methods of exact solutions. Comm. Pure Appl. Math., 27, 97-133.
Gel'fand, I.M. and Dikii, L.A. 1975. Asymptotic behavoir of the resolvent os Sturm-Liouville equations and the algebra of the KdV equations. Russ. Math. Surveys, 30, 77-113.
Gel'fand, I.M. and Dikii, L.A. 1976. A Lie algebra structure in formal variational calculus. Funct. Anal. Appl., 10, 16-22.
Gel'fand, I.M. and Dikii, L.A. 1976. Fractional powers of operators and Hamiltonian systems. Funct. Anal. Appl., 10, 259-273.
Gel'fand, I.M. and Dikii, L.A. 1977. The resolvent and Hamiltonian systems. Funct. Anal. Appl., 11, 93-104.
Gel'fand, I.M. and Dikii, L.A. 1978. The calculus of jets and nonlinear Hamiltonian systems. Funct. Anal. Appl., 12, 81-94.
Gel'fand, I.M. and Levitan, B.M. 1955. On the determination of a differential equation from its spectral function. Amer. Math. Soc. Transl., 1, 253-304.
Getmanov, B.S. 1977. New Lorentz-invariant system with exact multisoliton solutions. JETP Lett., 25, 119-122.
Gibbon, J.D., and Eilbeck, J.C. 1972. A possible N-soliton solution for a nonlinear optics equation. J. Phys. A: Gen. Phys., 5, L22-L24.
Gibbon, J.D., Caudrey, P.J., Bullough, R.K. and Eilbeck, J.C. 1973. An N-soliton solution of a nonlinear optics equation derived by a general inverse method. Lett. Nuovo Cim., 8, 775-779.
Gibbon, J.D., James, I.N. and Moroz, I.M. 1979. An example of soliton behaviour in a rotating baroclinic fluid. Proc. Roy. Soc. Lond., 367A, 219-237.
Gibbon, J.D., James, I.N., and Moroz, I.M. 1979. The sine-Gordon equation as a model for a rapidly rotating baroclinic fluid. Physica Scripta, 20, 402-408.
Gibbon, J.D., and McGuinness, M.J. 1980. Nonlinear focusing and the Kelvin-Helmholtz instability. Phys. Letts., 77A, 118-21.
Gibbon, J.D., and McGuinness, M.J. 1980. A derivation of the Lorenz equations

for some dispersively unstable flows. Phys. Letts., 77A, 295-299.
Gibbon, J.D., and McGuinness, M.J. 1981. Amplitude equations at the critical points of unstable dispersive physical systems. Proc. Roy. Soc. Lond., A377, 185-219.
Gibbon, J.D., and McGuinness, M.J. 1982. The real and complex Lorenz equations in rotating fluids and lasers. Physica D, 5, 108-122.
Gibbs, H.M. and Slusher, R.E. 1972. Sharp line self-induced transparency. Phys. Rev., 6, 2326-2334.
Ginsberg, V.L. and Landau, L.D. 1965. On the theory of superconductivity. In ter Haar, D. (Ed.), Collected papers of L.D. Landau, , pp 546-568. Oxford: Pergamon.
Goddard, P., Nuyts, J. and Olive, D. 1977. Gauge theories and magnetic charges. Nucl. Phys., B125, 1-28.
Goldberg, A. and Schey, H.M. 1967. Computer-generated motion pictures of one-dimensional quantum-mechanical transmission and reflection phenomena. Am. J. Phys., 35, 177-186.
Goldstein, H. 1980. Classical Mechanics (2nd. Ed.). New York: Addison Wesley.
Gor'kov, L.P. 1959. Microscopic derivation of the Ginzberg-Landau equations in the theory of superconductivity. JETP, 9, 1364-67.
Gor'kov, L.P. 1960. Theory of superconducting alloys in a strong magnetic field near the critical temperature. JETP, 10, 998-1004.
Gor'kov, L.P. and Eliashberg, G.M. 1969. Superconducting alloys in a strong alternating field. JETP, 29, 698-703.
Gottfried, K. 1966. Quantum Mechanics. New York: Benjamin.
Gottlieb, D., and Orszag, S.A. 1977. Numerical Analysis of Spectral Methods. Philadelphia: SIAM conference series in Applied Mathematics.
Gourlay, A.R. 1971. Some recent methods for the numerical solution of time-dependent partial differential equations. Proc. Roy. Soc. Lond. A, 323, 219-235.
Greenspan, H. 1968. Theory of rotating fluids. Cambridge: Cambridge University Press.
Greig, I.S., and Morris, J. Ll. 1976. A hopscotch method for the KdV equation. J. Comp. Phys., 20, 64-80.
Griffiths, D.F., Mitchell, A.R., and Morris, J.Ll. 1982. A numerical survey of the nonlinear Schrödinger equation. University of Dundee preprint, 1982, submitted to Comp. Methods in Appl. Mech. and Eng.
Grimshaw, R.H.J. 1977. The modulation of an internal gravity wave packet and the resonance with the mean motion. Studies in Appl. Math., 56, 241-266.
Grishin, V.E., Katyshev, Yu,V., Makhaldiani, N.V., and Shvachka, A.B. 1980. Interaction dynamics and nontopological soliton stability in an essentially nonlinear model of a complex scalar field. Phys. Lett., 78A, 423-428.
Guillemin, V. and Pollack, A. 1974. Differential topology. New York: Prentice Hall.
Haken, H. 1978. Synergetics: An Introduction. Berlin: Springer-Verlag.
Harrison, B.K. 1978. Bäcklund transformation for the Ernst equation of general relativity. Phys. Rev. Lett., 41, 1197-1200.
Hasegawa, A., and Tappert, F. 1973. Transmission of stationary nonlinear optical pulses in dispersive dielectric firbres II. Appl. Phys. Letts., 23, 171-2.
Hasegawa, A. 1975. Plasma Instabilities and Nonlinear Effects. Berlin: Springer.
Hasegawa, A. and Kodama, Y. 1981. Signal transmissions by optical solitons in monomode fibre. Proc. IEEE, 69, 1145-1150.
Hasimoto, H, and Ono, H. 1972. Nonlinear modulation of gravity waves. J. Phys. Soc. Japan, 33, 805-11.
Henon, M. 1974. Integrals of the Toda lattice. Phys. Rev. B, 9, 1921-3.

Hershkowitz, N., Romesser, T. and Montgomery, D. 1972. Multiple soliton production and the KdV equation. Phys. Rev. Lett., 29, 1586-9.

Hide, R. 1958. An experimental study of thermal convection in a rotating fluid. Phil. Trans. R. Soc. Lond. A, 250, 441-478.

Hide, R. 1969. Some laboratory experiments on free thermal convection in a rotating fluid. In Corby, G.A. (Ed.), Global Circulation of the Atmosphere, : Roy. Met. Soc. Lond.

Hide, R. and Mason, P.J. 1975. Sloping convection in a rotating fluid. Adv. Phys., 24, 47-100.

Hide, R., Mason, P.J. and Plumb, A. 1977. Thermal convection in a rotating fluid subject to a horizontal temperature gradient. J. Atmos. Sci., 34, 930-950.

Hilton, P. 1966. An introduction to homotopy theory. Cambridge: C.U.P.

Hirota, R. 1971. Exact solution of the Korteweg-de Vries equation for multiple collisions of solitons. Phys. Rev. Lett., 27, 1192-1194.

Hirota, R. 1972. Exact solution of the sine-Gordon equation for multiple collisions of solitons. J. Phys. Soc. Japan, 33, 1459-1463.

Hirota, R. 1973. Exact N-soliton solution of the wave equation of long waves in shallow-water and in nonlinear lattices. J. Math. Phys., 14, 810-815.

Hirota, R. 1974. A new form of Bäcklund transformation and its relation to the inverse scattering problem. Prog. Theor. Phys., 52, 1498-1512.

Hirota, R. 1976. Direct methods of finding exact solutions of nonlinear evolution equations. In Miura, R.M. (Ed.), Bäcklund Transformations, . Berlin: Springer.

Hirota, R. 1977. Nonlinear partial difference equations III; discrete sine-Gordon equation. J. Phys. Soc. Japan, 43, 2079-2086.

Hirota, R. 1980. Direct methods in soliton theory. In Bullough, R.K. and Caudrey, P.J. (Ed.), Solitons, Berlin: Springer.

Hirota, R. 1981. Bilinear forms of soliton equations. Tech. Rep. No. A-9, Dept. of Appl. Math., Hiroshima University.

Hocking, J.G. and Young, G.S. 1961. Topology. New York: Addison-Wesley.

Hocking, L., Stewartson, K. and Stuart, J.T. 1972. A nonlinear instability burst in plane parallel flow. J. Fluid Mech., 51, 705-735.

Holmberg, B. 1952. A remark on the uniqueness of the potential determined from the asymptotic phase. Nuovo Cim., 9, 597-604.

Hopf, E. 1950. The partial differential equation $u_t + uu_x = \delta u_{xx}$. Comm. Pure Appl. Math., 3, 201-230.

Howard, L.N. and Krishnamurti, R. 1980. A model illustrating a possible mechanism for large scale flow in convection. Bull. Amer. Phys. Soc., 25, 1080.

Hylleraas, E.A. 1948. Calculation of a perturbing central field of force from the elastic scattering phase shift. Phys. Rev., 74, 48-51.

Hyman, J.M., McLaughlin, D.W., and Scott, A.C. 1981. On Davydov's alpha-helix solitons. Physica D, 3, 23-44.

Ichikawa, Y.H., Mitsuhashi, T. and Konno, K. 1976. Contribution of higher order terms in the reductive perturbation theory I: A case of weakly dispersive wave. J. Phys. Soc. Japan, 41, 1382-86.

Ichikawa, Y.H. 1979. Topics on solitons in plasmas. Physica Scripta, 20, 296-305.

Ikezi, M., Taylor, R. and Baker, R.D. 1970. Formation and interaction of ion acoustic solitary waves. Phys. Rev. Lett., 25, 11-14.

Ince, E.L. 1956. Ordinary differential equations. New York: Dover.

Infeld, I. and Hull, T.E. 1951. The factorisation method. Rev. Mod. Phys., 23, 21-68.

Iooss, G. and Joseph, D.D. 1981. Bifurcation Theory. Berlin: Springer.

Jacobs, L. and Rebbi, C. 1979. Interaction energy of superconducting vortices. Phys. Rev., B19, 4486-94.

Jacobson, D.A. 1965. Ginzberg-Landau equations and the Josephson effect.

Phys. Rev., 138, 1066-70.
Jaffe, A. and Taubes, C. 1980. Vortices and Monopoles. Boston: Birkhauser.
Jeffrey, H. and Jeffrey, B.S. 1946. Methods of Mathematical Physics. Cambridge: C.U.P.
Jeffrey, A. and Kakutani, T. 1972. Weak nonlinear dispersive waves. SIAM Review, 14, 582-643.
Jimbo, M., Miwa, T. and Ueno, K. 1981. Monochromy preserving deformations of linear ordinary differential equations with rational coefficients. I - General theory and tau-functions. Physica D, 2, 306-351.
Jimbo, M. and Miwa, T. 1981. Monochromy preserving deformations of linear ordinary differential equations with rational coefficients. II. Physica Scripta D, 2, 407-448.
Jimbo, M. and Miwa, T. 1981. Monochromy preserving deformations of linear ordinary differential equations with rational coefficients. III. Physica Scripta D, 4, 26-46.
Johnson, R.S. 1973. On the development of a solitary wave moving over an uneven bottom. Proc. Cam. Phil. Soc., 73, 183-230.
Jordan, D., and Smith, P. 1977. Nonlinear Ordinary Differential Equations. Oxford: Applied Mathematics and Computing Series.
Jost, R. and Kohn, W. 1952. Construction of a potential from a phase shift. Phys. Rev., 87, 979-992.
Jost, R. and Kohn, W. 1952. Equivalent potentials. Phys. Rev., 88, 382-385.
Kac, M. and van Moerbeke, P. 1975. On an explicitly soluble system on nonlinear differential equations related to certain Toda lattices. Adv. in Math., 16, 160-9.
Kadomtsev, B.B., and Petviashvili, V.I. 1970. The stability of solitary waves in weakly dispersive media. Dokl. Akad. Nauk SSR, 192, 753-756.
Kako, M. and Rowlands, G. 1976. Two dimensional stability of ion acoustic solitons. Plasma Physics, 18, 165-70.
Kakutani, T., Ono, H., Taniuti, T. and Wei, C.C. 1968. Reductive perturbation method in nonlinear wave propagation II. Application to hydromagnetic waves in cold plasma. J. Phys. Soc. Japan, 24, 1159-66.
Kakutani, T., and Sugimoto, N. 1974. Krylov-Bogoliubov-Mitropolsky method for nonlinear wave modulation. Phys. of Fluids, 17, 1617-25.
Karpman, V.I. 1967. The structure of 2-dimensional flow round bodies in dispersive media. Sov. Phys. JETP, 25, 1102-11.
Karpman, V.I., and Krushkal, E.M. 1969. Modulated waves in nonlinear dispersive media. Sov. Phys. (JETP), 28, 277-81.
Karpman, V.I. 1975. Nonlinear waves in dispersive media. London: Pitman.
Kato, T. 1966. Perturbation theory for linear operators. Berlin: Springer.
Kaup, D. 1975. Exact quantisation of the nonlinear Schrödinger equation. J. Math. Phys., 16, 2036-41.
Kaup, D.J. 1976. The three wave interaction - A nondispersive phenomenon. Studies in App. Math., 55, 9-44.
Kaup, D.J. 1978. Simple harmonic generation: an exact method of solution. Studies in Appl. Math., 59, 25-35.
Kaup, D.J., Reiman, A.H., Bers, A. 1979. Space-time evolution of nonlinear three-wave interactions I. Rev. Mod. Phys., 51, 275-309.
Kaup, D.J. 1980. On the inverse scattering problem for cubic eigenvalue problems of the class $\psi_{xxx} + 6Q\psi_x + 6R\psi = \lambda\psi$. Studies in Appl. Math., 62, 189-216.
Kawahara, T. 1969. Oblique nonlinear hydromagnetic waves in a collision free plasma with isothermal electron pressure. J. Phys. Soc. Japan, 27, 1331-40.
Kawata, T. and Inoue, H. 1978. Inverse scattering method for the nonlinear evolution equations under nonvanishing conditions. J. Phys. Soc. Japan, 44, 1722-9.
Kawata, T., Sakai, J. and Kobayashi, N. 1980. Inverse method for the mixed

nonlinear Schrödinger equation and soliton solutions. J. Phys. Soc. Japan, 48, 1371-9.
Kay, I. and Moses ,H.E. 1955. The determination of the scattering potential from the spectral measure function I: Continuous spectrum. Nuovo Cimento, 2, 917-961.
Kay, I. and Moses ,H.E. 1956. The determination of the scattering potential from the spectral measure function II: Point eigenvalues and proper eigenfunctions. Nuovo Cimento, 3, 66-84.
Kay, I. and Moses ,H.E. 1956. The determination of the scattering potential from the spectral measure function III: Calculation of the scattering potential from the scattering operator for the one-dimensional Schrödinger equation. Nuovo Cimento, 3, 276-304.
Keller, J.B., Kay, I., and Shmoys, J. 1956. Determination of the potential from the scattering data. Phys. Rev., 102, 557-559.
Keller, J.B. 1967. Inverse problems. Am. Math. Monthly, 83, 107-118.
Kelley, P.L. 1965. Self-focussing of laser beams. Phys. Rev. Lett., 15, 1005-1008.
Kever, H. and Morikawa, G.K. 1969. KdV equation for nonlinear hydromagnetic waves in a warm collision free plasma. Physics of Fluids, 12, 2090-3.
Khalifa, A.K.A. 1979. Theory and application of the collocation method with splines for ordinary and partial differential equations. PhD thesis, Heriot-Watt University,
Kitchenside, P.W., Caudrey, P.J., Bullough, R.K. 1979. Soliton-like spin waves in He B. Physica Scripta, 20, 673-680.
Kodama, K. and Taniuti, T. 1978. Higher order approximation in the reductive perturbation method I and II. J. Phys. Soc. Japan, 45, 298-314.
Kodama, Y., Ablowitz, M.J. and Satsuma, J. 1982. Direct and scattering problems of the nonlinear intermediate long wave equation. J. Math. Phys., 23, 564-576.
Konno, K, Mitsuhashi, T. and Ichikawa, Y.H. 1977. Dynamical processes of the dressed ion acoustic solitons. J. Phys. Soc. Japan, 43, 669-74.
Konno, K., and Suzuki, H. 1979. Self-focussing of laser beam in nonlinear media. Physica Scripta, 20, 382-386.
Korepin, V.E. and Faddeev, L.D. 1975. Quantisation of solitons. Theor. Math. Phys., 25, 1039-49.
Korteweg, D.J., and de Vries, G. 1895. On the change of form of long waves advancing in a rectangular canal, and on a new type of long stationary waves. Phil. Mag., 39, 422-443.
Kramer, D. and Neugebauer, G. 1979. The superposition of two Kerr solutions. Phys. Letts., 75A, 259-261.
Krein, M.G. 1953. On the transfer function of a one-dimensional boundary problem of the second order. Doklady, Akad. Nauk., 88, 405-408.
Krein, M.G. 1955. On determination of the potential of a particle from its s-function. Doklady, Akad. Nauk., 105, 433-6.
Krichever, I.M. 1977. Methods of algebraic geometry in the theory of nonlinear equations. Russ. Math. Surveys, 32, 185-213.
Kruskal, M.D., Miura, R.M., and Gardner, C.S. 1970. Korteweg-de Vries equation and generalizations. V. Uniqueness and nonexistence of polynomial conservation laws. J. Math. Phys., 11, 952-960.
Kudryavtsev, A.E. 1975. Solitonlike solutions for a Higgs scalar field. JETP Lett., 22, 82-83.
Kuramoto, Y. 1978. Diffusion induced chaos in reaction systems. Suppl. Prog. Theor. Phys., 64, 364-367.
Lake, B.M., Yuen, H.C., Rungaldier, H., and Ferguson, W. 1977. Nonlinear deep-water waves: theory and experiment. Part 2 Evolution of a continuous wave train. J. Fluid Mech., 83, 49-74.
Lamb, H. 1932. Hydrodynamics. Cambridge: CUP.
Lamb, G.L. 1970. Higher conservation laws in ultrashort optical pulse propa-

gation. Phys. Lett., 32A, 251-252.
Lamb, G.L. 1971. Analytic descriptions of ultrashort optical pulse propagation in a resonant medium. Rev. Mod. Phys., 43, 99-124.
Lamb, G.L. 1973. Phase variation in coherent optical pulse propagation. Phys. Rev. Lett., 31, 196-199.
Lamb, G.L. 1973. On the connection between lossless propagation and pulse profile. Physica, 66, 298-314.
Lamb, G.R., Jr. 1976. Bäcklund transformations at the turn of the century. In Miura, R.M. (Ed.), Lecture Notes in Mathematics, 515: Bäcklund Transformations, Berlin: Springer.
Lamb, G.L. Jr. 1980. Elements of soliton theory. New York: J. Wiley.
Lamb, W.E. Jr. 1964. Theory of optical maser oscillators. In Miles, P.A. (Ed.), Proceedings of the Int. School of Phys. 'Enrico Fermi', Course XXXI, , pp 78-110. New York: Academic Press.
Landau, L.D. and Lifshitz, E. 1958. Quantum Mechanics. Oxford: Pergamon.
Landau, L.D. and Lifshitz, E.M. 1959. Fluid Mechanics. Oxford: Pergamon Press. Vol. 6 of Course of Theoretical Physics, 2nd. Ed.
Landau, L.D. and Lifshitz, E. 1960. Electrodynamics of continuous media. Oxford: Pergamon.
Landau, L.D. and Lifshitz, E.M. 1965. Quantum Mechanics. Oxford: Pergamon Press. Vol. 3 of Course of Theoretical Physics, 2nd. Ed.
Landau, L.D. 1969. On the theory of the dispersion of magnetic permability in ferromagnetic bodies. In ter Haar, D. (Ed.), Collected papers of L.D. Landau, . Oxford: Pergamon. pages 101-114.
Lang, S. 1973. Elliptic functions. New York: Addison Wesley.
Lange, C., and Newell, A.C. 1971. The post buckling problem for thin elastic shells. SIAM J. of Appl. Math., 21, 605-629.
Lange, C., and Newell, A.C. 1974. A stability criterion for envelope equations. SIAM J. of Appl. Math., 27, 441-456.
Lax, P.D. 1968. Integrals of nonlinear equations of evolution and solitary waves. Commun. Pure and Appl. Math., 21, 467-490.
Lax, P.D. 1975. Periodic solutions of the KdV equation. Comm Pure Appl. Math. 28, 141-188.
Lax, P.D. 1976. Almost periodic solutions of the KdV equation. SIAM Review, 18, 351-375.
Lax, P.D. 1978. A Hamiltonian approach to the KdV and other equations. In Crandall, M.G. (Ed.), Nonlinear Evolution Equations, London: Academic Press.
Lee, T.D. 1976. Examples of four dimensional soliton solutions and abnormal nuclear states. Phys. Rep., 23, 254-8.
Leibbrandt, G., Morf, R. and Wang, S. 1980. Solutions of the sine-Gordon equation in higher dimensions. J. Math. Phys., 21, 1613-1624.
Leibovich, S., and Seebass, A.R. (Eds.). 1974. Nonlinear waves. Ithaca: Cornell Univ. Press.
Leibovitch, S. 1970. Weakly nonlinear waves in rotating fluids. J. Fluid Mech., 42, 803-22.
Levinson, N. 1949. On the uniqueness of the potential in a Schrödinger equation for a given asymptotic phase. Danske. Vid. Selsk. Mat-Fys. Medd., 25(9), pp29.
Lighthill, M.J. 1978. Waves in fluids. Cambridge: CUP.
Lin, J. and Kahn, P. 1982. Phase and amplitude instability in delay-diffusion population models. J. Math. Biol., 13, 383-393.
Ljance, V.E. 1966. A differential operator with spectral singularities. Amer. Math. Soc. Trans., (2) 60, 185-225.
Ljance, V.E. 1967. An analog of the inverse problem of scattering theory for a non-self-adjoint operator. Math USSR -Sbornik, 1, 485-503.
Lomdahl, P.S., Olsen, O.H., and Christiansen, P.L. 1980. Return and collapse of solutions to the nonlinear Schrödinger equation in cylindrical sym-

metry. Phys. Lett., 78A, 125-128.
Lomdahl, P.S. 1982. Soliton dynamics in nonintegrable sine-Gordon systems. PhD thesis, LAMF, Technical University of Denmark, Lyngby,
Lomdahl, P.S., Soersen, O.H., Christiansen, P.L., Scott, A.C., and Eilbeck, J.C. 1981. Bunched multi-solitons in Josephson tunnel junctions. Phys. Rev. B, 24, 7460-7462.
Longuet-Higgins, M.S., and Cokelet, E.D. 1976. The deformation of steep surface waves on water I. A numerical method of computation. Proc. Roy. Soc. Lond., A350, 1-26.
Lonngren, K. and Scott, A. 1978. Solitons in Action. New York: Academic Press.
Lorenz, E.N. 1963. Deterministic nonperiodic flow. J. Atmos. Sci., 20, 130-141.
Lugovoi, V.N. and Prokhorov, A.M. 1974. Theory of high power laser radiation in a nonlinear medium. Sov. Phys. Uspekhi, 16, 658-678.
Lund, F. and Regge, T. 1976. Unified approach to strings and vortices with soliton solutions. Phys. Rev., D14, 1524-35.
Lund, F. 1977. Notes on the geometry of the nonlinear sigma-model in two dimensions. Phys. Rev., D15, 1540-3.
Lüscher, M. and Pohlmeyer, K. 1978. Scattering of massless lumps and nonlocal charges in the two dimensional classical sigma-model. Nuc. Phys., B137, 46-54.
Luther, A. 1976. Eigenvalue spectrum of interacting massive fermions in one dimension. Phys. Rev. B, 14, 2153-9.
Ma, Y.-C. 1978. The complete solution of the long wave - short wave resonance equations. Studies in Appl. Math., 59, 201-21.
Madore, J. 1981. Geometrical methods in classical field theory. Phys. Rep., 75, 125-204.
Madsen, O.S. and Mei, C.C. 1969. The transformation of a solitary wave over an uneven bottom. J. Fluid Mech., 39, 781-91.
Maison, D. 1979. On the complete integrability of the stationary axially symmetric Einstein equations. J. Math. Phys., 20, 871-7.
Makhankov,V.G. 1978. Dynamics of classical solitons (in non-integrable systems). Phys.Rep., 35, 1-128.
Makhankov, V.G., Kummer, G., and Shvachka, A.B. 1979. Many dimensional U(1) solitons, their interactions, resonances and bound states. Physica Scripta, 20, 454-461.
Makhankov, V.G. 1979. Computer and solitons. Physica Scripta, 20, 558-562.
Makhankov, V. 1980. Computer experiments in soliton theory. Comp. Phys. Comm. 21, 1-49.
Makhankov, V.G., Litvinenko, E.I., and Shvachka, A.B. 1981. Numerical investigation of the Kadomtsev-Petviashvili soliton stability. Comp. Phys. Comm. 2, 223-232.
Makhankov, V.G., Kummer, G. and Shvachka, A.B. 1981. Novel Pulsons (or stability from instability). Physica D, 3, 344-349.
Manakov, S.V. 1974. On the theory of two-dimensional stationary self-focusing of electromagnetic waves. Sov. Phys. JETP, 38, 248-253.
Manakov, S.V. 1975. Complete integrability and stochastization of discrete dynamical systems. Sov. Phys. JETP, 40, 269-274.
Manakov, S.V. 1976. Note on the integration of Euler's equations of the dynamics of an n-dimensional rigid body. Funct. Anal. Appl., 10, 93-??
Manakov, S.V., Zakharov, V.E., Bordag, L.A., Its, A.R. and Matveev, V.B. 1977. Two dimensional solitons of the Kadomtsev-Petviashvili equation and their interaction. Phys. Lett. A., 63, 203-6.
Manakov, S.V. and Zakharov, V.E. 1981. Three-dimensional model of relativistic-invariant field theory, integrable by the inverse scattering transform. Lett. Math. Phys., 5, 247-253.
Manakov, S.V. 1982. Pulse propagation in a long laser amplifier. Sov. Phys.

JETP Lett., 35, 237-240.
Manton, N.S. 1978. Complex structure of monopoles. Nucl. Phys., B135, 319-332.
Marchenko, V.A. 1950. Concerning the theory of a differential operator of the second order. Doklady Akad. Nauk., 72, 457-460.
Marchenko, V.A. 1952. Some questions of the theory of one dimensional differential operators of the second order, I. Trudy Moskov. Mat., 1, 327-420.
Marchenko, V.A. 1952. Some questions of the theory of one dimensional differential operators of the second order, II. Trudy Moskov. Mat., 2, 3-83.
Marchenko, V.A. 1955. On the reconstruction of the potential energy from phases of the scattered waves. Doklady Akad. Nauk., 104, 695-698.
Marchenko, V.A. 1960. Eigenfunction expansions for non-self-adjoint singular differential operators of the second order. Amer. Math. Soc. Trans., 25, 77-130.
Marsden, J. and McCracken, M. 1976. The Hopf Bifurcation and its Applications. Berlin: Springer series in Pure and Applied Sciences No. 19.
Maxon, S., and Viecelli, J. 1974. Cylindrical solitons. Phys. Fluid., 17, 1614-1616.
McCall, S.L., and Hahn, E.L. 1967. Self-induced transparency by pulsed coherent light. Phys. Rev. Lett., 18, 908-911.
McCall, S.L., and Hahn, E.L. 1969. Self-induced transparency. Phys. Rev., 183, 457-485.
Meis, T. and Marcowitz, U. 1978. Numerical solution of partial differential equations. Berlin: Springer.
Mermin, N.D. 1979. The topological theory of defects in ordered media. Rev. Mod. Phys., 51, 591-648.
Mertzbacher, E. 1961. Quantum mechanics. New York: John Wiley.
Messiah, A. 1961,1962. Quantum Mechanics. New York: North Holland.
Michel, L. 1980. Symmetry defects and broken symmetry. Rev. Mod. Phys., 52, 617-651.
Miles, J.W. 1977. Resonantly interacting solitary waves. J. Fluid. Mech, 79, 171-179.
Miles, J.W. 1981. The Korteweg-de Vries equation: a historical essay. J. Fluid. Mech., 106, 131-147.
Miller, J.J.H. 1971. On the location of zeros of certain classes of polynominals with applications for numerical analysis. J. Inst. Maths Applics, 8, 397-406.
Milnor, J. 1965. Topology from the differential viewpoint. Charlottesville: Univ. Virginia Press.
Mitchell, A.R., and Wait, R. 1977. The Finite Element Method in Partial Differential Equations. London: J. Wiley.
Mitchell, A.R., and Griffiths, D.F. 1980. The Finite Difference Method in Partial Differential Equations. London: John Wiley.
Mitchell, A.R. and Schoombie, S.W. 1981. Finite element studies of Solitons. To be published in the Proceedings of the Conference on Finite Element Methods in Coupled Problems, Swansea, September 1981.
Mittra, R.K. 1973. Inverse scattering and remote probing. In Computer techniques of electromagnetics, (Ch. 7), Oxford: Pergamon.
Miura, R.M. 1968. Korteweg-de Vries equation and generalizations. I. A remarkable explicit nonlinear transformation. J. Math. Phys, 9, 1202-1204.
Miura, R.M., Gardner, C.S., and Kruskal, M.D. 1968. Korteweg de Vries equation and generalizations. II. Existence of conservation laws and constants of the motion. J. Math. Phys, 9, 1204-1209.
Miura, R.M. 1974. The KdV equation: A model equation for nonlinear dispersive waves. In Leibovitch, S. and Seebass, A. (Ed.), Nonlinear Waves, Ithaca, N.Y.: Cornell University Press.
Miura, R.M. (Ed.). 1976. Bäcklund Transformations, the Inverse Scattering Method, Solitons, and Their Applications. Berlin: Springer. Lecture

Notes in Mathematics, Vol. 515.

Miura, R.M. 1976. The Korteweg-de Vries equation: a survey of results. SIAM Review, 18, 412-459.

Mohapatra, R.N. and Lai, C.H. 1981. Selected papers on gauge theory of fundamental interactions. Singapore: World Sci. Pub. Co.

Mollenauer, L., Stolen, R., and Gordon, J. 1980. Experimental observation of picosecond pulse narrowing and solitons in optical fibres. Phys. Rev. Letts., 45, 1095-8.

Montgomery, D., and Tidman, D. 1964. Secular and nonsecular behaviour for the cold plasma equations. Phys of Fluids, 7, 242-9.

Moroz, I.M. and Brindley, J. 1981. Evolution of baroclinic wave packets in a flow with continuous shear and stratification. Proc. Roy. Soc. Lond., 377A, 379-404.

Morris, H.C. and Dodd, R.K. 1980. A 2-connection and operator bundles for the Ernst equation for axially symmetric gravitational fields. Phys. Lett., 75A, 20-22.

Morton, K.W. 1964. Finite compression waves in a collision free plasma. Physics of Fluids, 7, 1801-15.

Moser, J. (ed.). 1975. Dynamical Systems. Theory and Applications. Berlin: Springer-Verlag. Lecture notes in physics no. 38.

Moser, J. 1975. Three Integrable Hamiltonian systems connected with isospectral deformations. Adv. in Math., 16, 197-220.

Murray, A.C. 1978. Solutions of the KdV equation evolving from irregular data. Duke Math. J., 45, 149-181.

Naimark, M.A. 1968. Linear differential operators in Hilbert spaces II: London: Harrap.

Nairboli, G.A. 1970. Nonlinear longitudinal dispersive waves in elastic rods. J. Math. Phys. Sci., 4, 64-73.

Narayanamurti, V. and Varma, C.M. 1970. Nonlinear propagation of heat pulses in solids. Phys. Rev. Lett., 25, 1105-8.

Nayfeh, A. 1973. Perturbation Methods. New York: Wiley Interscience.

Nayfeh, A., and Mook, D. 1979. Nonlinear Oscillations. New York: Wiley Interscience (Pure and Applied Mathematics).

Neugebaur, G. and Kramer, D. 1980. The superposition of two Kerr solutions. Phys. Lett., 75A, 259-261.

Newell, A.C. and Whitehead, J.A. 1969. Finite bandwidth, finite amplitude convection. J. Fluid Mech., 38, 279-303.

Newell, A.C., Lange, P.J. and Aucoin, P.J. 1970. Random convection. J. Fluid Mech., 40, 513-542.

Newell, A.C. (ed.). 1974. Nonlinear Wave Motion. Providence, R.I.: American Math. Soc.

Newell, A.C. 1977. Finite amplitude instabilities of partial difference equations. SIAM J. Appl. Math., 33, 133-160.

Newell, A.C. 1978. Soliton perturbations and Nonlinear Focusing. In A. Bishop and T. Schneider (Ed.), Solitons and Condensed Matter Physics, Springer Solid State Sciences Series, Vol. 8, Berlin: Springer.

Newell, A.C. 1978. Long waves - short waves, a solvable model. SIAM J. Appl. Math., 35, 650-64.

Newell, A.C. 1979. Bifurcation and nonlinear focusing. In Haken, H. (Ed.), Pattern Formation by Dynamical Systems and Pattern Recognition, Berlin: Springer.

Newell, A.C. 1979. The general structure of integtable evolution equations. Proc. Roy. Soc., A365, 283-311.

Nicholson, D.R., and Goldman, M.V. 1976. Damped Nonlinear Schrödinger Equation. Phys. Fluids, 19, 1621-5.

Nicholson, D.R. and Goldman, M.V. 1978. Virial theory of direct Langmuir collapse. Phys. Rev. Letts., 41, 406-410.

Novikov, S.P. 1974. A periodic problem for the KdV equation I: Funct. Anal.

Appl., 8, 236-246. translated from Funct. Anal. i Pril. 8 (1974) 54-66.
Olive, D., Cuito, S. and Crewther, R.J. 1979. Instantons in field theory. Riv. Nuovo Cim., 8, 1-117.
Olver, P.J. 1979. Euler operators and conservation laws of the BBM equation. Math. Proc. Camb. Phil. Soc., 85, 143-160.
Osborne, A.R. and Burch, T.L. 1980. Internal solitons in the Andaman Sea. Science, 208, 451-460.
Osborne, A.R. and Provenzale, A. 1982. Numerical methods for evaluation of the spectral transform of localized wave fields described by the KdV equation. to be published in the proceedings of the Recontre Interdisiplinaire Problemes Inverses L'Universite des Sciences et Techniques du Languedoc, Montpellier, France, December 1981.
Ostrovsky, L.A. and Sutin, A.M. 1977. Nonlinear elastic waves in rods. Appl. Math. and Mech., 41, 531-537.
Parkes, R.D. 1969. Superconductivity Vols. I & II. New York: Marcel Dekker Inc.
Parmentier, R. 1978. Fluxons on Long Josephson Junctions. In Scott, A.C. (Ed.), Solitons in Action, , pp 173-199. New York: Academic Press.
Patani, A., Scheindwein, M. and Shafi, Q. 1976. Topological charges in field theory. J. Phys. A., 9, 1513-20.
Pawlik, M. and Rowlands, G. 1975. The propagation of solitary waves in piezoelectric semiconductors. J.Phys.C., 8, 1189-1204.
Pedlosky, J. 1970. Finite amplitude baroclinic waves. J. Atmos. Sci., 27, 15-30.
Pedlosky, J. 1971. Finite amplitude baroclinic waves with small dissipation. J. Atmos. Sci., 28, 587-597.
Pedlosky, J. 1971. Geophysical fluid dynamics. In W. Reid (Ed.), Mathematical Problems in the Geophysical Sciences, . Providence: AMS.
Pedlosky, J. 1972. Finite amplitude baroclinic wave packets. J. Atmos. Sci., 29, 680-686.
Pedlosky, J. 1980. Geophysical fluid dynamcis. New York: Springer.
Pen-Yu, Kuo, and Sanz-Serna, J.M. 1981. Convergence of methods for the numerical solution of the KdV equation. IMA J Num. Anal., 1, 215-221.
Peregrine, D.H. 1966. Calculation of the development of an undular bore. J. Fluid Mech., 25, 321-330.
Perring, J.K., and Skyrme, T.H.R. 1962. A model unified field equation. Nucl. Phys., 31, 550-555.
Phillips, N.A. 1951. A simple 3-dim. model for the study of large scale extratropical flow patterns. J. Met., 8, 381-394.
Pines, D. 1962. The many body problem. New York: Benjamin.
Ranada, A.F. 1979. Nonlinear problems in theoretical physics. Berlin: Springer. Lecture notes in physics no.98.
Rebbi, C. 1980. Interaction of superconducting votices. In Harnad, J.P. and Shnider, S. (Ed.), Geometrical and topological methods in gauge theories, pp 96-113. Berlin : Springer.
Redekopp, L. 1977. On the theory of solitary Rossby waves. J. Fluid Mech., 82, 725-45.
Reid, B.K. and Walker, J.H. 1979. SCRIBE Introductory User's Manual (Second Edition). Pittsburg: Computing Science Department, Carnegie-Mellon University.
Reiman, A.H. 1979. Space-time evolution of nonlinear three wave interactions II. Rev. Mod. Phys., 51, 311-14.
Richtmyer, R.D., and Morton, K.W. 1967. Difference Methods for Initial-Value Problems (2nd Ed.). London: J. Wiley.
Rickayzen, G. 1965. Theory of superconductivity. New York: Wiley.
Romea, R. 1977. The effects of friction and bets on finite amplitude baroclinic waves. J. Atmos. Sci., 34, 1689-95.
Rosenberg, H.M. 1975. The solid state. Oxford: O.U.P.

Rubenstein, J. 1970. Sine-Gordon equation. J. Math. Phys., 11, 258-266.
Ruelle, D. and Takens, F. 1971. On the nature of turbulence. Comm. Math. Phys., 20, 167-192.
Ruelle, D. 1980. Strange attractors. Mathematical Intelligencer, 2, 126-137. Autumn Issue.
Santarelli, A.R. 1978. Numerical analysis of the regularized long-wave equation: anelastic collision of solitary waves. Nuovo Cim., 46B, 179-188.
Sanz-Serna, J.M. and Christie, I. 1981. Petrov-Galerkin methods for nonlinear dispersive waves. J. Comp. Phys., 39, 94-102.
Satsuma, J. 1976. N-soliton solution of the two-dimensional KdV equation. J. Phys. Soc. Japan, 40, 286-290.
Satsuma, J., Ablowitz, M.J. and Kodama, Y. 1979. On an internal wave equation describing a stratified fluid with finite depth. Phys. Lett. A, 73, 283-286.
Schamel, H., and Elsasser, K. 1976. The application of the spectral method to nonlinear wave propagation. J. Comp. Phys., 22, 501-516.
Schamel, H. 1979. Role of trapped particles and waves in plasma solitons - Theory and applications. Physica Scripta, 20, 306-316.
Schiff, L.I. 1949. Quantum mechanics. New York: McGraw-Hill.
Schlesinger, L. 1912. Reine Angewandte Math., 141, 96-145.
Schneider, T., and Stoll, E. 1980. Classical statistical mechanics of the sine-Gordon and Phi-four chains. II. Static properties. Phys. Rev. B, 22, 5317-5338.
Schneider, T., and Stoll, E. 1981. Classical statistical mechanics of the sine-Gordon and Phi-four chains. II. Dynamic properties. Phys. Rev. B, 23, 4631-4660.
Schoombie, S.W. 1982. Spline Petrov-Galerkin methods for the numerical solution of the KdV equation. IMA J. Num. Anal., 2, 95-109.
Scott, A.C. 1969. A nonlinear Klein-Gordon equation. Amer. J. Phys., 37, 52-61.
Scott, A.C. 1970. Active and nonlinear wave propagation in electronics. New York: Wiley-Interscience.
Scott, A.C., Chu, F.Y.F., and McLaughlin, D.W. 1973. The Soliton: a new concept in applied science. Proc. IEEE, 61, 1443-1483.
Scott Russell, J. 1840. Experimental researches into the laws of certain hydrodynamic phenomena that accompany the motion of floating bodies, and have not previously been reduced into conformity with the known laws of the Resistance of Fluids. Edinb. Roy. Soc. Trans., 14, 47-109, + 2 plates.
Scott Russell, J. 1844. Report on waves. In Rep. 14th Meeting of the British Assoc. for the Advancement of Science, London: John Murray.
Scott Russell, J. 1885. The Wave of Translation. : London.
Scott, A.C. 1981. The laser-raman spectrum of a Davydov soliton. Physics Letters, 86A, 60-62.
Scott, A.C. 1982. Dynamics of Davydov solitons. To be published in Physical Review B (1982).
Scott, A.C. 1982. The vibrational structure of Davydov solitons. to be published in Physica Scripta (1982).
Seeger, A., Donth, H. and Kochendorfer. 1953. Theorie der Versetzungen in eindimensionalen Atomreihen.III. Versetzungen, Eigenbewegungen und ihre Wechselwirkung. Zeitschrift fur Physik, 134, 173-193.
Seeger, A. and Schiller, P. 1966. Kinks and dislocation lines and their effects on internal friction in crystals. In Mason, P.A.W. (Ed.), Physical Acoustics, Vol. 3, , pp 361-495. New York: Academic Press.
Shiefman, J., and Kumar, P. 1979. Interaction between soliton pairs in a double sine-Gordon equation. Physica Scripta, 20, 435-439.
Siegel, C.L. and Moser, J. 1971. Lectures in celestial mechanics. Berlin:

Springer.
Skyrme, T.H.R. 1958. A non-linear theory of strong interactions. Proc. Roy. Soc. Lond., A247, 260-278.
Smirnov, V.A. 1964. A course of higher mathematics V: Oxford: Pergamon Press.
Smith, K., Solomon, D. and Wagner, S. 1977. Practical and mathematical aspects of the problem of reconstructing objects from radiographs. Bull. Am. Math. Soc., 83, 1227-70.
Smith, R.K. 1977. On a theory of amplitude vacillation in baroclinic waves. J. Fluid Mech., 79, 289-306.
Solymar, L. 1972. Superconducting tunnelling and applications. London: Chapman and Hall.
Spanier, E. 1966. Algebraic topology. New York: McGraw Hill.
Stewartson, K. and Stuart, J.T. 1971. Nonlinear instability of plane Poiseuille flow. J. Fluid. Mech., 1971, 529-545.
Stoker, J.J. 1957. Water Waves: The mathematical theory with applications. New York: Wiley Interscience.
Stokes, G.G. 1847. On the theory of oscillatory waves. Camb. Trans., 8, 441-473.
Strauss, W.A. 1978. Nonlinear Invariant Wave Equations. In Velo, G. and Wightman, A. (Ed.), Invariant Wave Equations, Berlin: Springer-Verlag.
Strauss, W.A. and Vazquez, L. 1978. Numerical solution of a nonlinear Klein-Gordon equation. J. Comp. Phys., 28, 271-278.
Stuart, J.T. 1960. On the nonlinear mechanics of wave disturbances in stable and unstable parallel flows I. J. Fluid Mech., 9, 353-370.
Stuart, J.T. 1971. Nonlinear stability theory. Ann. Rev. Fluid Mech., 3, 347-70.
Stuart, J.T., and DiPrima, R.C. 1978. The Eckhaus and Benjamin-Feir resonance mechanisms. Proc. Royal Soc. Lond., A362, 27-41.
Su, C.S., and Gardner, C.S. 1969. The KdV equation and generalizations. III. Derivation of the Korteweg-de Vries equation and Burgers' equation. J. Math. Phys, 10, 536-539.
Taha, T.R. and Ablowitz, M.J. 1982. On analytic and numerical aspects of certain nonlinear evolution equations, I-III. Clarkson College preprints I.F.N.S. #14-#16.
Takhtadzhyan, L.A. and Faddeev, L.D. 1974. Essantially nonlinear one-dimensional model of classical field theory. Theor. Math. Phys., 21, 1046-57.
Takhtadzhyan, L.A. and Faddeev, L.D. 1974. . Usp. Mat. Nauk, 29, 249.
Takhtadzhyan, L.A. 1977. Integration of the continuous Heisenberg spin chain through the inverse scattering method. Phys. Lett. A, 64, 235-7.
Tanaka, S. 1974. KdV equation: construction of solutions in terms of scattering data. Osaka J. Math., 11, 49-59.
Taniuti, T., and Wei, C.-C. 1968. Reductive perturbation method in nonlinear wave propagation propagation. J. Phys. Soc. Japan, 24, 941-6.
Taniuti, T. and Washimi, M. 1968. Self-trapping and instability of hydromagnetic waves along the magnetic field in a cold plasma. Phys. Rev. Lett. 21, 209-12.
Taniuti, T., and Yajima, N. 1969. Perturbation method for a nonlinear wave modulation I. J. Math. Phys., 10, 1369-1372.
Taniuti, T., and Yajima, N. 1973. Perturbation method for a nonlinear wave modulation III. J. Math. Phys., 14, 1389-97.
Tappert, F. and Varma, C.M. 1970. Asymptotic theory of self-trapping of heat pulses in solids. Phys. Rev. Lett., 25, 1108-11.
Tappert, F. and Zabusky, N.J. 1971. Gradient induced fission of solitons. Phys. Rev. Lett., 27, 1774-76.
Tappert, F. and Judice, C.N. 1972. Recurrence of nonlinear ion acoustic waves. Phys. Rev. Lett., 29, 1308-11.
Tappert, F. 1972. Improved KdV equation for ion acoustic waves. Physics of

Fluids, 15, 2446-7.
Tappert, F.D. 1974. Numerical solutions of the KdV equation and its generalisations by the split-step Fourier method. Lect. Appl. Math. Am. Math. Soc., 15, 215.
Tinkham, M. 1975. Introduction to superconductivity. New York: McGraw Hill.
Titchmarsh, E.C. 1948. Theory of Fourier Integrals. Oxford: O.U.P.
Tjon, J. and Wright, J. 1977. Solitons on the continuous Heisenberg spin chain. Phys. Rev., B15, 3470-6.
Toda, H. 1962. Composition methods in homotpy groups of spheres. Princeton: Princeton Univ. Press.
Toda, M. 1970. Waves in nonlinear lattices. Prog. Theor. Phys. Suppl., 45, 174-200.
Toda, M. 1976. Development of the theory of a nonlinear lattice. Prog. Theor. Phys. Suppl., 59, 1-35.
Toda, M. 1981. Theory of nonlinear lattices. Berlin: Springer.
Tran, M. Q. 1979. Ion acoustic solitons in a plasma: A review of their experimental properties and related theories. Physica Scripta, 20, 317-327.
Trubowitz, E. 1977. The inverse problem for periodic potentials. Comm. Pure and Appl. Math., 30, 321-337.
Trullinger, S. 1979. Dynamic polarizability of the double quadratic kink. J. Math. Phys., 21, 592-8.
Vliegenthart, A.C. 1971. On finite-difference methods for the KdV equation. J. Engrg. Math., 51, 137-155.
Wadati, M., and Toda, M. 1972. The exact n-soliton solution of the Korteweg-de Vries equation. J. Phys. Soc. Japan, 32, 1403-1411.
Wadati, M. 1972. The exact solution of the modified Korteweg-de Vries equation. J. Phys. Soc. Japan, 32, 1681.
Wahlquist, H.D. and Estabrook, F.B. 1973. Bäcklund transformation for solutions of the KdV equation. Phys. Rev. Lett., 31, 1386-1390.
Washimi, M. and Taniuti, T. 1966. Propagation of ion acoustic solitary waves of small amplitude. Phys. Rev. Lett, 17, 996-8.
Watson, J.N. 1960. Nonlinear mechanics of disturbances in parallel flows. J. Fluid Mech., 9, 371-389.
Weiland, J. and Wilhelmsson, H. 1977. Coherent nonlinear interaction of waves in plasmas. Oxford: Pergamon. International Series in Natural Philosophy, Vol. 88.
Weiland, J., Ichikawa, Y.H., and Wilhelmsson, H. 1978. A perturbation expansion for the NLS equation with application to the influence of nonlinear Landau damping. Physica Scripta, 17, 517-22.
Weissman, M. 1979. Nonlinear wave packets in the Kelvin-Helmholz instability. Phil. Trans. Roy. Soc., 290, 639-681.
White, R.M. 1971. Quantum theory of magnetism. New York: McGraw Hill.
Whitehead, J.H.C. 1947. An expression of Hopf's invariant as an integral. Proc. Nat. Acad. Sci. (USA), 33, 117-123.
Whitham, G.B. 1974. Linear and Nonlinear Waves. New York: John Wiley.
Whittaker, M.T. and Watson, J.N. 1962. A course in modern analysis. Cambridge: C.U.P.
Wijngaarden, L.A. 1968. On the equation of motion for mixtures of liquid and gas bubbles. J. Fluid Mech., 33, 465-74.
Williams, J.G. 1970. Topological analysis of a nonlinear field theory. J. Math. Phys., 11, 2611-6.
Williams, J.G. and Zvengrowski, P. 1977. Spin in kink type theories. Int. J. Theor. Phys., 16, 755-761.
Winther, . 1980. A conservation finite element method for the KdV equation. Math. Comp., 34, 23-43.
Yajima, N., and Oikawa, M. 1976. Formation and Interaction of Sonic-Langmuir Solitons - Inverse Scattering method. Progress of Theor. Phys., 56, 1719-39.

Yang, C.N. and Mills, R.L. 1954. Conservation of isotopic spin and isotopic gauge invariance. Phys. Rev., 96, 191-5.
Yuen, H.C., and Lake, B.M. 1975. Nonlinear deep water waves: theory and experiment. Physics of Fluids, 18, 906-960.
Zabusky, N.J., and Kruskal, M.D. 1965. Interaction of solitons in a collisionless plasma and the recurrence of initial states. Phys. Rev. Lett., 15, 240-243.
Zabusky, N.J. 1967. A synergetic approach to problems of nonlinear dispersive wave propagation and interaction. In Ames, W. (Ed.), Nonlinear Partial Differential Equations, New York: Academic Press.
Zabusky, N.J., Deem, G.S. and Kruskal, M.D. 1968. Formation, propagation and interaction of solitons. 16mm Cine Film.
Zabusky, N.J. 1968. Solitons and bound states of the time independent Schrödinger equation. Phys, Rev., 168, 124-28.
Zabusky, N.J. 1969. Nonlinear lattice dynamics and energy sharing. J. Phys. Soc. Japan, 26, 196-202.
Zabusky, N.J. and Galvin, C.J. 1971. Shallow water waves, the KdV equation and solitons. J. Fluid Mech., 47, 811-24.
Zabusky, N.J. 1973. Solitons and energy transport in nonlinear lattices. Computational Physics Comm., 5, 1-10.
Zabusky, N.J. 1981. Computational synergetics and mathematical innovation. J. Comp. Phys., 43, 195-249.
Zakharov, V.E., and Faddeev, L.D. 1971. Korteweg-de Vries equation: A complete integrable Hamiltonian system. Funct. Anal. Appl., 5, 280-287.
Zakharov, V.E., and Shabat, A.B. 1972. Exact theory of two-dimensional self-focussing and one-dimensional self-modulation of waves in nonlinear media. Sov. Phys.-JETP, 34, 62-69.
Zakharov, V.E. 1972. Collapse of Langmuir Waves. Sov. Phys. JETP, 72, 908-14.
Zakharov, V.E. and Faddeev, L.D. 1972. KdV equation: a completely integrable Hamiltonian system. Funct. Anal. Appl., 5, 280-287.
Zakharov,V.E., and Manakov, S.V. 1973. Resonant interaction of wave packets in nonlinear media. Sov. Phys. JETP Lett., 18, 243-5.
Zakharov, V.E. and Shabat, A.B. 1974. A scheme for integrating the nonlinear equations of mathematical physics by the method of the inverse scattering problem, I. Funct. Anal. Appl., 8, 226-235.
Zakharov, V.E. and Synakh, S. 1975. The nature of the self-focussing singularity. Sov. Phys. JETP, 41, 465-468.
Zakharov, V.E., Takhtadzhyan, L.A. and Faddeev, L.D. 1975. Complete description of the solution of the 'sine-Gordon' equation. Sov. Phys. Doklady, 19, 824-6.
Zakharov, V.E. 1976. Exact solutions to the problem of the parametric interaction of 3-dimensional wave packets. Soviet Physics Doklady, 21, 322-323.
Zakharov, V.E., and Manakov, S.V. 1976. Generalisation of the inverse scattering problem method. Theor. and Math. Phys., 27, 485-7.
Zakharov, V.E., and Manakov, S.V. 1976. Asymptotic behaviour of non-linear wave systems integrated by the inverse scattering method. Sov. Phys. JETP, 44, 106-112.
Zakharov, V.E., and Manakov, S.V. 1976. The theory of resonance interaction of wave packets in nonlinear media. Sov. Phys. JETP, 42, 842-50.
Zakharov, V.E. 1977. Kinetic equation for solitons. Sov. Phys. JETP, 33, 538-541.
Zakharov, V.E., and Belinskii, V.A. 1978. Stationary gravitational solitons with axial symmetry. JETP, 50, 1-9.
Zakharov, V.E. and Takhtajan, L.A. 1979. Equivalence of the nonlinear Schrödinger equation with the equation of the Heisenberg ferromagnet. Theor. and Math. Phys., 38, 17-23.

Zakharov, V.E. and Mikhailov, A.V. 1979. Relativistically invariant two-dimensional models of field theory which are integrable by means of the inverse scattering problem method. Sov. Phys. JETP, 47, 1017-27.

Zakharov, V.E. 1980. The inverse scattering method. In Bullough, R.K. and Caudrey, P.J. (Ed.), Solitons (Vol. 17 in Topics in Current Physics), , pp 243-285. Berlin": Springer.

Zakharov, V.E. and Shabat, A.B. 1980. Integration of nonlinear equations of mathematical physics by the method of inverse scattering, II. Funct. Anal. Appl., 13, 166-174.

Zakharov, V.E. and Manakov, S.V. 1974. On the complete integrability of the nonlinear Schrödinger equation. Theor. Math. Phys., 19, 551-9.

INDEX